Birkhäuser

Modeling and Simulation in Science, Engineering and Technology

José Antonio Carrillo • Eitan Tadmor
Editors

Active Particles, Volume 4

Theory, Models, Applications

Birkhäuser

Editors
José Antonio Carrillo
Mathematical Institute
University of Oxford
Oxford, UK

Eitan Tadmor
Department of Mathematics and Institute
for Physical Science & Technology
University of Maryland
College Park, MD, USA

ISSN 2164-3679 ISSN 2164-3725 (electronic)
Modeling and Simulation in Science, Engineering and Technology
ISBN 978-3-031-73422-9 ISBN 978-3-031-73423-6 (eBook)
https://doi.org/10.1007/978-3-031-73423-6

This book is published under the imprint Birkhäuser, www.birkhauser-science.com by the registered company Springer Nature Switzerland AG
The registered company address is: Gewerbestrasse 11, 6330 Cham, Switzerland

Preface

This edited book is a collection of contemporary surveys on modeling, qualitative analysis, and simulation of active particles, with applications in natural sciences. It is a follow-up to volumes 1–3 under the same title. As in the preceding volumes, this book blends diverse applications which indicate the interdisciplinary nature of the subject matter and the different mathematical tools to address these applications. Indeed, this new frontier of science offers a range of new challenging problems which require advanced mathematical tools and, in some cases, new mathematical theories.

The contributions have been selected with the aim of covering a variety of different viewpoints—from modeling the interactions in collective dynamics of animals and in population dynamics, through neural-networks, semi-supervised learning, and Monte Carlo methods in optimization, to kinetic and continuum theories with applications to aggregation and birth-and-death processes. The aim is to provide the scientific community with a timely, up-to-date overview of current research trends conducted by leading experts.

We have several contributions on rigorous derivation of mean-field macroscopic and continuum models from particle systems. In this context the reader will find relevant surveys on the following:

- In the chapter "Calculating Discrete Minimizers of the Interaction Energy in Collective Behavior," José Cañizo and Alejandro Ramos-Lora consider minimizers of the N-particle system governed by attractive-repulsive interaction energy potentials found in collective dynamics, including singular potentials. They illustrate 2D minimizers such as lattice structures, Wulff shapes, and the continuous large-Nlimit for locally integrable potentials.
- In the chapter "Large-Population Limits of Non-exchangeable Particle Systems," Nathalie Ayi and Nastassia Pouradier Duteil present novel approaches for the large-population limit of non-exchangeable particle systems, which do not admit the classical mean-field description. Instead, they describe a non-exchangeable mean-field limit on the product space of positions and labels in the frameworks of static and adaptive networks.

- In the chapter "Large Population Limit of Interacting Population Dynamics via Generalized Gradient Structures," Anastasiia Hraivoronska, Jasper Hoeksema, and Oliver Tse focus on a variational-based approach for the derivation of a doubly nonlocal Fisher-KPP model as a macroscopic description for the large population limit.

Analytical properties and numerical approaches of some of the macroscopic equations obtained are also covered in this volume with the following contributions:

- In the chapter "Aggregation-Diffusion Phenomena: From Microscopic Models to Free Boundary Problems," Inwon Kim, Antoine Mellet, and Jeremy Sheung-Him Wu explore the emergence of collective dynamics as a result of the competition between attraction and repulsion governed by aggregation-diffusion, such as those found in the context of attractive chemotaxis phenomena. Depending on the scale involved, one encounters phase transitions, free boundary problems, and surface tension among the different ingredients in the dynamics of active particles.
- In the chapter "Conservative Semi-Lagrangian Methods for Kinetic Equations," Seung-Yeon Cho, Maria Groppi, Jingmei Qiu, Giovanni Russo, and Seok-Bae Yun provide a detailed review of semi- Lagrangian methods for kinetic Boltzmann, BGK, and Vlasov-Poisson models. They discuss conservative non-oscillatory reconstructions, splitting methods, and spectral differentiation to allow accurate numerical solutions of collisional kinetic equations.

This volume also includes discussion on some of the most relevant applications nowadays. We have comprehensive surveys on applications in data science: sampling, inverse problems, machine learning, as well as collective dynamics in nature.

- In the chapter "Consistency of Semi-Supervised Learning, Stochastic Tug-of-War Games, and the p-Laplacian," Jeff Calder and Nadejda Drenska provide a broad overview of recent work on Partial Differential Equations (PDE) that arise as continuum limits of graph-based semi-supervised learning.
- In the chapter "Models of Animal Behavior as Active Particle Systems with Nonreciprocal Interactions," Dan Gorbonos, Amir Haluts, and Nir Gov discuss interactions observed in collective dynamics in nature, reflecting the complexity of animal movement and decision-making. These include the adversarial nature of "contestant particles" and the stabilization of cohesive swarms through long-range and adaptive gravity-like attraction.
- In the chapter "Bayesian Sampling Using Interacting Particles," Shi Chen, Zhiyun Ding, and Qin Li address the mean-field analysis for a large number of interactive Bayesian samples. They use coupling methods and compactness-based arguments to justify the validity of sampling strategies and establish a correspondence between coupled ODEs/SDEs and PDE descriptions.
- Finally, in the chapter "Adjoint Monte Carlo Methods," Russel Caflisch and Yunan Yang explore the development of adjoint Monte Carlo methods for solving kinetic-based optimization problems, a common challenge in plasma control

and device design. They describe various adjoint Monte Carlo methods that enable efficient gradient estimation in complex, high-dimensional optimization problems.

Oxford, UK José Antonio Carrillo
College Park, MD, USA Eitan Tadmor

Contents

Consistency of Semi-supervised Learning, Stochastic Tug-of-War Games, and the p-Laplacian

Jeff Calder and Nadejda Drenska

Abstract In this paper, we give a broad overview of the intersection of partial differential equations (PDEs) and graph-based semi-supervised learning. The overview is focused on a large body of recent work on PDE continuum limits of graph-based learning, which have been used to prove well-posedness of semi-supervised learning algorithms in the large data limit. We highlight some interesting research directions revolving around *consistency* of graph-based semi-supervised learning and present some new results on the consistency of p-Laplacian semi-supervised learning using the stochastic tug-of-war game interpretation of the p-Laplacian. We also present the results of some numerical experiments that illustrate our results and suggest directions for future work.

1 Introduction

Machine learning refers to algorithms that learn how to perform tasks, like image classification or text generation, from examples or experience, and are not explicitly programmed with step-by-step instructions in the way a human may be instructed to perform a similar task. The recent surge in machine learning and artificial intelligence is being driven by deep learning, which uses deep artificial neural networks and has found applications in nearly all areas of science, engineering, and everyday life [53]. Modern deep learning excels when provided with massive amounts of training data and computational resources. However, there are many applications, specifically with real-world problems, where labeled training data is hard to come by and number in the hundreds or thousands, instead of millions. For

J. Calder (✉)
School of Mathematics, University of Minnesota, Minneapolis, MN, USA
e-mail: jcalder@umn.edu

N. Drenska
Department of Mathematics, Louisiana State University, Baton Rouge, LA, USA
e-mail: ndrenska@lsu.edu

J. A. Carrillo, E. Tadmor (eds.), *Active Particles, Volume 4*, Modeling and Simulation in Science, Engineering and Technology,
https://doi.org/10.1007/978-3-031-73423-6_1

example, in medical image analysis, a human expert, i.e., a highly trained doctor, must annotate images in order to provide data for machine learning algorithms to train in. Obtaining labeled data is thus costly, and there is a tremendous interest in developing machine learning algorithms that perform well with as few labeled examples as possible.

There are many frameworks for learning from limited data. One effective method is *semi-supervised learning*, which makes use of both labeled and unlabeled data in the learning task. In contrast, the most common type of machine learning, called *fully supervised* learning, makes use of only labeled training data. The labeled training data for a fully supervised classification task includes data/label pairs $(x_1, y_1), \ldots, (x_n, y_n)$, where $x_i \in \mathbb{R}^d$ and $y_i \in \mathbb{R}^k$. The goal of fully supervised learning can generally be stated as finding, or "learning," a function $f : \mathbb{R}^d \to \mathbb{R}^k$ so that $f(x_i) \approx y_i$ for all i. This can be viewed as an *ill-posed* problem, especially when the number of labeled training points n is small, since there are many possible functions f that fit the training data. Furthermore, the ultimate goal is not just to fit the training data but to learn a function that *generalizes* well to new data that has not been seen before. Semi-supervised learning uses unlabeled data to improve the performance of classification algorithms in the context of small training sets. In many applications, unlabeled data is abundant and easy to obtain. In medical image analysis, for example, unlabeled data would correspond to medical images of a similar type and modality that have not been labeled or annotated by an expert.

An effective technique for exploiting unlabeled data in semi-supervised learning is to utilize a graph structure, which may be intrinsic to the data, or constructed based on similarities between data points. Graphs encode interdependencies between constituent data points that have proven useful for analyzing and representing high-dimensional data. There has been a surge of interest recently in graph-based semi-supervised learning techniques for problems where very few labeled examples are available, which is a setting that is challenging for existing techniques based on Laplacian regularization and harmonic extension. Various methods have been proposed, including p-Laplacian regularization, higher-order Laplacian methods, Poisson learning, and many others. Many of these algorithms are inspired by insights from the theory of partial differential equations (PDEs) or the calculus of variations, by examining the PDE-continuum limits of discrete graph-based learning algorithms, and their well-posedness properties, or lack thereof.

While there has been a substantial amount of work on PDE-continuum limits of graph-based learning, there has been relatively little work on the question of *consistency* of graph-based learning, using these well-developed PDE tools. The basic question of *consistency* is whether the machine learning algorithm is making the correct predictions, under a simplified model for the data. The current PDE-continuum limit results simply describe how the algorithms behave in the large data limit, but have not yet, with few exceptions, been used to prove that they work properly—that is, that they are consistent. It is arguably the case that consistency is more important than well-posedness in the continuum limit, yet the question has rarely been studied, which we suspect is due to the difficulty in defining what consistency means, and the difficulty in obtaining meaningful results outside of toy settings.

In this paper, we provide a broad overview of semi-supervised learning and its connections to PDEs, and we present some new consistency results for p-Laplacian-based semi-supervised learning. Our new results make use of the tug of war with noise interpretation of the p-Laplacian, for which we also provide a brief literature survey. In particular, one of our results uses the tug-of-war game on a stochastic block model graph, which does not have the geometric structure that is usually required for PDE-based analysis. Our consistency results for the p-Laplacian are preliminary results meant to spark new work, and they certainly leave many questions unanswered. Our overall goal in this paper is to highlight a number of open research problems that will benefit from fruitful collaboration between PDE analysts and more theoretically minded machine learning researchers. It would be interesting in future work to improve these results and extend them to other graph-based semi-supervised learning algorithms and other graph structures.

1.1 Outline

This paper is organized as follows. In Sect. 2, we provide a brief overview of the p-Laplacian and the stochastic tug-of-war game interpretation. In Sect. 3, we give a thorough survey of graph-based semi-supervised learning, and its connections to PDEs and tug-of-war games, with a particular emphasis on the p-Laplacian. Then in Sect. 4, we present some preliminary results on consistency properties of the p-Laplacian using the tug-of-war game interpretation. This includes results both on geometric graphs and stochastic block models, as well as some numerical results to illustrate the main theorems. Finally, we conclude and discuss directions for future work in Sect. 5.

2 Tug-of-War Games and the p-Laplacian

In this section, we provide a brief overview of the p-Laplacian and the connection to stochastic tug-of-war games.

2.1 The p-Laplacian

The p-Laplacian arises as the Euler-Lagrange equation, or necessary conditions, for the nonlinear potential problem in the calculus of variations

$$\min_{u\in W^{1,p}(\Omega)} \int_\Omega |\nabla u|^p \, dx, \tag{2.1}$$

where $p \geq 1$ and $\Omega \subset \mathbb{R}^d$, subject to some boundary conditions, such as a Dirichlet condition $u = g$ on $\partial\Omega$. The Euler-Lagrange equation [46] for (2.1) is

$$\Delta_p u := \operatorname{div}\left(|\nabla u|^{p-2}\nabla u\right) = 0 \quad \text{in} \quad \Omega, \tag{2.2}$$

and we call $\Delta_p u$ the p-Laplacian of u. We must take $p \geq 1$ to ensure (2.1) is convex and admits a minimizer. The case of $p = 2$ corresponds to the usual Laplacian $\Delta_p = \Delta$. When $1 \leq p < 2$ the diffusion is singular when $\nabla u = 0$, while for $p > 2$, the diffusion becomes degenerate when $\nabla u = 0$; both cases lead to drastically different properties and regularity theory compared to the uniformly elliptic case of $p = 2$ [45]. Functions that satisfy $\Delta_p u = 0$ are called *p harmonic*.

If we expand the divergence in (2.2), we find that any p-harmonic function u satisfies (provided $\nabla u \neq 0$ when $p < 2$)

$$\Delta_p u = |\nabla u|^{p-2}(\Delta u + (p-2)\Delta_\infty u) = 0 \quad \text{in} \quad \Omega, \tag{2.3}$$

where $\Delta_\infty u$ is the ∞-Laplacian, defined by

$$\Delta_\infty u = \frac{1}{|\nabla u|^2}\nabla u \cdot \nabla^2 u \nabla u = \frac{1}{|\nabla u|^2}\sum_{i,j=1}^{d} u_{x_i x_j} u_{x_i} u_{x_j}, \tag{2.4}$$

and $\nabla^2 u$ is the Hessian of u. The ∞-Laplacian is so named, because it is the $p \to \infty$ limit of the p-Laplacian in the sense that

$$\Delta_\infty u = \lim_{p\to\infty} \frac{1}{p-2}|\nabla u|^{2-p}\Delta_p u,$$

provided again that $\nabla u \neq 0$, which follows directly from (2.3). It is possible to interpret the ∞-Laplacian as the Euler-Lagrange equation for a variational problem like (2.1) with $p = \infty$; we refer the reader to [4] for more details. For a more detailed overview of the p-Laplacian, we also refer to [82].

2.2 *Tug-of-War Games*

When $p = 2$, there is a well-established classical connection between random walks, or Brownian motions, and harmonic functions. Indeed, any harmonic function u satisfies the mean value property [46]

$$u(x_0) = \fint_{B(x_0,\varepsilon)} u(y)\, dy, \tag{2.5}$$

where $\varepsilon > 0$ is any value for which $B(x_0, \varepsilon) \subset \Omega$, and the notation $\fint$ means

$$\fint_V u\,dx = \frac{1}{|V|}\int_V u\,dx.$$

We can interpret the right-hand side of (2.5) as the expectation of $u(X)$, where X is a random variable uniformly distributed on the ball $B(x_0, \varepsilon)$. Thus, if we define a random walk $X_1, X_2, \ldots$, on Ω, which is a sequence of random variables for which $X_0 = x_0 \in \Omega$ and, conditioned on X_k, X_{k+1} is uniformly distributed on $B(X_k, \varepsilon)$, we have

$$\mathbb{E}\left[u(X_{k+1}) \mid X_k\right] = \fint_{B(X_k,\varepsilon)} u(y)\,dy = u(X_k),$$

provided $B(X_k, \varepsilon) \subset \Omega$, provided we, for the moment, ignore the boundary $\partial\Omega$. Thus, since u is harmonic, we have that $Z_k = u(X_k)$ is a *martingale* [120]. This connection to probability theory allows simple alternative proofs of various estimates in harmonic function theory, such as Harnack's inequality and gradient estimates [79, 80], and has found applications in proving gradient estimates on graphs as well [26].

Over the past 15 years, there has been significant interest in extending these martingale techniques to the p-Laplacian. To do this, however, the definition (2.2) of the p-Laplacian is not very useful. Instead, provided that $\nabla u \neq 0$, we can drop the $|\nabla u|^{p-2}$ term in (2.3) to obtain the equation

$$\Delta u + (p-2)\Delta_\infty u = 0, \tag{2.6}$$

and restrict our attention to $p \geq 2$. Given a smooth function u, we can average the Taylor expansion for u about x_0 to obtain

$$\fint_{B(x_0,\varepsilon)} u(y)\,dy = u(x_0) + \frac{\varepsilon^2}{2(d+2)}\Delta u(x_0) + O(\varepsilon^3),$$

and so

$$\varepsilon^2 \Delta u(x_0) = 2(d+2)\fint_{B(x_0,\varepsilon)} u(y)\,dy - 2(d+2)u(x_0) + O(\varepsilon^3). \tag{2.7}$$

Noting that the ∞-Laplacian is the second derivative of u in the direction of the gradient $v = \frac{\nabla u(x_0)}{|\nabla u(x_0)|}$, we have

$$\begin{aligned}\varepsilon^2 \Delta_\infty u(x_0) &= \left[u\left(x_0 + \varepsilon v\right) - u(x_0)\right] - \left[u(x_0) - u\left(x_0 - \varepsilon v\right)\right] + O(\varepsilon^3) \qquad (2.8)\\ &= \max_{B(x_0,\varepsilon)} u + \min_{B(x_0,\varepsilon)} u - 2u(x_0) + O(\varepsilon^3).\end{aligned}$$

Inserting (2.7) and (2.8) into (2.6), we see that if u is a smooth p-harmonic function, i.e., satisfying (2.6), then

$$u(x_0) = \alpha \fint_{B(x_0,\varepsilon)} u(y)\,dy + \frac{1-\alpha}{2}\left(\max_{B(x_0,\varepsilon)} u + \min_{B(x_0,\varepsilon)} u\right) + O(\varepsilon^3). \tag{2.9}$$

as $\varepsilon \to 0$ where $\alpha = \frac{d+2}{d+p} \in [0,1]$ since $p \geq 2$. Thus, while p-harmonic functions do not satisfy a mean value property for any size ball, (2.9) gives an asymptotic version of a mean value property, with min and max terms arising from the ∞-Laplacian. In fact, the asymptotic mean value property (2.9) *characterizes* p-harmonic functions [89]. It is also an interesting equation to study in its own right; when the $O(\varepsilon^3)$ is dropped, the functions are called *p-harmonious* and studied in detail in [91].

The mean value property (2.9) suggests a way to adapt the random walk construction earlier to p-harmonic functions. We simply define a stochastic process $X_0, X_1, X_2, \ldots$ so that, given X_k, X_{k+1} is defined in the following way: With probability α, we take a random walk step, so X_{k+1} is uniformly distributed on $B(X_k, \varepsilon)$; with probability $\frac{1-\alpha}{2}$, we set $X_{k+1} = \operatorname{argmax}_{B(X_k,\varepsilon)} u$, and likewise with probability $\frac{1-\alpha}{2}$, we set $X_{k+1} = \operatorname{argmin}_{B(X_k,\varepsilon)} u$. When the argmax or argmin is not unique, we make a rule to break ties, which can be deterministic or random. By the definition of this stochastic process, for any continuous function u, we have

$$\mathbb{E}[u(X_{k+1}) \mid X_k] = \alpha \fint_{B(X_k,\varepsilon)} u(y)\,dy + \frac{1-\alpha}{2}\left(\max_{B(X_k,\varepsilon)} u + \min_{B(X_k,\varepsilon)} u\right).$$

In particular, if u is smooth and p-harmonic, then the discussion above shows that

$$\mathbb{E}[u(X_{k+1}) \mid X_k] = u(X_k) + O(\varepsilon^3).$$

Thus, we have recovered the martingale property, at least asymptotically as $\varepsilon \to 0$.

The stochastic process introduced above is often described as a two player *tug-of-war* game with noise. The game involves a token X_k that is moved by two players and by random noise. Player I is trying to move the token to locations that maximize u, while the goal of player II is to move the token to places that minimize u. The game is played by flipping two coins. The first comes up heads with probability α and tails with probability $1-\alpha$. If the first coin comes up heads, the token X_k is moved to a uniformly random point X_{k+1} in the ball $B(X_k, \varepsilon)$. If the first coin comes up tails, then the game switches to a *tug-of-war* game, where a second unbiased coin is flipped to decide which player gets to move the token to decide X_{k+1}. Whichever player wins the second coin flip is allowed to move the token wherever they like in the ball $B(X_k, \varepsilon)$; the idea being that player I will move the token to maximize u, while player II will minimize. The exact goal of each player depends on the boundary condition; they may want to maximize/minimize the value of u when the game stops by hitting the boundary $\partial\Omega$, in which case only the values of u on or near the boundary need to be specified in the game and the players are assumed to play optimal strategies to maximize or minimize the payoff—the value of u—at the end of the game. The random walk step, and the randomness in choosing between players, is interpreted as *noise*, hence the term tug of war with noise.

There is a close connection between tug-of-war games and nonlocal elliptic equations. Let us define the nonlocal 2 and ∞ Laplacians by

$$\Delta_2^\varepsilon u(x) = \fint_{B(x_0,\varepsilon)} u(y)\,dy - u(x), \text{ and } \Delta_\infty^\varepsilon u(x) = \frac{1}{2}\left(\max_{B(x,\varepsilon)} u + \min_{B(x,\varepsilon)} u\right) - u(x).$$

Then if we drop the error term in (2.9) and rearrange, we arrive at the equation

$$\Delta_p^\varepsilon u := \alpha \Delta_2^\varepsilon u + (1-\alpha)\Delta_\infty^\varepsilon u = 0. \tag{2.10}$$

The operator Δ_p^ε on the left above is a nonlocal approximation to the p-Laplacian that arises from the tug-of-war game perspective and is closely related to the graph p-Laplacian discussed in Sect. 3. We can also consider a corresponding nonlocal boundary value problem

$$\begin{cases} \Delta_p^\varepsilon u = 0, & \text{in } \Omega_\varepsilon \\ u = g, & \text{on } \partial_\varepsilon \Omega, \end{cases} \tag{2.11}$$

where g is given, $\partial_\varepsilon \Omega = \partial\Omega + B(0,\varepsilon)$ and $\Omega_\varepsilon = \Omega \setminus \partial_\varepsilon \Omega$. If we choose the stopping time τ to be the first time that the tug-of-war game hits the boundary strip $\partial_\varepsilon \Omega$, then the martingale property and the optional stopping theorem yield

$$u(x) = \mathbb{E}[u(X_\tau) \mid X_1 = x].$$

This gives a representation formula for solutions of the nonlocal p-Laplace equation (2.10) that is useful for studying properties of the solution through martingale techniques. Properties that are independent of the nonlocal scale $\varepsilon > 0$ are inherited by p-harmonic functions by sending $\varepsilon \to 0$.

Tug-of-war games for the p-Laplacian were originally introduced in [105, 106] with a version that holds for $1 \leq p \leq \infty$ using ideas from earlier work on deterministic games for the 1-Laplacian [69]. The version for $p \geq 2$ described above was introduced and studied in [90]. This work motivated a study of the game-theoretic p-Laplacian on graphs [92] as well as finite difference approaches for numerically approximating solutions [3, 102, 103]. The tug-of-war interpretation of the p-Laplacian has led to simple alternative proofs of regularity for p-harmonic functions, including the Harnack inequality and gradient estimates [5, 83]. In addition, many variants of tug-of-war games have been introduced, including games with bias [107], mixed Neumann/Dirichlet boundary conditions [32], obstacle problems [78], nonlocal tug-of-war for the fractional p-Laplacian [76], time-dependent equations [56], and variants on the core structure of the game [74]. We also refer the reader to the survey article [77] and two recent books on tug-of-war games [75, 104] for more details.

3 Semi-supervised Learning and PDEs

In this section, we overview graph-based semi-supervised learning and the recent connections to PDEs in the continuum limit, with a specific focus on the p-Laplacian and tug-of-war games.

3.1 Graph-Based Semi-supervised Learning

Semi-supervised learning uses both labeled and unlabeled data. As a toy example, in Fig. 1a, we show the famous two-moons data set with two labeled data points, the blue circle and green square. The black points are the *unlabeled data*, which are not used by fully supervised learning, and will be discussed momentarily. With only two labeled data points, it is difficult to learn a general function f that will correctly classify new data points. In Fig. 1b, we show the result of training a linear kernel support vector machine (SVM) with these two data points, which finds the linear decision boundary with maximal margin—in this case, the line equidistant[1] from the two training points. Given no other information, this may be a reasonable thing to do. However, suppose now that we also have access to the black points, which are *unlabeled* data points. That is, we have access to the coordinates $x_i \in \mathbb{R}^2$ but not the label y_i, which in this case is $y_i \in \{0, 1\}$ for binary classification, and these unlabeled data points are the *only* data points we will want to apply our classifier to in the future. In this case, the linear SVM decision boundary is a poor choice, since it cuts through a dense region of the unlabeled data—the lower moon—where we may not expect a true decision boundary to lie. Instead, we should seek to place the decision boundary in sparse regions between clusters. In Fig. 1c, we show the result of using a semi-supervised learning algorithm on the same data set, which correctly separates the two moons.[2]

There are many ways to incorporate unlabeled data into a machine learning algorithm. One common and successful approach is to utilize *graph-based* learning, where each node in the graph corresponds to a data point, and the edges in the graph correspond to either intrinsic relationships between data points or record some

[1] The line does not appear orthogonal to the vector between the training points, because the axes are scaled differently.

[2] To be precise, we applied graph-based Poisson learning, described below, to obtain label predictions for all unlabeled points, and then we trained a radial basis kernel SVM using the label predictions.

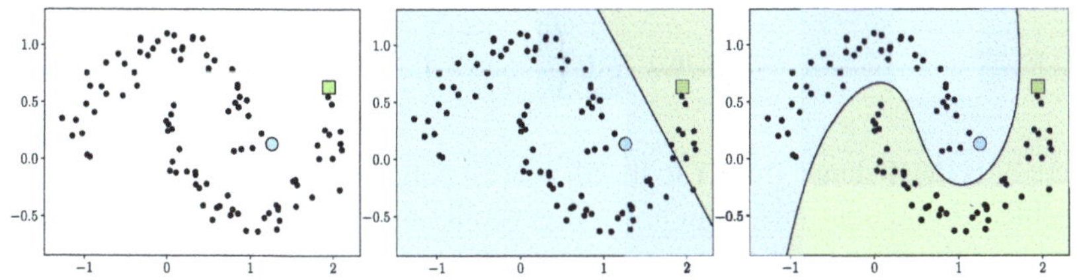

Fig. 1 A toy example comparing fully and semi-supervised learning. In (**a**), we show a data set with two labeled examples—the blue circle and green square—along with 98 unlabeled data points, the black dots. In (**b**), we show the decision regions from training a fully supervised classification algorithm, while in (**c**), we show the decision regions for a semi-supervised learning algorithm, which uses the unlabeled data to inform the decision boundary

notion of similarity between data points. Many types of data have intrinsic graph structures, like molecules in drug discovery problems, citation data sets, or networks like the internet. In other problems, like image classification, the graph structure is not intrinsic but can be constructed as a *similarity graph*, in which similar data points are connected by an edge with a weight that encodes the degree of similarity. Let $\mathcal{X} = \{x_1, \ldots, x_n\}$ be a set of data points, where $x_i \in \mathbb{R}^d$. Here, we assume we have a graph with vertex set $\mathcal{X}$ described by a symmetric weight matrix $W = W^T \in \mathbb{R}^{n\times n}$ in which each entry w_{xy}, for $x, y \in \mathcal{X}$, is nonnegative and encodes a notion of similarity between x and y. A zero weight $w_{xy} = 0$ indicates there is no edge between x and y, while a positive weight $w_{xy} > 0$ indicates the presence of an edge, and the larger the value the stronger the edge. We let $N_x \subset \mathcal{X}$ denote the graph neighbors of the vertex $x \in \mathcal{X}$, which is defined by

$$N_x = \{y \in \mathcal{X} : w_{xy} > 0\}. \tag{3.1}$$

A common way to construct a graph over a data set is the geometric graph construction

$$w_{xy} = \eta\left(\frac{|x - y|}{\varepsilon}\right), \tag{3.2}$$

where $\varepsilon > 0$ is the graph *bandwidth* and $\eta : [0, \infty) \to [0, \infty)$ is a nonnegative function that is typically decreasing with compact support. A common choice is Gaussian weights where $\eta(t) = e^{-t^2}$, with a possible truncation to zero at some distance $t = \tau$. One issue with the geometric graph construction is that one has to choose a very large value for the graph bandwidth $\varepsilon > 0$ to ensure the graph is connected, even in the sparsest regions of the data set, which leads to a very non-sparse weight matrix W that is difficult to work with computationally. To address this, it is common to adjust the bandwidth ε locally to reflect the density (or sparsity) of the data set, which results in various types of k-nearest neighbor graphs. One common construction is

$$w_{xy} = \eta\left(\frac{|x-y|}{\sqrt{\varepsilon_x \varepsilon_y}}\right),$$

where ε_i is the distance from x to its kth nearest neighbor, though other choices are possible. Throughout this section, we'll generally assume a graph with geometric weights of the form (3.2).

3.2 Laplacian Regularization

Suppose now that we have a graph structure over our data set, given by a weight matrix $W = (w_{xy}) \in \mathbb{R}^{n\times n}$ along with a subset of graph nodes $\Gamma \subset \mathcal{X}$ that are labeled, with corresponding labels $y = g(x) \in \mathbb{R}$ for $g : \Gamma \to \mathbb{R}$ a given labeling function, which we are taking to be scalar only for simplicity of discussion. The seminal approach in graph-based semi-supervised learning is *Laplacian regularization*, initially proposed in [135], which propagates the labels from Γ to the rest of the graph by solving the optimization problem

$$\min_{u:\mathcal{X}\to\mathbb{R}} \sum_{x,y\in\mathcal{X}} w_{xy}(u(x)-u(y))^2, \tag{3.3}$$

subject to the constraint that $u(x) = g(x)$ for $x \in \Gamma$. The real-valued solution $u(x)$ is then thresholded to the nearest label to make a class prediction. The energy in (3.3) is called the *graph Dirichlet energy*, and the minimizer $u : \mathcal{X} \to \mathbb{R}$ is exactly the harmonic extension of the labeled data, which satisfies the boundary value problem

$$\begin{cases} \mathcal{L}u(x) = 0, & \text{if } x \in \mathcal{X}\setminus\Gamma \\ u(x) = g(x), & \text{if } x \in \Gamma, \end{cases} \tag{3.4}$$

where $\mathcal{L}$ is the graph Laplacian defined by

$$\mathcal{L}u(x) = \sum_{y\in\mathcal{X}} w_{xy}(u(x)-u(y)). \tag{3.5}$$

In other words, we are seeking the *smoothest* function—in this case harmonic—that correctly classifies the labeled data points. In semi-supervised learning, this is often called the *semi-supervised smoothness assumption* [30], which stipulates that the labeling is smooth in high density regions of the data set. Since the initial development in [134, 135], graph-based learning using Laplacian regularization, and variations thereof, has grown into a wide class of useful techniques in machine learning [2, 8, 10–12, 15, 24, 25, 29, 55, 57, 58, 73, 117, 118, 121, 123, 128, 129, 131–133].

Graph harmonic functions satisfy a mean value property, similar to harmonic functions on $\mathbb{R}^d$. Indeed, if we rearrange the condition $\mathcal{L}u(x) = 0$, we obtain

$$u(x) = \frac{1}{d_x} \sum_{y \in \mathcal{X}} w_{xy} u(y), \tag{3.6}$$

where $d_x = \sum_{y \in \mathcal{X}} w_{xy}$ is the degree of vertex x. This is a local version of the continuous mean value property for harmonic functions. It says that the value $u(x)$ of a graph harmonic function at a node x is equal to the weighted average of the values $u(y)$ at neighboring nodes y, weighted by the graph edge weights w_{xy}. In fact, one way to solve equation (3.4) is by iterating the mean value property

$$u_{k+1}(x) = \frac{1}{d_x} \sum_{y \in \mathcal{X}} w_{xy} u_k(y), \tag{3.7}$$

for $x \in \mathcal{X} \setminus \Gamma$, fixing $u_k(x) = g(x)$ for $x \in \Gamma$. This is exactly the classical Jacobi iteration for solving a linear system, and while it is not the fastest technique—the preconditioned conjugate gradient method is much faster—it has a nice interpretation as *propagating* labels using the neighborhood structure of the graph and is thus called *label propagation* in the literature; see [134]. Equation (3.7) can also be viewed as a diffusion equation on the graph.

Laplacian regularized learning has an important connection to random walks on graphs. Let $X_1, X_2, \ldots$ be a random walk on the vertices $\mathcal{X}$ of the graph, with transition probabilities $P = D^{-1}W$, where $D = \operatorname{diag}(d_{x_1}, \ldots, d_{x_n})$. That is, the probability p_{xy} of the random walker moving from vertex x to vertex y is given by $p_{xy} = d_x^{-1} w_{xy}$. Given a function $u : \mathcal{X} \to \mathbb{R}$ on the vertices of the graph, we can compute

$$\mathbb{E}[u(X_{k+1}) - u(X_k) \mid X_k = x] = \sum_{y \in \mathcal{X}} p_{xy}(u(y) - u(x)) = -\frac{1}{d_x} \mathcal{L}u(x).$$

The object on the right-hand side is called the *random walk* graph Laplacian, and we denote it by $\mathcal{L}_{rw}$; that is,

$$\mathcal{L}_{rw} u(x) = \frac{1}{d_x} \mathcal{L}u(x).$$

The computation above thus yields

$$\mathbb{E}[u(X_{k+1}) \mid X_k] = u(X_k) - \mathcal{L}_{rw} u(X_k), \tag{3.8}$$

and thus, the random walk graph Laplacian is the generator for the random walk. In particular, if u is graph harmonic, so that $\mathcal{L}u(x) = 0$, then $u(X_k)$ is a *martingale*. Defining the stopping time τ as the first time the random walk hits the labeled data set Γ, the optional stopping theorem yields that the solution u of Laplace learning (3.4) satisfies

$$u(x) = \mathbb{E}[u(X_\tau) \mid X_1 = x] = \sum_{y\in\Gamma} \mathbb{P}(X_\tau = y \mid X_1 = x) g(y). \tag{3.9}$$

Thus, the solution of Laplacian regularized graph-based learning (3.4) can be interpreted as a weighted average of the given labels $g(y)$, weighted by the probability of the random walk hitting $y \in \Gamma$ first, before hitting any other labeled data point. This is a reasonable thing to do for semi-supervised learning at an intuitive level as well. Indeed, if each cluster in the data set is well separated from the others, then provided the label rate is not too low, the random walk is likely to stay in the cluster it started in long enough to hit a labeled data point in that cluster, giving the correct label and propagating labels well within clusters.

At moderately low label rates, Laplacian regularization performs very well for graph-based semi-supervised learning. However, the performance can become quite poor at extremely low label rates, which was first noted in [101] and later in [39]. When there are very few labels, the solution of the Laplace learning problem (3.4) prefers to be nearly constant over the whole graph, with sharp spikes at the labeled data points, as depicted in Fig. 2a. This is the configuration that gives the least graph Dirichlet energy, since the spikes are relatively inexpensive when the number of labeled data points is small. From the random walk perspective, when there are very few labels, the random walk takes very long to hit a labeled data point, so the stopping time τ is very large, and the distribution of the random walker approaches the limiting invariant distribution on the graph, which is proportional to the degree. Thus, by (3.9) we have

$$u(x) \approx \frac{\sum_{y\in\Gamma} d_y g(y)}{\sum_{y\in\Gamma} d_y} \tag{3.10}$$

at very low label rates. When the labels are binary $g(y) \in \{-1, 1\}$, the sign of the right-hand side of (3.10) is constant over the graph and simply chooses the class whose labeled points have the largest cumulative degree, which is one way to explain the very poor performance of Laplace learning at low label rates.

Even in settings where moderate amounts of labeled data will eventually be acquired, it is often necessary to consider, at least initially, problems with very few labels where Laplace learning is not useful. One of those areas is *active learning*, which refers to machine learning algorithms that address the problem of choosing the *best* data points to label in order to obtain superior model performance with as few labels as possible [110]. Active learning methods incorporate a human-in-the-loop that can be queried to label new data points as needed, and the goal is to achieve the highest accuracy with as few labels as possible by intelligently choosing the next point to label, often in a sequential manner. Active learning procedures normally start out with extremely small labeled sets that slowly grow during the acquisition of new labeled data points, which requires graph-based semi-supervised learning algorithms that perform well both with small and moderate amounts of labeled data. A number of graph-based active learning methods have been proposed [65, 85, 97, 100, 108, 136], including recent methods inspired by the PDE-continuum limits discussed below [98], and recent applications have been found in image classification problems [17, 31, 44, 99].

3.3 PDE-Inspired Insights and Algorithms

Another interpretation of the poor performance of Laplace learning at low label rates is through PDE continuum limits. To explain this, we need to place some additional assumptions on the graph construction. We assume the vertices $\mathcal{X} = \{x_1, \ldots, x_n\}$ of the graph are *i.i.d.* random variables distributed on a domain $\Omega \subset \mathbb{R}^d$ with a continuous and positive density $\rho : \Omega \to (0, \infty)$. We assume the random geometric graph construction, which uses weights (3.2) for a given bandwidth $\varepsilon > 0$. This is called a *random geometric graph*, and in this setting, the expectation of the graph Dirichlet energy in (3.3) is

$$\begin{aligned} n^2 \int_\Omega \int_\Omega \eta\left(\frac{|x-y|}{\varepsilon}\right) (u(x) - u(y))^2 \rho(x)\rho(y)\, dxdy \\ \approx C_\eta n^2 \varepsilon^{d+2} \int_\Omega \rho^2 |\nabla u|^2\, dx, \end{aligned} \tag{3.11}$$

provided $u : \Omega \to \mathbb{R}$ is a smooth function and ρ is Lipschitz, where C_η is a positive constant depending only on η. The asymptotics in (3.11) can be verified formally with Taylor expansions of u and ρ and can also be established rigorously in the language of Gamma convergence, as in [49]. Thus, the continuum version of Laplace learning (3.4) is the boundary value problem

$$\begin{cases} \operatorname{div}(\rho^2 \nabla u) = 0, & \text{in } \Omega \setminus \Gamma \\ u = g, & \text{on } \Gamma, \end{cases} \tag{3.12}$$

where $\Gamma \subset \Omega$ is purposely vaguely specified and should encode some notion of the continuum limit of the labeled data set, and $g : \Gamma \to \mathbb{R}$ the continuum limit of the values of the labels. When Γ does not contain the entire boundary $\partial\Omega$, we also must impose homogeneous Neumann conditions on $\partial\Omega \setminus \Gamma$.

The presence of the squared density ρ^2 as a diffusion coefficient in the PDE continuum limit illustrates how Laplace learning is able to diffuse labels quickly in high-density regions where ρ is large and place decision boundaries, which correspond to sharp transitions in the label function u, in sparse regions where ρ is small. However, the presence of the labeled set Γ as a boundary condition is

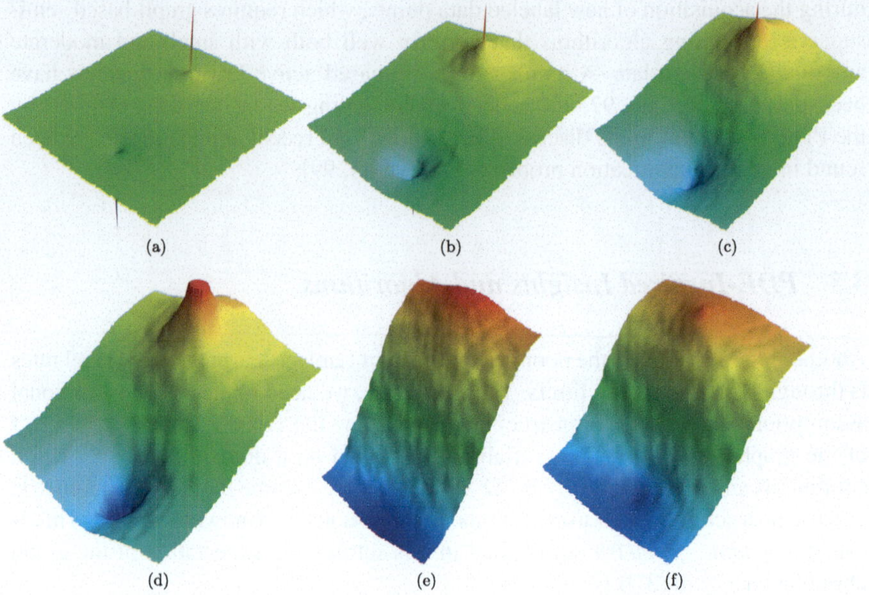

Fig. 2 Comparison of different semi-supervised learning methods on a toy example with two labeled data points with labels +1 and −1. The graph consists of $n = 20{,}000$ uniformly distributed random variables on the unit box $[0, 1]^2$ with geometric Gaussian kernel weights with $\varepsilon = 0.05$. Here WNLL stands for Weighted NonLocal Laplacian [111], and PWLL stands for Poisson Weighted Laplace Learning [27, 98]. (**a**) Laplace [135]. (**b**) p-Laplace, $p = 2.05$ [47]. (**c**) p-Laplace, $p = 2.25$ [47]. (**d**) WNLL [111]. (**e**) High-order Laplace, $m = 2$ [128]. (**f**) PWLL [27, 98]

problematic, since (3.12) is well-posed only when Γ is large enough, and regular enough. This is due to the fact that the Sobolev function space $H^1(\Omega)$, like the $L^p(\Omega)$ function spaces, consists of functions that are defined only up to sets of measure zero. So the values of $u(x)$ at particular points $x \in \Omega$ are not well-defined. This means, for example, that Γ cannot be a collection of isolated points, as in Fig. 2a; otherwise, (3.12) is ill-posed. In order for (3.12) to be well-posed, the boundary set Γ must be *large enough* and sufficiently regular so that we can interpret what it means to *evaluate* $u(x)$ for $x \in \Gamma$. Such results are called *trace theorems* in the PDE literature, and the restriction $u|_\Omega$ is the trace of u on Γ; we refer to [46] for more details.

The last decade has seen a surge of interest in developing graph-based learning algorithms that are well-posed with arbitrarily small labeled data sets by considering PDE continuum limits that are insensitive to the labeled data set. One natural idea proposed in [40] is to consider the p-Dirichlet energy on the graph, given by

$$\sum_{x,y\in\mathcal{X}} w_{xy}|u(x) - u(y)|^p. \tag{3.13}$$

In fact, the p-Laplacian was suggested in many other works as well [1, 16, 40–43, 54, 66, 130] though not in the low label rate context. By a similar argument as above, the continuum limit energy should be the weighted p-Dirichlet energy

$$\int_\Omega \rho^2 |\nabla u|^p \, dx.$$

When $p > d$, we have the Sobolev embedding $W^{1,p}(\Omega) \subset C^{0,\alpha}(\Omega)$ [46], which ensures that solutions are Hölder continuous, spikes cannot form, and the continuum boundary value problem

$$\begin{cases} \operatorname{div}(\rho^2 |\nabla u|^{p-2} \nabla u) = 0, & \text{in } \Omega \setminus \Gamma \\ u = g, & \text{on } \Gamma, \end{cases} \tag{3.14}$$

is well-posed without any conditions on Γ. This well-posedness was established rigorously in the continuum limit in [112] when $p > d$, and the number of labeled points is finite and fixed as the number of unlabeled data points tends to infinity. In this case, the authors of [112] identified an additional restriction on the length scale $\varepsilon > 0$ that arises through the nonlocal nature of the graph Laplacian. Namely, in addition to $p > d$, it is necessary that $n\varepsilon^p \ll 1$ to ensure that spikes do not form. Since $n\varepsilon^d \gg \log n$ is necessary to ensure connectivity of the graph,[3] this condition can only be satisfied when $p > d$. The authors of [112] also proposed a modification of p-Laplace learning for which the condition $n\varepsilon^p \ll 1$ is not required, by essentially extending the labels to all nearest neighbors on the graph.

It is worthwhile to pause and note that the condition $n\varepsilon^p \ll 1$ comes from a simple balancing of energy between "spiky" functions and smooth functions on the graph. If we write (3.11) for the p-Laplacian, we see that the graph p-Dirichlet energy of a smooth function scales like $n^2\varepsilon^{p+d}$. On the other hand, the energy of a constant function with spikes at the labeled nodes is

$$\sum_{x\in\Gamma} \sum_{y\in\mathcal{X}} \eta\left(\frac{|x-y|}{\varepsilon}\right) |g(y) - c|^p \approx C|\Gamma| S^p n\varepsilon^d,$$

where C is a constant depending on η, $c = u(x)$ is the constant value of u, S is the size of the spikes, and $|\Gamma|$ is the number of labeled points, all provided the label rate is sufficiently low and the labeled points are sufficiently far apart. In order to ensure that a smooth interpolating function has less energy than a spiky one, in the setting where $|\Gamma|$ is constant with n and ε, we require

$$n^2\varepsilon^{p+d} \ll n\varepsilon^d \iff n\varepsilon^p \ll 1.$$

[3] The quantity $n\varepsilon^d$ is the average number of neighbors of any nodes on the graph, up to a constant.

If we allow the number of labels $|\Gamma|$ to depend on n and ε, then the condition is

$$n^2\varepsilon^{p+d} \ll |\Gamma| n\varepsilon^d \iff \beta := \frac{1}{n}|\Gamma| \gg \varepsilon^p.$$

The quantity β above is the *label rate*—the ratio of labeled data points to all data points—and simply energy balancing arguments show that the label rate β should satisfy $\beta \gg \varepsilon^p$ to ensure convergence to a well-posed continuum problem. The case of finite labels as $n \to \infty$ corresponds to $\beta = n^{-1}$, which recovers the condition $n\varepsilon^p \ll 1$. In [28], it was shown that $\beta \sim \varepsilon^2$ for $p = 2$ is the threshold for a spiky versus smooth continuum limit in a setting that allowed the labeled data set to grow as $n \to \infty$. The proofs in [28] used the random walk interpretation of the 2-Laplacian and martingale arguments. A negative result was also established in [28] for $p > 2$ showing that spikes develop when $\beta \ll \varepsilon^p$. However, a positive result for $p > 2$—that $\beta \gg \varepsilon^p$ is sufficient to prevent spikes from forming—is currently lacking, outside of the finite label regime covered in [112].

As in Sect. 2, we can take the limit as $p \to \infty$ of the graph p-Dirichlet energy. It is more illustrative to do this with the necessary conditions for minimizing the graph p-Dirichlet energy (3.13), which is the graph p-Laplace equation

$$\sum_{y\in\mathcal{X}} w_{xy}|u(x) - u(y)|^{p-2}(u(x) - u(y)) = 0 \tag{3.15}$$

for each $x \in \mathcal{X} \setminus \Gamma$. To take the limit as $p \to \infty$, we separate the terms in the summation above by sign and take the pth root to obtain

$$\left(\sum_{y:g_{xy}>0}^{n} w_{xy}(u(x) - u(y))^{p-1}\right)^{1/p} = \left(\sum_{y:g_{yx}>0}^{n} w_{xy}(u(y) - u(x))^{p-1}\right)^{1/p}.$$

where $g_{xy} = w_{xy}(u(x) - u(y))$. The terms with $g_{xy} = 0$ do not contribute and can be neglected. We assume neither sum above is empty; otherwise, all terms in (3.15) are zero, which is trivial. Sending $p \to \infty$ yields the equation

$$\max_{y:g_{xy}>0}(u(x) - u(y)) = \max_{y:g_{yx}>0}(u(y) - u(x)).$$

The maximums above are unchanged by replacing the conditions $g_{xy} > 0$ and $g_{yx} > 0$ with $w_{xy} > 0$, as the graph is symmetric so $w_{xy} = w_{yx}$. Making this replacement, dividing by two on both sides, and rearranging yield

$$\mathcal{L}_\infty u(x) := u(x) - \frac{1}{2}\left(\max_{N_x} u + \min_{N_x} u\right) = 0, \tag{3.16}$$

where $\mathcal{L}_\infty$ is called the *graph ∞-Laplacian*, and we recall that N_x are the graph neighbors of x defined in (3.1). Notice the maximum and minimum above are over graph neighbors, which satisfy $w_{xy} > 0$, but that the weights in the graph do not enter directly into the operator otherwise. If we were to replace w_{xy} by w_{xy}^p in (3.15), then the weights would appear in the graph ∞-Laplacian, which is more informative of the graph structure and was the approach taken in [20]. We use the unweighted ∞-Laplacian here, since it connects more directly to the tug-of-war interpretation of the graph p-Laplacian, discussed below in this section. It turns out, similar to the continuum ∞-Laplace equation, that graph ∞-harmonic functions also solve L^∞-type variational problems on graphs (see, e.g., [71, 84]).

The graph ∞-Laplacian was first used for semi-supervised learning in [71, 84], where it is called *Lipschitz learning*. In [21], it was shown rigorously using PDE-continuum limits that solutions of the graph ∞-Laplace equation converged to ∞-harmonic functions in the continuum, even for finite isolated labeled data points. This established well-posedness of Lipschitz learning with arbitrarily few labeled examples. Subsequent work [18, 19] established convergence rates of graph ∞-harmonic functions to the continuum. As in Sect. 2, specifically (2.6), it is natural to define a p-Laplacian operator on the graph that is a combination of the 2-Laplacian and ∞-Laplacian. Here we use the definition

$$\mathcal{L}_p = \alpha \mathcal{L}_{rw} + (1-\alpha)\mathcal{L}_\infty, \tag{3.17}$$

where $\alpha = 1/(p-1)$ and $\alpha = 1$ when $p = \infty$. This type of graph p-Laplacian is often called the *game-theoretic* p-Laplacian due to its connection to tug-of-war games, as described in Sect. 2. The p-Laplace operator appearing earlier in (3.15), as the necessary conditions for the p-Dirichlet energy, is usually called the *variational* p-Laplacian and is a different operator on the graph, even though they agree in the continuum. The game-theoretic p-Laplacian on graphs originally appeared in [92] and was proposed for semi-supervised learning in [20], although the latter work considered a weighted version of the ∞-Laplacian. In [20], it was shown that the game-theoretic p-Laplacian is well-posed with arbitrarily few labeled examples when $p > d$,[4] by showing that the continuum limit was the well-posed continuum p-Laplace equation with $p > d$. An interesting and important detail is that the condition $n\varepsilon^p \ll 1$ is not required by the game-theoretic p-Laplacian due to the strong regularization provided by the ∞-Laplacian.

As in Sect. 2, we can develop a tug-of-war game interpretation for the game-theoretic graph p-Laplacian. Any function u on the graph satisfying $\mathcal{L}_p u = 0$ also satisfies, by rearranging (3.17), the equation

$$u(x) = \frac{\alpha}{d_x}\sum_{y\in\mathcal{X}} w_{xy}u(y) + \frac{1-\alpha}{2}\left(\max_{N_x} u + \min_{N_x} u\right), \tag{3.18}$$

[4] The definition of α differs by a constant in [20], so that p has the correct continuum interpretation.

which is the discrete graph version of the continuous identity (2.9), without the $O(\varepsilon)$ error term. As we did in Sect. 2, we can define a discrete stochastic process on the graph that produces the martingale property. We define $X_1, X_2, \ldots$ so that, given X_k, the choice of X_{k+1} is with probability α a random walk step from X_k, with probability $\frac{1-\alpha}{2}$ the vertex y maximizing $u(y)$ over neighbors y of X_k and with probability $\frac{1-\alpha}{2}$ the corresponding minimizing vertex. For the same reasons as in Sect. 2, any graph p-harmonic function, satisfying $\mathcal{L}_p u(x) = 0$ for all i, is a martingale when applied to the process X_k, in the sense that

$$\mathbb{E}[u(X_{k+1}) \mid X_k] = u(X_k).$$

This allows us to apply martingale techniques to study properties of the solution to the game-theoretic p-Laplacian on the graph. The variational graph p-Laplacian (3.15) does not have any such stochastic tug-of-war interpretation. To our knowledge, no existing works have used the tug-of-war interpretation of the graph p-Laplacian to study properties of semi-supervised p-Laplacian learning.

Figure 2b and c show the results of the game-theoretic p-Laplacian applied to the toy two labeled point problem, illustrating how the p-Laplacian corrects the spike phenomenon in Laplace learning. Note we only need $p > 2$ here since $d = 2$. Many other approaches have been proposed for correcting the degeneracy in Laplace learning. In [111], the authors proposed to reweight the edges of the graph that connect to labels more strongly to discourage spike formation and to use the ordinary graph Laplacian on the reweighted graph. The method is called the Weighted Nonlocal Laplacian (WNLL). Figure 2d shows an example of the WNLL on the two-point problem, where we can see that the method essentially just extends the labels to nearest neighbors on the graph. However, it was shown in [24] that the WNLL method remains ill-posed at arbitrarily low label rates. In [24], the authors proposed the *properly weighted* graph Laplacian, which reweights the graph in a nonlocal way that is *singular* near the labeled data, so that the continuum limit is well-posed with arbitrarily few labels. In [128], it was proposed to use a higher-order Laplace equation of the form $\mathcal{L}^m u = 0$ for problems with very few labels. The idea here is that in the continuum, the variational formulation of the equation would involve $u \in H^m(\Omega)$, and when $m > \frac{d}{2}$, the Sobolev embedding $H^m \subset C^{0,\alpha}(\Omega)$ ensures that spikes do not form and the continuum limit is well-posed. Figure 2e shows the two-point problem with higher-order Laplacian regularization. The well-posedness of higher-order Laplace learning with finite labeled data in the continuum limit was only very recently addressed in [119]. Finally, there is recent work on using Poisson equations on graphs for semi-supervised learning in [25, 27, 98], some of which connects to the earlier work on the properly weighted Laplacian [24]. In particular, the Poisson Weighted Laplace Learning (PWLL) method from [27, 98] involves reweighting the graph using the solution of a graph Poisson equation, whose singularities are sufficient to invoke the results in [24], though the theory for these methods is currently still under development. Figure 2f shows the behavior of PWLL on the two-point problem. The volume-constrained MBO method based on auction dynamics developed in [64] has also proven to be effective at low label rates, though we are not aware of any theory to explain this. Other methods at low label rates include Hamilton-Jacobi equations on graphs [22] and the *centered kernel method* [86–88]. We also mention here the work on MBO methods[5] for graph-based semi-supervised learning [14, 48, 62, 93–95] which consider approximations of the $p = 1$ Laplacian for graph-based learning. The MBO methods perform semi-supervised learning on graphs by alternating diffusion and thresholding to label vectors until convergence, which has the effect of finding the partition of the graph with the smallest perimeter that fits the labels correctly.

3.4 Consistency Versus Well-Posedness

Much of the work described above is aimed at establishing continuum PDE or variational descriptions of graph-based machine learning algorithms, which can, for example, prove that a method like p-Laplace learning provides a well-posed method for propagating labels on graphs that does not develop spikes or other degeneracies. Many other works have also considered the continuum limit of the graph Laplacian, its spectrum, and regularity properties [23, 26, 38, 50, 59, 60]. However, this abundance of work fails to address the problem of whether the machine learning methods work well or not; that is, they say little or nothing about whether the predicted label is correct! This is the question of *consistency*, which is more important and difficult, and thus less often studied.

As an example, in Fig. 3, we show the results of graph-based semi-supervised learning applied to the same two-moons example as in Fig. 1. The p-Laplace and higher-order Laplace methods are the only methods with rigorous well-posedness results with very few labels, and Poisson learning is expected to enjoy such results, which is the subject of a forthcoming paper by the first author. Laplace learning and WNLL are ill-posed in the low-label regime. The best performing method is Poisson learning [25], followed closely by high-order Laplace, which illustrates that well-posedness does not always go hand in hand with good consistency properties, or at least that the relation between consistency and having a well-posedness continuum limit is not clear or well-understood.

To address the consistency problem, one has to make an assumption on the underlying cluster structure of the graph, or the labeling function. Then the goal is to prove that the machine learning algorithm can identify the clusters or correct labels, under reasonable assumptions on the model and parameters. This kind of consistency analysis was carried out in the context of spectral clustering in [61], in which the probability density ρ was assumed to be highly concentrated on disjoint clusters. Spectral clustering uses the eigenvectors of the graph Laplacian

[5] MBO stands for Merriman-Bence-Osher, who originally proposed threshold dynamics approaches for numerically approximating mean curvature motion [96].

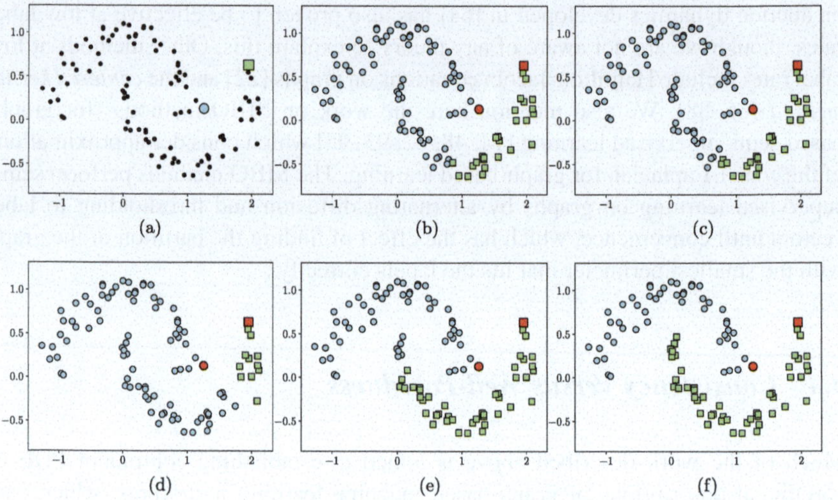

Fig. 3 Classification results using different graph-based semi-supervised learning algorithms on the two-moons data set. There are two classes, the upper moon and the lower moon, and the label function is binary $g(x) \in \{0, 1\}$. The red points are the only labeled data points. The p-Laplace method used $p = 3$, while high-order Laplace learning used order $m = 4$, both of which gave the best results over a reasonable search. **(a)** Data. **(b)** Laplace [135] (77%). **(c)** p-Laplace [47] (79%). **(d)** WNLL [111] (61%). **(e)** High-order Laplace [128] (94%). **(f)** Poisson [25] (100%)

for clustering [116] and are closely related to spectral methods in data science [9, 34, 37]. Another related work is [113], which assumes a well-separated mixture model for the data and shows that the spectral embedding reflects the clusters in the data [113]. Aside from these examples, there is a lack of consistency results for many graph-based semi-supervised learning methods, which should be viewed as an opportunity to utilize powerful PDE-continuum limit tools to establish important and interesting results about machine learning algorithms. We begin such a study in this paper.

3.5 *Deep Semi-supervised Learning*

There is a growing body of work on deep semi-supervised learning, which includes many methods based on graph neural networks; see [125] for a survey. The data science applications are somewhat different here; in addition to a graph, most problem formulations assume that each *node* $x \in \mathcal{X}$ in the graph has an associated feature vector $v(x) \in \mathbb{R}^k$. An example application would be a citation network, where each node in the graph is an academic paper and the edges represent citations between papers. The feature vector $v(x)$ for paper x may contain summary statistics of key words appearing in the abstract of the paper. An example task is to classify the subject of each paper, using both the graph structure and the feature vectors, which in general provide complimentary information.[6]

Let us briefly describe how some graph neural networks work for semi-supervised learning. A standard feed-forward neural network, or multilayer perceptron, iteratively composes linear functions with nonlinear activation function, $u_{k+1} = \sigma(\Theta_k u_k + b_k)$, where $u_k \in \mathbb{R}^{n_k}$ is the input to the kth layer, $\Theta_k \in \mathbb{R}^{n_{k+1} \times n_k}$ is the weight matrix for the kth layer, and $b_k \in \mathbb{R}^{n_{k+1}}$ the bias vector. A common choice for the activation σ is the rectified linear unit (ReLU) $\sigma(t) = \max\{t, 0\}$, but other choices are possible. The input to the network is u_1, and the output is u_L after L layers.

Graph neural networks incorporate the graph information into the neural network architecture. A very popular approach is the graph convolutional neural network (GCN) proposed in [68], which combines the standard feed-forward neural network with the graph structure as follows:

$$u_{k+1}(x) = \sigma\left(\frac{1}{d_x}\sum_{y\in\mathcal{X}} w_{xy}(\Theta_k u_k(y) + b_k)\right), \tag{3.19}$$

where $u_k : \mathcal{X} \to \mathbb{R}^{n_k}$ and Θ_k, b_k are as before.[7] The inputs are the node feature vectors $u_1(x) = v(x)$, and the output after L layers is, after thresholding, the network's prediction of the label for each node in the graph. This output is fed into a loss function that measures how well the network fits some training data, and the weights Θ_k and biases b_k are tuned by minimizing the loss function with gradient descent.

We note that the GCN layer (3.19) is essentially a combination of a feed-forward neural network with averaging over neighborhoods on the graph. These types of networks are also called *message passing* graph neural networks. The reader can compare this to label propagation, see (3.7), and notice that when σ is the identity $\sigma(t) = t$, $\Theta_k = I$, and $b_k = 0$, this appears to be exactly label propagation. The difference is that with graph neural networks, the operation acts on the *feature vectors* of the nodes, while label propagation acts directly on the labels, since feature vectors are not available. The graph convolutional layer (3.19) can be viewed as a special case of a more general class of of graph neural networks based on a spectral definition of convolution [35]. We also mention an important variant called graph attention networks [115], which allows the weights in (3.19) to be *learned* using an *attention mechanism*, in a similar way that transformers utilize the attention mechanism in large language models [114].

The analysis of graph neural network models like the GCN given in (3.19) would be substantially different from the analysis in this paper. The important problems

---

[6] A standard test data set called *PubMed* [124] uses exactly this setup.

[7] To be precise, the method in [68] divides by $\sqrt{d_x d_y}$ in the sum, instead of d_x, and adds self-loops to the graph.

concern how the *learned* weights Θ_k and biases b_k interact with the graph structure during training, especially for deeper networks, and whether the network relies more on the graph structure or the node features for classification. Although there is ostensibly a connection to label propagation (3.7), this is largely superficial, since GCNs (3.19) usually only have a handful of layers, say 2 or 3, while label propagation is often run for hundreds of steps or more, in order for the method to *converge* to the solution of Laplace's equation. Another difference with the setting of this paper is that many recent graph neural network approaches are aimed at tackling heterophily in graphs, in which adjacent nodes may very often belong to different classes [81, 122, 127, 137]. In this case, the semi-supervised smoothness assumption does not hold, and the question is how to combine information from both the feature vectors and graph structure, when neither on their own is highly informative.

There are, nevertheless, some settings where the analysis in this paper may be relevant in deep graph neural networks. Many recent works have shown that one can obtain effective methods by decoupling the graph diffusion or averaging in (3.19) from the neural network weights Θ_k and biases b_k [6, 36, 51, 63, 109]. One way to do this, proposed in [51], is to first run the feature vectors $v(x)$ for each node through a standard feed-forward neural network to make label predictions for each node x and then to diffuse those predictions over the graph before feeding them into a loss function. In the case of [51], the authors used the PageRank algorithm [52], which is essentially a graph reaction-diffusion equation [126], and the label predictions entered through the source term. The particular choice of PageRank is not essential, and other works have shown that various forms of graph diffusion can easily substitute [6, 109]. Thus, improvements in graph-based semi-supervised learning algorithms, especially theoretical guarantees, can have a direct impact on deep semi-supervised learning methods that decouple the graph from the neural network.

4 Consistency Results for the *p*-Laplacian

We present here a general framework for proving consistency results for graph-based semi-supervised learning with the game-theoretic p-Laplacian. The framework is based on the stochastic tug-of-war interpretation of the p-Laplacian on a graph that was introduced in Sect. 3. We illustrate how to use the framework to prove some preliminary consistency results on geometric graphs and stochastic block model graphs and then highlight some open research questions for future work.

As in Sect. 3, let $\mathcal{X} = \{x_1, \ldots, x_n\}$ denote our data points, which are the vertices of the graph, and let $W = (w_{xy})_{x,y\in\mathcal{X}}$ be a symmetric (i.e., $w_{xy} = w_{yx}$) weight matrix which encodes the graph structure. We assume the graph is connected. We let $d_x = \sum_{y\in\mathcal{X}} w_{xy}$ be the degree of vertex $x \in \mathcal{X}$, and we recall that $N_x \subset \mathcal{X}$

denotes the graph neighbors of x, as defined in (3.1). Let $\ell^2(\mathcal{X})$ denote the space of functions $u : \mathcal{X} \to \mathbb{R}$, and define the game-theoretic p-Laplacian on the graph G is the operator $\mathcal{L}_p : \ell^2(\mathcal{X}) \to \ell^2(\mathcal{X})$ defined in (3.17) in Sect. 3, where $\alpha = 1/(p-1)$ and $p \geq 2$. Let us denote by $\Gamma \subset \mathcal{X}$ the labeled data set and $g : \Gamma \to \mathbb{R}$ the labels. The game-theoretic p-Laplacian regularized semi-supervised learning problem is given by the boundary value problem

$$\begin{cases} \mathcal{L}_p u = 0, & \text{in } \mathcal{X} \setminus \Gamma \\ u = g, & \text{on } \Gamma. \end{cases} \tag{4.1}$$

Since the graph is connected, the p-Laplace equation (4.1) can be shown to have a unique solution via the Perron method [20].

In this section, we address the question of *consistency* of p-Laplacian regularization. That is, supposing that g is the restriction to Γ of a label function on the whole graph $g : \mathcal{X} \to \mathbb{R}$ that is in some sense smooth with respect to the geometry of the graph, we aim to give conditions under which the solution u of (4.1) is close to g at the unlabeled vertices $\mathcal{X} \setminus \Gamma$. In other words, consistency asks the question: Under what conditions on the underlying label function g, the graph structure W, and the label set Γ does the solution of p-Laplacian regularization make the correct label predictions? In this paper, we consider the *noise-free* setting, where we observe the true labels without any noise or corruption. It would be interesting to consider the setting of noisy labels in future work; we discuss this more in Sect. 5.

In consistency of graph-based learning, it is essential that the underlying label function g is in some way regular or smooth with respect to the graph structure. In other words, there should be some degree of correlation between the labels of vertices and their nearby neighbors in the graph topology. Otherwise graph-based learning is not useful and would not be expected to yield better results than fully supervised learning that ignores the graph structure. In this section, we present results for two different types of graph structures. In Sect. 4.2, we present results for geometric graphs, where the vertices are sampled from a Euclidean domain and nearby points are connected by edges. In this case, we assume that g has some degree of regularity with respect to the underlying Euclidean geometry. In Sect. 4.3, we present results for classification on stochastic block model graphs, where there is no continuous geometric structure (SBM graphs can be thought of as having a discrete geometric structure, but this is not directly related to continuous PDEs). In this case, the label function is concentrated on the blocks, and only changes between blocks where there are relatively few edges.

4.1 Results on General Graphs

Before presenting our main results in Sects. 4.2 and 4.3, we present our general stochastic tug-of-war framework on general graphs in Sect. 4.1.1, which amounts to selecting suboptimal strategies in the tug-of-war game to produce tractable upper and lower bounds on the solution u of the p-Laplace equation (4.1). It is interesting to remark that we do not explicitly use the symmetry assumption $w_{xy} = w_{yx}$ in our results, and our results hold with minor modifications to the quantities involved for directed (i.e., nonsymmetric) graphs.

4.1.1 The Tug-of-War Lemma

In this section, we present our main tug-of-war result (Lemma 4.3), which uses the stochastic tug-of-war game and martingale techniques to bound the difference $u(x) - g(x)$, where u solves (4.1). These techniques are later applied to geometric graphs in Sect. 4.2 and to stochastic block model (SBM) graphs in Sect. 4.3. The tug-of-war game for the p-Laplacian is based on the following dynamic programming principle associated with the p-Laplacian, which was discussed and established in Sect. 3.

Proposition 4.1 (Dynamic Programming Principle) *If $u \in \ell^2(\mathcal{X})$ and $x \in \mathcal{X}$ such that $\mathcal{L}_p u(x) = 0$ then*

$$u(x) = \frac{\alpha}{d_x} \sum_{y \in \mathcal{X}} w_{xy} u(y) + \frac{1-\alpha}{2} \left(\min_{N_x} u + \max_{N_x} u \right), \tag{4.2}$$

where we recall that $\alpha = 1/(p-1)$ and $p \geq 2$.

We call Eq. (4.2) a dynamic programming principle (DPP), because it expresses u at a point via u at nearby points and is closely related to the DPPs that appear in optimal control problems (see, e.g., [7]). In fact, the DPP arises from the stochastic two player tug-of-war game described in Sect. 3. The game is played between two players Paul and Carol. The game ends when the token of the game arrives at a point $x \in \Gamma$, and Paul pays Carol $g(x)$. Thus, Paul wants the game to end where g is smallest, while Carol wants the opposite. At each step of the game, with probability α, the token moves from its current position x to a neighbor $y \in N_x$ via a single step of a random walk on the graph (i.e., the next position is chosen randomly according to the distribution $\mathbb{P}(y) = w_{xy}/d_x$). With probability $(1-\alpha)/2$, Paul chooses the next position of the token, and with probability $(1-\alpha)/2$, Carol chooses the next position. We will use this interpretation in the proofs of our main results, without explicitly writing down the game or the appropriate spaces of strategies.

In order to use the tug-of-war game, we define a stochastic process that corresponds to a *suboptimal* strategy for Paul. We recall that $\alpha = 1/(p-1)$ and pick any $p \geq 2$. Let $u \in \ell^2(\mathcal{X})$ and $x \in \mathcal{X}$, and consider the following stochastic process defined on $\mathcal{X}$ associated with u and an initial point $x \in \mathcal{X}$. We define the sequence of random variables X_0, X_1, X_2 as follows. We first set $X_0 = x$. Then, conditioned on X_i, we choose X_{i+1} as follows: First, if $X_i \in \Gamma$, then we set $X_{i+1} = X_i$, so the process gets *stuck* when it hits Γ. If $X_i \notin \Gamma$, then we choose X_{i+1} by the following:

1. With probability α, we take an independent random walk step, that is

$$\mathbb{P}(X_{i+1} = y \mid X_i = x) = \frac{w_{xy}}{d_X}.$$

2. With probability $\frac{1}{2}(1-\alpha)$, we choose $X_{i+1} \in \underset{y \in N_{X_i}}{\operatorname{argmax}}\, u$.
3. With probability $\frac{1}{2}(1-\alpha)$, we choose $X_{i+1} \in \underset{y \in N_{X_i} \cap \Gamma}{\operatorname{argmin}}\, g$, when $N_{x_i} \cap \Gamma \neq \varnothing$, and $X_{i+1} \in \underset{y \in N_{X_i}}{\operatorname{argmin}}\, u$, when $N_{x_i} \cap \Gamma = \varnothing$.

All the probabilistic ingredients are chosen independently. When there are multiple choices for X_{i+1} in steps 2 and 3, the particular choice is irrelevant to the analysis, and any concrete choice will do (e.g., say, choose the vertex x_i with the smallest index). The stochastic process $X_0, X_1, X_2, \ldots,$ defined above can be interpreted as a realization of the two-player game where Carol plays optimally and our choice of strategy for Paul is to move to Γ as soon as possible, which is the *suboptimal* part of his strategy.

We now establish the sub-martingale property corresponding to our stochastic process, which is a key ingredient in our analysis.

Lemma 4.2 *Let u be the solution to* (4.1). *Then the random variables $Z_i = u(X_i)$ form a sub-martingale with respect to X_i, that is $\mathbb{E}[Z_{i+1} \mid X_i] \geq Z_i$.*

Proof Conditioned on $X_i \in \Gamma$, we have $\mathbb{E}[u(X_{i+1}) \mid X_i] = u(X_i)$, since $X_{i+1} = X_i$. Conditioned on $X_i \notin \Gamma$, we further condition on steps 1–3 in the definition of X_{i+1} and use $u = g$ on Γ to obtain

$$\mathbb{E}[u(X_{i+1}) \mid X_i] = \frac{\alpha}{d_{X_i}} \sum_{y \in \mathcal{X}} w_{X_i,y} u(y) + \frac{1-\alpha}{2}\left(\max_{N_{X_i}} u + \min_{V} u(y)\right), \tag{4.3}$$

where $V = N_{x_i}$ when $N_{X_i} \cap \Gamma = \varnothing$ and $V = N_{X_i} \cap \Gamma$ when $N_{X_i} \cap \Gamma \neq \varnothing$. Using Proposition 4.1, we have

$$\mathbb{E}[u(X_{i+1}) \mid X_i] \geq \frac{\alpha}{d_{X_i}} \sum_{y \in \mathcal{X}} w_{X_i,y} u(y) + \frac{1-\alpha}{2}\left(\max_{N_{X_i}} u + \min_{N_{X_i}} u\right) = u(X_i),$$

which completes the proof. □

We now give the main tug-of-war lemma, which uses the sub-martingale property to bound $u(x)$.

Lemma 4.3 *Let u be the solution to* (4.1). *Define the stopping time*

$$\tau = \inf\{i \geq 0 \, : \, X_i \in \Gamma\}. \tag{4.4}$$

Then for any $x \in \mathcal{X}$, it holds that

$$u(x) \leq \mathbb{E}[g(X_\tau) \,|\, X_0 = x]. \tag{4.5}$$

Proof By Lemma 4.2, $Z_i = u(X_i)$ is a sub-martingale, and so by Doob's Optional Stopping Theorem for sub-martingales, see, e.g., [120]; we have

$$\begin{aligned} u(x) &= \mathbb{E}[Z_0 \,|\, X_0 = x] \leq \mathbb{E}[Z_\tau \,|\, X_0 = x] = \mathbb{E}[u(X_\tau) \,|\, X_0 = x] \\ &= \mathbb{E}[g(X_\tau) \,|\, X_0 = x], \end{aligned} \tag{4.6}$$

since $X_\tau \in \Gamma$. □

Let us give a brief preview of how Lemma 4.3 is used. We can subtract $g(x)$ from both sides of (4.5) to obtain

$$|u(x) - g(x)| \leq \mathbb{E}[|g(X_\tau) - g(X_0)| \,|\, X_0 = x].$$

If g is Lipschitz continuous, which is a reasonable assumption on a *continuous geometric* graph, then the right-hand side is $O(|X_\tau - X_0|)$, and so bounding $|u(x) - g(x)|$ amounts to estimating the stopping time τ. In this context, it is not especially important that Paul chooses the point in $N_{x_i} \cap \Gamma$ that minimizes g, and the game could be modified so Paul picks, say, a random boundary point. However, for SBM graphs, the arguments are different, and the key is to allow the *noise* to exit the game, since the noise sees the block structure. In the SBM setting, it is important for Paul to choose the minimizer of g over $N_{x_i} \cap \Gamma$, since g cannot be Lipschitz in any sense on an SBM graph. The interested reader can skip to Sect. 4.3 for the SBM analysis, which is independent of much of the geometric results below.

Throughout the rest of the paper, we make the strong, but simplifying, assumption that every vertex $x \in \mathcal{X}$ has a neighbor $y \in N_x$ with $y \in \Gamma$, meaning:

(A1) We assume that for every $x \in \mathcal{X}$

$$\Gamma \cap N_x \neq \varnothing.$$

Assumption **(A1)** is a simplifying assumption that makes much of the analysis tractable. Essentially, it allows us to estimate the stopping time τ defined in (4.4) in many different settings by allowing Paul to end the game quickly. Depending on the graph model, assumption **(A1)** is a fairly strong assumption on both the *label rate* and the *graph structure*. In Sects. 4.2 and 4.3, we will give conditions under which **(A1)** holds for geometric and SBM graphs. The results in this paper are preliminary, and we discuss the importance of relaxing assumption **(A1)** in Sect. 5. This seems to us to be a nontrivial task that will require fine estimates on martingale stopping times.

4.1.2 General Consistency Results

Here, we use the martingale lemma, Lemma 4.3, to establish some general consistency results, which will rely on a Lipschitz or bounded derivative condition on the label function g. Thus, for $g : \mathcal{X} \to \mathbb{R}$, we need to construct a notion of gradient that is consistent with the gradient in the continuum when working with geometric graphs with a length scale ε. Given a graph described by a weight matrix W, we let $\mathrm{diam}(W)$ denote the *diameter* of the graph, which is defined as the largest number of hops required to get from any node to any other node in the graph. When the graph is a random geometric graph with bandwidth $\varepsilon > 0$, then each hop travels at most distance ε, and so we would expect to have $\mathrm{diam}(W) = O(\varepsilon^{-1})$. This motivates the definition of the *graph length scale*.

Definition 4.4 Given a weighted graph W, the *graph length scale* ε_W is defined as

$$\varepsilon_W = \frac{1}{\mathrm{diam}(W)}.$$

We can now define the gradient of g as

$$\nabla g(x, y) = \frac{g(x) - g(y)}{\varepsilon_W},$$

for any $x, y \in \mathcal{X}$ connected by an edge, so $w_{xy} > 0$. We also define

$$\|g\|_\infty = \max_{x \in \mathcal{X}} |g(x)|,$$

and

$$\|\nabla g\|_\infty = \max_{x,y \in X} \left\{ |\nabla g(x, y)| \, \mathbb{1}_{w_{xy} > 0} \right\}.$$

This is to say, the norm $\|\nabla g\|_\infty$ is the maximal absolute difference of g across all edges in the graph divided by the graph length scale ε_W.

Define the smallest *weighted* ratio of labeled neighbors to neighbors to be

$$\delta := \min_{x \in X} \frac{\Sigma_{y \in \mathcal{X}} w_{xy} \mathbb{1}_{y \in \Gamma}}{\Sigma_{z \in \mathcal{X}} w_{xz}}. \tag{4.7}$$

Note that if **(A1)** holds, then $\delta > 0$.

The following theorem gives our first consistency result on general graphs.

Theorem 4.5 *Assume (A1) holds. Then, for any $x \in \mathcal{X}$,*

$$|u(x) - g(x)| \le \left(\frac{2\log(\|g\|_\infty \|\nabla g\|_\infty^{-1} \varepsilon_W^{-1})}{1 - \alpha + 2\alpha\delta} + 2 \right) \|\nabla g\|_\infty \varepsilon_W. \tag{4.8}$$

Proof Throughout the proof, let us write $\zeta = \|\nabla g\|_\infty \varepsilon_W$ for simplicity. Define the stopping time τ as in (4.4). By Lemma 4.3, we have that

$$u(x) - g(x) \le \mathbb{E}[g(X_\tau) - g(x) \mid X_0 = x].$$

Let us estimate the right-hand side above by fixing some k (which we will choose shortly) and conditioning on $\tau > k$ or $\tau \le k$. For simplicity of notation, we will drop the conditioning $X_0 = x$ below. Then we have

$$\begin{aligned} u(x) - g(x) &\le \mathbb{E}[g(X_\tau) - g(x) \mid \tau \le k]\mathbb{P}(\tau \le k) \\ &\quad + \mathbb{E}[g(X_\tau) - g(x) \mid \tau > k]\mathbb{P}(\tau > k). \end{aligned} \tag{4.9}$$

First, we estimate the probability $\mathbb{P}(\tau > k)$. Since each vertex has a labeled neighbor (by Assumption (**A1**)), each step has probability at least $\frac{1}{2}(1-\alpha) + \alpha\delta$ of hitting Γ and exiting, so a general estimate is

$$\mathbb{P}(\tau > k) \le \left[1 - \tfrac{1}{2}(1-\alpha) - \alpha\delta\right]^k \le e^{-\frac{k}{2}(1-\alpha) - \alpha\delta k}. \tag{4.10}$$

For any k, we can use a telescoping series to obtain

$$g(X_k) - g(x) = \sum_{j=0}^{k-1} (g(X_{j+1}) - g(X_j)) \le \sum_{j=0}^{k-1} \|\nabla g\|_\infty \varepsilon_W = k\zeta.$$

It follows that

$$\mathbb{E}[g(X_\tau) - g(x) \mid \tau \le k] \le k\zeta. \tag{4.11}$$

Substituting (4.10) and (4.11) in (4.9), we have that

$$\begin{aligned} u(x) - g(x) &\le k\zeta \mathbb{P}(\tau \le k) + \mathbb{E}[g(X_\tau) - g(x) \mid \tau > k] e^{-\frac{k}{2}(1-\alpha) - \alpha\delta k} \\ &\le k\zeta + 2\|g\|_\infty e^{-\frac{k}{2}(1-\alpha) - \alpha\delta k}. \end{aligned} \tag{4.12}$$

We want to balance the two terms in the right-hand side of (4.12), so we choose k so that

$$\zeta = \|g\|_\infty e^{-\frac{(1-\alpha)k}{2} - \alpha\delta k},$$

or equivalently

$$k = \frac{2\log(\|g\|_\infty/\zeta)}{1-\alpha+2\alpha\delta}.$$

Therefore, we have

$$u(x) - g(x) \le \zeta(k+2) = \left(\frac{2\log(\|g\|_\infty/\zeta)}{1-\alpha+2\alpha\delta} + 2\right)\zeta. \tag{4.13}$$

To estimate $g - u$, we use a similar argument, where we replace u by $-u$ and g by $-g$. This concludes the proof. □

It turns out we can use the same argument to establish a gradient estimate on the solution u of the p-Laplace equation, which shows that the gradient of u is controlled by the gradient of g.

Theorem 4.6 *Assume (A1) holds, and let x and y be two neighbors on the graph. Then,*

$$\|\nabla u\|_\infty \le \left(\frac{4\log(\|g\|_\infty \|\nabla g\|_\infty^{-1} \varepsilon_W^{-1})}{1-\alpha+2\alpha\delta} + 5\right)\|\nabla g\|_\infty.$$

Proof We apply the triangle inequality

$$|u(x) - u(y)| \le |u(x) - g(x)| + |u(y) - g(y)| + |g(x) - g(y)|$$

and use the bound in Theorem 4.5 twice, noting that $g(x) - g(y) = \nabla g(x, y)\varepsilon_W$. □

4.1.3 Vertex Classification Consistency Results

We now turn to vertex classification results, where the label function $g : \mathcal{X} \to \mathbb{R}$ takes constant values, and hence it will have sharp transitions across edges in the graph. Thus, $\|\nabla g\|_\infty$ may not be small, and so Theorems 4.5 and 4.6 are not applicable. We will focus on binary classification, where every point $x \in \mathcal{X}$ has a label zero or a label one, so $g : \mathcal{X} \to \{0, 1\}$. To use the p-Laplace equation (4.1), we solve the equation with the binary values for the boundary condition $g : \Gamma \to \{0, 1\}$ and threshold the solution $u(x)$ at $\frac{1}{2}$ to obtain a binary label prediction for each vertex $x \in \mathcal{X} \setminus \Gamma$. Thus, we make the following definition.

Definition 4.7 Given $g : \mathcal{X} \to \{0, 1\}$, a point $x \in \mathcal{X}$ is classified *correctly* if $g(x) = \mathbb{1}_{u(x)\ge\frac{1}{2}}$, where u is the solution to (4.1).

In this section, we focus on identifying which points in the graph are classified correctly. This requires a few definitions.

Definition 4.8 A path on the graph is a sequence of edges which joins a sequence of vertices. The length of a path is the number of edges the path consists of.

Definition 4.9 The distance $\mathrm{dist}(x, y)$ between two vertices $x, y \in \mathcal{X}$ is the length of the shortest path connecting the vertices. The distance between a vertex x and a set of vertices $A \subset \mathcal{X}$, denoted $\mathrm{dist}(x, A)$, is the smallest distance between x and any of the vertices $y \in A$.

For $i = 0, 1$, we define

$$\mathcal{X}_i = \{x \in \mathcal{X} : g(x) = i\} \tag{4.14}$$

and note that $\mathcal{X} = \mathcal{X}_0 \cup \mathcal{X}_1$. For any integer $m \geq 0$ and $i = 0, 1$, we define

$$A^i_m = \{x \in \mathcal{X}_i : \mathrm{dist}(x, \mathcal{X}_{1-i}) \leq m\},$$

and note that $A_m = A^0_m \cup A^1_m$. That is to say, A^0_m is the set of points from class 0 that are at most m steps away from the set of class 1 points and vice versa for A^1_m. This means that when m is small, the set A_m contains points that are sufficiently close to the boundary between the two classes that we may expect data points in A_m to be misclassified.

Observe that for any $x \in \mathcal{X}$, there is a positive probability of at least $\frac{1-\alpha}{2} + \alpha\delta = \frac{p-2+2\delta}{2(p-1)}$ of hitting the boundary Γ.. Even when $p = 2$,, this probability is positive, since it is exactly $\delta > 0$. We use this to establish the classification bound below, which shows that whenever a vertex x is sufficiently far, in terms of graph distance, away from the true decision boundary, it will be classified correctly.

Theorem 4.10 *Assume (A1). Let k be the smallest integer strictly larger than*

$$\frac{2\log(2)}{1-\alpha+2\alpha\delta} = \frac{2(p-1)\log(2)}{p-2+2\delta}. \tag{4.15}$$

Then every $x \notin A_k$ is classified correctly.

Proof Let $x \in \mathcal{X} \setminus A_k$, and assume $X_0 = x$. Define the stopping time τ as in (4.4). For any point $x \in \mathcal{X}$, we have that

$$\begin{aligned} u(x) - g(x) \leq\; & \mathbb{E}[g(X_\tau) - g(x) \mid \tau \leq k]\mathbb{P}(\tau \leq k) \\ & + \mathbb{E}[g(X_\tau) - g(x) \mid \tau > k]\mathbb{P}(\tau > k). \end{aligned} \tag{4.16}$$

We want to classify x correctly whether it belongs to $\mathcal{X}_0$ or $\mathcal{X}_1$. Recall that the labels of vertices in $\mathcal{X}_0$ are zeros and in $\mathcal{X}_1$ are ones.

Since $x \notin A_k$, the labels of vertices reached by the stopping time $\tau \leq k$ can only be labels from the same class, so $g(X_\tau) = g(x)$. Thus, when $\tau \leq k$, the first term on the right-hand side of (4.16) is zero. Then, using that g is between zero and one, and that Eq. (4.10) holds, we obtain

$$u(x) - g(x) \leq \mathbb{E}[g(X_\tau) - g(x) \mid \tau > k]\mathbb{P}(\tau > k) \leq \mathbb{P}(\tau > k) \leq e^{-\frac{k}{2}(1-\alpha)-\alpha\delta k} \tag{4.17}$$

The analogous estimate on $g - u$ is similarly obtained, and so we have

$$|u(x) - g(x)| \leq e^{-\frac{k}{2}(1-\alpha)-\alpha\delta k}.$$

For *correct* classification of x, we need precisely $|u(x) - g(x)| < \frac{1}{2}$, which is equivalent to

$$e^{-\frac{k}{2}(1-\alpha)-\alpha\delta k} < \frac{1}{2},$$

and is satisfied by the choice of k in (4.15)

□

4.2 Geometric Graphs

In this section, we specialize our results to geometric graphs, which include the very commonly used random geometric ε-graphs and k-nearest neighbor (k-NN) graphs. The purpose of this section is to provide conditions under which ε-graphs and k-NN graphs, along with other graphs with geometric ε-scaling, fulfill the general assumption **(A1)** in Sect. 4.1.1, so that we can apply the theory developed in Sect. 4.1.1 to these graphs.

4.2.1 Main Assumptions

In order to construct ε-graphs and k-NN graphs and construct the labeled data set Γ, we need the following assumptions. Note that assumptions **(A2)**, **(A3)**, and **(A4)** also appear in [28].

(A2) Let $d \geq 2$ and let $\Omega \subset \mathbb{R}^d$ be an open, bounded, and connected domain with a Lipschitz boundary. Let $x_1, x_2, \ldots, x_n$ be *i.i.d.* with continuous density ρ satisfying

$$0 < \rho_{\min} := \inf_{x\in\Omega} \rho(x) \leq \sup_{x\in\Omega} p(x) =: \rho_{\max} < \infty.$$

We define $\mathcal{X} = \{x_1, x_2, \ldots, x_n\}$.

(A3) We construct Γ in the following way. Let $g \in \mathrm{Lip}(\overline{\Omega})$, where Ω is defined as in **(A2)**. Let $\beta \in (0, 1)$. For each $x \in \mathcal{X}$, define an independent random variable $z(x) \sim \mathit{Bernoulli}(\beta)$, and assign $x \in \Gamma$ whenever $z(x) = 1$. Each training data point $x \in \Gamma$ is assigned a label $g(x)$.

(A4) We define an interaction potential $\eta : [0, \infty) \to [0, \infty)$. The interaction potential $\eta(\cdot)$ has support on the interval $[0, 1]$, it is nonincreasing, and η restricted to $[0, 1]$ is Lipschitz continuous. For $t \geq 1$ we extend η so that $\eta(t) = 0$. Moreover, $\eta(\frac{1}{2}) > 0$ and η integrate to 1 :

$$\int_{\mathbb{R}^d} \eta(|x|)dx = 1.$$

We define $\eta_\varepsilon(t) = \frac{1}{\varepsilon^d}\eta(t/\varepsilon)$.

(A5) (Geometric graphs of length ε) There exist positive constants $c_1, c_2, \varepsilon_1, \varepsilon_2$ such that for every $x, y \in \mathcal{X}$ we have that

$$c_1\varepsilon_1^{-d}\mathbb{1}_{|x-y|<\varepsilon_1} \leq w_{xy} \leq c_2\varepsilon_2^{-d}\mathbb{1}_{|x-y|<\varepsilon_2}.$$

We note that assumptions **(A2)** and **(A4)** are used in the ε−graph construction (see Sect. 4.2.2). In Lemma 4.12, we prove that assumption **(A5)** holds for ε-graphs. Assumption **(A3)** has to do with how the labeled data points are chosen, and we note that the parameter $\beta \in (0, 1)$ is the *label rate.*

Assumption **(A2)** is satisfied for k-NN-graphs by construction. In Lemma 4.16, we prove that assumption **(A5)** holds with high probability for k-NN graphs. Applying **(A5)**, we prove **(A1)** holds with high probability, which means that Theorems 4.5 and 4.6 hold. For $x \in \mathcal{X}$ we define $d_n(x)$ and $p_n(x)$ as follows:

$$d_n(x) = \Sigma_{y\in\mathcal{X}} w_{xy}, \tag{4.18}$$

$$p_n(x) = \Sigma_{y\in\mathcal{X}} w_{xy}\mathbb{1}_{y\in\Gamma}. \tag{4.19}$$

Then, as before, we define

$$\delta = \min_{x\in\mathcal{X}} \frac{p_n(x)}{d_n(x)} = \min_{x\in\mathcal{X}} \frac{\Sigma_{y\in\mathcal{X}} w_{xy}\mathbb{1}_{y\in\Gamma}}{\Sigma_{z\in\mathcal{X}} w_{xz}}. \tag{4.20}$$

4.2.2 The ε-Graphs

Let $0 < \varepsilon \leq 1$. We construct our ε-graph as follows.

Definition 4.11 We define an ε-graph according to the following rules. The vertices $\mathcal{X} = \{x_1, \ldots x_n\}$ are constructed according to **(A2)**, and the labeled set Γ is chosen according to **(A3)**. We assume **(A4)**, and the symmetric edges $w_{xy} = w_{yx}$ are obtained as follows:

$$w_{xy} = \eta_\varepsilon(|x - y|) \tag{4.21}$$

Lemma 4.12 *ε-Graphs satisfy assumption **(A5)** with constants $c_1 = 2^{-d}\eta(\frac{1}{2})$, $c_2 = \sup \eta$, $\varepsilon_1 = \frac{\varepsilon}{2}$, and $\varepsilon_2 = \varepsilon$.*

Proof By their construction, $\varepsilon-$graphs have the following property:

$$w_{xy} = \varepsilon^{-d}\eta\left(\frac{|x-y|}{\varepsilon}\right) \in [\varepsilon^{-d}\eta(\tfrac{1}{2})\mathbb{1}_{|x-y|<\frac{\varepsilon}{2}}, \varepsilon^{-d}\sup\eta\mathbb{1}_{|x-y|<\varepsilon}] \tag{4.22}$$

$$\equiv [(\varepsilon/2)^{-d}(2^{-d}\eta(\tfrac{1}{2}))\mathbb{1}_{|x-y|<\frac{\varepsilon}{2}}, \varepsilon^{-d}\sup\eta\mathbb{1}_{|x-y|<\varepsilon}]. \tag{4.23}$$

So **(A5)** holds, with the constants stated in the Lemma. □

4.2.3 The k-Nearest Neighbors Graphs

We follow [23] in constructing a k-NN graph. We fix a number $k \in \mathbb{N}$, such that $1 \leq k \leq n-1$. The number of neighbors $N_\varepsilon(x)$ of x in an ε-ball is

$$N_\varepsilon(x) := \sum_{j:0<|x_j-x|\leq\varepsilon} 1$$

and the distance to the k-nearest neighbor is

$$\varepsilon_k(x) := \min\{\varepsilon > 0 \ : N_\varepsilon(x) \geq k\}.$$

Definition 4.13 We define a relation $\sim_k$ on $\mathcal{X} \times \mathcal{X}$ by declaring

$$x_i \sim_k x_j$$

if x_j is among the k nearest neighbors of x_i.

Definition 4.14 For the *symmetric* k-nearest neighbor graph, we construct the vertices according to **(A2)**. If $x_i \sim_k x_j$ **or** $x_j \sim_k x_i$, we place an edge between x_i and x_j. The edges are symmetric, meaning $w_{xy} = w_{yx}$ and

$$w_{xy} = \eta_{\max\{\varepsilon_k(x),\varepsilon_k(y)\}}(|x-y|).$$

Definition 4.15 In the *mutual* k-nearest neighbor graph, we construct the vertices according to **(A2)**. If $x_i \sim_k x_j$ **and** $x_j \sim_k x_i$, we place an edge between x_i and x_j. The edges are symmetric, meaning $w_{xy} = w_{yx}$ and

$$w_{xy} = \eta_{\min\{\varepsilon_k(x),\varepsilon_k(y)\}}(|x-y|).$$

In the following lemma, we verify that **(A5)** holds for k-NN graphs.

Lemma 4.16 *There exists a* $c > 0$, *such that symmetric k-NN graphs and mutual k-NN graphs satisfy assumption* **(A5)** *with probability* $1 - 4\exp(-\frac{c}{2}(k/n)^{2/d}k)$, *provided k is not too large, meaning*

$$1 \le k \le cn\varepsilon^d$$

holds.

Proof Consider $x \in \mathcal{X}$—a vertex in the $k-$NN graph. We set

$$\gamma = \frac{1}{\sqrt{2}}(k/n)^{1/d}.$$

Let α_d be the volume of the d-dimensional Euclidean unit ball.

We apply Lemma 3.8 from [23] to estimate the expected value of the number of neighbors: we have that

$$\mathbb{P}(|\alpha_d \rho(x) n \varepsilon_k(x)^d - k| \ge \tfrac{1}{2}k) \le 4\exp(-c\gamma^2 k).$$

Thus, with probability $1 - 4\exp(-c\gamma^2 k)$, we obtain

$$\alpha_d \rho(x) n \varepsilon_k^d \in \left[\tfrac{k}{2}, \tfrac{3k}{2}\right],$$

or equivalently

$$\left(\frac{k}{2\alpha_d \rho(x) n}\right)^{1/d} \le \varepsilon_k \le \left(\frac{3k}{2\alpha_d \rho(x) n}\right)^{1/d}.$$

Denote $m := \inf_{x \in X} (\frac{k}{2\alpha_d \rho(x) n})^{1/d}$ and $M := \sup_{x \in X} (\frac{3k}{2\alpha_d \rho(x) n})^{1/d}$. Thus, with probability at least $1 - 4\exp(-c\gamma^2 k)$, we have:

$$m \le \varepsilon_k \le M.$$

In order to obtain the constants in **(A5)**, we bound w_{xy} from below and above. Note that ε_k comes from the vertex x or the vertex y and that η is a nonincreasing function.

We start with the lower bound of w_{xy}:

$$\begin{aligned} w_{xy} = \frac{1}{\varepsilon_k^d}\eta\left(\frac{|x-y|}{\varepsilon_k}\right) &= \frac{1}{\varepsilon_k^d}\eta\left(\frac{|x-y|}{\varepsilon_k}\right)\mathbb{1}_{|x-y|\le\varepsilon_k} \\ &= \frac{1}{\varepsilon_k^d}\eta\left(\frac{|x-y|}{\varepsilon_k}\right)\mathbb{1}_{|x-y|\le m/2} \\ &\quad + \frac{1}{\varepsilon_k^d}\eta\left(\frac{|x-y|}{\varepsilon_k}\right)\mathbb{1}_{|x-y|\in[m/2,\varepsilon_k]} \end{aligned}$$

$$
\begin{aligned}
&\geq \frac{1}{\varepsilon_k^d}\eta\left(\frac{m/2}{\varepsilon_k}\right)\mathbb{1}_{|x-y|\leq m/2}\\
&\geq \frac{1}{\varepsilon_k^d}\eta\left(\frac{1}{2}\right)\mathbb{1}_{|x-y|\leq m/2}\\
&= \eta\left(\frac{1}{2}\right)\left(\frac{m}{2\varepsilon_k}\right)^d\left(\frac{m}{2}\right)^{-d}\mathbb{1}_{|x-y|\leq m/2}\\
&\geq \eta\left(\frac{1}{2}\right)\left(\frac{m}{2M}\right)^d\left(\frac{m}{2}\right)^{-d}\mathbb{1}_{|x-y|\leq m/2}.
\end{aligned}
$$

We choose $\varepsilon_1 = \frac{m}{2}$ and $c_1 = \eta(\frac{1}{2})(\frac{m}{2M})^d$.

Next, we compute the upper bound for the edge weight:

$$
\begin{aligned}
w_{xy} = \frac{1}{\varepsilon_k^d}\eta\left(\frac{|x-y|}{\varepsilon_k}\right) &= \frac{1}{\varepsilon_k^d}\eta\left(\frac{|x-y|}{\varepsilon_k}\right)\mathbb{1}_{|x-y|\leq\varepsilon_k}\\
&\leq \varepsilon_k^{-d}\max\eta\mathbb{1}_{|x-y|\leq\varepsilon_k}\\
&= M^{-d}\left(\frac{M}{\varepsilon_k}\right)^d\max\eta\mathbb{1}_{|x-y|\leq\varepsilon_k}\\
&\leq M^{-d}\left(\frac{M}{m}\right)^d\max\eta\mathbb{1}_{|x-y|\leq M}\\
&\leq M^{-d}\mathbb{1}_{|x-y|\leq M}\left\{\left(\frac{M}{m}\right)^d\max\eta\right\}.
\end{aligned}
$$

Now, we choose $c_2 = \left\{\left(\frac{M}{m}\right)^d\max\eta\right\}$ and $\varepsilon_2 = M$. This establishes all the constants in **(A5)**, which concludes the proof. □

4.2.4 General Geometric Graphs

Now that we have shown that ε-graphs and k-NN graphs satisfy **(A5)**, and we proceed to study general geometric graphs, which are essentially only assumed to satisfy **(A5)**. This covers a wide variety of geometrically constructed graphs, including, but not limited to, ε-graphs and k-NN graphs.

Theorem 4.17 *Assume that* ***(A2)**,**(A3)**, **(A4)**, **(A5)*** *hold. Consider the labeling rate β and δ defined as in* (4.20). *Then, there exist positive constants c_3, c_4, and c such that $\delta\geq c\beta$ and* ***(A1)*** *hold with probability at least*

$$1 - n\exp(-c_3 n\varepsilon_2^d) - n\exp(-c_4\beta n\varepsilon_1^d).$$

Here c_3, c_4, and c only depend on η, d, Ω, and ρ.

Proof Fix $x \in \mathcal{X}$. Because of **(A5)**, we have that:

$$d_n(x) = \sum_{y \in N_x} w_{xy} \leq \sum_{y \in \mathcal{X} \setminus x} c_2 \mathbb{1}_{|x-y| \leq \varepsilon_2} \varepsilon_2^{-d} := \Phi$$

Let us consider all the vertices in $\mathcal{X} \setminus x$ and treat them as random variables, calling them $Y_1, \ldots, Y_{n-1}$. Define $\psi_1(Y_i) = c_2 \mathbb{1}_{|x-Y_i| \leq \varepsilon_2} \varepsilon_2^{-d}$ and then it follows that

$$\Phi = \sum_{i=1}^{n-1} \psi_1(Y_i) = \sum_{i=1}^{n-1} c_2 \mathbb{1}_{|x-Y_i| \leq \varepsilon_2} \varepsilon_2^{-d}.$$

Thus,

$$\mathbb{E}[\Phi] = \sum_{Y_i \in \mathcal{X} \setminus x} \mathbb{E}[c_2 \mathbb{1}_{|x-Y_i| \leq \varepsilon_2} \varepsilon_2^{-d}] = (n-1)c_2 \int_{\Omega \cap B(x,\varepsilon_2)} \varepsilon_2^{-d} \rho(y) dy.$$

We apply Bernstein's inequality (30) in [20] to Φ obtaining that

$$\mathbb{P}(|\Phi - \mathbb{E}(\Phi)| > c_d(n-1)) \leq 2 \exp\left(\frac{-(n-1)c_d^2}{2(\sigma^2 + \frac{1}{3} b c_d)} \right)$$

for $\sigma^2 = Var(\psi_1(Y_i))$ and $b = \sup |\psi_1(Y_i)|$. We observe that $b = \sup |c_2 \mathbb{1}_{|x-Y_i| \leq \varepsilon_2} \varepsilon_2^{-d}| = c_2 \varepsilon_2^{-d}$ and also

$$\sigma^2 = Var(c_2 \mathbb{1}_{|x-Y_i| \leq \varepsilon_2} \varepsilon_2^{-d}) = c_2^2 \varepsilon_2^{-2d} Var(\mathbb{1}_{|x-Y_i| \leq \varepsilon_2}) \sim C \varepsilon_2^{-d},$$

because $Var(\mathbb{1}_{|x-Y_i| \leq \varepsilon_2})$ is of order ε_2^d as we integrate over a ball of radius ε_2. Thus, σ^2 and b depend on Ω, ρ, c_2, as well as on ε_2 explicitly.

We union-bound, obtaining that there exists two positive constants c_5 and c_3 for which

$$\Phi \leq c_d(n-1) + (n-1)c_2 \int_{\Omega \cap B(x,\varepsilon_2)} \varepsilon_2^{-d} \rho(y) dy = c_5(n-1)$$

holds with probability at least $1 - n \exp\{-c_3 n \varepsilon_2^d\}$. The same line of reasoning applies to $p_n(x)$ defined in (4.19). We define

$$\psi_2(Y_i) := c_1 \mathbb{1}_{|x-Y_i| \leq \varepsilon_1} \varepsilon_1^{-d} \mathbb{1}_{z(Y_i)=1}$$

and $\hat{\Phi} = \sum_{i=1}^{n} \psi_2(Y_i)$. We compute

$$\mathbb{E}[\hat{\Psi}] = \sum_{Y_i \in \mathcal{X}' \setminus x} \mathbb{E}[c_1 \mathbb{1}_{|x-Y_i| \le \varepsilon_1} \varepsilon_1^{-d} \mathbb{1}_{z(Y_i)=1}] = (n-1) c_1 \beta \int_{\Omega \cap B(x, \varepsilon_1)} \varepsilon_1^{-d} \rho(y) dy.$$

Thus there exists two positive constants c_6 and c_4 such that with probability $1 - n \exp(-c_4 \beta n \varepsilon_1^d)$ we have

$$p_n(x) \ge (n-1)\beta c_6.$$

Thus, there exist constants c_5, c_6 such that

$$d_n(x) < c_5(n-1) \tag{4.24}$$

and

$$p_n(x) \ge c_6 \beta (n-1) \tag{4.25}$$

hold for all all x with probability $1 - n\exp(-c_3 n \varepsilon_2^d) - n \exp(-c_4 \beta n \varepsilon_1^d)$.

We use the results we just obtained for (4.24) and (4.25) to estimate that

$$\delta = \min_{x \in \mathcal{X}} \frac{p_n(x)}{d_n(x)} \ge \min \frac{c_5(n-1)\beta}{c_6(n-1)} \ge c\beta. \tag{4.26}$$

We now focus on proving **(A1)**. Pick any vertex $x \in X$. Since by definition

$$p_n(x) = \Sigma_{y \in \mathcal{X}} w_{xy} \mathbb{1}_{z(y)=1}$$

is a sum of nonnegative quantities, and by (4.25), there exists a neighbor x_ℓ of x such that

$$w_{x x_\ell} \mathbb{1}_{z(x_\ell)=1} > 0.$$

Thus, $z_\ell = 1$ for some vertex x_ℓ, therefore x_ℓ is labeled, and so x has a neighbor in Γ. This is a restatement of **(A1)**. This concludes the proof. □

In particular, with probability at least $1 - n\exp(-c_3 n \varepsilon_2^d) - n\exp(-c_4 \beta n \varepsilon_1^d)$ assumption **(A1)** holds for k-NN graphs and ε-graphs. Therefore, Theorems 4.5 and 4.10 also hold (with the same probability) for ε-graphs and k-NN graphs.

Corollary 4.18 *For an ε-graph and $g \in Lip$, we have that*

$$|u(x) - g(x)| \le \varepsilon \operatorname{Lip}(g) \Big(\frac{2 \log \frac{\sup g}{\varepsilon \operatorname{Lip}(g)}}{1 - \alpha + 2\alpha\delta} + 2 \Big)$$

holds with with probability at least $1 - n\exp(-c_3 n \varepsilon^d) - n\exp(-c_4 \beta n \varepsilon^d)$.

Proof The statement of Corollary 4.18 holds from Theorem 4.5, the observation $|g(z) - g(y)| \leq \mathrm{Lip}(g)|z - y|$, and the fact that the distance between nearby nodes $|z - y|$ is of size ε. □

4.3 Stochastic Block Model Graphs

In this section, we apply our ideas and results from Sect. 4.1.1 to the stochastic block model (SBM) graph. Since SBM graphs have *discrete* geometric structure, as opposed to the *continuous* geometric structure of a random geometric graph, there is no obvious relationship to a continuum PDE on an SBM graph. Nevertheless, the tug-of-war lemma provides useful results in this setting.

We first define an SBM graph. The edge weights are binary $w_{xy} = 1$ if there is an edge between x and y and $w_{xy} = 0$ if there is no edge. We partition the vertices $\mathcal{X}$ of the SBM graph into two disjoint blocks $\mathcal{X} = \mathcal{X}_0 \cup \mathcal{X}_1$ according to an underlying binary label function $g : \mathcal{X} \to \{0, 1\}$, as in (4.14). We define $r, q \in (0, 1)$ as the two key probabilities for an SBM graph: r is the probability of an *intra-class* edge, and q is the probability of an *interclass* edge. In other words,

$$\mathbb{P}(w_{xy} = 1) = r \quad \text{if } x, y \in \mathcal{X}_i \text{ for } i = 0 \text{ or } i = 1$$

$$\mathbb{P}(w_{xy} = 1) = q \quad \text{if } x \in \mathcal{X}_i \text{ and } y \in \mathcal{X}_{1-i} \text{ for } i = 0 \text{ or } i = 1.$$

By **(A3)**, the probability that any vertex is labeled is β; thus, only a portion $\Gamma \in \mathcal{X}$ is labeled. The rest of $\mathcal{X}$ is unlabeled, and the goal is to label the points in $\mathcal{X}_0$ and $\mathcal{X}_1$ correctly. We also define $N_0 = |\mathcal{X}_0|$, $N_1 = |\mathcal{X}_1|$ and $n = |\mathcal{X}| = N_0 + N_1$. Let us set $\Gamma_i = \Gamma \cap \mathcal{X}_i$ so that $\Gamma = \Gamma_0 \cup \Gamma_1$.

Now, let u be the solution of the p-Laplace equation (4.1), and define the two sets $\mathcal{Y}_0$ and $\mathcal{Y}_1$ as follows: We assign x to $\mathcal{Y}_0$ whenever $u(x) \leq \frac{1}{2}$,, and we assign x to $\mathcal{Y}_1$ whenever $u(x) > \frac{1}{2}$. That is, $\mathcal{Y}_0$ is the set of vertices classified as class 0 by p-Laplace learning, and $\mathcal{Y}_1$ is the set classified as class 1. In order to classify correctly, we need $\mathcal{Y}_0 = \mathcal{X}_0$ and $\mathcal{Y}_1 = \mathcal{X}_1$.

Consider a vertex $x_i \in \mathcal{X}$. Let γ_i be the ratio of the number of other-class neighbors to the number of neighbors of vertex x_i, that is, if $x_i \in \mathcal{X}_j$, $j \in \{0, 1\}$, then

$$\gamma_i = \frac{\sum_{x \in \mathcal{X}_{1-j}} w_{x_i, x}}{\sum_{x \in \mathcal{X}} w_{x_i, x}}. \tag{4.27}$$

For a vertex x_i, the quantity γ_i is exactly the probability that a random walk step will jump to the opposite block. Let β_i be the fraction of neighbors from the same class that are labeled. Thus, if $x_i \in \mathcal{X}_j$ with $j \in \{0, 1\}$, then

$$\beta_i = \frac{\sum_{x \in \mathcal{X}_j \cap \Gamma} w_{x_i,x}}{\sum_{x \in \mathcal{X}_j} w_{x_i,x}}. \tag{4.28}$$

The term β_i is the probability that the random walk exits at at labeled node in the same class on the next step, conditioned on the event that the random walk remains in the same class.

We now derive upper and lower bounds for γ_i and β_i. The lower bound for β_i implies that **(A1)** holds with high probability, but we do not use **(A1)** directly in this section.

Lemma 4.19 *Let us consider an SBM graph with parameters N_0, N_1, r, q, and label rate β. Let $\sigma_1 \geq 0$ and $0 \leq \sigma_2 < 1$. Then, with probability of at least*

$$1 - \sum_{j=0}^{1} N_j \left\{ \exp\left(-\frac{qN_{1-j}\sigma_1^2}{2(1+\frac{1}{3}\sigma_1)} \right) - 3\exp\left(-\frac{1}{8}\beta r N_j \sigma_2^2 \right) \right\}. \tag{4.29}$$

for any $x_i \in \mathcal{X}_j$, $j \in \{0, 1\}$, we have

$$\gamma_i \leq \frac{(1+\sigma_1)qN_1}{(1+\sigma_1)qN_1 + (1-\sigma_2)r(N_0-1)}, \tag{4.30}$$

and

$$\beta_i \geq \frac{1-\sigma_2}{1+\sigma_2}\beta. \tag{4.31}$$

Proof Pick a vertex $x_i \in \mathcal{X}_0$. Define the following notation:

$$Y_j = \begin{cases} 1, & \text{if } x_i \text{ is connected to } x_j \in \mathcal{X}_1 \\ 0, & \text{otherwise.} \end{cases}$$

$$Z_j = \begin{cases} 1, & \text{if } x_i \text{ is connected to } x_j \in \mathcal{X}_0 \\ 0, & \text{otherwise.} \end{cases}$$

Denote $A = \sum_{j=1}^{N_1} Y_j$, $B = \sum_{j=1}^{N_0} Z_j$, and note that by (4.27), we have

$$\gamma_i = \frac{A}{A+B} = \frac{1}{1+C}, \quad C = \frac{B}{A}.$$

We now apply the Chernoff bounds (see, e.g., [13]) for A and B to obtain

$$\mathbb{P}(A \geq (1+\sigma_1)qN_1) \leq \exp\left(-\frac{qN_1\sigma_1^2}{2(1+\frac{1}{3}\sigma_1)}\right)$$

$$\mathbb{P}\left(|B - r(N_0-1)| \leq \sigma_2 r(N_0-1)\right) \leq 2\exp\left(-\frac{3}{8}r(N_0-1)\sigma_2^2\right),$$

for any $\sigma_1 > 0$ and $0 \leq \sigma_2 < 1$. Therefore,

$$C \geq \frac{(1-\sigma_2)r(N_0-1)}{(1+\sigma_1)qN_1}$$

holds with probability at least

$$1 - \exp\left(-\frac{qN_1\sigma_1^2}{2(1+\frac{1}{3}\sigma_1)}\right) - 2\exp\left(-\frac{3}{8}r(N_0-1)\sigma_2^2\right).$$

Assuming this event holds, we then have

$$\gamma_i \leq \frac{1}{1+C} \leq \frac{(1+\sigma_1)qN_1}{(1+\sigma_1)qN_1 + (1-\sigma_2)r(N_0-1)} = \frac{(1+\sigma)qN_1}{(1+\sigma)qN_1 + r(N_0-1)},$$

which establishes (4.30), upon union bounding over $\mathcal{X}_0$.

Now, let V_i be *i.i.d.* Bernoulli(β) for $i = 1, \ldots, n$ so that $x_i \in \Gamma$ if and only if $V_i = 1$. Then

$$\beta_i = \frac{1}{B}\sum_{j=1}^{n} Z_j V_j.$$

Since $Z_j V_j$ is Bernoulli(βr), we have

$$\mathbb{P}\left(\sum_{j=1}^{n} Z_j V_j \leq (1-\sigma_2)\beta r(N_0-1)\right) \leq \exp\left(-\frac{3}{8}\beta r(N_0-1)\sigma_2^2\right).$$

Assuming this event holds as well yields

$$\beta_i \geq \frac{(1-\sigma_2)\beta r(N_0-1)}{(1+\sigma_2)r(N_0-1)} = \frac{1-\sigma_2}{1+\sigma_2}\beta,$$

which establishes, with a union bound, (4.31). The argument is similar for $x_i \in \mathcal{X}_1$. The probabilities are simplified by combining like terms and using that $N_i - 1 \geq N_i/2$ since $N_i \geq 2$. □

Now we prove the main result for SBM graphs.

Theorem 4.20 *Let* $2 \leq p < \infty$, *and consider an SBM graph with parameters* r, q, N_0, N_1, *and label rate* β. *Let* $\sigma_1 \geq 0$, $0 \leq \sigma_2 < 1$, *and assume that*

$$\frac{r}{q}\beta > (1+\sigma)\max\left\{\frac{N_0}{N_1-1}, \frac{N_1}{N_0-1}\right\}, \tag{4.32}$$

where

$$1+\sigma = \frac{(1+\sigma_1)(1+\sigma_2)}{(1-\sigma_2)^2}. \tag{4.33}$$

Then the solution u *to* (4.1) *classifies all vertices correctly with probability at least* (4.29).

Remark 4.21 Note that the condition $p < \infty$ ensures that $\alpha > 0$ in (4.2), and thus there is always a positive probability of taking a random walk step. The proof of Theorem 4.20 relies entirely on the random walk and allows the two players to balance each other out. We expect this is suboptimal and is why the value of p does not explicitly enter the condition on $\frac{r}{q}$.

Proof of Theorem 4.20 We fix $\sigma_1 \geq 0$ and $0 \leq \sigma_2 \leq \frac{1}{2}$ and assume that the conclusions of Lemma 4.19 hold. We consider the stochastic tug-of-war process $X_0, X_1, \ldots$, starting from $X_0 = x \in \mathcal{X}_0$. We define the associated random variables W_k by

$$W_k = \begin{cases} 0, & \text{if } X_k \in \Gamma_0 \\ 1, & \text{if } X_k \in \mathcal{X}_1 \\ 2, & \text{if } X_k \in \mathcal{X}_0 \setminus \Gamma_0. \end{cases}$$

The random variables W_k indicate whether the process stops at Γ_0, moves to $\mathcal{X}_1$ or stays in $\mathcal{X}_0$ and does not exit. The probabilities of each of these transitions depend on the vertex the process is currently at. If $X_{k-1} = x_i \in \mathcal{X}_0$, then

$$\begin{aligned} p_{0,i} &:= \mathbb{P}(W_k = 0) = \frac{1-\alpha}{2} + \alpha(1-\gamma_i)\beta_i, \\ p_{1,i} &:= \mathbb{P}(W_k = 1) < \frac{1-\alpha}{2} + \alpha\gamma_i, \\ p_{2,i} &:= \mathbb{P}(W_k = 2) \geq \alpha(1-\gamma_i)(1-\beta_i), \end{aligned}$$

where γ_i is defined in (4.27), and $\beta_i > 0$ is fraction of same-class neighbors of x_i that are labeled. Let τ be the stopping time defined by

$$\tau = \inf\{k : W_k = 0 \text{ or } W_k = 1\}.$$

Let us define

$$p_0 = \min_{1\le i\le n} p_{0,i} = \frac{1-\alpha}{2} + \alpha(1-\gamma_{max})\beta_{min} \quad \text{and}$$

$$p_1 = \max_{1\le i\le n} p_{1,i} = \frac{1-\alpha}{2} + \alpha\gamma_{max},$$

where we write

$$\gamma_{max} = \max_{1\le i\le n} \gamma_i \quad \text{and} \quad \beta_{min} = \min_{1\le i\le n} \beta_i.$$

By Doob's optional stopping theorem and Lemma 4.2, we have $u(x) \le \mathbb{E}[u(X_\tau)]$. Note that the expected value of τ is finite, since the probability of stopping at each step is $p_{0,i} + p_{1,i} \ge p_0 > 0$.

Let i_k denote the vertex index of the stochastic tug-of-war process on the kth step, so that $X_k = x_{i_k}$. Then conditioned on $i_{\tau-1} = i$, the probability of $W_\tau = 1$ is exactly $p_{1,i}/(p_{0,i} + p_{1,i})$. Therefore by the law of conditional probability, and the fact that $x_{i_{\tau-1}} \in \mathcal{X}_0$, we have

$$\begin{aligned} u(x) \le \mathbb{E}[u(X_\tau)] \le \mathbb{P}(X_\tau \in \mathcal{X}_1) &= \sum_{i=1}^n \frac{p_{1,i}}{p_{0,i} + p_{1,i}} \mathbb{P}(i_{\tau-1} = i) \\ &= \sum_{x_i \in \mathcal{X}_0} \frac{p_{1,i}}{p_{0,i} + p_{1,i}} \mathbb{P}(i_{\tau-1} = i) \le \frac{p_1}{p_0 + p_1}. \end{aligned}$$

Since $x \in \mathcal{X}_0$, its correct label is 0, so in order to classify x correctly, we need to show that $u(x) < \frac{1}{2}$, which holds whenever $p_0/p_1 > 1$, or rather, $p_0 > p_1$. This is equivalent to

$$\beta_{min} > \frac{\gamma_{max}}{1-\gamma_{max}}. \tag{4.34}$$

By Lemma 4.19, we have

$$\gamma_{max} \le \frac{(1+\sigma_1)qN_1}{(1+\sigma_1)qN_1 + (1-\sigma_2)r(N_0-1)},$$

and so

$$1 - \gamma_{max} \ge \frac{(1-\sigma_2)r(N_0-1)}{(1+\sigma_1)qN_1 + (1-\sigma_2)r(N_0-1)}.$$

Therefore,

$$\frac{\gamma_{max}}{1-\gamma_{max}} \le \frac{(1+\sigma_1)qN_1}{(1-\sigma_2)r(N_0-1)}.$$

Also, by Lemma 4.19, we have

$$\beta_{min} \geq \frac{1-\sigma_2}{1+\sigma_2}\beta.$$

Thus, a sufficient condition for (4.34) to hold is

$$\frac{1-\sigma_2}{1+\sigma_2}\beta > \frac{(1+\sigma_1)qN_1}{(1-\sigma_2)r(N_0-1)},$$

which is equivalent to

$$\frac{r}{q}\beta > (1+\sigma)\frac{N_1}{N_0-1},$$

where σ is defined in (4.33). A similar argument holds for $x \in \mathcal{X}_1$, except with N_0 and N_1 reversed. This concludes its proof. □

4.4 Numerical Results

We present here some numerical results with synthetic and real data, in order to illustrate the main results established in this section, and the room for improvement in future work. The Python code to reproduce the results in this section is available on GitHub.[8]

Our first experiment is with a stochastic block model graph to illustrate Theorem 4.20. We take a graph with $n = 3000$ vertices and consider the case of equal block sizes $N_0 = N_1 = 1500$ and unbalanced block sizes $N_0 = 2000$, $N_1 = 1000$. We take the intraclass probability to be $r = 0.5$, and we vary the interclass probability $q \leq r$. In Fig. 4, we show the error rates for binary classification using a labeling rate of $\beta = 0.2$ as a function of the ratio r/q. For each value of r/q, we ran 100 randomized trials and averaged the error rates. For equal size blocks $N_0 = N_1$, we see that the error rate decreases rapidly when $r/q > 1$. Theorem 4.20 guarantees that all vertices are classified correctly when $r/q > \beta^{-1} = 5$, which is a pessimistic bound in this case, since all values of p classify correctly when $r/q = 2$. For unequal block sizes, $N_0/N_1 - 2$ in Fig. 4b, we see that for larger values of p, the error rate decreases quickly when $r/q > 2$, while $p = 2$ requires $r/q > 8$ to see a similar decrease and $r/q = 10$ to classify correctly. In this case, Theorem 4.20 guarantees correct classification with high probability when $r/q > 2\beta^{-1} = 10$, which agrees very closely with the $p = 2$ result but is pessimistic for $p > 2$. Thus, our results may be tight for $p = 2$, but there is clearly much room for improvement in the range $p > 2$. In particular, we currently cannot explain why the results in Fig. 4b

[8] https://github.com/jwcalder/p-Laplace-consistency

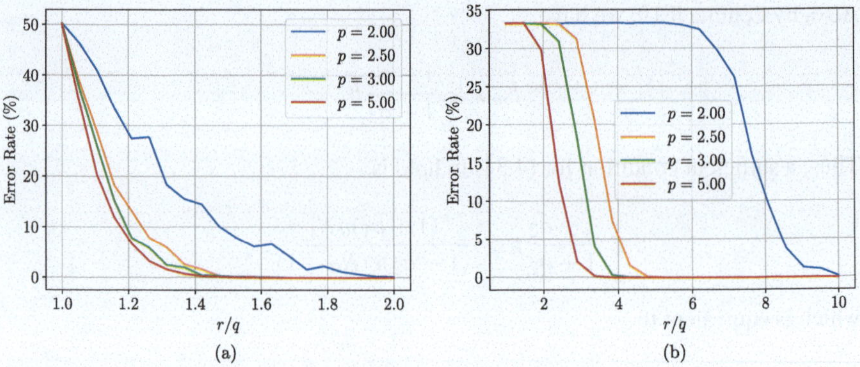

Fig. 4 Example of the effect of r/q on the classification performance on stochastic block model graphs for various values of p. (**a**) $N_0 = N_1 = 1500$. (**b**) $N_0 = 2000$, $N_1 = 1000$

(a) (b)

Fig. 5 Sample of MNIST 4s and 9s, and Cifar-10 deer and dogs. (**a**) MNIST. (**b**) Cifar-10

are so dramatically better as p increases. An analysis of this sort would presumably require a different approach to Theorem 4.20 that exploited the tug-of-war game as well.

To illustrate our results on geometric graphs, we turn to real data. We consider the MNIST data set which consists of 70,000 images of handwritten digits between 0 and 9 [72], with each image a 28×28 grayscale image. We also consider the Cifar-10 data set [70] which contains 60,000 natural images from ten classes (airplane, automobile, bird, cat, deer, dog, frog, horse, ship, truck). Each Cifar-10 image is a 32×32 pixel color image. To make the problem more similar to our setting, we restrict the data sets from ten classes down to two, so we have a binary classification problem. For MNIST, we use the 4s and 9s, which are one of the harder pairs of digits to separate, while for Cifar-10, we use the deer and dogs. Figure 5 shows some example images from each data set.

In order to obtain a good feature embedding for constructing the graph, we used a variational autoencoder [67] for MNIST and the contrastive learning SimCLR method [33] for Cifar-10. Both methods are unsupervised deep learning algorithm that learn feature embeddings, or representations, of image data sets that maps the

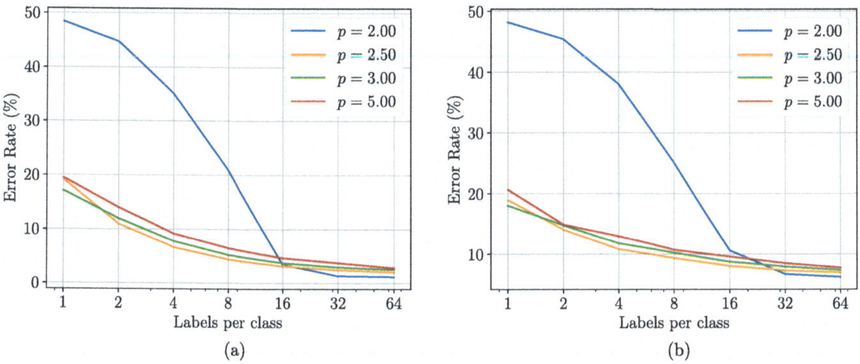

Fig. 6 Accuracy for binary classification of two classes in the MNIST and Cifar-10 data sets using p-Laplacian regularization with different values of p. We used the 4s and 9s from MNIST, which are one of the more difficult pairs to separate, and the deer and trucks from Cifar-10. (**a**) MNIST. (**b**) Cifar-10

images into a latent space identified with $\mathbb{R}^k$ where the similarity in latent features is far more informative of image similarity than pixel-wise similarity. After embedding the images into the feature space, we constructed a 10 nearest neighbor graph using the angular similarity, as in [25].

For our experiment, we used 1 up to 64 labeled examples per class for each data set, ran 100 trials at each label rate, and reported the average error rate. Figure 6 shows the error rates for different values of p as a function of label rate. We can see that larger values of p give better classification performance at very low label rates, as expected from previous work [47]. Once the label rate is sufficiently large, the error rate becomes small. As predicted in Theorem 4.10, the only points that are misclassified are those that are close to the boundary between classes in terms of graph distance, which are a small fraction of the data points—based on Fig. 6 about 2–4% for MNIST and 5–10% for Cifar-10.

5 Conclusions and Future Work

In this paper, we gave a thorough overview of the intersection between graph-based semi-supervised learning and PDEs and highlighted problems focused on consistency of classification that have not received significant attention yet in the community. We presented some preliminary results on consistency of p-Laplacian-based semi-supervised learning. Our results use the stochastic tug-of-war interpretation of the p-Laplacian on a graph, and we also provided a brief overview of this field. We proved consistency results for general graphs, geometric graphs, and stochastic block model graphs, the latter of which are not usually covered by continuum PDE-based arguments. One of our findings is that the tug-of-war game transfers nicely between different graph structures and does not require the geometric structure of the graph. We also presented numerical experiments on synthetic and real data that illustrated our results and suggested directions for future work.

We highlight below some open problems for future work.

1. **Relaxing the assumption (A1)**. The assumption **(A1)** is a strong simplifying assumption we made in order to obtain preliminary results. It asks that every vertex on the graph has a labeled neighbor and is used to make our results tractable. In practice, this means that, in the geometric graph setting, labels do not propagate very far on the graph. It would be very interesting, and more practically relevant, to extend these results to settings where **(A1)** does not hold. This would require a far more delicate martingale analysis than we provided in this paper.
2. **Stochastic block models**. Our proof technique in Theorem 4.20 for the stochastic block model graph only exploited the random walk and ignored the tug-of-war component of the game. The numerical results in Sect. 4.4 suggest that the classification accuracy improves dramatically as $p > 2$ increases, even at moderately large label rates—20% in this case. In order to explain this, it would seem necessary to improve Theorem 4.20 by utilizing the tug-of-war game as well, so that the condition (4.32) on r/q depends on p.
3. **Noise and corruption:** In real-world applications, we may observe noisy or corrupted labels $g = g^\dagger + \xi$, where $\xi(y)$ for $y \in \Gamma$ are independent and identically distributed random variables. For simplicity, assume they have mean zero and variance $\sigma^2 > 0$. The results in this paper would establish that we recover g, while the true goal is to recover the clean uncorrupted labels $g^\dagger$. Applying the sub-martingale estimate Lemma 4.3 in this case and subtracting $g^\dagger$ from both sides, we obtain

$$|u(x) - g^\dagger(x)| \leq \mathbb{E}[|g^\dagger(X_\tau) - g^\dagger(x) \mid X_0 = x] + |\mathbb{E}[\xi(X_\tau)|\ X_0 = x]|\,,$$

where the expectation is over the random walk, and not the independent noise ξ. To recover $g^\dagger$, we need to be able to show that the last term above is small, either in expectation or with high probability. Without the absolute values, it is given by

$$\mathbb{E}[\xi(X_\tau) \mid X_0 = x] = \sum_{y \in \Gamma} \mathbb{P}(X_\tau = y)\xi(y).$$

If we assume the distribution of the stopping vertex, X_τ is independent of the noise ξ, then the variance of this term with respect to the noise is given by

$$\operatorname{Var} \mathbb{E}[\xi(X_\tau) \mid X_0 = x] = \sigma^2 \sum_{y \in \Gamma} \mathbb{P}(X_\tau = y)^2. \tag{5.1}$$

Thus, understanding the noisy setting requires a finer analysis of the stochastic process, so that we can estimate the ℓ^2 norm of the probability distribution of X_τ. If X_τ is well-spread out on Γ, say $\mathbb{P}(X_\tau = y) = \frac{1}{\#\Gamma}$, then the variance (5.1) is small: $\sigma^2/\#\Gamma$. On the other hand, if the distribution concentrates on a single node, $\mathbb{P}(X_\tau = y) = \delta_{y=z}$ for some $z \in \Gamma$, then the variance is σ^2; the same as that of the noise $\xi(z)$. Essentially, the more labels we average over to compute $u(x)$, the smaller the effect of noise. Estimating the variance in (5.1) appears to be a nontrivial undertaking that we leave for future work. In particular, X_τ will generally only be independent of ξ when $p = 2$; for $p > 2$, the variance computation in (5.1) will also contain covariance terms that may be difficult to control.

On the other hand, we expect the SBM analysis in Theorem 4.20 to extend in a fairly direct way to the setting where some fraction $\theta \in (0, 1)$ of the labels are corrupted by flipping to the opposite class. In this case, we expect a similar result to hold where β_i decreases and γ_i increases, by an amount proportional to θ, and then the left-hand side of (4.32) is decreased by a factor like $1 - \theta$. We leave this to a future study as well.

4. **Extension to other random-walk models**. It would be interesting to extend these results to other models that have random walk interpretations, including Poisson learning [25], PWLL [27, 98], and the properly weighted Laplacian [24]. Some of the same high level ideas may work, but we expect many of the ingredients to be different. The case of Laplace learning was essentially already studied in [28].
5. **Similar results for other models**. There are a number of models that do not have random walk interpretations, such as the variational p-Laplacian [39, 112] and the MBO methods [14, 48, 62, 93–95]. It would be interesting to prove similar consistency results for these methods, though the techniques would be substantially different, since as far as we are aware, there are no representation formulas that express the solutions through stochastic processes in these works.

Acknowledgments

Funding: Calder was supported by NSF-DMS grant 1944925, the Alfred P. Sloan foundation, the McKnight foundation, and an Albert and Dorothy Marden Professorship.

Source Code: https://github.com/jwcalder/p-Laplace-consistency.

References

1. M. Alamgir, U.V. Luxburg, Phase transition in the family of p-resistances. in *Advances in Neural Information Processing Systems* (2011), pp. 379–387
2. R.K. Ando, T. Zhang, Learning on graph with Laplacian regularization, in *Advances in Neural Information Processing Systems* (2007), pp. 25–32
3. S. Armstrong, C. Smart, A finite difference approach to the infinity Laplace equation and tug-of-war games. Trans. Amer. Math. Soc. **364**(2), 595–636 (2012)

4. G. Aronsson, M. Crandall, P. Juutinen, A tour of the theory of absolutely minimizing functions. Bull. Amer. Math. Soc. **41**(4), 439–505 (2004)
5. A. Attouchi, H. Luiro, M. Parviainen, Gradient and Lipschitz estimates for tug-of-war type games. SIAM J. Math. Analy. **53**(2), 1295–1319 (2021)
6. A. Azad, Learning Label Initialization for Time-Dependent Harmonic Extension (2022). arXiv preprint arXiv:2205.01358
7. M. Bardi, I.C. Dolcetta, et al., *Optimal Control and Viscosity Solutions of Hamilton-Jacobi-Bellman Equations*, vol. 12 (Springer, Berlin, 1997)
8. M. Belkin, P. Niyogi, Using manifold stucture for partially labeled classification, in *Advances in Neural Information Processing Systems*, 15 (2002)
9. M. Belkin, P. Niyogi, Laplacian eigenmaps for dimensionality reduction and data representation. Neural Comput. **15**(6), 1373–1396 (2003)
10. M. Belkin, P. Niyogi, Semi-supervised learning on Riemannian manifolds. Mach. Learn. **56**, 209–239 (2004)
11. M. Belkin, I. Matveeva, P. Niyogi, Regularization and semi-supervised learning on large graphs, in *Learning Theory: 17th Annual Conference on Learning Theory, COLT 2004, Banff, July 1–4, 2004. Proceedings 17* (Springer, Berlin, 2004), pp 624–638
12. Y. Bengio, O. Delalleau, N. Le Roux, *Label Propagation and Quadratic Criterion*, Semi-supervised Learning edition (MIT Press, Cambridge, 2006), pp. 193–216
13. S. Boucheron, G. Lugosi, P. Massart, *Concentration Inequalities: A Nonasymptotic Theory of Independence* (Universities Press, Oxford , 2013)
14. Z.M. Boyd, E. Bae, X.-C. Tai, A.L. Bertozzi, Simplified energy landscape for modularity using total variation. SIAM J. Appl. Math. **78**(5), 2439–2464 (2018)
15. F. Bozorgnia, M. Fotouhi, A. Arakelyan, A. Elmoataz, Graph based semi-supervised learning using spatial segregation theory. J. Comput. Sci. **74**, 102153 (2023)
16. N. Bridle, X. Zhu, p-voltages: Laplacian regularization for semi-supervised learning on high-dimensional data, in *Eleventh Workshop on Mining and Learning with Graphs (MLG2013)* (2013)
17. J. Brown, R. O'Neill, J. Calder, A.L. Bertozzi, Utilizing contrastive learning for graph-based active learning of SAR data, in *SPIE Defense and Commercial Sensing: Algorithms for Synthetic Aperture Radar Imagery XXX* (2023)
18. L. Bungert, J. Calder, T. Roith, Uniform convergence rates for lipschitz learning on graphs. IMA J. Numer. Analy. **43**, 2445–2495 (2022)
19. L. Bungert, J. Calder, T. Roith, Ratio convergence rates for Euclidean first-passage percolation: applications to the graph infinity Laplacian. Ann. Appl. Probab. **34**(4), 3870–3910 (2024)
20. J. Calder, The game theoretic p-Laplacian and semi-supervised learning with few labels. Nonlinearity **32**(1), 301 (2018)
21. J. Calder, Consistency of Lipschitz learning with infinite unlabeled data and finite labeled data. SIAM J. Math. Data Sci. **1**(4), 780–812 (2019)
22. J. Calder, M. Ettehad, Hamilton-Jacobi equations on graphs with applications to semi-supervised learning and data depth. J. Mach. Learn. Res. **23**(318), 1–62 (2022)
23. J. Calder, N. García Trillos, Improved spectral convergence rates for graph Laplacians on ε-graphs and k-NN graphs. Appl. Comput. Harmonic Analy. **60**, 123–175 (2022)
24. J. Calder, D. Slepčev, Properly-weighted graph Laplacian for semi-supervised learning. Appl. Math. Optim. **82**, 1111–1159 (2020)
25. J. Calder, B. Cook, M. Thorpe, D. Slepčev, Poisson learning: graph based semi-supervised learning at very low label rates, in *Proceedings of the 37th International Conference on Machine Learning, PMLR*, vol. 119 (2020), pp. 1306–1316
26. J. Calder, N. García Trillos, M. Lewicka, Lipschitz regularity of graph Laplacians on random data clouds. SIAM J. Math. Analy. **54**(1), 1169–1222 (2022)
27. J. Calder, B. Cook, M. Thorpe, D. Slepčev, Y. Zhang, S. Ke, Graph-based semi-supervised learning with Poisson equations. In Preparation (2024)

28. J. Calder, D. Slepčev, M. Thorpe, Rates of convergence for Laplacian semi-supervised learning with low labeling rates. Res. Math. Sci. Special issue on PDE Methodsr Mach. Learn. **10**(1), 10 (2023)
29. J. Cervino, L.F. Chamon, B.D. Haeffele, R. Vidal, A. Ribeiro, Learning globally smooth functions on manifolds, in *International Conference on Machine Learning* (PMLR, London, 2023), pp. 3815–3854
30. O. Chapelle, B. Scholkopf, A. Zien, *Semi-supervised Learning* (MIT, Cambridge, 2006)
31. J. Chapman, B. Chen, Z. Tan, J. Calder, K. Miller, A.L. Bertozzi, Novel batch active learning approach and its application on the synthetic aperture radar datasets, in *SPIE Defense and Commercial Sensing: Algorithms for Synthetic Aperture Radar Imagery XXX (Best Student Paper)* (2023)
32. F. Charro, J. García Azorero, J.D. Rossi, A mixed problem for the infinity Laplacian via tug-of-war games. Calcul. Variat. Part. Differ. Equ. **34**(3), 307–320 (2009)
33. T. Chen, S. Kornblith, M. Norouzi, G. Hinton, A simple framework for contrastive learning of visual representations, in *International Conference on Machine Learning* (PMLR, London, 2020), pp. 1597–1607
34. R.R. Coifman, S. Lafon, Diffusion maps. Appl. Comput. Harmonic Analy. **21**(1), 5–30 (2006)
35. M. Defferrard, X. Bresson, P. Vandergheynst, Convolutional neural networks on graphs with fast localized spectral filtering, in *Advances in Neural Information Processing Systems*, 29 (2016)
36. H. Dong, J. Chen, F. Feng, X. He, S. Bi, Z. Ding, P. Cui, On the equivalence of decoupled graph convolution network and label propagation. Proc. Web Confer. **2021**, 3651–3662 (2021)
37. D.L. Donoho, C. Grimes, Hessian eigenmaps: locally linear embedding techniques for high-dimensional data. Proc. Natl. Acad. Sci. **100**(10), 5591–5596 (2003)
38. M.M. Dunlop, D. Slepčev, A.M. Stuart, M. Thorpe, Large data and zero noise limits of graph-based semi-supervised learning algorithms. Appl. Comput. Harmonic Analy. **49**(2), 655–697 (2020)
39. A. El Alaoui, X. Cheng, A. Ramdas, M.J. Wainwright, M.I. Jordan, Asymptotic behavior of ℓ_p-based Laplacian regularization in semi-supervised learning, in *Conference on Learning Theory* (2016), pp. 879–906
40. A. Elmoataz, M. Toutain, D. Tenbrinck, On the p-Laplacian and ∞-Laplacian on graphs with applications in image and data processing. SIAM J. Imag. Sci. **8**(4), 2412–2451 (2015)
41. A. Elmoataz, X. Desquesnes, M. Toutain, On the game p-Laplacian on weighted graphs with applications in image processing and data clustering. Eur. J. Appl. Math. **28**(6), 922–948 (2017)
42. A. Elmoataz, F. Lozes, M. Toutain, Nonlocal PDEs on graphs: from tug-of-war games to unified interpolation on images and point clouds. J. Math. Imag. Vis. **57**(3), 381–401 (2017)
43. H. Ennaji, Y. Quéau, A. Elmoataz, Tug of war games and PDEs on graphs with applications in image and high dimensional data processing. Sci. Rep. **13**(1), 6045 (2023)
44. J. Enwright, H. Hardiman-Mostow, J. Calder, A.L. Bertozzi, Deep semisupervised label propagation for SAR image classification, in *SPIE Defense and Commercial Sensing: Algorithms for Synthetic Aperture Radar Imagery XXX* (2023)
45. L.C. Evans, A new proof of local C1, α regularity for solutions of certain degenerate elliptic pde. J. Differ. Equ. **45**(3), 356–373 (1982)
46. L. Evans, *Partial Differential Equations (Graduate Studies in Mathematics, V. 19) GSM/19* (American Mathematical Society, Providence, 1998)
47. M. Flores, J. Calder, G. Lerman, Analysis and algorithms for Lp-based semi-supervised learning on graphs. Appl. Comput. Harmonic Analy. **60**, 77–122 (2022)
48. C. Garcia-Cardona, E. Merkurjev, A.L. Bertozzi, A. Flenner, A.G. Percus, Multiclass data segmentation using diffuse interface methods on graphs. IEEE Trans. Pattern Analy. Mach. Intell. **36**(8), 1600–1613 (2014)
49. N. García Trillos, D. Slepčev, Continuum limit of total variation on point clouds. Arch. Ration. Mech. Analy. **220**, 193–241 (2016)

50. N. García Trillos, M. Gerlach, M. Hein, D. Slepčev, Error estimates for spectral convergence of the graph Laplacian on random geometric graphs toward the Laplace–Beltrami operator. Foundat. Comput. Math. **20**(4), 827–887 (2020)
51. J. Gasteiger, A. Bojchevski, S. Günnemann, Predict then propagate: graph neural networks meet personalized PageRank, in *International Conference on Learning Representations* (2018)
52. D.F. Gleich, PageRank beyond the Web. SIAM Rev. **57**(3), 321–363 (2015)
53. I. Goodfellow, Y. Bengio, A. Courville, *Deep Learning* (MIT Press, Cambridge, 2016)
54. Y. Hafiene, J. Fadili, A. Elmoataz, Nonlocal p-Laplacian Variational problems on graphs (2018). arXiv:1810.12817
55. J. Ham, D. Lee, L. Saul, Semisupervised alignment of manifolds, in *International Workshop on Artificial Intelligence and Statistics* (PMLR, London, 2005), pp. 120–127
56. J. Han, Time-dependent tug-of-war games and normalized parabolic p-Laplace equations. Nonlinear Analy. **214**, 112542 (2022)
57. J. He, M. Li, H.-J. Zhang, H. Tong, C. Zhang, Manifold-ranking based image retrieval, in *Proceedings of the 12th Annual ACM International Conference on Multimedia* (ACM, New York, 2004), pp. 9–16
58. J. He, M. Li, H.-J. Zhang, H. Tong, C. Zhang, Generalized manifold-ranking-based image retrieval. IEEE Trans. Image Process. **15**(10), 3170–3177 (2006)
59. M. Hein, J.-Y. Audibert, U. Von Luxburg, From graphs to manifolds-weak and strong pointwise consistency of graph Laplacians, in *COLT*, vol. 3559 (Springer, Berlin, 2005), pp. 470–485
60. M. Hein, J.-Y. Audibert, U.V. Luxburg, Graph Laplacians and their convergence on random neighborhood graphs. J. Mach. Learn. Res. **8**(6), 1325–1368 (2007)
61. F. Hoffmann, B. Hosseini, A.A. Oberai, A.M. Stuart, Spectral analysis of weighted Laplacians arising in data clustering. Appl. Comput. Harmonic Analy. **56**, 189–249 (2022)
62. H. Hu, T. Laurent, M.A. Porter, A.L. Bertozzi, A method based on total variation for network modularity optimization using the MBO scheme. SIAM J. Appl. Math. **73**(6), 2224–2246 (2013)
63. Q. Huang, H. He, A. Singh, S.-N. Lim, A.R. Benson, Combining label propagation and simple models out-performs graph neural networks (2020). arXiv preprint arXiv:2010.13993
64. M. Jacobs, E. Merkurjev, S. Esedoḡlu, Auction dynamics: a volume constrained MBO scheme. J. Comput. Phys. **354**, 288–310 (2018)
65. M. Ji, J. Han, A variance minimization criterion to active learning on graphs, in *Artificial Intelligence and Statistics* (PMLR, London, 2012), pp. 556–564
66. A. Jung, A.O. Hero III, A. Mara, S. Jahromi, Semi-supervised learning via sparse label propagation (2016). arXiv preprint arXiv:1612.01414
67. D.P. Kingma, M. Welling, Auto-encoding variational bayes (2013). arXiv preprint arXiv:1312.6114
68. T.N. Kipf, M. Welling, Semi-supervised classification with graph convolutional networks (2016). arXiv:1609.02907
69. R. Kohn, S. Serfaty, A deterministic-control-based approach motion by curvature. Commun. Pure Appl. Math. J. Iss. Courant Instit. Math. Sci. **59**(3), 344–407 (2006)
70. A. Krizhevsky, G. Hinton, et al., Learning multiple layers of features from tiny images. *University of Toronto* (2009)
71. R. Kyng, A. Rao, S. Sachdeva, D.A. Spielman, Algorithms for Lipschitz learning on graphs, in *Conference on Learning Theory* (2015), pp. 1190–1223
72. Y. LeCun, L. Bottou, Y. Bengio, P. Haffner, Gradient-based learning applied to document recognition. Proc. IEEE **86**(11), 2278–2324 (1998)
73. W.-Y. Lee, L.-C. Hsieh, G.-L. Wu, W. Hsu, Graph-based semi-supervised learning with multi-modality propagation for large-scale image datasets. J. Visual Commun. Image Represent. **24**(3), 295–302 (2013)
74. M. Lewicka, Random Tug of War games for the p-Laplacian: $1 < p < \infty$ (2018). arXiv preprint arXiv:1810.03413

75. M. Lewicka, *A Course on Tug-of-War Games with Random Noise* (Springer, Berlin, 2020)
76. M. Lewicka, Non-local Tug-of-War with noise for the geometric fractional p-Laplacian. Adv. Differ. Equ. **27**(1–2), 31–76 (2022)
77. M. Lewicka, J.J. Manfredi, Game theoretical methods in PDEs. Bollettino dell'Unione Matematica Italiana **7**(3), 211–216 (2014)
78. M. Lewicka, J.J. Manfredi, The obstacle problem for the p-laplacian via optimal stopping of tug-of-war games. Probab. Theory Relat. Fields **167**, 349–378 (2017)
79. M. Lewicka, Y. Peres, The Robin mean value equation I: a random walk approach to the third boundary value problem. Potent. Analy. **59**, 1–32 (2022)
80. M. Lewicka, Y. Peres, The Robin mean value equation II: asymptotic Hölder regularity. Potent. Analy. **59**, 1–35 (2022)
81. X. Li, R. Zhu, Y. Cheng, C. Shan, S. Luo, D. Li, W. Qian, Finding global homophily in graph neural networks when meeting heterophily, in *International Conference on Machine Learning* (PMLR, London, 2022), pp. 13242–13256
82. P. Lindqvist, *Notes on the Stationary p-Laplace Equation* (Springer, Berlin, 2019)
83. H. Luiro, M. Parviainen, E. Saksman, Harnack's inequality for p-harmonic functions via stochastic games. Commun. Part. Differ. Equ. **38**(11), 1985–2003 (2013)
84. U.V. Luxburg, O. Bousquet, Distance-based classification with Lipschitz functions. J. Mach. Learn. Res. **5**(Jun), 669–695 (2004)
85. Y. Ma, R. Garnett, J. Schneider, Σ-optimality for active learning on gaussian random fields, in ed. by C.J.C. Burges, L. Bottou, M. Welling, Z. Ghahramani, K.Q. Weinberger, *Advances in Neural Information Processing Systems 26* (Curran Associates, San Jose, 2013), pp. 2751–2759
86. X. Mai, A random matrix analysis and improvement of semi-supervised learning for large dimensional data. J. Mach. Learn. Res. **19**(79), 1–27 (2018)
87. X. Mai, R. Couillet, Random matrix-inspired improved semi-supervised learning on graphs, in *International Conference on Machine Learning* (2018)
88. X. Mai, R. Couillet, Consistent semi-supervised graph regularization for high dimensional data. J. Mach. Learn. Res. **22**(1), 4181–4228 (2021)
89. J. Manfredi, M. Parviainen, J. Rossi, An asymptotic mean value characterization for p-harmonic functions. Proc. Am. Math. Soc. **138**(3), 881–889 (2010)
90. J.J. Manfredi, M. Parviainen, J.D. Rossi, Dynamic programming principle for tug-of-war games with noise. ESAIM Control Optim. Calcul. Variat. **18**(1), 81–90 (2012)
91. J.J. Manfredi, M. Parviainen, J.D. Rossi, On the definition and properties of p-harmonious functions. Annali della Scuola Normale Superiore di Pisa-Classe di Scienze **11**(2), 215–241 (2012)
92. J.J. Manfredi, A.M. Oberman, A.P. Sviridov, Nonlinear elliptic partial differential equations and p-harmonic functions on graphs. Differ. Integr. Equ. **28**(1–2), 79–102 (2015)
93. E. Merkurjev, T. Kostic, A.L. Bertozzi, An MBO scheme on graphs for classification and image processing. SIAM J. Imag. Sci. **6**(4), 1903–1930 (2013)
94. E. Merkurjev, C. Garcia-Cardona, A.L. Bertozzi, A. Flenner, A.G. Percus, Diffuse interface methods for multiclass segmentation of high-dimensional data. Appl. Math. Lett. **33**, 29–34 (2014)
95. E. Merkurjev, A.L. Bertozzi, F. Chung, A semi-supervised heat kernel pagerank MBO algorithm for data classification. Commun. Math. Sci. **16**(5), 1241–1265 (2018)
96. B. Merriman, J.K. Bence, S.J. Osher, Motion of multiple junctions: a level set approach. J. Comput. Phys. **112**(2), 334–363 (1994)
97. K. Miller, A.L. Bertozzi, Model-change active learning in graph-based semi-supervised learning (2021). arXiv preprint arXiv:2110.07739
98. K. Miller, J. Calder, Poisson reweighted Laplacian uncertainty sampling for graph-based active learning. SIAM J. Math. Data Sci. **5**, 1160–1190 (2023)
99. K. Miller, X. Baca, J. Mauro, J. Setiadi, Z. Shi, J. Calder, A. Bertozzi, Graph-based active learning for semi-supervised classification of SAR data, in *SPIE Defense and Commercial Sensing: Algorithms for Synthetic Aperture Radar Imagery XXIX* (2022), p. 12095

100. J.M. Murphy, M. Maggioni, Unsupervised clustering and active learning of hyperspectral images with nonlinear diffusion. IEEE Trans. Geosci. Remote Sens. **57**(3), 1829–1845 (2019)
101. B. Nadler, N. Srebro, X. Zhou, Semi-supervised learning with the graph Laplacian: the limit of infinite unlabelled data. Adv. Neural Inf. Process. Syst. **22**, 1330–1338 (2009)
102. A. Oberman, A convergent difference scheme for the infinity Laplacian: construction of absolutely minimizing Lipschitz extensions. Math. Comput. **74**(251), 1217–1230 (2005)
103. A.M. Oberman, Finite difference methods for the infinity Laplace and p-Laplace equations. J. Comput. Appl. Math. **254**, 65–80 (2013)
104. M. Parviainen, *Notes on Tug-of-War Games and the p-Laplace Equation*. Springer Briefs on PDEs and Data Science (Springer, Berlin, 2024)
105. Y. Peres, S. Sheffield et al., Tug-of-war with noise: a game-theoretic view of the p-Laplacian. Duke Math. J. **145**(1), 91–120 (2008)
106. Y. Peres, O. Schramm, S. Sheffield, D. Wilson, Tug-of-war and the infinity Laplacian. J. Amer. Math. Soc. **22**(1), 167–210 (2009)
107. Y. Peres, G. Pete, S. Somersille, Biased tug-of-war, the biased infinity Laplacian, and comparison with exponential cones. Calc. Var. Partial Differ. Equ. **38**(3–4), 541–564 (2010)
108. Y.-L. Qiao, C.X. Shi, C. Wang, H. Li, M. Haberland, X. Luo, A.M. Stuart, A.L. Bertozzi, Uncertainty quantification for semi-supervised multi-class classification in image processing and ego-motion analysis of body-worn videos, in *Image Processing: Algorithms and Systems* (2019)
109. P. Sellars, A.I. Aviles-Rivero, C.-B. Schönlieb, Laplacenet: a hybrid graph-energy neural network for deep semisupervised classification. IEEE Trans. Neural Netw. Learn. Syst. **35**(4), 5306–5318 (2022)
110. B. Settles, Active learning literature survey. University of Wisconsin-Madison Department of Computer Sciences (2009)
111. Z. Shi, S. Osher, W. Zhu, Weighted nonlocal Laplacian on interpolation from sparse data. J. Sci. Comput. **73**(2–3), 1164–1177 (2017)
112. D. Slepcev, M. Thorpe, Analysis of p-laplacian regularization in semisupervised learning. SIAM J. Math. Analy. **51**(3), 2085–2120 (2019)
113. N.G. Trillos, F. Hoffmann, B. Hosseini, Geometric structure of graph Laplacian embeddings. J. Mach. Learn. Res. **22**(1), 2934–2988 (2021)
114. A. Vaswani, N. Shazeer, N. Parmar, J. Uszkoreit, L. Jones, A.N. Gomez, L. Kaiser, I. Polosukhin, Attention is all you need, in *Advances in Neural Information Processing Systems, 30* (2017)
115. P. Veličković, G. Cucurull, A. Casanova, A. Romero, P. Lio, Y. Bengio, Graph attention networks (2017). arXiv preprint arXiv:1710.10903
116. U. von Luxburg, A tutorial on spectral clustering. Stat. Comput. **17**(4), 395–416 (2007)
117. B. Wang, Z. Tu, J.K. Tsotsos, Dynamic label propagation for semi-supervised multi-class multi-label classification, in *Proceedings of the IEEE International Conference on Computer Vision* (2013), pp. 425–432
118. Y. Wang, M.A. Cheema, X. Lin, Q. Zhang, Multi-manifold ranking: using multiple features for better image retrieval, in *Pacific-Asia Conference on Knowledge Discovery and Data Mining* (Springer, Berlin, 2013), pp. 449–460
119. A. Weihs, M. Thorpe, Consistency of Fractional Graph-Laplacian Regularization in Semi-Supervised Learning with Finite Labels (2023). arXiv preprint arXiv:2303.07818
120. D. Williams, *Probability with Martingales* (Cambridge University Press, Cambridge, 1991)
121. B. Xu, J. Bu, C. Chen, D. Cai, X. He, W. Liu, J. Luo, Efficient manifold ranking for image retrieval, in *Proceedings of the 34th International ACM SIGIR Conference on Research and Development in Information Retrieval* (ACM, New York, 2011), pp. 525–534
122. Y. Yan, M. Hashemi, K. Swersky, Y. Yang, D. Koutra, Two sides of the same coin: Heterophily and oversmoothing in graph convolutional neural networks, in *2022 IEEE International Conference on Data Mining (ICDM)* (IEEE, Piscataway, 2022), pp. 1287–1292
123. C. Yang, L. Zhang, H. Lu, X. Ruan, M.-H. Yang, Saliency detection via graph-based manifold ranking, in *Proceedings of the IEEE Conference on Computer Vision and Pattern Recognition* (2013), pp. 3166–3173

124. Z. Yang, W. Cohen, R. Salakhudinov, Revisiting semi-supervised learning with graph embeddings, in *International Conference on Machine Learning* (PMLR, London, 2016), pp. 40–48
125. X. Yang, Z. Song, I. King, Z. Xu, A survey on deep semi-supervised learning. IEEE Trans. Knowl. Data Eng. **35**, 8934–8954 (2022)
126. A. Yuan, J. Calder, B. Osting, A continuum limit for the PageRank algorithm. Eur. J. Appl. Math. **33**, 472–504 (2022)
127. X. Zheng, Y. Liu, S. Pan, M. Zhang, D. Jin, P.S. Yu, Graph neural networks for graphs with heterophily: A survey (2022). arXiv preprint arXiv:2202.07082
128. X. Zhou, M. Belkin, Semi-supervised learning by higher order regularization, in *Proceedings of the Fourteenth International Conference on Artificial Intelligence and Statistics*. JMLR Workshop and Conference Proceedings (2011), pp. 892–900
129. D. Zhou, B. Schölkopf, Learning from labeled and unlabeled data using random walks, in *Joint Pattern Recognition Symposium* (Springer, Berlin, 2004), pp. 237–244
130. D. Zhou, B. Schölkopf, Regularization on discrete spaces, in *Joint Pattern Recognition Symposium* (Springer, Berlin, 2005), pp. 361–368
131. D. Zhou, O. Bousquet, T.N. Lal, J. Weston, B. Schölkopf, Learning with local and global consistency, in *Advances in Neural Information Processing Systems* (2004), pp. 321–328
132. D. Zhou, J. Weston, A. Gretton, O. Bousquet, B. Schölkopf, Ranking on data manifolds, in *Advances in Neural Information Processing Systems* (2004), pp. 169–176
133. D. Zhou, J. Huang, B. Schölkopf, Learning from labeled and unlabeled data on a directed graph, in *Proceedings of the 22nd International Conference on Machine Learning* (ACM, New York, 2005), pp. 1036–1043
134. X. Zhu, Z. Ghahramani, Learning from labeled and unlabeled data with label propagation. *ProQuest number: information to all users* (2002)
135. X. Zhu, Z. Ghahramani, J.D. Lafferty, Semi-supervised learning using Gaussian fields and harmonic functions, in *Proceedings of the 20th International Conference on Machine Learning (ICML-03)* (2003), pp. 912–919
136. X. Zhu, J. Lafferty, Z. Ghahramani, Combining active learning and semi-supervised learning using Gaussian fields and harmonic functions, in *International Conference on Machine Learning (ICML) 2003 Workshop on The Continuum from Labeled to Unlabeled Data in Machine Learning and Data Mining* (2003), pp. 58–65
137. J. Zhu, Y. Yan, L. Zhao, M. Heimann, L. Akoglu, D. Koutra, Beyond homophily in graph neural networks: current limitations and effective designs. Adv. Neural Inf. Process. Syst. **33**, 7793–7804 (2020)

Discrete Minimizers of the Interaction Energy in Collective Behavior: A Brief Numerical and Analytic Review

José A. Cañizo and Alejandro Ramos-Lora

Abstract We consider minimizers of the N-particle interaction potential energy and briefly review numerical methods used to calculate them. We consider simple pair potentials which are repulsive at short distances and attractive at long distances, focusing on examples which are sums of two powers. The range of powers we look at includes the well-known case of the Lennard-Jones potential, but we are also interested in less singular potentials which are relevant in collective behavior models. We report on results using the software GMIN developed by Wales and collaborators for problems in chemistry. For all cases, this algorithm gives good candidates for the minimizers for relatively low values of the particle number N. This is well-known for potentials similar to Lennard-Jones, but not for the range which is of interest in collective behavior. Standard minimization procedures have been used in the literature in this range, but they are likely to yield stationary states which are not minimizers. We illustrate numerically some properties of the minimizers in 2D, such as lattice structure, Wulff shapes, and the continuous large-N limit for locally integrable (that is, less singular) potentials.

1 Introduction

Finding configurations with minimum potential energy for a set of particles is a common mathematical problem found across several fields. In its simplest version, we consider a potential $V : \mathbb{R}^d \to \mathbb{R} \cup \{+\infty\}$ and a number $N \geq 2$ of particles. The problem then consists in finding the configurations $X = (x_1, \ldots, x_N) \in \mathbb{R}^{Nd}$ such

J. A. Cañizo (✉) · A. Ramos-Lora
Departamento de Matemática Aplicada & IMAG, Universidad de Granada, Granada, Spain
e-mail: canizo@ugr.es; ramoslora@ugr.es

J. A. Carrillo, E. Tadmor (eds.), *Active Particles, Volume 4*, Modeling and Simulation in Science, Engineering and Technology,
https://doi.org/10.1007/978-3-031-73423-6_2

that the energy

$$E_N(X) = \sum_{i=1}^{N} \sum_{\substack{j=1 \\ j \neq i}}^{N} V(x_i - x_j) \tag{1.1}$$

is as small as possible. These configurations are called *minimizers*[1] or *ground states*. In a broader sense, the problem also involves understanding the structure of these minimizers: their shape, the distribution of their points, and their asymptotic properties for large N. The energy (1.1) is clearly invariant by a translation of the points $(x_1, \ldots, x_N)$, and consequently any translation of a minimizer is also a minimizer. One usually removes this invariance by looking for minimizers with some fixed property, most often by fixing the mean position $\sum_{i=1}^{N} x_i$ to be 0. If V is radially symmetric, then E_N is also invariant by rotations of the points $(x_1, \ldots, x_N$ and by their inversion $(-x_1, \ldots, -x_N)$ (and hence by any rigid motion of these points). In all cases, minimizers are the same if V is substituted by $V + C$ for some constant C, so some normalization on V is usually assumed (often, either $\lim_{|x| \to +\infty} V(x) = 0$ or $V(0) = 0$, but not necessarily).

In this paper, we give a short review of numerical methods to calculate these minimizers and report on some results in dimension $d = 2$ using the publicly available GMIN software developed by David J. Wales (see [68]). We carry out numerical simulations in 2D for potentials of the form

$$V(x) = \frac{|x|^a}{a} - \frac{|x|^b}{b}, \tag{1.2}$$

where $a > b$. Due to our choice of coefficients, the term with the a power always represents an attractive potential, and the term with the b power is repulsive. We will also give a short review of the available rigorous results to date.

In the previous statement of the problem, the particles are all identical: They interact with others in an identical way through the pair potential V. More complicated versions of the problem may allow for particles of different types, interacting in different ways with one another. One of the better known contexts where this problem is relevant is chemical physics, where one often wants to find the configuration of minimum possible energy for a given molecule. This is important, for example, in the study of protein folding, where the natural structure of a protein may be related to its minimum-energy configuration. In the context of molecular structure, the Lennard-Jones interaction potential

$$V(x) = \frac{A}{|x|^{12}} - \frac{B}{|x|^6}, \tag{1.3}$$

---

[1] In some contexts they are called *global minimizers*, but in the present paper, "minimizer" means "global minimizer" unless we explicitly mention local minimizers.

with $A, B > 0$ given constants, is often taken as a test case. The study of its minimizers (in dimension $d = 3$) has received a lot of attention and is a notable example of a hard global optimization problem.

In addition to these problems in physics, this paper is motivated by several collective behavior models where these minimizers are relevant. Among them, perhaps the simplest is the so-called *aggregation equation*,

$$\partial_t u = \operatorname{div}(u(\nabla u * V)), \tag{1.4}$$

which is a partial differential equation for $u = u(t, x)$, which represents the density of a certain population which interacts through the potential V. Equilibria of (1.4) (if any) must satisfy $\nabla u * V = 0$ on the support of u, which is a property shared by the minimizers of the *continuous interaction energy*

$$E(u) := \int_{\mathbb{R}^d} \int_{\mathbb{R}^d} u(x)u(y)V(x - y)\,\mathrm{d}x\,\mathrm{d}y \tag{1.5}$$

in the set of probability measures (or positive measures with a fixed total mass). Hence, probability minimizers of the energy (1.5) are particular examples of stationary solutions to (1.4). To avoid repetition, whenever we speak of minimizers of E, it is understood that we always refer to minimizers in the set of probability measures. Finding minimizers of E is a "continuous version" of the problem of minimizing E_N. The link between these problems is interesting, especially if we consider that numerical studies of the continuous problem must inevitably carry out some sort of discretization of the equation, which often leads to some version of the discrete minimization problem of E_N. Hence, a proper understanding of the large-N behavior of minimizers of E_N is also useful to justify numerical methods for calculating minimizers of the continuous energy E.

There are also many individual-based models whose "organized states" share similarities with minimizers of E_N. As an example, we mention a model from [29]: We consider N individuals with positions and velocities given by (x_i, v_i) for $i = 1, \ldots, N$, and the system of ordinary differential equations

$$\begin{aligned} \frac{\mathrm{d}}{\mathrm{d}t} x_i &= v_i, \\ \frac{\mathrm{d}}{\mathrm{d}t} v_i &= (\alpha - \beta |v_i|^2) v_i - \sum_{\substack{j=1 \\ j \neq i}}^{N} \nabla V(x_i - x_j). \end{aligned}$$

This models a set of individuals which interact through a potential V but also have a preferred movement speed $|v| = (\alpha/\beta)^{1/2}$. It is easy to see that if $X = (y_1, \ldots, y_N)$ is a critical point of the energy E_N, then $(x_i(t), v_i(t)) := (y_i + tv, v)$ is a solution to the previous system of equations. In particular, minimizers of E_N yield coherent states or *flocking states* for this model. Minimizers (or critical points) of E_N appear similarly in other models, and this has motivated interest in types of potentials different from the ones usually encountered in physics problems.

Existence of Minimizers of E_N For a very general family of potentials, it is easy to show that minimizers must exist. If one assumes that V is lower semicontinuous and radially strictly increasing whenever $|x| > R$ for some given $R > 0$, then one may carry out an argument like that in [13, Theorem 4.1] to show their existence: It is not hard to argue that there cannot be "gaps" of size larger than R between the particles of a minimizer, which gives a simple estimate on their diameter. A different argument is given in [19, Lemma 2.1] for power-law potentials and in [9, Section 1.2] for potentials satisfying $\lim_{|x|\to+\infty} V(x) = 0$.

2 Numerical Methods

In general, the energy E_N from (1.1), as a function $E_N \colon \mathbb{R}^{Nd} \to \mathbb{R} \cup \{+\infty\}$, is not convex, and the global optimization problem of finding its minima cannot use the comparatively simpler methods of convex optimization. As mentioned before, there's the problem of invariance by translations, easily avoided by fixing the mean position of particles. The problem of invariance by rotations is harder to avoid, and hence algorithms usually find one of any possible rotations of the minimizer. It can then be hard to decide whether two runs of the same algorithm have given a rotation of the same minimizer, or a fundamentally different minimizer. However, the deeper problem is that there are usually a huge number of stationary states for the energy E_N, that is, points $(x_1, \ldots, x_N)$ at which

$$\sum_{\substack{j=1\\ j\neq i}}^{N} \nabla V(x_i - x_j) = 0 \qquad \text{for all } i = 1, \ldots, N. \tag{2.1}$$

Many of these are local minima, and runs of standard local minimization algorithms will most often fall into one of these states, and not get to a minimizer. Hence simple descent algorithms such as variants of conjugate gradient descent, or Newton-type methods like LBFGS [48], will almost surely not find minimizers even for small N. Global minimization algorithms must somehow avoid these "traps" by more clever search techniques. We will give a short overview of numerical methods used for this kind of problems and then describe the algorithm we use, implemented in the program GMIN by D. J. Wales [68].

Most numerical methods found in the literature have been tested with the Lennard-Jones potential (1.3), and most often in 3D, which is a challenging problem of physical interest and has been the subject of many publications in chemistry journals. A database with all currently available minima of E_N for several potentials in 3D (and molecular minimization problems of other kinds) is available online [70], among them Lennard-Jones and Morse potentials. This database contains references of the first publication for each minimizer, and the reader may also check the unpublished paper [74] for a good list of historical references.

Early methods such as those in [40–43] are *biased* methods, in the sense that they work by "guessing" that minimizers have some kind of regular crystal structure. Initial guesses for a minimizer are obtained by placing particles at the sites of an appropriate lattice and then running local minimization algorithms to optimize the positions. Then one carries out some kind of random perturbation of these positions in order to find "nearby" local minima which may have a lower energy. In addition to guessing initial candidates by using the sites of a lattice, one may also start by finding minimizers for small N and add particles to build candidates for larger N. These methods were quite successful in finding many candidates for minimizers which are still standing today, but there are several "hard" values of N for which better (lower energy) configurations were found later [52, 69]. An important step of many of these methods, used for the "random perturbation" we mentioned, is *simulated annealing* [45], a global optimization method which has an appealing physical motivation for chemistry problems. Roughly speaking, in simulated annealing, a random perturbation X' of a given configuration X is considered, and then:

1. It is accepted if X' has lower energy than X, or
2. It is accepted with a probability proportional to $\exp(-\frac{E_N(X')-E_N(X)}{T})$ if X' has higher energy than X.

Here T is a parameter which has an analogy to temperature and governs how probable it is to jump "far" from the current state. In the literature on this, performing this algorithm any number of times for a fixed T is sometimes referred to as a *MONTE CARLO SIMULATION* and each step as a *Monte Carlo step*. Simulated annealing methods then work by running many steps of this random perturbation and gradually reducing T until departures from the current state become very unlikely. Methods focused on this technique have had some success [72, 73] and are also part of more recent methods.

Another family of methods is that of evolutionary algorithms, which have also been quite successful in finding minima for Lennard-Jones potentials, even for the values of N considered harder [27, 51]. Genetic algorithms are a particular kind of evolutionary algorithms, though the naming is not completely precise. These methods are based on an analogy with natural selection and are also an example of biased methods which make informed guesses on the structure of minimizers. We refer to the more recent paper by Dittner and Hartke [28] for a list of references on the use of this kind of algorithms for Lennard-Jones energy minimization.

The paper [5] contains a good presentation of related minimization problems and describes a simple algorithm for minimizing E_N for the Lennard-Jones potential, which is tested in 2D. It is also a biased method in the sense that particles are initially placed at the sites of a triangular lattice, and later suitable symmetrization methods are used.

As discussed, these biased methods have been tried mainly for potentials of Lennard-Jones type, motivated by problems in chemistry and crystallography. Since our aim is to use a good numerical algorithm for a wider range of potentials whose minimizers may behave very differently, an adaptation of these methods does not seem straightforward, especially for the cases which have a continuous limit such as those discussed in Sect. 4, where particles do not arrange as large-scale crystals. Instead, we use a technique known as *basin-hopping*, first reported for the Lennard-Jones potential in [69] and before used by Li and Scheraga [47] for a different problem in chemistry. This technique has been in particular quite successful for Lennard-Jones-type potentials. It is an *unbiased* algorithm in the sense that it does not assume any particular structure of the minimizers, and it seems a more appropriate algorithm for the range of potentials we are interested in. It works as follows: First, the energy $E_N(X)$ is substituted by an energy $\widetilde{E}_N(X)$, obtained by performing a suitable descent algorithm starting from the configuration X (the Polak-Ribiere variant of the conjugate gradient algorithm is mentioned in [69], although it currently uses a customized LBFGS method). That is: If we denote by $Y = M(X)$, the result of running a descent algorithm from the starting configuration X, then

$$\widetilde{E}_N(X) := E_N(M(X)).$$

Hence, the new *energy landscape* $\widetilde{E}_N$ is a piecewise constant function, which is constant in a region where our chosen descent algorithm stops at the same local minimum Y, and whose value in this region is $E_N(Y)$. This new landscape can be studied theoretically, taking $M(X)$ to be the point to which the descent algorithm converges, and it can be approximated numerically in practice by considering the point at which the numerical descent stops. Now several steps are followed:

1. A Monte Carlo exploration of the landscape $\widetilde{E}_N$ is carried out with a fixed temperature, chosen so that the acceptance rate of new points is about $1/2$ (as explained in our brief earlier discussion of simulated annealing).
2. During this exploration, we also observe the *individual energy* (or just energy) of each particle i, defined by

$$E_{N,i} = \sum_{j \neq i} V(x_i - x_j).$$

   We consider the particle i_{max} which has a maximum energy among all N particles, and the one i_{min} which has minimum energy. If this maximum energy is larger than a certain multiple of the lowest energy, then an *angular move* is performed for the particle i_{max}: This particle is removed from the cluster and placed at a random position close to the boundary of the cluster.

In addition to this, a further Monte Carlo exploration is carried out with the following special starting positions:

1. The best configuration found by the algorithm with $N - 1$ particles, adding one particle at the best position found by a fixed number of angular moves.

2. The best configuration found by the algorithm with $N + 1$ particles, removing the particle with highest energy.

We have applied this algorithm for different potentials, otherwise with no modifications. Some of the results are explored in the next sections, together with a short review of the expected behavior of minimizers. We would like to emphasize that while obtaining local minimizers is straightforward in almost all cases, obtaining good candidates for a *minimizer* is a hard numerical problem. As an example, the hexagonal-shaped minimizers for very singular potentials such as Lennard-Jones (see Fig. 1) require some kind of global search algorithm as described above. Or consider, for example, a minimizer for the locally integrable potential (1.2) with $a = 2$ and $b = 0.5$ (Fig. 2). Asymptotically as $N \to +\infty$, this density is known to converge to the density of a minimizer of the continuous energy (1.5) (see Sect. 4), but the way this happens shows several interesting phenomena. One sees that locally particles still seem to be arranged in triangular patterns, but since the overall shape must be round (since the support of the minimizer is strongly suspected to be a ball), these local triangular arrangements must undergo some kind of dislocation patterns in order to fit the global shape. In Fig. 2, one can clearly appreciate dislocation curves along which triangular patterns with different orientations meet. These patterns are quite sensitive to the minimization algorithm used, and we do not know whether the plot in Fig. 2 actually represents a minimizer. But an understanding of these patterns is certainly an interesting problem, and even its numerical study requires powerful algorithms which can handle a large number of particles and yield good approximations to the minimizer. For particle numbers over a few thousand, it is likely that this is still out of reach, even numerically.

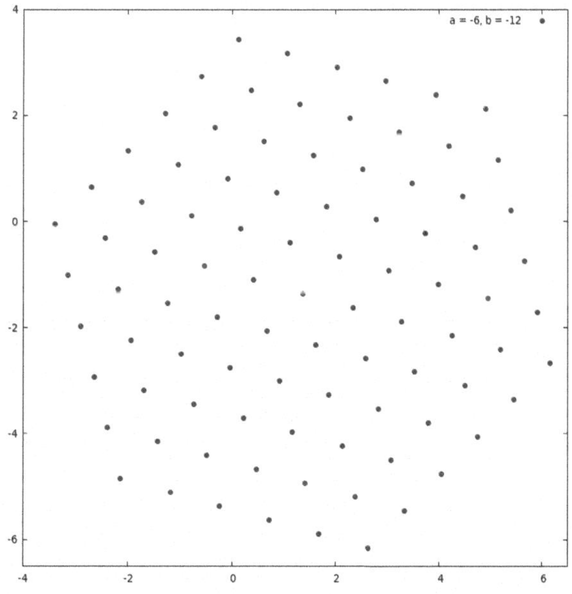

Fig. 1 Minimizer for the Lennard-Jones potential with 91 particles

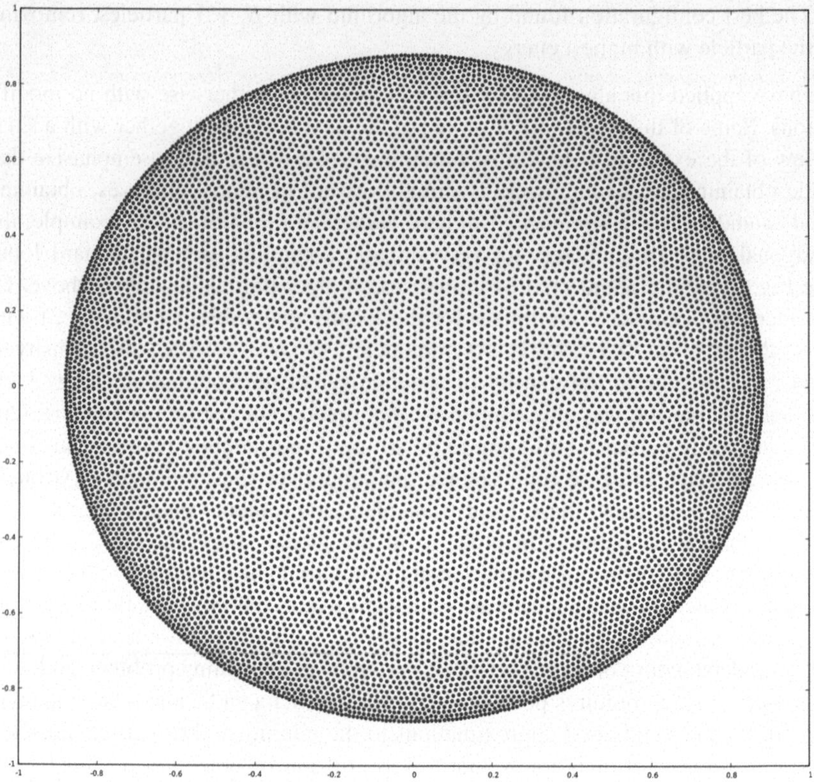

Fig. 2 Our best numerical candidate for a minimizer of the potential (1.2) with $a = 2$, $b = 0.5$ and 10^4 particles (We thank David J. Wales for providing this result)

In Figs. 2 and 3, we show minimizer candidates for 10,000 particles, which have required large computational power and several weeks of computer time.

3 Crystallization Phenomena

The wider problem of understanding the reason why matter is often arranged in crystalline patterns is known as the problem of crystallization. The reviews by Radin [57], Ventevogel [63] are a good introduction to this, giving a motivation and related mathematical problems. It happens that minimizers of E_N have a close relationship to the observed arrangement of atoms or molecules at low temperature, and hence an aspect of the crystallization problem which is particularly simple to state is *prove rigorously that (or whether) minimizers of E_N approach a periodic arrangement as $N \to +\infty$*. For some potentials V which have a singularity stronger than $|x|^{-d}$ at $x = 0$, minimizers of E_N numerically seem to do exactly this, but proofs are

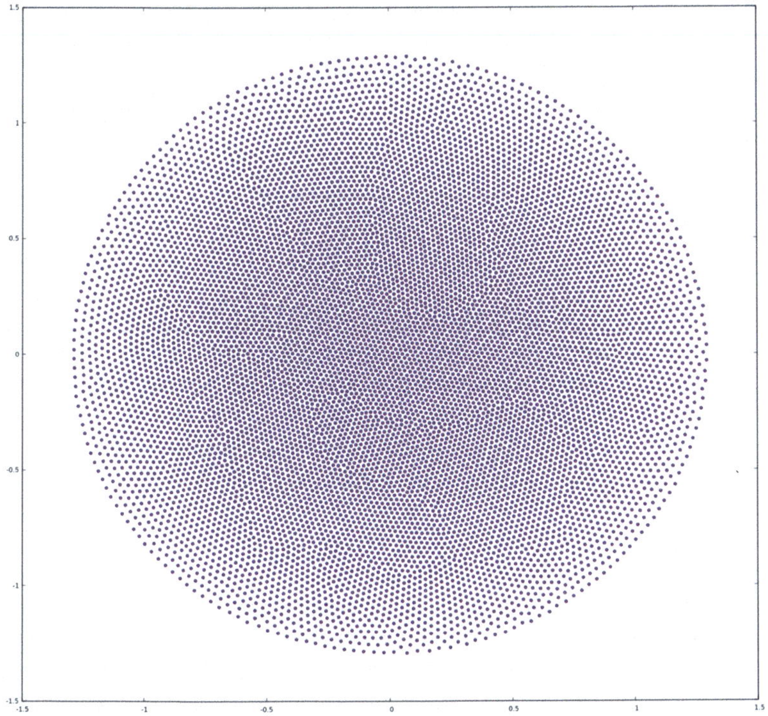

Fig. 3 Numerical candidate for minimizer of the potential (1.2) with $a = 2$, $b = -1$ and 10^4 particles. (We thank David J. Wales for providing this result.) This corresponds to the "strongly repulsive" regime considered in [20]. In this regime, it is conjectured that a sequence of minimizers $(X_N)_{N\geq 2}$ must converge to a measure ν which must have a smooth density with respect to Lebesgue measure. In fact, the most likely candidate measure is given in [15, eq. (38)] and is of the form

$$\nu(x) = C(R^2 - |x|^2)_+^{1-\frac{b+d}{2}},$$

with $(\cdot)_+$ denoting the positive part, with suitable constants $C, R > 0$ which ensure in particular that the total mass is 1. In the case corresponding to this figure, we have $d = 2$, $b = -1$, so the density of ν vanishes at the boundary. By contrast, in the case of Fig. 2, the exponent $1 - (b+d)/2$ becomes negative, and the density of ν should diverge at the boundary

surprisingly difficult even in dimension 1! The review by Blanc and Lewin [9] is a nice source of references on this problem, and we would like to give a short summary with our current understanding.

Let us first state the problem a bit more precisely, essentially following [9]. In the context of crystallization, we usually consider potentials $V = V(x)$ which are positive at $x = 0$ (or have a singularity where the potential diverges to $+\infty$ as $x \to 0$), have a strictly negative minimum value, and satisfy

$$\lim_{|x|\to+\infty} V(x) = 0. \tag{3.1}$$

Further conditions are needed for minimizers to show any kind of crystallization behavior; see more on this below. The canonical example of the potentials which exhibit this kind of phenomena is the Lennard-Jones potential from Eq. (1.3).

For a particle distribution $X = (x_1, \ldots, x_N) \in \mathbb{R}^{Nd}$, we consider the associated *empirical measure* given by

$$\mu_X := \sum_{i=1}^{N} \delta_{x_i}. \tag{3.2}$$

Notice that we are *not* dividing by N, so this measure has total mass equal to N. In many cases, we expect this measure to approach some periodic arrangement as $N \to +\infty$, and in order to define this precisely, we recall that a *lattice* in $\mathbb{R}^d$ is a subgroup of $\mathbb{R}^d$ of the form

$$L = \{k_1 v_1 + k_2 v_2 + \cdots + k_d v_d \mid k_1, \ldots, k_d \in \mathbb{Z}\}, \tag{3.3}$$

where $(v_1, \ldots, v_d)$ form a vector space basis of $\mathbb{R}^d$. Lattices are a simple form of periodic structures and are a basic concept in crystallography, where they are often referred to as *Bravais lattices*. The reason for this special naming is that in crystallography, the name *lattice* is often broadened to include periodic structures which are not lattices in the sense of the above definition; notable examples are the honeycomb arrangement in 2D or the hexagonal close packed arrangement in 3D. Here we will always use the word *lattice* in the mathematical sense stated in (3.3). Consistently with (3.2), for a lattice L, we denote by μ_L the measure

$$\mu_L := \sum_{p \in L} \delta_p. \tag{3.4}$$

Notice that μ_L has infinite mass (is an unbounded measure), but it is finite in compact sets. Here is a very strong formulation of the problem of crystallization:

Q1 (lattice crystallization) Does there exist a lattice $L \subseteq \mathbb{R}^d$ and a sequence of minimizers $(X_N)_{N \geq 2}$ of E_N such that (possibly up to a subsequence)

$$\mu_{X_N} \overset{*}{\rightharpoonup} \mu_L \qquad \text{as } N \to +\infty?$$

The convergence here is the weak-$*$ convergence of measures (convergence when tested against continuous, compactly supported functions). Or even stronger: Does this take place for *all* sequences of minimizers, up to rigid movements of these minimizers?

Crystalline structures may well be periodic structures which are not lattices. A very natural formalization of a periodic structure is a measure μ on $\mathbb{R}^d$ such that $\mu(\cdot + p) = \mu(\cdot)$ for all p in some lattice L (i.e., a measure which is invariant by translations of vector p, for all p in a given lattice). If this happens, we say that

μ is *invariant by the lattice* L. Following this, here is a weaker form of the above problem:

Q2 (periodic crystallization) Does there exist a lattice $L \subseteq \mathbb{R}^d$ and a sequence of minimizers $(X_N)_{N\geq 2}$ of E_N such that (possibly up to a subsequence)

$$\mu_{X_N} \overset{*}{\rightharpoonup} \mu \qquad \text{as } N \to +\infty,$$

where μ is a measure invariant by the lattice L? Or even stronger: Does this take place for *all* sequences of minimizers, up to rigid movements of these minimizers?

If these questions are too hard, one may consider the large-N behavior of some macroscopic quantities like the energy per particle. Given a lattice L in $\mathbb{R}^d$ and a potential V we define its *energy per particle* by

$$e(L) := \sum_{p\in L\setminus\{0\}} V(p), \tag{3.5}$$

when this sum converges absolutely. Also, given a configuration $X = (x_1, \ldots, x_N) \in \mathbb{R}^{Nd}$, we define its energy per particle by

$$e_N(X) := \frac{1}{N} E_N(X).$$

Here is a related, usually weaker, statement of the problems **Q1** and **Q2**:

Q3 (energy crystallization) Does there exist a lattice $L \subseteq \mathbb{R}^d$ and a sequence of minimizers $(X_N)_{N\geq 2}$ of E_N such that (possibly up to a subsequence)

$$e_N(X_N) \to e(L) \qquad \text{as } N \to +\infty?$$

Or even stronger: Does this take place for *all* sequences of minimizers?

In a related problem, one can try to give good estimates for the convergence of $e_N(X_N)$ to $e(L)$, for example, giving bounds on its rate of convergence. The fact that $e_N(X_N)$ has *some* limit is known for a wide range of potentials.

In the context of crystallization phenomena, one always assumes (3.1). As a consequence, a sequence of configurations with N particles such that the interparticle distances all diverge to $+\infty$ has energy as close to 0 as we wish. Hence, for a minimizer X_N, it must happen that $E_N(X_N) \leq 0$, so if **Q3** holds, one must have $e(L) \leq 0$. In any case, if **Q3** holds for the full sequence of minimizers (and not up to a subsequence), in particular, it implies that for some $C > 0$

$$E_N(X_N) \geq -CN, \qquad \text{for all } N \geq 2. \tag{3.6}$$

Potentials V satisfying (3.6) are called *H-stable*, or sometimes just *stable* in the literature; see, for example, [59, Section 3.2]. Often (3.6), though seemingly weaker, actually implies that $e_N(X_N)$ does have a limit as $N \to +\infty$. Since the terminology comes from statistical mechanics and thermodynamics, interest is placed on potentials for which thermodynamical quantities (such as the energy-per-particle) have a finite limit, which justifies the word *stable*. Minimizers for potentials which are *not* stable may still have different scaling limits which are also interesting, though their study has different motivations and is more recent; see Sect. 4.

Finally, there is the related question of the *formation of a macroscopic object*. The *diameter* of a minimizer is defined as the maximum distance between two of its points. If we rescale minimizers in order to fix their diameter, does the resulting measure converge to something? If we look at the regime where one expects that particles approach a periodic distribution, we would expect the diameter of a minimizer to be roughly proportional to $N^{1/d}$, and hence we pose the following question:

MO (formation of a macroscopic object) Does there exist a nontrivial measure μ on $\mathbb{R}^d$ and a sequence of minimizers $(X_N)_{N\geq 2}$ of E_N such that (possibly up to a subsequence)

$$\frac{1}{N}\mu_{X_N/N^{1/d}} \overset{*}{\rightharpoonup} \mu \qquad \text{as } N \to +\infty? \tag{3.7}$$

Or even stronger: Does this take place for *all* sequences of minimizers, up to rigid movements of these minimizers?

The empirical measure $\mu_{X_N/N^{1/d}}$ corresponds to the scaled minimizer

$$\frac{1}{N^{1/d}}X_N := \left(\frac{x_1}{N^{1/d}}, \dots, \frac{x_N}{N^{1/d}}\right),$$

and we divide the measure by N to ensure a total mass equal to 1. This question concerns behavior at a different scale from questions **Q1–Q3**. An answer to **MO** seems to not give much information on any of **Q1–Q3**, and reciprocally.

We have focused on the above questions in order to narrow the discussion to a representative case, since the field is too broad for a short review. However, there are many related aspects of these problems. Some of them involve infinite distributions of particles. Among these, we mention the following:

1. (*Lattice minimization*) Among all lattices in $\mathbb{R}^d$, which ones minimize the energy per particle (3.5)?
2. (*Infinite particle distributions*) Given a countably infinite set of points in $\mathbb{R}^d$, it is possible to define their *mean energy per particle* in a reasonable way, so that it is consistent with the energy per particle of a lattice given in (3.3). Among all infinite particle distributions, which ones minimize the energy per particle?

We do not elaborate on different mathematical interpretations of the problem, namely, minimization problems with fixed density, modified models with a background energy, and statistical mechanical or quantum mechanical models. We refer the reader to the reviews mentioned above for more on these matters [9, 57]. For minimization on *compact* submanifolds of $\mathbb{R}^d$, we refer to [10, 37].

Minimizers for a specific case of potential V have links to sphere packing problems. If we consider the following *sticky hard sphere* potential

$$V(x) := \begin{cases} +\infty & \text{if } |x| < 1, \\ -1 & \text{if } |x| = 1, \\ 0 & \text{if } |x| > 1, \end{cases} \tag{3.8}$$

then minimizers cannot have any two particles at a distance closer than 1 (otherwise their energy would be $+\infty$), and there is a reward for spheres which are touching. Hence, a minimizer of this potential is just a configuration of N hard spheres in d-dimensional space with the maximum total number of contact points among them. In dimension 2, this problem was solved by Heitmann and Radin [39]; see also [49] for a more recent proof. But the problem is far from understood in dimensions 3 and higher, with open questions even for low values of N (above 11). A popular review of the problem can be found in [38], and see [44, 50] for some recent results. Attempts have been made to relate the structure of these minimizers to those for the Lennard-Jones potential [62], but a precise link remains unclear. It is also tempting to make a connection of the sticky hard sphere problem with that of *sphere packing*, which is the problem of finding a distribution of identical hard spheres in $\mathbb{R}^d$ with maximum possible density. This link has often been remarked, but as far as we know, there are no rigorous results backing this up. There have been some surprising recent results in sphere packing by Cohn et al. [23], Viazovska [67], who give specific lattices in dimensions 8 and 24 which can be proved to be the densest possible sphere distributions in these dimensions. The proof relies on the existence of specially symmetric lattices in these dimensions, which are natural candidates for this densest packing.

Now that we have made an effort to state rigorously the results we are interested in, we can try to summarize the available rigorous results:

One-Dimensional Results In dimension 1, there were several early results by [64–66]. These results concern mainly infinite particle distributions (as mentioned in points 1,2 above), but specifically [64] contains a proof of the strong version of **Q3** for a wide class of power-law potentials of the form (1.2) with $b < 1$ and a certain restriction on a: The energy per particle of any sequence of minimizers converges to the energy per particle of a given lattice; this lattice is the unique one with minimum energy per particle. This is a rather complete answer to **Q3** for some power-law potentials. However, as far as we know, there are no satisfying general conditions on V which ensure a similar answer.

In contrast, problem **Q1** is not so well understood even in dimension 1. The best available result seems to be the one by Gardner and Radin [35], which does show the strong version of **Q1** for some potentials including the Lennard-Jones one.

The conditions for this to hold are more restrictive, and it seems a hard problem to generalize them to a wide class of potentials.

In dimension 1, the minimization problem for the sticky hard sphere potential (3.8) is trivial, and minimizers are distributions of N equally spaced particles at interparticle distance 1.

Two-Dimensional Results For the sticky hard sphere potential (3.8) in dimension 2, results are already interesting. A first proof that minimizers must be placed at the sites of a triangular lattice was given by Heitmann and Radin [39], giving an answer to **Q1** (and hence **Q2**) and **Q3**. For a specific piecewise linear potential not too far from (3.8), a positive answer to **Q3** was also given by Radin [56]. Then **MO** was answered by Au Yeung et al. [1] for a family of short-range potentials with V compactly supported and close enough to the sticky sphere potential (3.8): The convergence (3.7) takes place up to a subsequence, and the measure μ is $\frac{2}{\sqrt{3}}\mathbb{1}_E$, where $\mathbb{1}_E$ is the characteristic of a finite-perimeter set E of area $\sqrt{3}/2$. In [61], **Q3** is answered positively for a certain family of potentials which allow similar growth conditions as the Lennard-Jones one. In addition, it is proved that with periodic boundary conditions, the triangular lattice is a minimizer. Recently in [7], a computer-assisted method is used to show that the triangular lattice minimizes energy among lattices for the Lennard-Jones potential in 2D (hence giving an answer to the related lattice minimization problem we mentioned earlier).

Adding a three-body potential (and hence for an energy not strictly of the form (1.1)), and still with periodic boundary conditions, [71] followed the ideas in [61] to show that minimizers must have particles placed at the sites of a *hexagonal* periodic arrangement (the vertices of a honeycomb, which are not a lattice in the sense we are using in this paper). Hence in this case, a version of **Q2** holds, but not **Q1**. Problem **Q3** has been solved by Bétermin et al. [8] for some specific piecewise-constant potentials which give rise to a square lattice.

Three-Dimensional Results From the point of view of rigorous results, the situation in three dimensions is almost completely open. As far as we know, there are no results regarding any of **Q1**, **Q2**, **Q3**, or **MO**, even for the sticky hard sphere potential (3.8). The problem of optimal packing, however, was solved by Hales [36] with a nonconventional proof which requires several computer checks. One of the difficulties of the 3D problem is that there are many nonequivalent sphere distributions in $\mathbb{R}^3$ which maximize density, including nonperiodic structures.

Higher Dimensional Results In higher dimensions, there have been important breakthroughs on optimal sphere packing, for which the optimal lattices have been recently identified in dimension 8 [67] and 24 [23]. As a consequence of these ideas, in [24], it was proved that these optimal lattices are also *universally optimal*: They minimize the interaction energy among infinite configurations of points with fixed density 1, for any potentials V which are completely monotone functions of the squared distance. It is conjectured that the same holds in dimension 2 for the usual triangular lattice, but this is not proved as far as we know. In [55], it is shown that this universal optimality also implies a corresponding result for Riesz and Coulomb interaction energies, which require an appropriate renormalized definition for infinite particle distributions. Hence, this latter result is known in dimensions 8 and 24 and is a conjecture in dimension 2.

All of these results concern infinite particle configurations. As far as we know, no results are known concerning any of **Q1**, **Q2**, **Q3** or **MO** in dimensions $d \geq 3$.

4 Continuous Limits

Consider a sequence $(X_N)_{N\geq 2}$ of minimizers of E_N and a locally integrable potential V (in the case of potentials of the form (1.2) this holds if and only if $b > -d$). Then it may happen that

$$\nu_{X_N} := \frac{1}{N}\mu_{X_N} = \frac{1}{N}\sum_{i=1}^{N}\delta_{x_i^N} \overset{*}{\rightharpoonup} \nu \qquad \text{as } N \to +\infty,$$

for some measure ν, where $X_N = (x_1^N, \ldots, x_N^N)$. This type of limit is of *mean-field* type due to the factor $1/N$, and notice that there is no rescaling as there was in (3.7). In this case, the discrete interaction energy E_N is an approximation of the continuous one E,

$$\begin{aligned}\frac{1}{N^2}E_N(X) = \frac{1}{N^2}\sum_{i=1}^{N}\sum_{\substack{j=1\\ j\neq i}}^{N} V(x_i - x_j) &= \int_{\mathbb{R}^d}\int_{\mathbb{R}^d} V(x-y)\,\mathrm{d}\nu_{X_N}(x)\,\mathrm{d}\nu_{X_N}(y)\\ &\approx \int_{\mathbb{R}^d}\int_{\mathbb{R}^d} V(x-y)\,\mathrm{d}\nu(x)\,\mathrm{d}\nu(y) = E(\nu),\end{aligned} \tag{4.1}$$

when this can be rigorously justified. Then we expect the measure ν to be a minimizer of the continuous interaction energy (1.5). Proving this is another interesting question which we will state explicitly:

MF (Mean-field limit) Does there exist a nontrivial measure ν in $\mathbb{R}^d$ and a sequence of minimizers $(X_N)_{N\geq 2}$ of E_N such that (possibly up to a subsequence)

$$\frac{1}{N}\mu_{X_N} \overset{*}{\rightharpoonup} \nu \qquad \text{as } N \to +\infty?$$

Does this take place for *all* sequences of minimizers, up to rigid movements of these minimizers? In this case, is ν a minimizer of the continuous interaction energy (1.5)?

In analogy to **Q3**, one may also look for good asymptotics of the energy-per-particle $E_N(X_N)/N$ as $N \to +\infty$. For locally integrable potentials, if **MF** holds, we expect

(4.1) to hold, and hence we also expect that

$$E_N(X_N) \sim N^2 E(\nu).$$

It is interesting to study this approximation independently and improve it to more precise approximations. This problem was studied by Petrache and Serfaty [54] when the potential V is repulsive, and there is an additional external confining potential. For the particular case $-d < b < a = 2$ in (1.2), and when continuous minimizers are regular enough, this gives quite precise asymptotics for $E_N(X_N)$.

The only answer we know to question **MF** was given in [19] for power-law potentials (1.2) with $b \geq 1$, and in [13] for general potentials which include the power laws (1.2) with $b > 2 - d$. In this case, the answer to all questions in **MF** is positive for *strictly unstable* potentials V, that is, potentials V for which there exists a probability distribution ρ on $\mathbb{R}^d$ such that

$$E(\rho) < \lim_{|x| \to +\infty} V(x)$$

(assuming this limit exists or is $+\infty$). For a wide range of potentials, this concept is actually shown to imply that the classical H-stability condition (3.6) does *not* hold [13, 60]. We believe **MF** should also be true in the range $-d < b \leq 2 - d$, but as far as we know, there is no available proof. For the strategy in [13], two ingredients are needed:

1. One needs to show that the sequence of discrete energies E_N Gamma converges in a suitable sense to the continuum energy E.
2. One needs to show that the diameter of a minimizer X_N stays bounded as $N \to +\infty$. As a consequence, the empirical measures μ_{X_N}/N converge up to a subsequence, in the weak-$*$ sense, to a certain probability measure μ. By using the previous Gamma-convergence result (and possibly some additional uniform estimates on other properties of the minimizers), one may then deduce that μ must be a minimizer of E.

In relation to these questions, a description of the minimizers of E is also an interesting problem. As remarked in the introduction, these minimizers are of interest for collective behavior models and are also quite interesting from the point of view of calculus of variations. A general result on existence of these minimizers was given in [14, 60], but there remain many questions on their uniqueness, symmetry, and regularity properties. There has been a lot of recent activity regarding these, and we will try to give here a short account of the literature on this.

An early discussion on the problem of minimizing E can be found in [4, Section X]. In [58], for very regular potentials, it is proved that stationary states (and in particular minimizers, if any) must be sums of delta functions. A link between the repulsive singularity of the potential at $x = 0$ and the dimension of the support of minimizers was given by Balagué et al. [2], essentially by noticing that for any minimizer ρ, the convolution $\Delta V * \rho$ must be bounded. A consequence of this is that the support of a minimizer for V of the form (1.2) with $2 - d \le b < 2$ must have dimension at least $2 - b$. This is not optimal, since numerical simulations seem to indicate that, at least for "usual" potentials, the support always has integer dimension, and some effort has been devoted to understanding this effect. See however [16] for an example of fractal behavior for a specifically tailored potential V.

For "strongly repulsive" potentials with $-d < b < 2 - d$ (but still locally integrable), regularity of minimizers can be obtained via obstacle problems [20]. When $a = 2$ and $-d < b < 2 - d$, these minimizers correspond to the "fractional Barenblatt solutions" of [11, 12]; candidates for minimizers can be found for $-d < b < a$ with $b \le 2$, when a or b are even integers [15]. In dimension $d \ge 2$, when $a = 2$ and $2 - d < b < 4 - d$, minimizers are radially symmetric, unique, and were obtained in [16].

For $b = 2$ and $a \in (2, 4)$, the unique minimizer (up to translations) is the uniform measure on an $(n - 1)$-dimensional sphere of appropriate radius [25]. This has recently been extended by Frank and Matzke [34] to the case $d \ge 2$, $2 \le a \le 4$, $b_*(a) \le b \le 2$ with $b < a$, for a certain decreasing function $b_* = b_*(a)$ which is always between $-d + 3$ and $-d + 4$.

This is in contrast with minimizers for *mildly repulsive* potentials (potentials which are C^2 at 0), for which minimizers must be supported on a finite number of points under quite general conditions [21]. This result is true even for *local* minimizers in the topology of the ∞-Wasserstein transport distance. This applies to power-law potentials of the form (1.2) whenever $2 < b < a$. For $a \ge 4$ and $b \ge 2$ (excluding the case $(a, b) = (4, 2)$), the minimizer is unique up to rigid movements, and it must be a sum of equal delta functions supported at the vertices of a regular d-simplex [26]. In this same reference, the case $(a, b) = (4, 2)$ is studied. It is a special limit case in which minimizers are not unique (even up to rigid movements): Minimizers may be supported on a sphere, on a finite number of points, or may be convex combinations of these possibilities. In the 1D case, explicit minimizers are known for some power-law potentials of the form (1.2) [32].

In addition to this, the question of stability or instability of these minimizers for several dynamical models has been investigated in a few cases; see, for example, [30, 31] (for 1D stability of sums of Dirac masses for the aggregation equation (1.4)); [3, 46] (stability of spherical shells and rings); and [6] (behavior of patch solutions for $b = 2 - d$).

We will also mention some results on non-isotropic potentials: See [22] and [17, 18].

There are also related minimization problems which would take us too far from the topic; we refer the reader to [9] and [33] for an introduction to similar problems and many references on them.

We will illustrate the above properties in 2D with a few numerical results for power-law potentials of the form (1.2). Let us first fix $a = 2$ and look at the appearance of numerical minimizers for several values of b. As the exponent b increases, the repulsion becomes progressively weaker. It is clearly seen in Figs. 4 and 5 that minimizers become more concentrated as b increases, since repulsion

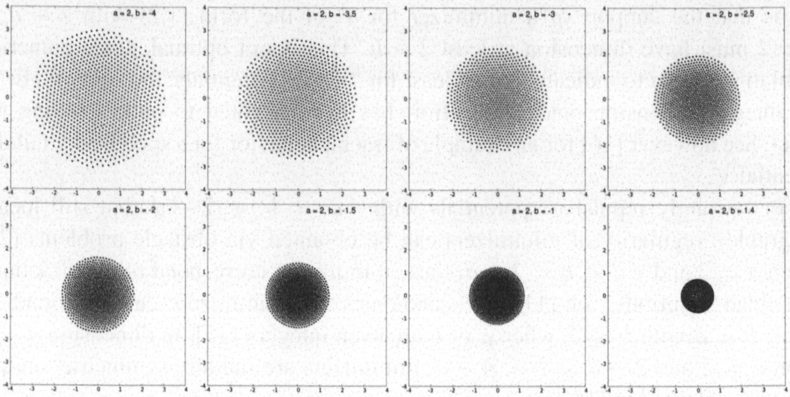

Fig. 4 Minimizer for the power-law potential (1.2) with $a = 2$ and several values of b, always with 10^3 particles

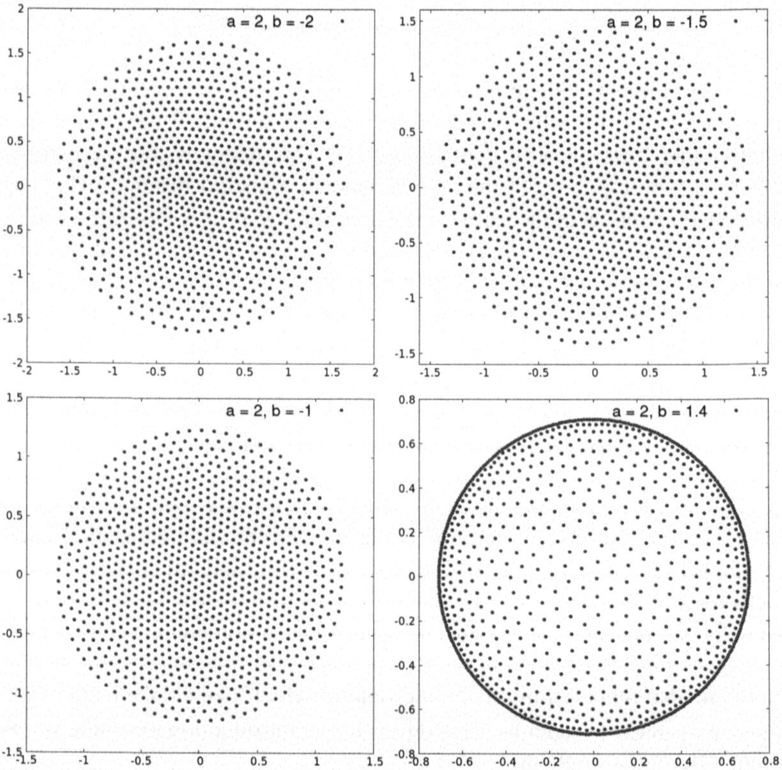

Fig. 5 Zoomed-in view of the last four minimizers in Fig. 4

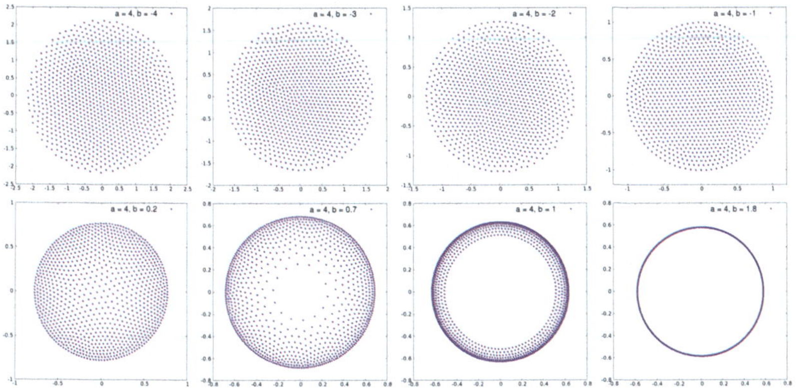

Fig. 6 Minimizer for the power-law potential (1.2) with $a = 4$ and several values of b, always with 10^3 particles

becomes locally weaker. Also, they become more concentrated on the boundary of a ball for higher b. For $b < 0$, we are in the regime studied in [20], so we know minimizers of E should be smooth, but it is not proved that N-particle minimizers approach minimizers of E. In contrast, for $b > 0$, we know from [13] that N-particle minimizers must approach minimizers of the continuous energy E, but we do not know the specific shape of these minimizers; for example, it is not proved in this range that minimizers must be regular functions on their support, or even L^p functions for some $p \geq 1$. Reasonable candidates for minimizers of E are given in [15], and they are all smooth functions on their support.

For $a = 4$, the situation is similar, but now there is a range of exponents with $b \geq 2$ where the potential V is $\mathcal{C}^2$. This is the range studied in [25] and [21]. In Fig. 6, we observe how the increase of b causes the appearance of empty space in the middle of the particle configuration, and its further increase implies the accumulation of all particles in a ring.

5 Between Crystallization and Continuous Limits

We will also point out that there is a range of exponents for power-law potentials which has been left almost untouched in the literature. For potentials V which are not locally integrable (that is, with $b \leq -d$), there are ranges of a for which crystallization (in the sense of **Q1** to **Q3**) is not expected to take place or has not been considered. The main unexplored range is $b \leq -d$, $a > 0$, since in the crystallization literature and in problems motivated by statistical mechanics, it is always assumed that $V(x)$ decays to 0 as $|x| \to +\infty$. But there is also a range of exponents $b < a < 0$ for which crystallization does not seem to happen, and we are left with some kind of limiting behavior which is not yet understood. We conjecture

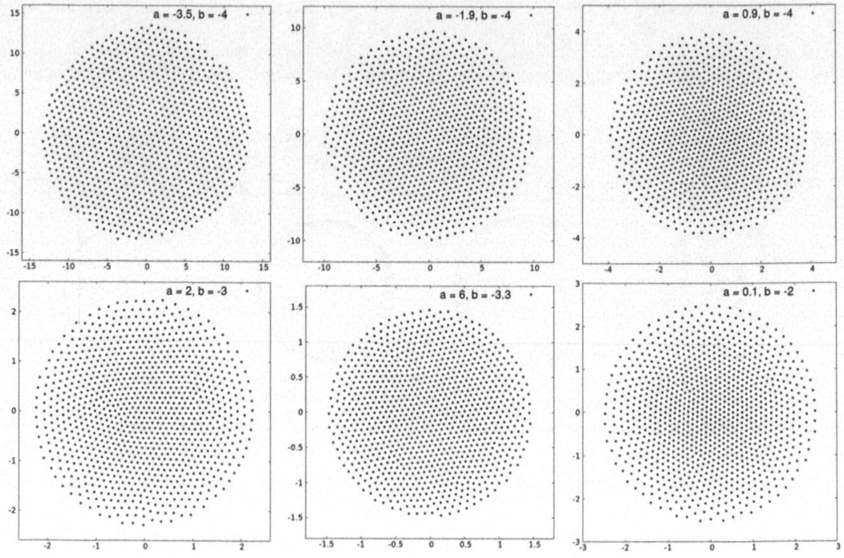

Fig. 7 Some minimizers with $b \leq -2$

that in this case, there is still a scaling which leads to a well-defined limiting profile along the lines of **MO**:

MS (macroscopic scaling) Do there exist $s > d$, a nontrivial measure μ on $\mathbb{R}^d$ and a sequence of minimizers $(X_N)_{N\geq 2}$ of E_N such that (possibly up to a subsequence)

$$\frac{1}{N}\mu_{X_N/N^{1/s}} \overset{*}{\rightharpoonup} \mu \qquad \text{as } N \to +\infty? \tag{5.1}$$

Or even stronger: Does this take place for *all* sequences of minimizers, up to rigid movements of these minimizers?

There are numerical results in 1D which seem to indicate that this is the case [19], but this has not been systematically explored, much less the subject of any rigorous proofs. Figure 7 shows 2D configurations of the kind $b < a < 0$ and $b < -d, a > 0$. In some of them, we appreciate certain similarities with crystallization cases (see Fig. 1), but we are not aware of any works exploring their behavior. This regime may be related to scaling limits of dynamical problems with non-integrable potentials such as [53], but the link is far from clear to us.

Acknowledgments We would like to thank David J. Wales for several interesting discussions and help with the software GMIN, used for the main numerical calculations in this paper. We also thank J. A. Carrillo and R. Frank for some interesting discussions. The authors acknowledge support from grant PID2020-117846GB-I00, the research network RED2022-134784-T, and the María de Maeztu grant CEX2020-001105-M from the Spanish government.

References

1. Y. Au Yeung, G. Friesecke, B. Schmidt, Minimizing atomic configurations of short range pair potentials in two dimensions: crystallization in the Wulff shape. Calcul. Variat. Part. Differ. Equ. **44**(1–2), 81–100 (2011) http://doi.org/10.1007/s00526-011-0427-6
2. D. Balagué, J.A. Carrillo,T. Laurent, G. Raoul, Dimensionality of local minimizers of the interaction energy. Arch. Ration. Mech. Analy. **209**(3),1055–1088 (2013). arXiv:1210.6795. http://doi.org/10.1007/s00205-013-0644-6
3. D. Balagué, J.A. Carrillo, T. Laurent, G. Raoul, Nonlocal interactions by repulsive–attractive potentials: Radial ins/stability. Phys. D Nonlinear Phenom. **260**, 5–25 (2013). arXiv:1109.5258 http://doi.org/10.1016/j.physd.2012.10.002
4. F. Bavaud, Equilibrium properties of the Vlasov functional: The generalized Poisson-Boltzmann-Emden equation. Rev. Modern Phys. **63**(1), 129 (1991). https://doi.org/10.1103/revmodphys.63.129
5. A. Ben Haj Yedder, X. Blanc, C. Le Bris, A numerical investigation of the 2-dimensional crystal problem. Preprint CERMICS (2003). Available at http://www.ann.jussieu.fr/publications/2003/R03003.html
6. A.L. Bertozzi, T. Laurent, F. Léger, Aggregation and spreading vie the Newtonian potential: the dynamics of patch solutions. Math. Models Methods Appl. Sci. **22**(supp01), 1140005 (2012). http://doi.org/10.1142/s0218202511400057
7. L. Bétermin, Optimality of the triangular lattice for Lennard–Jones type lattice energies: a computer-assisted method. J. Phys. A Math. Theor. **56**(14), 145204 (2023). http://dx.doi.org/10.1088/1751-8121/acc21d
8. L. Bétermin, L. De Luca, M. Petrache, Crystallization to the square lattice for a two-body potential. Arch. Ration. Mech. Analy. **240**(2), 987–1053 (2021). https://doi.org/10.1007/s00205-021-01627-6
9. X. Blanc, M. Lewin, The crystallization conjecture: a review. EMS Surv. Math. Sci. **2**(2), 225–306 (2015). https://doi.org/10.4171/emss/13
10. S.V. Borodachov, D.P. Hardin, E.B. Saff, *Discrete Energy on Rectifiable Sets*. (Springer, New York, 2019). http://dx.doi.org/10.1007/978-0-387-84808-2
11. L. Caffarelli, J.L. Vázquez, Asymptotic behaviour of a porous medium equation with fractional diffusion. Discrete Contin. Dyn. Syst. Ser. A **29**(4), 1393–1404 (2011). arXiv:1004.1096. https://doi.org/10.3934/dcds.2011.29.1393
12. L.A. Caffarelli, J.L. Vázquez, Nonlinear porous medium flow with fractional potential pressure. Arch. Ration. Mech. Analy. **202**(2), 537–565 (2011). arXiv:1001.0410. https://doi.org/10.1007/s00205-011-0420-4
13. J.A. Cañizo, F.S. Patacchini, Discrete minimisers are close to continuum minimisers for the interaction energy. Calcul. Variat. Partial Differ. Equ. **57**(1), (2018). https://doi.org/10.1007/s00526-017-1289-3
14. J.A. Cañizo, J.A. Carrillo, F.S. Patacchini, Existence of compactly supported global minimisers for the interaction energy. Arch. Ration. Mech. Analy. **217**(3), 1197–1217 (2015).https://doi.org/10.1007/s00205-015-0852-3. arXiv:1405.5428
15. J.A. Carrillo, Y. Huang, Explicit equilibrium solutions for the aggregation equation with power-law potentials. Kinetic Related Models **10**(1), 171–192 (2017). https://doi.org/10.3934/krm.2017007
16. J.A. Carrillo, R. Shu, From radial symmetry to fractal behavior of aggregation equilibria for repulsive–attractive potentials. Calcul. Variat. Part. Differ. Equ. 62(1) (2021). https://doi.org/10.1007/s00526-022-02368-4
17. J.A. Carrillo, R. Shu, Minimizers of 3D anisotropic interaction energies. Adv. Calcul. Variat. (2022). https://doi.org/10.1515/acv-2022-0059
18. J.A. Carrillo, R. Shu, Global minimizers of a large class of anisotropic attractive-repulsive interaction energies in 2D. Commun. Pure Appl. Math. **77**(2), 1353–1404 (2023). https://doi.org/10.1002/cpa.22162

19. J.A. Carrillo, M. Chipot, Y. Huang, On global minimizers of repulsive–attractive power-law interaction energies. Philosoph. Trans. R. Soc. A Math. Phys. Eng. Sci. **372** (2028), 20130399 (2014). https://doi.org/10.1098/rsta.2013.0399
20. J.A. Carrillo, M.G. Delgadino, A. Mellet, Regularity of local minimizers of the interaction energy via obstacle problems. Commun. Math. Phys. **343**(3), 747–781 (2016). arXiv:1406.4040. https://doi.org/10.1007/s00220-016-2598-7
21. J. Carrillo, A. Figalli, F. Patacchini, Geometry of minimizers for the interaction energy with mildly repulsive potentials. Annales de l'Institut Henri Poincaré C, Analyse non linéaire **34**(5),1299–1308 (2017). https://doi.org/10.1016/j.anihpc.2016.10.004
22. J.A. Carrillo, J. Mateu, M.G. Mora, L. Rondi, L. Scardia, J. Verdera, The equilibrium measure for an anisotropic nonlocal energy. Calcul. Variat. Part. Differ. Equ. **60**(3) (2021). https://doi.org/10.1007/s00526-021-01928-4
23. H.Cohn, A. Kumar, S. Miller, D. Radchenko, M. Viazovska, The sphere packing problem in dimension 24. Ann. Math. **185**(3), 1017–1033 (2017). https://doi.org/10.4007/annals.2017.185.3.8
24. H. Cohn, A. Kumar, S. Miller, D. Radchenko, M. Viazovska, Universal optimality of the E_8 and Leech lattices and interpolation formulas. Ann. Math. **196**(3), 983–1082 (2022). https://doi.org/10.4007/annals.2022.196.3.3
25. C. Davies, T. Lim, R.J. McCann, Classifying minimum energy states for interacting particles: Spherical shells. SIAM J. Appl. Math. **82**(4), 1520–1536 (2021). https://doi.org/10.1137/21m1455309
26. C. Davies, T. Lim, R.J. McCann, Classifying minimum energy states for interacting particles: Regular simplices. Commun. Math. Phys. **399**(2), 577–598 (2022). https://doi.org/10.1007/s00220-022-04564-x
27. D.M. Deaven, N. Tit, J.R. Morris, K.M. Ho, Structural optimization of Lennard-Jones clusters by a genetic algorithm. Chem. Phys. Lett. **256**(1–2), 195–200 (1996). https://doi.org/10.1016/0009-2614(96)00406-x
28. M. Dittner, B. Hartke, Conquering the hard cases of Lennard-Jones clusters with simple recipes. Comput. Theoret. Chem. **1107**, 7–13 (2017). https://doi.org/10.1016/j.comptc.2016.09.032
29. M.R. D'Orsogna, Y.L. Chuang, A.L. Bertozzi, L.S. Chayes, Self-propelled particles with soft-core interactions: Patterns, stability, and collapse. Phys. Rev. Lett. **96**(10), 104302 (2006). https://doi.org/10.1103/physrevlett.96.104302
30. K. Fellner, G. Raoul, Stability of stationary states of non-local equations with singular interaction potentials. Math. Comput. Modell. **53**, 1436–1450 (2010). https://doi.org/10.1016/j.mcm.2010.03.021
31. K. Fellner, G. Raoul, Stable stationary states of non-local interaction equations. Math. Models Methods Appl. Sci. **20**(12), 2267–2291 (2010). https://doi.org/10.1142/s0218202510004921
32. R.L. Frank, Minimizers for a one-dimensional interaction energy. Nonlinear Analy. **216**, 112691 (2022). https://doi.org/10.1016/j.na.2021.112691
33. R.L. Frank, *Some Minimization Problems for Mean Field Models with Competing Forces*. (EMS Press, Helsinki, 2023), pp. 277–294. ISBN 978-39-85475-51-3. https://doi.org/10.4171/8ecm/06
34. R.L. Frank, R.W. Matzke, Minimizers for an aggregation model with attractive-repulsive interaction. (2023). arXiv:2307.13769
35. C.S. Gardner, C. Radin, The infinite-volume ground state of the Lennard-Jones potential. J. Stat. Phys. **20**(6), 719–724 (1979). https://doi.org/10.1007/bf01009521
36. T. Hales, A proof of the Kepler conjecture. Ann. Math. **162**(3), 1065–1185 (2005). https://doi.org/10.4007/annals.2005.162.1065
37. D.P. Hardin, E.B. Saff, Minimal riesz energy point configurations for rectifiable d-dimensional manifolds. Adv. Math. **193**(1), 174–204 (2005). https://doi.org/10.1016/j.aim.2004.05.006
38. B. Hayes, The science of sticky spheres. Am. Sci. **100**(6), 442 (2012). https://doi.org/10.1511/2012.99.442

39. R.C. Heitmann, C. Radin, The ground state for sticky disks. J. Stat. Phys. **22**(3), 281–287 (1980). https://doi.org/10.1007/bf01014644
40. M.R. Hoare, Structure and dynamics of simple microclusters, in ed. by I. Prigogine, S.A. Rice, *Advances in Chemical Physics*, vol. 40 (Wiley, Hoboken, 1979), pp. 49–135
41. M.R. Hoare, P. Pal, Physical cluster mechanics: statics and energy surfaces for monatomic systems. Adv. Chem. Phys. **20**, 161–196 (1971). https://doi.org/10.1080/00018737100101231
42. M.R. Hoare, P. Pal, Statics and stability of small cluster nuclei. Nat. (Phys. Sci.) **230**, 5–8 (1971). https://doi.org/10.1038/physci230005a0
43. M.R. Hoare, P. Pal, Geometry and stability of "spherical" f.c.c. microcrystallites. Nat. (Phys. Sci.) **236**, 35–37 (1972). https://doi.org/10.1038/physci236035a0
44. R.S. Hoy, J. Harwayne-Gidansky, C.S. O'Hern, Structure of finite sphere packings via exact enumeration: Implications for colloidal crystal nucleation. Phys. Rev. E **85**(5), 051403 (2012). https://doi.org/10.1103/physreve.85.051403
45. S. Kirkpatrick,, C.D. Gelatt, M.P. Vecchi, Optimization by simulated annealing. Science **220**(4598), 671–680 (1983). https://doi.org/10.1126/science.220.4598.671
46. T. Kolokolnikov,, H. Sun,, D. Uminsky, A.L. Bertozzi, Stability of ring patterns arising from two-dimensional particle interactions. Phys. Rev. E **84**, 015203 (2011). https://doi.org/10.1103/physreve.84.015203
47. Z. Li, H.A. Scheraga, Monte Carlo-minimization approach to the multiple-minima problem in protein folding. Proc. Natl. Acad. Sci. **84**(19), 6611–6615 (1987). https://doi.org/10.1073/pnas.84.19.6611
48. D.C. Liu, J. Nocedal, On the limited memory BFGS method for large scale optimization. Math. Programm. **45**(1), 503–528 (1989)
49. L.D. Luca, G. Friesecke, Crystallization in two dimensions and a discrete Gauss–Bonnet theorem. J. Nonlinear Sci. **28**(1), 69–90 (2018). https://doi.org/10.1007/s00332-017-9401-6
50. G. Meng,, N. Arkus,, M.P. Brenner, V.N. Manoharan, The free-energy landscape of clusters of attractive hard spheres. Science **327**(5965), 560–563 (2010). https://doi.org/10.1126/science.1181263
51. J.A. Niesse, H.R. Mayne, Minimization of small silicon clusters using the space-fixed modified genetic algorithm method. Chem. Phys. Lett. **261**(4–5), 576–582 (1996). https://doi.org/10.1016/0009-2614(96)01000-7
52. J.A. Northby, Structure and binding of Lennard-Jones clusters: $13 \leq N \leq 147$. J. Chem. Phys. **87**(10), 6166–6177 (1987). https://doi.org/10.1063/1.453492
53. K. Oelschläger, Large systems of interacting particles and the porous medium equation. J. Differ. Equ. **88**(2), 294–346 (1990). https://doi.org/10.1016/0022-0396(90)90101-t
54. M. Petrache, S. Serfaty, Next order asymptotics and renormalized energy for Riesz interactions. J. Instit. Math. Jussieu **16**(3), 501–569 (2017). https://doi.org/10.1017/s1474748015000201
55. M. Petrache, S. Serfaty, Crystallization for Coulomb and Riesz interactions as a consequence of the Cohn-Kumar conjecture. Proc. Am. Math. Soc. **148**(7), 3047–3057 (2020). https://doi.org/10.1090/proc/15003
56. C. Radin, The ground state for soft disks. J. Stat. Phys. **26**(2), 365–373 (1981). https://doi.org/10.1007/bf01013177
57. C. Radin, Low temperature and the origin of crystalline symmetry. Int. J. Modern Phys. B **1**(5/6), 1157–1191 (1987). https://doi.org/10.1142/S0217979287001675
58. G. Raoul, Nonlocal interaction equations: Stationary states and stability analysis. Differ. Integr. Equ. **25**(5/6), 417–440 (2012). https://doi.org/10.57262/die/1356012673
59. D. Ruelle, *Statistical Mechanics: Rigorous Results*. (W.A. Benjamin, Los Angeles, 1969). ISBN 978-08-05383-60-7. http://www.worldcat.org/isbn/9780805383607
60. R. Simione, D. Slepčev, I. Topaloglu, Existence of ground states of nonlocal-interaction energies. J. Stat. Phys. **159**(4), 972–986 (2015). `arXiv:1405.5146`. https://doi.org/10.1007/s10955-015-1215-z
61. F. Theil, A proof of crystallization in two dimensions. Commun. Math. Phys. **262**(1), 209–236 (2006). https://doi.org/10.1007/s00220-005-1458-7

62. L. Trombach, R.S. Hoy, D.J. Wales, P. Schwerdtfeger, From sticky-hard-sphere to Lennard-Jones-type clusters. Phys. Rev. E **97**(4), 043309 (2018). https://doi.org/10.1103/physreve.97.043309
63. W.J. Ventevogel, Why do crystals exist? Phys. Lett. A **64**(5), 463–464 (1978). https://doi.org/10.1016/0375-9601(78)90685-0
64. W.J. Ventevogel, On the configuration of a one-dimensional system of interacting particles with minimum potential energy per particle. Phys. A Stat. Mech. Appl. **92**(3–4), 343–361 (1978). https://doi.org/10.1016/0378-4371(78)90136-x
65. W.J. Ventevogel, B.R.A. Nijboer, On the configuration of systems of interacting particles with minimum potential energy per particle. Phys. A Stat. Mech. Appl. **99**(3), 569–580 (1979). https://doi.org/10.1016/0378-4371(79)90072-4
66. W.J. Ventevogel, B.R.A. Nijboer, On the configuration of systems of interacting particle with minimum potential energy per particle. Phys. A Stat. Mech. Appl. **98**(1–2), 274–288 (1979). https://doi.org/10.1016/0378-4371(79)90178-x
67. M. Viazovska, The sphere packing problem in dimension 8. Ann. Math. **185**(3), 991–1015 (2017). https://doi.org/10.4007/annals.2017.185.3.7
68. D.J. Wales, GMIN optimisation software. Available at https://www-wales.ch.cam.ac.uk/GMIN/. https://www-wales.ch.cam.ac.uk/software.html
69. D.J. Wales, J.P.K. Doye, Global optimization by basin-hopping and the lowest energy structures of Lennard-Jones clusters containing up to 110 atoms. J. Phys. Chem. A **101**(28), 5111–5116 (1997). https://doi.org/10.1021/jp970984n
70. D.J. Wales, J.P.K. Doye, A. Dullweber, M.P. Hodges, F.Y. Naumkin, F. Calvo, J. Hernández-Rojas, T.F. Middleton, The Cambridge Cluster Database. http://www-wales.ch.cam.ac.uk/CCD.html
71. E. Weinan, D. Li, On the crystallization of 2D hexagonal lattices. Commun. Math. Phys. **286**(3), 1099–1140 (2008). https://doi.org/10.1007/s00220-008-0586-2
72. L.T. Wille, Searching potential energy surfaces by simulated annealing. Nature **325**, 374–374 (1987). https://doi.org/10.1038/325374c0
73. L.T. Wille, Minimum-energy configurations of atomic clusters: new results obtained by simulated annealing. Chem. Phys. Lett. **133**(5), 405–410 (1987). https://doi.org/10.1016/0009-2614(87)87091-4
74. J. Zhang, A brief review on results and computational algorithms for minimizing the Lennard-Jones potential. (2010), `arXiv:1101.0039 [physics.comp-ph]`

Large-Population Limits of Non-exchangeable Particle Systems

Nathalie Ayi and Nastassia Pouradier Duteil

Abstract A particle system is said to be non-exchangeable if two particles cannot be exchanged without modifying the overall dynamics. Because of this property, the classical mean-field approach fails to provide a limit equation when the number of particles tends to infinity. In this review, we present novel approaches for the large-population limit of non-exchangeable particle systems, based on the idea of keeping track of the identities of the particles. These can be classified in two categories. The non-exchangeable mean-field limit describes the evolution of the particle density on the product space of particle positions and labels. Instead, the continuum limit allows to obtain an equation for the evolution of each particle's position as a function of its (continuous) label. We expose each of these approaches in the frameworks of static and adaptive networks.

1 Introduction

1.1 Setting

Interacting particle systems refer to a large class of coupled differential equations modeling populations of interacting agents ("particles") susceptible to exhibit global organizational patterns without any centralized intelligence. They are used by various scientific communities to model many different phenomena, such as opinion formation [31], organization within animal groups [1, 5, 41], synchronization of

N. Ayi
Sorbonne Université, CNRS, Université Paris Cité, Inria, Laboratoire Jacques-Louis Lions (LJLL), Paris, France

Institut Universitaire de France, Paris, France

N. Pouradier Duteil (✉)
Sorbonne Université, CNRS, Université Paris Cité, Inria, Laboratoire Jacques-Louis Lions (LJLL), Paris, France
e-mail: nastassia.pouradier_duteil@sorbonne-universite.fr

J. A. Carrillo, E. Tadmor (eds.), *Active Particles, Volume 4*, Modeling and Simulation in Science, Engineering and Technology,
https://doi.org/10.1007/978-3-031-73423-6_3

coupled oscillators [39], and biological tissue formation [6]. These models have also been of great interest to the mathematical community. In particular, the study of their long-time behavior can reveal remarkable *global* patterns, although the particles interact only *locally* (see, for instance, [13, 15, 29, 30, 34]). Another direction of work consists of studying the control or optimal control of such systems to drive the population to a target configuration (see [18] and references within). A third area of interest concerns their so-called *large-population limit*, which is the focus of the present review.

The need for large-population limits of interacting particle systems becomes evident when one considers the size of some of the populations of interest. For instance, starlings are known to form flocks of several thousands of individuals [5], and neural networks in mammals contain up to 10^{11} neurons [33]. In this context, simulating such large populations via a system of N-coupled equations (with up to N^2 pairwise interactions) rapidly becomes computationally too expensive as the number of particles N increases. To resolve this issue, a crucial idea introduced in the pioneering works [17, 22, 28, 32, 50, 59, 60] consists of focusing on the evolution of the population density, instead of the individual trajectories.

Mean-Field Limit of Exchangeable Particle Systems Consider a system of N interacting particles characterized by their positions $(x_i^N)_{i\in\{1,\cdots,N\}} \in C([0,T];(\mathbb{R}^d)^N)$, whose evolution is given by the general system of coupled differential equations:

$$\begin{cases} \dfrac{d}{dt}x_i^N(t) = \dfrac{1}{N}\displaystyle\sum_{j=1}^{N} \phi(x_i^N(t), x_j^N(t)) \quad \text{for all } i \in \{1,\cdots,N\}, \quad t \in [0,T] \\ x_i^N(0) = x_i^{N,0} \text{ for all } i \in \{1,\cdots,N\}, \end{cases} \tag{1}$$

where

- $\{1,\cdots,N\}$ denotes the set of labels of the interacting particles, which represents their individual identities;
- $(x_i^N)_{i\in\{1,\cdots,N\}} \in C([0,T];(\mathbb{R}^d)^N)$ denotes the state of the interacting particles, and can, for instance, represent positions [23], opinions [31], velocities [20], phases [39], and angles [62]. The dimension d of the state space is then determined by the model;
- $\phi : \mathbb{R}^d \times \mathbb{R}^d \to \mathbb{R}^d$ is the so-called *interaction function*, encoding the pairwise interaction between any two positions x_i and x_j.

As N tends to infinity, one can show that the system (1) is well approximated by its so-called *mean-field limit* $\tilde{\nu} \in C([0,T];\mathcal{P}(\mathbb{R}^d))$ which satisfies the following Vlasov equation:

$$\begin{cases} \partial_t \tilde{\nu}_t(x) + \nabla_x \cdot \left(\left(\displaystyle\int_{\mathbb{R}^d} \phi(x,y) d\tilde{\nu}_t(y) \right) \tilde{\nu}_t(x) \right) = 0, \\ \tilde{\nu}_{t=0} = \tilde{\nu}_0. \end{cases} \tag{2}$$

For reasons that will become clear, in this review, we will refer to (2) as the **exchangeable mean-field limit** of the particle system. The link between the particle system (1) and its mean-field limit equation (2) becomes clear when introducing the so-called *empirical measure* $\tilde{\nu}^N$ constructed from the solution $(x_i^N)_{i\in\{1,\cdots,N\}}$ to the microscopic system (1):

$$\tilde{\nu}_t^N := \frac{1}{N}\sum_{i=1}^{N} \delta_{x_i^N(t)}, \tag{3}$$

and which satisfies the mean-field equation (2). Convergence of the solution $(x_i^N)_{i\in\{1,\cdots,N\}}$ of the microscopic system (1) toward the solution $\tilde{\nu}$ to the mean-field equation (2) can then be obtained as the result of a stability argument for the solution to (2) of the form:

$$W_1(\tilde{\nu}_t, \tilde{\nu}_t^N) \leq C(T) W_1(\tilde{\nu}_0, \tilde{\nu}_0^N), \tag{4}$$

where W_1 is the 1-Wasserstein (also known as Rubinstein-Kantorovich) distance [22]. An important condition to establish the stability estimate (4) is the Lipschitz continuity of the interaction function ϕ with respect to its two arguments.

Exchangeable vs. Non-exchangeable Particle Systems Interacting particle systems can be broadly classified into two categories: *exchangeable* (or indistinguishable) and *non-exchangeable* (or non-indistinguishable). A particle system is said to be exchangeable if any two particles can be exchanged without modifying the dynamics of the other particles. More precisely,

Definition 1 Let $(x_i^N)_{i\in\{1,\cdots,N\}} \in C([0,T];(\mathbb{R}^d)^N)$ denote the trajectories of N interacting particles satisfying

$$\begin{cases} \dfrac{d}{dt}x_i^N(t) = F_i(t, x^N(t)), & i \in \{1,\cdots,N\}, \quad t \in [0,T], \\ x_i^N(0) = x_i^{N,0}, & i \in \{1,\cdots,N\}, \end{cases}$$

and let $x_i^N(t) := \Psi_t \# x_i^{N,0}$ denote their push-forward by the flow of F. The particle system is said to be *exchangeable* (or, equivalently, *indistinguishable*) if, for any permutation function $\sigma : \{1,\cdots,N\} \to \{1,\cdots,N\}$ of the sets of particle labels, it holds

$$\begin{aligned} &\forall i \in \{1,\cdots,N\},\ y_i^N(0) = x_{\sigma(i)}^N(0) \\ &\Rightarrow \forall i \in \{1,\cdots,N\},\ \forall t \in [0,T] \Psi_t \# y_i^{N,0} = \Psi_t \# x_{\sigma(i)}^{N,0}, \end{aligned}$$

In other words, an interacting particle system is exchangeable if one can relabel the particles without modifying the dynamics. Trivially, the particle system (1) is exchangeable, since the right-hand side does not depend explicitly on the particle labels.

Exchangeable particle systems provide a good modeling framework for many applications in which the *labels*, which represent the *identities* of the particles, do not influence the dynamics. In this case, considering all particles to be identical is a good approximation and has been validated experimentally [1, 5, 41].

An example of system in which the particles are a priori *non-exchangeable* is provided by the Kuramoto model, used to describe the evolution of coupled oscillators' phases $(x_i^N)_{i\in\{1,\cdots,N\}} \in C([0,T];(\mathbb{T}^1)^N)$:

$$\frac{d}{dt}x_i^N(t) = u_i^N + \frac{C}{N}\sum_{j=1}^N \sin(x_j^N(t) - x_i^N(t)), \quad i \in \{1,\ldots,N\}. \tag{5}$$

In (5), the evolution of each oscillator's phase x_i^N depends on an intrinsic frequency u_i^N, which itself is determined by the particles' index i, so at first glance, this system is a non-exchangeable particle systems. However, note that by defining a new compound variable $y_i^N := (x_i^N, u_i^N)$, System (5) can be rewritten as $\dot{y}_i = \frac{1}{N}\sum_{i=1}^N \phi(y_i^N, y_j^N)$, where $\phi(y, y') := Nu + C\sin(x' - x)$. The new function ϕ is Lipschitz-continuous with respect to y and y', which allows to use the stability argument (4) for exchangeable systems, and derive its mean-field limit in the classical sense.

Why the Classical Mean-Field Limit Fails for Non-exchangeable Particle Systems In this review, we will focus on *non-exchangeable* particle systems of the form:

$$\begin{cases} \frac{d}{dt}x_i^N(t) = \frac{1}{N}\sum_{j=1}^N w_{ij}^N \phi(x_i^N(t), x_j^N(t)) \text{ for all } i \in \{1,\cdots,N\},\ t \in [0,T], \\ x_i^N(0) = x_i^{N,0}. \end{cases} \tag{6}$$

In system (6), the effect of the particles' labels $\{1,\cdots,N\}$ on the dynamics is decoupled from that of the particles' positions $(x_i^N)_{i\in\{1,\cdots,N\}}$, as it is introduced as a family of multiplicative weights $(w_{ij}^N)_{i,j\in\{1,\cdots,N\}} \in \mathbb{R}^N$. Particle systems of the form (6) can be seen as posed on an underlying weighted graph, in which the set of nodes corresponds to the set of labels $\{1,\cdots,N\}$ and to each edge (i,j) is attributed a weight $w_{ij}^N \in \mathbb{R}$.

Finding a good approximation of non-exchangeable particle systems (6) when N is large has been the subject of many recent works [3, 19, 25, 32, 35, 37, 46, 47, 53]. One main difficulty comes from the fact that the classical mean-field approach is no longer applicable. Indeed, studying the population density instead of the individual particles' trajectories entails an irreversible information loss, as one loses track of the particles' labels. In particular, notice that the empirical measure (3) is blind to any permutation $\sigma : \{1,\cdots,N\} \to \{1,\cdots,N\}$ of the sets of labels, since

$$\tilde{\nu}_t^N = \frac{1}{N}\sum_{i=1}^{N}\delta_{x_i^N(t)} = \frac{1}{N}\sum_{i=1}^{N}\delta_{x_{\sigma(i)}^N(t)}.$$

On the other hand, permuting the labels of the particle system (6) evidently modifies the individual trajectories, due to the presence of the weights w_{ij}^N. Thus, one cannot hope to be able to capture the dynamics of system (6) with a mean-field equation of the type (2) posed only on the space of positions.

Importantly, as mentioned above for the specific case of the Kuramoto model (5), if the interaction weights can be written as $w_{ij}^N := w(\xi_i^N, \xi_j^N)$ for a Lipschitz function w and some variables $(\xi_i^N)_{i\in\{1,\cdots,N\}}$, the particle system (6) can be rewritten for the new variables $y_i^N := (x_i^N, \xi_i^N)$, and a new empirical measure can be introduced on the product space of labels and positions. One can then apply the classical theory of Dobrushin [22] and leverage the Lipschitz continuity of w and ϕ to establish a stability estimate for solutions to the transport equation satisfied by this new empirical measure and proceed to prove the convergence of the particle system in this extended space. To summarize, in this Lipschitz framework, the particle system (6) can be treated as an *exchangeable* particle system. This approach hence fails if the interaction weights are not of the form $w_{ij}^N := w(\xi_i^N, \xi_j^N)$ for a Lipschitz function w, that is, if the particle system is truly *non-exchangeable*.

New Results in Limits of Graphs Non-exchangeable particle systems of the form (6) can be seen as posed on an underlying graph. For instance, in [63], a modified version of the Kuramoto model is introduced, in which the coupled oscillators are positioned in a one-dimensional ring, so that each oscillator interacts only with k oscillators on each side (for $k < \frac{N}{2}$):

$$\frac{d}{dt}x_i^N(t) = u + \frac{C}{N}\sum_{j=i-k}^{i+k}\sin(x_j^N(t) - x_i^N(t)). \tag{7}$$

System (7) can be interpreted as a system of differential equations on the graph $G_N = \langle V(G_N), E(G_N)\rangle$ whose vertex set is $V(G_N) = \{1, \ldots, N\}$ and edge set is

$$E(G_N) = \{(i, j) \in \{1, \ldots, N\} : 1 \leq \text{dist}(i, j) \leq k\}$$

where $\text{dist}(i, j) = \min\{|i - j|, N - |i - j|\}$.

The graph G_{10} for $k = 2$ and its adjacency matrix $(w_{ij}^N)_{i,j\in\{1,\cdots,N\}}$ are represented in Fig. 1.

Notice that the adjacency matrix of G_N belongs to $M_N(\mathbb{R})$, the space of N-by-N matrices with real coefficients, whose dimension changes with each value of N. One can instead introduce the piecewise-constant function $w_{G_N} : I^2 \to \{0, 1\}$, where $I := [0, 1]$, such that

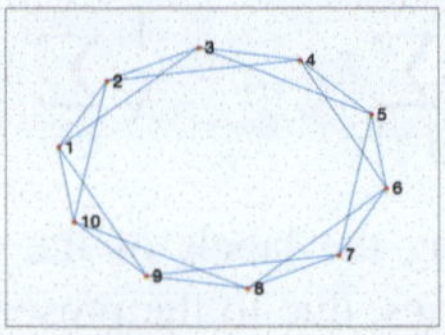

$$\begin{pmatrix} 0&1&1&0&0&0&0&0&1&1\\ 1&0&1&1&0&0&0&0&0&1\\ 1&1&0&1&1&0&0&0&0&0\\ 0&1&1&0&1&1&0&0&0&0\\ 0&0&1&1&0&1&1&0&0&0\\ 0&0&0&1&1&0&1&1&0&0\\ 0&0&0&0&1&1&0&1&1&0\\ 0&0&0&0&0&1&1&0&1&1\\ 1&0&0&0&0&0&1&1&0&1\\ 1&1&0&0&0&0&0&1&1&0 \end{pmatrix}$$

Fig. 1 Graph and adjacency matrix associated with system (7), for $N = 10$ and $k = 2$

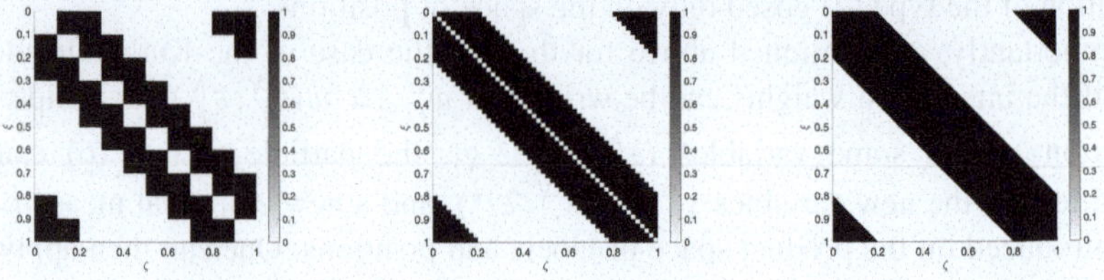

Fig. 2 Left and center: Pixel matrices of the graphs associated with (5) for $N = 10$ and $N = 50$, with $k = \frac{N}{5}$. Right: Plot of the limit graphon w

$$w_{G_N}(\xi, \zeta) = \begin{cases} 1 & \text{if } (\xi, \zeta) \in \left[\frac{i-1}{N}, \frac{i}{N}\right) \times \left[\frac{j-1}{N}, \frac{j}{N}\right) \text{ and } (i, j) \in E(G_N), \\ 0 & \text{otherwise.} \end{cases}$$

The plot of w_{G_N}'s support is nothing else than a pixel representation of the adjacency matrix of G_N (see Fig. 2). Moreover, with this new characterization, the space $L^\infty(I^2)$ to which w_{G_N} belongs no longer varies with N, which allows to consider its limit in the same space. In example (7), if k is proportional to N, one can show that w_{G_N} converges as N goes to infinity to a $[0, 1]$-valued function denoted w with respect to the cut distance, which we describe below. This function w is called a *graphon* in Graph Theory. This central object appears when considering limits of sequences of convergent *dense* graphs $(G_N)_{N\geq 1}$ (see [43]), that is, sequences in which $|E(G_N)| \approx C|V(G_N)|^2$ for some constant $C \in (0, 1)$. In other words, dense graph sequences are sequences of graphs whose average degree (i.e., average number of edges per node) stays proportional to the total number of possible edges $|V(G_N)|^2$. The other extreme situation, in which the average degree of the graph stays bounded although the number of nodes tends to infinity, is referred to as *sparse* graph sequence. The intermediate cases, in which the average degree grows slower than $|V(G_N)|^2$, are referred to as having *intermediate density* [42].

Remark 1.1 Note the graph G_N associated with example (7) is an undirected, unweighted graph (i.e., the associated adjacency matrix contains either zeros or ones). The theory applies similarly to directed and weighted graphs, for which the adjacency matrix $(w_{ij}^N)_{i,j\in\{1,\cdots,N\}}$ takes values in $\mathbb{R}$ and is not necessarily symmetric.

The limit of the particle system (6) when N goes to infinity then naturally reveals objects from Graph Theory, which provides powerful tools to tackle the question of the large-population limit (see, e.g., [4, 43]). For instance, the convergence of a graph sequence is linked to the convergence of w_{G_N} to w in the so-called *cut-norm*, which is defined, for $w \in L^1(I^2)$, as

$$\|w\|_{\square} := \sup_{\substack{S,T \text{ measurable} \\ \text{subsets of } I}} \left| \int_{S\times T} w(\xi,\zeta)d\xi d\zeta \right|. \tag{8}$$

The cut-norm can be used to define the corresponding metric

$$d_{\square}(\tilde{w}, w) = \|\tilde{w} - w\|_{\square}.$$

However, to compare graphs (or graphons), it seems natural to identify two graphs (respectively, graphons) which are the same up to a relabeling of the nodes. To allow for such comparisons, the cut distance is defined by

$$\delta_{\square}(\tilde{w}, w) = \inf_{\Psi \in S_I} d_{\square}(\tilde{w}, w^{\Psi})$$

where S_I denote the set of all invertible measure-preserving maps and $w^{\Psi}(\xi,\zeta) := w(\Psi(\xi), \Psi(\zeta))$. Notice that $\delta_{\square}$ is only a pseudometric, since it can be zero for two different graphons, as long as they are identical up to relabeling. Importantly, it holds

$$\delta_{\square}(\tilde{w}, w) \leq \|\tilde{w} - w\|_{L^1(I^2)}, \tag{9}$$

so that convergence in L^1-norm implies convergence in cut distance of the graph sequence $(G_N)_{N\geq 1}$. This explains why the L^p- setting is adopted in many frameworks (as will be seen in Sects. 2.1.1 and 2.2). One of the drawbacks of this setting is that it is suited for *dense* graphs sequences. Indeed, in sparse graph sequences (in which a node has a bounded number of neighbors), the limiting graphon is trivial, meaning $w = 0$. Thus, this setting does not allow to capture the limiting object associated with this particular class of density.

However, recent results in [4] allow to revisit this question in order to address sparse graphs or intermediate densities. For instance, a population which has a strong community structure can be represented by a locally dense while globally sparse network. In [4], Backhaus and Szegedy provide a general framework unifying dense and sparse graph limit theory. Their approach relies on the fact that graphs can actually be represented as operators called *graphops*. More precisely, a graphop is a bounded and positivity-preserving operator $A : L^{\infty}(I) \to L^1(I)$, which is self-adjoint in the sense that for any $f, g \in L^{\infty}(I)$, it holds $\langle Af, g\rangle = \langle g, Af\rangle$, with

$$\langle f, g\rangle := \int_I f(\xi)g(\xi)d\xi.$$

Graphops can be seen as a generalization of graphons by defining for each graphon $w \in L^\infty(I^2)$ an associated graphop

$$A_w f(\zeta) := \int_I w(\xi, \zeta) f(\zeta) d\zeta.$$

Note that there exists alternative objects more general than graphons but less than graphops such as digraph measures [37] or extended graphons [33]. We will introduce them when presenting the associated large-population results (Sect. 2.1.2).

The Non-exchangeable Mean-Field Limit and the Continuum Limit The graphon framework allows to characterize the convergence of the weighted graphs defined by the interaction weights $(w_{ij}^N)_{i,j\in\{1,\cdots,N\}}$ as N tends to infinity, by considering the corresponding piecewise-constant graphons $w_{G_N} \in L^\infty([0,1]^2)$ defined for all $(\xi, \zeta) \in [0,1]^2$ by

$$w_{G_N}(\xi, \zeta) = \sum_{i,j=1}^{N} w_{ij}^N \mathbb{1}_{[\frac{i-1}{N}, \frac{i}{N})\times[\frac{j-1}{N}, \frac{j}{N})}(\xi, \zeta).$$

Convergence of the sequence $(w_{G_N})_{N\in\mathbb{N}}$ to a limit graphon $w \in L^\infty([0,1]^2)$ can be quantified in terms of the natural cut-distance $\delta_\square$ or one of the L^p-norms. If the interaction weights converge as N tends to infinity to a limit graphon $w \in L^\infty([0,1]^2)$, then the microscopic system can be shown to converge weakly toward a measure $\nu \in C([0,T], \mathcal{P}([0,1]\times\mathbb{R}^d))$, solution to a Vlasov equation of the form:

$$\begin{cases} \partial_t \nu_t(\xi, x) + \nabla_x \cdot \left(\left(\int_{[0,1]} \int_{\mathbb{R}^d} w(\xi, \zeta)\phi(x, y) d\nu_t(\zeta, y) d\zeta \right) \nu_t(\xi, x) \right) = 0, \\ \nu_{t=0} = \nu_0. \end{cases} \tag{10}$$

As we will see, the limit of the graph can be more general (such as a so-called extended graphon, a graphop, or a digraph measure), and the limit equation will be written accordingly. To differentiate this limit from the classical mean-field limit (2), we will refer to Eq. (10) as the **non-exchangeable mean-field limit** of System (6). Here, the variable $x \in \mathbb{R}^d$, as in the microscopic model (6), represents the position (or phase, opinion, etc.), and the newly introduced variable $\xi \in [0,1]$ is a continuous representation of the particles' labels, or identities. Notice in particular the asymmetric roles of of the two variables x and ξ. The limit equation is a transport equation, in which the probability measure ν is transported only in the direction of the variable x, whereas the variable ξ plays the role of a structure variable. Due to its similarity with the previously mentioned mean-field limit, we will refer to Eq. (10) as the *non-exchangeable mean-field limit* of system (6). The non-exchangeable mean-field limit has been derived in various frameworks in [19, 25, 32, 35, 37, 53], presented in Sect. 2.1.

As shown in [3, 46, 47, 53], when the interaction weights $(w_{ij}^N)_{i,j\in\{1,\cdots,N\}}$ converge as N tends to infinity to a graphon $w \in L^\infty([0,1]^2)$, the microscopic system (6) can also be shown to converge pointwise toward the solution $x \in C([0,T]; L^\infty(\mathbb{R}^d))$ to an integrodifferential Euler-type equation (also denoted *nonlinear heat equation* in [46] or *graph limit* in [12, 46, 47]):

$$\begin{cases} \partial_t x(t,\xi) = \displaystyle\int_{[0,1]} w(\xi,\zeta)\phi(x(t,\xi),x(t,\zeta))d\zeta \\ x(0,\cdot) = x_0. \end{cases} \tag{11}$$

In this review, we will refer to this approach as the *continuum limit*, as in [26]. Notice that the integrodifferential equation (11) is nothing but the particle system (6) written on an infinite graph, whose nodes are labeled by the continuous variable $\xi \in [0,1]$ instead of the discrete labels $\{1,\cdots,N\}$, hence the name "continuum limit." This approach will be presented in Sect. 2.2.

The relations between the two limit equations (10) and (11) are presented in Sect. 2.3.

Adaptive Dynamical Networks Adaptive dynamical networks represent a broad class of interconnected dynamical systems, useful to cover a wide range of real-life applications. Their main feature is that the connectivity of the network evolves over time and that this evolution can depend on the states of the system itself. For instance, in the context of opinion dynamics, nodes symbolize individual agents, while links mirror the myriad connections that we maintain in our social spheres, be it with friends, family, or colleagues. In that framework, the adaptive nature is to be understood as follows: not only are relationships likely to influence our opinions, but our opinions also exert a reciprocal effect, inducing alterations in the network structure (our relationships). An illustrative instance is found in the adaptive voter model (see, for instance, [64]), where agents update their opinions and connections. With a certain probability, each agent may adopt the viewpoint of interacting agents or, alternatively, shift their connections toward those who share more similar opinions. These intricate models offer a nuanced representation of reality, acknowledging that in numerous scenarios, networks are far from static. This not only holds true as mentioned above for social interactions [2, 7, 8, 44, 52, 54, 56] but extends to diverse domains such as neural and neuronal networks [55, 57], machine learning applications [40, 58], power grid models [51], and beyond. For a comprehensive exploration of various domains and examples, we direct readers to the review [9].

As in [9], we will say that a network is *adaptive* if the evolution of the edge (i,j) explicitly depends on the states of the nodes i and j. A counterexample is the blinking system in [3] for which every edge of the graph is reset at regular time intervals, independently of the states of the nodes at this time. To summarize, as mentioned in [9], there are two main classes for adaptive networks:

- *event-based adaptation*, where the network structure changes at certain discrete points in time and the triggers for the changes depend on the system itself,
- *continuous adaptation*.

In this review, we will exclusively focus on the second class. It can be presented in its general form as

$$\begin{cases} \dfrac{d}{dt}x_i^N(t) = f_i(x_i^N(t), t) + \displaystyle\sum_{j=1}^{N} w_{ij}^N(t)\phi\left(x_i^N(t), x_j^N(t), t\right) & \text{for all } i \in \{1, \cdots, N\} \\ \dfrac{d}{dt}w_{ij}^N(t) = h_{ij}(w^N(t), x^N(t), t) \end{cases} \tag{12}$$

where $w^N := (w_{ij}^N)_{i,j\in\{1,\cdots,N\}}$, where w_{ij}^N represents the (now time-evolving) weight of the connection from node j to i, and $x^N = (x_1^N, \ldots, x_N^N)$.

The question of the large-population limit, whether through the mean-field approach or the continuum limit one, is challenging in its generality. We choose to focus on two specific frameworks that have successfully addressed this problem. The first one is the setting of Kuramoto-type models on N oscillators [39]. The second one can be seen as a variant of the Hegselmann-Krause dynamics [31] where, additionally to the opinions, we also are interested in the evolving-in-time weights of agents which represent their charisma.

We will present mean-field limits for those models in Sect. 3.1 and continuum limits in Sect. 3.2. As in the static case, we will exhibit the links between the different limit equations in Sect. 3.3.

1.2 Notations and Preliminary Remarks

In the last 10 years, the surge of interest in large-population limits for non-exchangeable particle systems has led to a wealth of publications. Consequently, due to the rapid growth of the field and to the variety of models considered, no unifying set of notations has emerged within the community. In this review, we have decided to use the following notations:

- x will denote a particle's state (which can represent an opinion, position, velocity or phase, as mentioned above). Note that it is also denoted by x in [2, 33], but it is denoted by ξ in [53], by u in [3, 19, 25, 35, 46, 47], by ϕ in [37], and by θ in [10].
- Ω will denote the state space in which the particles evolve. We can distinguish between two main classes of results. Articles focusing on the Kuramoto model (5) consider that x represents an oscillator phase and thus take the state space to be the torus $\Omega = \mathbb{T} := \mathbb{R}/(2\pi\mathbb{Z})$ [19, 25–27, 35]. On the other hand, when the state is assumed to be a position, the state space is assumed to be $\Omega = \mathbb{R}^d$, for instance, in [2, 33, 53]. For presentation simplicity or technical reasons, one can also find $\Omega = \mathbb{R}$ in [3, 46, 47].

- ξ will denote a particle's label or identity. It is also denoted by ξ in [33] but is denoted by s in [2, 12] and by x in [10, 25, 37, 46, 47, 53].
- I will denote the probability space to which the (continuous) label ξ belongs. Most often, the discrete label $i \in \{1, \cdots, N\}$ will be mapped to the continuous label ξ by the transformation $i \mapsto \frac{i}{N}$, so that $I = [0, 1]$ [2, 3, 19, 35, 46, 47]. In this case, the probability space I is endowed with the Lebesgue measure $d\xi$. In other works, the label set is taken to be a more general multidimensional set, denoted X [37] or Ω [25, 53]. Choosing a different reference measure allows to give more or less weight to different parts of the population. For instance, the mixed model composed of finitely many leaders and infinitely many followers in [24] can be written using the set $I = [0, 1]$ endowed with the probability measure $\frac{1}{2}d\xi + \frac{1}{2N}\sum_{k=1}^{m}\delta_{\frac{k}{m}}$. To keep the presentation as clear as possible, we chose to state all results in the framework $([0, 1], d\xi)$ but will mention the possible extensions to more general sets when applicable.
- ϕ will denote the interaction function, as in [2]. It is also denoted by K in [33], by g in [37], and by D in [25, 35, 46, 47]. Note that in many applications, $\phi : (x, y) \mapsto \tilde{\phi}(y - x)$ for some function $\tilde{\phi}$.
- $(w_{ij}^N)_{i,j\in\{1,\cdots,N\}}$ will denote the interaction weights as in [33, 46]. The weight w_{ij}^N is denoted by $W_{N,ij}$ in [35], by a_{ij} in [37], by A_{ij} in [25], and by ξ_{ij}^N in [47].

Moreover, we will consistently use the following conventions for probability measures (where VE stands for the Vlasov equation):

	Empirical measures	Continuous empirical measures	Solution to the VE
Static	$\tilde{\nu}_t^N(x) := \frac{1}{N}\sum_{i=1}^{N}\delta_{x_i^N(t)}(x)$	$\bar{\tilde{\nu}}_t(x) = \int_I \delta_{x(t,\xi)}(x)d\xi$	$\tilde{\nu}_t(x)$
Static	$\nu_t^N(\xi, x) := \frac{1}{N}\sum_{i=1}^{N}\delta_{x_i^N(t)}(x)\delta_{\xi_i^N}(\xi)$	$\overline{\nu}_t(\xi, x) = \int_I \delta_{x(t,\zeta)}(x)\delta_\zeta(\xi)d\zeta$	$\nu_t(\xi, x)$
Adaptive	$\mu_t^N(k, x, m) := \frac{1}{N}\sum_{i=1}^{N}\delta_{x_i^N(t)}(x) \times \delta_{m_i^N(t)}(m)\delta_i(k)$	$\overline{\mu}_t(\xi, x, m) = \int_I \delta_{x(t,\zeta)}(x)\delta_{m(t,\zeta)}(m) \times \delta_\zeta(\xi))d\zeta$	$\mu_t(\xi, x, m)$
Adaptive	$\tilde{\mu}_t^N(x, m) := \frac{1}{N}\sum_{i=1}^{N}\delta_{x_i^N(t)}(x)\delta_{m_i^N(t)}(m)$	$\bar{\tilde{\mu}}_t(x, m) = \int_I \delta_{x(t,\xi)}(x)\delta_{m(t,\xi)}(m)d\xi$	$\tilde{\mu}_t(x, m)$
Adaptive	$\hat{\mu}_t^N(x) := \frac{1}{N}\sum_{i=1}^{N}m_i^N(t)\delta_{x_i^N(t)}(x)$	$\bar{\hat{\mu}}_t(x) = \int_I m(t, \xi)\delta_{x(t,\xi)}(x)d\xi$	$\hat{\mu}_t(x)$

In order to keep our presentation as clear as possible given the variety of existing frameworks, we have made the choice to present some results in a simplified formalism. We encourage the reader to consult the more general statements in the corresponding original articles.

2 Large-Population Limits of Non-exchangeable Particle Systems on Static Graphs

Let $T > 0$ and $x_i^{N,0} \in \Omega^N$. Let $(x_i^N)_{i\in\{1,\cdots,N\}} \in C([0,T];\Omega^N)$ represent a system of non-exchangeable interacting particles satisfying the system of coupled equations

$$\begin{cases} \frac{d}{dt}x_i^N(t) = \frac{1}{N}\sum_{j=1}^N w_{ij}^N \phi(x_j^N(t) - x_i^N(t)) & \text{for all } i \in \{1,\cdots,N\},\ t\in[0,T] \\ x_i^N(0) = x_i^{N,0} & \text{for all } i \in \{1,\cdots,N\}. \end{cases} \tag{13}$$

- The set $\{1,\cdots,N\}$ represents the set of *labels* of the interacting particles, i.e., their identities. Two distinct particles cannot have the same label.
- For each $t \in [0,T]$, $x_i^N(t) \in \Omega$ represents the state of the particle with label i at time t. Depending on the model, it might represent an opinion, a position (in which case $\Omega = \mathbb{R}^d$), a phase ($\Omega = \mathbb{T}$), etc. For simplicity, from here onward, we will refer to $x_i^N(t)$ as the particle's *position*.
- The interaction function $\phi \in \mathrm{Lip}(\Omega^2)$ models the spatial interaction between any two particles. It will be considered to be a Lipschitz function.
- $(w_{ij}^N)_{i,j\in\{1,\cdots,N\}}$ are interaction weights, modeling an underlying interaction network. We suppose that for every $(i,j) \in \mathbb{N}^2$, $w_{ij}^N \in \mathbb{R}$. Depending on the framework, it will also sometimes be assumed that the matrix $(w_{ij}^N)_{i,j\in\{1,\cdots,N\}}$ is symmetric, corresponding to an undirected graph, and that $w_{ij}^N \in \mathbb{R}_+$. These assumptions will be made clear when necessary.

Under these hypotheses, there exists a unique solution $(x_i^N)_{i\in\{1,\cdots,N\}} \in C^1([0,T];\Omega^N)$ to (13) (see [53] for the proof under more general hypotheses).

2.1 *Mean-Field Limit*

The question of the mean-field limit of system (13) has been the focus of several works in the recent years [19, 33, 35, 36, 53]. The main requirement is that as N tends to infinity, $(w_{ij}^N)_{i,j\in\{1,\cdots,N\}}$ has a limit, which can be either a graphon $w \in L^\infty(I\times I)$, or a more general object. Then, the microscopic system can be shown to converge (in a certain sense to be specified) toward a measure ν, solution to (10), which by disintegration of the measure ν, can be rewritten as follows: for almost every ξ,

$$\begin{cases} \partial_t \nu_t^\xi(x) + \nabla_x \cdot \left(\left(\int_I \int_\Omega w(\xi,\zeta)\phi(x,y) d\nu_t^\zeta(y) d\zeta \right) \nu_t^\xi(x) \right) = 0, \\ \nu_{t=0} = \nu_0. \end{cases} \tag{14}$$

As in the microscopic model (13), the variable $x \in \Omega$ represents the position (or phase, or opinion), and the newly introduced variable $\xi \in I$ is a continuous representation of the particles' labels, or identities. As announced in the introduction, we refer to this limit process as the **non-exchangeable mean-field limit** of System 13.

The interpretation of ν can be twofold. Let $\Delta\xi \subset I$ and $\Delta x \subset \mathbb{R}^d$.

- In a probabilistic setting, $\nu_t(\Delta\xi \times \Delta x) = \int_{\Delta\xi \times \Delta x} d\nu_t(\xi, x) d\xi$ represents the probability of finding an agent with label in $\Delta\xi$ and position in Δx at time t.
- In a deterministic setting, $\nu_t(\Delta\xi \times \Delta x)$ represents the mass of agents with labels in $\Delta\xi$ and positions in Δx at time t.

Existence and uniqueness of the solution to (14) can be proven using a fixed-point argument similar to the classical Neunzert's fixed point for the exchangeable Vlasov equation [49], where the Lipschitz continuity of the interaction function ensures well-posedness of the fibered characteristics system associated with (14). However, since it is not the main focus of this review, we do not go into details and refer the reader to the cited articles for the proof in each of the frameworks that we mention below (see, for instance, Sect. 2.1.1).

The first results for non-exchangeable mean-field limit theory appeared in the framework of *dense* graphs (Sect. 2.1.1). These results were then extended to graphs with intermediate or sparse densities Sect. 2.1.2.

2.1.1 Dense Graphs

For $N \in \mathbb{N}$, let $\langle G_N, V(G_N), E(G_N) \rangle$ denote a graph of N nodes, whose set of vertices and set of edges are, respectively, given by $V(G_N)$ and $E(G_N)$. A sequence of graphs $(G_N)_{N\in\mathbb{N}}$ is called "dense" [42, 43] if the number of edges each graph contains is proportional to the maximal number of edges it can contain, i.e., if

$$|E(G_N)| \approx C|V(G_N)|^2$$

where $C \in (0, 1)$. The limit of a convergent dense graph sequence is given by a so-called *graphon* [42, 43]. Although definitions of graphons vary depending on the context in which they are introduced, in this review, we will use the following notion:

Definition 2 A *graphon* is a function $w \in L^\infty(I^2; \mathbb{R})$.

Lipschitz Graphon In the seminal paper [19], Chiba and Medvedev derive the mean-field limit of the microscopic system (13) with the following regularity assumption:

Hypothesis 1 (Graphon Regularity) *The graphon* $w \in \mathrm{Lip}(I^2; \mathbb{R})$ *is a Lipschitz function.*

This assumption allows to treat the mean-field equation (14) as the usual Vlasov equation associated with exchangeable particle systems. The proof is provided in the specific case of the Kuramoto model, in which the variables $(x_i)_{i\in\{1,\cdots,N\}}$ represent oscillator phases and evolve on the torus $\Omega = \mathbb{T}$. However, it can easily be extended to the case of particles in $\mathbb{R}^d$. For each $N \in \mathbb{N}$, we introduce a discretization $(\xi_1^N, \cdots, \xi_N^N) \in I^N$ of the unit interval, with the requirement that for all continuous function $f \in C(I)$, the Riemann sum of f evaluated at $(\xi_i^N)_{i\in\{1,\cdots,N\}}$ converge to the integral of f on I, i.e.,

$$\lim_{N\to\infty} \frac{1}{N}\sum_{i=1}^{N} f(\xi_i^N) = \int_I f(\xi)d\xi.$$

Given such a discretizations, each discrete weight w_{ij}^N is defined by evaluating the continuous graphon w at the corresponding discretization points:

$$w_{ij}^N := w(\xi_i^N, \xi_j^N). \tag{15}$$

This specific choice along with the continuity assumption Hypothesis 1 endows the non-exchangeable system (13) with a hidden exchangeable structure. Similarly to the transformation of the Kuramoto model (5) (see Sect. 1.1), by denoting $y_i^N := (\xi_i^N, x_i^N)$, one can rewrite the particle system (13) as

$$\begin{cases} \dfrac{d}{dt} y_i^N(t) = \dfrac{1}{N}\displaystyle\sum_{j=1}^{N} \psi(y_i^N(t), y_j^N(t)) \quad \text{for all } i \in \{1, \cdots, N\},\ t \in [0, T] \\ y_i^N(0) = (\xi_i^N, x_i^{N,0}) \quad \text{for all } i \in \{1, \cdots, N\}, \end{cases}$$

where $\psi(y, y') := (0, w(\xi, \xi')\phi(x' - x))$. Then, the empirical measure associated with a solution $(x_i^N)_{i\in\{1,\cdots,N\}}$ to the microscopic system (13) with such interaction weights is defined for all $t \in [0, T]$ by

$$\nu_t^N(\xi, x) := \frac{1}{N}\sum_{i=1}^{N} \delta_{x_i^N(t)}(x)\delta_{\xi_i^N}(\xi). \tag{16}$$

Importantly, probability measures in $\mathcal{P}(I \times \Omega)$ are compared via the *Bounded Lipschitz distance* d_{BL} *on the product space* $I \times \Omega$, defined for all $\mu_0, \nu_0 \in \mathcal{P}(I \times \Omega)$

by

$$d_{\mathrm{BL}}(\mu_0, \nu_0) := \sup \left\{ \int_{I\times\Omega} f(\xi, x) d(\mu_0(\xi, x) - \nu_0(\xi, x)) \mid f \in \mathrm{Lip}(I \times \Omega), \|f\|_{\mathrm{Lip}} \leq 1, \|f\|_{L^\infty} \leq 1 \right\}.$$

One can show that given Hypothesis 1, for any $T \in \mathbb{R}_+$, there exists a unique weak solution $\nu \in C([0, T], \mathcal{P}(I \times \Omega))$ to the mean-field equation (14). The proof relies on the theory of Neunzert developed for the Vlasov equation [48, 49], consisting in writing ν_t as the push-forward of ν_0 via the flow of the vector field $\tilde{V}[\nu_t](\xi, x) := (0, V[\nu_t](\xi, x))$, where

$$V[\nu_t](\xi, x) := \int_I \int_{\mathbb{R}^d} w(\xi, \zeta)\phi(x, y) d\nu_t^\zeta(y) d\zeta, \tag{17}$$

and in showing the existence and uniqueness of the solution to the associated fixed-point equation. Moreover, the Lipschitz continuity of the function $((\xi, x), (\zeta, y)) \mapsto w(\xi, \zeta)\phi(x, y)$ allows to obtain continuity of the solution to (10) with respect to the initial data in the Bounded Lipschitz distance. Given two solutions ν_t^1, ν_t^2 to (10) associated with the initial data $(\nu_0^1, \nu_0^2) \in \mathcal{P}(I \times \Omega)$, it holds for all $t \in [0, T]$:

$$d_{\mathrm{BL}}(\nu_t^1, \nu_t^2) \leq e^{CT} d_{\mathrm{BL}}(\nu_0^1, \nu_0^2),$$

for some constant $C > 0$. Then, as in the classical exchangeable case [22], the proof of convergence relies on two main arguments:

1. The empirical measure (16) built from the solution to the particle system (13) is a solution to the limit equation (14)
2. Solutions to (14) satisfy the previous stability estimate.

This implies the following convergence result:

$$d_{\mathrm{BL}}(\nu_0^N, \nu_0) \xrightarrow[N\to\infty]{} 0 \Rightarrow d_{\mathrm{BL}}(\nu_t^N, \nu_t) \xrightarrow[N\to\infty]{} 0,$$

where ν_t is the solution to (10).

Non-Lipschitz Graphon In [35], Kaliuzhnyi-Verbovetsky and Medvedev extend the previous result by revising Neunzert's fixed-point theory to adapt it to non-Lipschitz limit graphons. The main idea for this improvement is to exploit the fundamental asymmetry in the roles played by the variables x and ξ. More specifically, if ν_0 is absolutely continuous with respect to the variable ξ, one can show that the measure ν_t remains absolutely continuous with respect to ξ for all time (as shown in [19]). This allows to rewrite $d\nu_0(\xi, x) = d\nu_0^\xi(x)d\xi$, where for all

$\xi \in [0, 1]$, $\nu_0^\xi \in \mathcal{P}(\Omega)$. In this way, we consider the set of $\mathcal{P}(\Omega)$-valued functions $\bar{\mathcal{M}} := \{\mu : I \to \mathcal{P}(\Omega)\}$ equipped with the L^1-Bounded-Lipschitz distance

$$\bar{d}(\mu_0, \nu_0) = \int_I d_{\mathrm{BL}}(\mu_0^\xi, \nu_0^\xi) d\xi,$$

which is a complete metric space. The Lipschitz assumption on the graphon w can then be relaxed to the much weaker condition

Hypothesis 2 (Graphon Regularity) *The graphon $w \in L^\infty(I^2)$ satisfies*

$$\lim_{\delta \to 0} \int_I |w(\xi + \delta, \zeta) - w(\xi, \zeta)| d\zeta = 0.$$

With this assumption, one can show that the vector field V defined by (17) is Lipschitz continuous in x, continuous in ξ, and Lipschitz continuous with respect to ν_t. This implies that the equation of characteristics $\dot{x}(t) = V[\nu_t](\xi, x(t))$ is well-posed, and so that it generates a flow $\Psi_{t,0}$ defined for almost all $\xi \in I$ by $\Psi_{t,0}^\xi[w, \nu]x(0) = x(t)$. Then, it can be shown that the fixed-point equation $\nu_t = \Psi_{0,t}[w, \nu]\#\nu_0$ has a unique solution $\nu \in C([0, T], \bar{\mathcal{M}})$, which in turn implies existence and uniqueness of the solution to the mean-field equation (14). Moreover, the solution to (14) is continuous with respect to its initial data in the $L^1 - \mathrm{BL}$ distance $\bar{d}$ and is continuous with respect to the graphon w in the L^1-norm.

The main result of [35] is then to show the convergence to the mean-field equation (14) of the solution $(x_i^N)_{i \in \{1, \cdots, N\}}$ to the microscopic particle system (13) in which the discrete weights $(w_{ij}^N)_{i,j \in \{1, \cdots, N\}}$ are given by an L^1-approximation of the graphon w:

$$w_{ij}^N := N^2 \int_{\frac{i-1}{N}}^{\frac{i}{N}} \int_{\frac{j-1}{N}}^{\frac{j}{N}} w(\xi, \zeta) d\xi d\zeta. \tag{18}$$

Importantly, this construction implies the L^1 convergence of the graph sequence, as seen in Sect. 1.1, which is a stronger requirement than the convergence in cut distance (see Eq. (9)).

The link between the microscopic system (13) and the mean-field equation (14) is done via the empirical measure $\nu_{n,m} \in C([0, T], \bar{M})$, defined for all $N := nm$, for all $\xi \in [\frac{i-1}{n}, \frac{i}{n})$, for all $t \in [0, T]$ by

$$\nu_{n,m,t}^\xi(x) := \frac{1}{m} \sum_{j=1}^m \delta_{x_{(i-1)m+j}^N(t)}(x). \tag{19}$$

Contrarily to the Lipschitz case, this new empirical measure $\nu_{n,m,t}$ does not solve directly the mean-field equation (14). Therefore, the argument does not follow simply from Dobrushin's stability estimate but consists of introducing an auxiliary system which does solve the mean-field equation while remaining close to the empirical measure $\nu_{n,m,t}$ as m tends to infinity. A suitable stability estimate then allows to prove the convergence of the auxiliary system toward the solution to the mean-field equation (14). More precisely, for any $\varepsilon > 0$, there exists $n_1 \in \mathbb{N}$ such that for all $n \geq n_1$, there exists $m_1 \in \mathbb{N}$, such that for all $m \geq m_1$, for all $t \in [0, T]$,

$$\bar{d}(\nu^{\xi}_{n,m,t}, \nu_t) \leq \varepsilon,$$

where ν_t is the solution to the mean-field equation (14).

General Interaction Function In [53], Paul and Trélat provide a general result for particle systems of the form

$$\begin{cases} \frac{d}{dt}x_i^N(t) = \frac{1}{N}\sum_{j=1}^{N} G(t, \frac{i}{N}, \frac{j}{N}, x_i^N(t), x_j^N(t)) \text{ for all } i \in \{1, \cdots, N\},\ t \in [0, T] \\ x_i^N(0) = x_i^{N,0} \quad \text{for all } i \in \{1, \cdots, N\}, \end{cases} \tag{20}$$

in which the effect of the particles' labels is no longer decoupled from that of the particles' positions. The function

$$\begin{aligned} G : \mathbb{R} \times I \times I \times \mathbb{R}^d \times \mathbb{R}^d &\to \mathbb{R}^d \\ (t, \xi, \zeta, x, y) &\mapsto G(t, \xi, \zeta, x, y) \end{aligned}$$

is assumed to be locally Lipschitz with respect to (x, y) uniformly with respect to (t, ξ, ζ) on any compact subset of $\mathbb{R} \times I \times I$. Convergence can be proven to the following Vlasov equation:

$$\begin{cases} \partial_t \nu_t(\xi, x) + \nabla_x \cdot \left(\left(\int_{I\times\mathbb{R}^d} G(t, \xi, \zeta, x, y) d\nu_t(\zeta, y) d\zeta \right) \nu_t(\xi, x) \right) = 0, \\ \nu_{t=0} = \nu_0. \end{cases} \tag{21}$$

Existence and uniqueness of the weak solution $\nu \in C([0, T]; \mathcal{P}_c(I \times \mathbb{R}^d))$ to (21) is proven for all compactly supported initial data $\nu_0 \in \mathcal{P}_c(I \times \mathbb{R}^d)$. Moreover, the link between the discrete and the continuous system is provided by introducing the empirical measure ν_t^N defined in (16) with $\xi_i = \frac{i}{N}$ for all $i \in \{1, \ldots, N\}$. Then, denoting by W_p the p-Wasserstein distance on the product space $I \times \mathbb{R}^d$, one has the following result for all $p \geq 1$:

$$W_p(\nu_0, \nu_0^N) \xrightarrow{N\to\infty} 0 \quad \Rightarrow \quad W_p(\nu_t, \nu_t^N) \xrightarrow{N\to\infty} 0$$

on all compact time interval $[0, T]$.

Moreover, if G is assumed to be locally Lipschitz with respect to all four arguments (ξ, ζ, x, y), a stability estimate for (21) allows to write

$$\forall t \in [0, T], \quad W_p(\nu_t, \nu_t^N) \leq C_{\nu,\nu^N}(t) W_p(\nu_0, \nu_0^N),$$

where the constant $C_{\nu,\nu^N}(t)$ depends on the Lipschitz norm of G with respect to x and y on the supports of ν_t and ν_t^N. Note that in that framework, introducing the new variable $y_i^N = (x_i^N, \frac{i}{N})$ reduces the system to an exchangeable one.

Moreover, if the system is posed on a graph, i.e., $G(t, \xi, \zeta, x, y) := w(\xi, \zeta)\phi(x, y)$, the Lipschitz condition on G implies that the graphon w is Lipschitz, as in [19].

Remark 2.1 In its full generality, the result from [53] is stated for a general n-dimensional set I, and the Lebesgue measure $d\xi$ in (21) can be replaced by a more general measure $d\nu(\zeta)$. The final time of existence T is not necessarily finite.

Unlabeled Graphon The previous three results [19, 35, 53] are all based on a same general idea: A limit graphon w is given and used to build a graph sequence converging to w in L^1-norm by discretizing w in an appropriate way (see, for instance, (15) or (18)). The convergence of the microscopic system (13) toward the limit Vlasov equation is obtained as a consequence of this discretization procedure. This "top-down" process can be argued to be somewhat artificial, since the graph sequence is built using the *a priori* knowledge of the limit graphon. Another downside is the assumption that the graph sequence converges in L^1-norm, which is a stronger requirement than convergence in the cut-distance $\delta_\square$.

In [10], Bet, Coppini, and Nardi propose a different approach, with no *a priori* knowledge of the limit object, for the random system on a graph G_N

$$\begin{cases} dX_t^{i,N} = \dfrac{1}{N} \displaystyle\sum_{i=1}^{N} w_{ij}^N \phi(X_t^{i,N}, X_t^{j,N}) dt + dB_t^i \\ X_0^{i,N} = X_0^i, \end{cases} \tag{22}$$

where $(B^i)_{i\in\{1,\cdots,N\}}$ is a sequence of independent and identically distributed (i.i.d.) Brownian motions on $\Omega := \mathbb{T}$, the initial conditions $(X_0^i)_{i\in\{1,\cdots,N\}}$ are i.i.d. sampled from some probability distribution $\boldsymbol{\nu}_0$, and the weights (w_{ij}^N) are the edge weights of the graph G_N.

The main tool of this approach is the cut norm $\|\cdot\|_\square$ (see Equation (8)), well-known to the Graph Theory community [42, 43]. To prove convergence of the microscopic system with noise (22), one needs to consider the cut distance between two graphons, taking into consideration all possible relabelings, defined by

$$\delta_\square(\tilde{w}, w) = \inf_{\Psi \in S_I} \|\tilde{w} - w^\Psi\|_\square,$$

where S_I denotes the space of invertible measure-preserving maps Ψ from I to I and where $w^\Psi : (\xi, \zeta) \mapsto w(\Psi(\xi), \Psi(\zeta))$. Importantly, in this setting, the cut distance between two graphons w and $\tilde{w}$ can be zero for two different graphons. The important point is that the graphons be equal up to relabeling.

In this framework, the microscopic system is linked to its mean-field limit by the traditional empirical measure $\tilde{\boldsymbol{\nu}}_t^N \in \mathcal{P}(\Omega)$ defined as in (3) by

$$\tilde{\boldsymbol{\nu}}_t^N := \frac{1}{N} \sum_{i=1}^{N} \delta_{X_i^N(t)}. \tag{23}$$

Notice that so defined, the empirical measure $\tilde{\boldsymbol{\nu}}_t^N$ does not contain the information of the labels i and is blind to any relabeling $X_{\sigma(i)}^N$ of the particle system.

The main result can then be stated as follows. Assume that $\phi \in C^{1+\varepsilon}(\Omega)$ for some $\varepsilon > 0$. Consider a sequence of graphs G_N, whose associated graphons w_{G_N} defined by

$$w_{G_N} : (\xi, \zeta) \mapsto \sum_{i=1}^{N} \sum_{j=1}^{N} w_{ij}^N \mathbf{1}_{[\frac{i-1}{N}, \frac{i}{N})}(\xi) \mathbf{1}_{[\frac{j-1}{N}, \frac{j}{N})}(\zeta)$$

converge in cut norm to an a priori unknown limit graphon in probability, in the sense that

$$\lim_{N \to +\infty} \mathbb{E}[\delta_\square(w_{G_N}, w)] = 0.$$

Note that the limit graphon w is not unique, since any relabeling of w would also be a limit of w_{G_N} in the cut distance. For this reason, w is referred to as an *unlabeled graphon* in [10]. Then, the empirical measures $\tilde{\boldsymbol{\nu}}_t^N$ converge to a limit measure $\tilde{\boldsymbol{\nu}} \in \mathcal{P}(C([0, T], \Omega))$, which is the weak solution to the nonlinear Fokker-Plank equation

$$\begin{cases} \partial_t \tilde{\boldsymbol{\nu}}_t(x) + \nabla_x \cdot \left(\int_I \left(\int_I \int_\Omega w(\xi, \zeta) \phi(x, y) d\check{\nu}_t^\zeta(y) d\zeta \right) \check{\nu}_t^\xi(x) d\xi \right) = \frac{1}{2} \Delta_x \tilde{\boldsymbol{\nu}}_t(x), \\ \tilde{\boldsymbol{\nu}}_{t=0} = \tilde{\boldsymbol{\nu}}_0, \end{cases} \tag{24}$$

in which the measure $\check{\nu}_t^\xi$ is defined by the nonlinear process

$$\begin{cases} X_t = X_0 + \int_0^t \int_I \int_\Omega w(U, \zeta) \phi(X_s, y) d\check{\nu}_s^\zeta(y) d\zeta ds + B_t \\ \check{\nu}_t^\zeta = \mathcal{L}(X_t | U = \zeta) \quad \text{for } \zeta \in I, t \in [0, T], \end{cases} \tag{25}$$

where $\mathcal{L}(X_0) = \tilde{\boldsymbol{\nu}}_0$, U is a uniform random variable on I and B_t is a Brownian motion.

Importantly, this convergence result for the marginal $\tilde{\boldsymbol{\nu}}_t^N$ provides less information than the convergence results for the associated full empirical measure $\boldsymbol{\nu}_t^N$ (see (16)) defined on the product space $I \times \Omega$, since the marginal $\tilde{\boldsymbol{\nu}}_t^N$ does not contain the information of the labels. We refer the reader to [10, 11, 21] for more details on such probabilistic approaches.

2.1.2 Sparse Graphs

As seen in the Introduction, the graph sequence corresponding to Example (7) admits a graphon limit when k increases proportionally to N as N tends to infinity. What can be said when k is not proportional to N?

Recent results in Graph Theory have provided multiple ways of defining the limit of a graph sequence which is not dense. These new limit objects include L^p-graphons, graphops, graphings, and measures [4, 14, 38]. We refer to [33] for an illustration of the relationships between the various limit objects.

Graphop The result of [35] is further extended by Gkogkas and Kuehn in [25] to a class of sparse graphs defined using so-called graphops (graph operators). Graphops were introduced in [4] by Backhausz and Szegedy in the aim of providing a general framework unifying dense and sparse graph theories. The main idea consists of moving away from the object of graphons (i.e., functions $w \in L^\infty(I^2)$) and of considering instead the action of graphs as operators from $L^\infty(I)$ to $L^1(I)$.

Definition 3 A graphop is a linear operator $A : L^\infty(I) \to L^1(I)$ which has finite operator norm $\|A\|_{\infty\to 1}$, is positivity-preserving, and is self-adjoint, in the sense that for any $f, g \in L^\infty(I)$, it holds $\langle Af, g\rangle = \langle g, Af\rangle$, with

$$\langle f, g\rangle := \int_I f(\xi)g(\xi)d\xi.$$

Importantly, there exists a family $(\nu_A^\xi)_{\xi\in I}$ of finite fiber measures so that the action of the graphop A on a function $f \in L^\infty(I)$ is given by:

$$Af : \xi \mapsto \int_I f(\zeta)d\nu_A^\xi(\zeta).$$

This object generalizes the concept of symmetric graphons in the following way: For every symmetric graphon $w \in L^\infty(I^2)$, one can define a graphop $A_w : L^\infty(I) \to L^1(I)$ such that for all $f \in L^\infty(I)$,

$$A_w f : \xi \mapsto \int_I w(\xi, \zeta)f(\zeta)d\zeta.$$

Following a similar strategy to that of [35], the authors of [25] introduce an auxiliary system which approximates the empirical measure, and, importantly, which does solve the mean-field equation. In the stability estimate, the authors found that they needed a stronger notion of convergence than the so-called action convergence in the theory of graphops by Backhausz and Szegedy [4]. More precisely, consider a sequence of graphons $(w_K)_{K\in\mathbb{N}}$ such that their associated graphops $(A_{w_K})_{K\in\mathbb{N}}$ converge as K goes to infinity to a limit graphop A, in the sense that for almost all $\xi \in I$, $\nu^{\xi}_{A_K} \rightharpoonup \nu^{\xi}_{A}$. As in [35], the authors of [25] propose to derive the mean-field limit of the solution $x^{N,K}$ of the microscopic system (13) in which the weights $w^{N,K}_{ij}$ are given by the L^1 approximation of w_K:

$$w^{N,K}_{ij} := N^2 \int_{\frac{i-1}{N}}^{\frac{i}{N}} \int_{\frac{j-1}{N}}^{\frac{j}{N}} w^K(\xi,\zeta)d\xi d\zeta.$$

To this microscopic system is associated an empirical measure

$$\nu^{\xi}_{n,m,K,t}(x) := \frac{1}{m}\sum_{j=1}^{m} \delta_{x^{N,K}_{(i-1)m+j}(t)}(x)\mathbf{1}_{[\frac{i-1}{n},\frac{i}{n})}(\xi),$$

which is piecewise constant with respect to the variable ξ. As in [35], the empirical measure is shown to approximate as N tends to infinity the unique solution ρ^K to the equation

$$\partial_t \rho^K(t,\xi,x) + \nabla_x \cdot \left(\rho^K(t,\xi,x)\left(\int_{\Omega} \phi(x,y)(A_{w_K}\rho^K)(t,\xi,y)dy\right)\right) = 0,$$

given the necessary regularity condition:

Hypothesis 3 (Graphop Regularity) *For all* $\xi, \xi_0 \in I$, *if* $\xi \to \xi_0$, *then* $\nu^{\xi}_{A} \rightharpoonup \nu^{\xi_0}_{A}$.

The empirical measure $\nu_{n,m,K,t}$ defined as in (19) from $x^{N,K}$ can then be shown to converge toward ν^{ξ}_t, defined for all $S \in \mathcal{B}(\mathbb{R})$ by $\nu^{\xi}_t(S) := \int_S \rho(t,\xi,x)dx$, where ρ is the solution to the mean-field equation

$$\partial_t \rho(t,\xi,x) + \nabla_x \cdot \left(\rho(t,\xi,x)\left(\int_{\Omega} \phi(x,y)(A\rho)(t,\xi,y)dy\right)\right) = 0,$$

in the L^1-Bounded-Lipschitz distance.

Remark 2.2 In [25], the proof is done on $\Omega = \mathbb{T}$ for the Kuramoto model but can be easily extended to $\mathbb{R}$. Moreover, the space of nodes can be extended from $(I, d\xi)$ with $I = [0, 1]$ as exposed here to a more general $(\tilde{I}, dm(\xi))$, with $\tilde{I} \subset \mathbb{R}^n$.

Remark 2.3 As explained above, this approach requires the converging sequence of graphons to be symmetric (i.e., the corresponding graphs to be undirected).

Digraph Measures In [37], Kuehn and Xu further generalize the result of [25] to sparse directed graphs (also called "digraphs"), using the framework of *digraph* *measures*. Denoting by $\mathcal{M}_+$ the set of finite Borel positive measures, one can define digraph measures as everywhere defined in the first variable, bounded measures in the second variable:

Definition 4 Any measure-valued function $\eta \in \mathcal{B}(I, \mathcal{M}_+(I))$ is a *digraph measure*.

Convergence is shown in the *uniform bounded Lipschitz metric* d_∞, defined for all $\mu_1, \mu_2 \in \mathcal{B}(I, \mathcal{M}_+(\mathbb{R}^d))$ by

$$d_\infty(\mu_1, \mu_2) := \sup_{\xi \in I} d_{\mathrm{BL}}(\mu_1^\xi, \mu_2^\xi).$$

As in the works [35] and [25], the strategy in [37] involves an auxiliary system solving the mean-field equation, which approximates the empirical measures not solving the mean-field equation. In the stability estimate, convergence in the uniform bounded-Lipschitz distance d_∞ is needed for the digraph measures, which requires suitable approximations by finite graphs. These can be obtained under an additional continuity assumption with respect to the first variable for the limit digraph measure ν to which the sequence of graphs converge:

Hypothesis 4 $\eta \in \mathcal{C}(I, \mathcal{M}_+(I))$.

Using this framework, a *graphon* can be viewed as the digraph limit of a sequence of dense graphs, and a *graphop* is a symmetric digraph measure.

The results presented in [37] apply to general compact label sets I (not necessarily $[0, 1]$), with a reference measure that can differ from the usual Lebesgue one, and in particular that can be discrete or singular. However, in order to be consistent with what has been presented above, we present the mean-field limit equation that is obtained in this simplified framework, which can be written as

$$\partial_t \rho(t, \xi, x) + \nabla_x \cdot \left(\rho(t, \xi, x) \left(\int_I \int_{\mathbb{R}^d} \phi(x, y)\rho(t, \zeta, y) dy d\eta^\xi(\zeta)\right)\right) = 0.$$

Extended Graphon The approach proposed by Jabin, Poyato, and Soler in [33] differs from the approaches in [35], [25], and [37], in that it requires no continuity of the limit object (in this new framework, called *extended graphon*), which is defined by

Definition 5 An extended graphon is a measure $w \in L^\infty_\xi \mathcal{M}_\zeta \cap L^\infty_\zeta \mathcal{M}_\xi$.

Importantly, similarly to the approach of [10] exposed in the last paragraph of Sect. 2.1.1, this limit procedure requires no a priori knowledge of the limit of the discrete coupling weights $(w_{ij}^N)_{i,j \in \{1,\cdots,N\}}$. Consequently, instead of making assumptions on the limit extended graphon, assumptions are made on the discrete weights:

Hypothesis 5 *The discrete weights* $(w_{ij}^N)_{i,j \in \{1,\cdots,N\}}$ *satisfy:*

- $\max_{1\le i\le N} \frac{1}{N}\sum_{j=1}^{N} |w_{ij}^N| = O(1)$ *and* $\max_{1\le j\le N} \frac{1}{N}\sum_{i=1}^{N} |w_{ij}^N| = O(1)$
- $\max_{1\le i,j\le N} \frac{|w_{ij}^N|}{N} = o(1)$.

In [33], the authors introduce the additional hypothesis that initial data $x_{i,0}^N$ are independent random variables distributed according to absolutely continuous laws and that $\mathrm{Law}(x_{i,0}^N) = f_{i,0}^N(x)dx$ with

$$\sup_N \max_i \|f_{i,0}^N(x)\|_{W^{1,1}\cap W^{1,\infty}} < \infty.$$

Given weights satisfying Hypothesis 5, the microscopic system (13) is shown to converge to a limit function $f \in L^\infty([0,T]\times I, W^{1,1}\cap W^{1,\infty}(\Omega))$ for any $T > 0$, solution to

$$\partial_t f(t,x,\xi) + \nabla_x \cdot \left(\left(\int_I w(\xi, d\zeta) \int_\Omega \phi(x,y) f(t,y,\zeta) \right) f(t,x,\xi) \right) = 0,$$

where $w \in L_\xi^\infty \mathcal{M}_\zeta \cap L_\zeta^\infty \mathcal{M}_\xi$ is an *extended graphon*. More precisely, convergence is obtained in the following sense: up to the extraction of a subsequence,

$$\lim_{N\to\infty} \sup_{0\le t\le T} \mathbb{E} W_1 \left(\int_I f(t,\cdot,\xi)d\xi, \tilde{\nu}_t^N \right) = 0,$$

where W_1 is the 1-Wasserstein distance on Ω and the empirical measures $\tilde{\nu}_t^N$ are defined by (3) from the solution x^N to the discrete system (13). Notice that symmetry is required neither of the discrete weights $(w_{ij}^N)_{i,j\in\{1,\cdots,N\}}$ nor of the extended graphon w, unlike in the *graphop* approach [25]. However, contrarily to the digraph measure framework, a symmetric role is given to ξ and ζ in the definition of the extended graphon $w \in L_\xi^\infty \mathcal{M}_\zeta \cap L_\zeta^\infty \mathcal{M}_\xi$.

As in [10], this approach comes with a cost, as it loses some information by providing convergence of $\tilde{\nu}_t^N$ only toward the marginal $\int_I f(t,\cdot,\xi)d\xi$.

2.2 Continuum Limit

The mean-field limit presented in Sect. 2.1 provides *weak* convergence of the microscopic system (13) toward the solution to a transport equation (32). Another approach, denoted *continuum limit* or *graph limit*, provides a *pointwise* convergence of the solution to (13) toward the solution $x \in C([0,T]; L^\infty(\mathbb{R}^d))$ to an integrodifferential Euler-type equation:

$$\begin{cases} \partial_t x(t,\xi) = \displaystyle\int_I w(\xi,\zeta)\phi(x(t,\xi),x(t,\zeta))d\zeta \\ x(0,\cdot) = x_0. \end{cases} \tag{26}$$

In the case of the mean-field limit, the link between the solution x^N to the microscopic system (13) and the solution ν to a limit equation is done via a so-called *empirical measure* (given, e.g., in Eqs. (16), (19)). In the continuum limit framework, the link between the solution x^N to (13) and the solution x to (26) is provided by constructing a piecewise-constant function $x_N \in C([0,T]; L^\infty(\mathbb{R}^d))$ from x^N:

$$\forall \xi \in I, \quad x_N(t,\xi) := \sum_{i=1}^{N} x_i^N(t)\mathbf{1}_{[\frac{i-1}{N},\frac{i}{N})}(\xi). \tag{27}$$

The crucial point consists of noticing that x_N is solution to (26) if and only if x^N is solution to the microscopic equation (13) with the graphon defined by $w^N(\xi,\zeta) = w_{ij}^N$ for $(\xi,\zeta) \in I_i^N \times I_j^N$.

Another key difference with the mean-field limit approach is that most results currently published are in the context of *dense graphs*, i.e., graphs whose limits are graphons $w \in L^\infty(I^2)$ (see Definition 2 in Sect. 2.1.1). The available results can be divided into results for deterministic graphs and results for random graphs.

2.2.1 Continuum Limit on Deterministic Graphs

A seminal paper providing the first proof of convergence of the solution to the microscopic equation (13) toward its continuum limit (26) was published in 2014 by Medvedev [46]. The main result can be stated as follows. Let $x \in C([0,T]; L^\infty(\mathbb{R}^d))$ denote the solution to the integrodifferential equation (26) on a given graphon $w \in L^\infty(I^2)$ and with an initial condition $x_0 \in L^\infty(I)$. For each $N \in \mathbb{N}$, consider the solution x^N to the microscopic system (13) on the underlying graph whose adjacency matrix w^N is given by

$$\forall (i,j) \in \{1,\cdots,N\}^2, \quad w_{ij}^N := \int_{\frac{i-1}{N}}^{\frac{i}{N}} \int_{\frac{j-1}{N}}^{\frac{j}{N}} w(\xi,\zeta)d\zeta d\xi. \tag{28}$$

and the initial condition by

$$\forall i \in \{1,\cdots,N\}, \quad x_i^{N,0} := \int_{\frac{i-1}{N}}^{\frac{i}{N}} x_0(\xi)d\xi.$$

Let $x_N \in C([0,T]; L^\infty(\mathbb{R}^d))$ denote the piecewise-constant function built from the vector x^N by (27). Then, x_N converges to x in L^2−norm, satisfying

$$\lim_{N\to\infty} \sup_{t\in[0,T]} \|x(t,\cdot) - x_N(t,\cdot)\|_{L^2(I)} = 0.$$

A more detailed explanation of this limit process is presented in Sect. 3.2, in the case of a particle system with time-evolving weights.

Remarkably, this result requires no continuity of the graphon w. However, as shown in [53] (Theorem 4), if both the initial data and the graphon are regular enough, one can obtain a quantitative convergence result. More precisely, let $\alpha \in (0, 1]$ such that $x_0 \in C^{0,\alpha}(I; \mathbb{R}^d)$. Let $G : [0, T] \times I^2 \times (\mathbb{R}^d)^2$ be locally Lipschitz with respect to its last two arguments uniformly with respect to the first three, and suppose in addition that G is locally α-Hölder with respect to its last four arguments. Then, the solution x to the integrodifferential equation

$$\begin{cases} \partial_t x(t,\xi) = \displaystyle\int_I G(t,\xi,\zeta,x(t,\xi),x(t,\zeta))d\zeta \\ x(0,\cdot) = x_0, \end{cases}$$

satisfies $x(t,\cdot) \in C^{0,\alpha}(I; \mathbb{R}^d)$ for all $t \in [0, T]$. Moreover, let x^N denote the solution to the particle system

$$\begin{cases} \dfrac{d}{dt}x_i^N(t) = \dfrac{1}{N}\displaystyle\sum_{i=1}^{N} G(t,\frac{i}{N},\frac{j}{N},x_i^N(t),x_j^N(t)) \quad \text{for all } i \in \{1,\cdots,N\} \\ x_i^N(0) = x_0(\dfrac{i}{N}). \end{cases}$$

Then, for every $N \in \mathbb{N}$, it holds

$$\max_{i\in\{1,\cdots,N\}} |x(t,\frac{i}{N}) - x_i^N(t)| \le \frac{1}{N^\alpha}(1 + \mathrm{Hol}_\alpha(x_0))e^{2tL_x^N(t)},$$

where $\mathrm{Hol}_\alpha(x_0)$ denotes the Hölder constant of x_0 and $L_x^N(t)$ depends on the Hölder constant of $G(\tau,\cdot,\cdot,\cdot,\cdot)$ and on the Lipschitz constant of $G(\tau,\xi,\zeta,\cdot,\cdot)$ for all $\tau \in [0, t]$.

In addition, letting x_N be the piecewise-constant function defined by (27) from the solution x^N to the microscopic system, it holds

$$\|x(t,\cdot) - x_N(t,\cdot)\|_{L^\infty(I)} \le \frac{2}{N^\alpha}(1 + \mathrm{Hol}_\alpha(x_0))e^{2tL_x^N(t)}.$$

Remark 2.4 In order to insist on the role of the regularity, we stated the result from Theorem 4 in [53] in a simplified form. The full result, as stated in its most general form, only requires I to be a general compact smooth n-dimensional manifold, and the final time $T \in \mathbb{R}_+ \cup \{+\infty\}$ corresponds to the uniform maximal time until which the solution to the microscopic system is well-defined. We encourage the reader to consult [53] for a complete statement.

2.2.2 Continuum Limit on Random Graphs

Random Unweighted Graphs In [42, 43], Lovasz and Szegedy introduce a method to construct random graphs from a graphon w. Such random graphs, named w-random graphs, are unweighted, meaning that each edge is either present or absent, and consequently, the corresponding adjacency matrix $(w_{ij})_{i,j\in\{1,\cdots,N\}}$ contains values in $\{0, 1\}$. Given a graphon $w \in L^\infty(I^2; [0, 1])$, a w-random graph can be defined in two different ways, as proposed in [47]: either from a sequence of i.i.d. random variables or from a sequence of deterministic evenly spaced variables. Although these differences may seem subtle, they are indeed fundamental, and the type of convergence one may hope to obtain depends on the degree of randomness ("random-random" or "random-deterministic") introduced in the graph construction.

Definition 6 Let $w \in L^\infty(I^2; [0, 1])$. A w-random graph can be constructed either from a sequence of random variables (r-r) or from a sequence of deterministic variables (r-d):

(r-r) Let $Z = (Z_i)_{i\in\mathbb{N}}$ be a sequence of i.i.d. random variables, uniformly distributed in $I = [0, 1]$. Let $N \in \mathbb{N}$. A random (unweighted) graph G_N generated by the random sequence Z is constructed by inserting each edge (i, j) with the following conditional probability:

$$\mathbb{P}[(i, j) \in E(G_N) \mid Z] = w(Z_i, Z_j).$$

The corresponding adjacency matrix is given by $(w_{ij})_{i,j\in\{1,\cdots,N\}}$ with

$$w_{ij} = \begin{cases} 1 \text{ if } (i, j) \in E(G_N) \\ 0 \text{ otherwise.} \end{cases}$$

(r-d) Let $N \in \mathbb{N}$. Let $\tilde{Z} = (\tilde{Z}_i^N)_{i\in\{1,\cdots,N\}}$ be a determistic sequence satisfying $\tilde{Z}_i^N \in [\frac{i-1}{N}, \frac{i}{N})$ for all $i \in \{1, \cdots, N\}$. A random (unweighted) graph $\tilde{G}_N$ generated by the deterministic sequence $\tilde{Z}$ is constructed by inserting each edge (i, j) with probability $w(\tilde{Z}_i^N, \tilde{Z}_j^N)$:

$$\mathbb{P}[(i, j) \in E(\tilde{G}_N)] = w(\tilde{Z}_i^N, \tilde{Z}_j^N).$$

The corresponding adjacency matrix is given by $(w_{ij})_{i,j\in\{1,\cdots,N\}}$ with

$$w_{ij} = \begin{cases} 1 \text{ if } (i, j) \in E(\tilde{G}_N) \\ 0 \text{ otherwise.} \end{cases}$$

The approach (r-r) (standing for random-random) requires randomness in both steps: that of randomly attributing a realization of the random variables Z_i to each node

i and that of randomly selecting the weights ξ_{ij} with probability $w(Z_i(\omega), Z_j(\omega))$ where $Z_i(\omega), Z_j(\omega)$ stands for a realization of the random variable Z_i, Z_j. The approach (r-d) (for random-deterministic) requires randomness only in the second step. Two convergence results can then be given, one in each of the settings (r-r) and (r-d).

Firstly, in [47], Medvedev proves that given a symmetric graphon $w \in L^\infty(I^2, [0,1])$, a Lipschitz function $\phi : \mathbb{R} \to \mathbb{R}$ and $x_0 \in L^\infty(I)$, if the solution $x \in C([0,T]; L^\infty(I))$ to (26) satisfies the inequality

$$\sup_{t\in[0,T]} \int_I \left(\int_I w(\xi,\zeta)\phi(x(t,\xi), x(t,\zeta))^2 d\zeta \right.$$
$$\left. - \left(\int_I w(\xi,\zeta)\phi(x(t,\xi), x(t,\zeta)) d\zeta \right)^2 \right) d\xi \geq C_1$$

for some constant $C_1 > 0$, then the solution x^N to the microscopic system (13) posed on the w-random graph generated by a random sequence (r-r) constructed as in Definition 6 converges toward x. The convergence is obtained in the following sense: for some $C > 0$,

$$\lim_{N\to+\infty} \mathbb{P}\left[\sup_{t\in[0,T]} \|x^N(t) - \mathbf{P}_{Z^N} x(t,\cdot)\|_{2,N} \leq \frac{C}{\sqrt{N}} \right] = 1,$$

where for all $a \in \mathbb{R}^N$, $\|a\|_{2,N} := \frac{1}{N}\sum_{i=1}^N a_i$, and $\mathbf{P}_{Z^N} x(t,\cdot) := (x(t,Z_1), \cdots, x(t,Z_N))$ denotes the evaluation of the function $x(t,\cdot) \in L^\infty(I)$ at the discretization points $(Z_1, \cdots, Z_N)$.

On the other hand, the convergence of the microscopic system on a w-random graph generated by a deterministic sequence (r-d) requires more regularity from the graphon w. More specifically, in [47], Medvedev proves that given a symmetric graphon $w \in L^\infty(I^2; [0,1])$ almost everywhere continuous, a Lipschitz function $\phi : \mathbb{R} \to \mathbb{R}$ and $x_0 \in L^\infty(I)$, if the solution $x \in C([0,T]; L^\infty(I))$ to (26) satisfies the inequality

$$\sup_{t\in[0,T]} \int_{I^2} w(\xi,\zeta)(1 - w(\xi,\zeta))\phi(x(t,\xi) - x(t,\zeta)) d\xi d\zeta > 0$$

for some $T > 0$, then defining x_N from the solution x^N to the microscopic system (13) on the w-random graph (r-d) as by the relation (27), it holds

$$\|x_N - x\|_{C([0,T];L^2(I))} \xrightarrow[N\to+\infty]{\mathbb{P}} 0,$$

where the convergence is in probability.

The proofs of both results rely on several applications of the Central Limit Theorem and, in the case (r-d), on the introduction of an intermediate deterministic system, known to converge due to results in [46] (See Sect. 2.2.1).

Random Weighted Graphs A generalization of this convergence result to general directed weighted random graphs is provided in [3]. In that aim, the concept of w-random graphs is generalized to that of q-weighted random graphs as follows:

Definition 7 Let $q : I^2 \to \mathcal{P}(\mathbb{R}_+)$. A q-weighted random graph can be constructed either from a sequence of random variables (r-r) or from a sequence of deterministic variables (r-d).

(r-r) Let $Z = (Z_i)_{i\in\mathbb{N}}$ be a sequence of i.i.d. random variables, uniformly distributed in $I = [0, 1]$. Let $N \in \mathbb{N}$. A q-weighted random graph generated by the random sequence Z is constructed by randomly attributing to each edge (i, j) a weight $w_{ij} \in \mathbb{R}_+$ with law $q(Z_i, Z_j, \cdot)$.

(r-d) Let $N \in \mathbb{N}$. Let $\tilde{Z} = (\tilde{Z}_i^N)_{i\in\{1,\cdots,N\}}$ be a determistic sequence satisfying $\tilde{Z}_i^N \in [\frac{i-1}{N}, \frac{i}{N})$ for all $i \in \{1, \cdots, N\}$. A q-weighted random graph generated by the deterministic sequence $\tilde{Z}$ is constructed by randomly attributing to each edge (i, j) a weight $w_{ij} \in \mathbb{R}_+$ with law $q(\tilde{Z}_i, \tilde{Z}_j, \cdot)$.

Definition 7 is indeed a generalization of Definition 6, since for any graphon $w \in L^\infty(I^2; [0, 1])$, one can define the law

$$q_w(\xi, \zeta, du) := (1 - w(\xi, \zeta))\delta_0(du) + w(\xi, \zeta)\delta_1(du).$$

Again, the convergence results depend heavily on the degree of randomness ((r-r) or (r-d)) of the q-weighted random graph.

In the case of a q-weighted random graph generated by a random sequence, given a Lipschitz function $\phi : \mathbb{R} \to \mathbb{R}$ and an initial condition $x_0 \in L^\infty(I; \mathbb{R})$, the solution x^N to the discrete system posed on the q-weighted random graph converges to the solution x to the continuous equation (26), where the limit graphon is the first moment of the weighted random graph law q:

$$\forall(\xi, \zeta) \in I^2, \quad w(\xi, \zeta) = \int_{\mathbb{R}_+} u \, q(\xi, \zeta; du). \tag{29}$$

The convergence is obtained quantitatively in the following sense:

$$\mathbb{P}\left[\sup_{t\in[0,T]} \|x^N(t) - \mathbf{P}_{Z^N} x(t, \cdot)\|_{2,N} \geq \frac{C_1(T)}{\sqrt{N}}\right] \leq \frac{\tilde{C}_1}{N}$$

for some constants $C_1(T)$ and $\tilde{C}_1$.

In the case of a q-weighted random graph generated by a deterministic sequence, more regularity is required both of the initial data and of the first moment of q, the weighted random graph law. More specifically, if $x_0 \in C^{0,\frac{1}{2}}(I)$ and $(\xi, \zeta) \mapsto wq(\xi, \zeta; dw)$ is $\frac{1}{2}$–Hölder on I^2, then the solution x^N to the microscopic system (13) posed on the q-weighted random graph converges to the solution x to the continuous equation (26), where the limit graphon is the first moment of the weighted random graph law q, as given by Eq. (29). Denoting by x_N the projection of the vector x^N onto $C([0,T]; L^\infty(I))$ given by (27), it holds

$$\mathbb{P}\left[\|x_N - x\|_{C([0,T];L^2(I))} \geq \frac{C_2(T)}{\sqrt{N}}\right] \leq \frac{\tilde{C}_2}{N}$$

for some constants $C_2(T)$ and $\tilde{C}_2$.

Both proofs of convergence rely on the Bienaymé-Chebyshev inequality and, in the case (r-d), on the convergence of an intermediate deterministic system given by Theorem 4 in [53] (see Sect. 2.2.1).

2.3 Links Between Continuum Limit and Mean-Field Limits

Let $N \in \mathbb{N}$. Denoting by $x_i^N(t) \in \mathbb{R}^d$ the position of particle i at time t, a general particle system can be described by $(x_i^N)_{i\in\{1,\cdots,N\}} \in C([0,T],\mathbb{R}^d)^N$, whose evolution is given by:

$$\begin{cases} \dfrac{d}{dt}x_i^N = \dfrac{1}{N}\displaystyle\sum_{i=1}^{N} \phi\left(\frac{i}{N}, \frac{j}{N}, x_i^N, x_j^N\right) \quad \text{for all } i \in \{1, \cdots, N\} \\ x_i^N(0) = x_i^{N,0}. \end{cases} \tag{30}$$

As seen in Sect. 2.2, its limit as N goes to infinity can be written as the solution $x \in C([0,T], L^2(I;\mathbb{R}^d))$ to the following integrodifferential equation, provided that $x_i^{N,0}$ converges to x^0 in a suitable sense:

$$\begin{cases} \partial_t x(t,\xi) = \displaystyle\int_I \phi(\xi, \zeta, x(t,\xi), x(t,\zeta))d\zeta \\ x(0,\cdot) = x^0. \end{cases} \tag{31}$$

We refer to x as the **continuum limit** of x^N. Within this framework, the infinitely numerous agents are assumed to be labeled by the variable ξ, which spans the set $I = [0, 1]$. Thus, $x(t, \xi)$ denotes the position of agent with label ξ at time t.

Secondly, as seen in Sect. 2.1, taking the **non-exchangeable mean-field** limit of the microscopic system (30) yields the following equation on the probability density

$\nu \in C([0,T], \mathcal{P}(I \times \mathbb{R}^d))$:

$$\begin{cases} \partial_t \nu_t(\xi, x) + \nabla_x \cdot \left(\left(\int_{I \times \mathbb{R}^d} \phi(\xi, \zeta, x, y) d\nu_t(\zeta, y) d\zeta \right) \nu_t(\xi, x) \right) = 0 \\ \nu_{t=0} = \nu_0. \end{cases} \tag{32}$$

Thirdly, if the agents are indistinguishable, i.e., if $\phi(\xi, \zeta, x(t,\xi), x(t,\zeta)) = \phi(x(t,\xi), x(t,\zeta))$, we can derive the classical **exchangeable mean-field limit** giving the evolution of the probability measure $\tilde{\nu} \in C([0,T], \mathcal{P}(\mathbb{R}^d))$:

$$\begin{cases} \partial_t \tilde{\nu}_t(x) + \nabla_x \cdot \left(\left(\int_{\mathbb{R}^d} \phi(x, y) d\tilde{\nu}_t(y) \right) \tilde{\nu}_t(x) \right) = 0 \\ \tilde{\nu}_{t=0} = \tilde{\nu}_0. \end{cases} \tag{33}$$

Here, $\tilde{\nu}_t(\Omega)$ denotes the mass of agents in the space region Ω at time t. We refer to Sect. 1.1 for a reminder of the different notations used consistently in this review.

All three Eqs. (31), (32), and (33) are obtained as the limit of (30) as N goes to infinity (remembering that (33) is valid only if the exchangeable case), as shown in Sects. 2.1 and 2.2. Here, we explain briefly and formally how these descriptions are related to one another (see Fig. 3 for a visual summary). A complete and detailed overview of how all models are related can be found in [53].

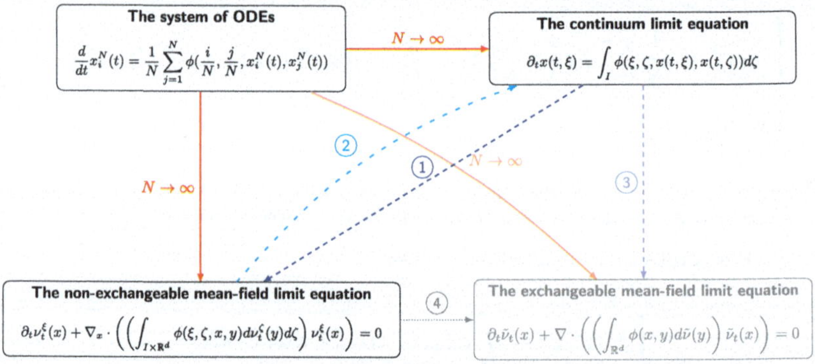

Fig. 3 Links between the different equations. The red arrows show the large-population limits described in Sects. 2.1 and 2.2. The dashed arrows 1, 2, and 3 are, respectively, explained in Sects. 2.3.1, 2.3.2 and 2.3.3. Arrow 4 corresponds to Remark 2.6. The dark blue arrows 1 and 3 correspond to links from the continuum limit equation to the non-exchangeable or exchangeable mean-field equations. The light blue arrow 2 corresponds to the reverse link from the non-exchangeable mean-field equation to the continuum limit equation, in the specific case $\phi(\xi, \zeta, x, y) = w(\xi, \zeta)(\lambda_1 x + \lambda_2 y)$ (Sect. 2.3.2). Note that all arrows leading to the exchangeable mean-field limit equation (bottom right corner) are valid only in the exchangeable case, i.e., $\phi = \phi(x, y)$

2.3.1 From Continuum Limit to Non-exchangeable Mean-Field Limit

Let $x(t,\xi)$ denote the solution to (31), and let $\overline{\nu}_t$ denote a modified "continuous" empirical measure defined by

$$\overline{\nu}_t(\xi, x) = \int_I \delta_{x(t,\zeta)}(x)\delta_\zeta(\xi)d\zeta.$$

For all test functions $f \in C^\infty(I \times \mathbb{R}^d)$, it holds

$$\begin{aligned}\frac{d}{dt}\int_{I\times\mathbb{R}^d} f(\xi,x)d\overline{\nu}_t(\xi,x)d\xi &= \frac{d}{dt}\int_I f(\xi, x(t,\xi))d\xi \\ &= \int_I \nabla_x f(\xi, x(t,\xi)) \cdot \left(\int_I \phi(\xi,\zeta,x(t,\xi),x(t,\zeta))d\zeta\right)d\xi \\ &= \int_{I\times\mathbb{R}^d} \nabla_x f(\xi,x) \cdot \left(\int_{I\times\mathbb{R}^d} \phi(\xi,\zeta,x,y)d\overline{\nu}_t(\zeta,y)d\zeta\right)d\overline{\nu}_t(\xi,x)d\xi,\end{aligned}$$

which shows that the continuous empirical measure $\overline{\nu}$ built from the solution x to (31) is itself a solution to (32). This computation corresponds to the arrow 1 in Fig. 3.

Note that the set of measures $\nu_t \in \mathcal{P}(I \times \mathbb{R}^d)$ that can be written as a continuous empirical measure $\overline{\nu}_t$ is the measures supported on curves $\xi \mapsto x(t,\xi)$.

2.3.2 From Non-exchangeable Mean-Field Limit to Continuum Limit

In general, the exchangeable mean-field equation (32) and the continuum equation (31) cannot be obtained from one another. However, as shown in [53], if the interaction kernel is assumed to be linear, a solution to (31) can be recovered from a solution to the non-exchangeable mean-field limit equation (32) by taking its first moment with respect to the space variable.

In this aim, we suppose that the marginal of ν_0 with respect to its first variable is the Lebesgue measure, i.e., that

$$\pi_I\#\nu_0 = d\xi.$$

As observed in [53] (Remark 1), the marginal of ν_t with respect to its first variable is constant in time, since the transport term acts only on the space variable. It then holds $\pi_I\#\nu_t = d\xi$ for all $t \in [0,T]$. Let us then write the disintegration of ν_t with respect to its marginal $d\xi$ as

$$\nu_t = \int_I \nu_t^\xi\, d\xi,$$

and further assume the following

Hypothesis 6 *$\int_{\mathbb{R}^d} d\nu_t^\xi(x) = 1$ for almost all $\xi \in I$.*

We then define the first moment of ν_t with respect to the space variable as

$$\bar{x}(t,\xi) := \int_{\mathbb{R}^d} x\, d\nu_t^\xi(x).$$

Decomposing $\bar{x}$ on its coordinates using an orthonormal basis $(e_k)_{k\in\{1,\cdots,d\}}$ of $\mathbb{R}^d$, it then holds for all $k \in \{1, \cdots, d\}$

$$\begin{aligned}
\frac{d}{dt}\bar{x}(t,\xi)\cdot e_k &= \frac{d}{dt}\int_{\mathbb{R}^d} (x\cdot e_k)\, d\nu_t^\xi(x) \\
&= \int_{\mathbb{R}^d} \nabla_x(x\cdot e_k)\cdot\left(\int_{I\times\mathbb{R}^d} \phi(\xi,\zeta,x,y)d\nu_t^\zeta(y)d\zeta\right) d\nu_t^\xi(x) \\
&= \int_{\mathbb{R}^d} e_k\cdot\left(\int_{I\times\mathbb{R}^d} \phi(\xi,\zeta,x,y)d\nu_t^\zeta(y)d\zeta\right) d\nu_t^\xi(x).
\end{aligned}$$

At this point, we are led to make a strong assumption on the form of the interaction function ϕ.

Hypothesis 7 *We suppose that $(\xi,\zeta,x,y) \mapsto \phi(\xi,\zeta,x,y)$ can be written as the product of a function of ξ, ζ and of a linear function of x, y, i.e.,*

$$\phi(\xi,\zeta,x,y) = w(\xi,\zeta)(\lambda_1 x + \lambda_2 y),$$

with $w : I^2 \mapsto \mathbb{R}$ and $(\lambda_1,\lambda_2) \in \mathbb{R}^2$.

This form is actually common in models for opinion dynamics with linear interaction of the type Hegselmann-Krause (see [31]), for which the interation is given by $w(\xi,\zeta)(y-x)$. We obtain

$$\begin{aligned}
\frac{d}{dt}\bar{x}(t,\xi) &= \int_{\mathbb{R}^d}\left(\int_{I\times\mathbb{R}^d} w(\xi,\zeta)(\lambda_1 x+\lambda_2 y)d\nu_t^\zeta(y)d\zeta\right) d\nu_t^\xi(x) \\
&= \int_I w(\xi,\zeta)\left(\lambda_1\int_{\mathbb{R}^d} x d\nu_t^\xi(x) + \lambda_2\int_{\mathbb{R}^d} y d\nu_t^\zeta(y)\right) d\zeta \\
&= \int_I w(\xi,\zeta)\left(\lambda_1\bar{x}(t,\xi)+\lambda_2\bar{x}(t,\zeta)\right) d\zeta
\end{aligned}$$

which is the continuum limit equation (31) for this specific choice of ϕ. This computation corresponds to the arrow 3 in Fig. 3.

The fact that one can recover a closed equation on the first moment of ν_t with respect to the space variable is remarkable and can be intuitively explained as follows. In the general case, the evolution of $\bar{x}(t,\xi)$ depends on all remaining particles labeled by all $\zeta \in I$ and located at all $y \in \mathbb{R}^d$. The linear condition on ϕ implies that the combined effect of all the particles with label ζ is equivalent to that of their first moment $\bar{x}(t, \xi)$. The problem of obtaining a closed equation in the general (nonlinear) case is still open, and we refer the reader to [53] (Section 3.1.2.) for further comments on this issue.

Remark 2.5 The assumption on the marginal of ν_0 with respect to its first variable can be lifted without loss of generality. If $\pi_I \# \nu_0 := \mu$ for a general $\mu \in \mathcal{P}(I)$, one can recover the general graph-limit equation

$$\partial_t x(t, \xi) = \int_I \phi(\xi, \zeta, x(t, \xi), x(t, \zeta)) d\mu(\zeta).$$

2.3.3 Subordination of the Mean-Field Limit to the Continuum Limit Equation (Indistinguishable Case)

In the case in which the particles are indistinguishable, i.e., $\phi(\xi, \zeta, x, y) = \phi(x, y)$, one can derive the classical mean-field limit equation (33) as the limit as N tends to infinity of the particle system (30). As shown in [12], Eq. (33) can also be obtained directly from the continuum limit (31) using the "continuous" empirical measure

$$\bar{\tilde{\nu}}_t(x) = \int_I \delta_{x(t,\xi)}(x) d\xi$$

where $x(t, \xi)$ is a solution to (31). Indeed, for all test functions $f \in C^\infty(\mathbb{R}^d)$, it holds

$$\begin{aligned} \frac{d}{dt} \int_{\mathbb{R}^d} f(x) d\bar{\tilde{\nu}}_t(x) =& \frac{d}{dt} \int_I f(x(t, \xi)) d\xi = \int_I \nabla_x f(x(t, \xi)) \\ & \cdot \left(\int_I \phi(x(t, \xi), x(t, \zeta)) d\zeta \right) d\xi \\ =& \int_{\mathbb{R}^d} \nabla_x f(x) \cdot \left(\int_{\mathbb{R}^d} \phi(x, y) d\bar{\tilde{\nu}}_t(y) \right) d\bar{\tilde{\nu}}_t(x), \end{aligned}$$

which is the weak formultion of (33). This computation corresponds to the arrow 2 in Fig. 3. Thus, deriving the continuum limit of (30) as N goes to infinity and doing the transformation above is an alternative way of obtaining the classical mean-field limit equation (33).

Two important remarks need to be made.

Firstly, all initial measures $\tilde{\nu}_0 \in \mathcal{P}(\mathbb{R}^d)$ can be approximated by empirical measures of the form $\bar{\tilde{\nu}}_t(x) = \int_I \delta_{x(t,\xi)}(x) d\xi$, as shown in the following proposition.

Proposition 1 *The set of measures*

$$\left\{ \int_I \delta_{x_0(\xi)}(x) d\xi \quad | \quad x_0 : I \to \mathbb{R}^d \text{ measurable} \right\}$$

is dense in $\mathcal{P}(\mathbb{R}^d)$ for the weak topology.

Moreover, if $d = 1$, for any $\tilde{\nu}_0 \in \mathcal{P}(\mathbb{R})$, there exists a measurable function $x_0 : I \to \mathbb{R}$ such that $\tilde{\nu}_0(x) = \int_I \delta_{x_0(\xi)}(x) d\xi$.

Proof It is well known that the set of measures that can be written as finite sums of Dirac masses is dense in $\mathcal{P}(\mathbb{R}^d)$. We claim that for any $\tilde{\nu}_0^n := \sum_{i=1}^n a_i \delta_{x_i}$ where $a_i \in [0, 1]$ for all $i \in \{1, \cdots, n\}$ and $\sum_{i=1}^n a_i = 1$, there exists a function $x_0 : I \to \mathbb{R}^d$ such that

$$\tilde{\nu}_0^n(x) = \int_I \delta_{x_0(\xi)}(x) d\xi.$$

We prove this claim constructively. Let x_0 be defined by the piecewise-constant function

$$\forall i \in \{1, \cdots, n\}, \; \forall \xi \in [b_i, b_{i+1}), \; x_0(\xi) = x_i$$

where $b_i := \sum_{k=1}^{i-1} a_k$. Then

$$\int_I \delta_{x_0(\xi)}(x) d\xi = \sum_{i=0}^n \int_{b_i}^{b_i + a_i} \delta_{x_i}(x) d\xi = \sum_{i=1}^n a_i \delta_{x_i}.$$

Moreover, if $\tilde{\nu}_0 \in \mathcal{P}(\mathbb{R})$, defining the nondecreasing function

$$x_0 : \xi \mapsto \min\{x \in \mathbb{R} \mid \nu_0((-\infty, x]) \geq \xi\}$$

implies $\tilde{\nu}_0(x) = \int_I \delta_{x_0(\xi)}(x) d\xi$. □

Secondly, the choice of x_0 is in general not unique. Any measure-preserving rearrangement of x_0 gives the same mesure $\tilde{\nu}_0$. This implies that in passing from the solution $x(t, \xi)$ to the continuum limit equation to the solution $\tilde{\nu}_t(x)$ to the mean-field equation, there is an irreversible information loss. In this indistinguishable case, the agents can be relabeled without impacting the dynamics.

Remark 2.6 In the indistinguishable case, the exchangeable mean-field equation (33) can be recovered from the non-exchangeable mean-field equation (32) by integrating the solution ν_t^ξ in ξ.

3 Large-Population Limits of Particle Systems on Adaptive Dynamical Networks

As mentioned in the introduction, the setting of adaptive dynamical networks is of great interest, since it allows to build more realistic models. Indeed, in many real-life situations, the graph involved is not static, and its evolution can depend on the states of the system itself. The natural question is to see if and how the large-population limit considered above, i.e., the non-exchangeable and exchangeable mean-field limits and the continuum limit can be extended to this setting. Actually, it is a challenging problem in its generality, and up to recently, two specific frameworks have successfully addressed it.

The first one is the setting of Kuramoto-type models on N oscillators [26]. It can be written in the following form

$$\begin{cases} \frac{d}{dt} x_i^N = \omega_i^N(x_i^N, t) + \frac{1}{N}\sum_{j=1}^{N} w_{ij}^N \phi\left(x_i^N, x_j^N\right) & \text{for all } i \in \{1, \cdots, N\} \\ \frac{d}{dt} w_{ij}^N = -\varepsilon\left(w_{ij}^N + H(x_i^N, x_j^N)\right) \end{cases} \tag{34}$$

where $x_i^N = x_i^N(t) \in \mathbb{T} = \mathbb{R}/(2\pi\mathbb{Z})$ represents the phase of the i-th oscillator for $i \in \{1, \ldots, N\}$, $\omega_i^N : \mathbb{T} \times \mathbb{R} \to \mathbb{R}$ is the vector field describing the intrinsic frequency, $\phi : \mathbb{T}^2 \to \mathbb{R}$ is a coupling function, $w^N(t) = (w_{ij}^N(t))_{1\le i,j\le N}$ is the time-evolving weight matrix of the network of oscillators, which takes into account the local information of two interacting oscillators via the function $H : \mathbb{T}^2 \to \mathbb{R}$ and $\varepsilon > 0$ is a parameter which controls the time scale between the dynamics on the network of the phases and the dynamics of the network weights.

The second one can be seen as a variant of the Hegselmann-Krause dynamics [31] where the particle system is composed not only of agents' opinions but also of time-evolving weights of influence, which represent individual charisma [2]. This model can also be viewed as a system of ODEs on a evolving-weighted (on-symmetric) graph (see Fig. 4). It can be written in the following way

$$\begin{cases} \frac{d}{dt} x_i^N(t) = \frac{1}{N}\sum_{j=1}^{N} m_j^N(t)\,\phi(x_j^N(t) - x_i^N(t)) \\ \frac{d}{dt} m_i^N(t) = \psi_i^{(N)}(x^N(t), m^N(t)) \end{cases} \tag{35}$$

where $x^N = (x_1^N, \ldots, x_N^N) : [0, T] \to (\mathbb{R}^d)^N$ represent the opinions of N agents, and $m^N = (m_1^N, \ldots, m_N^N) : [0, T] \to \mathbb{R}^N$ represent their individual weights of influence. Each opinion's time-evolution is affected by the opinion of each neighboring agent via the interaction function ϕ, proportionally to the neighboring

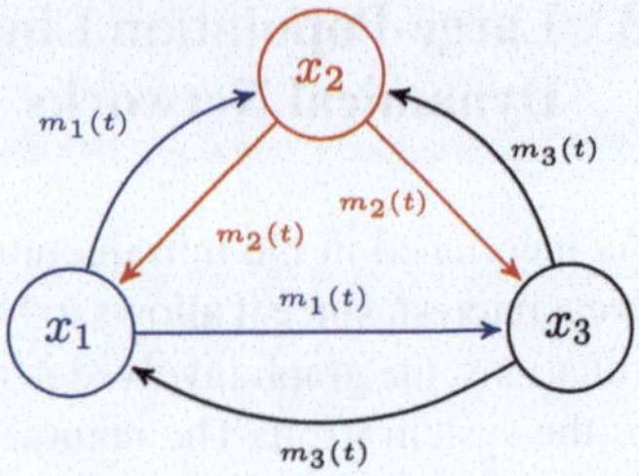

Fig. 4 Evolving-weighted graph associated with (35) in the case of three agents [2]

agent's weight of influence. Moreover, the agents' weights are assumed to evolve in time, and their dynamics may depend on the opinions and weights of all the other agents, via functions $\psi_i^{(N)} : (\mathbb{R}^d)^N \times \mathbb{R}^N \to \mathbb{R}$.

More recently, in [61], inspired by the approach developed for the particular model of evolving-in-time weights in [2], some results have been obtained for a more general Kuramoto-type model of the form

$$\begin{cases} \frac{d}{dt} x_i^N = \omega_i^N(x^N, t) + \frac{1}{N} \sum_{j=1}^{N} w_{ij}^N \phi\left(x_i^N, x_j^N\right) & \text{for all } i \in \{1, \cdots, N\} \\ \frac{d}{dt} w_{ij}^N = \psi_{ij}^{(N)}(x^N(t), w^N(t)) \end{cases} \tag{36}$$

As in the static case, when considering large-population limit, several approaches are possible: mean-field and continuum limits. We start with the mean-field one.

3.1 *Mean-Field Limits*

3.1.1 The Non-exchangeable Mean-Field Limit

Let us start by recalling that the question of the derivation of a non-exchangeable mean-field limit is very difficult if we want to deal with general adaptive networks. So far, there exists only partial results on very specific cases.

The Kuramoto-Type Model In [27], Gkogkas, Kuehn, and Xu are interested in a particular form of (34):

$$\begin{cases} \frac{d}{dt} x_i^N = \omega_i^N + \frac{1}{N} \sum_{j=1}^{N} w_{ij}^N \varphi\left(x_j^N - x_i^N\right) & \text{for all } i \in \{1, \cdots, N\} \\ \frac{d}{dt} w_{ij}^N = -\varepsilon\left(w_{ij}^N + h(x_j^N - x_i^N)\right). \end{cases} \tag{37}$$

Note that the authors explain that their result can actually extend to the more general form (34) easily. Going back to the framework of digraph measures defined in Sect. 2.1.2, we recall that digraph measures are measure-valued function $\eta \in \mathcal{B}(I, \mathcal{M}_+(I))$. In this framework, the mean-field limit equation is a generalized Vlasov equation:

$$\partial_t \mu_t^\xi(x) + \partial_x \left(\mu_t^\xi(x) \left(\omega(t,\xi) + e^{-\varepsilon(t-t_0)} \int_I \int_{\mathbb{T}} \varphi(y-x) d\mu_t^\zeta(y) d\eta_0^\xi(\zeta) \right.\right.$$
$$\left.\left. + \int_I \int_{\mathbb{T}} \varphi(y-x)\mu_t^\zeta(dy) \left(\varepsilon \int_{t_0}^t e^{-\varepsilon(t-s)} \int_{\mathbb{T}} h(z-x) d\mu_s^\zeta(z) ds \right) d\zeta \right)\right) = 0 \tag{38}$$

where $\mu \in \mathcal{C}([0,T]; \mathcal{C}(I, \mathcal{M}_+(I))$ for $T > 0$.

In order to study (38), the following generalized adaptive Kuramoto network is introduced:

$$\begin{cases} \partial_t x(t,\xi) = \omega(t,\xi) + \displaystyle\int_I \int_{\mathbb{T}} \varphi(y - x(t,\xi)) d\mu_t^\zeta(y) d\eta_t^\xi(\zeta), \\ \partial_t \eta_t^\xi(\cdot) = -\varepsilon \eta_t^\xi(\cdot) - \varepsilon \displaystyle\int_{\mathbb{T}} h(y - x(t,\xi)) d\mu_t^\cdot(y) \lambda(\cdot) \end{cases} \tag{39}$$

which is interpreted in a weaker sense in integral form and where λ is the Lebesgue measure. The equation on η^ξ is to be understood in the weak sense, meaning that for any bounded continuous test function $f \in \mathcal{C}_b(I)$, η_t^ξ satisfies

$$\int_I f(\zeta) d\eta_t^\xi(\zeta) = \int_I f(\zeta) d\eta_0^\xi(\zeta) - \varepsilon \int_{t_0}^t \left(\int_I f(\zeta) d\eta_\tau^\xi(\zeta) \right) d\tau$$
$$- \varepsilon \int_{t_0}^t \int_I \left(f(\zeta) \int_{\mathbb{T}} h(y - x(\tau,\xi)) d\mu_\tau^\zeta(y) d\zeta \right) d\tau. \tag{40}$$

The key idea of the article is that, thanks to a variation of constants formula for η, one can prove that being a solution to (39) in its integral form is equivalent to being a solution to

$$\begin{cases} x(t,\xi) = x_0(\xi) + \displaystyle\int_{t_o}^t \left(\omega(s,\xi) + e^{-\varepsilon(s-t_0)} \int_I \int_{\mathbb{T}} \varphi(y - x(s,\xi)) d\mu_s^\zeta(y) d\eta_0^\xi(\zeta) \right., \\ \qquad - \varepsilon \displaystyle\int_{t_0}^s e^{-\varepsilon(t-\tau)} \int_I \left(\int_{\mathbb{T}} \varphi(y - x(s,\xi)) d\mu_s^\zeta(y) \right. \\ \qquad\qquad \left.\left. \cdot \displaystyle\int_{\mathbb{T}} h(y - x(\tau,\xi)) d\mu_\tau^\zeta(y) \right) d\zeta \right) \\ \eta_t^\xi(\cdot) = e^{-\varepsilon(t-t_0)} \eta_0^\xi(\cdot) - \left(\varepsilon \displaystyle\int_{t_0}^t e^{-\varepsilon(t-s)} \int_{\mathbb{T}} h(y - x(s,\xi)) d\mu_s^\cdot(y) ds \right) \lambda(\cdot) \end{cases} \tag{41}$$

Under this form, the dynamics of the oscillators are decoupled from the dynamics of the digraph measures. One can then reduce the hybrid system to a one-dimensional integral equation indexed by the vertex variable coupled on the prescribed initial graph measures as well as prescribed time-dependent measure valued functions. Well-posedness of Eq. (38), which is referred to as *equation of mean-field characteristics*, is proven by a fixed-point argument on the semi-flow of $(41)_1$.

Moreover, regarding the mean-field limit, the result consists in establishing approximations of the solution to the mean-field limit (38) by empirical distributions

$$\mu^{\xi}_{n,m,t}(x) := \sum_{i=1}^{m} \mathbf{1}_{A_i^m}(\xi)\frac{a_{m,i}}{n}\sum_{j=1}^{n}\delta_{x^{m,n}_{(i-1)n+j}(t)}(x), \tag{42}$$

as $n, m \to \infty$. Here,

- A_i^m is a partition of I,
- $\{a_{m,i} : i = 1, \ldots, m\}_{m\in\mathbb{N}} \subseteq \mathbb{R}_+$ is a sequence,
- $\{x^{m,n,0}_{(i-1)n+j} : i = 1, \ldots, m,\ j = 1, \ldots, n\}_{n,m\in\mathbb{N}} \subseteq \mathbb{T}$ is a sequence

such that

$$\mu^{\xi}_{n,m,0} = \sum_{i=1}^{m} \mathbf{1}_{A_i^m}(\xi)\frac{a_{m,i}}{n}\sum_{j=1}^{n}\delta_{x^{m,n,0}_{(i-1)n+j}}(x).$$

The measures $\mu_{n,m,0}$ converge to the initial data μ_0 associated with (38) (see [37, Theorem 5.4] for the existence of such partition and sequences), and $(x_i)_{i=1,\ldots,mn}$ satisfy a sequence of ODEs which has a similar form as (37). The proof relies on stability estimates and the discretization of a given initial digraph measure and of the initial condition of the generalized Vlasov equation.

The Evolving-in-Time Weight Model So far, there is no result of non-exchangeable mean-field limit for the general system (35). However, for weight dynamics of the form

$$\psi_i(x^N, m^N) = \int_{\{1,\ldots,N\}\times\mathbb{R}^d\times\mathbb{R}} S(i, x_i^N, m_i^N, k, y, n)d\mu_t^N(k, y, n) \tag{43}$$

where S is locally Lipschitz in (x_i^N, m_i^N, y, n) uniformly in (i, k) and

$$\mu_t^N(k, x, m) = \frac{1}{N}\sum_{i=1}^{N}\delta(x - x_i^N(t))\delta(m - m_i^N(t))\delta(k - i),$$

we conjecture that it should converge to the probability density $\mu \in \mathcal{C}([0, T], \mathcal{P}(I \times \mathbb{R}^d \times \mathbb{R}))$ solution to the following equation:

$$\partial_t \mu_t(\xi, x, m) + \nabla_x \cdot \left(\left(\int_{I\times\mathbb{R}^d\times\mathbb{R}} n\phi(x, y) d\mu_t(\zeta, y, n) d\zeta \right) \mu_t(\xi, x, m) \right)$$

$$+ \partial_m \left(\left(\int_{I\times\mathbb{R}^d\times\mathbb{R}} S(\xi, x, m, \zeta, y, n) d\mu_t(\zeta, y, n) d\zeta \right) \mu_t(\xi, x, m) \right) = 0. \tag{44}$$

Note that, if S is locally Lipschitz in $(i, x_i^N, m_i^N, k, y, n)$, the convergence can be established applying the result proven in [53]. Indeed, in that case, by setting $z_i^N = (x_i^N, m_i^N)$, we can rewrite the system as

$$\begin{cases} \dfrac{d}{dt} z_i^N = \dfrac{1}{N} \displaystyle\sum_{i=1}^{N} \Phi(\frac{i}{N}, \frac{j}{N}, z_i^N, z_j^N) \quad \text{for all } i \in \{1, \cdots, N\} \\ z_i(0) = z_i^{N,0} \end{cases} \tag{45}$$

where Φ satisfies the necessary regularity assumptions.

3.1.2 The Exchangeable Mean-Field Limit

In this subsection, we will only focus on the evolving-in-time weight model, which was first introduced in [44], with a focus on its long-time behavior for specific choices of weight dynamics. The first considerations regarding the large-population limit have been addressed in [54] for a more general model and rigorously proven in [2, 56].

In this review, we denote by *exchangeable mean-field limit* the classical mean-field limit, which initially appeared in the context of gas dynamics (see [16], for instance), and which only applies to systems that preserve *indistinguishibility* (or *exchangeability*, see Definition 1). Recall that for classical particle systems, the empirical measure $\tilde{\nu}_t^N \in \mathcal{P}(\mathbb{R}^d)$ is defined by

$$\tilde{\nu}_t^N(x) = \frac{1}{N} \sum_{i=1}^{N} \delta_{x_i^N(t)}(x).$$

Importantly, $\tilde{\nu}_t^N$ stays unchanged for any relabeling of the particles, which justifies that it can be used as a link between a particle system and a Vlasov-type equation only if the particle system preserves exchangeability.

In the case of System (35), due to the specific role played by the weights, it has been chosen to work with a modified empirical measure $\hat{\mu}_t^N \in \mathcal{P}(\mathbb{R}^d)$, defined by

$$\hat{\mu}_t^N(x) := \frac{1}{N} \sum_{i=1}^{N} m_i^N(t) \delta_{x_i^N(t)}(x). \tag{46}$$

Notice that in addition to being blind to any relabeling of the agents, this new empirical measure stays unchanged if the weights of agents at the same location x are redistributed between them. Thus, the notion of indistinguishibility in that case needs to include the preservation of "grouping of the agents": The dynamics must not be influenced by redistribution of the weights of grouped agents.

For this reason, the mean-field limit has been proven for a particular class of weight dynamics which preserves this new notion of indistinguishability, given by

$$\psi_i^{(N)}(x^N, m^N) = m_i^N(t)\frac{1}{N^k}\sum_{j_1=1}^{N}\cdots\sum_{j_k=1}^{N} m_{j_1}^N(t)\cdots m_{j_k}^N(t) \times S_k(x_i^N(t), x_{j_1}^N(t), \cdots x_{j_k}^N(t)), \quad (47)$$

where $k \in \mathbb{N}$ and $S_k : (\mathbb{R}^d)^{k+1} \to \mathbb{R}$ is globally bounded and Lipschitz. Another assumption is that the total weight of the system is conserved, i.e., $\sum_i \psi_i^{(N)} = 0$, in order for $\hat{\mu}_t^N$ to be a probability measure for all $t \geq 0$.

Remark 3.1 This last assumption is formulated as a skew-symmetry property on S_k in [2, 56], which indeed leads to the conservation of the total weight.

Thus, for this special class of weights, it can be proven that the empirical measure converges to a measure $\hat{\mu}_t \in \mathcal{P}(\mathbb{R}^d)$ solution to the following transport equation with source:

$$\partial_t \hat{\mu}_t(x) + \nabla \cdot (V[\hat{\mu}_t](x)\hat{\mu}_t(x)) = h[\hat{\mu}_t](x) \quad (48)$$

with the nonlocal vector field given by

$$\forall \mu \in \mathcal{P}(\mathbb{R}^d), \quad \forall x \in \mathbb{R}^d,$$
$$V[\mu](x) = \int_{\mathbb{R}^d} \phi(y-x)d\mu(y),$$

the nonlocal source term given by

$$\forall \mu \in \mathcal{P}(\mathbb{R}^d), \quad \forall x \in \mathbb{R}^d,$$
$$h[\mu](x) = \left(\int_{(\mathbb{R}^d)^k} S_k(x, y_1, \cdots, y_k)d\mu(y_1)\cdots d\mu(y_k)\right)\mu(x).$$

Regular Interaction Function For $\phi \in \mathrm{Lip}(\mathbb{R}^d; \mathbb{R})$, this result is actually proven by two different approaches. In [56], the proof consists of noticing that the empirical measure is already a solution to (48) and in establishing continuity with respect to the initial data, implying that

$$d_{\mathrm{BL}}(\hat{\mu}_0^N, \hat{\mu}_0) \xrightarrow[N\to\infty]{} 0 \Rightarrow d_{\mathrm{BL}}(\hat{\mu}_t^N, \hat{\mu}_t) \xrightarrow[N\to\infty]{} 0,$$

where $\hat{\mu}_t$ is the solution to (48) with initial data $\hat{\mu}_0$. The other proof, in [2], is actually based on exploiting the continuum limit (see Sect. 3.3).

Singular Interaction Function In [8], the authors are concerned with analyzing the mean-field limit for (35) in the particular case of the attractive 1D Coulomb interaction where $\phi(x) = V'(x)$ with $V(x) = |x|$, i.e., $\phi(x) = \text{sgn}(x)$ with the convention $\text{sgn}(0) = 0$. More precisely, they study the following system

$$\begin{cases} \frac{d}{dt}x_i^N = \frac{1}{N}\sum_{j=1}^N m_j^N \,\text{sgn}(x_j^N - x_i^N) \\ \frac{d}{dt}m_i^N(t) = \frac{1}{N}m_i^N \sum_{j=1}^N m_j^N S_1(x_j^N - x_i^N) \end{cases} \tag{49}$$

where $S_1 \in C_0^\infty(\mathbb{R})$ is assumed to be odd.

The mean-field equation for this ODE system is then the following

$$\partial_t \hat{\mu}_t(x) - \partial_x \left(\left(\int_{-\infty}^x \hat{\mu}_t(dy) - \int_x^{+\infty} \hat{\mu}_t(dy) \right) \hat{\mu}_t(x) \right) = \hat{\mu}_t(x)(S_1 \star \hat{\mu}_t)(x).$$

We start by noticing that the odd character of S_1 implies that $\hat{\mu}_t$ stays a probability density for all times. Thus, the above equation can be rewritten as

$$\partial_t \hat{\mu}_t(x) - \partial_x \left(\left(2\int_{-\infty}^x \hat{\mu}_t(dy) - 1 \right) \hat{\mu}_t(x) \right) = \hat{\mu}_t(x)(S_1 \star \hat{\mu}_t)(x). \tag{50}$$

The main idea developed in [8] is to shift the focus and instead of studying (50) to analyze the equation for the primitive of $\hat{\mu}_t$. Let us set

$$F(t,x) := -\frac{1}{2} + \int_{-\infty}^x \hat{\mu}_t(dy),$$

then integrating (50) yields the following nonlocal Burgers-type equation for F

$$\partial_t F + \partial_x(A(F)) = \mathbf{S}[F](t,x) \tag{51}$$

where $A(F) = -F^2$ and

$$\mathbf{S}[F](t,x) := F(t,x)(\varphi \star F)(t,x) - \int_{-\infty}^x F(t,z)(\partial_z\varphi \star F)(t,z)dz$$

with $\varphi := \partial_x S$.

One of the advantages of this new formulation is that the flux term in the new Burgers equation is local. The strategy is to consider a primitive of the empirical measure associated with (49)

$$F_N(t,x) := -\frac{1}{2} + \frac{1}{N}\sum_{j=1}^{N} m_j(t) H(x - x_j(t))$$

where H is the Heaviside function. One can then establish that F_N is an entropy solution to

$$\partial_t F_N + \partial_x (A_N(t, F_N)) = \mathbf{S}[F_N](t,x) \tag{52}$$

where A_N is a time-dependent approximation of the time-independent flux A. This proof relies on classical arguments like the use of Rankine-Hugoniot and Oleink conditions adapted to this setting (see [8, Appendix A] for more details). The mean-field limit is proven by extracting a converging subsequence and establishing some stability estimates. More precisely, provided that F_N^0 converges to F^0 in $L^1(\mathbb{R})$, F_N can be proven to converge to F in $\mathcal{C}([0,T], L^1(\mathbb{R}))$ where F_N and F are, respectively, the solutions to (52) and (51) associated with the respective initial data F_N^0 and F^0.

3.2 *Continuum Limit*

As mentioned in the static case, the study of continuum limits is of great interest. For instance, in [45], the analysis of the steady states on small-world graphs was carried out by using those continuum limits. Thus, because of the relevance of adaptive dynamical networks from a modeling point of view, it seems natural to extend those limits to that setting.

Regular Weight Dynamics The first result of this type was established in [2], where the following limit equation

$$\begin{cases} \partial_t x(t,\xi) = \int_I m(t,\zeta)\phi(x(t,\zeta) - x(t,\xi)) d\zeta \\ \partial_t m(t,\xi) = \psi(\xi, x(t,\cdot), m(t,\cdot)), \end{cases} \tag{53}$$

was obtained as the continuum limit of the particle system with time-varying weights (35). Here, ξ represents the continuous index variable taking values in $I := [0,1]$, as introduced in the previous sections. The proof requires some regularity assumptions and some bounds.

Hypothesis 8 *The function* $\psi : I \times L^\infty(I;\mathbb{R}^d) \times L^\infty(I;\mathbb{R}) \to \mathbb{R}$ *is assumed to satisfy the following Lipschitz properties: There exists* $L_\psi > 0$ *such that for all*

$(x_1, x_2, m_1, m_2) \in L^2(I)^4$,

$$\begin{cases} \|\psi(\cdot, x_1, m_1) - \psi(\cdot, x_2, m_1)\|_{L^2(I)} \leq L_\psi \|x_1 - x_2\|_{L^2(I)} \\ \|\psi(\cdot, x_1, m_1) - \psi(\cdot, x_1, m_2)\|_{L^2(I)} \leq L_\psi \|m_1 - m_2\|_{L^2(I)}. \end{cases} \tag{54}$$

Assume also that there exists $C_\psi > 0$ such that for all $(x, m) \in L^\infty(I, \mathbb{R}^d \times \mathbb{R})$, for all $\xi \in I$,

$$|\psi(\xi, x, m)| \leq C_\psi(1 + \|m\|_{L^\infty(I)}). \tag{55}$$

while $\phi \in \mathrm{Lip}(\mathbb{R}^d; \mathbb{R})$.

The sublinear growth assumption (55) may seem restrictive, and it is actually necessary in order to prevent the blowup in finite time of the weight function m. Moreover, it provides a framework which is coherent with the one developed in [46] on graphs with L^∞ weights. Indeed, one can view system (53) as the evolution of the opinions x on a time-dependent weighted nonsymmetric graph with weights $w(t, \xi, \zeta) = m(t, \zeta)$.

The result follows the same path as in the static case. Let us define

$$\begin{cases} x_N(t, \xi) := \sum_{i=1}^{N} x_i^N(t) \mathbf{1}_{[\frac{i-1}{N}, \frac{i}{N})}(\xi) \\ m_N(t, \xi) := \sum_{i=1}^{N} m_i^N(t) \mathbf{1}_{[\frac{i-1}{N}, \frac{i}{N})}(\xi). \end{cases} \tag{56}$$

Consider the weight dynamics (35) with ψ_i defined using the functional ψ appearing in (53) as follows:

$$\forall i \in \{1, \cdots, N\}, \qquad \psi_i^{(N)}(x^N(t), m^N(t)) = N \int_{\frac{i-1}{N}}^{\frac{i}{N}} \psi(\xi, x_N(t, \xi), m_N(t, \xi)) d\xi. \tag{57}$$

The main observation relies on the fact that there is an equivalence between being a solution to (35) and the fact that the associated x_N and m_N defined in (56) satisfy the following system of integrodifferential equations

$$\begin{cases} \partial_t x_N(t, \xi) = \int_I m_N(t, \zeta) \phi(x_N(t, \zeta) - x_N(t, \xi)) \, d\zeta \\ \partial_t m_N(t, \xi) = N \int_{\frac{1}{N}\lfloor \xi N \rfloor}^{\frac{1}{N}(\lfloor \xi N \rfloor + 1)} \psi(\zeta, x_N(t, \cdot), m_N(t, \cdot)) \, d\zeta. \end{cases} \tag{58}$$

Through some fixed-point argument, one can prove that there exists a unique solution (x^N, m^N) to (35) and (x, m) to (53) that, respectively, belong to

$\mathcal{C}([0,T];\mathbb{R}^{dN}\times\mathbb{R}^N)$ and $\mathcal{C}([0,T];L^\infty(I;\mathbb{R}^{dN}\times\mathbb{R}^N))$. Then, as in [46], the proof relies on some L^2-estimates of x_N-x and m_N-m. More precisely, using the regularity assumptions, some bounds, Cauchy-Schwarz and Gronwall inequalities, it holds

$$\sup_{t\in[0,T]}\left(\|(x_N-x)(t)\|_{L^2(I)}+\|(m_N-m)(t)\|_{L^2(I)}\right)$$
$$\leq\left(\|(x_N-x)(0)\|_{L^2(I)}+\|(m_N-m)(0)\|_{L^2(I)}+\int_0^T\|g_N(\tau)\|_{L^2(I)}d\tau\right)e^{KT} \tag{59}$$

for some $K>0$ with

$$g_N:(t,\xi)\mapsto N\int_{\frac{1}{N}\lfloor\xi N\rfloor}^{\frac{1}{N}(\lfloor\xi N\rfloor+1)}\psi(\zeta,x(t),m(t))\,d\zeta-\psi(\xi,x(t),m(t)).$$

Thus, defining the initial conditions for the microscopic dynamics as

$$\begin{cases} x^{0,N}:=\left(N\int_{\frac{i-1}{N}}^{\frac{i}{N}}x_0(s)ds\right)_{i\in\{1,\dots,N\}} & \in(\mathbb{R}^d)^N\\ m^{0,N}:=\left(N\int_{\frac{i-1}{N}}^{\frac{i}{N}}m_0(s)ds\right)_{i\in\{1,\dots,N\}} & \in(\mathbb{R})^N,\end{cases} \tag{60}$$

the result is obtained using Lebesgue's differentiaton theorem and the dominated convergence theorem for the term containing g_N. Indeed, it implies the convergence in, respectively, $\mathcal{C}([0,T];L^2(I;\mathbb{R}^{dN}))$ and $\mathcal{C}([0,T];L^2(I;\mathbb{R}^N))$ of x_N and m_N to x and m, solutions to (53) with initial data $x(0,\cdot)=x_0$ and $m(0,\cdot)=m_0$.

Regular Edge Dynamics The second result of this type was established in [26] for a particular adaptive Kuramoto-type model (34). Although the result is only valid for dense graphs and therefore involves working with the graphon framework just as in [2, 46], it is stated in a slightly more general framework which, if the theory allowed, would make it possible to deal with a more general class of graphs.

We start by recalling some concepts of Graph Theory mentioned in Sects. 1 and 2.1.2. As seen previously, a graphop is a bounded self-adjoint and positivity-preserving operator $A:L^\infty(I)\to L^1(I)$. It can be viewed as a generalized concept for the adjacency matrix. Importantly, the Riesz representation theorem allows to define graphops using a family of finite measures $(\eta^\xi)_{\xi\in I}$ called fiber measures as follows: In the case where η^ξ has a density w as

$$d\eta^\xi(\zeta)=w(\xi,\zeta)d\zeta,$$

then w plays the role of the usual graphon (see Def. 2 and Sect. 2.1.1). In [26], the continuum limit of the adaptive Kuramoto-type model (34) is proven to be

$$\begin{cases} \partial_t x(t,\xi) = \omega(\xi, x(t,\xi), t) + \int_I \phi(x(t,\xi), x(t,\zeta)) d\eta_t^\xi(\zeta), \\ \partial_t \eta_t^\xi(\zeta) = -\varepsilon \eta_t^\xi(\zeta) + \varepsilon H(x(t,\xi), x(t,\zeta))\lambda(\zeta) \end{cases} \tag{61}$$

where λ is the Lebesgue measure. Equation (61) is to be interpreted in the integral form and the equation on η^ξ in the weak sense, meaning that for any bounded continuous test-function $f \in \mathcal{C}_b(I)$,

$$\int_I f(\zeta) d\eta_t^\xi(\zeta) = \int_I f(\zeta) d\eta_0^\xi(\zeta) - \varepsilon \int_{t_0}^t \left(\int_I f(\zeta) d\eta_\tau^\xi(\zeta) \right) d\tau$$
$$- \varepsilon \int_{t_0}^t \left(\int_I f(\zeta) H(x(\tau,\xi), x(\tau,\zeta)) d\zeta \right) d\tau. \tag{62}$$

As in the previous case, the proof heavily relies on the regularity assumptions. The functions $\phi, H : \mathbb{T}^2 \to \mathbb{R}$ are Lipschitz continuous and $\omega : I \times \mathbb{T} \times \mathbb{R} \to \mathbb{R}$ is continuous in t and Lipschitz continuous in x and ξ. Provided some additional continuity and absolute continuity hypotheses on (x_0, η_0) (see [26, Assumptions (A5)-(A6)]), one can prove the convergence in $\mathcal{C}([t_0, t_0+T], L^\infty(I;\mathbb{T})) \times \mathcal{C}([t_0, t_0+T], \mathcal{B}(I, \mathcal{M}(I)))$ of the piecewise-constant functions (x_N, η_N) defined as

$$\begin{cases} x_N(t,\xi) := \sum_{i=1}^N x_i^N(t) \mathbf{1}_{[\frac{i-1}{N}, \frac{i}{N})}(\xi) \\ w_N(t,\xi,\zeta) := \sum_{i=1}^N w_{ij}^N(t) \mathbf{1}_{[\frac{i-1}{N}, \frac{i}{N})}(\xi) \mathbf{1}_{[\frac{j-1}{N}, \frac{j}{N})}(\zeta) \\ d(\eta_N)_t^\xi(\zeta) := w_N(t,\xi,\zeta) d\zeta \end{cases} \tag{63}$$

to (x, η_t) solution to (61) for the appropriate initial data (they are build similarly as for the previous proof) with $\mathcal{B}(I, \mathcal{M}(I))$ being the space of bounded measurable functions from I to $\mathcal{M}(I)$. As before, the existence and uniqueness of solutions relies on the use of a fixed-point theorem, while the convergence uses, among other things, the continuous dependence on the initial data and on the function ω that is established through some stability estimates.

More recently, in the graphon framework, following the spirit of the proofs in [3, 46], the convergence of (36) to the continuum limit equation

$$\begin{cases} \partial_t x(t,\xi) = \omega(t,\xi, x(t,\cdot)) + \int_I m(t,\zeta) \phi(t, x(t,\xi), x(t,\zeta)) d\zeta \\ \partial_t w(t,\xi,\zeta) = \psi(\xi,\zeta, x(t,\cdot), w(t,\cdot,\cdot)), \end{cases} \tag{64}$$

was established under some regularity assumptions and some bounds similar to those of Hypothesis 8.

Singular Weight Dynamics In [7], the authors proposes to deal with the model (35) in the case where the weight dynamics contain a singularity. This approach is motivated by the "pairwise competition" model introduced in [44], which can be written as

$$\begin{cases} \frac{d}{dt}x_i^N = \frac{1}{N}\sum_{j=1}^{N} m_j^N \phi(x_j^N - x_i^N) \\ \frac{d}{dt}m_i^N = \frac{1}{2N^2}m_i^N \sum_{k,j=1}^{N} m_k^N m_j^N \left(\phi(x_k^N - x_i^N) + \phi(x_k^N - x_j^N)\right) \mathbf{s}(x_i^N - x_j^N). \end{cases} \tag{65}$$

The aim of [7] is to extend the result established in [2] to the case where $\mathbf{s}$ presents a jump discontinuity at the origin. More precisely, one considers the following one-dimensional setting:

Hypothesis 9 *The restriction* $\mathbf{s}_{|(0,\infty)}$ *and* $\mathbf{s}_{|(-\infty,0)}$ *are Lipschitz-continuous, and* $\mathbf{s}$ *is odd and bounded.*

Under these assumptions, convergence is obtained to the following system of equations

$$\begin{cases} \partial_t x(t,\xi) = \int_I m(t,\zeta)\phi(x(t,\zeta) - x(t,\xi))d\zeta \\ \partial_t m(t,\xi) = m(\xi)\int\int_{I^2} m(\zeta)m(\zeta_*)\left(\phi(x(\zeta_*) - x(\xi)) + \phi(x(\xi) - x(\zeta_*))\right) \\ \times \mathbf{s}(x(\xi) - x(\zeta))d\zeta d\zeta_*. \end{cases} \tag{66}$$

An important necessary condition for the well-posedness of the microscopic dynamics is the fact that the opinions remain separated for all times provided that they are separated initially. Similarly, the solution to the continuum limit equation is expected to satisfy an analogue property: In dimension 1, this is contained in the assumption that x^0 is one to one. Then, a pointwise evaluation of x^0 is needed, and therefore, unlike the previous frameworks, a more natural assumption is $x^0 \in \mathcal{C}(I)$ rather than $x^0 \in L^\infty(I)$. Note that, as for the microscopic dynamics, this assumption propagates over time.

The proofs for the existence and uniqueness of solutions to (66) and of the convergence to the continuum limit equation of (65) are quite similar to those in [2], the main difference being of course in the handling of the equation for the weight dynamics. Because of the singularity in 0, one introduces the sets

$$A_N(t,\xi) := \{\zeta \in I \mid x_N(t,\zeta) - x_N(\xi,t) > 0\}, \quad B_N(t,\xi) := A_N^c(t,\xi)$$

and

$$A(t,\xi) := \{\zeta \in I \mid x(t,\zeta) - x(\xi,t) > 0\}, \quad B(t,\xi) := A^c(t,\xi)$$

and uses them to split the integrals appearing in the computations into several terms. One can then conclude using the convergence of $\mathbf{1}_{A_N(0,\xi)}(\zeta)$ to $\mathbf{1}_{A(0,\xi)}(\zeta)$ and of $\mathbf{1}_{B_N(0,\xi)}(\zeta)$ to $\mathbf{1}_{B(0,\xi)}(\zeta)$ as $N \to \infty$.

Remark 3.2 This result is also extended to any dimension $d > 1$ with ξ varying in the d-dimensional unit cube I^d, but we will not expand on this point and refer the reader to [7, Section 5] for more details.

3.3 Links Between Continuum Limit and Mean-Field Limits for the Case of Evolving-in-Time Weights

This section is devoted to bridging all the different limit equations obtained for the opinion model with time-evolving weights. As previously, for $N \in \mathbb{N}$, we denote $x_i^N(t) \in \mathbb{R}^d$ the opinion of agent i at time t and $m_i^N(t) \in \mathbb{R}$ its weight (representing its charisma). We consider the system

$$\begin{cases} \dfrac{d}{dt} x_i^N(t) = \dfrac{1}{N} \displaystyle\sum_{j=1}^{N} m_j^N(t)\, \phi(x_i^N(t), x_j^N(t)) \\ \dfrac{d}{dt} m_i^N(t) = \psi_i^{(N)}(x^N(t), m^N(t)). \end{cases} \tag{67}$$

For initial data such that m_0 is strictly positive, one can actually prove that $m^N(t)$ remains positive for all $t \in [0,T]$. Thus, from here onward, the weights are assumed to belong to $\mathbb{R}_+^*$. Taking the continuum limit leads to the convergence in $\mathcal{C}([0,T]; L^2(I;\mathbb{R}^{dN})) \times \mathcal{C}([0,T]; L^2(I;(\mathbb{R}_+^*)^N))$ to (x,m) which is a solution to the integrodifferential equation

$$\begin{cases} \partial_t x(t,\xi) = \displaystyle\int_I m(t,\zeta)\phi(x(t,\xi), x(t,\zeta))d\zeta \\ \partial_t m(t,\xi) = \psi(\xi, x(t,\cdot), m(t,\cdot)), \end{cases} \tag{68}$$

provided that the initial data converges in a suitable sense.

The Non-exchangeable Mean-Field Limit As mentioned in Sect. 3.1.1, for weight dynamics of the form

$$\psi_i(x^N, m^N) = \int_{\{1,\dots,N\}\times\mathbb{R}^d\times\mathbb{R}_+^*} S_i(x_i^N, m_i^N, k, y, n)\, d\tilde{\mu}_t^N(k,y,n)$$

with the appropriate regularity assumptions, the non-exchangeable mean-field limit satisfied by the limit probability density $\mu \in \mathcal{C}([0,T], \mathcal{P}(I \times \mathbb{R}^d \times \mathbb{R}_+^*))$ is

$$\begin{aligned}\partial_t \mu_t^\xi(x,m) + \nabla_x \cdot \left(\left(\int_{I \times \mathbb{R}^d \times \mathbb{R}_+^*} n\phi(x,y) d\mu_t^\zeta(y,n) d\zeta \right) \mu_t^\xi(x,m) \right) \\ + \partial_m \left(\left(\int_{I \times \mathbb{R}^d \times \mathbb{R}_+^*} S(\xi, x, m, \zeta, y, n) d\mu_t^\zeta(y,n) d\zeta \right) \mu_t^\xi(x,m) \right) = 0 \qquad (69)\end{aligned}$$

The Exchangeable Mean-Field Limits In the case where the agents are indistinguishable, in the sense mentioned in Sect. 3.1, we can be interested in two different probability densities: the one for weights and opinions belonging to $\mathcal{C}([0,T], \mathcal{P}(\mathbb{R}^d \times \mathbb{R}_+^*))$ and the one only for opinions belonging to $\mathcal{C}([0,T], \mathcal{P}(\mathbb{R}^d))$. In both cases, the mean-field limit can actually be obtained for a class, slightly different from the one presented previously, with a straightforward adaptation of the proof.

1. For weight dynamics of the form

$$\psi(\xi, x(t,\cdot), m(t,\cdot)) = \int_{\mathbb{R}^d \times \mathbb{R}_+^*} \tilde{S}(x(t,\xi), m(t,\xi), y, n) d\overline{\tilde{\mu}}_t(y,n) \qquad (70)$$

   with

$$\overline{\tilde{\mu}}_t(x,m) = \int_I \delta_{x(t,\xi)}(x) \delta_{m(t,\xi)}(m) d\xi, \qquad (71)$$

   the classical mean-field equation can be derived for the probability measure $\tilde{\mu} \in \mathcal{C}([0,T], \mathcal{P}(\mathbb{R}^d \times \mathbb{R}_+^*))$, obtaining a transport equation in the variables x and m:

$$\begin{aligned}\partial_t \tilde{\mu}_t(x,m) + \nabla_x \cdot \left(\left(\int_{\mathbb{R}^d \times \mathbb{R}_+^*} n\phi(x,y) d\tilde{\mu}_t(y,n) \right) \tilde{\mu}_t(x,m) \right) \\ + \partial_m \left(\left(\int_{\mathbb{R}^d \times \mathbb{R}_+^*} \tilde{S}(x,m,y,n) d\tilde{\mu}_t(y,n) \right) \tilde{\mu}_t(x,m) \right) = 0 \qquad (72)\end{aligned}$$

2. As discussed in Sect. 3.1, for the general class of weight dynamics

$$\psi(\xi, x(t,\cdot), m(t,\cdot)) = m(t,\xi) \int_{\mathbb{R}^d} \hat{S}(x(t,\xi), y) d\hat{\mu}_t(y) \qquad (73)$$

with

$$\overline{\tilde{\mu}}_t(x) := \int_I m(t,\xi)\delta_{x(t,\xi)}(x)d\xi, \tag{74}$$

the probability measure $\hat{\mu} \in \mathcal{C}([0,T], \mathcal{P}(\mathbb{R}^d))$ obtained in the limit satisfies the following transport equation with source:

$$\partial_t \hat{\mu}_t(x) + \nabla_x \cdot \left(\int_{\mathbb{R}^d} \phi(x,y) d\hat{\mu}(y) \hat{\mu}_t(x) \right) = \hat{\mu}_t(x) \int_{\mathbb{R}^d} \hat{S}(x,y) d\hat{\mu}_t(y). \tag{75}$$

Thus, (68), (69), (72), and (75) can be obtained as the limit of (67) as N goes to infinity. The derivations of (68) and (75) have been done rigorously in [2]. The derivation of (44) can be obtained, thanks to the results in [53]. The derivation of (72) will be mentioned in the following. As in Sect. 2.3, let us try to clarify how all theses different descriptions relate to each other (see Fig. 5 for a visual summary).

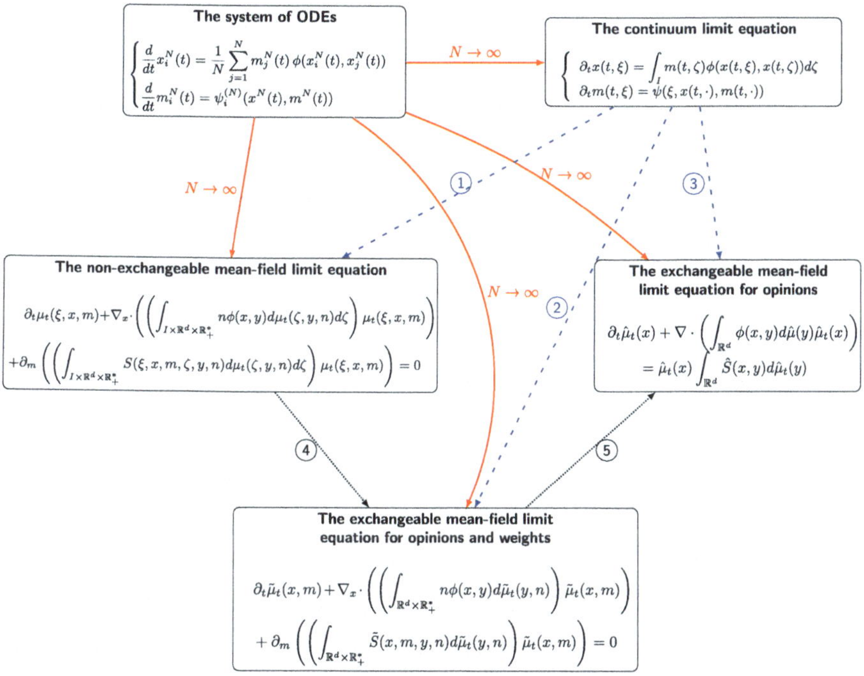

Fig. 5 Links between the different equations. The arrow 1 corresponds to the weight dynamics $\psi_1(\xi, x(t,\cdot), m(t,\cdot)) = \int_{I\times\mathbb{R}^d\times\mathbb{R}_+^*} S(\xi, x(t,\xi), m(t,\xi), \zeta, y, n) d\overline{\mu}_t(\zeta, y, n) d\zeta$, the arrow 2 to the weight dynamics $\psi_2(\xi, x(t,\cdot), m(t,\cdot)) = \int_{\mathbb{R}^d\times\mathbb{R}_+^*} \tilde{S}(x(t,\xi), m(t,\xi), y, n) d\overline{\tilde{\mu}}_t(y,n)$ and the arrow 3 to the weight dynamics $\psi_3(\xi, x(t,\cdot), m(t,\cdot)) = m(t,\xi)\int_{\mathbb{R}^d} \hat{S}(x(t,\xi), y) d\overline{\hat{\mu}}_t(y)$. The arrows 4 and 5 correspond to Remark 3.3

3.3.1 From Continuum Limit to Non-exchangeable Mean-Field Limit

We denote $\overline{\mu}_t$ the following continuous empirical measure

$$\overline{\mu}_t(\xi, x, m) = \int_I \delta_{x(t,\zeta)}(x)\delta_{m(t,\zeta)}(m)\delta_\zeta(\xi)d\zeta. \tag{76}$$

where $(x(t,\xi), m(t,\xi))$ is a solution to (68) for the weight dynamics

$$\psi(\xi, x(t,\cdot), m(t,\cdot)) = \int_{I\times\mathbb{R}^d\times\mathbb{R}_+^*} S(\xi, x(t,\xi), m(t,\xi), \zeta, y, n)d\overline{\mu}_t(\zeta, y, n)d\zeta. \tag{77}$$

For all test functions $f \in C^\infty(I \times \mathbb{R}^d \times \mathbb{R}_+^*)$, it holds

$$\begin{aligned}
&\frac{d}{dt}\int_{I\times\mathbb{R}^d\times\mathbb{R}_+^*} f(\xi, x, m)d\overline{\mu}_t(\xi, x, m)d\xi = \frac{d}{dt}\int_I f(\xi, x(t,\xi), m(t,\xi))d\xi \\
&= \int_I \nabla_x f(\xi, x(t,\xi), m(t,\xi)) \cdot \left(\int_I m((t,\zeta)\phi(x(t,\xi), x(t,\zeta))d\zeta\right) d\xi \\
&\quad + \int_I \partial_m f(\xi, x(t,\xi), m(t,\xi)) \\
&\quad \times \left(\int_{I\times\mathbb{R}^d\times\mathbb{R}_+^*} S(\xi, x(t,\xi), m(t,\xi), \zeta, y, n)d\overline{\mu}_t(\zeta, y, n)d\zeta\right) d\xi \\
&= \int_{I\times\mathbb{R}^d\times\mathbb{R}_+^*} \nabla_x f(\xi, x, m) \cdot \left(\int_{I\times\mathbb{R}^d\times\mathbb{R}_+^*} n\phi(x, y)d\overline{\mu}_t(\zeta, y, n)d\zeta\right) \\
&\quad \times d\overline{\mu}_t(\xi, x, m)d\xi \\
&\quad + \int_{I\times\mathbb{R}^d\times\mathbb{R}_+^*} \partial_m f(\xi, x, m) \\
&\quad \times \left(\int_{I\times\mathbb{R}^d\times\mathbb{R}_+^*} S(\xi, x, m, \zeta, y, n)d\overline{\mu}_t(\zeta, y, n)d\zeta\right) d\overline{\mu}_t^\xi(x, m)d\xi
\end{aligned}$$

Thus, the continuous empirical measure $\overline{\mu}_t$ built from the solution to (68) for the weight dynamics (77) is a solution to the non-exchangeable mean-field limit equation (44). This computation corresponds to the arrow 1 in Fig. 5.

3.3.2 From Continuum Limit to Exchangeable Mean-Field Limits

For weight dynamics of the form (70), similar computations against test functions $f \in C^\infty(\mathbb{R}^d \times \mathbb{R}_+^*)$ show that the continuous empirical measure $\overline{\tilde{\mu}}_t(x, m)$ defined in (71) satisfies the exchangeable mean-field equation for opinions and weights (72). Indeed, we get, for all test functions $f \in C^\infty(\mathbb{R}^d \times \mathbb{R}_+^*)$,

$$
\begin{aligned}
\frac{d}{dt}\int_{\mathbb{R}^d\times\mathbb{R}_+^*} f(x,m)d\overline{\tilde{\mu}}_t(x,m)d\xi &= \frac{d}{dt}\int_I f(x(t,\xi),m(t,\xi))d\xi \\
&= \int_I \nabla_x f(x(t,\xi),m(t,\xi))\cdot\left(\int_I m((t,\zeta)\phi(x(t,\xi),x(t,\zeta))d\zeta\right)d\xi \\
&\quad + \int_I \partial_m f(x(t,\xi),m(t,\xi))\left(\int_{\mathbb{R}^d\times\mathbb{R}_+^*} \tilde{S}(x(t,\xi),m(t,\xi),y,n)d\overline{\tilde{\mu}}_t(y,n)\right)d\xi \\
&= \int_{\mathbb{R}^d\times\mathbb{R}_+^*} \nabla_x f(x,m)\cdot\left(\int_{\mathbb{R}^d\times\mathbb{R}_+^*} n\phi(x,y)d\overline{\tilde{\mu}}_t(y,n)\right)d\overline{\tilde{\mu}}_t(x,m) \\
&\quad + \int_{\mathbb{R}^d\times\mathbb{R}_+^*} \partial_m f(x,m)\left(\int_{\mathbb{R}^d\times\mathbb{R}_+^*} \tilde{S}(x,m,y,n)d\overline{\tilde{\mu}}_t(y,n)\right)d\overline{\tilde{\mu}}_t(x,m).
\end{aligned}
$$

This computation corresponds to the arrow 2 in Fig. 5.

Lastly, similarly to [2], for weight dynamics of the form (73), the continuous empirical measure $\overline{\hat{\mu}}_t(x)$ defined in (74) satisfies the exchangeable mean-field equation for opinions (75). Indeed, for all test functions $f \in C^\infty(\mathbb{R}^d)$,

$$
\begin{aligned}
\frac{d}{dt}\int_{\mathbb{R}^d} f(x)d\overline{\hat{\mu}}_t(x)d\xi &= \frac{d}{dt}\int_I m(t,\xi)f(x(t,\xi))d\xi \\
&= \int_I m(t,\xi)\nabla_x f(x(t,\xi))\cdot\left(\int_I m((t,\zeta)\phi(x(t,\xi),x(t,\zeta))d\zeta\right)d\xi \\
&\quad + \int_I m(t,\xi)f(x(t,\xi))\left(\int_{\mathbb{R}^d} \hat{S}(x(t,\xi),y)d\overline{\hat{\mu}}_t(y)\right)d\xi \\
&= \int_{\mathbb{R}^d} \nabla_x f(x)\cdot\left(\int_{\mathbb{R}^d} \phi(x,y)d\overline{\hat{\mu}}_t(y)\right)d\overline{\hat{\mu}}_t(x) \\
&\quad + \int_{\mathbb{R}^d} f(x)\left(\int_{\mathbb{R}^d} \hat{S}(x,y)d\overline{\hat{\mu}}_t(y)\right)d\overline{\hat{\mu}}_t(x).
\end{aligned}
$$

This computation corresponds to the arrow 3 in Fig. 5.

Remark 3.3 We notice that $\overline{\tilde{\mu}}_t(x, m) = \int_I \overline{\mu}(\xi, x, m)d\xi$ and

$$\overline{\hat{\mu}}_t(x) = \int_{\mathbb{R}} m\overline{\tilde{\mu}}_t(x, dm).$$

Thus, for the appropriate choice for the weight dynamics, this gives the path to obtain (72) from (44) and (75) from (72) and corresponds to the arrows 4 and 5 in Fig. 5.

Remark 3.4 Because of the presence of nonlinearity, we cannot adapt the argument developed in Sect. 2.3.2, and we cannot exhibit a path to get the continuum limit (68) from the non-exchangeable mean-field limit (44).

The continuum limit can actually provide a tool for an alternative proof for the exchangeable mean-field limit. Indeed, noticing that we can rewrite $\hat{\mu}_t^N$ defined in (46) as

$$\hat{\mu}_t^N = \int_I m_N(t, \xi)\delta_{x_N(t,\xi)}(x)$$

with (x_N, m_N) defined in (56), we can easily show that, for all $t \in [0, T]$,

$$W_1(\hat{\mu}_t^N, \overline{\hat{\mu}}_t) \xrightarrow[N \to +\infty]{} 0$$

where we recall

$$\overline{\hat{\mu}}_t(x) = \int_I m(t, \xi)\delta_{x(t,\xi)}(x)d\xi,$$

provided the appropriate choice for the initial data using the convergence of (x_N, m_N) to (x, m) with W_1 being the 1-Wasserstein distance. Moreover, we just showed that $\overline{\hat{\mu}}_t$ happens to be a solution to (75), hence the conclusion.

All the results mentioned in that section require some regularity assumptions on both the interaction function and the weight dynamics. However, in [7], by similar computations, the authors are able to obtain the derivation of the exchangeable mean-field limit for the probability density of opinions associated with their singular weight dynamics. It is worth noticing that the continuum limit turns out to be a powerful tool since, in that case, there is no alternative proof through the classical approach. Indeed, all the classical approaches use some regularity of the source term which is no longer true for singular weight dynamics.

Finally, noticing that can rewrite $\tilde{\mu}_t^N$ defined in (71) as

$$\tilde{\mu}_t^N = \int_I \delta_{x_N(t,\xi)}(x)\delta_{m_N(t,\xi)}(m),$$

we can similarly prove that, for all $t \in [0, T]$,

$$W_1(\tilde{\mu}_t^N, \overline{\tilde{\mu}}_t) \xrightarrow[N \to +\infty]{} 0$$

where we recall

$$\overline{\tilde{\mu}}_t(x, m) = \int_I \delta_{x(t,\xi)}(x)\delta_{m(t,\xi)}(m)d\xi,$$

provided the appropriate choice for the initial data. Thus, this provides the derivation of (72) from (67) as N goes to infinity.

Acknowledgments NA has received funding from the Emergence grant of Sorbonne University. NPD has received funding from the Emergence grant of Sorbonne University.

References

1. I. Aoki, A simulation study on the schooling mechanism in fish. Nippon Suisan Gakkaishi **48**(8), 1081–1088 (1982)
2. N. Ayi, N. Pouradier Duteil, Mean-field and graph limits for collective dynamics models with time-varying weights. J. Diff. Equ. **299**, 65–110 (2021)
3. N. Ayi, N. Pouradier Duteil, Graph limit for interacting particle systems on weighted random graphs (2023). arXiv preprint arXiv:2307.12801
4. A. Backhausz, B. Szegedy, Action convergence of operators and graphs. Canadian J. Math. **74**(1), 72–121 (2022)
5. M. Ballerini, N. Cabibbo, R. Candelier, A. Cavagna, E. Cisbani, I. Giardina, V. Lecomte, A. Orlandi, G. Parisi, A. Procaccini et al., Interaction ruling animal collective behavior depends on topological rather than metric distance: evidence from a field study. Proc. Natl. Acad. Sci. **105**(4), 1232–1237 (2008)
6. J. Barré, P. Degond, D. Peurichard, E. Zatorska, Modelling pattern formation through differential repulsion. Netw. Heterogen. Media **15**(3), 307–352 (2020)
7. I. Ben-Porat, J.A. Carrillo, P.-E. Jabin, The graph limit for a pairwise competition model. J. Differ. Equ. **413**, 329–369 (2024)
8. I. Ben Porat, J.A. Carrillo, S.T. Galtung. Mean field limit for one dimensional opinion dynamics with coulomb interaction and time dependent weights. Nonlinear Analy. **240**, 113462 (2024)
9. R. Berner, T. Gross, C. Kuehn, J. Kurths, S. Yanchuk, Adaptive dynamical networks. Phys. Rep. **1031**, 1–59 (2023)
10. G. Bet, F. Coppini, F. R. Nardi, Weakly interacting oscillators on dense random graphs, J. Appl. Probab. **61**(1), 255–278 (2024)
11. S. Bhamidi, A. Budhiraja, R. Wu, Weakly interacting particle systems on inhomogeneous random graphs. Stochastic Processes and their Applications **129**(6), 2174–2206 (2019)
12. U. Biccari, D. Ko, E. Zuazua, Dynamics and control for multi-agent networked systems: a finite-difference approach. Math. Models Methods Appl. Sci. **29**(04), 755–790 (2019)
13. B. Bonnet, N. Pouradier Duteil, M. Sigalotti, Consensus formation in first-order graphon models with time-varying topologies. Math. Models Methods Appl. Sci. **32**(11), 2121–2188 (2022)

14. C. Borgs, J. Chayes, H. Cohn, Y. Zhao, An l^p theory of sparse graph convergence i: limits, sparse random graph models, and power law distributions. Trans. Am. Math. Soc. **372**(5), 3019–3062 (2019)
15. L. Boudin, F. Salvarani, E. Trélat, Exponential convergence towards consensus for non-symmetric linear first-order systems in finite and infinite dimensions. SIAM J. Math. Analy. **54**(3), 2727–2752 (2022)
16. W. Braun, K. Hepp, The Vlasov dynamics and its fluctuations in the $1/N$ limit of interacting classical particles. Comm. Math. Phys. **56**(2), 101–113 (1977)
17. W. Braun, K. Hepp, The vlasov dynamics and its fluctuations in the 1/n limit of interacting classical particles. Commun. Math. Phys. **56**(2), 101–113 (1977)
18. F. Bullo, J. Cortés, S. Martinez, *Distributed Control of Robotic Networks: A Mathematical Approach to Motion Coordination Algorithms*, vol. 27 (Princeton University Press, Princeton, 2009)
19. H. Chiba, G.S. Medvedev, The mean field analysis of the Kuramoto model on graphs i. the mean field equation and transition point formulas. Discrete Contin. Dyn. Syst. **39**(1), 131–155 (2019)
20. F. Cucker, S. Smale, Emergent behavior in flocks. IEEE Trans. Autom. Control **52**, 852–862 (2007)
21. S. Delattre, G. Giacomin, E. Luçon, A note on dynamical models on random graphs and fokker-planck equations. J. Stati. Phys. **165**, 785–798 (2016)
22. R.L. Dobrushin, Vlasov equations. Funct. Analy. Appl. **13**(2), 115–123 (1979)
23. S. Faure, B. Maury, Crowd motion from the granular standpoint. Math. Models Methods Appl. Sci. **25**(03), 463–493 (2015)
24. M. Fornasier, B. Piccoli, F. Rossi, Mean-field sparse optimal control. Philosoph. Trans. R. Soc. A Math. Phys. Eng. Sci. **372**(2028), 20130400 (2014)
25. M.A. Gkogkas, C. Kuehn, Graphop mean-field limits for Kuramoto-type models. SIAM J. Appl. Dyn. Syst. **21**(1), 248–283 (2022)
26. M.A. Gkogkas, C. Kuehn, C. Xu, Continuum limits for adaptive network dynamics. Commun. Math. Sci. **21**, 83–106 (2023)
27. M.A. Gkogkas, C. Kuehn, C. Xu, Mean field limits of co-evolutionary heterogeneous networks (2023)
28. F. Golse, On the dynamics of large particle systems in the mean field limit, in *Macroscopic and Large Scale Phenomena: Coarse Graining, Mean Field Limits and Ergodicity* (Springer, Cham, 2016), pp. 1–144
29. S.-Y. Ha, J.-G. Liu, A simple proof of the Cucker-Smale flocking dynamics and mean-field limit. Commun. Math. Sci. **7**(2), 297–325 (2009)
30. S.-Y. Ha, J. Kim, X. Zhang, Uniform stability of the cucker-smale model and its application to the mean-field limit. Kinetic Related Models **11**(5), 1157–1181 (2018)
31. R. Hegselmann, U. Krause, Opinion dynamics and bounded confidence models, analysis and simulation. J. Artif. Societ. Soc. Simul. **5**, 07 (2002)
32. P.-E. Jabin, Z. Wang, Quantitative estimates of propagation of chaos for stochastic systems with $w^{-1,\infty}$ kernels. Invent. Math. **214**, 523–591 (2018)
33. P.-E. Jabin, D. Poyato, J. Soler, Mean-field limit of non-exchangeable systems. Working paper or preprint (2022)
34. A. Jadbabaie, N. Motee, M. Barahona, On the stability of the kuramoto model of coupled nonlinear oscillators, in *Proceedings of the 2004 American Control Conference*, vol. 5 (2004), pp. 4296–4301
35. D. Kaliuzhnyi-Verbovetskyi, G.S. Medvedev, The mean field equation for the Kuramoto model on graph sequences with non-lipschitz limit. SIAM J. Math. Analy. **50**(3), 2441–2465 (2018)
36. C. Kuehn, Network dynamics on graphops. New J. Phys. **22**(5), 053030 (2020)
37. C. Kuehn, C. Xu, Vlasov equations on digraph measures. J. Diff. Equ. **339**, 261–349 (2022)
38. D. Kunszenti-Kovács, L. Lovász, B. Szegedy, Measures on the square as sparse graph limits. J. Combinat. Theory Ser. B **138**, 1–40 (2019)

39. Y. Kuramoto, Self-entrainment of a population of coupled nonlinear oscillators, in ed. by H. Araki, *International Symposium on Mathematical Problems in Theoretical Physics, Lecture Notes in Physics, Vol. 39* (Springer, New York, 1975), pp. 420–422
40. Y. LeCun, Y. Bengio, G. Hinton, Deep learning. Nature **521**, 436–44 (2015)
41. U. Lopez, J. Gautrais, I.D. Couzin, G. Theraulaz, From behavioural analyses to models of collective motion in fish schools. Interface Focus **2**(6), 693–707 (2012)
42. L. Lovász, *Large Networks and Graph Limits*. Colloquium Publications, vol. 60 (American Mathematical Society, Providence, 2012)
43. L. Lovász, B. Szegedy, Limits of dense graph sequences. J. Combinat. Theory Series B **96**(6), 933–957 (2006)
44. S. McQuade, B. Piccoli, N. Pouradier Duteil. Social dynamics models with time-varying influence. Math. Models Methods Appl. Sci. **29**(4),681–716 (2019)
45. G.S. Medvedev, Small-world networks of kuramoto oscillators. Phys. D Nonlinear Phenomena **266**, 13–22 (2014)
46. G.S. Medvedev, The nonlinear heat equation on dense graphs and graph limits. SIAM J. Math. Analy. **46**, 2743–2766 (2014)
47. G.S. Medvedev. The nonlinear heat equation on W-random graphs. Arch. Ration. Mech. Analy. **212**, 781–803 (2014)
48. H. Neunzert, Mathematical investigations on particle-in-cell methods. Fluid Dyn. Trans. **9**, 229–254 (1978)
49. H. Neunzert. An introduction to the nonlinear boltzmann-vlasov equation, in ed. by C. Cercignani, *Kinetic Theories and the Boltzmann Equation* (Springer, Berlin, 1984) , pp. 60–110
50. H. Neunzert, J. Wick, Die approximation der lösung von integro-differentialgleichungen durch endliche punktmengen, in ed. by R. Ansorge, W. Törnig, *Numerische Behandlung nichtlinearer Integrodifferential-und Differentialgleichungen* (Springer, Berlin, 1974), pp. 275–290
51. T. Nishikawa, A.E. Motter, Comparative analysis of existing models for power-grid synchronization. New J. Phys. **17**(1), 015012 (2015)
52. A. Nugent, S.N. Gomes, M.-T. Wolfram, On evolving network models and their influence on opinion formation. Phys. D Nonlinear Phenomena **456**, 2023, 133914 (2024)
53. T. Paul, E. Trélat, From microscopic to macroscopic scale equations: mean field, hydrodynamic and graph limits (2022)
54. B. Piccoli, N. Pouradier Duteil, Control of collective dynamics with time-varying weights, in ed. by F. Salvarani, *Recent Advances in Kinetic Equations and Applications* (Springer International Publishing, Cham, 2021), pp. 289–308
55. O. Popovych, M. Xenakis, P. Tass, The spacing principle for unlearning abnormal neuronal synchrony. PLoS ONE **10**, 02 (2015)
56. N. Pouradier Duteil, Mean-field limit of collective dynamics with time-varying weights. Netw. Heterog. Media **17**(2), 129–161 (2022)
57. V. Rohr, R. Berner, E. Lameu, O. Popovych, S. Yanchuk, Frequency cluster formation and slow oscillations in neural populations with plasticity. PLOS ONE **14**, e0225094 (2019)
58. D.E. Rumelhart, R.J. Williams, G.E. Hinton, Learning representations by back-propagating errors. Nature **323**, 533–536 (1986)
59. S. Serfaty, Mean field limit for Coulomb-type flows. Duke Math. J. **169**(15), 2887–2935 (2020)
60. A.-S. Sznitman. Topics in propagation of chaos, in *Lecture Notes in Mathematics* (Springer, Berlin, 1991), pp. 165–251
61. S. Throm, Continuum limit for interacting systems on adaptive networks (2023)
62. T. Vicsek, A. Czirók, E. Ben-Jacob, I. Cohen, O. Shochet, Novel type of phase transition in a system of self-driven particles. Phys. Rev. Lett. **75**(6), 1226 (1995)
63. D.A. Wiley, S.H. Strogatz, M. Girvan, The size of the sync basin. Chaos An Interdiscipl. J. Nonlinear Sci. **16**(1), 015103 (2006)
64. G. Zschaler, G.A. Bohme, M. Seissinger, C. Huepe, T. Gross, Early fragmentation in the adaptive voter model on directed networks. Phys. Rev. E **85**(4), 046107 (2012)

Models of Animal Behavior as Active Particle Systems with Nonreciprocal Interactions

Amir Haluts, Dan Gorbonos, and Nir S. Gov

Abstract Active particle systems of interacting self-propelled particles offer a versatile framework for modeling complex systems. When employed to describe aspects of animal behavior, the complexity of animal movement and decision-making often requires the use of unique types of effective interactions between the particles—notably nonreciprocal effective forces that do not obey the usual conservation laws of Newtonian mechanics. Here, we review two recent empirically motivated models of two very different types of animal behavior, where the behavior is described in terms of active particles which interact through nonreciprocal effective forces. The first model describes the dynamics of animal contests, wherein typically two rivals fight over a localized resource. The uniquely shaped effective potentials between the model's "contestant particles" manifest the adversarial nature of contest interactions and capture the dynamical essence of contest behavior in space and time. The second model describes the stabilization of cohesive swarms through long-range and adaptive gravity-like attraction. This "adaptive gravity" model explains the observed mass and velocity profiles of laboratory midge swarms. These examples demonstrate that theoretical models that use the framework of active particles to describe animal behavior can expand the scope of active-particle research as well as explain complex phenomena in animal behavior.

A. Haluts · N. S. Gov (✉)
Department of Chemical and Biological Physics, Weizmann Institute of Science, Rehovot, Israel
e-mail: amir.haluts@weizmann.ac.il; nir.gov@weizmann.ac.il

D. Gorbonos
Department of Collective Behaviour, Max Planck Institute of Animal Behavior, Konstanz, Germany
e-mail: dgorbonos@ab.mpg.de

J. A. Carrillo, E. Tadmor (eds.), *Active Particles, Volume 4*, Modeling and Simulation in Science, Engineering and Technology,
https://doi.org/10.1007/978-3-031-73423-6_4

1 Introduction

Active particle systems are composed of interacting, self-propelled particles. In theoretical models, implemented using computer simulations, interactions between particles can take a variety of forms [1]. These interactions can take the form of the (reciprocal) forces that arise from the potentials between physical particles, for example, of the Lennard-Johns-type, but they could also be nonphysical, such as those leading to velocity alignment between neighboring particles in Vicsek-type models [2]. The latter types of interactions are motivated by biological systems and are often nonreciprocal, for example, due to alignment [3, 4] or due to velocity-aligned anisotropic interaction potentials [5]. The resulting behavior of such nonequilibrium systems is determined by the self-propulsion properties of the particles, the type of interactions, and noise.

Since animals are self-propelled, modeling the movement-related behaviors of animals as individuals and in groups is naturally approached by active particle models [6]. In these systems, the interactions between the animals, which control their movements, are not easy to extract from experimental data. This difficulty stems from these interactions being driven by animal communication, which can involve a complex set of signals, in addition to each animal's interpretation of the movements and intentions of others. The mapping between the biological system and an active particle system is therefore nontrivial.

Here we review two recent models of animal behavior that describe the dynamics in their respective systems in terms of interacting active particles. These models demonstrate that the effective interactions which arise within the context of animal behavior can have distinct properties that do not appear in typical physical systems: Forces between animals can be nonreciprocal and adaptive.

Nonreciprocal interactions have garnered increasing attention in recent research [7–12]. They arise naturally in the context of animal contests (short-range fights), which we discuss in Sect. 2 of this chapter. We review a general spatiotemporal model that describes the dynamics of contestants as they escalate their interaction. In the model, the contestants are active particles which interact through effective interaction potentials. This model was originally developed to describe a system of male spider contestants which compete on the two-dimensional orb web of the female [13] and subsequently extended to describe contests between animals in general [14]. The dynamics of these agonistic interactions is driven by each contestant's assessment of its own and its rival's "strength" and the response to these assessments [14]. The asymmetric scenario where the animals have different "strengths" therefore naturally gives rise to nonreciprocal interactions, which can be described by each contestant responding to a different effective potential landscape. The most noticeable effect of these nonreciprocal forces during contests is the transition from bounded tussle dynamics (of closely matched rivals) to chase dynamics (between highly unmatched rivals).

In Sect. 3 of this chapter, we demonstrate the role of "adaptivity" in the effective interactions between animals. Adaptivity is the property of animals to respond to the relative fold change in a signal that reaches their sensory organs [15]. This means that a constant persistent change in the background signal does not have a lasting effect on the behavior, as the sensory modality adapts to the new background. This property appears in all animals, from bacteria to humans, and in all sensory systems. We review the modeling of adaptive interactions in the context of midge swarms, which form cohesive yet disordered swarms of flying insects [16]. The model explores long-range attractive interactions between particles, which are adaptive, resulting in an "adaptive gravity" form [17–20]. The adaptivity introduces a nonequilibrium property by violating the conservation of energy and momentum of the particles and giving rise to nonreciprocal interactions. Note that we focus here on adaptivity of the animal's sensory sensitivity (and the effective interactions associated with it) due to an external signal, over short timescales that are relevant for the animals' motion, as in adaptive chemotaxis [21]. We therefore do not refer to adaptivity on the timescale of genetic mutations, as notably occurs in growing bacterial systems [22].

In both parts of this chapter, we demonstrate the novel theoretical aspects of the biologically inspired interactions between the particles as well as compare to experimental data. This chapter emphasizes how modeling animal behavior provides us with motivation to introduce and explore unique types of interactions and emergent phenomena in the context of active particle systems.

2 Nonreciprocally Interacting Particles Model of Animal Contests

It is commonplace for animals to engage in agonistic behavior to obtain and defend food, mates, and territory [23]. These fights over resources, widely termed "animal contests," inherently entail the resolution of an interplay between risk and reward. Animal contests have therefore attracted extensive research from ecologists and biologists who sought to understand the evolutionary logic behind such nontrivial trade-offs [24–32]. From a theoretical standpoint, these risk-reward trade-offs have also made animal contests particularly amenable to the modeling tools of game theory, which, ever since the seminal works of the 1970s [33–37], has been the central analytical framework used to model and understand them. But while game-theoretic contest models make predictions that can, in principle, be tested empirically [38], mapping these models to real contest scenarios has proven difficult and often lead to ambiguous conclusions [38–41]. An intrinsic hurdle for such mapping has been the omission of within-contest dynamics by game-theoretic models, which typically meet the observable dynamics of contest behavior only in their endpoint predictions (such as who won and how long was the contest) and disregard the spatiotemporal intricacies of the contestants' trajectories.

While detailed tracking of spatiotemporal trajectories was out of reach for the original contest theorists, it is now a standard practice in animal behavior research

[42, 43]. This motivated the development of a new theoretical framework for the dynamics of animal contests, which describes them based on their spatiotemporal properties [13, 14]. In this physics-inspired framework, the real-space dynamics of contestants is explicitly described in terms of (active) "contestant particles," which interact through effective interaction potentials. The adversarial nature of the inter-contestant interactions makes their properties distinct from those of the typical pairwise potentials that govern the interactions between physical particles. Notably, since each contestant's behavior is ultimately driven by the assessment of its own and its rival's "strength,", contest interactions are directional and nonreciprocal. This is handled by assigning the dynamics of each contestant with the influence of a distinct potential landscape that cannot be simply deduced from a global potential. In the following, we discuss the motivation behind this framework, its construction, and some of its implications—both in general and in a specific experimental system. We review the model presented in refs. [13] and [14] as a cohesive body of work, to which we add retrospective insights.

2.1 Summary

The ubiquitous scenario that motivates our modeling approach is that of a dyadic contest, in which two contestants compete over a localized resource. In this generic setup, their mutual attraction to the resource brings the contestants into contest range, and once within this range, they typically become engaged in a short-range contest interaction until they resolve the encounter (disengage)—as illustrated by the trajectories in Fig. 3c.

The fundamental assumption behind our theoretical treatment is that the spatiotemporal properties of contest behavior can be effectively described by some archetypal interaction potentials (which give rise to effective interaction forces) and that these potentials predominantly depend on the contestants' positions. Each contestant is treated as an active Brownian particle that moves due to deterministic (interaction-induced) and stochastic effective forces—so that its spatiotemporal dynamics is governed by a Langevin equation, as described in Sect. 2.2. The attraction toward the resource is encoded by an effective "resource potential" that can be thought of as the effective potential landscape on which (and due to which) contests take place. The inter-contestant interactions, including the "contest" itself, are encoded by effective "contestant interaction potentials"—the model's most essential building block, which are constructed based on generic features of contest behavior. These potentials are described in Sect. 2.3.

This framework enables a well-defined notion of a pairwise "contest" according to the relative motion between the contestants, as described in Sect. 2.4. Furthermore, the framework can account for different rival assessment strategies, which have been extensively studied in game-theoretic contest models, as described in Sect. 2.5. The detailed description of the contestants' motion accounts for spatiotemporal features of asymmetric contests. Notably, our model's contestants exhibit chase dynamics as an emergent property of broken symmetry in their nonreciprocal interaction forces, as described in 2.6. Finally, explicit time dependence of our model's interaction potentials (e.g., due to accumulation of fighting costs) is discussed briefly in Sect. 2.7.

2.2 *Dynamics of Contestant Particles*

To describe the spatiotemporal dynamics of our model's contestants—notably in order to simulate a dyadic contest over a localized resource (as visualized in Fig. 3c)—we treat each contestant as an active Brownian particle that moves due to deterministic (interaction-induced) and stochastic effective forces. Although the motion characteristics of real animals are typically much richer than those of active particles, such models have proven useful in capturing average properties of animal motion across species, contexts, and timescales [44], and here we will demonstrate this usefulness for modeling contests. The spatiotemporal dynamics of a "contestant particle" i (where $i \in \{1, 2, ..., N\}$ in a system of N contestants) is modeled to be governed by the following (overdamped) Langevin equation (see also [45] for a comprehensive review of such models for active particle dynamics)

$$\dot{\mathbf{r}}_i = -\eta \nabla V_{\mathrm{tot}\to i}(\mathbf{r}_i, \{\mathbf{r}_j\}) + \sqrt{2D}\,\boldsymbol{\xi}_i + \text{activity}. \tag{1}$$

Note that we denote vectors by boldface letters and scalars by regular letters throughout this part of the chapter. On the right-hand side of Eq. (1), the first term accounts for the deterministic driving force felt by contestant i at position $\mathbf{r}_i$ due to the (total) effective potential $V_{\mathrm{tot}\to i}$, where η is the contestant's "mobility." $V_{\mathrm{tot}\to i}$ generally depends on $\mathbf{r}_i$ and on the positions $\{\mathbf{r}_j\}$ of all other rival contestants, where these positions are defined with respect to a single resource at the origin (as illustrated in Fig. 1a). The construction of $V_{\mathrm{tot}\to i}$, which combines all the fundamental building blocks of the model, is detailed in the next section; the second term is the usual translational noise term, where D is an effective translational

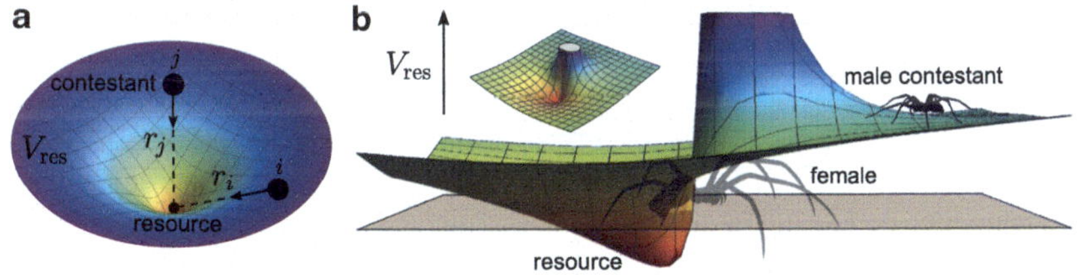

Fig. 1 Effective resource potentials. (**a**) The landscape of an effective "resource potential" with a simple radially symmetrical form, as in Eq. (4). (**b**) The landscape of a non-isotropic effective resource potential, which encodes the influence of a female *T. clavipes* spider and its orb web on the motion of male spider contestants [13] (Adapted from refs. [14] and [13])

diffusion coefficient and $\boldsymbol{\xi}_i$ is a vector of mutually uncorrelated sources of standard Gaussian white noise for each of the spatial dimensions (i.e., $\boldsymbol{\xi}_i = (\xi_{x_i}, \xi_{y_i})$, with $\xi_{x_i}, \xi_{y_i} \sim \mathcal{N}(0, 1)$, in two dimensions); the third term accounts for the internal (active) directional persistence of the contestant (which is independent from the directional persistence driven by $V_{\text{tot}\to i}$) and is commonly governed by its own Langevin equation. While the activity term can be used to better match motion characteristics of particular animal contestants, such as the male spiders studied in ref. [13], for timescales sufficiently longer than the persistence time it mostly acts as an additional source of noise [46].

In general, the noise in the model manifests the contribution of 'behavioral noise'—which is integral to biological decision-making—to the motion of the contestants. The practical role of this noise is to allow our simulated contestant particles to dynamically explore the effective potential landscape of $V_{\text{tot}\to i}$. Notably, this allows us to simulate the fast engage-and-disengage dynamics of typical animal contests [13, 23], as in Fig. 3c. In what follows, we assume for simplicity that the behavioral properties which make a contestant "active" can be sufficiently encoded into $V_{\text{tot}\to i}$, drop the activity term in Eq. (1), and thereby also leave the translational diffusion term (driven by Brownian, thermal noise) to solely account for the noise. For the practical implementation of Eq. (1) to simulate the dynamics of interacting contestant particles, yielding contest trajectories as in Figs. 3 and 6, we used the first-order (Euler) integration scheme—as detailed in the supporting information of ref. [14].

2.3 Effective Interaction Potentials and Nonreciprocity

The total effective potential felt by contestant i combines the influence of a resource and of all other contestants $\{j\}$ (which compete over this particular resource) on the motion of contestant i. Observations suggest that when contestants become strongly engaged in a contest interaction, their attention is predominantly given to their rival(s). This can be modeled by an "attention switch"—where the contestants' motion is affected by their interaction with the resource (which brings them into contest range) only when they are relatively far apart and is otherwise dominated by their interactions with each other. Therefore, we write $V_{\text{tot}\to i}$ as follows

$$V_{\text{tot}\to i}(\mathbf{r}_i, \{\mathbf{r}_j\}) = \sum_{j\neq i} V_{j\to i}(\mathbf{r}_i, \mathbf{r}_j) + \theta(\mathrm{x}_{ij} - \hat{\mathrm{x}}_{ij})\, V_{\text{res}\to i}(\mathbf{r}_i) \tag{2}$$

where the $V_{j\to i}$'s are the "contestant interaction potentials"—each encoding the influence of a rival contestant j on the motion of contestant i, $V_{\text{res}\to i}$ is the effective "resource potential" felt by contestant i—which encodes the influence of the resource on the motion of contestant i, and $\theta(x)$ is a Heaviside step function—where $\mathrm{x}_{ij} \equiv |\mathbf{r}_i - \mathbf{r}_j|/\mathrm{x}_0$ is the dimensionless (scaled) distance between contestants

i and j (x_0 is a typical length scale), and $x_{ij}^{\wedge}$ is the "contest onset" distance between i and j (which is defined in Sect. 2.4). For now, we will treat these effective potentials as time-independent but note that they could also vary with (interaction) time—notably due to accumulation of fighting costs and learning—as discussed briefly in Sect. 2.7.

Before discussing explicit functional forms for $V_{\text{res}\to i}$ and $V_{j\to i}$, note that Eqs. (1) and (2) can be used to describe the interactions between any number of contestants that compete over a single resource (where the motion of each contestant is governed by its own Langevin equation) and that one can further generalize Eq. (2) to include the influence of more than one resource (such that the contestants move in a "field" of resources). Nevertheless, although many-contestant and/or many-resource scenarios exist in nature (a specific many-contestant scenario is discussed in Sect. 2.6), the simplest scenario of a dyadic contest, in which two contestants fight over a resource, is by far the most common and well-studied [32, 38]. For this reason, and since a "contest" (in our context) is inherently a pairwise concept, we will mostly deal with the dyadic version of Eq. (2),

$$V_{\text{tot}\to i}(\mathbf{r}_i, \mathbf{r}_j) = V_{j\to i}(\mathbf{r}_i, \mathbf{r}_j) + \theta(x_{ij} - x_{ij}^{\wedge})\, V_{\text{res}\to i}(\mathbf{r}_i), \tag{3}$$

where contestants 'i' and 'j' are, henceforth, the only two contestants (henceforth, we use the indices 'i' and 'j' as the dummy identifiers of the contestants but note that $N = 2$ contestants for all of the remaining text, equations, and figures). We proceed by describing the spatiotemporal features common to contest dynamics in many animal species and map these generic rules into the shapes of the effective potentials $V_{\text{res}\to i}$ and $V_{j\to i}$. The finer details that these potentials have in specific animals and setups can be extracted empirically from contestant trajectories, as described in [13] and [14]. In the following, we often visualize and apply the model in a two-dimensional space but note that the model is equally applicable in three dimensions.

A typical dyadic contest is initiated when two animals seek access to a spatially localized resource, commonly a territory or a mate [47–49]. In terms of their physical influence, these resources act as attractors that drive contestants toward each other until conflict becomes inevitable and can thus be thought of as the effective potential landscape on which (and due to which) contests take place. The shape of this potential landscape depends on the properties of the resource. For most resources, an adequate model might be a simple radially symmetrical "sink," for example

$$V_{\text{res}\to i}(\mathbf{r}_i) = p_i \ln(|\mathbf{r}_i| + \epsilon), \tag{4}$$

where $p_i, \epsilon > 0$ are parameters (p_i sets the attractiveness of the resource, as perceived by contestant i). The potential landscape created by Eq. (4) is visualized in Fig. 1a. For some animate resources, however, behavior and setup could render the shape of the resource potential's landscape less trivial. For example, in the natural contest arena that is the orb web of a female *Trichonephila clavipes* spider, the resource landscape on which male contestants are moving comprises of the female and the architecture of its web. In this system, males are observed to approach the female from the back and avoid its front—where the potentially cannibalistic female is most reactive [13]. This resource should therefore be modeled by a non-isotropic effective potential. Specifically, in [13], the influence of this resource is described by a potential with a landscape as in Fig. 1b. Regardless of these system-specific details, the global qualitative effect of resource landscapes is generic: to bring the contestants into contest range due to their mutual attraction to the resource. This generic property implies that effective resource potentials have a global minimum, as in Fig. 1a, b.

The inter-contestant interactions, including the "contest" itself, are encoded by the contestant interaction potential $V_{j\to i}$—the model's most essential building block. The archetypal shape of this potential can be constructed based on the following generic (and in a sense minimal) features of contest behavior, which are stated in terms of effective attraction or repulsion between the contestants depending on the inter-contestant distance: (1) long-range repulsion due to mutual avoidance (which can be surmounted due to the attracting resource), (2) medium- to short-range attraction when the contestants reach a separation distance in which conflict escalation is inevitable (and hence move toward each other), (3) strong repulsion at contact, and (4) the strength of the interaction decays to zero when the contestants are far apart. Importantly, conflict escalation is associated with the tendency to decrease the inter-contestant distance (effective attraction), while conflict de-escalation is associated with the tendency to increase this distance (effective repulsion).

The above rules amount to a contestant interaction potential with a qualitative shape as in Fig. 2a, b, which can be satisfied by various functional forms. One such particular potential is a combination of a logarithmic repulsion and an attractive Gaussian well,

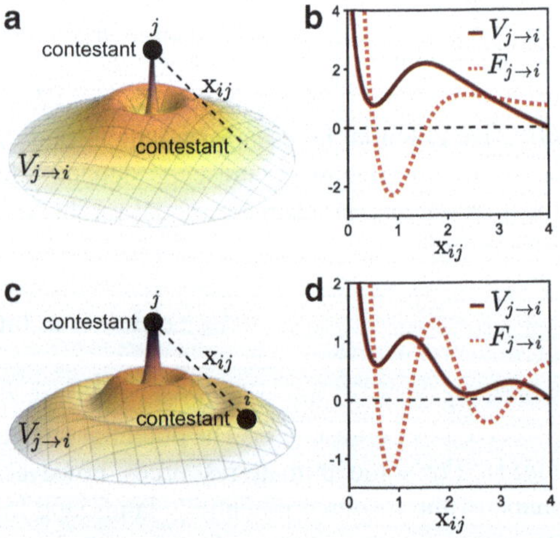

Fig. 2 Effective contestant interaction potentials. (**a**) The landscape of an effective "contestant interaction potential" $V_{j\to i}$, as given in Eq. (5). (**b**). The profiles of $V_{j\to i}$ and its corresponding force $F_{j\to i} = -dV_{j\to i}/d\mathrm{x}_{ij}$ as a function of the inter-contestant distance x_{ij} (Eq. (5) with $\alpha_{j\to i} = 7$, $\delta_{j\to i} = 3$, and $\beta = 1$). (**c**, **d**) A more elaborate $V_{j\to i}$ that encodes a two-stage contest escalation scheme (Adapted from ref. [14])

$$V_{j\to i}(\mathbf{r}_i, \mathbf{r}_j) = -\alpha_{j\to i}\exp(-\beta \mathrm{x}_{ij}{}^2) - \delta_{j\to i}\ln(\mathrm{x}_{ij}), \qquad \mathrm{x}_{ij} = \frac{|\mathbf{r}_i - \mathbf{r}_j|}{\mathrm{x}_0} \tag{5}$$

where $\alpha_{j\to i}$ and $\delta_{j\to i}$ are positive interaction parameters that set the strengths of effective attraction and repulsion (experienced by contestant i when interacting with a rival j), $\beta > 0$ determines the range of effective attraction (assumed to be the same for both contestants), and x_0 is a typical length scale. This potential has two *distinct* local extrema (as in Fig. 2b), at

$$\mathrm{x}_{ij}^{\min} = \sqrt{-\frac{1}{\beta}W_0(\Gamma_{j\to i})} \quad \text{and} \quad \mathrm{x}_{ij}^{\max} = \sqrt{-\frac{1}{\beta}W_{-1}(\Gamma_{j\to i})}, \quad \text{with}$$

$$\Gamma_{j\to i} = -\frac{\delta_{j\to i}}{2\alpha_{j\to i}}, \tag{6}$$

(where $W_{-1}(x)$ $(-1/e \le x < 0)$ and $W_0(x)$ $(-1/e \le x)$ are the two real branches of the Lambert W function [50]) provided that

$$\frac{\alpha_{j\to i}}{\delta_{j\to i}} > \frac{e}{2}, \tag{7}$$

where e is Euler's number. The values of $\alpha_{j\to i}$ and $\delta_{j\to i}$ are related to contestant i's assessment of itself and of its rival and therefore depend on the "strength," or (henceforth) effective "size," of each contestant. In the animal contest literature, this effective size is typically termed "resource holding potential" (RHP) [24–32]. Not to be confused with our model's effective interaction potentials, the RHP is a species-dependent measure for a contestant's ability to win fights and is commonly correlated (but not strictly interchangeable) with size-related features, such as claw size in crabs or antler size in deer [32, 38]. In our case, $\alpha_{j\to i}$ reflects the motivation of contestant i to escalate the interaction (i.e., how aggressive i is) and is therefore associated with the (absolute or relative) effective size of i, while $\delta_{j\to i}$ reflects how intimidating (repulsive) the rival j is perceived by i and is therefore associated with the (absolute or relative) effective size of j. Under this interpretation of $\alpha_{j\to i}$ and $\delta_{j\to i}$, the case where Eq. (7) is not satisfied corresponds to contestant i being "not motivated enough," or equivalently, "too intimidated," to escalate the interaction with the rival j. In Sect. 2.5, we discuss explicit models for the dependence of $\alpha_{j\to i}$ and $\delta_{j\to i}$ on the contestants' effective sizes and the implications of these models in the context of nonreciprocal interactions.

The interaction potential of Eq. (5) describes a rather minimal contest escalation scheme, in the sense that it features a single escalation barrier (local maximum, see Fig. 2b) toward a short-range interaction. Proposing this functional form as the "archetypal" representation of contest interactions is motivated by the premise that contestant dynamics in the vicinity of this ultimate escalation stage account for the qualitative essence of contests in many animal species. Nevertheless, system-specific details that diverge from this minimal description can be included in $V_{j\to i}$, as long as they correspond to escalation stages that occur at typical inter-contestant distances. For example, such stages exist in contests that escalate through an intermediate evaluation or display stage [31, 51–53], where this stage precedes the ultimate escalation into a short-range interaction. These two-stage contest escalation schemes can be encoded by a contestant interaction potential with a qualitative shape as in Fig. 2c, d. In Ref. [13], the interaction between male spider contestants—which mostly interact through a single modality of vibratory cues transmitted through the web—was modeled by a similar two-stage potential.

Completing our basic pairwise treatment of contests, a respective interaction potential $V_{i\to j}$ encodes the influence of contestant i on the motion of the rival j. Importantly, since these potentials manifest the (behavioral and perceptual) asymmetries between the contestants, the adversarial interactions encoded by $V_{j\to i}$ and $V_{i\to j}$ are nonreciprocal (they violate Newton's third law). This means that, for a general pair of contestants i and j, $V_{j\to i} \neq V_{i\to j}$ (as illustrated in Fig. 3a). The influence of $V_{j\to i}$ and $V_{i\to j}$ on the motion of the contestants is governed by the nonreciprocal effective forces that they generate:

$$\mathbf{F}_{j\to i}(\mathrm{x}_{ij}) = -\frac{dV_{j\to i}(\mathrm{x}_{ij})}{d\mathrm{x}_{ij}}\hat{\mathbf{x}}_{ij} \quad \text{and} \quad \mathbf{F}_{i\to j}(\mathrm{x}_{ij}) = -\frac{dV_{i\to j}(\mathrm{x}_{ij})}{d\mathrm{x}_{ij}}(-\hat{\mathbf{x}}_{ij}), \tag{8}$$

where $\hat{\mathbf{x}}_{ij} = (\mathbf{r}_i - \mathbf{r}_j)/|\mathbf{r}_i - \mathbf{r}_j|$ denotes the unit vector in the $j \to i$ direction, and again, in general, $\mathbf{F}_{j\to i} \neq -\mathbf{F}_{i\to j}$.

2.4 Definition and Duration of a Pairwise Contest

While the adversarial inter-contestant interactions that govern contest dynamics are nonreciprocal, to clearly define a "contest" in our model (in a sense that simultaneously applies to both contestants), it is necessary to define a pairwise contest onset between our model's contestant particles. For this purpose, we consider the combined contributions of the effective forces $\mathbf{F}_{j\to i}$ and $\mathbf{F}_{i\to j}$ to the relative motion between i and j along the inter-contestant axis. Taking the position of contestant j as a fixed point of reference, contestant i appears to be driven along $\hat{\mathbf{x}}_{ij}$ by a relative "contest force" $\mathbf{F}_{\text{contest}}$, given by

$$\mathbf{F}_{\text{contest}}(\mathrm{x}_{ij}) = \mathbf{F}_{j\to i}(\mathrm{x}_{ij}) - \mathbf{F}_{i\to j}(\mathrm{x}_{ij}) = -\frac{d}{d\mathrm{x}_{ij}}\left[V_{j\to i}(\mathrm{x}_{ij}) + V_{i\to j}(\mathrm{x}_{ij})\right]\hat{\mathbf{x}}_{ij} \tag{9}$$

where note that $\mathbf{F}_{i\to j}$, which is applied by i on j in the $-\hat{\mathbf{x}}_{ij}$ direction, appears to drive i relative to j in the opposing $+\hat{\mathbf{x}}_{ij}$ direction. Equation (9), which describes an effectively reciprocal force (it acts on i and j with the same magnitude and in opposing directions), motivates the definition of a corresponding relative "contest potential" V_{contest} as the sum of the individual interaction potentials,

$$V_{\text{contest}}(\mathrm{x}_{ij}) = V_{j\to i}(\mathrm{x}_{ij}) + V_{i\to j}(\mathrm{x}_{ij}) \tag{10}$$

such that $F_{\text{contest}} = -dV_{\text{contest}}/d\mathrm{x}_{ij}$. Note that according to the particular contestant interaction potential of Eq. (5), the relative contest potential is given by

$$V_{\text{contest}}(\mathrm{x}_{ij}) = -(\alpha_{j\to i} + \alpha_{i\to j})\exp(-\beta \mathrm{x}_{ij}{}^2) - (\delta_{j\to i} + \delta_{i\to j})\ln(\mathrm{x}_{ij}), \tag{11}$$

and its two local extrema are at

$$\mathrm{x}_{ij}^{\vee} = \sqrt{-\frac{1}{\beta}W_0(\Gamma_{ij})} \quad \text{and} \quad \mathrm{x}_{ij}^{\wedge} = \sqrt{-\frac{1}{\beta}W_{-1}(\Gamma_{ij})}, \quad \text{with}$$

$$\Gamma_{ij} = -\frac{\delta_{j\to i} + \delta_{i\to j}}{2\left(\alpha_{j\to i} + \alpha_{i\to j}\right)}, \tag{12}$$

where $\mathrm{x}_{ij}^{\vee}$ and $\mathrm{x}_{ij}^{\wedge}$ denote the locations of the minimum and maximum, respectively, as indicated in Fig. 3b. The contest potential of Eq. (11) has these two *distinct* extrema when

$$\frac{\alpha_{j\to i} + \alpha_{i\to j}}{\delta_{j\to i} + \delta_{i\to j}} > \frac{e}{2}, \tag{13}$$

and becomes repulsive for all x_{ij} when this inequality is not satisfied.

Regardless of the particular functional form of Eq. (11), but assuming that V_{contest} has a qualitative shape as in Fig. 3b (with corresponding local extrema at some $\mathrm{x}_{ij}^{\vee}$ and $\mathrm{x}_{ij}^{\wedge}$), we now define the onset of a contest according to the direction of relative motion between the contestants—governed by the sign of F_{contest} (as illustrated in Fig. 3b). When the contestants reach a separation x_{ij} that is shorter than $\mathrm{x}_{ij}^{\wedge}$ (driven by their mutual attraction to the resource and the stochastic components of their motion), the relative force becomes attractive ($F_{\text{contest}} < 0$) toward $\mathrm{x}_{ij}^{\vee}$, and the inter-contestant dynamics is characterized by a transient bounded state. In Fig. 3c, d, the dynamics of two contestant particles getting in and out of this transient bounded state is illustrated by simulated contestant trajectories. We identify this bounded state with the ultimate escalation into a short-range contest, in which the contestants are completely engaged with each other, and define $\mathrm{x}_{ij}^{\wedge}$ as the contest onset distance—such that two of our model's contestant particles are considered to be "engaged in a contest" when $\mathrm{x}_{ij} < \mathrm{x}_{ij}^{\wedge}$ (in Eqs. (2) and (3), this is the regime where $\theta = 0$). Note that even if the contest potential has a more elaborate shape than in Fig. 3b (as in Fig. 2d and ref. [13]), the above principles can still be used to define an analogous pairwise contest onset—notably by taking the onset of the most short-range potential well to play the role of $\mathrm{x}_{ij}^{\wedge}$.

From our definition of a contest, it immediately follows that contest duration (t_c)—a common measurable of interest in studies of animal contests [38]—is

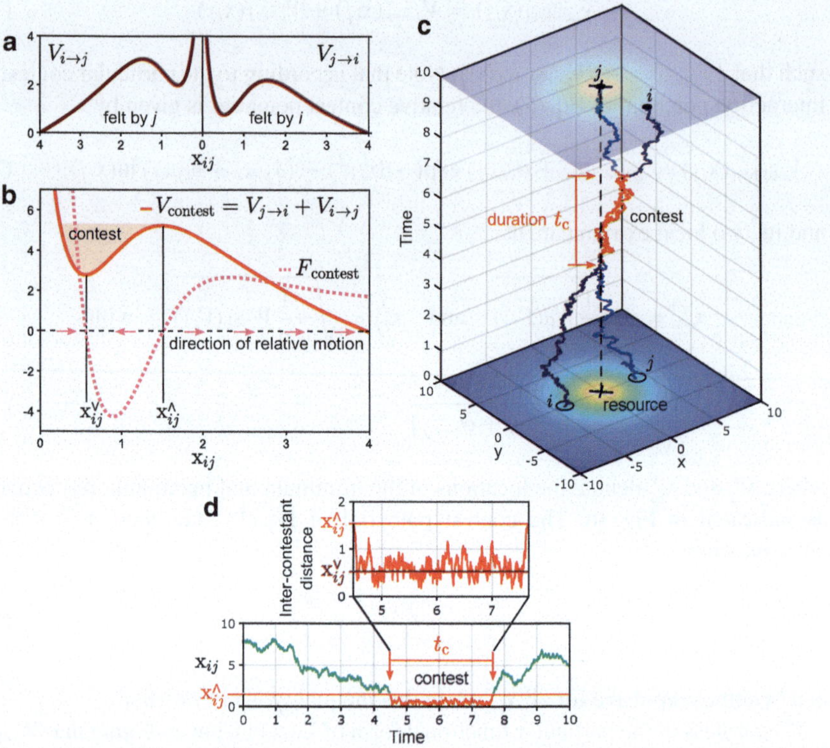

Fig. 3 Nonreciprocity and definition of a contest. (**a**) The potentials $V_{j\to i}$ and $V_{i\to j}$ encode nonreciprocal interactions that manifest asymmetries between contestants (in general, $V_{j\to i} \neq V_{i\to j}$). Here, these asymmetries are described by variations in the interaction parameters of Eq. (5), where $V_{j\to i}$ is shown with $\alpha_{j\to i} = 7$ and $\delta_{j\to i} = 3$, and $V_{i\to j}$ with $\alpha_{j\to i} = 8$ and $\delta_{j\to i} = 4$ ($\beta = 1$ for both). (**b**) The relative "contest potential" $V_{\text{contest}} = V_{j\to i} + V_{i\to j}$ governs the relative motion between the contestants along the inter-contestant axis (arrows). The distance $\mathrm{x}_{ij} = \mathrm{x}_{ij}^{\wedge}$ is defined as the contest onset. (**c**) Typical simulated trajectories of two identical contestants ($V_{j\to i} = V_{i\to j}$) in the vicinity of a resource. The dynamics of each contestant particle was simulated based on Eq. (1), with the relevant parameter values as in Table 1 (see SI of ref. [14] for further details). Their mutual attraction to the resource brings the contestants into contest range. Segments of the trajectories in which the contestants were engaged in a contest ($\mathrm{x}_{ij} < \mathrm{x}_{ij}^{\wedge}$) are shown in red. (**d**) The distance between the contestants throughout the simulation. As evident in the close-up view, the contestants spend the majority of the contest near the minimum of V_{contest} (Adapted from Ref. [14])

equivalent to the escape time of our contestant particles from their transient bounded state within the potential well of V_{contest} (as illustrated in Fig. 3d). Then, with an effective "contest bounding energy" of

$$U = V_{\text{contest}}(\mathrm{x}_{ij}^{\wedge}) - V_{\text{contest}}(\mathrm{x}_{ij}^{\vee}), \tag{14}$$

Table 1 Model parameters. A list of all parameters used in the model's equations. The column "Equation" indicates the equation in which each parameter is first used. The column "Role" provides a brief description of each parameter's role in the model. The column "Value used" indicates the numerical value (or range of values) assigned to each parameter, in all or in most cases, and the parameter's dimensions (L = Length, F = Force, and T = Time)

Equation	Parameter	Role	Value used
(1)	η	Mobility. Sets contribution of deterministic effective forces to the motion of contestant particles	1.0 $\left[\mathrm{L}\cdot\mathrm{F}^{-1}\,\mathrm{T}^{-1}\right]$
(1)	D	Diffusivity. Sets contribution of stochastic effective forces to the motion of contestant particles	0.5 $\left[\mathrm{L}^2\,\mathrm{T}^{-1}\right]$
(4)	p_i	Sets attractiveness of resource, as perceived by contestant i	4.0 $[\mathrm{F}\cdot\mathrm{L}]$
(4)	ε	Prevents divergence of Logarithmic resource potential	1.0
(5)	$\alpha_{j\to i}$	Reflects the motivation of contestant i to escalate the interaction with the rival j (effective attraction of i towards j)	7.0 $[\mathrm{F}\cdot\mathrm{L}]$
(5)	β	Determines the range of effective attraction set by $\alpha_{j\to i}$	1.0
(5)	$\delta_{j\to i}$	Reflects how intimidating the rival j is perceived by contestant i (effective repulsion of i from j)	3.0 $[\mathrm{F}\cdot\mathrm{L}]$
(5)	x_0	Typical length scale for inter-contestant interaction	1.0 $[\mathrm{L}]$
(17)	m_i	Effective 'size' (or 'strength') of contestant i	0.8–1.2
(17)	α_0	Sets the basal value of $\alpha_{j\to i}$	7.0 $[\mathrm{F}\cdot\mathrm{L}]$
(17)	δ_0	Sets the basal value of $\delta_{j\to i}$	3.0 $[\mathrm{F}\cdot\mathrm{L}]$
(17)	s	Power-law exponent representing absolute self-assessment	1 [in Eq. (18)]
(17)	r	Power-law exponent representing absolute rival assessment	1 [in Eq. (20)]
(17)	s_Q	Power-law exponent representing relative self-assessment	1 [in Eq. (19)]
(17)	r_Q	Power-law exponent representing relative rival assessment	2 [in Eq. (19)]
(21)	K_{self}	Rate at which 'self-inflicted' costs are accrued	0.01 $\left[\mathrm{T}^{-1}\right]$
(21)	K_{ij}	Rate at which 'rival-inflicted' costs are accrued	0.1 $\left[\mathrm{T}^{-1}\right]$

the mean contest duration is analogous to the Karmer's escape time [54] and thus follows an Arrhenius' law

$$\langle t_{\mathrm{c}}\rangle \simeq \Pi \exp\left(\frac{U}{T_{\mathrm{eff}}}\right), \tag{15}$$

where the pre-exponential factor Π mostly depends on the potential's curvature at its extrema, and the effective temperature T_{eff} mostly depends on the properties of the active particle—namely, on the diffusivity D and the activity [54, 55]. Equation (15) assumes that, after they first cross the contest onset $\mathrm{x}_{ij}^{\wedge}$, the contestants reach the minimum $\mathrm{x}_{ij}^{\vee}$ quickly—within an average time that is significantly shorter than t_{c}—and spend the rest of the contest near $\mathrm{x}_{ij}^{\vee}$ until they escape, as in Fig. 3d. Note that this implies that both the translational diffusion and the activity terms in Eq. (1) should be taken to be sufficiently small compared to the contest bounding energy U. Deriving analytical expressions for Π and T_{eff} in the context of our active particles setup is possible in some limits (see [55] for a rather instructive derivation for active particles obeying the same Langevin Equation as ours). But importantly, one fundamental assumption in Kramer's analytical derivation [54], that the potential well is very deep compared to the stochastic fluctuations, is generally not satisfied for our contestant particles—as they have to describe the fast engage-and-disengage dynamics of typical animal contests [13, 23]. Nevertheless, even without exact expressions for Π and T_{eff}, Eq. (15) can be used to predict the qualitative trends of $\langle t_{\mathrm{c}} \rangle$ with respect to variations in the parameters of $V_{j\to i}$ and $V_{i\to j}$ (which determine U), as shown in Ref. [14].

2.5 Assessment Strategies and Interaction Asymmetries

The type of RHP-related information used by contestants in their decision-making, commonly termed the "assessment strategy," has been extensively studied in game-theoretic contest models—especially in order to describe asymmetric contests [32, 36, 56] [recall the definition of RHP, given after Eq. (7)]. In [14], we proposed that different assessment strategies can be expressed in our framework in terms of an "assessment function" $\mathbf{A}_{j\to i}$ (of contestant i with respect to a rival j), which defines the functional relations between the interaction parameters of $V_{j\to i}$ and $V_{i\to j}$ and the contestants' effective sizes—where the effective size of contestant i is defined as a dimensionless proxy for its RHP,

$$m_i = \frac{\text{RHP of contestant } i}{\text{RHP of reference}}, \tag{16}$$

with the "RHP of reference" being, for example, the population average. For the contestant interaction potential $V_{j\to i}$ of Eq. (5), with interaction parameters $\alpha_{j\to i}$ and $\delta_{j\to i}$, we write $\mathbf{A}_{j\to i}$ as

$$\mathbf{A}_{j\to i}(m_i, m_j) \equiv \begin{pmatrix} \alpha_{j\to i}(m_i, m_j) \\ \delta_{j\to i}(m_i, m_j) \end{pmatrix} = \begin{pmatrix} \alpha_0\, {m_i}^{s} (m_i/m_j)^{s_Q} \\ \delta_0\, {m_j}^{r} (m_j/m_i)^{r_Q} \end{pmatrix}, \tag{17}$$

where α_0 and δ_0 are scaling parameters, and the different power laws represent: ${m_i}^{s}$—absolute self-assessment; ${m_j}^{r}$—absolute rival assessment; $(m_i/m_j)^{s_Q}$ and $(m_j/m_i)^{r_Q}$—relative self- and rival-assessment, respectively. These relations are best understood by recalling that $\alpha_{j\to i}$ reflects the motivation of contestant i to escalate the interaction with the rival j and therefore increases with m_i but decreases with m_j, while $\delta_{j\to i}$ reflects how intimidating the rival j is perceived by i and therefore decreases with m_i but increases with m_j. Note that the rival's assessment function, $\mathbf{A}_{i\to j} = \left(\alpha_{i\to j},\, \delta_{i\to j}\right)^T$, is obtained by swapping the functional roles of m_i and m_j in Eq. (17) [i.e., $\mathbf{A}_{i\to j}(m_i,\, m_j) = \mathbf{A}_{j\to i}(m_j,\, m_i)$]. The assessment strategy, which is now defined in terms of how $\alpha_{j\to i}$ and $\delta_{j\to i}$ vary with m_i and m_j, is represented by the (nonnegative) values of the exponents s, r, s_Q, r_Q. In particular, with $s > 0$ and r, s_Q, $r_Q = 0$, we obtain our model's representation of the so-called "pure self-assessment" [38],

$$\mathbf{A}_{j\to i}(m_i,\, m_j) = \begin{pmatrix} \alpha_0\, {m_i}^{s} \\ \delta_0 \end{pmatrix} \quad \text{for pure self-assessment,} \tag{18}$$

where contestants only assess their own (absolute) effective size and disregard their rival's effective size, and with s, $r = 0$ and s_Q, $r_Q > 0$, we obtain the representation for pure "mutual assessment" [38],

$$\mathbf{A}_{j\to i}(m_i,\, m_j) = \begin{pmatrix} \alpha_0\, (m_i/m_j)^{s_Q} \\ \delta_0\, (m_j/m_i)^{r_Q} \end{pmatrix} \quad \text{for pure mutual assessment,} \tag{19}$$

where the contestants conduct a purely relative assessment of their own effective size with respect to their rival's effective size. Note that, by construction, for $m_i = m_j = 1$ ('RHP of contestant i' = 'RHP of reference' in Eq. (16)), any mode of assessment described by Eq. (17) yields the same interaction potentials.

Figure 4 illustrates the differences between the assessment functions of Eqs. (18) and (19) in terms of how $V_{i\to j}$ and $V_{j\to i}$ vary with m_i and m_j; it shows these potentials, based on Eq. (5), under pure self-assessment (Eq. (18) with $s = 1$) and pure mutual assessment (Eq. (19) with $s_Q = 1$ and $r_Q = 2$), for the interactions of small size-matched contestants, a small contestant with a large contestant, and large size-matched contestants—where "small" and "large" refer to the contestants' effective sizes. Notably, compared to pure self-assessment—which detects absolute changes in m_i and m_j (compare Fig. 4a–c), pure mutual assessment is scale-invariant, as Eq. (19) satisfies $\mathbf{A}_{i\to j}(c\, m_i,\, c\, m_j) = \mathbf{A}_{i\to j}(m_i,\, m_j)$ for any scalar $c > 0$. This property is illustrated in Fig. 4d, f by the fact that, under pure mutual assessment, the interaction potentials of two size-matched small contestants and two size-matched large contestants—are identical. In addition, detection of asymmetries between m_i and m_j is much stronger in mutual assessment, as evident by the pronounced broken symmetry between $V_{j\to i}$ and $V_{i\to j}$ for the interaction of unmatched contestants (compare Fig. 4b, e).

It should be noted that even a highly asymmetric inter-contestant interaction as in Fig. 4e (where notably, $V_{i\to j}$ is strictly repulsive while $V_{j\to i}$ has a deep attractive well with a lowered escalation barrier), can still describe a pairwise "contest" as defined in Sect. 2.4—as long as the potential well of $V_{\text{contest}} = V_{j\to i} + V_{i\to j}$ exists (i.e., as long as V_{contest} supports a transient bounded state for some $\mathrm{x}_{ij} < \mathrm{x}_{ij}^{\wedge}$). This is

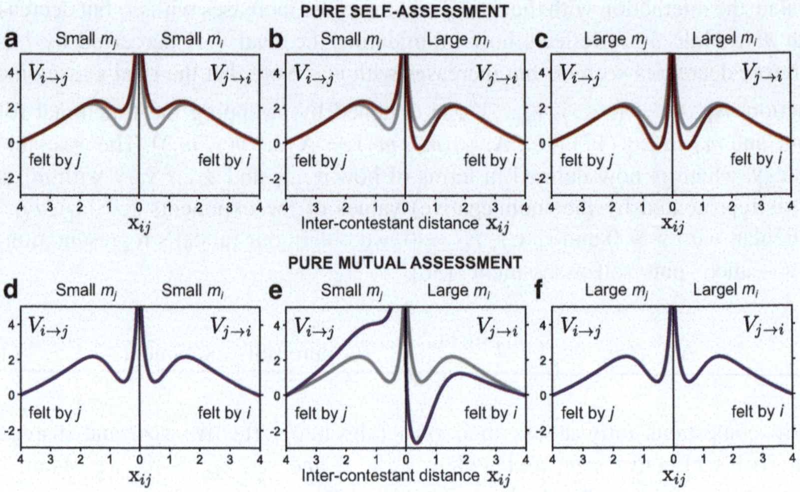

Fig. 4 Assessment strategies in terms of contestant interaction potentials. Under pure self-assessment (Eq. (18) with $s = 1$) and pure mutual assessment (Eq. (19) with $s_Q = 1$ and $r_Q = 2$), the interaction potentials $V_{i\to j}$ and $V_{j\to i}$ are shown for the interaction of (**a, d**) small size-matched contestants ($m_j = m_i = 0.8$), (**b, e**) a small contestant with a large contestant ($m_j = 0.8$, $m_i = 1.2$ for pure self-assessment, and $m_j = 0.9$, $m_i = 1.1$ for pure mutual assessment [different values merely for graphing purposes]), and (**c, f**) large size-matched contestants ($m_j = m_i = 1.2$). For reference, each graph also shows the potentials of medium size-matched contestants ($m_j = m_i = 1$) in grey. Equations (18) and (19) were used with $\alpha_0 = 7$ and $\delta_0 = 3$ (Adapted from ref. [14])

illustrated in Fig. 5 by very different contest interactions (in terms of the asymmetry between $V_{j\to i}$ and $V_{i\to j}$) that were chosen to yield the same contest potential. As discussed in Sect. 2.6, the qualitative properties of contest trajectories with symmetric interactions (as in Fig. 5a) will clearly differ from those of contests with strongly asymmetric interactions (as in Fig. 5b). But interestingly, the mean contest duration predicted by Eqs. (14) and (15) is the same for both of these interactions (in [14], this property is shown in simulations), as they share the same contest potential of Fig. 5c. From a behavioral perspective, this can be understood through the heuristic that both interactions have the same "total" amounts of motivation ($\alpha_{j\to i} + \alpha_{i\to j}$) and intimidation ($\delta_{j\to i} + \delta_{i\to j}$) and therefore persist for the same amount of time, although in Fig. 5a, these are split equally between the contestants, while in Fig. 5b, contestant i has most of the motivation and contestant j has most of the intimidation.

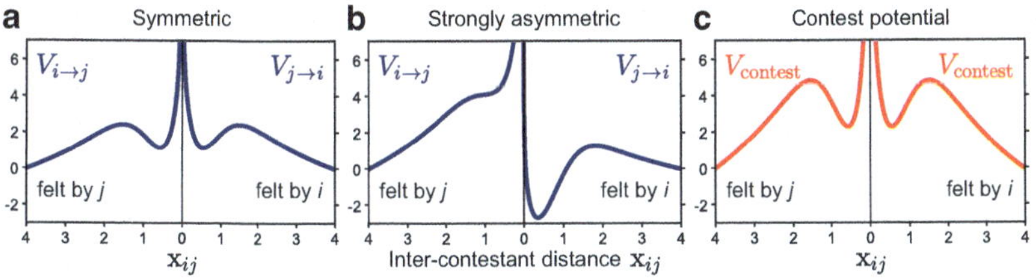

Fig. 5 Very different contest interactions with the same contest potential. (**a**) The contestant interaction potentials of a symmetric interaction between two identical contestants ($V_{j\to i} = V_{i\to j}$, with $\alpha_{j\to i} = \alpha_{i\to j} = 7.14$, $\delta_{j\to i} = \delta_{i\to j} = 3.25$, and $\beta = 1$). (**b**) The contestant interaction potentials of a strongly asymmetric interaction ($\alpha_{j\to i} = 8.55$, $\alpha_{i\to j} = 5.73$, $\delta_{j\to i} = 2.02$, $\delta_{i\to j} = 4.48$, and $\beta = 1$), in which $V_{i\to j}$ is strictly repulsive. (**c**) These very different interactions have the same contest potential $V_{\text{contest}} = V_{j\to i} + V_{i\to j}$, since $\alpha_{j\to i} + \alpha_{i\to j}$ and $\delta_{j\to i} + \delta_{i\to j}$ (and β) are the same for both

2.6 *Chase Dynamics Emerges from Interaction Asymmetry*

Chasing behavior is evidently a ubiquitous feature of agonistic interactions in animals and is intuitively associated with some form of asymmetry between the chasing individual and the one being chased. The essence of this behavior is captured by our model's contestant particles, which exhibit chase dynamics as an emergent property of broken symmetry in their nonreciprocal interaction forces. As illustrated in Fig. 6a for strongly asymmetric contestant interaction potentials, within the "chase" range the larger contestant is attracted by a deep potential well which drives it toward its smaller rival, while the smaller contestant is repelled in the same direction—away from its larger rival. Together, these effects give rise to directed chase behavior during the contest. This "microscopic" chase—at the single-particle level—is analogous, both in qualitative nature and in the mechanism that creates it, to the macroscopic chase and traveling wave phenomena that occur between (chemical) species in active mixtures in the presence of nonreciprocal interactions [8, 57].

Figure 6b, c compare typical trajectories of strongly asymmetric and symmetric contests under "pure mutual assessment" in simulations (namely, the dependence of $\alpha_{j\to i}$ and $\delta_{j\to i}$ on the contestants' effective sizes is given by Eq. (19), with $s_Q = 1$ and $r_Q = 2$). While symmetric contests are characterized by scrambled and relatively localized trajectories, the trajectories of strongly asymmetric contests feature substantial directional alignment and persistence, and consequently much greater displacement, due to chase dynamics. The extent to which chase dynamics governs the contestants' trajectories can be quantified by considering the direction correlation between the velocity of the contestants' midpoint—measured by $\hat{\mathbf{v}}_{\text{m}}$, and the inter-contestant direction $\hat{\mathbf{x}}_{ij}$, as defined in Fig. 6b (*Inset*). The temporal mean of this "chase correlator," $\langle \hat{\mathbf{v}}_{\text{m}} \cdot \hat{\mathbf{x}}_{ij} \rangle$, is zero for symmetric interactions, and $\left| \langle \hat{\mathbf{v}}_{\text{m}} \cdot \hat{\mathbf{x}}_{ij} \rangle \right| \sim 1$ if the interaction is dominated by the chase phase.

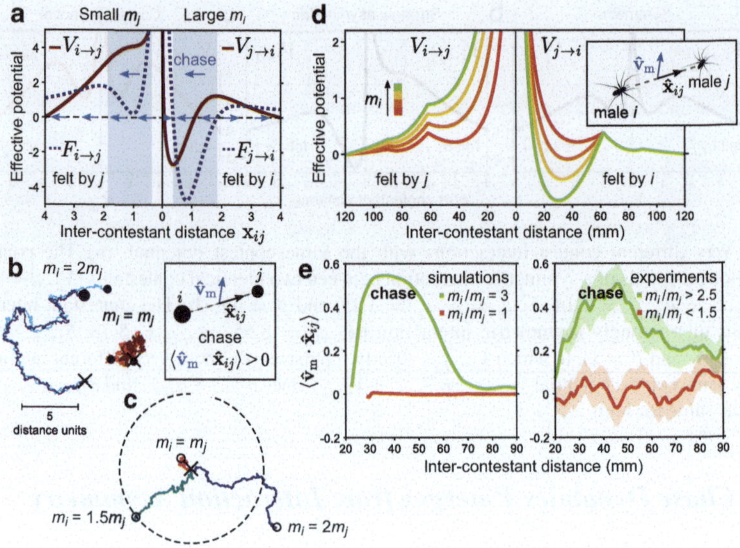

Fig. 6 Chase dynamics emerges from interaction asymmetry. (**a**) Strongly asymmetric interaction potentials illustrate how interaction asymmetry leads to chase dynamics. Shaded rectangles mark the chase range, within which the larger contestant is attracted by a deep minimum, while the smaller contestant is repelled in the same direction. Arrows indicate the direction of motion for each contestant, as dictated by the signs of the forces $F_{j\to i}$ and $F_{i\to j}$. (**b**) Typical trajectories of strongly asymmetric and symmetric contests in simulations. Chase dynamics is quantified by the direction correlation between the velocity of the contestants' midpoint—measured by $\hat{\mathbf{v}}_{\mathrm{m}}$, and the inter-contestant direction $\hat{\mathbf{x}}_{ij}$ (see *Inset*). (**c**) Trajectories of the contestants' midpoint during contests under pure mutual assessment (the dependence of $\alpha_{j\to i}$ and $\delta_{j\to i}$ on m_i and m_j is given by Eq. (19), with $s_Q = 1$ and $r_Q = 2$) for different effective size ratios. These trajectories, of equal durations, demonstrate the effect of chase dynamics on the contestants' displacement during the contest. In (**b**) and (**c**), "X" marks the contestants' midpoint at contest onset and circles mark the final positions. (**d**) Increasing asymmetry in the contestant interaction potentials of Ref. [13] (used to model the interaction between male spider contestants) as a function of increasing $m_i \geq m_j$ (at constant m_j). (**e**) Profiles of $\langle \hat{\mathbf{v}}_{\mathrm{m}} \cdot \hat{\mathbf{x}}_{ij} \rangle$ in contests between two male spiders (from Ref. [13]), in simulations and experiments, for symmetric and asymmetric contests. In experiments, m_i and m_j denote the actual body weights of the spiders. Error bars show SEM (Adapted from Refs. [14] and [13]. See these works for further details)

Accounting for the dynamical aspects of asymmetric contests was particularly important in the natural system of *T. clavipes* male spider contestants studied in Ref. [13]—as mature specimen of these male spiders can vary greatly in size [58, 59]. The observed coexistence of sexually mature *T. clavipes* males that differ in their weights by as much as an order of magnitude is puzzling, as one would expect stabilizing evolution to prevent such extraordinary size variability [60]. Figure 6d shows the experimentally calibrated, two-stage interaction potentials used to model the interactions between the spider contestants of Ref. [13]. The observation-based model for assessment in this system (if it were to be mapped to the one-stage interaction potential of Eq. (5)) corresponds to the following assessment function

$$\mathbf{A}_{j\to i}(m_i, m_j) = \begin{pmatrix} \alpha_0 \, m_i \\ \delta_0 \, m_j \end{pmatrix}. \tag{20}$$

The way in which Eq. (20) detects effective size asymmetries is qualitatively similar to the pure mutual assessment of Eq. (19), as evident by how $V_{j\to i}$ and $V_{i\to j}$ vary with an increasing m_i in Fig. 6d. Accordingly, it predicts substantial chase dynamics for large asymmetries in the contestants' effective sizes, as shown in Fig. 6e *Left*. This prediction is in good agreement with the dynamics of real spider contests, as shown in Fig. 6e *Right*.

The ability of the model in Ref. [13] to account for these details motivated its application to a many-contestant scenario taking place in this system. It was found that local male density, defined as the number of males occupying a single female's web, is highly variable in wild populations of *T. clavipes* [60]. Some studies suggested that this could resolve the puzzle of high size variability, if large males have a significant reproductive advantage when male density is high, but this advantage is diminished when male density is low—allowing for a wide distribution of sizes to be stable [60]. Using simulations, we provided in Ref. [13] a mechanistic explanation for the hypothesized competitive advantage of large males at high male densities. We have shown that the repulsive effective interaction potentials inflicted by large males on smaller males enable them to resolve the interactions with their smaller rivals swiftly and at a distance—and thereby reach the female-resource faster and retain it better, while small males form long-lasting contests with other small males (which are more abundant, [60]) and therefore suffer disproportionately from a high density of rivals, in which the frequency of contests is high. Conversely, when male density is low, we have found in simulations that large males have a less significant advantage at reaching and retaining the female-resource [13]. On such female webs, which coexist in nature with high-density webs [60], the disadvantages of being large (e.g., higher detectability by the potentially cannibalistic female or by other predators) may outweigh its benefits.

2.7 *Time Dependence of Contest Interactions*

Until now we have treated inter-contestant interactions as if they do not (explicitly) depend on time: The effective potentials that describe these interactions were time-independent, and so the only "time dependence" in our framework was due to the motion of the contestant particles (driven by their governing Langevin equations), which continuously alters the potential landscapes that they experience. However, contestant interaction potentials reflect the contestants' current effective sizes, and these could vary during (and due to) the interaction—notably as fighting costs accumulate (and decrease the effective sizes compared to their initial states) and as the contestants update their perception of their own and of their rival's status (learning). The dependence of m_i and m_j on interaction time due to such effects introduce into $V_{j\to i}$ and $V_{i\to j}$ an explicit time dependence.

As an illustrative example for the notable consequences of such effects on the properties of the inter-contestant interaction, consider the simple model from Ref. [14] for the dependence of the effective sizes on the accumulated contest time t_σ, given by

$$m_i(t_\sigma) = \frac{\mu_i}{1 + \left(K_{\text{self}} + K_{ij}\,\dfrac{\mu_j}{\mu_i}\right) t_\sigma} \quad \text{and} \quad m_j(t_\sigma) = \frac{\mu_j}{1 + \left(K_{\text{self}} + K_{ij}\,\dfrac{\mu_i}{\mu_j}\right) t_\sigma}, \tag{21}$$

where μ_i and μ_j are the effective sizes of contestants i and j when the contest starts, K_{self} is the rate at which "self-inflicted" costs are accrued (e.g., average rate of energy expenditure), $K_{ij}\,\mu_j/\mu_i$ is the rate at which "rival-inflicted" costs are accrued (e.g., average rate of injuries), and t_σ is accumulated only during the contest (i.e., when $\mathrm{x}_{ij} < \mathrm{x}_{ij}^{\wedge}$). Equation (21) naively assumes that costs are accrued continuously and deterministically and that the rates of accruing them are independent of the instantaneous values of m_i and m_j. Nevertheless, it includes a simple form of feedback, since the rate of incurring costs from the rival is proportional to the initial effective size ratio—such that the larger contestant inflicts costs faster but incurs them slower. Note that according to Eq. (21), while both m_i and m_j decrease monotonically with t_σ, their ratio m_i/m_j increases with t_σ when $\mu_i > \mu_j$ and $K_{ij} > 0$. It approaches a constant as $t_\sigma \to \infty$, where

$$\lim_{t_\sigma \to \infty} \frac{m_i}{m_j} = \frac{\mu_i}{\mu_j} \cdot \frac{K_{\text{self}}\,\mu_j + K_{ij}\,{\mu_i}^2}{K_{\text{self}}\,\mu_i + K_{ij}\,{\mu_j}^2}. \tag{22}$$

The dynamics of m_i, m_j, and m_i/m_j according to Eq. (21) is shown in Fig. 7a for a case where $\mu_i > \mu_j$.

Now, through the assessment function of Eq. (17), $V_{j\to i}$ and $V_{i\to j}$ also depend on t_σ. In Fig. 7b, this dependence is visualized under pure mutual assessment (Eq. (19) with $s_Q = 1$ and $r_Q = 2$). The increasing asymmetry (in terms of the growing ratio m_i/m_j) is manifested by an increasingly asymmetric inter-contestant interaction—as evident by the opposing trends of the potentials with respect to t_σ in Fig. 7b. An interesting consequence of this interaction-induced amplification of asymmetry is that it offers a generic mechanism for the emergence of strong chase dynamics during the contest, even if the initial asymmetry is small. Moreover, note that under Eq. (19) with $s_Q = 1$ and $r_Q = 2$, the bounding well that defines the contest regime becomes shallower with m_i/m_j, and therefore with t_σ (as shown in Fig. 7c)—which in turn reduces the mean contest duration (recall Eq. (15)). This trend can eventually lead to a cost-driven contest termination when the bounding well completely disappears and V_{contest} becomes strictly repulsive (recall that this happens when Eq. (13) is not satisfied), as explored in detail in ref. [14]. These effects, we propose, could model a ubiquitous mechanism for contest resolution in real animal contests.

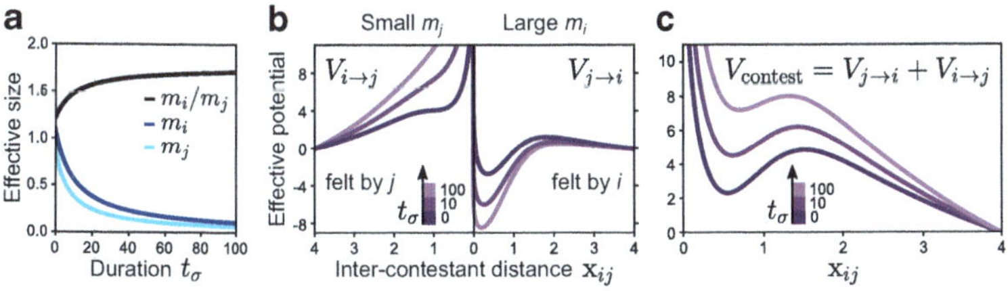

Fig. 7 Time dependence of contest interactions. (**a**) The dynamics of the effective sizes m_i and m_j and of their ratio m_i/m_j, according to Eq. (21) with $K_{\text{self}} = 0.01$, $K_{ij} = 0.1$, and with initial effective sizes $\mu_i = 1.1$ and $\mu_j = 0.9$. (**b**) The dependence of m_i and m_j on the accumulated contest duration t_σ due to costs means that $V_{i\to j}$ and $V_{j\to i}$ themselves depend on t_σ. Here, these potentials vary with the values of m_i and m_j in A as t_σ increases, according to pure mutual assessment (Eq. (19) with $s_Q = 1$ and $r_Q = 2$). (**c**) Under this assessment strategy, the bounding well that defines the contest regime becomes shallower with m_i/m_j, and therefore with t_σ

3 Effective Gravity Model of Swarming Midges

A special type of collective animal behavior is the formation of cohesive swarms [16, 61]. These swarms are characterized by having the individual animals moving in an uncoordinated manner, unlike flocks, which are characterized by strong directional alignment, but nevertheless maintaining a cohesive structure that does not disperse. There have been various theoretical models proposed to describe this phenomenon, mainly based on velocity alignment that is mitigated by large noise [62, 63].

Here we describe a theoretical model that is based on adaptive long-range interactions [17]. For swarming midges, the effective interactions are attractive and mediated by acoustics due to the sound they emit while flying. Male midges form large swarms at dusk, thereby attracting females for mating. Within our model, midges form such cohesive swarms by accelerating toward each other in proportion to the intensity of the sound received. For pure acoustics, the leading term (monopole) gives a functional form of the acceleration which is similar to the force of gravitational attraction, decaying as $1/r^2$. It was shown in [64, 65] that in general, power law decaying attraction and nonlinear diffusion are sufficient for the formation of local stable swarms from initially dispersed population profiles.

An additional, and crucial, component of this model is adaptivity. Any sensory mechanism in nature is subject to a modification due to adaptivity, which means that the sensory mechanism adapts itself to the background stimuli—its sensitivity is higher when the background is lower and vice versa. Exact adaptation means that the steady-state output is independent of the steady-state level of input. This is part of a fold-change detection mechanism [66], which is ubiquitous in nature, and involves a response whose entire shape, including amplitude and duration, depends only on fold change and not on the absolute levels of the input. The adaptivity is described by a scalar symmetry of the sensory mechanism whereby multiplication of the input fields or stimuli by a scalar does not change the reaction of the system.

Table 2 The parameters of the adaptive gravity model. A list of all parameters used in the model's equations. The column "Equation" indicates the equation in which each parameter is first used. The column "Role" provides a brief description of each parameter's role in the model. The column "Dimensions" gives the dimensions in terms of (l—length, m—mass, t—time)

Equation	Parameter	Role	Dimensions
(23)	c_g	Strength of gravitational potential	ml^3t^{-2}
(26)	R_{ad}	Length scale of adaptivity	l
(35)	R_s	The radius of the swarm—the mean distance of a particle from the center of mass	l
(59)	I_{bgd}	The leading order of the expansion of the "background noise" term	mlt^{-2}
(61) (63)	$\gamma, \gamma_1, \gamma_2$	The background sound parameter	–

Adding the two components together we reach the formulation of "adaptive gravity" which is described below. The adaptive-gravity model was shown to provide a good description of the observed mass and velocity profiles of laboratory midge swarms, which differ from those produced by regular gravity observed in star clusters [19]. Here we summarize some of the main theoretical results that describe the special features of the adaptive-gravity and appeared already in [17–20], focusing on the special properties that emerge due to these nonreciprocal long-range interactions. The parameters of the model are summarized in Table 2.

3.1 *Adaptive Gravity*

Let us consider N identical particles, whose positions are denoted by $\vec{x}_i$ $(i = 1, .., N)$, interacting by adaptive gravity. The contribution of a particle at $\vec{x}_j$ to the (regular) gravitational potential at $\vec{x}_i$ is given by

$$V^{ij}(\vec{x}_i) = \begin{cases} \frac{c_g}{|\vec{x}_i - \vec{x}_j|}, & i \neq j \\ 0, & i = j. \end{cases} \tag{23}$$

where c_g is a positive constant with dimensions of $mass \cdot length^3/time^2$.

The contribution of the particle at $\vec{x}_j$ to the gravitational force at the point $\vec{x}_i$ is given by the gradient of the potential

$$\vec{F}^g_{ij}(\vec{x}_i) = -\nabla V^{ij}(\vec{x}_i), \tag{24}$$

so that total gravitational (regular) linear force that acts on the particle at $\vec{x}_i$ is

$$\vec{F}^g(\vec{x}_i) = \sum_{j=1}^{n} \vec{F}^g_{ij}(\vec{x}_i) = -\sum_{j=1}^{n} \nabla V^{ij}(\vec{x}_i) \tag{25}$$

In adaptive gravity, the regular gravitational force is "renormalized" by a factor that depends on the total background force (scalar sum in the denominator), resulting in a highly nonlinear form of the interactions between the particles

$$\vec{F}^{ad}(\vec{x}_i) = -\frac{\sum_{j=1}^{n} \nabla V^{ij}(\vec{x}_i)}{1 + (c_g^{-1} R_{ad}^2) \sum_{j=1}^{n} |\nabla V^{ij}|} \tag{26}$$

where R_{ad} is the length scale over which adaptivity occurs. This way to introduce adaptivity first appeared in the context of alignment interactions [3, 67]. In the context of the midge swarm the effective interactions are mediated by acoustic (and also visual) signals, such that the background force represents the local background noise produced by the swarm.

Since $|\nabla V^{ij}| \sim |\vec{x}_i - \vec{x}_j|^{-2}$, when the distances between pairs are large $|\vec{x}_i - \vec{x}_j| \gg \sqrt{N} R_{ad}$ the adaptivity does not play a role, and the interaction approaches regular gravity in this limit

$$\vec{F}^{ad}(\vec{x}_i) \sim \vec{F}^{g}(\vec{x}_i). \tag{27}$$

On the other hand, when $|\vec{x}_i - \vec{x}_j| < \sqrt{N} R_{ad}$ adaptivity is strong and then the following expression is a good approximation, which we term "perfect adaptvity"

$$\vec{F}^{ad}(\vec{x}_i) \sim \frac{c_g}{R_{ad}^2 \, N_{tot}(\vec{x}_i)} \sum_{j=1}^{n} \nabla V^{ij}(\vec{x}_i), \tag{28}$$

where $N_{tot}(\vec{x}_i)$ is a factor that represents the total background "noise" (or the total stimuli) at the point $\vec{x}_i$ as a result of all the interactions that involve the particle at this point

$$N_{tot}(\vec{x}_i) = \sum_{j=1}^{n} |\nabla V^{ij}(\vec{x}_i)|. \tag{29}$$

The appearance of this factor in the denominator of the expressions in Eqs. (26) and (28) of the effective force, due to adaptivity, is responsible for the nonlinear and nonreciprocal nature of the interactions in this model. It is also a good example of the scalar symmetry in adaptation mechanisms [66]. We refer to $V^{ij}(i, j = 1, .., N)$ as our input fields and multiplication of the input fields by a scalar does not change the reaction of the system $\vec{F}^{ad}$.

For two particles at positions $\vec{x}_1$ and $\vec{x}_2$, this factor is reciprocal

$$N_{tot}(\vec{x}_1) = N_{tot}(\vec{x}_2) = \frac{c_g}{|\vec{x}_1 - \vec{x}_2|}, \tag{30}$$

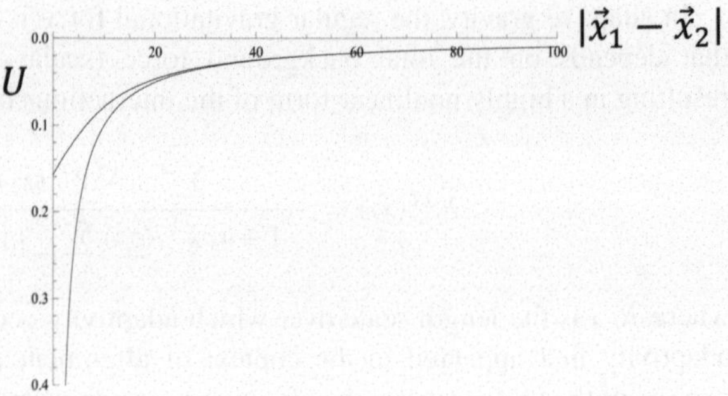

Fig. 8 The potential energy for two particles in adaptive gravity $U_{\text{pair}}(|\vec{x}_1 - \vec{x}_2|)$ calculated according to Eq. (31) using $R_{ad}/c_g = 10$ (blue line). The red line is regular gravity potential for comparison with $c_g = 1$ (Adapted from [17])

and therefore the interaction is reciprocal. Note that the effective force between two particles under perfect adaptivity becomes a constant, independent of the separation between them. Also, energy is conserved and the force from Eq. (26) can be integrated to give the effective potential between the pair of particles

$$U_{\text{pair}}(|\vec{x}_1 - \vec{x}_2|) = \frac{c_g}{R_{\text{ad}}}\left[\arctan\left(|\vec{x}_1 - \vec{x}_2|/R_{\text{ad}}\right) - \pi/2\right]. \tag{31}$$

Unlike regular gravity, where the potential energy diverges when the two particles approach each other, the adaptivity prevents this and the potential in Eq. (31) approaches a finite value with a "cusp" shape, so that it is linear in $|\vec{x}_1 - \vec{x}_2|$. For $|\vec{x}_1 - \vec{x}_2| \geq R_{\text{ad}}$, the effective potential approaches the long-range regular gravitational behavior of $U_{\text{pair}} = c_g/|\vec{x}_1 - \vec{x}_2|$ (Fig. 8).

In the case of $N \geq 3$, the interactions are in general nonreciprocal as we can see for three particles at $\vec{x}_i$, $(i = 1, .., 3)$, which have different normalization factors

$$N_{tot}(\vec{x}_1) = \frac{c_g}{|\vec{x}_1 - \vec{x}_2|} + \frac{c_g}{|\vec{x}_1 - \vec{x}_3|}, \tag{32}$$

$$N_{tot}(\vec{x}_2) = \frac{c_g}{|\vec{x}_2 - \vec{x}_1|} + \frac{c_g}{|\vec{x}_2 - \vec{x}_3|} \tag{33}$$

$$N_{tot}(\vec{x}_3) = \frac{c_g}{|\vec{x}_1 - \vec{x}_3|} + \frac{c_g}{|\vec{x}_2 - \vec{x}_3|} \tag{34}$$

We therefore can not write an effective potential for this system. When Newton's third law is not satisfied, there is no conservation of momentum. In addition, nonreciprocal interactions imply that the system is out of equilibrium and time reversibility is broken [68], and the energy cannot be conserved. Next we explore some of the notable consequences of the adaptive interactions on the dynamics and properties of the swarms.

3.2 Enhanced Stability of Adaptive Spherical-Uniform Swarms

We now show that due to adaptivity, a swarm of particles interacting through adaptive gravity does not suffer from the famous Jeans instability that is responsible for collapse of groups of particles under normal gravitational or other attractive power-law long-range forces [69, 70]. Jeans instability means that any self-gravitating system that is dense enough is unstable to random fluctuations, which will result in the collapse of the system. In the case of biological matter, we expect that at short distances, the particles will repel each other, and the meaning of this instability could be a tearing off of the structure of the swarm. Nevertheless, adaptivity prevents it. Adaptivity thus provides swarms with a natural mechanism for self-stabilization, which entails preserving their structure and preventing such instabilities. The argument that we present below is heuristic. A more rigorous derivation in terms of a critical wave number for a wavelike periodic perturbation appeared in [18] and reached the same expressions up to numerical constants.

For the purpose of the argument, let us consider a spherical uniform swarm whose radius is $R_s = \langle r \rangle$ (defined as the mean distance of a particle from the center of the mass of the swarm, which is $R_s = (3/4)R$ for a uniform density spherical swarm of radius R). We want to compare the response to a local fluctuation in the density for regular vs. adaptive gravity. In the case of a random fluctuation in the density, the random movement of the particles could stabilize the swarm if their random velocities are high enough. The typical time for stabilization is the typical time it takes to move across the swarm: $t_{esc} = R_s/\sigma_v$, where $\sigma_v = \sqrt{\langle v^2 \rangle}$ is the root-mean-square velocity. This velocity can arise from thermal motion, or from the chaotic motion due to the attractive forces themselves (that depend on initial conditions), as well as due to the noisy active propulsion forces of the living (active) particles. The question is whether σ_v is high enough, so that the typical time to stabilize (or equivalently the time for the density fluctuation to escape) t_{esc} is shorter than the typical time to collapse t_{col}, which is given by the typical time a particle falls to the center of mass due to the overall gravitational attraction: $t_{col} = 2\pi/\sqrt{K}$, where K is the effective spring constant for the linear restoring force of the form $\vec{F} = -K\vec{r}$ that acts toward the center of the swarm. This linear force appears for a uniform density spherical swarm and is calculated in the appendix for both regular gravity K^g (Eq. (68) in the Appendix) and adaptive gravity K^{ad} (Eq. (72) in the Appendix).

The criterion for instability is therefore $t_{esc} > t_{col}$. For regular gravity, we get from it the following inequality

$$\rho > \frac{3\pi \, \langle v^2 \rangle}{c_g \, R_s^2}, \tag{35}$$

which gives the famous critical Jeans density for collapse in regular gravity [69]

$$\rho_{Jeans} = \frac{3\pi \, \langle v^2 \rangle}{c_g \, R_s^2}, \tag{36}$$

which means that for any density higher than this value, the swarm will be unstable and collapse as a result of any arbitrarily small density fluctuation.

In the case of adaptive gravity, we get the following inequality:

$$\langle v^2 \rangle < \frac{c_g R_s}{16\pi^2 R_{ad}^2}, \tag{37}$$

which shows that the collapse does not depend on the density, and there is a threshold on the value of the thermal velocity fluctuations below which the swarm will collapse. This means that any swarm that is "hot" enough is perfectly stable against density fluctuations. The adaptivity makes the mutual interactions weaker in the dense regions, and as a result, the dependence of stability on the density disappears, and this way it contributes to the stability of the swarm.

The calculation given above for the stability condition of a uniform density swarm applies also for a real swarm (as well as a galaxy or a star cluster), even though its density is not uniform [19].

It was found in [18] that for attractive forces with higher-order power laws and adaptivity, the stabilization is even stronger. Instead of a minimal density for collapse, there is a maximal density for collapse, and above it, the swarm becomes stable, which therefore protects the swarms from collapse.

3.3 Generalized Virial Relation for Active Systems

We now develop virial equations that relate the kinetic energy to a generalized analogue of potential energy when the system is stationary, based only on the conservation of mass. The equations are of second order, since they relate second-order moments. In their derivation, we follow closely Chandrasekhar [71] that derived them for the gravitational potential, and we write them in a more general form for any force assuming only mass conservation. We derive here the scalar virial equation, while the more general tensorial version appeared in [17].

Let us start from the following version of Newton's second law governing the velocity $\vec{v}$ of an element of the swarm whose density is ρ:

$$\rho \frac{d\vec{v}}{dt} = -\nabla p + \rho \vec{f}, \tag{38}$$

where $p(\vec{r}, t)$ is an isotropic pressure whose gradient creates a force and $\vec{f}$ stands for the rest of the forces per unit mass. We multiply the equation by the position vector $\vec{r}$ (scalar product) and integrate over the entire volume V. Then from the left-hand side of Eq. (38), we obtain from mass conservation

$$\int \rho\vec{r} \cdot \frac{d\vec{v}}{dt} dV = \frac{d^2 I}{dt^2} - 2K, \tag{39}$$

where I is the scalar moment of inertia defined as

$$I \equiv \frac{1}{2} \int \rho \vec{r}^2 \, dV, \tag{40}$$

and

$$K \equiv \frac{1}{2} \int \rho \vec{v}^2 \, dV. \tag{41}$$

The first term on the right-hand side of Eq. (38) gives us:

$$\int \vec{r} \cdot \nabla p \, dV = \int \nabla \cdot (\vec{r} p) \, dV - 3 \int p \, dV. \tag{42}$$

Then using Gauss's theorem for a volume V and its boundary ∂V, we get a surface term for the pressure:

$$\int \vec{r} \cdot \nabla p \, dV = S - 3\Pi, \tag{43}$$

where the surface term is

$$S \equiv \oint_{\partial V} p \vec{r} \cdot d\vec{s}, \tag{44}$$

and we define the total pressure over the volume of the swarm to be:

$$\Pi \equiv \int p \, dV. \tag{45}$$

The second term in the right-hand side of Eq. (38) gives the analogue of potential energy of the swarm:

$$W \equiv \int \rho \vec{r} \cdot \vec{f} \, dV. \tag{46}$$

Then combining Eqs. (38), (39) and (43), we obtain the following equation:

$$\frac{d^2 I}{dt^2} = 2K - S + 3\Pi + W. \tag{47}$$

Assuming that the system is stationary and defining the total analogue of kinetic energy T as a sum of the kinematic motion and the internal thermal motions in the form of isotropic pressure:

$$T \equiv K + \frac{3}{2} \Pi, \tag{48}$$

we arrive to a generalized virial theorem that relates (with a surface term) the kinetic energy with the "potential energy" W:

$$2T - S + W = 0. \tag{49}$$

This relation is very general for any force $\vec{f}$ that acts on the particles without a requirement for any conservation laws except for the mass conservation. We can obtain average values for the analogues of kinetic and potential energies T and W by calculating the average of $\vec{v}^2$ and $\vec{r} \cdot \vec{f}$ (where $\vec{f}$ is taken as the acceleration toward the center of the swarm with the calculation being averaged per midge over the swarm volume and a long period of time).

Comparing this virial expression to data of real midge swarms in lab experiments, the following mean values per midge were found [17]:

$$\begin{aligned} \langle T \rangle &= (3.42 \pm 0.08) \cdot 10^2 \quad cm^2/s^2, \\ \langle -W/2 \rangle &= (2.80 \pm 0.08) \cdot 10^2 \quad cm^2/s^2. \end{aligned} \tag{50}$$

which were found to be approximately constant as a function of the swarm size R_s (which increases with the number of midges in the swarm). Then according to Eq. (49), the difference between the two gives the mean surface pressure in the swarm:

$$S = (1.24 \pm 0.14) \cdot 10^2 \quad cm^2/s^2. \tag{51}$$

This pressure acts in the opposite direction to the kinetic energy in Eq. (49), indicating that the swarm is experiencing a stabilizing inward effective pressure on its surface. The origin of this pressure could arise from interactions of the swarm midges with midges outside the swarm or the environment. Such external stabilizing pressures are commonly found in astrophysical stellar systems, such as globular clusters [72].

For a spherical swarm of uniform density, we can calculate the "potential energy" W, find its scaling with the swarm size R_s, and compare with the experimental data. We expect that near the center of the swarm, the adaptivity will be the strongest and the "perfect adaptivity" regime will be the best approximation for the effective force between the particles. The "potential energy" (Eq. (46)) in a uniform spherical symmetric swarm is given by:

$$W = \frac{27}{64\, R_s^3} \int_0^{\frac{4}{3} R_s} r^3\, f(r)\, dr. \tag{52}$$

If it were a regular gravity then $f(r) = F^g(r)$ according to Eq. (67) and

$$W^g = -\frac{9\, c_g}{20\, R_s} \propto -\frac{1}{R_s}. \tag{53}$$

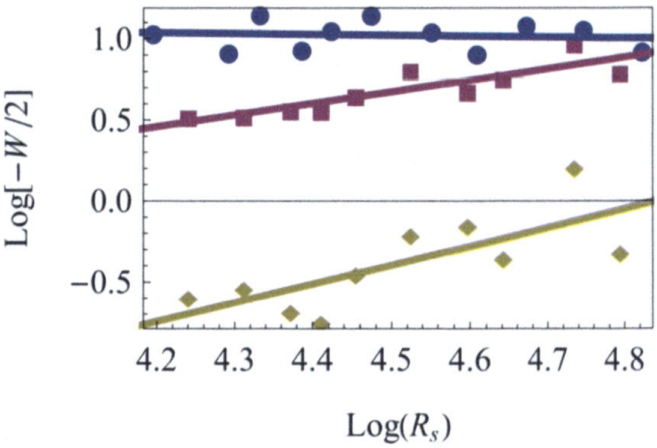

Fig. 9 The "potential energy" W in adaptive gravity. Log-Log plot for the "potential energy" W in the center of the swarm with a linear fit, constrained for $r < R_s/2, R_s, \infty$ (yellow, purple, and blue, respectively). The power-law slopes are (yellow and purple, respectively): $1.16 \pm 0.12, 0.71 \pm 0.05$ (Adapted from [17])

However, in adaptive gravity in the "perfect adaptivity" regime, we have to divide $F^g(r)$ by $N_{tot}(r)$ according to the continuum limit of Eq. (28)

$$f(r) = \frac{c_g\, F^g(r)}{R_{ad}^2\, N_{tot}(r)}. \tag{54}$$

Substituting into Eq. (52) and taking the expression for $N_{tot}(r)$ from Eq. (69) and $F^g(r)$ from Eq. (67), we get:

$$W^{ad} = -\frac{16\, c_g\, R_s}{3\, R_{ad}^2} \int_0^1 \frac{x^5\, dx}{2x + (x^2 - 1)\ln\left(\frac{1-x}{1+x}\right)} \sim -0.4 \frac{c_g}{R_{ad}^2} R_s \propto -R_s, \tag{55}$$

and this linear scaling with R_s is indeed observed in the swarm data when we constrain the data closer to the center of the swarm (where the density is rather uniform and the effect of adaptivity is strongest), as we see from the slopes in Fig. 9.

Another way to look at Eq. (47) (when the system is stationary) is as a definition of the pressure:

$$\Pi = -\frac{1}{3}\,(2\,K + W) + \frac{S}{3}, \tag{56}$$

that was used to characterize the phases of midge swarms [73] and to write the equation of state for the swarms [74].

3.4 *Spontaneous Formation of Bounded Pairs in Adaptive-Gravity Swarms*

Besides the quantitative modifications as a result of adaptivity, which gives rise to modified global swarm behavior and scaling as shown above, there is a striking new

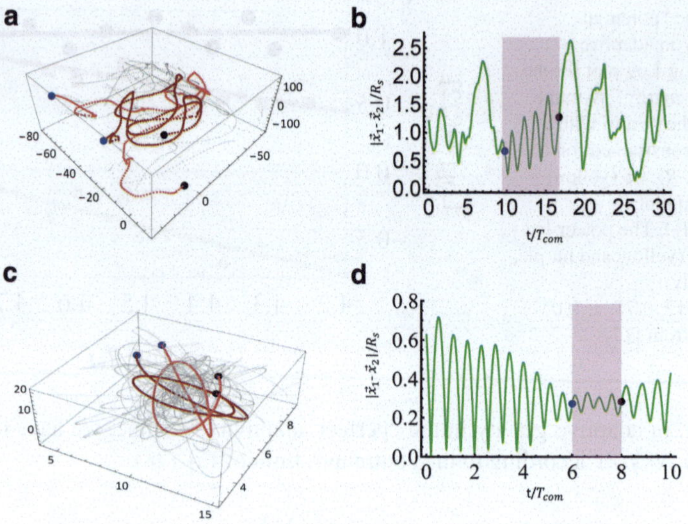

Fig. 10 Pair formation in laboratory observations of midge swarms. (**a**) Trajectories of two midges in a laboratory swarm with 21 midges that exhibited pairing (identified as being above a threshold value for the oscillation frequency, see [20, 76]). The midges were identified as belonging to a pair between the blue and black points. Paired parts of the trajectories are colored in red, while unpaired parts are in gray. Distances are in mm. (**b**) Distance between the members of the laboratory pair in (**a**) as a function of time. Distances are normalized by the swarm size R_s. The time is normalized by the typical orbit time around the center of mass T_{com}. (**c**) Pairing event in a simulation with 30 particles, $c_g = 1$, $R_{ad} = 15$, and $R_s = 5.08$ (defined as a pair in a bound orbit). Distances are in simulation unit lengths. Here the segments between the blue and the black points are only a part of a longer pairing event. The red and the gray colors are as in (**a**). (**d**) Distance between the members of the pair in (**c**) (simulation) as a function of time. Distances are normalized by the swarm size R_s. The time is normalized by the typical orbit time around the center of mass T_{com} (Adapted from [20])

local phenomenon of pair formation that can be qualitatively explained to arise from adaptivity. It has been observed that midges form transient pairs with synchronized relative motion while moving through the swarm [75]. An example from real data is shown in Fig. 10a, b and from simulations in Fig. 10c, d. In the simulation, we integrated the following equation (assuming a unit mass per particle)

$$\ddot{\vec{x}}_i = \vec{F}^{ad}(\vec{x}_i) \tag{57}$$

for all the particles in order to obtain an individual trajectory for each one of them, where $\vec{F}^{ad}(\vec{x}_i)$ is given in Eq. (26). The definition of a pair in the data is according to the frequency of their relative movement being high enough, above a threshold value, and in the simulation the pair is defined by the energy ratio that forms a bound pair. We can see in the example of Fig. 10 that the bound pair in the simulation shows higher oscillation frequency in their mutual orbits and it is consistent with

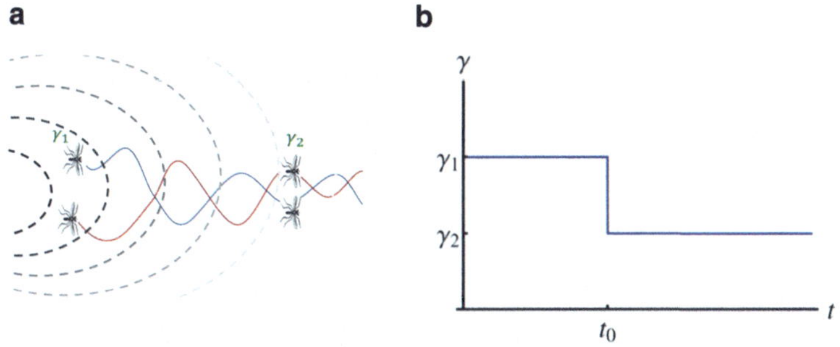

Fig. 11 The background sound parameter. (**a**) Illustration of the proposed pair formation mechanism. When two interacting midges leave the dense region of the swarm (darker dashed lines), where the background sound parameter γ is high, and move to a lower density region (such that $\gamma_1 > \gamma_2$), the mutual pull between them becomes stronger, their orbit gets tighter, and they become bound. (**b**) The background sound parameter γ is taken to be a step function. For $t \leq t_0$, the background sound level is higher than the background sound for $t > t_0$ (Adapted from [20])

the definition of the pair in the laboratory data. It was also shown in [20] that this identification is statistically robust.

Within the adaptive-gravity model, these pairing events occur whenever two midges happen to move together from the center of the swarm (where the total background noise is high) toward the swarm periphery (where the total background noise is low). As a result of adaptivity, during this movement, the attraction in the pair increases as the total noise decreases, thereby forming a bound state (see Fig. 11a). The full analysis and comparison with experimental data appeared in [20]. Here we give a simple analytical sketch of the formation of a bound pair as a result of adaptivity.

Let us consider the interaction of two midges that happen by chance to come very close to each other, in the background of the rest. We assume that the separation between the two midges is small compared to their average distance to the rest of the midges in the swarm, and therefore the interactions with the rest of the swarm will be negligible except for a contribution to the total noise in Eq. (26), so that the effective force felt by one member of the pair (at $\vec{x}_1$) is

$$\vec{F}^{ad}(\vec{x}_1) = -\frac{\nabla V^{12}(\vec{x}_1)}{1 + c_g^{-1} R_{ad}^2 I_{background} + c_g^{-1} R_{ad}^2 |\nabla V^{12}|}, \tag{58}$$

where

$$I_{background} \equiv \sum_{i=3}^{n} |\nabla V^{1i}| \sim I_{bgd} + \mathcal{O}.(|\vec{x}_1 - \vec{x}_i|). \tag{59}$$

The leading order of the expansion at infinity gives us a constant background contribution. Integrating this force, we get the effective two-body potential (compare with Eq. (31))

$$U_{\text{pair}}(|\vec{x}_1 - \vec{x}_2|) = \frac{c_g}{\gamma R_{\text{ad}}}\left[\arctan\left(\gamma\, |\vec{x}_1 - \vec{x}_2|/R_{\text{ad}}\right) - \pi/2\right], \tag{60}$$

where

$$\gamma \equiv \sqrt{1 + c_g^{-1} R_{ad}^2\, I_{bgd}} \tag{61}$$

is the background sound parameter. Hence, we effectively have two-body motion under the influence of a mutual central force. In the case of two bodies, not only the conservation laws are valid but also the additivity of the effective force, so we can use all the conservation laws of a central force such as energy and angular momentum conservation. In particular, this two-body system can be reduced to an equivalent one-dimensional motion in the effective potential

$$U_{\text{eff},12}(|\vec{x}_1 - \vec{x}_2|) = \frac{\tilde{l}^2}{2\,|\vec{x}_1 - \vec{x}_2|^2} + \frac{c_g}{\gamma\, R_{\text{ad}}}\left(\arctan\left(\frac{\gamma\, |\vec{x}_1 - \vec{x}_2|}{R_{\text{ad}}}\right) - \frac{\pi}{2}\right), \tag{62}$$

where $\tilde{l}$ is the angular momentum per unit mass (and we take the reduced mass of the pair).

When the background sound is reduced (as the pair move away from the swarm center), the sensitivity of each pair member increases, their effective mutual force becomes stronger, and an unbound pair may become bound. Let us analyze it for simplicity when the background sound parameter decreases as a step function (Fig. 11):

$$\gamma = \begin{cases} \gamma_1 & t \leq t_0 \\ \gamma_2 & t > t_0, \end{cases} \tag{63}$$

where $\gamma_1 > \gamma_2$.

When the mutual attractive force is stronger, even when we start from an elliptical bound orbit, it becomes tighter, and therefore the particles approach closer to each other. The total energy, which is conserved, is given by the following expression:

$$E_i = E_k(|\vec{x}_1 - \vec{x}_2|) + \frac{\tilde{l}^2}{2\,|\vec{x}_1 - \vec{x}_2|^2} + \frac{c_g}{\gamma_i\, R_{\text{ad}}}\left(\arctan\left(\frac{\gamma_i\, |\vec{x}_1 - \vec{x}_2|}{R_{\text{ad}}}\right) - \frac{\pi}{2}\right), \tag{64}$$

where $E_k(|\vec{x}_1 - \vec{x}_2|)$ is the kinetic energy of the radial movement. At $t = t_0$, let us denote the radial distance of the reduced mass by r_0 and the energy of the pair is changing due to the change in the background sound. The energy difference is given by:

$$\Delta E = \Delta U_{\text{eff},12} = \frac{c_g}{\gamma_2 R_{\text{ad}}}\left(\arctan\left(\frac{\gamma_2 |\vec{x}_1 - \vec{x}_2|}{R_{\text{ad}}}\right) - \frac{\pi}{2}\right) - \frac{c_g}{\gamma_1 R_{\text{ad}}}\left(\arctan\left(\frac{\gamma_1 |\vec{x}_1 - \vec{x}_2|}{R_{\text{ad}}}\right) - \frac{\pi}{2}\right) \quad (65)$$

When $\gamma_2 < \gamma_1$ in this case $\Delta E < 0$ since $U_{\text{eff},12}(r)$ which is given in Eq. (62) is monotonically increasing (becomes less negative) as a function of γ for any positive value of R_{ad}

$$\partial_\gamma U_{\text{eff},12} = c_g \frac{2\gamma R_{ad} |\vec{x}_1 - \vec{x}_2| + (R_{ad}^2 + \gamma^2 |\vec{x}_1 - \vec{x}_2|^2)(\pi - 2 \arctan\left(\frac{\gamma |\vec{x}_1 - \vec{x}_2|}{R_{ad}}\right))}{2\gamma^2 R_{ad} \left(R_{ad}^2 + \gamma^2 |\vec{x}_1 - \vec{x}_2|^2\right)} > 0. \quad (66)$$

The "tightness" of an orbit can be regarded as the absolute value of the ratio of the kinetic to potential energy $|E_k/U_{\text{eff},12}|$. When it is smaller than one $|E_k/U_{\text{eff},12}| < 1$, the orbit is bound, and that is how a pair is defined, if the two midges move in bound orbits (see Fig. 10). Since the effective potential energy is lowered as a result of the reduction of the background sound and the kinetic energy remains the same, this process can indeed produce bound orbits out of unbound ones. In addition, it can be shown [20] that the maximal distance between the members of the pairs is larger when the background sound parameter is higher. The proposed mechanism for pair formation is supported by experimental data [20], where it is shown that the background sound gradient tends to be negative where the pairs form (upon leaving the swarm center) and tend to be positive where they dissociate (entering the swarm center).

This phenomenon of separated particles forming bounded pairs can be explained by the adaptive-gravity model, while in regular gravitational attraction, it is extremely unlikely due to constraints of conservation of momentum and energy [77]. Due to momentum conservation, two particles in regular gravity can form a bound pair only if a third particle removes the excess momentum.

Note that the pairing mechanism that we describe is general to any many-body system with long-range adaptive interactions, and is not limited to the gravity-like functional form that we used here.

3.5 *Summary*

The fundamental ingredient of the adaptive gravity model is an ansatz for the effective force between particles. This force is a nonlinear function of the pairwise Newtonian gravitational interaction, inspired by the long-range acoustic interactions between midges in a swarm (Eq. (26)):

$$\vec{F}^{ad}(\vec{x}_i) = -\frac{\sum_{j=1}^{n} \nabla V^{ij}(\vec{x}_i)}{1 + (c_g^{-1} R_{ad}^2) \sum_{j=1}^{n} |\nabla V^{ij}|}.$$

This force is nonreciprocal for three particles or more. Among the various properties of the swarm in this model (that were compared to experimental results in [17–20]), we chose to highlight here the following:

- Stabilization of a swarm, based on the absence of gravitational Jeans instability, as a result of the adaptivity of the interactions.
- Generalized virial theorem which relates generalized kinetic energy T with generalized potential energy W (up to a surface term S) and is a consequence of only mass conservation (Eq. (49)):

  $$2T - S + W = 0.$$

  In addition we can predict the scaling relation between the effective potential energy and the swarm size (Fig. 9).
- Spontaneous formation of pairs as a result of the adaptivity of the interactions is another emergent feature of the long-range adaptive interactions.

4 Discussion

Nonreciprocal interactions between active particle have recently become a topic of growing interest. In this chapter, we have demonstrated that nonreciprocal forces, which arise from effective potentials that do not obey Newton's third law, and do not allow to formulate conservation of momentum and energy, arise naturally within the context of interactions between animals. We presented two very different examples for such effective interactions, in the form of (mostly) pairwise animal contests [13, 14], and in the context of the many-body long-range interactions within a cohesive swarm [17–20]. These examples expand the types and scope of application of nonreciprocal interactions within the wider field of active matter.

The nonreciprocally interacting particles model of animal contests represents a departure from conventional approaches by explicitly addressing the spatiotemporal dynamics of contest behavior. Unlike traditional game-theoretic models [24–32], which often overlook these dynamics, this model focuses on emergent spatiotemporal phenomena such as chase dynamics. By using effective interaction potentials distinct from traditional pairwise potentials, it captures the directional and nonreciprocal nature of contest interactions, highlighting the significance of asymmetric interactions in shaping agonistic behavior dynamics.

The effective gravity model of swarming midges introduces a unique perspective on collective behavior in animal groups. Moving beyond traditional velocity alignment-based models [62, 63], this model leverages adaptive long-range interactions mediated by acoustics to explain swarm cohesion. The incorporation of adaptivity ensures sensitivity adjustment to background stimuli, leading to stable swarm formations and the spontaneous formation of pairs within the swarm. These findings underscore the importance of long-range interactions and adaptive mechanisms in maintaining cohesion within animal groups.

These models demonstrate that the study of active-particle systems with unconventional interactions enriches our understanding of animal behavior, while in turn, the study of animal systems enriches and expands the scope of active-particle theory.

Appendix

Explicit Calculation of the Linear Restoring Force of Adaptive Gravity

Under the assumptions of a spherical and uniform swarm, there is a linear restoring force toward the center of the swarm of the form $\vec{F} = -K\vec{r}$ in the leading order in the deviation from it (where $\vec{r} = 0$ is the center of the swarm). Let us calculate explicitly the effective spring constant of the linear adaptive-gravitational field at a point inside a spherical swarm with radius $R_s = (3/4)R$ (the mean distance from the center is 3/4 of the radius R) and uniform density ρ, according to Eq. (26) where we take the continuum limit of the sums. We will use cylindrical coordinates (r, z, φ) and calculate the field at $(r = 0, z = z_0)$ without loss of generality (the point A in Fig. 12). The symmetry of the problem implies that the field is along the z axis. The contribution of a point at (r', z') to the gravitational force at $(0, z_0)$ is

$$\frac{c_g}{r'^2 + (z' - z_0)^2},$$

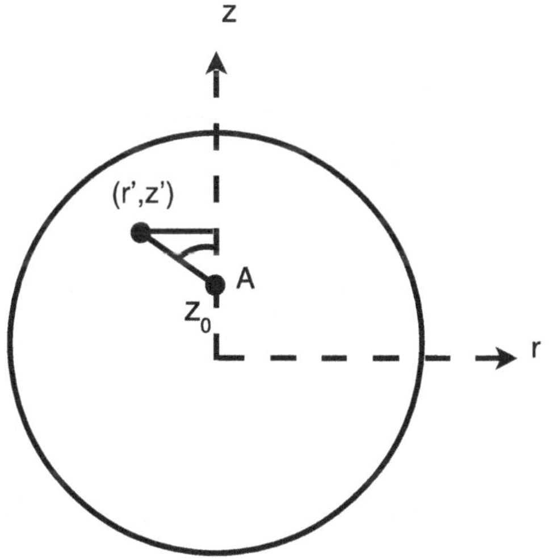

Fig. 12 The cylindrical coordinates (r, z) (and φ) that we use for the calculation of the effective gravitational field at a point A in a uniform-density spherical swarm

and the angle is

$$\cos\varphi = \frac{z' - z_0}{\sqrt{r'^2 + (z' - z_0)^2}}.$$

Hence the regular gravitational force at z_0 is

$$F^g(z_0) = 2\pi\,\rho\,c_g \int_{-R}^{R} dz' \int_0^{\sqrt{R^2 - z'^2}} r'dr' \frac{z' - z_0}{\left[r'^2 + (z' - z_0)^2\right]^{\frac{3}{2}}} = -\frac{4\pi\,\rho\,c_g}{3} z_0, \tag{67}$$

and then the effective spring constant of this restoring force is

$$K^g = \frac{4\pi\,\rho\,c_g}{3}. \tag{68}$$

This result can be also obtained from Gauss's law (as it was derived in [17]). The "total noise" factor in the denominator (also in Eq. (29)), which is given by the summation of the absolute values of the contributions to the point $(0, z_0)$, reads

$$\begin{aligned} N_{\text{tot}}(z_0) &= 2\pi\,c_g\,\rho \int_{-R}^{R} dz' \int_0^{\sqrt{R^2 - z'^2}} \frac{r'dr'}{r'^2 + (z' - z_0)^2} \\ &= \pi\,c_g\,\rho[2R - \frac{(R^2 - z_0^2)}{z_0} \ln\left(\frac{R - z_0}{R + z_0}\right)]. \end{aligned} \tag{69}$$

To leading order in z_0, we have

$$N_{\text{tot}}(z_0) = 4\pi\rho\,c_g\,R + \mathcal{O}(z_0^2), \tag{70}$$

which after substitution into Eq. (26) gives to leading order in z_0 the following restoring force toward the center:

$$F^{ad}(z_0) = -\frac{4\pi\,\rho\,c_g}{3 + 16\pi\,\rho\,R_s\,R_{ad}^2} z_0. \tag{71}$$

The number of particles N in a uniform and spherical swarm is given by $N = 4\pi\,\rho\,R_s^3/3$, and then we can write the denominator in Eq. (71) in the following way

$$3 + 16\pi\,\rho\,R_s\,R_{ad}^2 = 3 + 16\,N\,R_{ad}^2/R_s^2.$$

Assuming the adaptivity range is approximately the size of the swarm $R_{ad} \sim R_s$ and $N >> 1$. Then the first constant term in the denominator is negligible compared to the second. Then within this approximation, we get the following expression for the effective spring constant:

$$K^{ad} = \frac{c_g}{4\,R_s\,R_{ad}^2}. \tag{72}$$

References

1. A. Zöttl, H. Stark, Modeling active colloids: from active brownian particles to hydrodynamic and chemical fields. Ann. Rev. Condens. Matter Phys. **14**, 109–127 (2023)
2. T. Vicsek, A. Zafeiris, Collective motion. Phys. Rep. **517**(3–4), 71–140 (2012)
3. S. Motsch, E. Tadmor, A new model for self-organized dynamics and its flocking behavior. J. Stat. Phys. **144**, 923–947 (2011)
4. J. Haskovec, Flocking dynamics and mean-field limit in the cucker–smale-type model with topological interactions. Phys. D: Nonlin. Phenom. **261**, 42–51 (2013)
5. L. Barberis, F. Peruani, Phase separation and emergence of collective motion in a one-dimensional system of active particles. J. Chem. Phys. **150**(14), 144905 (2019)
6. N. Bellomo, M. Esfahanian, V. Secchini, P. Terna, What is life? Active particles tools towards behavioral dynamics in social-biology and economics. Phys. Life Rev. **43**, 189–207 (2022)
7. M. Durve, A. Saha, A. Sayeed, Active particle condensation by non-reciprocal and time-delayed interactions. Eur. Phys. J. E **41**, 1–9 (2018)
8. S. Saha, J. Agudo-Canalejo, R. Golestanian, Scalar active mixtures: the nonreciprocal cahn-hilliard model. Phys. Rev. X **10**(4), 041,009 (2020)
9. M. Fruchart, R. Hanai, P.B. Littlewood, V. Vitelli, Non-reciprocal phase transitions. Nature **592**(7854), 363–369 (2021)
10. K.L. Kreienkamp, S.H. Klapp, Clustering and flocking of repulsive chiral active particles with non- reciprocal couplings. New J. Phys. **24**(12), 123,009 (2022)
11. M. Kneževic´, T. Welker, H. Stark, Collective motion of active particles exhibiting non-reciprocal orientational interactions. Sci. Rep. **12**(1), 19,437 (2022)
12. A. Dinelli, J. O'Byrne, A. Curatolo, Y. Zhao, P. Sollich, J. Tailleur, Non-reciprocity across scales in active mixtures. Nat. Commun. **14**(1), 7035 (2023)
13. A. Haluts, S.F.G. Reyes, D. Gorbonos, R.I. Etheredge, A. Jordan, N.S. Gov, Spatiotemporal dynamics of animal contests arise from effective forces between contestants. Proc. Natl. Acad. Sci. **118**(49), e2106269118 (2021)
14. A. Haluts, A. Jordan, N.S. Gov, Modelling animal contests based on spatio-temporal dynamics. J. R. Soc. Interface **20**(202), 20,220,866 (2023)
15. M. Adler, U. Alon, Fold-change detection in biological systems. Curr. Opin. Syst. Biol. **8**, 81–89 (2018)
16. N.T. Ouellette, A physics perspective on collective animal behavior. Phys. Biol. **19**(2), 021,004 (2022)
17. D. Gorbonos, R. Ianconescu, J.G. Puckett, R. Ni, N.T. Ouellette, N.S. Gov, Long-range acoustic interactions in insect swarms: an adaptive gravity model. New J. Phys. **18**(7), 073,042 (2016)
18. D. Gorbonos, N.S. Gov, Stable swarming using adaptive long-range interactions. Phys. Rev. E **95**, 042,405 (2017)
19. D. Gorbonos et al., Similarities between insect swarms and isothermal globular clusters. Phys. Rev. Res. **2**, 013,271 (2020)
20. D. Gorbonos, J.G. Puckett, K. van der Vaart, M. Sinhuber, N.T. Ouellette, N.S. Gov, Pair formation in insect swarms driven by adaptive long-range interactions. J. R. Soc. Interface **17**(171), 20,200,367 (2020)

21. A. Shklarsh, G. Ariel, E. Schneidman, E. Ben-Jacob, Smart swarms of bacteria-inspired agents with performance adaptable interactions. PLoS Comput. Biol. **7**(9), e1002177 (2011)
22. E. Ben-Jacob, H. Shmueli, O. Shochet, A. Tenenbaum, Adaptive self-organization during growth of bacterial colonies. Phys. A: Stat. Mech. Appl. **187**(3–4), 378–424 (1992)
23. M. Briffa, I.C. Hardy, Introduction to animal contests, in *Animal Contests* (Cambridge University Press Cambridge, 2013), pp. 1–4
24. R.J. Payne, M. Pagel, Escalation and time costs in displays of endurance. J. Theor. Biol. **183**(2), 185–193 (1996)
25. M. Mesterton-Gibbons, J.H. Marden, L.A. Dugatkin, On wars of attrition without assessment. J. Theor. Biol. **181**(1), 65–83 (1996)
26. M. Enquist, O. Leimar, Evolution of fighting behaviour: decision rules and assessment of relative strength. J. Theor. Biol. **102**(3), 387–410 (1983)
27. M. Enquist, O. Leimar, T. Ljungberg, Y. Mallner, N. Segerdahl, A test of the sequential assessment game: fighting in the cichlid fish nannacara anomala. Anim. Behav. **40**(1), 1–14 (1990)
28. G.A. Parker, D.I. Rubenstein, Role assessment, reserve strategy, and acquisition of information in asymmetric animal conflicts. Anim. Behav. **29**(1), 221–240 (1981)
29. P. Hammerstein, G.A. Parker, The asymmetric war of attrition. J. Theor. Biol. **96**(4), 647–682 (1982)
30. R.J. Payne, M. Pagel, Why do animals repeat displays? Anim. Behav. **54**(1), 109–119 (1997).
31. R.J. Payne, Gradually escalating fights and displays: the cumulative assessment model. Anim. Behav. **56**(3), 651–662 (1998)
32. H. Kokko et al., Dyadic contests: modelling flights between two individuals, in *Animal Contests* (Cambridge University Press, Cambridge, 2013)
33. J. Smith, G.R. Price, The logic of animal conflict. Nature **246**(5427), 15–18 (1973)
34. J.M. Smith, The theory of games and the evolution of animal conflicts. J. Theor. Biol. **47**(1), 209–221 (1974)
35. G.A. Parker, Assessment strategy and the evolution of fighting behaviour. J. Theor. Biol. **47**(1), 223–243 (1974)
36. J.M. Smith, G.A. Parker, The logic of asymmetric contests. Anim. Behav. **24**(1), 159–175 (1976)
37. G.A. Parker et al., Sexual selection and sexual conflict. Sex. Sel. Reprod. Competition Insects **123**, 166 (1979)
38. G. Arnott, R.W. Elwood, Assessment of fighting ability in animal contests. Anim. Behav. **77**(5), 991–1004 (2009)
39. P. Taylor, R.W. Elwood, The mismeasure of animal contests. Anim. Behav. **65**(6), 1195–1202 (2003)
40. N.S. Pinto, A.V. Palaoro, P.E. Peixoto, All by myself? Meta-analysis of animal contests shows stronger support for self than for mutual assessment models. Biol. Rev. **94**(4), 1430–1442 (2019)
41. K.J. Chapin, P.E.C. Peixoto, M. Briffa, Further mismeasures of animal contests: a new framework for assessment strategies. Behav. Ecol. **30**(5), 1177–1185 (2019)
42. A.I. Dell et al., Automated image-based tracking and its application in ecology. Trends Ecol. Evol. **29**(7), 417–428 (2014)
43. R. Nathan et al., Big-data approaches lead to an increased understanding of the ecology of animal movement. Science **375**(6582), eabg1780 (2022)
44. I. Giardina, Collective behavior in animal groups: theoretical models and empirical studies. HFSP J. **2**(4), 205–219 (2008)
45. P. Romanczuk, M. Bär, W. Ebeling, B. Lindner, L. Schimansky-Geier, Active brownian particles. Euro. Phys. J. Spec. Top. **202**(1), 1–162 (2012)
46. G. Volpe, S. Gigan, G. Volpe, Simulation of the active brownian motion of a microswimmer. Am. J. Phys. **82**(7), 659–664 (2014)
47. L.A. Jordan, H. Kokko, M. Kasumovic, Reproductive foragers: male spiders choose mates by selecting among competitive environments. Am. Nat. **183**(5), 638–649 (2014)

48. S.T. Emlen, L.W. Oring, Ecology, sexual selection, and the evolution of mating systems. Science **197**(4300), 215–223 (1977)
49. M.J. Fuxjager, X. Zhao, N.S. Rieger, C.A. Marler, Why animals fight: uncovering the function and mechanisms of territorial aggression, in *APA Handbook of Comparative Psychology: Basic Concepts, Methods, Neural Substrate, and Behavior*, vol. 1 (American Psychological Association, Washington, 2017), pp. 853–875
50. R.M. Corless, G.H. Gonnet, D.E. Hare, D.J. Jeffrey, D.E. Knuth, On the lambert w function. Adv. Comput. Math. **5**, 329–359 (1996)
51. J.W. Popp, Risk and effectiveness in the use of agonistic displays by American goldfinches. Behaviour **103**(1–3), 141–156 (1987)
52. R.A. Martin, A review of shark agonistic displays: comparison of display features and implications for shark–human interactions. Mar. Freshwater Behav. Physiol. **40**(1), 3–34 (2007)
53. D. Paton, P. Caryl, Communication by agonistic displays. Behaviour **98**(1–4), 213–239 (1986)
54. P. Hänggi, P. Talkner, M. Borkovec, Reaction-rate theory: fifty years after kramers. Rev. Mod. Phys. **62**(2), 251 (1990)
55. A. Geiseler, P. Hänggi, G. Schmid, Kramers escape of a self-propelled particle. Euro. Phys. J. B **89**, 1–7 (2016)
56. G. Arnott, R.W. Elwood, Information gathering and decision making about resource value in animal contests. Anim. Behav. **76**(3), 529–542 (2008)
57. Z. You, A. Baskaran, M.C. Marchetti, Nonreciprocity as a generic route to traveling states. Proc. Natl. Acad. Sci. **117**(33), 19,767–19,772 (2020)
58. F. Vollrath, Male body size and fitness in the web-building spider nephila clavipes. Zeitsch-rift für Tierpsy- Chologie **53**(1), 61–78 (1980)
59. J. Schneider, M. Herberstein, F.C. De Crespigny, S. Ramamurthy, M. Elgar, Sperm competition and small size advantage for males of the golden orb-web spider nephila edulis. J. Evol. Biol. **13**(6), 939–946 (2000)
60. C.C. Rittschof, Male density affects large-male advantage in the golden silk spider, Nephila clavipes. Behav. Ecol. **21**(5), 979–985 (2010)
61. I.D. Couzin, Collective cognition in animal groups. Trends Cognit. Sci. **13**(1), 36–43 (2009)
62. A. Cavagna et al., Dynamic scaling in natural swarms. Nat. Phys. **13**(9), 914–918 (2017)
63. A. Cavagna, I. Giardina, T.S. Grigera, The physics of flocking: correlation as a compass from experiments to theory. Phys. Rep. **728**, 1–62 (2018)
64. C.M. Topaz, A.L. Bertozzi, Swarming patterns in a two-dimensional kinematic model for biological groups. SIAM J. Appl. Math. **65**(1), 152–174 (2004)
65. C.M. Topaz, A.L. Bertozzi, M.A. Lewis, A nonlocal continuum model for biological aggregation. Bull. Math. Biol. **68**, 1601–1623 (2006)
66. O. Shoval, L. Goentoro, Y. Hart, A. Mayo, E. Sontag, U. Alon, Fold-change detection and scalar symmetry of sensory input fields. Proc. Natl. Acad. Sci. USA **107**, 15,995–16,000 (2010)
67. S. Motsch, E. Tadmor, Heterophilious dynamics enhances consensus. SIAM Rev. **56**(4), 577–621 (2014)
68. S.A.M. Loos, S.H.L. Klapp, Irreversibility, heat and information flows induced by non-reciprocal interactions. New J. Phys. **22**(12), 123,051 (2020)
69. J.H. Jeans, The stability of a spherical nebula. Philos. Trans. R. Soc. Lond. Ser. A **199**, 1–53 (1902)
70. J. Binney, S. Tremaine, Galactic Dynamics (Princeton University Press, Princeton, 2011)
71. S. Chandrasekhar, *Ellipsoidal Figures of Equilibrium*. Silliman Foundati Lect. (Yale University Press, New Haven, 1969)
72. P.R. Shapiro, I.T. Iliev, H. Martel, K. Ahn, M.A. Alvarez, The Equilibrium structure of CDM halos (2004). arXiv: astro-ph/0409173 (astro-ph)
73. M. Sinhuber, N.T. Ouellette, Phase coexistence in insect swarms. Phys. Rev. Lett. **119**(17), 178,003 (2017)
74. M. Sinhuber, K. van der Vaart, Y. Feng, A.M. Reynolds, N.T. Ouellette, An equation of state for insect swarms. Sci. Rep. **11**(1), 3773 (2021)

75. J.G. Puckett, R. Ni, N.T. Ouellette, Time-frequency analysis reveals pairwise interactions in insect swarms. Phys. Rev. Lett. **114**, 258,103 (2015)
76. J.G. Puckett, R. Ni, N.T. Ouellette, Time-frequency analysis reveals pairwise interactions in insect swarms. Phys. Rev. Lett. **114**, 258,103 (2015). https://doi.org/10.1103/PhysRevLett.114.258103 [Online]. Available: https://link.aps.org/doi/10.1103/PhysRevLett.114.258103
77. P. Bodenheimer, *Principles of Star Formation*. Astronomy and Astrophysics Library (Springer, Berlin, 2011). ISBN: 9783642150623 [Online]. Available: https://books.google.co.il/books?id=ZlfANAEACAAJ

Bayesian Sampling Using Interacting Particles

Shi Chen, Zhiyan Ding, and Qin Li

Abstract Bayesian sampling is an important task in statistics and machine learning. Over the past decades, many ensemble-type sampling methods have been proposed. In contrast to the classical Markov chain Monte Carlo methods, these new methods deploy a large number of interactive samples, and the communication between these samples is crucial in speeding up the convergence. To justify the validity of these sampling strategies, the concept of interacting particles naturally calls for the mean-field theory. The theory establishes a correspondence between particle interactions encoded in a set of coupled ODEs/SDEs and a PDE that characterizes the evolution of the underlying distribution. This bridges numerical algorithms with the PDE theory used to show convergence in time. Many mathematical machineries are developed to provide the mean-field analysis, and we showcase two such examples: the coupling method and the compactness argument built upon the martingale strategy. The former has been deployed to show the convergence of the ensemble Kalman sampler and the ensemble Kalman inversion, and the latter will be shown to be immensely powerful in proving the validity of the Vlasov-Boltzmann simulator.

1 Introduction

Sampling is a fundamental task in the field of machine learning and serves as a crucial building block in various applications [1]. For example, sampling plays a significant role in Bayesian statistics, data assimilation [82], generative modeling [90], volume computation [98], and bandit optimization [73, 74, 86].

S. Chen · Q. Li (✉)
Mathematics Department, University of Wisconsin-Madison, Madison, WI, USA
e-mail: schen636@wisc.edu; qinli@math.wisc.edu

Z. Ding
Department of Mathematics, University of California, Berkeley, CA, USA
e-mail: zding.m@math.berkeley.edu

J. A. Carrillo, E. Tadmor (eds.), *Active Particles, Volume 4*, Modeling and Simulation in Science, Engineering and Technology,
https://doi.org/10.1007/978-3-031-73423-6_5

Sampling methods have been widely used in various scientific domains, including atmospheric science, petroleum engineering, and epidemiology [41, 66, 75].

The goal of sampling is to obtain approximately independent, identically distributed (i.i.d.) samples that come from a target distribution. Given a target distribution with density

$$\rho_{\text{target}}(x) \propto e^{-f(x)}, \quad \text{with} \quad f(x) : \mathbb{R}^d \to \mathbb{R}, \tag{1.1}$$

we are to find

$$\{x_i\}_{i=1}^N \sim \rho_{\text{target}}\,.$$

It is typically assumed $d \gg 1$. A classical origin of the sampling problem is Bayesian inference, in which the function f takes the form of

$$f(x; y) = \Phi(x; y) + \frac{1}{2}\,|x - x_0|^2_{\Gamma_0}, \quad \text{with} \quad \Phi(x; y) = \frac{1}{2}\,|y - \mathcal{G}(x)|^2_{\Gamma}, \tag{1.2}$$

where $\mathcal{G} : \mathbb{R}^d \to \mathbb{R}^{d'}$ denotes the forward map that maps input x to the output $\mathcal{G}(x)$ that hopefully agrees with the data $y \in \mathbb{R}^{d'}$, and Φ is to penalize the measuring error. Γ and $\Gamma_0 \in \mathbb{R}^d$ are positive semidefinite matrices that serve as the covariance matrix in the domain and range of the map with $|\cdot|_{\Gamma_0} := |\Gamma_0^{-\frac{1}{2}} \cdot|$ and $|\cdot|_{\Gamma} := |\Gamma^{-\frac{1}{2}} \cdot|$.

This formulation stems from the Bayesian theorem that states

$$\rho_{\text{target}}(x) = \rho_{\text{post}}(x) \propto \exp(-f(x)) = \rho_{\text{prior}}(x) l(y; x) \tag{1.3}$$

with $\rho_{\text{prior}}(x) \propto \exp\left(-\frac{1}{2}\,|x - x_0|^2_{\Gamma_0}\right)$ and the likelihood function $l(y; x) = \exp(-\Phi(x; y))$ that depends on the forward map $\mathcal{G}$. Since y is the given data and is not subject to change, when the context is clear, we drop its dependence in l, f, and Φ.

Sampling from a target distribution has a rich literature. Dating back to 1953 with the development of the Metropolis algorithm, for several decades, Markov chain Monte Carlo (MCMC) algorithms have been the primary tool for generating samples [43, 48, 83]. Different MCMC methods adopt different dynamics, but they share the same mathematical structure. Each MCMC method is run according to a transition matrix, and the target distribution is the unique invariant measure of such transition matrix. Subsequently, as a sample evolves according to this transition matrix, and its final distribution approaches the invariant measure, meaning that the sample can be viewed as a drawing from the target distribution. Over the decades, MCMC methods have attracted many investigations, and we have obtained a collection of results on MCMC convergence, including the convergence rate on the continuous and discrete level, both asymptotically and non-asymptotically with explicit rates [12, 19–22, 24, 25, 32–35, 64, 88, 99, 102].

In contrast to the classical single-particle MCMC methods, a new line of techniques surrounding "ensemble method" emerged over the past decade and have garnered a lot of interest. Instead of moving a single particle, many particles evolve simultaneously and interactively. The concept of utilizing a many-particle system instead of one particle traces back to the Ensemble Kalman Filter [38, 39, 47, 53], an ensemble-based version of the Kalman filter [58, 59] that stores Kalman matrices through sample representations. In the early 2010s, various ensemble-based methods were developed, including Affine Invariant Ensemble Sampler [49], Ensemble Kalman Inversion [40, 54] and, more recently, Consensus Based Sampling [15, 46], Stein Variational Gradient Descent [67, 69], Langevin Sampling with Birth-death [70], and Ensemble Kalman Sampler [44], and their derivatives such as the Weighted Ensemble Kalman Inversion [29].

It is believed that ensemble methods have some advantages over single-particle MCMC methods. The argument is that if many particles are exploring the full landscape together, the chance of them covering the full domain and discovering all local and global minima is higher and thus can potentially capture the target distribution faster. Yet, this promise has not been fully realized. The first step toward this is to spell out explicitly the convergence rate of the ensemble: With what convergence rate in time and in the number of samples can the ensemble converge to the target distribution?

Unfortunately, unlike single-particle MCMC solvers, the numerical analysis for ensemble-type algorithms is considerably more intricate. Many particles interact in a nonlinear manner. It is not immediate that such an interaction speeds up the convergence. Unlike MCMC, which typically requires one to trace the evolution of the distribution of the single sample through the application of the transition matrix, for an ensemble, since many samples are involved, the underlying distribution lives on a tensorized space that is significantly higher dimensional. This perspective necessitates the deployment of the mean-field limit to compress the tensorized probability space down to the original space. Upon such compression, the many-particle system is altogether described by an ensemble (empirical) distribution, providing a fair platform to compare with the classical MCMC evolution.

A variety of approaches can be used to construct a sampling algorithm involving many particles, but the general procedure is usually the same. Generally speaking, there are three steps:

1. Identify an evolutional PDE whose long time behavior recovers an invariant measure that coincides with the target distribution;
2. Design its particle method and translate the evolutional PDE for the underlying distribution to a coupled ODE/SDE system for each particle;
3. Discretize the ODE/SDE system by implementing the Euler-Maruyama or high-order discretization methods.

Accordingly, the analysis of the numerical error is divided into three sections.

- Show that the solution to the PDE indeed converges to the target distribution efficiently. This is mostly PDE analysis.

- Prove that the PDE can be solved by the particle method. This is to show that the ensemble distribution of the N-particle accurately recovers the PDE solution. This step equates the PDE and the coupled ODE/SDE system and is typically called deriving the mean-field limit.
- Demonstrate that the SDE discretization accurately recovers the dynamics of the underlying SDE system. This corresponds to analyzing the discretization error and showing that it is small when the stepsize h is small.

We summarize the above three steps in Fig. 1, and the associated strategy deployed to validate the algorithms. We should stress that these steps are ordered, and the order cannot be switched. Mathematically analyzing the algorithm amounts to taking the following limits:

$$\lim_{t\to\infty}\lim_{N\to\infty}\lim_{h\to 0}\,.$$

The three limits do not commute, and the analysis has to be done accordingly. In particular, for a preset accuracy level ϵ, running the analysis in Step 1 amounts to pin a time t, so that the PDE solution at this time is within $\frac{\epsilon}{3}$ to the target distribution, and for this t, we run analysis in Step 2 to identify the N and to ensure the N coupled SDE solution approximates the PDE within $\frac{\epsilon}{3}$, and finally, for the picked (t, N), we run analysis in Step 3 to identify a small enough h that ensures the discrete solution is within $\frac{\epsilon}{3}$ to the continuous-in-time SDE solution. This ordered constraint comes from the techniques that we deploy, such as the Grönwall inequality. It is possible for the limits to commute, if other machinery gets deployed. We do not discuss these options in this paper.

Each of these three steps of justification carries its own challenges and attracts a great deal of investigation. In this paper, we only discuss the mean-field analysis component of it (Step 2), using the Ensemble Kalman Sampler, Ensemble Kalman Inversion, and Boltzmann simulator as examples. The other two steps are equally

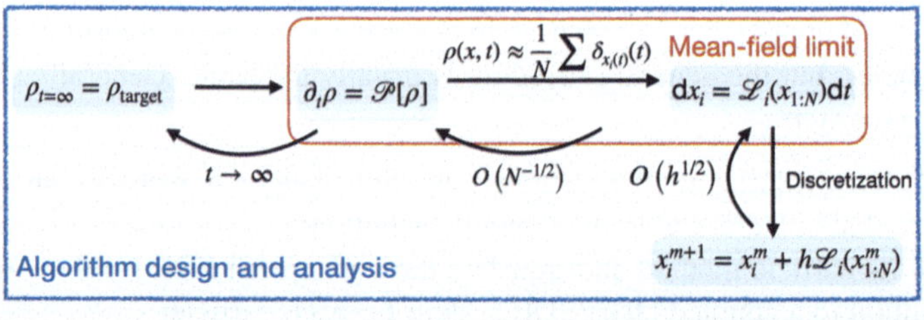

Fig. 1 The process of algorithm design (left to right) typically involves three steps: designing a PDE, designing its particle method, discretization. The analysis procedure (right to left) is consistent with the design process in reverse. This paper only focuses on the second step of algorithm analysis, which involves applying mean-field analysis to validate the mean-field limit of multiple-particle algorithms

important in the whole algorithm justification process but are not the main focus of the paper. We refer the interested readers to [13, 16, 26, 29, 60].

To a large extent, the goal of proving "mean-field limit" and justifying particle methods for simulating PDEs very much align. Essentially one claims that a distribution density function can be represented as an ensemble of many particles drawn from it. Accordingly, the evolution of this density function encoded in a dynamical PDE can be translated to the evolution of these representative particles. The question then becomes quantifying the differences between these two systems when the number of representative particles is large.

Multiple approaches have been taken to achieve mean field derivation. The most straightforward machinery is the coupling method [7, 11, 23, 42, 93]. This amounts to designing an auxiliary particle system with all particles passively pushed by the underlying "field." This auxiliary system serves as a bridge to link the underlying distribution and the ensemble system. When the system demonstrates good properties, such as Lipschitz forces, the method becomes very handy, as will be summarized in Sect. 2.1. Often in practice, forces, though regular enough, are not globally Lipschitz, preventing the direct application of the coupling method. Depending on the context, a mathematical treatment specifically designed for the SDEs at hand is required. This is a case-by-case discussion, and we will dive in detail in Sect. 2, using the ensemble Kalman inversion (EKI)[54] and the ensemble Kalman sampler (EKS)[44] as examples.

Another line of mean-field analysis is termed the martingale method. Within this framework, one can translate the PDE problems into their associated martingale problems that look for solutions as probability measures over the space of path measures. In this path measure space, one first demonstrates the tightness for a family of path measures. The tightness allows one to pass to a limit, upon which one can show that this limit happens to solve the mean-field PDE. Since the martingale method translates the mean-field convergence to the weak convergence of a sequence of measures over stochastic processes, the notion of convergence is quite weaker. Weak convergence is deployed to accommodate systems that are not so smooth, and thus the approach can handle a much larger class of interacting particle systems. In particular, we will use it to show the Vlasov-Boltzmann simulator in Sect. 3. The Boltzmann collision operator is significantly more complex, and mean-field analysis is beyond what the coupling method can handle. However, while the method is much more versatile and relaxes the assumptions on systems' coefficients, it lacks a precise description of the convergence rate. In Sect. 3.4, we present some numerical evidence.

The mean-field limit results that deploy the coupling method (Sect. 2) have appeared in literature, and we specifically mention [27] and [28]. The machinery is rather standard, and the set of theory is complete. Section 3 discusses the Boltzmann simulator to perform Bayesian sampling. For this new design of sampling method, we adopt the martingale strategy utilized in showing the mean-field limit for the classical Boltzmann equation and transition it into our new artificial system.

We finally stress that Bayesian sampling and optimization are two machine learning domains that are closely tied to each other, and they share mathematical strategies. Mean-field limit and interactive particles are also heavily relied on in the design and analysis of optimization solvers, typically under the name of "consensus-based" or "swarm-based." This is a vast domain of research, and we can only list a few recent studies [2, 8, 14, 17, 30, 46, 71, 78, 94].

2 Mean-Field Limit for Ensemble Based Sampling Methods

Ensemble-based sampling methods are immensely popular in areas related to atmospheric science and weather prediction. The concept originates from Ensemble Kalman Filter, an algorithmic pipeline for implementing Kalman Filter. Kalman Filter, as a standard technique for data assimilation, allows one to gradually assimilate information from collected data to dynamical models that have unknown parameters, so to gradually refine our estimates for these parameters. In the most brute-force simulation, the Kalman filter calls for the computation of the Kalman gain matrix, a term strongly tied to the covariance of the parameters. This is a very expensive numerical step when the problem is high dimensional. Ensemble Kalman Filter replaces the computation of the covariance by the ensemble covariance that can be extracted from a large quantity of samples. This way, one avoids direct computing the covariance, and it significantly reduces the computational cost [38–40, 47, 53, 62, 63].

The idea of using the ensemble covariance to replace the true covariance was later adopted by the field of Bayesian sampling; see [10, 82] and the references therein, e.g., [3, 4]. Starting in the early 2000s, the idea is at the core of a long list of exciting algorithm developments [29, 44, 54]. A similar perspective-shifting took place in optimization as well. Ensemble-based or swarm-based methods were designed to specifically tackle optimization problems that are non-convex and have since generated tremendous excitement [14, 71, 94].

Two distinctive and representative solvers in this research direction are the Ensemble Kalman Inversion (EKI) [54] and the Ensemble Kalman Sampler (EKS) [44]. The two methods are widely used to predict geophysical turbulence [31, 87], modeling reservoirs [55], and modeling fluid flow from the cardiovascular system [81] and generated excitement in various scientific domains.

We are interested in understanding the validity of the algorithms. As discussed in Sect. 1, we are particularly interested in showing the "mean-field" component of the method justification. Both systems deploy covariance matrices for samples to communicate with each other, and the flow fields for the samples are rather smooth. These properties allow us to implement the coupling method to conduct the mean-field analysis [27, 28] and demonstrate the convergence rate of the mean-field limit.

Below, we will first present in Sect. 2.1 a unified framework that discusses a typical use of the coupling method. Sections 2.2 and 2.3 are designated for EKI and EKS, respectively.

2.1 Unified Coupling Method Framework

The coupling method was beautifully shown in [93, 101], and this framework can handle a large class of systems that have nice properties. We take Fokker-Planck type PDE for the illustration:

$$\begin{aligned}\partial_t \rho(t,x) &= \mathcal{P}[\rho(t,x))] \\ &= \nabla_x \cdot (F(t,x;\rho)\rho(t,x))) + \frac{1}{2}\nabla_x \cdot (D(t,x;\rho)^\top D(t,x;\rho)\nabla_x \rho(t,x)),\end{aligned} \tag{2.1}$$

where $\rho(t)$ is the PDE solution. The operator $\mathcal{P}$ is composed of two components, the drift term with the coefficient $F(t,x;\rho)$ and the diffusion term with the coefficient $D(t,x;\rho)$. Both F and D can depend on ρ, and when they do, this equation is nonlinear. Throughout the section, we term the equation System A.

This system is closely related to the following two types of SDEs.

System B: The particle system pushed by the underlying characteristics. Denote $\{\tilde{x}_i\}$ a collection of samples i.i.d. drawn from $\rho(t=0,\cdot)$ at the initial time $t=0$. These samples evolve according to:

$$\mathrm{d}\tilde{x}_i(t) = F(t,\tilde{x}_i(t);\rho)\mathrm{d}t + D(t,\tilde{x}_i(t);\rho)\mathrm{d}W_t^i\,, \tag{2.2}$$

where W_t^i are independent Brownian motions. We further denote the ensemble distribution

$$\widetilde{\rho}_N(t,x) := \frac{1}{N}\sum_{n=1}^{N}\delta_{\tilde{x}_i(t)}\,. \tag{2.3}$$

The system is driven by the true underlying characteristics and thus honestly reflects the flow field. We expect $\widetilde{\rho}_N$ provided by System B to approximate ρ. However, we should note that (2.2) is not a system that can be numerically realized, since its parameters depend on $\rho(t,\cdot)$, the solution to the PDE, and is unknown.

System C: The particle system driven by the self-generated field. Let $\{x_i(t)\}_{i=1}^N$ be a collection of N samples i.i.d. drawn from $\rho(t=0,\cdot)$ at time $t=0$. We evolve them according to:

$$\mathrm{d}x_i(t) = F(t,x_i(t);\rho_N)\mathrm{d}t + D(t,x_i(t);\rho_N)\mathrm{d}W_t^i\,, \tag{2.4}$$

where drift and diffusion coefficient F and D depend on the self-induced ensemble distribution:

$$\rho_N(t,x) = \frac{1}{N}\sum_i \delta_{x_i(t)}(x)\,. \tag{2.5}$$

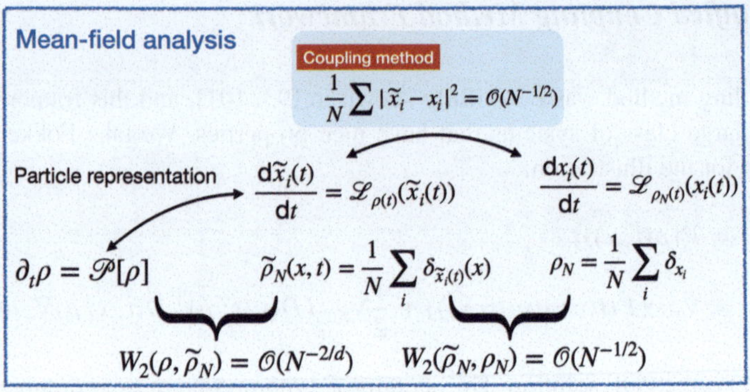

Fig. 2 The general routine of the mean-field analysis from coupling perspective. The auxiliary particles $\tilde{x}_i(t)$ evolve according to a coupled ODE/SDE with $\rho(t)$ generated from the mean-field limiting PDE. The coupling method shows that x_i is close to $\tilde{x}_i$, which implies that $\tilde{\rho}_N$ is close to ρ_N. This, combined with the fact that $\tilde{\rho}_N \approx \rho$ from [42], ultimately justifies $\rho_N \approx \rho$

The system looks almost identical to System B in (2.2). However, this new system deploys a self-generated ρ_N instead of relying on an external ρ and thus can be implemented numerically.

The mean-field derivation is to show that the solution of the underlying PDE (2.1), in System A, can be well approximated by the ensemble ρ_N in (2.5) in System C. This way, one translates the computation of the underlying PDE to the particle simulation in (2.4). This procedure is illustrated in Fig. 2.

The coupling method amounts to showing the similarity between System A and System C by equating both with the middle agent: System B. Indeed, according to [42], System A and B are already close:

$$\mathrm{dist}(\rho, \tilde{\rho}_N) \to 0\,, \quad \text{as} \quad N \to \infty\,, \tag{2.6}$$

for some proper definition of the dist function.[1] Most of the technical component of the proof goes to showing that System B and C are also close, meaning that, for all $0 \le t < \infty$,

$$\frac{1}{N}\sum_j \mathbb{E}\left|\tilde{x}_j - x_j\right|^2 \xrightarrow{N\to\infty} 0\,, \quad \Rightarrow \quad \mathrm{dist}(\tilde{\rho}_N, \rho_N) \xrightarrow{N\to\infty} 0\,. \tag{2.7}$$

If this can be proved, one can run the triangle inequality for:

[1] According to [42], the W_2-distance between the measures $\rho\mathrm{d}x$ and $\tilde{\rho}_N\mathrm{d}x$ converges to 0 when N approaches to infinity. We omit the symbol $\mathrm{d}x$ for simplicity of the notation when the context is clear.

$$\mathrm{dist}(\rho, \rho_N) \leq \mathrm{dist}(\rho, \tilde{\rho}_N) + \mathrm{dist}(\widetilde{\rho}_N, \rho_N) \to 0\,, \quad \text{as} \quad N \to \infty\,,$$

concluding the mean-field derivation.

To show (2.7), we note the similarity in form between (2.2) and (2.4). The two sets of SDEs are presented in almost identical form, and mathematically it is straightforward to compute the difference. A typical strategy is to subtract the two systems and evolve their differences.

$$\begin{aligned} \mathrm{d}\left(\tilde{x}_i(t) - x_i(t)\right) = &\left[F(t, \tilde{x}_i(t); \rho) - F(t, x_i(t); \rho_N)\right] \mathrm{d}t \\ &+ \left[D(t, \tilde{x}_i(t); \rho) - D(t, x_i(t); \rho_N)\right] \mathrm{d}W_t^i\,. \end{aligned} \tag{2.8}$$

Considering $x_i(t = 0) = \tilde{x}_i(t = 0)$, we need to show that the two terms on the right-hand side are controllable. This is indeed true when some proper assumptions are made. We now give a quick overview of the general recipe for doing so.

Deploying Itô's formula, we obtain

$$\begin{aligned} &\frac{\mathrm{d}}{\mathrm{d}t}\mathbb{E}\left(|\tilde{x}_i(t) - x_i(t)|^2\right) \\ =&2\,\underbrace{\mathbb{E}\left((\tilde{x}_i(t) - x_i(t))\left[F(t, \tilde{x}_i(t); \rho) - F(t, x_i(t); \rho_N)\right]\right)}_{\text{Term I}} \\ &+ \underbrace{\mathbb{E}\left(\|D(t, \tilde{x}_i(t); \rho) - D(t, x_i(t); \rho_N)\|_F^2\right)}_{\text{Term II}}\,. \end{aligned} \tag{2.9}$$

The cross terms vanish because $\mathbb{E}(\mathrm{d}W_t^i) = 0$. The two terms are dealt with separately:

- To deal with Term I, we first notice

$$\begin{aligned} &|F(t, \tilde{x}_i; \rho) - F(t, x_i; \rho_N)| \\ \leq &|F(t, \tilde{x}_i; \rho) - F(t, \tilde{x}_i; \widetilde{\rho}_N)| + |F(t, \tilde{x}_i; \widetilde{\rho}_N) - F(t, x_i; \rho_N)| \\ \leq &\underbrace{|F(t, \tilde{x}_i; \rho) - F(t, \tilde{x}_i; \widetilde{\rho}_N)|}_{\leq L W_2(\rho, \widetilde{\rho}_N)} + \underbrace{|F(t, \tilde{x}_i; \widetilde{\rho}_N) - F(t, x_i; \widetilde{\rho}_N)|}_{\leq L|\tilde{x}_i - x_i|} \\ &+ \underbrace{|F(t, x_i; \widetilde{\rho}_N) - F(t, x_i; \rho_N)|}_{\leq L W_2(\rho_N, \widetilde{\rho}_N)} \end{aligned} \tag{2.10}$$

  In this inequality, we repeatedly used the triangle inequality, and the last line assumes the Lipschitz condition of F on both ρ and x, with L denoting the Lipschitz constant over both arguments. Plugging (2.10) into Term I, we obtain

$$\text{Term I} \leq C\mathbb{E}\left(|\tilde{x}_i(t) - x_i(t)|^2 + W_2^2(\rho_N, \widetilde{\rho}_N) + W_2^2(\rho, \widetilde{\rho}_N)\right)\,, \tag{2.11}$$

where C is a constant depending on the Lipschitz constants, and Young's inequality is used.

- To deal with Term II, we use similar strategy. Assuming D is Lipschitz with respect to both ρ and x with constant L, we have

$$\begin{aligned}
&\|D(t,\tilde{x}_i;\rho)-D(t,x_i;\rho_N)\|_F \\
&\leq \|D(t,\tilde{x}_i;\rho)-D(t,\tilde{x}_i;\widetilde{\rho}_N)\|_F+\|D(t,\tilde{x}_i;\widetilde{\rho}_N)-D(t,x_i;\rho_N)\|_F \\
&\leq \underbrace{\|D(t,\tilde{x}_i;\rho)-D(t,\tilde{x}_i;\widetilde{\rho}_N)\|_F}_{\leq LW_2(\rho,\widetilde{\rho}_N)}+\underbrace{\|D(t,\tilde{x}_i;\widetilde{\rho}_N)-D(t,x_i;\widetilde{\rho}_N)\|_F}_{\leq L|\tilde{x}_i-x_i|} \\
&\quad+\underbrace{\|D(t,x_i;\widetilde{\rho}_N)-D(t,x_i;\rho_N)\|_F}_{\leq LW_2(\rho_N,\widetilde{\rho}_N)},
\end{aligned}$$

which in turn bound Term II as

$$\text{Term II} \leq C\mathbb{E}\left(|\tilde{x}_i(t)-x_i(t)|^2+W_2^2(\rho_N,\widetilde{\rho}_N)+W_2^2(\rho,\widetilde{\rho}_N)\right), \tag{2.12}$$

where C is a constant depending on the Lipschitz constants.

Plugging (2.11) and (2.12) in (2.9) and averaging over i, we derive:

$$\begin{aligned}
&\frac{\mathrm{d}}{\mathrm{d}t}\left(\frac{1}{N}\sum_{i=1}^{N}\mathbb{E}\left(|\tilde{x}_i(t)-x_i(t)|^2\right)\right) \\
\leq &C\left(\frac{1}{N}\sum_{i=1}^{N}\mathbb{E}\left(|\tilde{x}_i(t)-x_i(t)|^2\right)+\mathbb{E}\left(W_2^2(\rho_N,\widetilde{\rho}_N)\right)+\mathbb{E}\left(W_2^2(\rho,\widetilde{\rho}_N)\right)\right) \\
\leq &C'\left(\frac{1}{N}\sum_{i=1}^{N}\mathbb{E}\left(|\tilde{x}_i(t)-x_i(t)|^2\right)+\mathbb{E}\left(W_2^2(\rho,\widetilde{\rho}_N)\right)\right) \\
\leq &C'\left(\frac{1}{N}\sum_{i=1}^{N}\mathbb{E}\left(|\tilde{x}_i(t)-x_i(t)|^2\right)\right)+\mathcal{O}\left(N^{-\gamma}\right).
\end{aligned}$$

In the second inequality, we absorb $W_2^2(\rho_N,\widetilde{\rho}_N)$ because $W_2^2(\rho_N,\widetilde{\rho}_N) \leq \frac{1}{N}\sum_{i=1}^{N}\mathbb{E}\left(|\tilde{x}_i(t)-x_i(t)|^2\right)$, and in the last, we bounded $\mathbb{E}\left(W_2^2(\rho,\widetilde{\rho}_N)\right)$ with $\gamma > 0$ being a constant that depends on the dimension. This estimate comes from an off-the-shelf result [42] and will be presented in Lemma 2.3.

Finally, applying the Grönwall inequality, we conclude that $\tilde{x}_i(t)$ and $x_i(t)$ are close for all finite times when N is sufficiently large, thereby confirming (2.7).

The above intuitive derivation largely depends on the Lipschitz condition of the drift coefficients F and diffusion coefficients D. Unfortunately, most ensemble methods only have locally Lipschitz coefficients, and for such methods, care has to be given in calculations in different contexts. In particular, in the mean-field proof for EKI, a boot-strapping argument is called, and for EKS, the Ando-Hemmen lemma is used. The choice of these arguments is not unique. A new technique was recently introduced in a beautiful set of work in [97] that utilizes the concept of stopping time to enforce a Lipschitz condition before the stopping time, and this condition is matched together with the application of the Burkholder–Davis–Gundy inequality to pass to use of the Grönwall inequality. It seems that the framework is drastically more general than the bootstrapping technique, but more investigation is still needed to understand the mechanism. Nevertheless, the coupling method strategy remains the backbone of all these proofs.

Throughout the section, various statistical quantities appeared multiple times. We unify the notation here. For a given probability density $\rho(t,x)$, we define expectations

$$\mathbb{E}_{\rho_t}(x) = \int_{\mathbb{R}^d} x\rho(t,x)\mathrm{d}x, \quad \mathbb{E}_{\rho_t}(\mathcal{G}(x)) = \int_{\mathbb{R}^d} \mathcal{G}(x)\rho(t,x)\mathrm{d}x\,,$$

and the covariance matrices

$$\mathrm{Cov}_{\rho_t} = \int_{\mathbb{R}^d} \left(x - \mathbb{E}_{\rho_t}(x)\right) \otimes \left(x - \mathbb{E}_{\rho_t}(x)\right) \rho\mathrm{d}x\,, \tag{2.13}$$

$$\mathrm{Cov}_{\rho_t,\mathcal{G}} = \mathrm{Cov}^{\top}_{\mathcal{G},\rho_t} = \int_{\mathbb{R}^d} \left(x - \mathbb{E}_{\rho_t}(x)\right) \otimes \left(\mathcal{G}(x) - \mathbb{E}_{\rho_t}(\mathcal{G}(x))\right) \rho(t,x)\mathrm{d}x\,. \tag{2.14}$$

Similarly, for an ensemble of particles $\{x_n\}_{n=1}^N$, we define the empirical means and covariance matrices

$$\bar{x} = \frac{1}{N}\sum_{i=1}^N x_i, \quad \bar{\mathcal{G}} = \frac{1}{N}\sum_{i=1}^N \mathcal{G}(x_i)\,,$$

$$\mathrm{Cov}_{x,x} = \frac{1}{N}\sum_{i=1}^N (x_i - \bar{x}) \otimes (x_i - \bar{x})\,, \quad \text{and}$$

$$\mathrm{Cov}_{x,\mathcal{G}} = \frac{1}{N}\sum_{i=1}^N (x_i - \bar{x}) \otimes (\mathcal{G}(x_i - \bar{\mathcal{G}})\,. \tag{2.15}$$

2.2 *Ensemble Kalman Inversion*

This section reviews the EKI method [54] and summarizes its results of the mean-field analysis [27].

2.2.1 Algorithm

Ensemble Kalman Inversion (EKI) is an algorithm that designs a dynamics aiming at driving the prior distribution to the posterior distribution in a finite time horizon. More specifically, a PDE will be designed, so that its solution has the following form.

$$\rho(t,x) \propto \rho_{\text{prior}}(x) \exp\left(-t\Phi(x;y)\right) . \tag{2.16}$$

Recall Bayes' theorem (1.3); it is immediate that

$$\rho(t=0,x) = \rho_{\text{prior}}(x) , \quad \rho(t=1,x) = \rho_{\text{target}}(x) .$$

One proposal to achieve this is to implement the following PDE:

$$\begin{aligned}\partial_t \rho = \mathcal{L}[\rho] = & -\nabla_x \cdot \left((y - \mathcal{G}(x))^\top \Gamma^{-1} \text{Cov}_{\mathcal{G},\rho_t} \rho \right) \\ & + \frac{1}{2} \text{Tr} \left(\text{Cov}_{\rho_t,\mathcal{G}} \Gamma^{-1} \text{Cov}_{\mathcal{G},\rho_t} \mathcal{H}_x(\rho) \right)\end{aligned} \tag{2.17}$$

where $\mathcal{H}_x$ denotes the hessian in x.

Therefore, as a consequence, a Monte Carlo type solver that simulates (2.17) provides samples drawn from the target distribution ρ_{target}.

Remark 2.1 It is typically overlooked in the literature that in the most general setting, (2.16) does not solve (2.17) unless the forward map is linear. Namely, only when $\mathcal{G}(x) = Ax$ and thus $\Phi(x;y) = \frac{1}{2}\left(x^\dagger - x\right)^\top A^\top \Gamma^{-1} A(x^\dagger - x)$ where we denote $y = Ax^\dagger$, (2.16) solves (2.17), and thus the PDE solver generates samples from ρ_{target} at $t = 1$. For general nonlinear $\mathcal{G}$, this is only an approximation.

Indeed, as proved in [29], the correct PDE, besides having the $\mathcal{L}$ operator defined as in (2.17), should also make use of a correction term:

$$\partial_t \rho(t,x) = \mathcal{L}\left[\rho(t,x)\right] + \mathcal{R}(t,x;\rho)\rho(t,x) ,$$

where $\mathcal{R}(t,x;\rho)$ has a complicated nonlinear form that gets expanded out in [29]. This form strongly suggests that a sampling strategy should have all particles weighted with the weight adjusted according to $\mathcal{R}$.

Depending on the size of $\mathcal{R}$, EKI may produce samples that are $O(1)$ away from the target distribution ρ_{target}. It is still an open problem to estimate and control this weight term $\mathcal{R}$, but some discussions can be found in [29, 36, 62].

According to [77], the SDE representation of (2.17), the associated System B is:

$$\mathrm{d}\tilde{x}_i(t) = \text{Cov}_{\rho_t,\mathcal{G}} \Gamma^{-1} \left(y - \mathcal{G}(\tilde{x}_i(t))\right) \mathrm{d}t + \text{Cov}_{\rho_t,\mathcal{G}} \Gamma^{-\frac{1}{2}} \mathrm{d}W_t^i . \tag{2.18}$$

As discussed in Sect. 2.1, System B cannot be realized numerically. To turn it into an algorithm, the groundtruth covariance $\mathrm{Cov}_{\rho_t,\mathcal{G}}$ is replaced by the ensemble covariance matrix $\mathrm{Cov}_{x_t,\mathcal{G}_t}$, and we obtain the new system, the associated System C:

$$\mathrm{d}x_i(t) = \mathrm{Cov}_{x_t,\mathcal{G}_t}\Gamma^{-1}\left(y - \mathcal{G}(x_i(t))\right)\mathrm{d}t + \mathrm{Cov}_{x_t,\mathcal{G}_t}\Gamma^{-\frac{1}{2}}\mathrm{d}W_t^i\,, \tag{2.19}$$

where the covariance matrices are self-generated defined in (2.15).

Both System B and System C set the samples at the initial time $\{x_i(t=0)\}$ i.i.d. drawn from the prior distribution $\rho(t=0,\cdot) = \rho_{\text{prior}}$.

The algorithm of EKI is to further discretize system C using the Euler-Maruyama method with time step $h > 0$, as summarized in Algorithm 1. The algorithm is terminated at $M = 1/h$ time step to ensure that the system runs to the pseudo-time $t = 1$.

Algorithm 1 Ensemble Kalman Inversion

Preparation:

1. Input: $N \gg 1$; h (Stepsize); $M = 1/h$ (stopping index); Γ; and y (data).
2. Initial: $\{x_i^0\}_{i=1}^N$ sampled from initial distribution ρ_{ini}.

Run: Set time step $m = 0$;

While $m < M$:

1. Define empirical means and covariance:

$$\overline{x}^m = \frac{1}{N}\sum_{i=1}^N x_i^m\,,\ \overline{\mathcal{G}}^m = \frac{1}{N}\sum_{i=1}^N \mathcal{G}(x_i^m) \text{ and } \mathrm{Cov}_{x^m,\mathcal{G}^m} = \frac{1}{N}\sum_{i=1}^N (x_i^m - \overline{x}^m)\otimes\left(\mathcal{G}(x_i^m) - \overline{\mathcal{G}}^m\right).$$

2. Update ensemble particles ($\forall 1 \le i \le N$)

$$x_{*,i}^{m+1} = x_i^m + h\mathrm{Cov}_{x^m,\mathcal{G}^m}\Gamma^{-1}(y - \mathcal{G}(x_i^m))\,,$$

$$x_i^{m+1} = x_{*,i}^{m+1} + \sqrt{h}\mathrm{Cov}_{x^m,\mathcal{G}^m}\Gamma^{-1/2}\xi_i^m, \quad \text{with} \quad \xi_i^m \sim \mathcal{N}(0,\mathrm{I})\,.$$

3. Set $m \to m+1$

end

Output: Ensemble particles $\{x_i^M\}_{i=1}^N$.

2.2.2 Mean-Field Analysis for EKI

The mean-field derivation of EKI amounts to equating (2.19) with its underlying PDE (2.17). The following theorem summarizes the result.

Theorem 2.2 (Mean-Field Limit for EKI) *Suppose that $\mathcal{G}$ satisfies the weak nonlinear assumption.*[2] *Denote $\rho_N(t,x)$ the ensemble distribution of $\{x_i(t)\}$ as defined in* (2.5) *and $\rho(t,x)$ the weak solution to* (2.17) *with $\rho(0,x) = \rho_{\text{prior}}$, then $\rho_N \to \rho$ in the Wasserstein sense and the weak sense for all finite time t:*

- *For any $\epsilon > 0$, there is a constant $C_\epsilon(t)$ independent of N such that:*

$$\mathbb{E}\left(W_2(\rho_N(t,x), \rho(t,x))\right) \le C_\epsilon(t) \begin{cases} N^{-\frac{1}{4}}, & d \le 4 \\ N^{-1/d}, & d > 4 \end{cases} . \tag{2.20}$$

- *For any l-Lipschitz test function g, for any $\epsilon > 0$, there is a constant C_ϵ depending on ϵ, l and t and independent of N such that:*

$$\left(\mathbb{E}\left|\int g(x)\left[\rho_N(t,x)\mathrm{d}x - \rho(x,t)\mathrm{d}x\right]\right|^2\right)^{\frac{1}{2}} \le C_\epsilon N^{-\frac{1}{2}+\epsilon} . \tag{2.21}$$

We note the convergence is in expectation sense. Different realizations of Brownian motion provides different configuration of ρ_N. In expectation, ρ_N convergences to ρ, and the expectation sign takes expectation over the randomness from Brownian motions (2.5).

The theorem clearly suggests that the Wasserstein convergence rate sees the curse of dimensionality. For $d \ge 4$, the rate is slower than the classical Monte Carlo rate of $\frac{1}{\sqrt{N}}$. We believe that the rate is optimal within the current machinery; see [42]. On the other hand, the convergence in the weak sense (tested by a Lipschitz function) is always as strong as one would expect from an MC sampling. It should be noted that when $g(x) = x^p$ for any $p \in \mathbb{N}_+$, we expect convergence of the moments.[3]

The main strategy is already discussed in Sect. 2.1: the proof boils down to showing both (2.6) and (2.7), as presented below.

Firstly, we note that (2.6) is proved by calling [42, Theorem 1]. In the current context,

Lemma 2.3 *Let $\{\tilde{x}_i(t)\}$ solve* (2.18) *with $\rho(t,x)$ being the solution to* (2.17). *Further denote $\tilde{\rho}_N(t,x)$ the ensemble distribution of $\{\tilde{x}_i(t)\}$ as defined in* (2.3), *then, under the condition in Theorem 2.2, there is a constant $C(t)$ independent of N such that for all finite t:*

---

[2] For technical reasons, we can only prove the mean-field under the weak nonlinearity assumption. In particular, we assume that $\mathcal{G}$ is weak nonlinear, meaning that there is a matrix $A \in \mathbb{R}^{d\times d}$ such that $\mathcal{G}(x) = Ax + \mathrm{b}(x)$ where $\mathrm{b}(x) : \mathbb{R}^d \to \mathbb{R}^d$ is a smooth bounded function satisfying $\mathrm{Range}(\mathrm{b}) \perp_{\Gamma^{-1}} \mathrm{Range}(A)$, $|\mathrm{b}(x)| + |\nabla_x \mathrm{b}(x)| \le B$ with some constant $B > 0$ in $\mathbb{R}^d$, and $a \perp_{\Gamma^{-1}} b$ means $a^\top \Gamma^{-1} b = 0$ and $a^\top$ is to take transpose of a.

[3] $g(x) = x^p$ is not global Lipschitz, so a strategy that regularizes the test function through truncation is necessary to deploy (2.21). We leave out details.

$$\mathbb{E}\left(W_2^2(\widetilde{\rho}_N(t,x),\rho(t,x))\right) \le C(t)N^{-\min\{1/2,2/d\}}\,. \tag{2.22}$$

Now, we prove (2.7). It should be noted that the calculation in (2.10) does not apply directly: We do not have the Lipschitz continuity against ρ. In [27], a bootstrapping argument was deployed. To be more precise, we define the differences

$$\Delta_i(t) = x_i(t) - \tilde{x}_i(t)\,.$$

By subtracting the two Eqs. (2.18) and (2.19), a conditional convergence can be shown, namely:

Lemma 2.4 *If there is $\gamma \ge 0$ so that*

$$\mathbb{E}(|\Delta_i(t)|^2) \le N^{-\gamma}\,, \quad \forall i\,,$$

then for any $\epsilon > 0$, there is a constant C_ϵ so that

$$\mathbb{E}(|\Delta_i(t)|^2) \le C_\epsilon N^{-1/2-\gamma/2+\epsilon}\,, \quad \forall i\,.$$

Note that

$$\frac{1}{2}+\frac{\gamma}{2}-\epsilon > \gamma\,, \quad \text{for} \quad \gamma < \frac{1}{2}-\epsilon\,,$$

so conditioned on a lower convergence rate, the lemma can be deployed to push the convergence to a faster rate. To kick off this bootstrapping, we note that by setting $\gamma = 0$, one needs the boundedness of the second moment, which can be proved. Then Lemma 2.4 gets repeatedly deployed to finally push the results to:

$$\mathbb{E}(|\Delta_i(t)|^2) \le C_\epsilon N^{-1+\epsilon} \quad \Rightarrow \quad \mathbb{E}(W_2(\tilde{\rho}_N\,,\rho_N)) \le C_\epsilon N^{-1/2+\epsilon/2}\,, \tag{2.23}$$

for any $\epsilon > 0$.

Combining it with Lemma 2.3 and choosing $\epsilon = 1/2$ in (2.23), we obtain:

$$\begin{aligned}&\mathbb{E}\left(W_2(\rho_N,\rho)\right) \le \mathbb{E}\left(W_2(\rho_N,\tilde{\rho}_N)\right) + \mathbb{E}\left(W_2(\tilde{\rho}_N,\rho)\right)\\ \le&\mathbb{E}\left(W_2(\rho_N,\tilde{\rho}_N)\right) + \left(\mathbb{E}\left(W_2^2(\tilde{\rho}_N,\rho)\right)\right)^{1/2} \le C\begin{cases} N^{-\frac{1}{4}}, & d\le 4\\ N^{-1/d}, & d>4\end{cases},\end{aligned}$$

concluding (2.20).

The weak convergence in (2.21) is a classical practice. First, noting the triangle inequality, we have

$$
\begin{aligned}
&\left(\mathbb{E}\left|\int g(x)\left[\rho_N \mathrm{d}x-\rho \mathrm{d}x\right]\right|^2\right)^{\frac{1}{2}} \\
\leq &\left(\mathbb{E}\left|\int g(x)\left[\rho_N \mathrm{d}x-\tilde{\rho}_N \mathrm{d}x\right]\right|^2\right)^{\frac{1}{2}}+\left(\mathbb{E}\left|\int g(x)\left[\tilde{\rho}_N \mathrm{d}x-\rho \mathrm{d}x\right]\right|^2\right)^{\frac{1}{2}} .
\end{aligned}
\tag{2.24}
$$

These two terms can be controlled separately. To control the first term, we note

$$
\begin{aligned}
\mathbb{E}\left|\int g(x)\left[\rho_N \mathrm{d}x-\tilde{\rho}_N \mathrm{d}x\right]\right|^2 &= \mathbb{E}\left|\frac{1}{N}\sum_{i=1}^N g(x_i(t))-g(\tilde{x}_i(t))\right|^2 \\
\leq &\frac{l^2}{N^2}\mathbb{E}\left(\sum_{i=1}^N |x_i(t)-\tilde{x}_i(t)|^2\right) \leq C_\epsilon l^2 N^{-1+\epsilon} ,
\end{aligned}
\tag{2.25}
$$

where the l-Lipschitz and Hölder's inequality are used in the first inequality and (2.23) is used in the second inequality. The control of the second term is essentially the law of large numbers:

$$
\mathbb{E}\left|\int g(x)\left[\tilde{\rho}_N \mathrm{d}x-\rho(t,x)\mathrm{d}x\right]\right|^2 = \mathbb{E}\left|\frac{1}{N}\sum_{i=1}^N g(\tilde{x}_i(t))-\mathbb{E}_{\rho_t}(g)\right|^2 = \frac{1}{N}\mathrm{Var}_\rho[g]
$$

where $\mathrm{Var}_{\rho_t}[g] = \mathbb{E}_\rho[g^2] - \left(\mathbb{E}_\rho[g]\right)^2$. The term is bounded due to the l-Lipschitz property of g.

$$
\mathbb{E}\left|\int g(x)\left[\widetilde{\rho}_N \mathrm{d}x-\rho(t,x)\mathrm{d}x\right]\right|^2 \leq C\,(l)\,N^{-1} . \tag{2.26}
$$

Combining the two terms into (2.24), (2.21) is proved.

2.3 *Ensemble Kalman Sampler*

This section is dedicated to the mean-field analysis of the algorithm Ensemble Kalman sampler (EKS). As for EKI, we first present the algorithm and then show the application of the procedure as demonstrated in Sect. 2.1.

2.3.1 Algorithm

EKS was proposed in [44] and to some extent can be viewed as a modification to the overdamped Langevin dynamics. Unlike Langevin, which induces a single-particle MCMC type algorithm, EKS introduces the communications between particles and induces an ensemble method.

There are many approaches to justify the success of the overdamped Langevin. One approach is to view it as a particle method for the Fokker-Planck equation, and the beautiful discovery [57] reformulates it as the gradient flow of relative entropy. More specifically, the Fokker-Planck PDE writes as:

$$\frac{\mathrm{d}\rho}{\mathrm{d}t} = \nabla_x \cdot (\nabla_x f(x;y)\rho + \nabla_x \rho) = -\nabla_{W_2}\mathsf{KL}(\rho|\rho_{\text{target}})\,. \tag{2.27}$$

Here $\mathsf{KL}(\mu|\nu) = \int \log(\mathrm{d}\mu/\mathrm{d}\nu)\mathrm{d}\mu$ is termed relative entropy, or sometimes also called the Kullback-Leibler divergence. In the equation, $\nabla_{W_2}\mathsf{KL}$ stands for the gradient of ρ of the functional KL against the Wasserstein metric. Since it is a gradient flow-type PDE, the KL monotonically decreases in time. Moreover, under proper assumptions on the function $f(x;y)$ (e.g., log-concave), when $t \to \infty$, $\rho(t,x)$ converges to $\rho_{\text{target}} \propto \exp(-f(x;y))$ [72].

On the discrete level, to simulate the equation, one can deploy the particle method. It is then run according to the following Langevin dynamics:

$$\mathrm{d}x(t) = -\nabla_x f(x;y)\mathrm{d}t + \sqrt{2}\mathrm{d}W_t\,. \tag{2.28}$$

As a natural consequence, with long enough time, the particle produced by (2.28) can be seen as a sample drawn from the target distribution ρ_{target}. This classical sampling algorithm is called overdamped Langevin Monte Carlo [80, 84, 85].

The Ensemble Kalman Sampler (EKS) can be seen as an ensemble modification of (2.28). Indeed, it rewrites (2.27) into the following form:

$$\partial_t \rho = \nabla \cdot (\rho \mathrm{Cov}_{\rho_t} \nabla_x f(x;y)) + \mathrm{Tr}\left(\mathrm{Cov}_{\rho_t}\mathcal{H}_x \rho\right)\,, \tag{2.29}$$

where Cov_{ρ_t} is defined in (2.13).

One can further formulate it as a gradient flow over a specially designed metric [44, Section 3.3] to analytically quickly obtain convergence in time. We omit the details here.

The stark difference between this modified Fokker-Planck equation (2.29) and the classical Fokker-Planck equation (2.27) is that this new equation is nonlinear. When discretized using particle methods, one needs to differentiate System B and System C, as discussed in Sect. 2.1. System B is a collection of particles that are passively pushed by the underlying flow and in this context becomes:

$$\mathrm{d}\tilde{x}_i = -\mathrm{Cov}_{\rho_t}\nabla_x f(\tilde{x}_i;y)\mathrm{d}t + \sqrt{2\mathrm{Cov}_{\rho_t}}\mathrm{d}W_t^i\,.$$

In the Bayesian setting, $f(x; y)$ adopts the form in (1.2), and the formulation rewrites as:

$$d\tilde{x}_i = -\mathrm{Cov}_{\rho_t} \nabla_x \mathcal{G}(x_i)\Gamma^{-1} (\mathcal{G}(\tilde{x}_i(t)) - y)\, dt - \mathrm{Cov}_{\rho_t} \Gamma_0^{-1} (\tilde{x}_i(t) - x_0)\, dt + \sqrt{2\mathrm{Cov}_{\rho_t}} \mathrm{d}W_t^i\,. \tag{2.30}$$

It is clear that the covariance in the above formulation is the groundtruth covariance computed from the underlying flow ρ and is not available, so in real computation, the ensemble covariance is deployed as an approximation. This leads to System C:

$$dx_i = -\mathrm{Cov}_{x_t,x_t} \nabla_x \mathcal{G}(x_i)\Gamma^{-1} (\mathcal{G}(x_i(t)) - y)\, dt - \mathrm{Cov}_{x_t,x_t} \Gamma_0^{-1} (x_i(t) - x_0)\, dt + \sqrt{2\mathrm{Cov}_{x_t,x_t}} \mathrm{d}W_t^i\,. \tag{2.31}$$

where Cov_{x_t,x_t} is defined in (2.15).

There are two nice features of this modified Fokker-Planck equation and the ensemble discretization pair, in comparison to the original Fokker-Planck equation and the Langevin dynamics pair. As discovered in [13], the modified Fokker-Planck (2.29) is more stable against the conditioning of the landscape f. More specifically, when ρ_{target} is a Gaussian distribution, the convergence of $\rho(t)$ to ρ_{target} in the Wasserstein-2 distance is exponential, and the rate of convergence is independent of the conditioning of the target Gaussian.

The second advantage of deploying this new equation is that, when the forward map $\mathcal{G}$ is almost linear, we are in a very good position to transition the problem into a gradient-free formulation. More specifically, by formulating

$$\mathrm{Cov}_{\rho_t} \nabla_x \mathcal{G}(x(t)) \quad \rightarrow \quad \mathrm{Cov}_{\rho_t,\mathcal{G}_t}\,,$$

then System B as in written in (2.30) takes a gradient-free form:

$$d\tilde{x}_i = -\mathrm{Cov}_{\rho_t,\mathcal{G}_t} \Gamma^{-1} (\mathcal{G}(\tilde{x}_i) - y) - \mathrm{Cov}_{\rho_t} \Gamma_0^{-1} (\tilde{x}_i - x_0)\, dt + \sqrt{2\mathrm{Cov}_{\rho_t}} \mathrm{d}W_t^i\,, \tag{2.32}$$

and System C, as in (2.31), becomes:

$$dx_i = -\mathrm{Cov}_{x_t,\mathcal{G}_t} \Gamma^{-1} (\mathcal{G}(x_i) - y) dt - \mathrm{Cov}_{x_t,x_t} \Gamma_0^{-1} (x_i(t) - x_0) dt + \sqrt{2\mathrm{Cov}_{x_t,x_t}} \mathrm{d}W_t^i\,, \tag{2.33}$$

where $\mathrm{Cov}_{x_t,\mathcal{G}_t}$ is the ensemble covariance between $\{x_i(t)\}$ and $\{\mathcal{G}(x_i(t))\}$ defined in (2.15). Note that the computation does not require the access of $\nabla\mathcal{G}$, a desirable feature in many scientific domains [31]. We summarize in Algorithm 2 the numerical implementation of Eq. (2.33).

For mathematical justification of the algorithm, we provide the mean-field analysis in the theorem below.

Algorithm 2 Ensemble Kalman sampling

Preparation:

1. Input: $N \gg 1$; h (stepsize); M (stopping index); Γ; Γ_0; and y (data).
2. Initial: $\{x_i^0\}_{i=1}^N$ sampled from an initial distribution ρ_{ini}.

Run: Set time step $m = 0$;

While $m < M$:

1. Define empirical means and covariance:

$$\overline{x}^m = \frac{1}{N}\sum_{i=1}^N x_i^m\,, \text{ and } \overline{\mathcal{G}}^m = \frac{1}{N}\sum_{i=1}^N \mathcal{G}(x_i^m)\,,$$

$$\mathrm{Cov}_{x^m,x^m} = \frac{1}{N}\sum_{i=1}^N \left(x_i^m - \overline{x}^m\right)\otimes\left(x_i^m - \overline{x}^m\right) \text{ and } \mathrm{Cov}_{x^m,\mathcal{G}^m} = \frac{1}{N}\sum_{i=1}^N \left(x_i^m - \overline{x}^m\right)$$
$$\otimes\left(\mathcal{G}(x_i^m) - \overline{\mathcal{G}}^m\right)\,.$$

2. Update ensemble particles ($\forall 1 \le i \le N$)

$$x_{*,i}^{m+1} = x_i^m - h\mathrm{Cov}_{x^m,\mathcal{G}^m}\Gamma^{-1}\left(\mathcal{G}(x_i^m) - y\right) - h\mathrm{Cov}_{x^m,x^m}\Gamma_0^{-1}\left(x_{*,i}^{m+1} - x_0\right)\,,$$

$$x_n^{m+1} = x_{*,i}^{m+1} + \sqrt{2h\mathrm{Cov}_{x^m,x^m}}\xi_i^m\,, \quad \text{with} \quad \xi_i^m \sim \mathcal{N}(0,\mathrm{I})\,.$$

3. Set $m \to m+1$.

end

Output: Ensemble particles $\{x_i^M\}_{i=1}^N$.

Theorem 2.5 (Mean-Field Limit of EKS) *Suppose $\mathcal{G}(x) = Ax$ with $A \in \mathbb{R}^{d\times d}$. Let $\rho(t,x)$ be the weak solution to the Fokker-Planck equation and $\{x_i(t)\}$ be a solution to* (2.33) *with $x_i(0)$ i.i.d. drawn from the distribution ρ_{prior}. Assume:*

$$\lambda_{\min}\left(\mathrm{Cov}_{\rho_t}\right) > \lambda_{\max}\left(\mathrm{Cov}_{\rho_{target}}\right), \quad \forall 0 \le t \le T\,, \tag{2.34}$$

then for any $0 < \epsilon < 1/2$, there exits C, depending on T and ϵ, but independent of N such that

$$\mathbb{E}\left(W_2(\rho_N(t,x),\rho(t,x))\right) \le C\begin{cases} N^{-\frac{1}{4}}, & d \le 4 \\ N^{-1/d}, & d > 4 \end{cases}.$$

In addition, given any l-Lipschitz function g, for any $\epsilon > 0$, there is a constant C_ϵ such that for any $t < \infty$

$$\left(\mathbb{E}\left|\int g(x)\left[\rho_N(t,x)\mathrm{d}x - \rho(t,x)\mathrm{d}x\right]\right|^2\right)^{\frac{1}{2}} \le C_\epsilon N^{-\frac{1}{2}+\epsilon}\,.$$

where the constant C_ϵ depends on l, ϵ and time t and is independent of N

The proof of the above theorem is similar to that of Theorems 2.2, so we will not include it here. We refer interested readers to [28].

We note that the assumption (2.34) is for a technical purpose that we hope to be able to remove. In the proof, we deploy the Ando-Hemmen inequality that studies the differences between two matrices when square roots are taken. As long as some other reasonable estimates, which avoid direct use of Ando-Hemmen, can be found to control the sensitivity of taking square roots of matrices, we see the possibility of removing this assumption. In addition, when $\mathcal{G}$ is weakly nonlinear, we anticipate that convergence to the mean-field limit is still valid.

3 Boltzmann Simulator

We switch gears to a new set of sampling strategy called the Boltzmann simulator. In contrast to global interactions, the Boltzmann simulator deploys local interactions, drawing inspiration from the Vlasov-Boltzmann equation.

To facilitate our discussion, we begin by providing an overview of the Vlasov-Boltzmann equation in Sect. 3.1. This gives us the motivation to design sampling algorithms. Following this, in Sect. 3.2, we propose two sampling methods derived from the Vlasov-Boltzmann equation, namely, the Nanbu simulator and the Bird simulator. The mean-field justification of these two methods will then be discussed in Sect. 3.3. Numerical results are presented in Sect. 3.4.

3.1 Vlasov-Boltzmann and Its Properties

The Boltzmann simulator should be viewed as a particle method for the Vlasov-Boltzmann equation. In the domain $(x, v) \in \mathbb{R}^{2d}$:

$$\partial_t \mu_t(x, v) + \mathcal{L}\mu_t(x, v) = \mathcal{Q}\mu_t(x, v)\,, \tag{3.1}$$

where the transport operator $\mathcal{L}$ is defined by

$$\mathcal{L}\mu_t(x, v) = v \cdot \nabla_x \mu_t(x, v) - \nabla_x f(x) \cdot \nabla_v \mu_t(x, v)\,, \tag{3.2}$$

and the collision operator $\mathcal{Q}$ is defined by

$$\mathcal{Q}\mu_t(x, v) = \iint_{\mathbb{S}^{d-1}\times\mathbb{R}^d} q(v, w, n)(\mu_t(x, v^*)\mu_t(x, w^*) - \mu_t(x, v)\mu_t(x, w))\, \mathrm{d}n\mathrm{d}w\,. \tag{3.3}$$

The transport operator $\mathcal{L}$ is the Vlasov component of the equation driven by the Hamiltonian:

$$\dot{x} = v\,, \qquad \dot{v} = -\nabla_x f \tag{3.4}$$

where $f(x)$ is a given potential function. The Boltzmann collision operator $\mathcal{Q}$ describes the interaction between two particles located at the same position x and having pre-collisional velocities v and w. During this interaction, they exchange momentum and energy, resulting in new velocities v^* and w^* after the collision. The vector n is a unit vector that describes in which direction the particles are scattered off to:

$$v^* = v + ((w-v)\cdot n)n\,, \qquad w^* = w + ((v-w)\cdot n)n\,. \tag{3.5}$$

It can be verified that the relative velocity before and after collision is unchanged:

$$|v-w| = |v^* - w^*|\,. \tag{3.6}$$

The term $q(v, w, n)$ is called the cross-section. It quantifies the likelihood of such interactions. We have assumed that the medium is homogeneous, meaning q is independent of x, and only depends on $|v-w|$ and $(v-w)\cdot n$. This assumption is useful when we derive the weak form PDE in Sect. 3.3.1. We usually say that the Boltzmann operator is local because all particle interactions take place when their spatial coordinates agree—thus local in x.

This classical local Boltzmann collision operator in Eq. (3.1) can sometimes be relaxed to allow particles at different spatial locations to interact as well. We consider the mollified Vlasov-Boltzmann equation [51, 52]

$$\partial_t \mu_t(x, v) + \mathcal{L}\mu_t(x, v) = \widetilde{\mathcal{Q}}\mu_t(x, v)\,, \tag{3.7}$$

where the collision operator

$$\begin{aligned}\widetilde{\mathcal{Q}}\mu_t(x, v) = \iint_{\mathbb{S}^{d-1}\times\mathbb{R}^{2d}} & \widetilde{q}(x, v, y, w, n)(\mu_t(x, v^*)\mu_t(y, w^*) \\ & - \mu_t(x, v)\mu_t(y, w))\,\mathrm{d}n\mathrm{d}w\mathrm{d}y\,. \end{aligned} \tag{3.8}$$

The new cross-section $\widetilde{q}(x, v, y, w, n) = I(|x-y|)q(v, w, n)$ is delocalized, with $I(|x-y|)$ serving as a mollifier. It is nonnegative and normalized, i.e., $\int_{\mathbb{R}^d} I(|x|)\mathrm{d}x = 1$. This equation allows a particle at location x with velocity v to interact with a particle at location y with velocity w to obtain new velocities v^* and w^*. The intensity of such interaction is upper bounded by I. Note that the classical Boltzmann (3.1) can be recovered if $I = \delta$.

The equation can also be interpreted with a probabilistic viewpoint, dating back to Tanaka [95, 96], where he connected the solution of the Boltzmann equation to a Piecewise Deterministic Markov Process (PDMP). In this context, we view the solution of the Boltzmann equation as the distribution of particles that evolve according to the force field $-\nabla f(x)$. Each particle is equipped with a Poisson clock, and when the clock rings, the particle, with coordinate (x, v), chooses another particle (y, w) to interact with. The particle being chosen is selected at random from an ocean of infinitely many independent particles distributed according to the same distribution. The particle (x, v) collides with the particle (y, w) by randomly sampling a normal direction $n \sim \widetilde{q}(x, v, y, w, n)dn$. After the collision, the particle (x, v) adopts a new coordinate (x, v^*) and forgets all knowledge about the particle (y, w). It then continues to evolve under the force field. For mathematical convenience, it is further assumed the cross-section is bounded:

Assumption 3.1 (Boundedness of Cross-Section) There exists $\Lambda > 0$ such that

$$\sup_{x,y,v,w\in\mathbb{R}^d} \int_{\mathbb{S}^{d-1}} \widetilde{q}(x, v, y, w, n)dn \leq \Lambda < \infty . \tag{3.9}$$

According to the PDE analysis perspective, there are two nice features of this modified Boltzmann equation.

- Equilibrium state is the target distribution. As will be shown in Theorem 3.2, the equilibrium of the equation is the following:

$$\mu^* \propto \exp\left(-f(x) - \frac{1}{2}|v|^2\right), \tag{3.10}$$

  meaning its x-marginal distribution truly recovers the target distribution (1.1).
- Long-time limit convergences to the equilibrium. With proper conditions, as discussed in Theorem 3.3, one can also show that, in a weighted L_∞-norm, the PDE drives the initial distribution to the equilibrium state (3.10).

The goal of sampling is to design a dynamics whose invariant measure is the target distribution, so that samples following this dynamics, in long time, can be viewed as a drawing from the target distribution. The two features above suggest that the dynamics provided by the Vlasov-Boltzmann system (3.7) truly drives some arbitrary initial data to the target distribution and thus is a feasible choice. If one can design a particle method to simulate (3.7), the particles, in the long time limit, can be viewed as samples drawn from the target distribution.

We should also note that under the boundedness assumption of the cross-section, the well-posedness of the mollified Vlasov-Boltzmann equation is known [50], so the existence and uniqueness of the solution are not a concern.

We now present the mathematically rigorous description of the two features discussed above. Firstly, the target distribution is an invariant measure of the Vlasov-Boltzmann equation.

Theorem 3.2 *The mollified Vlasov-Boltzmann equation* (3.7) *admits an equilibrium state* μ^* *defined in* (3.10).

Proof To show the theorem, we are to plug in μ^* into the equation and demonstrate that both the left-hand side and the right-hand side vanishes.

The left-hand side transport term vanishes can be shown by direct computation:

$$v \cdot \nabla_x \mu^*(x, v) - \nabla_x f(x) \cdot \nabla_v \mu^*(x, v) = (-v \cdot \nabla_x f(x))\mu^*(x, v) \\ + (\nabla_x f(x) \cdot v)\mu^*(x, v) = 0\,. \quad (3.11)$$

The right-hand side collision term also vanishes. From the conservation of energy upon collision:

$$|v^*|^2 + |w^*|^2 = |v|^2 + |w|^2\,, \quad (3.12)$$

then

$$\iint_{\mathbb{S}^{d-1}\times\mathbb{R}^{2d}} \tilde{q}(x, v, y, w, n)(\mu^*(x, v^*)\mu^*(y, w^*) - \mu^*(x, v)\mu^*(y, w))\,\mathrm{d}n\mathrm{d}w\mathrm{d}y$$
$$\propto \iint_{\mathbb{S}^{d-1}\times\mathbb{R}^{2d}} \tilde{q}(x, v, y, w, n)\exp(-f(x) - f(y)) \\ \left[\exp(-|v^*|^2 - |w^*|^2) - \exp(-|v|^2 - |w|^2)\right]\mathrm{d}n\mathrm{d}w\mathrm{d}y = 0\,. \quad (3.13)$$

□

The stability of the equilibrium state μ^* to the Vlasov-Boltzmann equation (3.1) has also been studied, see [60, 65]. In particular, it is proved in [60] that for an potential function f with degenerate coordinate dependence, an initial state converges to μ^* exponentially in a weighted-L_∞ norm:

Theorem 3.3 (Theorem 1 in [60]) *Assume the potential function $f(x)$ has degenerate coordinate dependence, that is, there exists an $0 < n < d$, so that $f(x) = f(x_{n+1}, \ldots, x_d)$ for any $x = (x_1, \ldots, x_d)$. Assume the initial condition $\mu(0, x, v) = \mu_0(x, v)$ satisfies*

$$\int_{\mathbb{R}^{2d}} \mu_0(x, v) - \mu^*(x, v)\mathrm{d}x\mathrm{d}v = 0\,, \quad \textit{(Mass condition)}$$
$$\int_{\mathbb{R}^{2d}} \left(\frac{1}{2}|v|^2 + f(x)\right)(\mu_0(x, v) - \mu^*(x, v))\mathrm{d}x\mathrm{d}v = 0\,, \quad \textit{(Energy condition)}$$
$$\int_{\mathbb{R}^{2d}} v_i(\mu_0(x, v) - \mu^*(x, v))\mathrm{d}x\mathrm{d}v{=}0\,, \quad i{=}1, \ldots, n, \quad \textit{(Momentum condition)}\,, \quad (3.14)$$

and is close enough to the equilibrium state $\mu^*(x, v)$. Then under mild regularity conditions and periodic boundary conditions, the Vlasov-Boltzmann equation has a unique global solution $\mu_t(x, v)$. Furthermore, there exists a constant $\lambda = \lambda(q, \|f\|_\infty) > 0$ such that

$$\|\Phi \cdot (\mu_t - \mu^*)\|_\infty \le \mathcal{O}(e^{-\lambda t}), \quad \forall t \ge 0, \tag{3.15}$$

with the weight function $\Phi(x, v) = \left(\frac{1}{2}|v|^2 + f(x, v)\right)^{\beta/2}$ *for* $\beta > d/2$.

Remark 3.4 Before we dive in to use it as a sampling strategy, a couple of comments are in place.

- The convergence theorem has no requirement on the convexity of the potential, and hence it holds promise for applications in non-logconcave sampling problems. Indeed, in our numerical section, we will demonstrate one example where we have a double-well potential.
- There is still one step before turning it directly into a sampling method. To be more precise, Theorem 3.3 states the exponential convergence to the classical Vlasov-Boltzmann (3.1). However, this equation cannot be simulated using the particle method. The particle methods can only simulate the modified Boltzmann equation (3.7). We do not yet have a convergence-in-time result that shows the solution converging to the equilibrium state for the modified Boltzmann equation.
- The mathematical reason underpinning the Boltzmann simulator is the fact that the Bhatnagar-Gross-Krook (BGK) operator is typically deployed as an approximation to the collision operator. The linearized Vlasov-BGK equation writes:

  $$\partial_t \mu_t(x, v) + v \cdot \nabla_x \mu_t(x, v) - \nabla_x f(x) \cdot \nabla_v \mu_t(x, v) = \rho_t M - \mu_t$$

  where $M \propto e^{-|v|^2/2}$ is the normalized Maxwellian, and $\rho_t(x) = \int \mu_t(x, v)\mathrm{d}v$ is the density. The particle simulation of this equation is to have the particle running according to the Hamiltonian flow (3.4) and scatters off with a new velocity drawn according to Gaussian M upon a Poison clock. This formulation corresponds exactly to the randomized Hamiltonian Monte Carlo algorithm [9, 68].

These nice features suggest that the particle simulator of the Vlasov-Boltzmann equation should give a good sampling method.

3.2 Nanbu and Bird Algorithm

Both the Nanbu sampler and the Bird sampler are termed Direct Simulation Monte Carlo (DSMC) methods to simulate the mollified Vlasov-Boltzmann equation (3.7), see [79].

According to the probabilistic interpretation mentioned in Sect. 3.1, the Poisson clock determines the time for the particle at hand to "collide" with peers. However, this interpretation does not say what happens to the collision partner. Depending on what happens to this collision partner, different algorithms have been developed by Nanbu [76] and Bird [6].

We first introduce the Bird Algorithm. To start, we generated several samples $\{(x_i(0), v_i(0))\}_{i=1}^N$ at the initial time. For each ordered pair (i, j), we attach an independent Poisson process with rate Λ/N. Between two jump times, the particles evolve independently according to the free transport (3.4). When the clock of the ordered pair (i, j) rings at time T_{ij}, the state of the pair (i, j) is reevaluated. There are two possibilities:

- With probability $\Delta_{ij} = \Lambda^{-1} \int_{\mathbb{S}^{d-1}} \widetilde{q}(x_i(T_{ij}), v_i(T_{ij}), x_j(T_{ij}), v_j(T_{ij}), n)\mathrm{d}n$, the pair is updated. They adopt new velocities according to (3.5) with n drawn from:

$$n \sim \Lambda^{-1}\widetilde{q}(x_i(T_{ij}), v_i(T_{ij}), x_j(T_{ij}), v_j(T_{ij}), n)\mathrm{d}n\,.$$

- With probability $1 - \Delta_{ij}$, the state of the (i, j) pair is unchanged.

We summarize the Bird method in Algorithm 3.

Algorithm 3 Bird sampling method

Preparation:
1. Input: $N \gg 1$; T (terminal time);Λ (upper bound for total cross-section); $f(x)$.
2. Initial: $\{(x_i(0), v_i(0))\}_{i=1}^N$ sampled from an initial distribution μ_{ini}.

Run: Set system time $t = 0$; Sample initial collision time $t_{ij} \sim \mathrm{Exp}(\Lambda/N)$, $i, j = 1, \ldots, N$.
While $t < T$:
1. Pick $\tau_0 = \min_{i,j=1,\ldots,N} t_{ij}$ and $(i_c, j_c) = \operatorname{argmin}_{i,j=1,\ldots,N} t_{ij}$; set $\tau_1 = \tau_0 \wedge T$.
2. Evolve over $[t, \tau_1]$ Hamilton's equations

$$\dot{x}_i = v_i\,, \quad \dot{v}_i = -\nabla_x f(x_i)\,, \quad i = 1, \ldots, N$$

with initial condition $(x_i(t), v_i(t))$.
3. If $\tau_0 \leq T$, implement the collision stage; else jump to Step 4. Sample a number a uniformly over $[0, 1]$. If

$$a \leq \frac{1}{\Lambda} \int_{\mathbb{S}^{d-1}} \widetilde{q}(x_i(\tau_1), v_i(\tau_1), x_j(\tau_1), v_j(\tau_1), n)\mathrm{d}n\,,$$

then perform collision: sample the normal direction $n \in \mathbb{S}^{d-1}$ by

$$n \sim \widetilde{q}(x_{i_c}(\tau_1), v_{i_c}(\tau_1), x_j(\tau_1), v_j(\tau_1), n)\mathrm{d}n\,,$$

then update the velocity of particle i_c and j_c according to the rule

$$v_{i_c} \leftarrow v_{i_c} + ((v_{j_c} - v_{i_c}) \cdot n)n$$
$$v_{j_c} \leftarrow v_{j_c} + ((v_{i_c} - v_{j_c}) \cdot n)n\,;$$

else jump to Step 4.
4. Update system time $t \leftarrow \tau_1$ and update the next collision time of the pair (i_c, j_c) by $t_{i_c j_c} \leftarrow t_{i_c j_c} + \delta t$ with $\delta t \sim \mathrm{Exp}(\Lambda/N)$.

end
Output: Ensemble particles $\{(x_i(T), v_i(T))\}_{i=1}^N$.

The distinctive feature of the Nanbu method, compared to the Bird method, is that its Poisson clock is attached to a single particle, and it more views each particle to interact with a "mean-field" ocean. Each particle has an independent Poisson clock and evolves according to the Hamiltonian dynamics (3.4) until the clock rings. When one clock rings, the associated particle finds, uniformly among its peers, another particle with which to collide with according to the rate described in $\widetilde{q}$. However, only the specific particle whose clock rings undergoes an update.

We summarize the Nanbu method in Algorithm 4.

Algorithm 4 Nanbu sampling method

Preparation:
1. Input: $N \gg 1$; T (terminal time);Λ (upper bound for total cross-section); $f(x)$.
2. Initial: $\{(x_i(0), v_i(0))\}_{i=1}^N$ sampled from an initial distribution μ_{ini}.

Run: Set system time $t = 0$; Sample initial collision time $t_i \sim \text{Exp}(\Lambda)$, $i = 1, \ldots, N$.

While $t < T$:

1. Pick $\tau_0 = \min_{i=1,\ldots,N} t_i$ and $i_c = \operatorname{argmin}_{i=1,\ldots,N} t_i$; set $\tau_1 = \tau_0 \wedge T$.
2. Evolve over $[t, \tau_1]$ Hamilton's equations

$$\dot{x}_i = v_i\,, \quad \dot{v}_i = -\nabla_x f(x_i)\,, \quad i = 1, \ldots, N$$

with initial condition $(x_i(t), v_i(t))$.

3. If $\tau_0 \leq T$, implement the collision stage; else jump to Step 4. Sample a number a uniformly over $[0, 1]$. If

$$a \leq \frac{1}{N\Lambda} \sum_{j=1}^{N} \int_{\mathbb{S}^{d-1}} \widetilde{q}(x_i(\tau_1), v_i(\tau_1), x_j(\tau_1), v_j(\tau_1), n)\mathrm{d}n\,,$$

then perform collision: uniformly sample an index $j \in \{1, \ldots, N\}$, and sample $n \in \mathbb{S}^{d-1}$ according to the density $\widetilde{q}(x_{i_c}(\tau_1), v_{i_c}(\tau_1), x_j(\tau_1), v_j(\tau_1), n)\mathrm{d}n$, then update the velocity of particle i_c according to the rule

$$v_{i_c} \leftarrow v_{i_c} + ((v_j - v_{i_c}) \cdot n)n\,;$$

else jump to Step 4.

4. Update system time $t \leftarrow \tau_1$, and update the next collision time of particle i_c by $t_{i_c} \leftarrow t_{i_c} + \delta t$ with $\delta t \sim \text{Exp}(\Lambda)$.

end

Output: Ensemble particles $\{(x_i(T), v_i(T))\}_{i=1}^N$.

3.3 *Mean-Field Analysis for Boltzmann Simulator*

To provide the justification of these methods simulating the Boltzmann equation, one recognizes the algorithm calls for interactions between samples. The procedure naturally invites the theory of the mean-field derivation. We dedicate this section to provide a mean-field proof for the Nanbu and the Bird method.

It is clear that the coupling method of Sect. 2 can hardly be used. We switch our strategy and adopt the compactness argument for the associated martingale problems [18, 51, 52, 92]. In Sect. 3.3.1, we lay out the weak formulation of the Vlasov-Boltzmann equation and introduce the associated martingale problem. In Sect. 3.3.2, we follow a similar approach to introduce the weak formulation and the martingale problem for the two particle methods, the Nanbu sampler and the Bird sampler. A mean-field proof of the two particle methods is discussed in Sect. 3.3.3.

3.3.1 Weak Formulation and the Martingale Problem

Throughout the section, we assume that the PDE solution is absolutely continuous, so we rewrite $\mu_t(\mathrm{d}x, \mathrm{d}v)$ as $\mu_t(x, v)\mathrm{d}x\mathrm{d}v$. To obtain weak formulation, continuous test function $\phi \in C_{\mathrm{b}}^{\infty}(\mathbb{R}^{2d})$, we integrate it against (3.7) and obtain that

$$\frac{\mathrm{d}}{\mathrm{d}t}\int \phi(x, v)\mu_t(x, v)\mathrm{d}x\mathrm{d}v = \int \mathcal{L}\phi(x, v)\mu_t(x, v)\mathrm{d}x\mathrm{d}v + \int \phi(x, v)\widetilde{Q}\mu_t(x, v)\mathrm{d}x\mathrm{d}v\,. \tag{3.16}$$

The integral with respect to the collision term can be simplified through a pre-post-collisional change of variables [100]. According to (3.5), the change of velocities has a Jacobian whose determinant is 1. Making use of (3.6) and that $\widetilde{q}$ depends only on $|v-w|$ and $(v-w)\cdot n$,

$$\begin{aligned}&\int \phi(x, v)\widetilde{Q}\mu_t(x, v)\mathrm{d}x\mathrm{d}v\\ &= \iint_{\mathbb{S}^{d-1}\times\mathbb{R}^{4d}} (\phi(x, v^*) - \phi(x, v))\widetilde{q}(x, v, y, w, n)\mu_t(y, w)\mu_t(x, v)\,\mathrm{d}n\mathrm{d}y\mathrm{d}w\mathrm{d}x\mathrm{d}v\,.\end{aligned}$$

Let $\mathcal{K}$ denote the generator of the stochastic jump process

$$\mathcal{K}\phi(x, v, y, w) = \int_{\mathbb{S}^{d-1}} (\phi(x, v^*) - \phi(x, v))\widetilde{q}(x, v, y, w, n)\mathrm{d}n\,, \tag{3.17}$$

and denote $\mathcal{C}[\mu_t]$ the collision operator acting on the test function ϕ:

$$\mathcal{C}[\mu_t]\phi(x, v) = \langle \mathcal{K}\phi(x, v, \cdot, \cdot), \mu_t\rangle = \iint_{\mathbb{R}^{2d}} \mathcal{K}\phi(x, v, y, w)\mu_t(y, w)\,\mathrm{d}y\mathrm{d}w\,. \tag{3.18}$$

Then (3.16) has its weak formulation:

$$\frac{\mathrm{d}}{\mathrm{d}t}\langle\phi, \mu_t\rangle = \langle\mathcal{L}\phi, \mu_t\rangle + \langle\mathcal{C}[\mu_t]\phi, \mu_t\rangle = \langle(\mathcal{L} + \mathcal{C}[\mu_t])\,\phi, \mu_t\rangle\,. \tag{3.19}$$

Throughout the calculation, we use the bracket notation $\langle \cdot, \cdot \rangle$ to denote the integral of two functions in the full $(x, v) \in \mathbb{R}^{2d}$ domain.

This weak formulation (3.19) can then be translated to a martingale problem. For any $T > 0$, we denote by $\mathcal{D}([0, T], \mathbb{R}^{2d})$ the Skorohod space[4] of all càdlàg functions defined over the interval $[0, T]$ and take values in $\mathbb{R}^{2d}$. Furthermore, denote $\mathcal{P}(\mathcal{D}([0, T], \mathbb{R}^{2d}))$ the collection of probability path measures over $\mathcal{D}([0, T], \mathbb{R}^{2d})$. The mollified Vlasov-Boltzmann equation (3.19) can be translated to the following martingale problem:

Problem 3.5 (Martingale Problem Associated with the Mollified Vlasov-Boltzmann) *Find a path measure $P^* \in \mathcal{P}(\mathcal{D}([0, T], \mathbb{R}^{2d}))$ such that for any $\phi \in C_b^\infty(\mathbb{R}^{2d})$ and for any initial condition $P_{\text{ini}} \in \mathcal{P}_2(\mathbb{R}^{2d})$, the process defined by*

$$M_t^\phi(\omega) = \phi(Z_t(\omega)) - \phi(Z_0(\omega)) - \int_0^t \left(\mathcal{L} + \mathcal{C}[P_s^*]\right) \phi(Z_s(\omega)) \mathrm{d}s\,, \qquad \forall \omega \in \mathcal{D}([0, T], \mathbb{R}^{2d})\,, \tag{3.20}$$

is a P^*-martingale and $(Z_0)_\sharp P^* = P_{\text{ini}}$. Here we define $Z_t : \mathcal{D}([0, T], \mathbb{R}^{2d}) \to \mathbb{R}^{2d}$ to be the canonical process such that $Z_t(\omega) = \omega(t)$ for any $\omega \in \mathcal{D}([0, T], \mathbb{R}^{2d})$, and $P_t^* = (Z_t)_\sharp P^* \in \mathcal{P}(\mathbb{R}^{2d})$ is the pushforward of the measure P^* under the canonical map Z_t. In other words, P_t^* is the t-marginal of the path measure.

A weak form of the PDE is usually connected to the associated Martingale Problem; see, e.g., [61]. In our context, the solution to the Martingale Problem 3.5 provides a solution to the weak form of the mollified Vlasov-Boltzmann equation (3.19). In particular, let $P^* \in \mathcal{P}(\mathcal{D}([0, T], \mathbb{R}^{2d}))$ be the solution to the Martingale Problem 3.5, then one can check that its t-marginal satisfies the weak form equation of the mollified Vlasov-Boltzmann equation. To see this, we take the expectation on both sides of (3.20) and obtain

$$\begin{aligned} 0 = \mathbb{E}_{P^*}[M_{t=0}^\phi] = \mathbb{E}_{P^*}[M_t^\phi] &= \mathbb{E}_{P^*}[\phi \circ Z_t] - \mathbb{E}_{P^*}[\phi \circ Z_0] \\ &- \int_0^t \mathbb{E}_{P^*}\left[(\mathcal{L}\phi + \mathcal{C}[P_s^*]\phi) \circ Z_s\right] \mathrm{d}s\,, \end{aligned} \tag{3.21}$$

where we used $\circ$ to denote function composition, the martingale property to propagate $\mathbb{E}_{P^*}[M_t^\phi]$ back to $\mathbb{E}_{P^*}[M_{t=0}^\phi]$, and that, according to the definition, $M_{t=0}^\phi(\omega) = 0$.

Note that the t-marginal of P^* is defined by $P_t^* = (Z_t)_\sharp P^*$, and hence for all t,

[4] It is a convention to equip $\mathcal{D}([0, T], \mathbb{R}^{2d})$ with the J_1 topology [5].

$$\mathbb{E}_{P^*}[\phi \circ Z_t] = \int \phi(Z_t(\omega))\mathrm{d}P^*(\omega) = \langle \phi, P_t^* \rangle ,$$

$$\mathbb{E}_{P^*}\left[(\mathcal{L} + \mathcal{C}[P_s^*])(\phi \circ Z_s)\right] = \int (\mathcal{L}\phi + \mathcal{C}[P_s^*]\phi)(Z_s(\omega))\mathrm{d}P^*(\omega)$$

$$= \langle \mathcal{L}\phi, P_s^* \rangle + \langle \mathcal{C}[P_s^*]\phi, P_s^* \rangle .$$

Plug these back in (3.21), and identify $\mu_t = P_t^*$; we recover the weak form of the mollified Vlasov-Boltzmann equation (3.19).

We do not delve into the details, but we do point out that just as one needs to show the well-posedness of the PDE, Problem 3.5 also requires the well-posedness theory to ensure these is a unique path measure satisfying (3.20). Uniqueness can be proved under the mild regularity condition and the growth condition of the cross-section kernel $\widetilde{q}(x, v, y, w, n)$. We refer interested readers to Theorem 2.2 in [50].

3.3.2 Weak Formulation and the Martingale Problems for the N-Particle Systems

We switch gears to study the discrete setting: the Nanbu method and the Bird method. In this section, we lay out the weak form PDEs, as Piecewise Deterministic Markov Processes for both methods and their associated martingale problems.

To do so, we first lift ourselves up first to the N-fold coordinate system, with each coordinate keeping information of one particle: $x^N = (x_1, \ldots, x_N) \in (\mathbb{R}^d)^{\otimes N}$. Similarly, the velocity coordinate is also lifted up to the N-fold dimension by $v^N = (v_1, \ldots, v_N) \in (\mathbb{R}^d)^{\otimes N}$. We denote $\mu_t^N \in \mathcal{P}((\mathbb{R}^{2d})^{\otimes N})$ the distribution over the N-fold phase space.

Both Nanbu and Bird methods provide descriptions on how to propagate particles, so we can trace the evolution and translate them into the PDE language. For any given $\phi^N \in C_b^\infty((\mathbb{R}^{2d})^{\otimes N})$,

$$\frac{\mathrm{d}}{\mathrm{d}t}\left\langle \phi^N, \mu_t^N \right\rangle = \sum_{i=1}^{N} \left\langle \mathcal{L}_i \phi^N, \mu_t^N \right\rangle + \sum_{i,j=1}^{N} \left\langle \mathcal{K}_{ij} \phi^N, \mu_t^N \right\rangle , \tag{3.22}$$

where the free transport operator $\mathcal{L}_i$ acts only on the i-th coordinate pairs:

$$\mathcal{L}_i \phi^N(x^N, v^N) = v_i \cdot \nabla_{x_i} \phi^N(x^N, v^N) - \nabla_{x_i} f(x_i) \cdot \nabla_{v_i} \phi^N(x^N, v^N) , \quad i=1, \ldots, N , \tag{3.23}$$

and the form of the binary collision operator $\mathcal{K}_{ij}$ depends on the specific method in hand. For example, the Nanbu method only updates one coordinate, and it uses the following collision operator, for $i, j = 1, \ldots, N$:

$$\mathcal{K}_{ij}^{(\mathrm{N})}\phi^N(x^N,v^N) = \frac{1}{N}\int_{\mathbb{S}^{d-1}}(\phi^N(x^N,v^{N,i,*}) - \phi^N(x^N,v^N))\widetilde{q}(x_i,v_i,x_j,v_j,n)\mathrm{d}n\,, \tag{3.24}$$

where

$$v^{N,i,*} = (v_1,\ldots,v_i + ((v_j - v_i)\cdot n)n,\ldots,v_N)\,,$$

is the post-collision velocity. We note that only the i-th coordinate is changed. The superindex (N) stands for Nanbu. The Bird method, on the other hand, picks a pair and updates them simultaneously. The associated collision operator then reads:

$$\mathcal{K}_{ij}^{(\mathrm{B})}\phi^N(x^N,v^N) = \frac{1}{N}\int_{\mathbb{S}^{d-1}}(\phi^N(x^N,v^{N,i,j,*}) - \phi^N(x^N,v^N))\widetilde{q}(x_i,v_i,x_j,v_j,n)\mathrm{d}n\,, \tag{3.25}$$

where

$$v^{N,i,j,*} = (v_1,\ldots,v_i + ((v_j - v_i)\cdot n)n,\ldots,v_j + ((v_i - v_j)\cdot n)n,\ldots,v_N)$$

is the post-collision velocity, and both (i,j)-coordinates are changed. The superindex (B) stands for Bird.

Similar to the Vlasov-Boltzmann equation, these weak formulations (3.22) can be translated into martingale problems for the N-particle system. Similar to deriving the weak solution form, all terms are lifted up to the N-fold space. In particular, the N-fold path $\omega^N = (\omega_1,\ldots,\omega_N) \in \mathcal{D}([0,T],(\mathbb{R}^{2d})^{\otimes N})$, and the path measure P^N lives in the product space $\mathcal{P}(\mathcal{D}([0,T],(\mathbb{R}^{2d})^{\otimes N}))$, with the initial condition living in $P_{\mathrm{ini}}^N \in \mathcal{P}_2((\mathbb{R}^{2d})^{\otimes N})$. The martingale problem is to find a path on the product space:

Problem 3.6 (Martingale Problem Associated with the N-Particle System) *Find a path measure $P^N \in \mathcal{P}(\mathcal{D}([0,T],(\mathbb{R}^{2d})^{\otimes N}))$ such that for any test function $\phi^N \in C_b((\mathbb{R}^{2d})^{\otimes N})$ and initial condition $P_{\mathrm{ini}}^N \in \mathcal{P}_2((\mathbb{R}^{2d})^{\otimes N})$, the process defined by*

$$\begin{aligned} M_t^{\phi^N}(\omega^N) &= \phi^N(Z_t^N(\omega^N)) - \phi^N(Z_0^N(\omega^N)) \\ &\quad - \int_0^t \sum_{i=1}^N \mathcal{L}_i\phi^N(Z_s^N(\omega^N)) + \sum_{i,j=1}^N \mathcal{K}_{ij}\phi^N(Z_s^N(\omega^N))\,\mathrm{d}s\,, \\ &\quad \forall \omega^N \in \mathcal{D}([0,T],(\mathbb{R}^{2d})^{\otimes N})\,, \end{aligned} \tag{3.26}$$

is a P^N-martingale and $(Z_0^N)_\sharp P^N = P_{\mathrm{ini}}^N$. Here we define $Z_t^N : \mathcal{D}([0,T],(\mathbb{R}^{2d})^{\otimes N}) \to (\mathbb{R}^{2d})^{\otimes N}$ to be the canonical process such that $Z_t^N(\omega^N) = \omega^N(t)$ for any N-fold path ω^N.

The well-posedness of Problem 3.6 has also been investigated in [37, 56, 89].

3.3.3 The Mean-Field Analysis

The mean-field analysis concerns connecting the two martingale problems and shows the convergence of empirical path measures. We now define the empirical path measure for the N-particle system

$$Q_N(\omega^N) = \frac{1}{N}\sum_{i=1}^N \delta_{\omega_i} \in \mathcal{P}(\mathcal{D}([0,T],\mathbb{R}^{2d}))\,,$$
$$\forall \omega^N = (\omega_1,\ldots,\omega_N) \in \mathcal{D}([0,T],(\mathbb{R}^{2d})^{\otimes N})\,. \tag{3.27}$$

We note that Q_N itself is a path measure, and due to randomness in ω^N, Q_N is not deterministic either, and it is a random path measure whose law lives in a probability measure space over the probability path measure. More specifically, $\mathrm{Law}(Q_N) \in \mathcal{P}(\mathcal{P}(\mathcal{D}([0,T],\mathbb{R}^{2d})))$. We want to prove that this random path measure, in the large N limit, converges to a deterministic path measure. Hence, its law is a delta measure in the probability over probability path measure space, centered at the solution that solves Problem 3.5.

Theorem 3.7 (Theorem 4.5 in [51]) *Suppose the initial condition P^N_{ini} is exchangeable in the sense that P^N_{ini} is invariant under the permutation of coordinates. Under Assumption 3.1 and mild regularity assumptions on the cross-section $\widetilde{q}$ and the potential f, there is*

$$\mathrm{Law}(Q_N) := P^N \circ Q_N^{-1} \xrightarrow{N\to\infty} \delta_{P^*} \quad \textit{weak-}* \textit{ in } \mathcal{P}(\mathcal{P}(\mathcal{D}([0,T],\mathbb{R}^{2d})))\,, \tag{3.28}$$

where $P^* \in \mathcal{P}(\mathcal{D}([0,T],\mathbb{R}^{2d}))$ is the path measure solution to Problem 3.5.

Remark 3.8 The detailed assumption for this theorem to hold is technical. Generally speaking, according to [51], these assumptions can be categorized into three groups: The gradient of f is Lipschitz, the kernel $\widetilde{q}$ is smooth enough in its variables (x, y, v, w), and the kernel $\widetilde{q}$ has suitable decay in its variable (y, w). The same regularity assumptions are required in proving the well-posedness of the Martingale Problem (3.5) (see Theorem 2.2 in [50]). In the Bayesian sampling setting, the assumption on f typically hold, and the assumptions on $\widetilde{q}$ can be satisfied through a proper algorithmic design.

This convergence result leads to the mean-field limit. To be more precise, one can derive that the ensemble distribution living in $\mathcal{P}(\mathbb{R}^{2d})$ converges in law to the solution of the Vlasov-Boltzmann equation. More precisely:

Corollary 3.9 *Consider the N-particle ensemble distribution, defined by*

$$Q_{N,t}(\omega^N) := \frac{1}{N}\sum_{i=1}^N \delta_{(x_i(t),v_i(t))} \in \mathcal{P}(\mathbb{R}^{2d})\,, \tag{3.29}$$

where $\{(x_i(t), v_i(t))\}_{0\le t\le T} \in \mathcal{D}([0,T],\mathbb{R}^{2d})$ *denotes the path of the i-th particle. Then* $Q_{N,t}$ *is a random measure on* $\mathbb{R}^{2d}$*. Under the same assumption as Theorem 3.7, and taking further assumptions that* $Z_t : \mathcal{D}([0,T],\mathbb{R}^{2d}) \to \mathbb{R}^{2d}$ *is a continuous map, we know that* $Q_{N,t}$ *converges in distribution to* P_t^**, which we identify with* μ_t*, the solution of the mollified Vlasov-Boltzmann equation* (3.7)*. In other words, we have the convergence of the law of* $Q_{N,t}$*:*

$$\mathrm{Law}(Q_{N,t}) := P^N \circ Q_{N,t}^{-1} \xrightarrow{N\to\infty} \delta_{\mu_t} \quad \textit{weak-}* \textit{ in } \mathcal{P}(\mathcal{P}(\mathbb{R}^{2d}))\,. \tag{3.30}$$

Proof Note that the N-particle ensemble distribution at time t can be written as the pushforward of the empirical path measure:

$$Q_{N,t}(\omega^N) = (Z_t)_\sharp Q_N(\omega^N) \in \mathcal{P}(\mathbb{R}^{2d})\,. \tag{3.31}$$

For any test function $g \in C_b(\mathcal{P}(\mathbb{R}^{2d}))$, by making use of (3.31) and the pushforward definition:

$$\int_{\mathcal{P}(\mathbb{R}^{2d})} g(\mu)\,\mathrm{d}P^N \circ Q_{N,t}^{-1}(\mu) = \int_{\mathcal{P}(\mathcal{D}([0,T],\mathbb{R}^{2d}))} g((Z_t)_\sharp P)\,\mathrm{d}P^N \circ Q_N^{-1}(P)\,. \tag{3.32}$$

Assuming the continuity of the canonical map Z_t, we know that the functional $h(P) = g((Z_t)_\sharp P)$ is bounded and continuous, that is, $h \in C_b(\mathcal{P}(\mathcal{D}([0,T],\mathbb{R}^{2d})))$.

Noting the measure P_t^* is a pushforward of P^*: $P_t^* = (Z_t)_\sharp P^* \in \mathcal{P}(\mathbb{R}^{2d})$, we draw the conclusion from Theorem 3.7 to rewrite (3.32) for the convergence of the law of the ensemble distribution:

$$\begin{aligned}\int_{\mathcal{P}(\mathbb{R}^{2d})} g(\mu)\,\mathrm{d}P^N \circ Q_{N,t}^{-1}(\mu) &\xrightarrow{N\to\infty} \int_{\mathcal{P}(\mathcal{D}([0,T],\mathbb{R}^{2d}))} g((Z_t)_\sharp P)\,\mathrm{d}\delta_{P^*}(P)\\ &= \int_{\mathcal{P}(\mathbb{R}^{2d})} g(\mu)\,\mathrm{d}\delta_{\mu_t}(\mu)\,.\end{aligned} \tag{3.33}$$

□

In general, the map Z_t is not a continuous map from $\mathcal{D}([0,T],\mathbb{R}^{2d})$ to $\mathbb{R}^{2d}$, but the conclusion still holds. We refer the readers to [51, 52] for a rigorous justification of the convergence.

Remark 3.10 We stress a few remarks.

- There are many layers of randomness. The algorithm encodes Poisson randomness, so even with fixed initial data, each run of the algorithm provides a different configuration of $Q_{N,t}$, and thus it presents a probability measure over $\mathbb{R}^{2d}$, which means $Q_{N,t} \in \mathcal{P}(\mathbb{R}^{2d})$. At the initial time, N samples are drawn at random, and putting all $Q_{N,t}$ equipped with different initial data together, one forms another probability measure over $\mathcal{P}(\mathbb{R}^{2d})$. Theorem 3.7 and consequently Corollary 3.9

claim the convergence of the law of $Q_{N,t}$, in the space of probability over probability on $\mathbb{R}^{2d}$, to a Dirac delta centered on the solution to the Boltzmann equation (3.7).

- As spelled out in (3.33), the convergence is rather weak: The weak-$*$ convergence on $\mathcal{P}(\mathcal{P}(\mathbb{R}^{2d}))$ means that the test function g in actually a *functional* over $\mathcal{P}(\mathbb{R}^{2d})$ itself. However, this weak convergence is strong enough to provide convergence of all moments. To see this, we set

$$g(\mu) = \int (|x|^p + |v|^p)\mu(\mathrm{d}x\mathrm{d}v)\,,$$

a moment-taking functional. It is clearly a continuous bounded functional over $\mathcal{P}_p(\mathbb{R}^{2d})$. In the limit $N \to \infty$, $\int_{\mathcal{P}(\mathbb{R}^{2d})} g(\mu)\,\mathrm{d}\delta_{\mu_t}(\mu) = g(\mu_t)$ provides the moments of the solutions to the Vlasov-Boltzmann equation. The left-hand side in (3.30) when tested by moments, or equivalently (3.33), can be interpreted as

$$\mathbb{E}_{P^N_{\text{ini}}}\mathbb{E}_\omega\left(\frac{1}{N}\sum_i(|x_i(t)|^p + |v_i(t)|^p)\right)$$

with the two expectations taking over different initial configurations and different Poisson clock ringings, and hence all randomness involved in the algorithms.

- Although we do not dive into details, the proof of Theorem 3.7 follows a fixed set of machinery, see [51, 52]. The proof strategy heavily relies on the compactness argument and is composed of two steps.

Step 1. The law of Q_N is tight so that there exist a limit point $\mathfrak{Q}_\infty \in \mathcal{P}(\mathcal{P}(\mathcal{D}([0,T],\mathbb{R}^{2d})))$;

Step 2. For $\mathfrak{Q}_\infty$-almost surely $P \in \mathcal{P}(\mathcal{D}([0,T],\mathbb{R}^{2d}))$, P solves the mollified Vlasov-Boltzmann Martingale Problem.

These two steps combined, and calling the uniqueness to the Martingale Problem 3.5, we have $\mathfrak{Q}^\infty = \delta_{P^*}$ with P^* providing the weak solution to the mollified Vlasov-Boltzmann equation (3.19).

3.4 Numerical Experiments

In this section, we report on numerical experiments. We consider problems with $d = 1$ and $d = 2$ in the following. It is worth noting that for $d = 1$, the collision operator in the Vlasov-Boltzmann equation (3.1) becomes zero, whereas the mollified collision operator remains nontrivial and continues to play a role in the convergence to the target distribution.

To set up the problem, we consider a target probability measure $\rho^* \in \mathcal{P}(\mathbb{R}^d)$, where $\rho^* \propto \exp(-f(x))$. Initially, we sample N particles $\{(x_i(0), v_i(0))\}_{i=1}^N$ independently from a distribution $\mu_0 \in \mathcal{P}(\mathbb{R}^{2d})$, defined by

$$\mu_0 \propto \mathbb{1}_{|x|_\infty \le L} \cdot \exp\left(-\frac{|v|^2}{2\sigma^2}\right), \quad \forall (x, v) \in \mathbb{R}^{2d}. \tag{3.34}$$

This initial condition is chosen to fit the energy condition (3.14) required for achieving the convergence of the PDE, Theorem 3.3. This translates to the following configuration for σ when L is fixed.

$$\sigma^2 = 1 + \frac{2}{d}\left(\frac{\int_{\mathbb{R}^d} f(x)\exp(-f(x))\mathrm{d}x}{\int_{\mathbb{R}^d}\exp(-f(x))\mathrm{d}x} - \frac{1}{(2L)^d}\int_{|x|_\infty \le L} f(x)\mathrm{d}x\right).$$

We take the following regularized collision kernel

$$\widetilde{q}(x, v, y, w, n) = \frac{1}{(\epsilon\sqrt{\pi})^d}\exp\left(-\frac{|x-y|^2}{\epsilon^2}\right), \tag{3.35}$$

so that the total cross-section $\Lambda = \frac{2}{\epsilon\sqrt{\pi}}$ if $d = 1$ and $\Lambda = \frac{2}{\epsilon^2\pi}$ if $d = 2$. Since the regularized collision kernel is independent of n, upon collision, the vector n is sampled uniformly in the set $\{-1, 1\}$ when $d = 1$, and uniformly in the circle $\mathbb{S}^1$ when $d = 2$.

Remark 3.11 We choose the form of $\widetilde{q}$ for its simplicity. We do observe that the numerical performance of the algorithm depends on ϵ. Preliminary numerical test suggests larger ϵ leads to a faster convergence of the PDE and requires a smaller number of particles to simulate the PDE. This observation suggests a better algorithm performance for a larger ϵ. We do not yet have a theoretical justification to support this observation. More numerical examples are also needed to support this finding.

To measure the sampling performance, we consider two different metrics using the Kullback–Leibler (KL) divergence: the KL-divergence between the x-marginal distribution, defined by

$$KL_X(\mu_X(t)||\mu_X^*) = \int_{\mathbb{R}^d} \mu_X(t, x)\ln\frac{\mu_X(t, x)}{\mu_X^*(t, x)}\mathrm{d}x = \int_{\mathbb{R}^d}\mu_X(t, x)\ln\mu_X(t, x)\mathrm{d}x$$
$$+ \int_{\mathbb{R}^d} f(x)\mu_X(t, x)\mathrm{d}x\,, \tag{3.36}$$

where μ_X denotes the x-marginal of μ and the KL divergence between the phase space distribution, defined by

$$
\begin{aligned}
KL(\mu(t)||\mu^*) &= \int_{\mathbb{R}^{2d}} \mu(t,x,v) \ln \frac{\mu(t,x,v)}{\mu^*(t,x,v)} \mathrm{d}x \, \mathrm{d}v \\
&= \int_{\mathbb{R}^{2d}} \mu(t,x,v) \ln \mu(t,x,v) \, \mathrm{d}x \, \mathrm{d}v \\
&+ \int_{\mathbb{R}^{2d}} \left(f(x) + \frac{v^2}{2} \right) \mu(t,x,v) \, \mathrm{d}x \, \mathrm{d}v \,.
\end{aligned} \tag{3.37}
$$

In both the Nanbu sampler and the Bird sampler, the measure $\mu(t, x, v)$ is approximated by the ensemble distribution, defined by $\mu(t, x, v) = \frac{1}{N}\sum_{i=1}^{N} \delta_{(x_i(t), v_i(t))}(x, v)$. Neither of the KL divergence is defined for the ensemble distribution. To circumvent this difficulty, we evaluate the following mollified KL divergence, defined by

$$
\begin{aligned}
KL_X^\delta &= \frac{1}{N}\sum_{i=1}^{N} \ln\left(\frac{1}{N}\sum_{j=1}^{N} \rho^\delta(x_i(t) - x_j(t)) \right) + \frac{1}{N}\sum_{i=1}^{N} f(x_i(t)) \,. \\
KL^\delta &= \frac{1}{N}\sum_{i=1}^{N} \ln\left(\frac{1}{N}\sum_{j=1}^{N} \rho^\delta(x_i(t) - x_j(t))\rho^\delta(v_i(t) - v_j(t)) \right) \\
&+ \frac{1}{N}\sum_{i=1}^{N} \left[f(x_i(t)) + \frac{v_i^2(t)}{2} \right] .
\end{aligned} \tag{3.38}
$$

Here, we take the mollifier $\rho^\delta(x) = (8\pi\delta^2)^{-d/2}\exp(-|x|^2/(8\delta^2))$.

In all the examples below, to generate reference solutions, we use the Inverse Transform Sampling (see, e.g., page 102 in [45]) to obtain $N = 1000$ samples from $\mu_X^* \propto e^{-f(x)}$. In the simple settings like ours ($d = 1, 2$), Inverse Transform Sampling can still be deployed relatively easily [45]. The obtained samples are then used to compute a baseline value for the regularized KL divergences as the reference data (we term it the baseline solution in the plots).

Both algorithms are run with $N = 1000$ particles. For simulating the free transport term (3.4), Verlet algorithm [91] is deployed as the discretization step. In (3.35), we take the parameter $\epsilon = 1$ when $d = 1$ and $\epsilon = 4$ when $d = 2$ to ensure the time interval between collision is not too large. In (3.38), we take the regularization constant $\delta = 0.3$ across all examples. We take the support of initial condition $L = 2$ in (3.34).

In the first example, we set $f(x) = \frac{x^2}{2}$. Figure 3 shows the decrease in the relative entropy KL_X^δ, KL^δ and the histogram of samples obtained by the Nanbu and Bird sampler. Both methods provide ensemble distribution that quickly converges to the target distribution. The phase space relative entropy decays monotonically at the initial stage, which is aligned with the H-theorem [100]. This is because the phase space entropy can be decomposed into the sum of the absolute entropy and the total

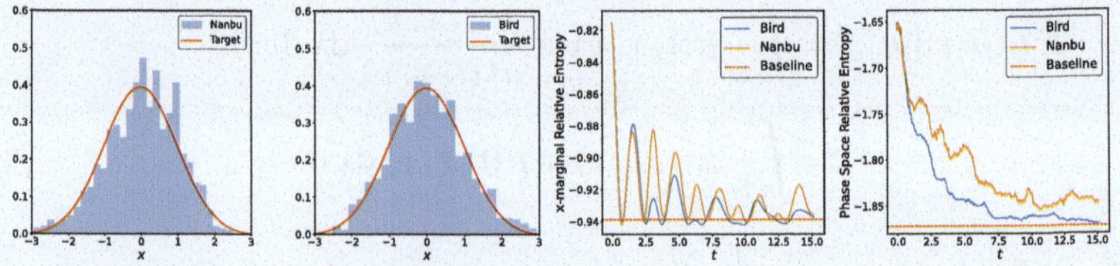

Fig. 3 **Left two plots**: The target distribution $\rho^* \propto \exp(-\frac{x^2}{2})$ and the histogram of samples obtained from the Nanbu sampler (leftmost) and the Bird sampler (second-left). **Right two plots**: The regularized relative entropy KL_X^δ (second-right) and KL^δ (rightmost) of the Bird sampler and the Nanbu sampler. The baseline value is computed by using $N = 1000$ samples obtained by the Inverse Transform method

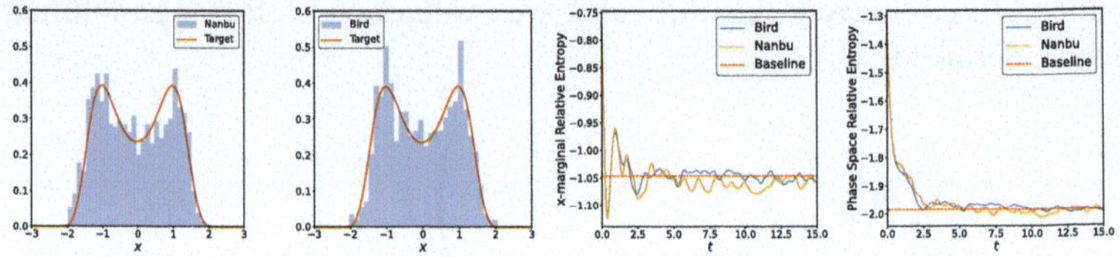

Fig. 4 **Left two plots:** The target distribution $\rho^* \propto \exp(-(x-1)^2(x+1)^2)$ and the histogram of samples obtained from the Nanbu sampler (leftmost) and the Bird sampler (second-left). **Right two plots:** The regularized relative entropy KL_X^δ (second-right) and KL^δ (rightmost) of the Bird sampler and the Nanbu sampler. The baseline value is computed by using $N = 1000$ samples obtained by the Inverse Transform method

energy, where the former monotonically decreases, and the latter remains constant. Following the initial stage, the phase space entropy oscillates near the equilibrium due to the randomness of the particles. We observe that the Nanbu method exhibits larger fluctuations and reaches saturation earlier. This is because the energy is not conserved upon collisions in the Nanbu method. In comparison, the Bird method demonstrates better stability due to the energy conservation.

In the second example, we set $f(x) = (x-1)^2(x+1)^2$, so the target distribution is clearly not log-concave. In Fig. 4, we nevertheless show the relative entropy KL_X^δ and KL^δ decreases over time. This is a strong indication in this strongly non-convex case; the Boltzmann simulator can nevertheless achieve global convergence.

In the third and fourth example, we consider two-dimensional problems with potential $f(x, y) = 0.5(x^2 + y^2)$ and $f(x, y) = 0.1((x-1)^2 + (y-1)^2)((x+1)^2 + (y+1)^2)$. While the former one corresponds to the Gaussian distribution, the latter is a double-well potential, and the target distribution is not log-concave. In Figs. 5 and 6, both the relative entropy KL_X^δ and KL^δ decrease over time, regardless of the convexity of the potential.

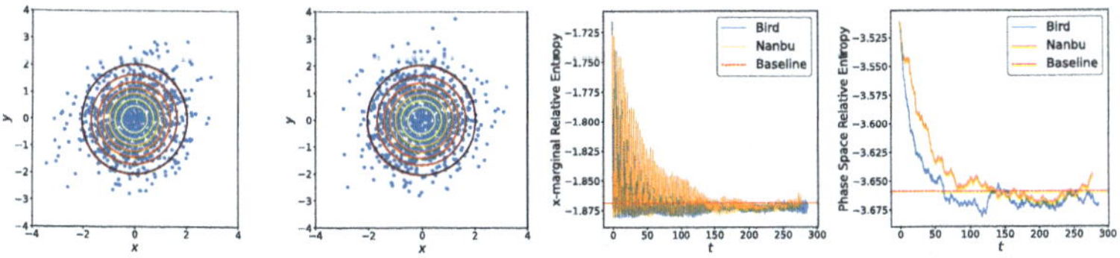

Fig. 5 **Left two plots:** The contour of the target distribution $\rho^* \propto \exp(-0.5(x^2 + y^2))$ and the samples obtained from the Nanbu sampler (leftmost) and the Bird sampler (Second-left). **Right two plots:** The regularized relative entropy KL^δ_X (Second-right) and KL^δ (rightmost) of the Bird sampler and the Nanbu sampler. The baseline value is computed by using $N = 1000$ samples obtained by the Inverse Transform method

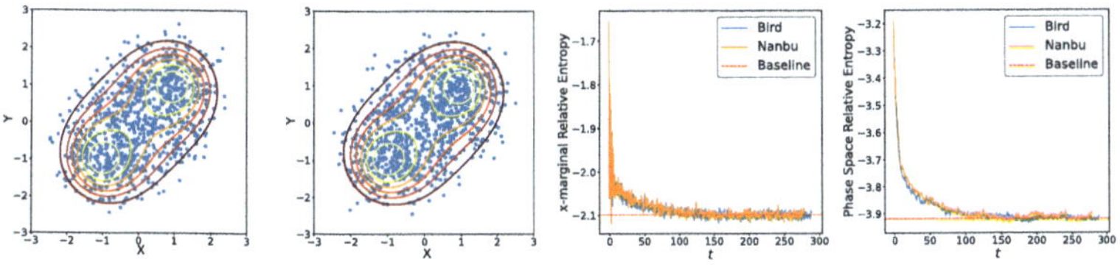

Fig. 6 **Left two plots**: The contour of the target distribution $\rho^* \propto \exp(-0.1((x-1)^2 + (y-1)^2)((x+1)^2 + (y+1)^2))$ and the samples obtained from the Nanbu sampler (leftmost) and the Bird sampler (second-left). **Right two plots**: The regularized relative entropy KL^δ_X (second-right) and KL^δ (rightmost) of the Bird sampler and the Nanbu sampler. The baseline value is computed by using $N = 1000$ samples obtained by the Inverse Transform method

References

1. C. Andrieu, N. de Freitas, A. Doucet, M.I. Jordan, An introduction to MCMC for machine learning. Mach. Learn. **50**, 5–43 (2003)
2. A. Benfenati, G. Borghi, L. Pareschi, Binary interaction methods for high dimensional global optimization and machine learning. Appl. Math. Optim. **86**(1), 9 (2022)
3. K. Bergemann, S. Reich, A localization technique for ensemble kalman filters. Q. J. R. Meteorolog. Soc. **136**(648), 701–707 (2010)
4. K. Bergemann, S. Reich, A mollified ensemble kalman filter. Q. J. R. Meteorolog. Soc. **136**(651), 1636–1643 (2010)
5. P. Billingsley, *Convergence of Probability Measures* (John Wiley & Sons, Hoboken, 2013)
6. G. Bird, Direct simulation and the Boltzmann equation. Phys. Fluids **13**(11), 2676–2681 (1970)
7. F. Bolley, J.A. Cañizo, J.A. Carrillo, Stochastic mean-field limit: non-Llipschitz forces and swarming. Math. Models Methods Appl. Sci. **21**(11), 2179–2210 (2011)
8. G. Borghi, L. Pareschi, Kinetic description and convergence analysis of genetic algorithms for global optimization. Preprint. arXiv:2310.08562 (2023)
9. N. Bou-Rabee, J.M. Sanz-Serna, Randomized hamiltonian Monte Carlo. Ann. Appl. Probab. **27**(4), 2159–2194 (2017)
10. E. Calvello, S. Reich, A.M. Stuart, Ensemble kalman methods: a mean field perspective. Preprint. arXiv:2209.11371 (2022)
11. J.A. Cañizo, J.A. Carrillo, J. Rosado, A well-posedness theory in measures for some kinetic models of collective motion. Math. Models Methods Appl. Sci. **21**(03), 515–539 (2011)

12. Y. Cao, J. Lu, L. Wang, On explicit L^2-convergence rate estimate for Underdamped Langevin dynamics. Arch. Ration. Mech. Anal. **247**(5), 90 (2023). https://doi.org/10.1007/s00205-023-01922-4
13. J.A. Carrillo, U. Vaes, Wasserstein stability estimates for covariance-preconditioned fokker–planck equations. Nonlinearity **34**(4), 2275–2295 (2021)
14. J.A. Carrillo, Y.-P. Choi, C. Totzeck, O. Tse, An analytical framework for consensus-based global optimization method. Math. Models Methods Appl. Sci. **28**(06), 1037–1066 (2018)
15. J.A. Carrillo, F. Hoffmann, A.M. Stuart, U. Vaes, Consensus-based sampling. Stud. Appl. Math. **148**(3), 1069–1140 (2022)
16. J.A. Carrillo, F. Hoffmann, A.M. Stuart, U. Vaes, The mean field ensemble kalman filter: near-gaussian setting. arXiv/2212.13239 (2023)
17. J.A. Carrillo, S. Jin, H. Zhang, Y. Zhu, An interacting particle consensus method for constrained global optimization. Preprint. arXiv:2405.00891 (2024)
18. L.-P. Chaintron, A. Diez, Propagation of chaos: a review of models, methods and applications. II. Applications. Kinet. Rel. Models **15**(6), 1017–1173 (2022)
19. X. Cheng, N.S. Chatterji, P.L. Bartlett, M.I. Jordan, Underdamped Langevin MCMC: a non-asymptotic analysis, in *Proceedings of the 31st Conference On Learning Theory* (2018), pp. 300–323
20. S. Chewi, T. Le Gouic, C. Lu, T. Maunu, P. Rigollet, A. Stromme, Exponential ergodicity of mirror-langevin diffusions, in *Proceedings of the 34th International Conference on Neural Information Processing Systems* (2020)
21. S. Chewi, C. Lu, K. Ahn, X. Cheng, T.L. Gouic, P. Rigollet, Optimal dimension dependence of the metropolis-adjusted langevin algorithm, in *Annual Conference Computational Learning Theory* (2020)
22. S. Chewi, M.A. Erdogdu, M. Li, R. Shen, S. Zhang, Analysis of langevin monte carlo from poincare to log-Sobolev, in *Proceedings of Thirty Fifth Conference on Learning Theory*, vol. 178. *Proceedings of Machine Learning Research*. PMLR, 02–05 Jul (2022), pp. 1–2
23. K. Craig, A. Bertozzi, A blob method for the aggregation equation. Math. Comput. **85**, 1681–1717 (2016). ISSN: 00255718, 10886842, https://www.jstor.org/stable/mathcomp.85.300.1681
24. A. Dalalyan, Theoretical guarantees for approximate sampling from smooth and log-concave densities. J. R. Stat. Soc. Ser. B (Stat. Methodol.) **79**(3), 651–676 (2017)
25. A. Dalalyan, A. Karagulyan, User-friendly guarantees for the Langevin Monte Carlo with inaccurate gradient. Stoch. Proces. Appl. **129**(12), 5278–5311 (2019)
26. L. Desvillettes, C. Villani, On the trend to global equilibrium for spatially inhomogeneous kinetic systems: the Boltzmann equation. Invent. Math. **159**(2), 245–316 (2005)
27. Z. Ding, Q. Li, Ensemble Kalman inversion: mean-field limit and convergence analysis. Stat. Comput. **31**, article number 9 (2021). https://doi.org/10.1007/s11222-020-09976-0
28. Z. Ding, Q. Li, Ensemble kalman sampler: mean-field limit and convergence analysis. SIAM J. Math. Anal. **53**(2), 1546–1578 (2021)
29. Z. Ding, Q. Li, J. Lu, Ensemble kalman inversion for nonlinear problems: weights, consistency, and variance bounds. Found. Data Sci. **3**(3), 371–411 (2021)
30. Z. Ding, M. Guerra, Q. Li, E. Tadmor, Swarm-based gradient descent meets simulated annealing. Preprint. arXiv:2404.18015 (2024)
31. K. Duraisamy, G. Iaccarino, H. Xiao, Turbulence modeling in the age of data. Ann. Rev. Fluid Mech. **51**, 357–377 (2019)
32. A. Durmus, É. Moulines, Non-asymptotic convergence analysis for the Unadjusted Langevin algorithm. Ann. Appl. Probab. **27**(3), 1551–1587 (2017)
33. A. Durmus, S. Majewski, B. Miasojedow, Analysis of Langevin Monte Carlo via convex optimization. J. Mach. Learn. Res. **20**(73), 1–46 (2019)
34. R. Dwivedi, Y. Chen, M.J. Wainwright, B. Yu, Log-concave sampling: metropolis-hastings algorithms are fast! in *Proceedings of the 31st Conference On Learning Theory*, ed. by S. Bubeck, V. Perchet, P. Rigollet, vol. 75. *Proceedings of Machine Learning Research*. PMLR, 06–09 Jul (2018), pp. 793–797

35. A. Eberle, A. Guillin, R. Zimmer, Couplings and quantitative contraction rates for Langevin dynamics. Ann. Probab. **47**(4), 1982–2010 (2019)
36. O.G. Ernst, B. Sprungk, H.-J. Starkloff, Analysis of the ensemble and polynomial chaos Kalman filters in bayesian inverse problems. SIAM/ASA J. Uncertainty Quantif. **3**(1), 823–851 (2015)
37. S.N. Ethier, T.G. Kurtz, *Markov Processes: Characterization and Convergence* (John Wiley & Sons, Hoboken, 2009)
38. G. Evensen, Sequential data assimilation with a nonlinear quasi-geostrophic model using Monte Carlo methods to forecast error statistics. J. Geophys. Res. Oceans **99**(C5), 10143–10162 (1994)
39. G. Evensen, The ensemble kalman filter: theoretical formulation and practical implementation. Ocean Dyn. **53**(4), 343–367 (2003)
40. G. Evensen, *Data Assimilation - The Ensemble Kalman Filter* (Springer-Verlag New York, Inc., Secaucus, 2006)
41. P. Fabian, Atmospheric sampling. Adv. Space Res. **1**(11), 17–27 (1981)
42. N. Fournier, A. Guillin, On the rate of convergence in wasserstein distance of the empirical measure. Probab. Theory Related Fields **162**(3), 707–738 (2015)
43. D. Gamerman, H.F. Lopes, *Markov Chain Monte Carlo: Stochastic Simulation for Bayesian Inference* (CRC Press, Boca Raton, 2006)
44. A. Garbuno-Inigo, F. Hoffmann, W. Li, A.M. Stuart, Interacting Langevin diffusions: gradient structure and ensemble Kalman sampler. SIAM J. Appl. Dyn. Syst. **19**(1), 412–441 (2020)
45. J.E. Gentle, *Random Number Generation and Monte Carlo Methods*, vol. 381 (Springer, Berlin, 2003)
46. N.J. Gerber, F. Hoffmann, U. Vaes, Mean-field limits for consensus-based optimization and sampling. Preprint. arXiv:2312.07373 (2023)
47. M. Ghil, S. Cohn, J. Tavantzis, K. Bube, E. Isaacson, *Applications of Estimation Theory to Numerical Weather Prediction* (Springer, New York, 1981)
48. W.R. Gilks, S. Richardson, D. Spiegelhalter, *Markov Chain Monte Carlo in Practice.* (CRC Press, Boca Raton, 1995)
49. J. Goodman, J. Weare, Ensemble samplers with affine invariance. Commun. Appl. Math. Comput. Sci. **5**(1), 65–80 (2010)
50. C. Graham, Nonlinear diffusion with jumps, in *Annales de l'IHP Probabilités et statistiques* (1992)
51. C. Graham, S. Méléard, Stochastic particle approximations for generalized Boltzmann models and convergence estimates. Ann. Probab. **25**(1), 115–132 (1997)
52. C. Graham, T.G. Kurtz, S. Méléard, P.E. Protter, M. Pulvirenti, D. Talay, S. Méléard, Asymptotic behaviour of some interacting particle systems; Mckean-Vlasov and Boltzmann models, in *Probabilistic Models for Nonlinear Partial Differential Equations: Lectures given at the 1st Session of the Centro Internazionale Matematico Estivo (CIME) held in Montecatini Terme, May 22–30, 1995* (1996), pp. 42–95
53. P. Houtekamer, H.L. Mitchell, A sequential ensemble kalman filter for atmospheric data assimilation. Mon. Weather Rev. **129**(1), 123–137 (2001)
54. M.A. Iglesias, K. Law, A.M. Stuart, Ensemble Kalman methods for inverse problems. Inverse Prob. **29**(4), 045001 (2013)
55. M.A. Iglesias, K. Law, A.M. Stuart, Evaluation of Gaussian approximations for data assimilation in reservoir models. Comput. Geosci. **17**, 851–885 (2013)
56. J. Jacod, A. Shiryaev, *Limit Theorems for Stochastic Processes*, vol. 288 (Springer Science & Business Media, Berlin, 2013)
57. R. Jordan, D. Kinderlehrer, F. Otto, The variational formulation of the fokker–planck equation. SIAM J. Math. Anal. **29**(1), 1–17 (1998)
58. R.E. Kalman, A new approach to linear filtering and prediction problems. J. Basic Eng. **82**(1), 35–45 (1960)
59. R.E. Kalman, R.S. Bucy, New results in linear filtering and prediction theory. J. Basic Eng. **83**(1), 95–108 (1961)

60. C. Kim, Boltzmann equation with a large potential in a periodic box. Commun. Partial Differ. Equ. **39**(8), 1393–1423 (2014)
61. T.G. Kurtz, Equivalence of stochastic equations and martingale problems, in *Stochastic Analysis 2010* (2011), pp. 113–130
62. K. Law, H. Tembine, R. Tempone, Deterministic mean-field ensemble kalman filtering. SIAM J. Sci. Comput. **38**(3), A1251–A1279 (2016)
63. F. Le Gland, V. Monbet, V.-D. Tran, Large sample asymptotics for the ensemble kalman filter. *Handbook on Nonlinear Filtering* (2011)
64. Y. T. Lee, R. Shen, K. Tian, Logsmooth gradient concentration and tighter runtimes for metropolized hamiltonian Monte Carlo, in *Conference on Learning Theory, COLT 2020, 9–12 July 2020, Virtual Event [Graz]*, vol. 125. *Proceedings of Machine Learning Research*. PMLR (2020), pp. 2565–2597
65. L. Li, On the trend to equilibrium for the Vlasov-Poisson-Boltzmann equation. J. Differ. Equ. **244**(6), 1467–1501 (2008)
66. R. Li, S. Pei, B. Chen, Y. Song, T. Zhang, W. Yang, J. Shaman, Substantial undocumented infection facilitates the rapid dissemination of novel coronavirus (SARS-CoV-2). Science **368**(6490), 489–493 (2020)
67. Q. Liu, D. Wang, Stein variational gradient descent: a general purpose Bayesian inference algorithm. Adv. Neural Inf. Proces. Syst. **29** (2016)
68. J. Lu, L. Wang, On explicit l^2-convergence rate estimate for piecewise deterministic Markov processes in MCMC algorithms. Ann. Appl. Probab. **32**(2), 1333–1361 (2022)
69. J. Lu, Y. Lu, J. Nolen, Scaling limit of the Stein variational gradient descent: the mean field regime. SIAM J. Math. Anal. **51**(2), 648–671 (2019)
70. Y. Lu, J. Lu, J. Nolen, Accelerating Langevin sampling with birth-death. Preprint. arXiv:1905.09863 (2019)
71. J. Lu, E. Tadmor, A. Zenginoglu, Swarm-based gradient descent method for non-convex optimization. Preprint. arXiv:2211.17157 (2022)
72. P. Markowich, C. Villani, On the trend to equilibrium for the Fokker-Planck equation: an interplay between physics and functional analysis, in *Physics and Functional Analysis, Matematica Contemporanea (SBM) 19* (1999), pp. 1–29
73. B.C. May, N. Korda, A. Lee, D.S. Leslie, Optimistic Bayesian sampling in contextual-bandit problems. J. Mach. Learn. Res. **13**(1), 2069–2106 (2012)
74. E. Mazumdar, A. Pacchiano, Y.-A. Ma, P.L. Bartlett, M.I. Jordan, On approximate Thompson sampling with Langevin algorithms, in *Proceedings of the 37th International Conference on Machine Learning* (2020), pp. 6797–6807
75. N.R. Nagarajan, M.M. Honarpour, K. Sampath, Reservoir-fluid sampling and characterization – key to efficient reservoir management. J. Pet. Technol. **59**(8), 80–91 (2007)
76. K. Nanbu, Direct simulation scheme derived from the Boltzmann equation. I. Monocomponent gases. J. Phys. Soc. Jpn. **49**(5), 2042–2049 (1980)
77. B. Øksendal, *Stochastic Differential Equations* (Springer, Berlin, 2003), pp. 65–84
78. L. Pareschi, Optimization by linear kinetic equations and mean-field langevin dynamics. Preprint. arXiv:2401.05553 (2024)
79. L. Pareschi, G. Russo, An introduction to Monte Carlo method for the Boltzmann equation, in *ESAIM: Proceedings*, vol. 10 (EDP Sciences, Les Ulis, 2001), pp. 35–75
80. G. Parisi, Correlation functions and computer simulations. Nucl. Phys. B **180**(3), 378–384 (1981)
81. A. Quarteroni, A. Manzoni, C. Vergara, The cardiovascular system: mathematical modelling, numerical algorithms and clinical applications. Acta Numer. **26**, 365–590 (2017)
82. S. Reich, A dynamical systems framework for intermittent data assimilation. BIT Numer. Math. **51**(1), 235–249 (2011)
83. C.P. Robert, G. Casella, G. Casella, *Monte Carlo Statistical Methods*, vol. 2 (Springer, Berlin, 1999)
84. G.O. Roberts, R.L. Tweedie, Exponential convergence of Langevin distributions and their discrete approximations. Bernoulli **2**(4), 341–363 (1996)

85. P.J. Rossky, J.D. Doll, H.L. Friedman, Brownian dynamics as smart Monte Carlo simulation. J. Chem. Phys. **69**(10), 4628–4633 (1978)
86. D. Russo, B.V. Roy, A. Kazerouni, I. Osband, Z. Wen, A tutorial on Thompson sampling. Found. Trends Mach. Learn. **11**(1), 1–96 (2018)
87. T. Schneider, S. Lan, A. Stuart, J. Teixeira, Earth system modeling 2.0: a blueprint for models that learn from observations and targeted high-resolution simulations.Geophys. Res. Lett. **44**(24), 12–396 (2017)
88. R. Shen, Y.T. Lee, The randomized midpoint method for log-concave sampling, in *Advances in Neural Information Processing Systems 32: Annual Conference on Neural Information Processing Systems 2019, NeurIPS 2019, 8–14 December 2019, Vancouver, BC* (2019), pp. 2098–2109
89. T. Shiga, H. Tanaka, Central limit theorem for a system of Markovian particles with mean field interactions. Zeitschrift für Wahrscheinlichkeitstheorie und verwandte Gebiete **69**(3), 439–459 (1985)
90. Y. Song, J. Sohl-Dickstein, D.P. Kingma, A. Kumar, S. Ermon, B. Poole, Score-based generative modeling through stochastic differential equations. Preprint. arXiv:2011.13456 (2020)
91. W.C. Swope, H.C. Andersen, P.H. Berens, K.R. Wilson, A computer simulation method for the calculation of equilibrium constants for the formation of physical clusters of molecules: application to small water clusters. J. Chem. Phys. **76**(1), 637–649 (1982)
92. A. Sznitman, Équations de type de Boltzmann, spatialement homogenes. Zeitschrift für Wahrscheinlichkeitstheorie und verwandte Gebiete **66**, 559–592 (1984)
93. A. Sznitman, Topics in propagation of chaos, in *Ecole d'Eté de Probabilités de Saint-Flour XIX — 1989*. Springer, Berlin (1991), pp. 165–251
94. E. Tadmor, A. Zenginoglu, Swarm-based optimization with random descent. Preprint. arXiv:2307.12441 (2023)
95. H. Tanaka, Probabilistic treatment of the Boltzmann equation of Maxwellian molecules. Zeitschrift für Wahrscheinlichkeitstheorie und Verwandte Gebiete **46**(1), 67–105 (1978)
96. H. Tanaka, Stochastic differential equation corresponding to the spatially homogeneous Boltzmann equation of Maxwellian and non-cutoff type, in *Stochastic Processes: Selected Papers of Hiroshi Tanaka* (World Scientific, Singapore, 2002), pp. 292–310
97. U. Vaes, Sharp propagation of chaos for the ensemble langevin sampler. Preprint. arXiv:2404.06456 (2024)
98. S. Vempala, Recent progress and open problems in algorithmic convex geometry, in *IARCS Annual Conference on Foundations of Software Technology and Theoretical Computer Science (FSTTCS 2010)*, vol. 8 (2010), pp. 42–64
99. S. Vempala, A. Wibisono, Rapid convergence of the unadjusted langevin algorithm: isoperimetry suffices, in *Advances in Neural Information Processing Systems*, vol. 32 (2019)
100. C. Villani, A review of mathematical topics in collisional kinetic theory. Handb. Math. Fluid Dyn. **1**(71-305), 3–8 (2002)
101. C. Villani et al., *Optimal Transport: Old and New*, vol. 338 (Springer, Berlin, 2009)
102. K. Wu, S. Schmidler, Y. Chen, Minimax mixing time of the metropolis-adjusted langevin algorithm for log-concave sampling. J. Mach. Learn. Res. (2022)

Aggregation-Diffusion Phenomena: From Microscopic Models to Free Boundary Problems

Inwon Kim, Antoine Mellet, and Jeremy Sheung-Him Wu

Abstract This chapter reviews (and expands) some recent results on the modeling of aggregation-diffusion phenomena at various scales, focusing on the emergence of collective dynamics as a result of the competition between attractive and repulsive phenomena—especially (but not exclusively) in the context of attractive chemotaxis phenomena.

At microscopic scales, particles (or other agents) are represented by spheres of radius $\delta > 0$ and we discuss both soft-sphere models (with a pressure term penalizing the overlap of the particles) and hard-sphere models (in which overlap is prohibited). The first case leads to so-called blob models which have received some attention recently as a tool to approximate nonlinear diffusion by particle systems. The hard-sphere model is similar to a classical model for congested crowd motion. We will review well-posedness results for these models and discuss their relationship to classical continuum description of aggregation-diffusion phenomena in the limit $\delta \to 0$: the classical nonlinear drift diffusion equation and its incompressible counterpart.

In the second part of the chapter, we discuss recent results on the emergence and evolution of sharp interfaces when a large population of particles is considered at appropriate space and timescales: At some intermediate timescale, phase separation occurs and a sharp interface appears which evolves according to a Stefan free boundary problem (and the density function eventually relaxes to a characteristic function—metastable steady state for the original problem). At a larger timescale

I. Kim was partially supported by NSF grant DMS-2153254. A. Mellet was partially supported by NSF grant DMS-2009236 and DMS-2307342.

I. Kim · J.S.-H. Wu
Department of Mathematics, UCLA, Los Angeles, CA, USA
e-mail: ikim@math.ucla.edu; jeremywu@math.ucla.edu

A. Mellet (✉)
Department of Mathematics, University of Maryland, College Park, MD, USA
e-mail: mellet@umd.edu

J. A. Carrillo, E. Tadmor (eds.), *Active Particles, Volume 4*, Modeling and Simulation in Science, Engineering and Technology,
https://doi.org/10.1007/978-3-031-73423-6_6

the attractive forces lead to surface tension phenomena and the evolution of the sharp interface can be described by a Hele-Shaw free boundary problem with surface tension. At that same timescale, we will also discuss the emergence of contact angle conditions for problems set in bounded domains.

1 Introduction

1.1 Aggregation-Diffusion Phenomena

The study of emergent collective dynamics in large systems of interacting agents has applications in many areas of applied sciences and has been the source of many interesting mathematical problems for decades. Classical examples in mathematical biology include the collective dynamics of microorganisms such as cells or amoeba via chemotaxis [83, 118] or that of larger living organisms (insect swarms, the motion of animal herds or human crowds) [11, 102, 134]. Throughout this chapter, we will refer to these organisms as active particles and assume that they are experiencing two competing effects: repulsion, modeled by nonlinear diffusion (or even density constraint), and attraction, modeled by nonlocal interactions. The simple mathematical model at the center of this chapter is the following aggregation-diffusion equation:

$$\partial_t \rho + \operatorname{div}(\rho \nabla (G * \rho)) = \operatorname{div}(\rho \nabla f'(\rho)) \qquad \text{in } [0,\infty) \times \mathbb{R}^d, \quad \rho(t=0) = \rho_{in}, \tag{1.1}$$

which describes the evolution of the density of particles $\rho(t, x)$ and provides a continuum description of the particle systems at some macroscopic scale (see [24] and references therein).

The nonlinearity $\rho \mapsto f(\rho)$ describes the repulsive force which models a natural volume exclusion principle (or incompressibility of the particles) and will play a central role throughout this chapter. The standard linear diffusion corresponds to $f(\rho) = \rho \log \rho$, but we will consider pressure laws which penalize values of the density above a critical threshold ρ^* and lead to degenerate diffusion for small values of ρ. Taking $\rho^* = 1$ (to simplify the notations), we will focus on the following power law:

$$f_m(\rho) := \frac{\rho^m}{m-1}, \qquad \forall \rho \geq 0, \qquad m > 1, \tag{1.2}$$

and its limit $m \to \infty$ (which enforces the congestion constraint $\rho \leq 1$):

$$f_\infty(\rho) := \begin{cases} 0 & \text{when } \rho \in [0,1], \\ \infty & \text{when } \rho > 1, \end{cases} \tag{1.3}$$

(in that case (1.1) should be written as (1.10)). More general convex functions can be considered as well.

The interaction kernel G in (1.1) describes long-range interactions. We will focus our analysis on kernels that are radially symmetric, **nonnegative**, and **integrable**:

$$G(x) = G(|x|) \geq 0, \qquad G \in L^1(\mathbb{R}^d).$$

In many applications, the interaction kernels will be radially decreasing[1] at least for large $|x|$ though we do not need such an assumption. One can think of the potential $\phi = G * \rho$ as a chemical sensing function and the drift term in (1.1) models the motion of the agents toward higher values of that function. An important example of an attractive kernel is G given by the equation

$$\sigma G - \eta \Delta G = \delta_{x=0} \qquad \text{in } \mathbb{R}^d, \qquad \lim_{|x|\to\infty} G(x) = 0, \tag{1.4}$$

which satisfies $G(x) \sim |x|^{2-d}$ when $|x| \to 0$ and $\int_{\mathbb{R}^d} G(x)\,dx = \sigma^{-1}$. When f is given by (1.2) and G by (1.4), Eq. (1.1) is the classical (parabolic-elliptic) Patlak-Keller-Segel model (PKS) for chemotaxis [83, 118] which describes the collective motion of cells that are attracted by a self-emitted chemical substance (we refer to Hillen and Painter [71] for a review on various modeling aspects):

$$\begin{cases} \partial_t \rho + \operatorname{div}(\rho \nabla \phi) = \Delta \rho^m, \\ \sigma\phi - \eta\Delta\phi = \rho. \end{cases} \tag{1.5}$$

While we will not exclusively focus on this model, all the results presented here will hold in particular for (1.5) when $m > 2$.

Purpose of this Chapter Aggregation-diffusion equations such as (1.1) have been an active area of mathematical research for the past few decades. Well-posedness, blowups, long-time behavior, steady states, and energy minimizers have all been studied intensely and we will not attempt to present a complete picture of the existing theory (see [7, 8, 21, 23, 24, 46, 91, 132] and references therein). Instead, our focus will be on two recent trends in the analysis of these phenomena at both smaller and larger scales than the continuum model (1.1).

In the first part of the chapter, we will describe a microscopic model for aggregation phenomena that includes a repulsive term taking into account the finite size of the particles, modeled by balls of radius $\delta > 0$. When f is given by (1.2), this model is similar to the blob method introduced for approximating nonlinear diffusion equations [19, 30, 31, 36, 37, 95]. Since this model allows for the overlap

[1] With the sign convention used in (1.1), radially decreasing kernels model an attractive force between particles. While the opposite convention is used in some of the references mentioned in this chapter, this convention allows us to work with a nonnegative kernel.

of the balls, we will refer to it as a soft-sphere model. We will also consider a hard-sphere microscopic model, which includes a strict nonoverlapping condition and which has been used to model congested crowd motion [58, 85, 100–103]. At the macroscopic scale, this hard-sphere model formally corresponds to (1.1) with f given by (1.3) and is related to Hele-Shaw free boundary problems. The purpose of this first part of the chapter is thus to review these microscopic models (their derivation and their properties) and to discuss the connection with the continuum model (1.1).

In the second part of this chapter, we will consider these same phenomena at a larger scale: When $m \in (2, \infty]$, we will show that Eq. (1.5) leads to phase separation (or phase segregation) and to the emergence of sharp interfaces. Phase separation, or the ability of a given population to organize itself into segregated groups with sharp edges, is an important feature of biological aggregation. Examples include animals moving as a homogeneous patch [96, 117] (e.g., insect swarms), pattern formation in cell tissues [3, 28], or the formation of membraneless organelles in eukaryotic cells[2] [16, 72, 107]. The fact that attractive nonlocal interactions can explain this phenomena has been previously pointed out [57, 134], and we present here a rigorous approach based on the recent work [86, 104] to derive these phenomena from (1.1) in regimes corresponding to a very large population of particles observed from far away over large timescales (or, equivalently, small attraction range). Characterizing the dynamics of these interfaces, and thus describing the collective behavior of the particles after aggregation, is a fascinating and challenging problem and we will present some recent results in this direction: At some appropriate timescale, the evolution of this interface is described by free boundary problems (Stefan and Hele-Shaw), which include surface tension effects as well as contact angle conditions and which are reminiscent of classical models for the evolution of interfaces in fluid dynamics.

In the rest of this introduction, we recall some classical facts about aggregation-diffusion equations (1.1) and then present all the models (microscopic, macroscopic, and geometric) that will be studied in the chapter as well as their relationships. Figure 1 summarizes that presentation.

1.2 *Finite Time Blowup vs Global Well-Posedness*

Before presenting these various models, we recall that since the attractive potential G leads to concentration, the well-posedness of (1.1) globally in time is not obvious. Preventing blowups (the formation of Dirac masses) in the mathematical model is

[2] Membraneless organelles are mesoscopic structures found in the nucleus and cytoplasm of cells, characterized by higher protein density and weaker molecular motion than the surrounding medium, but not separated from it by a membrane. They play a crucial role in the cell's operations by allowing for increased rates of biochemical reactions.

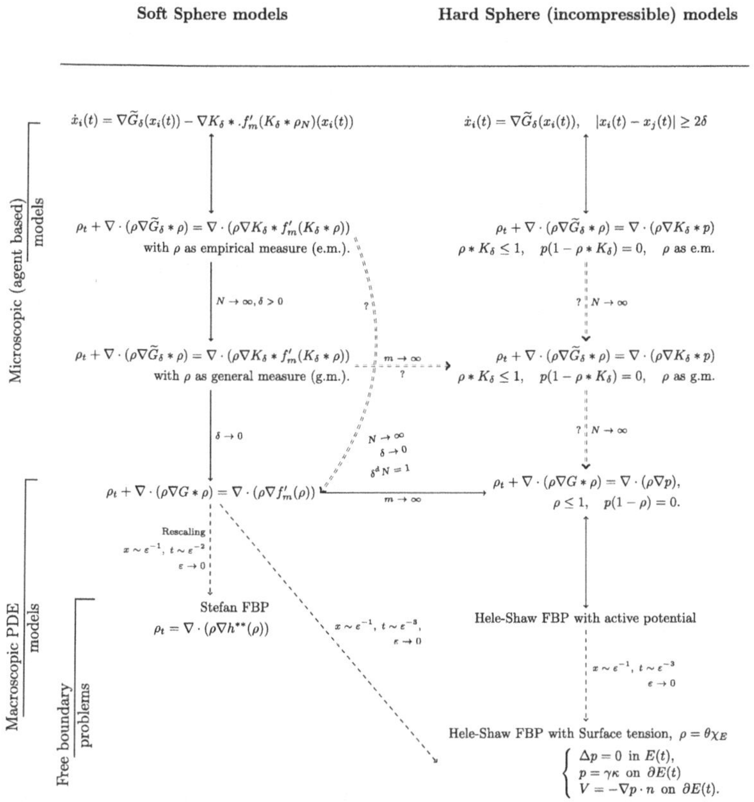

Fig. 1 The various models discussed in this chapter and their relationships. The dashed arrows indicate conditional convergence results. The double dashed arrows indicate open problems

relatively natural from a modeling point of view since one would expect that the particles (if they represent living organisms) have some built-in incompressibility which prevents overlap and thus limits the maximum density. For the classical PKS model (1.5) with $\sigma \geq 0$, it is known that the standard **linear diffusion** ($m = 1$) is not enough to prevent finite time blowup, but for $m > 2 - \frac{2}{d}$, solutions remain bounded in L^∞ uniformly in time (see [7, 8, 21, 132]) and global well-posedness holds for this equation (when $m = 2 - \frac{2}{d}$, well-posedness depends on the initial mass). On the other hand, when $\sigma > 0$ (so that the interaction kernel G is integrable), the power $m_c = 2$ is known to be critical for the existence of stationary states and energy minimizers for (1.5). In particular, while stationary states exist for all $\sigma > 0$ when $m > 2$ (see [6]), they exist if and only if $\sigma < \sigma_0$ when $m = 2$ (see [18, 81]). We will see in the second half of the chapter that the condition $m > 2$ is also the natural condition for solutions of (1.5) to experience phase separation phenomena.

1.3 Micro, Macro, and Geometric Models

We now present the various models that will be discussed in this chapter. In order to keep this introduction relatively short, we did not include references below. Relevant references for each of these models will be discussed in the corresponding sections of the chapter. Figure 1 (at the end of this section) summarizes the relationship between the various models.

Microscopic "agent-based" Models (Soft Sphere and Hard Sphere) At the microscopic level with $N \gg 1$ particles, each particle is represented by a ball $B_\delta(x_i(t))$ with radius $\delta > 0$, whose center $x_i(t)$ evolves according to a first-order ODE for $i = 1, \dots, N$. In the soft-sphere model, the overlap of each ball $B_\delta(x_i(t))$ is allowed, but penalized, resulting in a repulsive force depending on the local density. This leads to the following microscopic model (see Sect. 3.1):

$$\dot{x}_i(t) = \frac{1}{N}\sum_{j=1}^{N} \nabla \widetilde{G}_\delta(x_j(t) - x_i(t))$$
$$- \nabla \int K_\delta(x_i(t) - y) f'\left(\frac{1}{N}\sum_{j=1}^{N} K_\delta(y - x^j(t))\right) dy, \quad \forall i = 1, \dots, N,$$

where $K_\delta = \frac{1}{|B_\delta|}\chi_{B_\delta(0)}$ (or another approximation of unity supported in $B_\delta(0)$) and $\widetilde{G}_\delta = K_\delta * G * K_\delta$. The corresponding empirical distribution $\rho(t,x) = \frac{1}{N}\sum_{i=1}^{N} \delta(x - x_i(t))$ solves the nonlocal PDE:

$$\partial_t \rho + \operatorname{div}(\rho \nabla(\widetilde{G}_\delta * \rho)) = \operatorname{div}(\rho \nabla K_\delta * f'(K_\delta * \rho)) \qquad \text{in } [0,\infty) \times \mathbb{R}^d \tag{1.6}$$

with initial condition

$$\rho(0,x) = \frac{1}{N}\sum_{i=1}^{N} \delta(x - x_i(0)). \tag{1.7}$$

Similar equations and corresponding systems of ODEs have been introduced and used in particular to develop numerical methods for nonlinear diffusion models. In Sect. 3.1, we will recall several results concerning the well-posedness of (1.6) for general initial data (not necessarily empirical distributions) and review convergence results as $N \to \infty$ (mean-field limit—see Theorem 3.1) and $\delta \to 0$ (Theorem 3.3).

The microscopic hard-sphere model is a similar model in which overlap is not permitted: The constraint

$$|x_i(t) - x_j(t)| \geq 2\delta, \qquad \forall i, j = 1, \dots, N, \qquad i \neq j,$$

is strictly enforced, leading to a well-posed system of ODE with constraints (see Sect. 3.2.1). A corresponding PDE can be obtained formally by taking the limit $m \to \infty$ in (1.6), leading to the following equation (see Sect. 3.2):

$$\partial_t \rho + \operatorname{div}(\rho \nabla(\widetilde{G}_\delta * \rho)) = \operatorname{div}(\rho \nabla K_\delta * p), \quad p \in \partial f_\infty(K_\delta * \rho) \qquad \text{in } [0, \infty) \times \mathbb{R}^d, \tag{1.8}$$

which includes a pressure function $p(t, x)$ playing the role of a Lagrange multiplier for the nonlocal constraint $K_\delta * \rho \leq 1$. The condition $p \in \partial f_\infty(K_\delta * \rho)$[3] is a shorthand notation for the three conditions:

$$K_\delta * \rho(t, x) \leq 1, \qquad p(t, x) \geq 0 \qquad p(t, x)(1 - K_\delta * \rho(t, x)) = 0$$
$$\text{a.e. } x \in \mathbb{R}^d, \ t > 0$$

(note that (1.6) can be written in a similar form with $p = \partial f(K_\delta * \rho)$). However, while (1.8) is indeed related to a well-posed microscopic model when the initial condition is an empirical distribution (1.7), we will see that its well-posedness for general initial conditions is far from clear (see Sect. 3.2.5).

Macroscopic Models (Soft Sphere and Hard Sphere) In the limit $\delta \to 0$, Eq. (1.6) leads to the classical aggregation-diffusion equation which we recall here:

$$\partial_t \rho + \operatorname{div}(\rho \nabla(G * \rho)) = \operatorname{div}(\rho \nabla f'(\rho)) \qquad \text{in } [0, \infty) \times \mathbb{R}^d, \tag{1.9}$$

while the hard-sphere model (1.8) yields

$$\partial_t \rho + \operatorname{div}(\rho \nabla(G * \rho)) = \operatorname{div}(\rho \nabla p), \qquad p \in \partial f_\infty(\rho) \qquad \text{in } [0, \infty) \times \mathbb{R}^d. \tag{1.10}$$

A similar equation was introduced as a model for congested crowd motion and some variations of it were derived in many frameworks, in particular as mechanical models for tumor growth. Interpretations and properties of this model are recalled in Sects. 3.2.2–3.2.4. It is closely related to the classical Hele-Shaw free boundary problem with active potential (see Sect. 3.3).

Geometric Models: Phase Separation and Free Boundary Problems The second half of this chapter is devoted to phase separation, sharp interface limits, and the derivation of free boundary problems from (1.9) when f is given by (1.2) with $m > 2$ or from (1.10). The *sharp interface limit* corresponds to the regime in which the size of the support of ρ is very large compared to the interfacial region, and we will derive mathematical models for the evolution of this interface. The results of Sects. 4.1–4.4 can be summarized as follows: When the total mass $\int \rho\, dx =: \varepsilon^{-d}$ is

[3] ∂f_∞ denotes the subdifferential of the convex function f_∞ and is sometimes called the *Hele-Shaw graph*. It is a multivalued function given by

$$\partial f_\infty(\rho) := \begin{cases} 0 & \text{if } \rho < 1 \\ [0, \infty] & \text{if } \rho = 1 \\ \infty & \text{if } \rho > 1 \end{cases}$$

very large ($\varepsilon \ll 1$), we observe the development of an interface separating regions of low and high densities, which happens at a timescale of order ε^{-2}. After that time, the evolution of the interface can be described by a *Stefan free boundary problem*. In that regime, the motion of the interface is driven by the evolution of the density in the bulk. This bulk density eventually relaxes toward a constant (stable) value θ. At that point, the motion of the interface stops, at least at this timescale. But such a state is only metastable and the slower evolution of the interface, at timescale ε^{-3}, can be described by a one-phase Hele-Shaw free boundary problem with surface tension.

To justify this analysis, we consider a large number of particles, $\int \rho_{in}(x)\,dx = \varepsilon^{-d} \gg 1$. Rescaling the variables $x \mapsto \varepsilon x$ and $t \mapsto \varepsilon^2 t$ leads to the equation

$$\partial_t \rho + \operatorname{div}(\rho \nabla (G_\varepsilon * \rho)) = \operatorname{div}(\rho \nabla f'(\rho)) \qquad \text{in } [0,\infty) \times \mathbb{R}^d, \tag{1.11}$$

with $G_\varepsilon(x) := \varepsilon^{-d} G(\varepsilon^{-1}x)$ and $\int \rho_{in}(x)\,dx = 1$ (we can also consider the corresponding hard-sphere model). Recalling that $G \in L^1(\mathbb{R}^d)$ and denoting

$$\int_{\mathbb{R}^d} G(x)\,dx = \frac{1}{\sigma} < \infty,$$

(which is satisfied for (1.4)), we rewrite (1.11) as follows:

$$\partial_t \rho + \operatorname{div}\left(\rho \nabla \int G_\varepsilon(x-y)(\rho(y)-\rho(x))\,dy\right) = \operatorname{div}(\rho \nabla h'(\rho)) \quad \text{in } [0,\infty) \times \mathbb{R}^d, \tag{1.12}$$

where

$$h(\rho) := f(\rho) - \frac{1}{2\sigma}\rho^2 + a\rho \tag{1.13}$$

for some constant $a \in \mathbb{R}$—the addition of the term $a\rho$ in the function h does not affect the equation but will be important for the energy. When f is given by (1.2) with $m > 2$, the function $h''(\rho) = f''(\rho) - \frac{1}{\sigma}$ is negative for small ρ and positive for large ρ. This is a fundamental property which implies that the limiting equation $\partial_t \rho = \operatorname{div}(\rho \nabla h'(\rho))$ is an ill-posed forward-backward diffusion equation and leads to phase separation. In fact, the function h is a **double-well potential**: When f is given by (1.2) with $m > 2$, the function h in (1.13) becomes

$$h_m(\rho) = \frac{\rho}{m-1}\left(\rho^{m-1} - (m-1)\theta_m^{m-2}\rho + (m-2)\theta_m^{m-1}\right), \quad \theta_m = \left(\frac{1}{2\sigma}\right)^{\frac{1}{m-2}}, \tag{1.14}$$

where we took $a = \frac{m-2}{m-1}\theta_m^{m-1}$. This function is a double-well potential with a (singular) well at $\rho = 0$ and a (smooth) well at $\rho = \theta_m$ (see Fig. 2). When $f = f_\infty$, we take $a = \frac{1}{2\sigma}$ in (1.13) and find

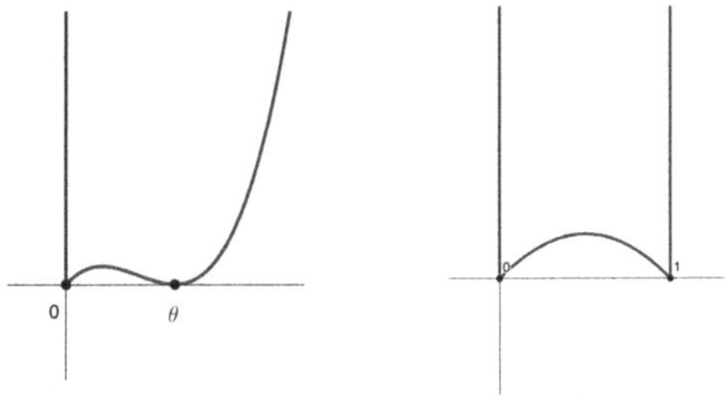

Fig. 2 The double-well potential $h_m(\rho)$ when $m = 3$ (left) and $h_\infty(\rho)$ (right)

$$h_\infty(\rho) := \begin{cases} \frac{1}{2\sigma}\rho(1-\rho) & \text{if } 0 \le \rho \le 1, \\ \infty & \text{otherwise,} \end{cases} \tag{1.15}$$

which is a double-well (or double-obstacle) potential, with wells at $\rho = 0$ and $\rho = 1$ (see Fig. 2).

We do not need to restrict ourselves to f given by (1.2) or (1.3) for the results below. All we need is that the function $\rho \mapsto h(\rho)$ is a double-well potential with stable region $\{0\} \cup (\theta, \infty)$ for some $\theta > 0$. Under this condition, (1.12) is a nonlocal approximation of a Cahn-Hilliard equation whose sharp interface limit $\varepsilon \to 0$ is a classical problem. We will show (in terms of Γ-convergence of the energy—see Theorem 4.5) that the limit $\varepsilon \to 0$ of (1.11) or (1.12) leads to a generalized Stefan problem:

$$\partial_t \rho = \operatorname{div}(\rho \nabla h^{**\prime}(\rho)) \tag{1.16}$$

where h^{**} denotes the convex hull of h. When the initial density does not take values in the unstable phase, that is, if $|\{\rho_{in}(x) \in (0, \theta)\}| = 0$, then (1.16) is a weak formulation of

$$\begin{cases} \rho \ge \theta, \quad \partial_t \rho = \operatorname{div}(\rho \nabla h^{**\prime}(\rho)) & \text{in } E(t) \\ \rho = 0 & \text{in } \mathbb{R}^d \setminus E(t) \\ V = -\nabla h^{**\prime}(\rho) \cdot n & \text{on } \partial E(t), \end{cases} \tag{1.17}$$

where V denotes the normal velocity of $\partial E(t)$. The evolution of ρ in the set $E(t)$ (described by a nonlinear diffusion equation) is thus responsible for the motion of the interface $\partial E(t)$ and will eventually lead to constant density, that is, $\rho(t) \to \theta \chi_E$ for some (constant) set E.

If $\rho_{in}(x) = \theta \chi_{E_{in}}$, then the solution of (1.16) is constant in time which indicates that the corresponding solution of (1.11) evolves very slowly when $\varepsilon \ll 1$ (characteristic functions are metastable for (1.11)). In order to characterize the evolution of such solutions over larger timescale (of order ε^{-3} compared to the microscopic timescale), we consider (in Sect. 4.3) the equation

$$\varepsilon \partial_t \rho^\varepsilon + \operatorname{div}(\rho^\varepsilon \nabla (G_\varepsilon * \rho^\varepsilon)) = \operatorname{div}(\rho^\varepsilon \nabla f'(\rho^\varepsilon)) \qquad \text{in } [0, \infty) \times \mathbb{R}^d, \tag{1.18}$$

with initial data $\rho_{in}(x) = \theta\chi_{E_{in}}$ for some set E_{in} with finite perimeter. When G is the Newtonian kernel (1.4), the evolution of ρ^ε, solution of (1.18), is then asymptotically described by the Hele-Shaw free boundary problem with surface tension (see Theorem 4.7):

$$\begin{cases} \Delta p = 0 & \text{in } E(t), \\ p = \gamma\kappa & \text{on } \partial E(t), \\ V = -\nabla p \cdot \nu & \text{on } \partial E(t), \end{cases} \tag{1.19}$$

where $\kappa(t, x)$ denotes the mean curvature of $\partial E(t)$ (with the convention that $\kappa \geq 0$ when E is convex) and V denotes the normal velocity of the interface $\partial E(t)$. The constant γ depends on m and σ (see (4.10)).

This shows that while the Stefan problem (1.16) describes the phase separation phenomena that takes place for the solutions of (1.11) when $\varepsilon \ll 1$, the resulting clusters (regions when $\rho = \theta$) will continue to evolve over a much larger timescale. This evolution, due to the attractive interactions between the particles, is akin to the evolution of an interface separating two immiscible fluids under the effect of surface tension phenomenon. This is not completely surprising: Surface tension phenomena can be explained by stating that the bond between two molecules of the same fluid (cohesion) is stronger than that between two molecules of different fluids (adhesion). The same can be said about our particles, which are happier when surrounded by other particles than by vacuum.

Another important phenomenon that often comes together with surface tension is the notion of **contact angle**, which appears when the free surface comes in contact with a fixed boundary. In most of the chapter we will look at problems set in the whole space $\mathbb{R}^d$, but in Sect. 4.4 we look at problems set in a bounded domain and show that contact angle conditions are a natural byproduct of the sharp interface limit discussed above. More precisely, we will consider the initial boundary value problem:

$$\begin{cases} \varepsilon\partial_t\rho + \operatorname{div}(\rho\nabla\phi_\varepsilon) = \operatorname{div}(\rho\nabla f'(\rho)) & \text{in } [0, \infty) \times \Omega, \\ \rho\nabla[-f'(\rho) + \phi_\varepsilon] \cdot n = 0 & \text{on } [0, \infty) \times \partial\Omega, \\ \rho(x, 0) = \rho_{in} & \text{in } \Omega. \end{cases} \tag{1.20}$$

In the context of chemotaxis, ϕ_ε will be the solution of an elliptic boundary value problem in Ω (see (4.16)). In that case we will derive a contact angle condition which depends on the boundary conditions for ϕ_ε (see (4.19)). In some experimental settings, it can also be interesting to keep $\phi_\varepsilon = G_\varepsilon * \rho$ (where ρ is extended by 0 to $\mathbb{R}^d$ to make sense of the convolution), modeling interactions that only depend on the distance between the particles. In that case, Ω acts as an obstacle and the limiting free boundary problem is (1.19) supplemented with contact angle condition $\alpha = \pi$ (i.e., the contact is tangential). But general contact angle conditions can be recovered as well by taking into account the interactions of our particles with the fixed boundary (Sect. 4.4.2).

Longer Timescale As a final note, we observe that stationary solutions for (1.9) and global minimizers of the corresponding energy have been the subject of intense research in the past two decades. We will not attempt to give a full review of these results here, but we recall that in many settings it is known that stationary solutions must be radially decreasing. For example, when G is the Newtonian attractive kernel and f is given by (1.2) with $m > 1$, the unique stationary state in two dimensions is radially symmetric and decreasing (see, for instance, [23, 46]). In contrast, both (1.16) and (1.19) have stationary solutions that are not radially decreasing, but consist of clusters separated by vacuum. This suggests that when $\varepsilon \ll 1$, the solutions of (1.11) and (1.18) may first approach some metastable equilibrium, close to some steady states of the limiting free boundary problems, but eventually these clusters will coalesce and converge as $t \to \infty$ to a radially symmetric stationary state (which consists of a unique cluster). Such dynamics are indeed observed numerically (see, for instance, [24]).

2 Energies and Gradient Flows

Formally, all the models described in this chapter have a gradient flow structure on the space of probability measures with finite second moment:

$$\mathcal{P}_2(\mathbb{R}^d) := \left\{ \rho \in \mathcal{P}(\mathbb{R}^d)\,;\ \int_{\mathbb{R}^d} |x|^2 d\rho(x) < \infty \right\},$$

endowed with the 2-Wasserstein metric which we denote by d_W. This structure was first formalized for the Fokker-Planck equation in the seminal work of Jordan, Kinderlehrer, and Otto [80] and extended to other PDEs in various work; see, for instance, [5, 22, 27]. Every equation presented in the introduction can be written in the form

$$\partial_t \rho + \operatorname{div}(\rho v) = 0, \qquad v = -\nabla \frac{\delta \mathscr{E}}{\delta \rho}(\rho) \text{ in } (0,\infty) \times \mathbb{R}^d, \tag{2.1}$$

for some energy functional $\mathscr{E} : \mathcal{P}_2 \to \mathbb{R} \cup \{+\infty\}$, which in turn provides a natural Lyapunov functional for the evolution equation: Weak solutions of (2.1) satisfy the energy dissipation inequality:

$$\mathscr{E}[\rho(t)] + \int_0^t \int_{\mathbb{R}^d} |v|^2 d\rho\, ds \le \mathscr{E}[\rho(0)].$$

We briefly present below the energy functionals associated to our models. Throughout the chapter we will rely on these functionals and their asymptotic behavior to develop heuristic arguments and to justify singular limits of the associated PDEs. For most of this chapter, we will work with measures ρ which are absolutely continuous with respect to Lebesgue measure. This is denoted by $d\rho \ll dx$ and we will use ρ to denote both the measure and its density with respect to Lebesgue measure.

The (macroscopic soft-sphere) aggregation-diffusion equation (1.9) is classically associated with the energy:

$$\mathscr{E}_f[\rho] := \mathscr{F}_f[\rho] - \frac{1}{2}\iint_{\mathbb{R}^d\times\mathbb{R}^d} G(x-y)\,d\rho(x)\,d\rho(y),$$

$$\mathscr{F}_f[\rho] := \begin{cases} \int_{\mathbb{R}^d} f(\rho(x))\,dx, & \text{if } \rho \in \mathcal{P}_2(\mathbb{R}^d),\ d\rho \ll dx, \\ +\infty, & \text{otherwise.} \end{cases} \tag{2.2}$$

We use the notation $\mathscr{E}_m$ (resp. $\mathscr{E}_\infty$) and $\mathscr{F}_m$ (resp. $\mathscr{F}_\infty$) when f is given by (1.2) (resp. (1.3)). In particular, for $\rho \in \mathcal{P}_2(\mathbb{R}^d)$ such that $d\rho \ll dx$, we have

$$\mathscr{F}_m[\rho] = \int_{\mathbb{R}^d} \frac{\rho(x)^m}{m-1}\,dx, \qquad \mathscr{F}_\infty[\rho] = \begin{cases} 0 & \text{if } \rho \le 1 \text{ a.e.} \\ +\infty & \text{otherwise,} \end{cases}$$

and the energy $\mathscr{E}_\infty$, given by

$$\mathscr{E}_\infty[\rho] = \mathscr{F}_\infty[\rho] - \frac{1}{2}\iint G(x-y)\,d\rho(x)\,d\rho(y), \tag{2.3}$$

is associated with the hard-sphere model (1.10).

Similarly, the microscopic model (1.6) is a gradient flow for the regularized energy:

$$\mathscr{E}_\delta[\rho] := \mathscr{F}_f[\rho * K_\delta] - \frac{1}{2}\iint_{\mathbb{R}^d\times\mathbb{R}^d} \widetilde{G}_\delta(x-y)\,d\rho(x)\,d\rho(y). \tag{2.4}$$

Notice that $\mathscr{E}_\delta[\rho] = \mathscr{E}_f[K_\delta * \rho]$. In particular, thanks to the convolution with K_δ, the measure ρ does not have to be absolutely continuous with respect to the Lebesgue measure for $\mathscr{E}_\delta$ to be finite. These energies play an important role in our proofs, and the Γ-convergence of $\mathscr{E}_\delta$ to $\mathscr{E}_f$ as $\delta \to 0$ and that of $\mathscr{E}_m$ to $\mathscr{E}_\infty$ as $m \to \infty$ are good indicators of the relationship between the corresponding PDEs.

Such Γ-convergence results play an even more central role in the second part of the chapter and the derivation of the free boundary problems (1.16) and (1.19). First, we note that the rescaled equation (1.11) is a gradient flow for the rescaled energy:

$$\mathscr{F}_f[\rho] - \frac{1}{2}\iint_{\mathbb{R}^d\times\mathbb{R}^d} G_\varepsilon(x-y)\,d\rho(x)\,d\rho(y), \qquad G_\varepsilon(x) = \varepsilon^{-d} G(\varepsilon^{-1}x). \tag{2.5}$$

Up to a constant, this energy can also be written as

$$\mathscr{J}_\varepsilon[\rho] = \begin{cases} \int_{\mathbb{R}^d} h(\rho)\,dx + \frac{1}{4}\iint_{\mathbb{R}^d\times\mathbb{R}^d} G_\varepsilon(x-y)[\rho(x)-\rho(y)]^2\,dx\,dy & \text{if } \rho \in \mathcal{P}_2(\mathbb{R}^d),\ d\rho \ll dx, \\ \infty & \text{otherwise,} \end{cases} \tag{2.6}$$

where h is the double-well potential defined by (1.13) (with wells at $\rho = 0$ and $\rho = \theta$). This functional is a nonlocal approximation of the Allen-Cahn energy:

$$\int_{\mathbb{R}^d} h(\rho(x))\,dx + \varepsilon^2\beta \int_{\mathbb{R}^d} \frac{1}{2}|\nabla\rho|^2\,dx, \tag{2.7}$$

commonly used in problems that exhibit phase separation. We will prove in particular that as long as $\int_{\mathbb{R}^d}(1+|x|^2)G(x)\,dx < \infty$, $\mathscr{J}_\varepsilon$ Γ-converges to (see Proposition 4.2)

$$\mathscr{J}^*[\rho] = \int_{\mathbb{R}^d} h^{**}(\rho)\,dx, \tag{2.8}$$

whose corresponding gradient flow is the generalized Stefan free boundary problem (1.16). Characteristic functions $\theta\chi_E$ are stationary solutions of (1.16) and $\mathscr{J}_\varepsilon(\theta\chi_E) \to \mathscr{J}^*(\theta\chi_E) = 0$ as $\varepsilon \to 0$. But the rescaled functional $\mathscr{G}_\varepsilon[\rho] := \varepsilon^{-1}\mathscr{J}_\varepsilon[\rho]$ Γ-converges to the perimeter functional (see Theorem 4.6):

$$\mathscr{G}_0(\rho) = \begin{cases} \gamma\theta P(E) & \text{if } \rho \in \mathcal{P}_2(\mathbb{R}^d), \quad \rho = \theta\chi_E \in \mathrm{BV}(\mathbb{R}^d), \\ \infty & \text{otherwise,} \end{cases}$$

for some constants γ depending on σ and θ where the perimeter $P(E)$ of the set $E \subset \mathbb{R}^d$ is defined by

$$P(E) = \sup\left\{ \int_E \operatorname{div} g\,dx\,;\ g \in C_0^1(\mathbb{R}^d;\,\mathbb{R}^d), \quad |g(x)| \le 1,\ \forall x \in \mathbb{R}^d \right\}. \tag{2.9}$$

Such results, for general kernels G and smooth double-well potentials, were first derived in [1]. This convergence is also proved by a different approach in [104] when G is given by (1.4) and f is given by (1.2) with $m > 2$ and in [86, 106] when $f = f_\infty$. In these papers, boundary conditions (when the problem is set in a bounded domain) play a crucial role (see Theorem 4.8) and led to the derivation of contact angle conditions.

The final observation is that the Wasserstein gradient flow for the perimeter functional is the Hele-Shaw free boundary problem with surface tension (1.19) (see [114]). The Γ-convergence of $\mathscr{G}_\varepsilon$ to $\mathscr{G}_0$ is thus consistent with the convergence of (1.11) to (1.19). However, the Γ-convergence of the energy alone does not guarantee the convergence of the corresponding gradient flows. Additional sufficient conditions for such a convergence were established by Sandier-Serfaty [129] and used, for example, to justify the convergence of (1.5) to (1.10) in the limit $m \to \infty$ in [39]. These additional conditions can be difficult to establish in other frameworks and we will not review or extend such results in this chapter. For the derivation of the free boundary problems (1.16) and (1.19) we will take a different approach and state conditional convergence results (see Theorems 4.5 and 4.7), which imply the convergence of the solutions under an assumption on the convergence of the energy (see (4.7)). Such conditional convergence results have a long history with free boundary problems involving mean curvature (see [55, 78, 94, 114]), and proving rigorous convergence results without this assumption remains an important and challenging problem.

3 Part 1: Microscopic/Macroscopic Soft-Sphere and Hard-Sphere Models

3.1 *Soft-Sphere Models: The Blob Method*

The main goal of this section is to derive the microscopic model (1.6) from particle dynamics and review some of the recent literature on this and related models. The convolution with the kernel K_δ is the key feature of this model. It accounts for the finite size ($\delta > 0$) of the particles, as opposed to treating them as points. We will see that (1.6) has solutions in the form of empirical distributions (sums of Dirac masses centered at points $x_i(t)$), a fact that allows us to connect (1.6) to the corresponding microscopic model describing the motion of individual particles via a system of coupled ODEs.

From a mathematical perspective, (1.6) is the gradient flow of a λ-convex energy functional (under appropriate assumptions on K_δ and f). This classical framework yields the well-posedness of (1.6) in the set of probability measures. Finally, we will discuss the connection between (1.6) and the classical aggregation-diffusion equation (1.9) in the limit $\delta \to 0$.

3.1.1 Transport Equation with Interaction Potential

Transport equations without any diffusion and with sufficiently smooth interaction potential $\widetilde{G} : \mathbb{R}^d \to \mathbb{R}$ (satisfying $\nabla \widetilde{G}(0) = 0$) are naturally associated to deterministic particle systems via the empirical distribution: Denoting by $\{x_i(t)\}_{i=1,\dots,N}$ the positions of N particles, solutions to the following system of ODEs,

$$\dot{x}_i(t) = \frac{1}{N} \sum_{j=1}^{N} \nabla \widetilde{G}(x_j(t) - x_i(t)) \qquad \forall i = 1, \dots, N, \tag{3.1}$$

the corresponding empirical distribution

$$\rho_N(t,x) = \frac{1}{N}\sum_{j=1}^{N} \delta(x - x_j(t))$$

is the solution (in the sense of distribution) of the continuity equation:

$$\partial_t \rho + \operatorname{div}(\rho \nabla \widetilde{G} * \rho) = 0 \tag{3.2}$$

with initial data $\rho_N(0,x) = \frac{1}{N}\sum_{j=1}^{N} \delta(x - x_j(0))$ (see, for instance, [22, 29]).

Equation (3.2) appears in many settings, particularly in the mathematical modeling of collective behaviors. We refer to [22] and the many references therein for a detailed discussion of the associated well-posedness theory. We recall that (3.2) is a gradient flow for the energy:

$$-\frac{1}{2}\int_{\mathbb{R}^d}\int_{\mathbb{R}^d} \widetilde{G}(x-y)d\rho(x)\,d\rho(y)$$

defined on $\mathcal{P}_2(\mathbb{R}^d)$ (this was formalized in [27] following ideas that were first introduced in [115]). In that context, a key assumption for the well-posedness of (3.2) is the λ-convexity of the kernel $-\widetilde{G}$ for some $\lambda \leq 0$ (see [22]). Under this assumption, we also get the following stability estimate:

$$d_W(\rho^1(t), \rho^2(t)) \leq d_W(\rho^1(0), \rho^2(0))e^{-\lambda t} \tag{3.3}$$

for any two weak measure solutions $\rho^1(t)$ and $\rho^2(t)$ of (3.2) in $C([0,\infty); \mathcal{P}(\mathbb{R}^d))$. Stability estimates have a long history and a similar inequality was first proved by Dobrushin [49] (see also Golse [65]) with the 1-Wasserstein distance. Deriving such estimates (which can lead to useful numerical approximations) when the interaction potential $\widetilde{G}$ is less regular and possibly in different topologies is still an active area of research. Inequality (3.3) implies in particular that if $\rho_N(0)$ converges to ρ_{in} (with respect to d_W), the measure $\rho_N(t,x)$ converges to the corresponding solution of (3.2). We will say that (3.2) is the mean-field limit of the ODE system (3.1).

The mean field limit (but not (3.3)) can also be justified for some more singular potentials (see, for instance, [29] when $\widetilde{G}(x) \sim |x|^b$ with $b > 2-d$), but not, to our knowledge, with the Newtonian kernel as in the PKS model (in dimension 1, some results have been obtained by Bonashi-Carrillo-Di Francesco-Peletier [14] for the attractive Newtonian kernel). Other classical references for the mean-field limit of a related model include Sznitman [133] and Golse [65] as well as the recent reviews with both theoretical and numerical perspectives [32, 33].

3.1.2 Linear/Nonlinear Repulsion

When diffusion (linear or nonlinear) is added to (3.2), solutions starting with empirical measures are instantly regularized by the diffusion. This makes the approximation of the corresponding PDE by a deterministic ODE system challenging (see [44, 128]). A classical way to deal with this issue is to consider instead a system of stochastic ODEs (adding a Brownian motion to (3.1)) as is done with the discretization of the linear Fokker-Planck equation by Langevin dynamics. Some nonlinear diffusion equations have also been handled via such methods [59, 110, 112, 122]. We also refer to some recent work treating singular potentials via the so-called relative-entropy method [17, 76, 77] and modulated energy method [126] (the result of [17] applies to the Patlak-Keller-Segel model in dimension 2).

We will now describe a different approach, based on the so-called blob method , which is a deterministic approach in which the (nonlinear) diffusion is approximated by a nonlocal interaction term which preserves the empirical distribution (see [19, 30, 31, 36, 37, 95]). This approach was also used by Motsch and Peurichard in [111] to derive a model for tumor growth.

3.1.3 Nonlocal Nonlinear Repulsion

The *blob method* amounts to taking into account the finite size $\delta > 0$ of the particles (at the microscopic level) and some volume exclusion principle. A key feature of the microscopic models we have in mind is a repulsive effect with finite radius: Particles that are within a certain distance of each other feel the effect of a strong repulsion, which prevents the density of particles from becoming too large.

Given a compactly supported function $\bar{K}_\delta : \mathbb{R}^d \to [0, \infty)$, we introduce the counting function:

$$\mu_N(x) = \sum_{j=1}^{N} \bar{K}_\delta(x - x_j(t)).$$

When $\bar{K}_\delta = \chi_{B_\delta(0)}$ (an important example we will come back to regularly), $\mu_N(x)$ counts how many particles are located in $B_\delta(x)$. If we think of the particles as *hard spheres* of radius δ, it is natural to impose the constraint $\mu_N \leq 1$. We will discuss this setting, and its many challenges, in the next section, but for now we will consider a *soft-sphere* model which instead of this hard constraint includes a repulsion force that drives the particles away from crowded regions with a strength that depends on μ_N.

Before introducing the model, we note that we can interpret the quantity $\int \bar{K}_\delta(x)\,dx = |B_\delta| = \omega_d \delta^d$ as the volume occupied by one particle, so the total volume occupied by all particles is equal to $m_0 := N \int \bar{K}_\delta(x)\,dx$. Introducing the normalized kernel $K_\delta = \frac{1}{\int \bar{K}_\delta(x)\,dx} \bar{K}_\delta$, we can write

$$\mu_N(x) = N\bar{K}_\delta * \rho_N(x) = m_0 K_\delta * \rho_N(x).$$

In order to simplify the notations, we will take $m_0 = 1$ below so that the counting function μ_N can be written as

$$\mu_N(x) = K_\delta * \rho_N(x), \qquad \int K_\delta(x)\,dx = 1, \quad \mathrm{Supp}\, K_\delta = B_\delta.$$

Note that we can choose $K_\delta(x) = \frac{1}{\omega_d \delta^d} K(\frac{x}{\delta})$ with $K(x) = \chi_{B_1}(x)$, but when convenient we will replace this kernel with a smooth radially symmetric decreasing function supported in B_1. In what follows N and δ will be treated as independent parameters, but we should remember that this interpretation of the microscopic model imposes the constraint $N \sim \delta^{-d}$.

The soft-sphere model is obtained by adding a repulsive term to (3.1) as follows:

$$\dot{x}_i(t) = \frac{1}{N}\sum_{j=1}^{N} \nabla\widetilde{G}(x_j(t)-x_i(t)) - \int K_\delta(x_i(t)-y)\nabla f'(\mu_N)(y)\,dy, \quad i = 1, \dots N, \tag{3.4}$$

for a convex function f (for instance, given by (1.2)). Equation (3.4) can also be written as

$$\dot{x}_i(t) = \nabla\widetilde{G} * \rho_N(x_i(t)) - \nabla K_\delta * f'(K_\delta * \rho_N(x))(x_i(t)), \qquad i = 1, \dots N. \tag{3.5}$$

The particular structure of this repulsive term, with the double convolution, is convenient from a mathematical viewpoint (it leads to an interpretation of the corresponding PDE as a gradient flow). From a modeling viewpoint, it can be interpreted as follows (when $K(x) = \chi_{B_1}(x)$): Since a particle is represented by the ball $B_\delta(x_i(t))$, the repulsion force exerted on its center $x_i(t)$ is the average over $B_\delta(x_i(t))$ of the gradient of the pressure $f'(\mu_N(x))$—which depends on the counting function μ_N.

System (3.5) is similar to the deterministic system used, for example, in [30, 31, 36, 37] to approximate the nonlinear diffusion equation via what has been called the blob method. One advantage of this approach is the fact that given a solution $\{x_i(t)\}_{i=1,\dots,N}$ of (3.4), the empirical distribution $\rho_N(t,x)$ is a solution (in the sense of distribution) of the nonlinear transport equation:

$$\partial_t \rho + \mathrm{div}\,(\rho\nabla\widetilde{G} * \rho) = \mathrm{div}\,(\rho\nabla K_\delta * f'(K_\delta * \rho)). \tag{3.6}$$

However, it has been observed (see, for example, [111]) that the long-time behavior of (3.6) exhibits discrepancies with that of the microscopic dynamics (3.5): When $\widetilde{G}$ is zero and K_δ is compactly supported, numerical simulations show that solutions of (3.6) with absolutely continuous densities spread and converge (locally) to zero (while preserving the mass) as $t \to \infty$ (which is consistent with the behavior of solutions of the porous media equation), whereas for a finite number of particles N, the solutions to (3.5) converge to a finite nonzero value (in particular, the dynamics stops when all the points $x_i(t)$ are at distance greater than 2δ from one another). So, while we have a one-to-one correspondence between the solutions of (3.5) and empirical measures solutions of (3.6), these empirical measures are *unstable* solutions of (3.6) (in the sense that perturbations of the Dirac masses diffuse in space under the dynamics of (3.6), hence departing further from their original Dirac structure). This discrepancy between the microscopic (ODE) and macroscopic (PDE) dynamics is not unique to this model and we shall see this again for the hard-sphere (or hard congestion) models in the next section.[4]

Note that (3.6) is reminiscent of the transport equation with Brinkman's law, which can be written as

$$\partial_t \rho = \operatorname{div}(\rho \nabla \widetilde{G} * \rho) + \operatorname{div}(\rho \nabla K_\delta * f'(\rho)), \qquad K_\delta - \delta \Delta K_\delta = \delta_{x=0}. \tag{3.7}$$

This model has a similar behavior as (3.6) when $f'(\rho) = \rho$, but when $f(\rho)$ is given by (1.2) with $m \gg 1$, (3.7) imposes a height constraint on ρ while (3.6) imposes a constraint on $K_\delta * \rho$. Both the dynamics and the mathematical theory are quite different in that case.

3.1.4 A Nonlocal, Nonlinear Keller-Segel Model

In the parabolic-elliptic approximation of the Patlak-Keller-Segel model, the attractive potential $\phi = G * \rho$ solves $\sigma\phi - \eta\Delta\phi = \rho$ and represents the concentration of a self-emitted chemoattractant substance. At the microscopic scale, we should also take into account the finite size of the particles B_δ (here, we take $K = \chi_{B_1}$ again): First, the production of the chemoattractant occurs over the whole ball $B_\delta(x_i)$ instead of being concentrated at the center x_i, leading to

$$\sigma\phi_\delta - \eta\Delta\phi_\delta = K_\delta * \rho_N . \tag{3.8}$$

Next, the particle sensors are also located over the whole ball $B_\delta(x_i)$, rather than only in the center x_i, so that the velocity is given by $\nabla K_\delta * \phi_\delta(x_i(t))$:

$$\dot{x}_i(t) = \nabla K_\delta * \phi_\delta(x_i(t)) - \nabla K_\delta * f'(K_\delta * \rho_N(x))(x_i(t)) . \tag{3.9}$$

This system is exactly (3.4) if we replace $\widetilde{G}$ with $\widetilde{G}_\delta := K_\delta * \widetilde{G} * K_\delta$. The good news is that while G is too singular to apply the existing well-posedness theory, the

[4] Motsch and Peurichard [111] proposed a stabilizing method which consists in assuming that $f(\rho) = 0$ for $\rho \le \rho^*$. This is to ensure that no repulsion occurs when the particle density ρ falls below some threshold value ρ_*.

regularized kernel $\widetilde{G}_\delta$ is much nicer. In particular,[5] we have $\widetilde{G}_\delta \in W^{2,\infty}$ as soon as $K_\delta \in L^2$.

Combining the results mentioned above (for instance, [22] and [31]), we deduce that when $\{x_i(t)\}_{i=1,\dots,N}$ solves (3.8)–(3.9), the empirical distribution $\rho_N(t,x)$ solves (1.6) which is the gradient flow in $\mathcal{P}_2(\mathbb{R}^d)$ for the energy functional (2.4). This gradient flow is studied in [22] when $f = 0$ (in particular $\widetilde{G}_\delta$ satisfies the assumptions of [22] when $K_\delta \in L^2$) and in [31] when $\widetilde{G}_\delta = 0$ and f satisfies some assumptions satisfied in particular when $f(\rho) = \frac{1}{m-1}\rho^m$ (but with additional assumption on K_δ). We also refer to earlier work in a similar spirit by Carrillo, Craig, and Patacchini [30]. The special case $m = 2$ has been given detailed attention by Burger and Esposito [19] and Craig, Elamvazhuthi, Haberland, and Turanova [37] (with applications in cross-diffusion and sampling, respectively). Adapting the method developed in these papers (in particular [22, 30, 37]), we get:

Theorem 3.1 *Fix $\delta > 0$; assume that $K \in C^2(\mathbb{R}^d)$, $K \in W^{2,\infty}(\mathbb{R}^d)$; and take $f = f_m$ for some $m > 1$. For all $\rho_{in} \in \mathcal{P}_2(\mathbb{R}^d)$ with $\mathscr{E}_\delta[\rho_{in}] < \infty$ (where E_δ is defined by (2.4)), there exists a unique gradient flow solution $\rho^\delta(t)$ in $AC_{loc}([0,\infty); \mathcal{P}_2(\mathbb{R}^d))$ of (1.6). This solution is characterized by*

$$\begin{cases} \partial_t \rho^\delta + \operatorname{div}(\rho^\delta \nabla \widetilde{G}_\delta * \rho^\delta) = \operatorname{div}(\rho^\delta \nabla K_\delta * f_m'(K_\delta * \rho^\delta)) & \text{in } \mathcal{D}'((0,\infty)\times\mathbb{R}^d) \\ \lim_{t\to 0^+} d_W(\rho^\delta(t), \rho_{in}) = 0. \end{cases} \tag{3.10}$$

Furthermore, given two such solutions $\rho^1(t)$ and $\rho^2(t)$, the following stability estimate holds:

$$d_W(\rho^1(t), \rho^2(t)) \le d_W(\rho^1(0), \rho^2(0)) e^{-\lambda_0 t} \tag{3.11}$$

with

$$\lambda_0 := -C(\|K\|_{W^{2,\infty}(\mathbb{R}^d)})\delta^{-d(m-1)-2} + C(\|K\|_{L^2})\delta^{-d}. \tag{3.12}$$

Inequality (3.11) implies the mean-field limit for fixed δ: If the empirical measure converges to a density ρ_{in}, in the sense that $d_W(\rho_N(t=0), \rho_{in}) \to 0$, then the empirical distribution $\rho_N(t,x)$ converges weakly to the solution ρ of (3.10). The constant λ_0 given by (3.12) is a (lower bound for the) modulus of convexity for the energy functional $\mathscr{E}_\delta$ (defined by (2.4)) with respect to the underlying optimal transport geodesic structure on $\mathcal{P}_2$. Importantly, we note that when $\delta \to 0$, we have $\lambda_0 \to -\infty$ as $\delta \to 0$. On the other hand, it is known that the limiting energy $\mathscr{E}$ is convex. This discrepancy suggests that this (3.12) is far from optimal.

---

[5] Indeed, the function $u = G * K_\delta$ solves $\sigma u - \eta \Delta u = K_\delta \in L^2$ and is thus in H^2. We deduce $\|D^2 \widetilde{G}_\delta\|_{L^\infty} = \|K_\delta * D^2 u\|_{L^\infty} \le \|K_\delta\|_{L^2} \|D^2 u\|_{L^2} \le C\|K_\delta\|_{L^2}^2$.

Sketch of the Proof of Theorem 3.1 Our assumptions on K_δ and the properties of $\widetilde{G}_\delta$ ensures that $\mathscr{E}_\delta : \mathcal{P}_2(\mathbb{R}^d) \to (-\infty, +\infty]$ is proper, lower-semicontinuous, and λ-convex with modulus of convexity given by (3.12) (see [22, 30]). The differentiability of $\mathscr{E}_\delta$ can be proved using the same arguments as in [22] (for the $\widetilde{G}_\delta$ term) and [30] (for the f_m-term) and leads to

$$\frac{\delta \mathscr{E}_\delta}{\delta \rho}[\rho] = -\widetilde{G}_\delta * \rho + f_m'(K_\delta * \rho) * K_\delta.$$

Existence and uniqueness of the gradient flow ρ^δ, as well as the fact that the gradient flow is a curve of maximal slope, follows from [5, Theorem 11.2.1]. Furthermore, ρ^δ solves the continuity equation:

$$\partial_t \rho^\delta + \operatorname{div}(\rho^\delta v^\delta) = 0, \quad \text{in } \mathcal{D}'(\mathbb{R}^d \times (0, \infty)),$$

with velocity

$$v^\delta(t) = \nabla \widetilde{G}_\delta * \rho^\delta(t) - \nabla K_\delta * f_m'(K_\delta * \rho^\delta(t)) \qquad \text{a.e. } t > 0,$$

and satisfies the energy inequality

$$\mathscr{E}_\delta[\rho^\delta(t)] + \int_0^t \int_{\mathbb{R}^d} |v^\delta|^2 d\rho^\delta(s)\, ds \leq \mathscr{E}_\delta[\rho_{\text{in}}]. \tag{3.13}$$

We note that the presence of the convolution and the assumption on K_δ implies that v^δ is uniformly bounded (and C^1), so ρ^δ solves (3.10). Conversely, this also implies (see [37, Proposition 3.12]) that any solution of (3.10) is in fact in $AC_{loc}([0, \infty); \mathcal{P}_2(\mathbb{R}^d))$ and is a gradient flow of $\mathscr{E}_\delta$.

Finally, we get (3.11) as a consequence of the usual stability estimate for λ-gradient flows [5, Theorem 11.2.1] (see also [37, Theorem 1.4]).

For future references, we also recall that ρ^δ satisfies the propagation of the second moment:

$$\int_{\mathbb{R}^d} |x|^2 \rho^\delta(t, x)\, dx \leq \left(\int_{\mathbb{R}^d} |x|^2 \rho_{\text{in}}(x)\, dx + \mathcal{E}_\delta[\rho_{\text{in}}] \right) e^t, \tag{3.14}$$

which follows from (3.13) since we have

$$\frac{d}{dt} \int_\Omega |x|^2 \rho^\delta(t, x)\, dx = 2 \int_\Omega x \cdot v^\delta \rho^\delta\, dx \leq 2 \left(\int_\Omega |x|^2 \rho^\delta\, dx \right)^{1/2} \left(\int_\Omega |v^\delta|^2 \rho^\delta\, dx \right)^{1/2}.$$

3.1.5 Convergence to the Macroscopic Model: The Limit $\delta \to 0$

We now turn to the question of the convergence of the nonlocal model (1.6) to the classical aggregation-diffusion equation (1.9) when $\delta \to 0$. We note that such a convergence cannot hold unless (1.9) has a solution, and it is well known that aggregation diffusion equations can experience finite time blowup. When f is given by (1.2) and G is the chemotaxis potential (1.4), then global in time (bounded) solutions exist in the subcritical regime $m > m_c := 2 - \frac{2}{d}$ (more generally, if $G(x) \sim \frac{1}{|x|^k}$, then the critical power is given by $m_c := 1 + \frac{k}{d}$; see [24]). In fact, we have:

Theorem 3.2 *Let $f = \frac{1}{m-1}\rho^m$ with $m > 2 - \frac{2}{d}$ and G be given by* (1.4). *Given $\rho_{in} \in \mathcal{P}_2(\mathbb{R}^d)$ such that $\|\rho_{in}\|_{L^\infty(\mathbb{R}^d)} < \infty$ and* $\operatorname{Supp} \rho_{in} \in B_R$, *Eq.* (1.9) *has a unique global solution $\rho(t,x)$ with initial data $\rho_{in}(x)$.*

Furthermore, for all $T > 0$ there exist $C(T)$ and $R(T)$ such that

$$\|\rho\|_{L^\infty((0,T)\times\mathbb{R}^d)} \le C(T), \qquad \operatorname{Supp} \rho(t) \subset B_{R(T)} \quad \forall t \in [0,T].$$

We refer, for example, to [8, 21, 91, 132] for the existence of a global bounded solution. The propagation of the L^∞ norm and compact support can also be proved as in [68, Appendix A and B] and uniqueness is proved in [8].

While this result only requires $m > 2 - \frac{2}{d}$, we will now make the stronger assumption $m > 2$ to simplify the analysis (the assumption $m > 2$ will be crucial in the second part of this chapter but the result below should hold for all $m > 2 - \frac{2}{d}$). Indeed, since $\int_{\mathbb{R}^d} G(x)\,dx = \frac{1}{\sigma}$, we can rewrite the energy (2.2) as

$$\begin{aligned}
\mathcal{E}_m[\rho] &= \int_{\mathbb{R}^d} f_m(\rho)\,dx - \frac{1}{2}\iint_{\mathbb{R}^d\times\mathbb{R}^d} G(x-y)\,\rho(x)\,\rho(y)\,dx\,dy \\
&= \int_{\mathbb{R}^d} f_m(\rho) - \frac{1}{2\sigma}\rho^2\,dx + \frac{1}{2}\iint_{\mathbb{R}^d\times\mathbb{R}^d} G(x-y)\left[\rho(x)^2 - \rho(x)\,\rho(y)\right]dx\,dy \\
&= \int_{\mathbb{R}^d} f_m(\rho) - \frac{1}{2\sigma}\rho^2\,dx + \frac{1}{4}\iint G(x-y)[\rho(x)-\rho(y)]^2\,dx\,dy.
\end{aligned}$$

When $m > 2$, the function $\rho \mapsto f_m(\rho) - \frac{1}{2\sigma}\rho^2$ is bounded below. Since the equation preserves $\int \rho(t,x)\,dx$ over time, we can add a term $a\rho$ to make it nonnegative (see (1.13)). This leads to the function $\rho \mapsto h_m(\rho)$ defined by (1.14), which satisfies $h_m(\rho) \ge 0$ for all $\rho \ge 0$. We can then work with the energy functional:

$$\overline{\mathcal{E}_m}[\rho] := \begin{cases} \displaystyle\int h_m(\rho(x))\,dx + \frac{1}{4}\iint G(x-y)[\rho(x)-\rho(y)]^2\,dx\,dy & \\ \qquad\qquad\qquad \text{if } \rho \in \mathcal{P}_2(\mathbb{R}^d),\ d\rho(x) \ll dx, \\ \infty & \\ \qquad\qquad\qquad \text{otherwise,} \end{cases}$$

which differs from $\mathscr{E}_m$ by a constant. Similarly, we can rewrite the energy (2.4), up to a constant, as

$$\overline{\mathscr{E}_\delta}[\rho] := \int h_m(K_\delta * \rho(x))\, dx + \frac{1}{4} \iint G(x-y)[K_\delta * \rho(x) - K_\delta * \rho(y)]^2\, dx\, dy,$$
$$\rho \in \mathcal{P}_2(\mathbb{R}^d).$$

In particular, the energy inequality (3.13) holds with $\mathscr{E}_\delta$ replaced by $\overline{\mathscr{E}_\delta}$.

We can now prove:

Theorem 3.3 *Assume that G is given by* (1.4) *and that* $f(\rho) = \frac{1}{m-1}\rho^m$ *with* $m > 2$ *and let* ρ_{in} *be as in Theorem 3.2. Let* $\rho^\delta(t,x)$ *be the solution of* (1.6) *given by Theorem 3.1 and* $\rho(t,x)$ *be the solution of* (1.9) *given by Theorem 3.2. Then, the entire sequence* $\rho^\delta(t)$ *converges narrowly to* $\rho(t)$ *(uniformly in* $[0,T]$*).*

We provide a proof of this result in Appendix section "Appendix A From Micro to Macro: Convergence $\delta \to 0$" by adapting the arguments developed in [31] to prove a similar results when $G = 0$.

Theorem 3.3 does not address the convergence of the microscopic model (3.5) when $\delta \to 0$ since the condition $\overline{\mathscr{E}_m}[\rho_{in}] < \infty$ excludes empirical measures. As explained in [31], this result can however be combined with (3.11): If we approximate the measure ρ_{in} by a sequence of empirical measures such that $d_W(\rho_{N,\delta}(t=0), \rho_{in}) \leq \frac{C}{N}$ and $N = o\left(e^{-\frac{1}{\delta^{2+d(m-1)}}}\right)$, then using (3.11) and the convergence of ρ^δ to ρ, we can show that the empirical distribution $\rho_{N,\delta}(t,x)$ converges to $\rho(t,x)$ solution of (1.9) with initial condition ρ_{in}. This restriction on N is very far from the scaling $N \sim \delta^{-d}$ necessary to preserve the total mass in our microscopic interpretation of the blob model. Numerical evidence (see [36, 37]) suggests that a good level of approximation holds for much smaller number of particles but justifying this limit is a challenging open problem.

To summarize, Eq. (3.6) preserves the empirical measure and thus describes precisely the evolution of a system of particles evolving according to the microscopic model (3.5). When $\delta \to 0$, this Eq. (1.6) is an approximation of (1.9) (which does not preserve the Dirac structure of ρ_N in its evolution).

To the best of our knowledge, this approach to approximating nonlocal diffusion equations by particle methods originates from the works of Mas-Gallic [99] for kinetic equations and Oelschläger [113] for the quadratic porous medium equation. We also mention [25, 44, 59, 95] in this direction. The convolution by K_δ alters the diffusion in the sense that particles remain particles: $\rho_N(t) = \frac{1}{N}\sum_{i=1}^N \delta_{x_i(t)}$ is a solution to the regularized equation where each particle $x_i(t)$ evolves according to the system of characteristics.

3.1.6 Other Microscopic Models with Repulsion

A similar regularization approach can be used, at least formally, to justify other macroscopic models. For example, Karper, Lindgren, and Tadmor [82] followed a similar approach and proposed the following microscopic model for chemotaxis:

$$\dot{x}_i(t) = \nabla\phi(x_i(t)) - \frac{\beta(\mu_N(x_i))}{\mu_N(x_i(t))}\nabla\mu_N(x_i(t)),$$

with $\beta(\mu) := \frac{\mu}{1-\mu}$. The corresponding empirical distribution then solves

$$\partial_t \rho_N + \operatorname{div}(\rho_N \nabla\phi) = \operatorname{div}\left(\frac{\rho_N}{K_\delta * \rho_N}\beta(K_\delta * \rho_N)\nabla(K_\delta * \rho_N)\right), \tag{3.15}$$

and in the limit $N \to \infty$ and $\delta \to 0$, we formally obtain

$$\partial_t \rho + \operatorname{div}(\rho\nabla\phi) = \operatorname{div}\left(\frac{\rho}{1-\rho}\nabla\rho\right) \tag{3.16}$$

which corresponds to (1.9) with $f(\rho) = (1-\rho)\ln(1-\rho)$. The use of such singular pressure is a classical way to enforce a congestion constraint $\rho \leq 1$ (see, for instance, [116]) and a model similar to (3.16) was derived in [98] from a stochastic particle system with volume effect, by enforcing the fact that two particles (which are assumed to be rigid squares in [98]) cannot overlap. Justifying the limit from (3.15) to (3.16) rigorously, in the spirit of Theorem 3.3, is, to the best of our knowledge, an interesting open problem.

Other microscopic models have been proposed to describe short-range repulsion. For example, Fischer, Kanzler, and Schmeiser [60] consider a one-dimensional model of particles $x_0(t) < x_1(t) < \cdots < x_N(t)$ in which each agent x_i only interacts with its two immediate neighbors x_{i-1} and x_{i+1} via a repulsive force which depends only on their distance. They derive a nonlinear diffusion model depending on the repulsive force (in [84] we use a similar approach to study the hard-sphere model in dimension 1). The extension of this model to higher dimensions is highly nontrivial and is very much related to similar difficulties faced in the so-called "upwind" scheme for hyperbolic equations [47, 56] for continuity equations with nonlinear mobilities. Other topological models, in which interactions depend on the "rank" of a neighbor rather than its distance, have been studied, for example, by Blanchet and Degond [12, 13]. Finally, let us point out that there are many models of pedestrian flows that also include the congestion constraint (see, for instance, [35, 74, 93]).

3.2 *Hard-Sphere (Incompressible) Models*

We now turn our attention to the hard-sphere (or incompressible) model. At the microscopic scale, the particles, still represented by balls of radius $\delta > 0$, are prevented from overlapping (congestion constraint). At the macroscopic scale, this will lead to the assumption that the population density cannot increase above a prescribed value.

On the one hand, we will see that the relationship between microscopic and macroscopic models is a lot more complex in this case than in the soft-sphere case of the previous section. On the other hand, we will show that the macroscopic hard-sphere model can easily be derived as the singular limit $m \to \infty$ of the (macroscopic) soft-sphere models. In the first part of the discussion, we assume that the "desired" velocity field is given a priori, but we also discuss the well-posedness properties of this macroscopic model when this velocity field is the gradient of the attractive chemotaxis potential.

3.2.1 The Microscopic Hard-Sphere Model

The microscopic model we now introduce has been extensively studied in the context of congested crowd motion, in particular by Maury et al. (see [58, 101–103]). In this context the domain Ω and the boundary condition (existence of "exits") play a very important role; this aspect will not be discussed in this chapter.

Unlike the soft-sphere model, in which the velocity is modified by a repulsive force, we now impose the hard-sphere constraint:

$$D_{ij}(\mathbf{x}(t)) := |x_i(t) - x_j(t)| \geq 2\delta \qquad \text{for all } i \neq j, \text{ for all } t \geq 0. \tag{3.17}$$

In terms of the empirical measure $\rho_N(t,x) = \frac{1}{N}\sum_{j=1}^{N} \delta(x - x_j(t))$ and using the notation of the previous section with $K = \chi_{B_1}$, we can also write this constraint as

$$\mu_N = K_\delta * \rho_N \leq 1. \tag{3.18}$$

As a consequence, the actual velocity of a particle will be the projection of the desired velocity onto the set of admissible velocity, namely, velocities that preserve the constraint (3.17).

Following [102], we assume that the particle i has a desired velocity v_i which might depend on the positions of all the particles. Below, we use the notation $\mathbf{x} = (x_1, \ldots, x_N) \in (\mathbb{R}^d)^N$ to denote the position of the N particles and $\mathbf{v}(\mathbf{x}) = (v_1(\mathbf{x}), \ldots, v_N(\mathbf{x}))$ for their desired velocity. The constrained microscopic model can then be written (see [102]):

$$\dot{\mathbf{x}}(t) = \mathbf{u}(\mathbf{x}), \quad \mathbf{u} = P_{C(\mathbf{x})}\mathbf{v}, \tag{3.19}$$

where $P_{C(\mathbf{x})}\mathbf{v}$ denotes the projection of $\mathbf{v}$ onto the set $C(\mathbf{x})$ of admissible velocities. If we define $D_{ij}(\mathbf{x}) := |x_i - x_j|$, then admissible velocities must be such that $D_{ij}(\mathbf{x})$ cannot decrease if the i and j particles are already touching. This leads to

$$C(\mathbf{x}) := \{\mathbf{u} \in (\mathbb{R}^d)^N \ : \ \nabla D_{ij}(\mathbf{x}) \cdot \mathbf{u} \geq 0 \text{ for all } i, j \text{ such that } D_{ij}(\mathbf{x}) = 2\delta\}.$$

Alternatively, using the fact that $\frac{d}{dt}|x_i(t) - x_j(t)|^2 = 2(x_i(t) - x_j(t)) \cdot (u_i(\mathbf{x}(t)) - u_j(\mathbf{x}(t)))$, we can write

$$C(\mathbf{x}) = \{\mathbf{u} \in (\mathbb{R}^d)^N \ : \ (u_i - u_j) \cdot (x_i - x_j) \geq 0 \text{ for all } i, j \text{ such that } |x_i - x_j| = 2\delta\}. \tag{3.20}$$

Following [102, 103], one can show that the projection $\mathbf{u} = P_{C(\mathbf{x})}\mathbf{v}$ in $(\mathbb{R}^d)^N$ can be characterized as follows: Using the notation $i \sim j$ to indicate that the particle i is in contact with the particle j and $e_{ij} = \frac{x_j - x_i}{|x_j - x_i|}$, we have

$$u_i = v_i - \sum_{j \sim i} p_{ij} e_{ij}, \tag{3.21}$$

where the pressure $p_{ij} = p_{ji}$ satisfies

$$p_{ij} \geq 0, \quad u_i \cdot e_{ij} + u_j \cdot e_{ji} \leq 0, \qquad \forall i \sim j,$$

(this last condition expresses the nonoverlapping constraint) and

$$u_i \cdot e_{ij} + u_j \cdot e_{ji} = 0 \text{ if } p_{ij} \neq 0.$$

This condition expresses the fact that if $p_{ij} \neq 0$, then the particles will stay in contact, i.e., the pressure prevents overlapping but does not push the particles away from each other.

The existence of a unique solution to (3.19) (for $\mathbf{v}$ Lipschitz and bounded) is proved in [103] (see also [101, Theorem 3.2]). Except in the case of $d = 1$, this does not fall into standard ODE theory because the set of admissible particle configurations

$$Q = \{\mathbf{x} = (x_1, x_2, \ldots, x_N) \in (\mathbb{R}^d)^N \ : \ |x_i - x_j| \geq 2\delta, \quad \forall i \neq j\}$$

is not convex in general. But it satisfies the weaker property of *uniform prox-regularity* which is sufficient for well-posedness.

As a final comment, we recall that in this microscopic model, it is possible to consider that each particle x_i has its own desired velocity v_i (see [102]). This distinction is not possible in the macroscopic model, which does not distinguish particles from each other. Thus, in consideration of passing to the macroscopic model it makes sense to assume that the velocity of the ith particles $v_i(\mathbf{x})$ is equal to

$\nabla\phi(x_i)$ for some potential $\phi : \mathbb{R}^d \to \mathbb{R}$. Given the corresponding projection $\mathbf{u}$, we denote by $\mathbb{P}_C(\nabla\phi)$ a vector field $\mathbb{R}^d \to \mathbb{R}^d$ such that $\mathbb{P}_C(\nabla\phi)(x_i) = u_i(\mathbf{x})$ (only the values of this vector field at the points x_i are uniquely defined). We can then write that the empirical distribution satisfies

$$\partial_t \rho_N + \operatorname{div}(\rho_N \mathbb{P}_C(\nabla\phi)) = 0, \qquad \mu_N(t) = K_\delta * \rho_N \leq 1. \tag{3.22}$$

3.2.2 The Macroscopic Hard-Sphere Model

Formally, the same principle can be applied to the macroscopic system: Here, we consider the formulation (3.22) of (3.19) and we write that the density $\rho(t, x)$ solves the continuity equation:

$$\partial_t \rho + \operatorname{div}(\rho\, \mathbb{P}_{C_\rho}(\nabla\phi)) = 0, \qquad \rho \leq 1.$$

The set of admissible velocities, which preserves the local constraint $\rho \leq 1$, can now be defined by duality as

$$C_\rho = \left\{ U \in L^2(\mathbb{R}^d;\, \mathbb{R}^d);\ \int_{\mathbb{R}^d} U \cdot \nabla q \leq 0, \quad \forall q \in H^1_\rho(\mathbb{R}^d) \right\},$$

where

$$H^1_\rho(\mathbb{R}^d) = \left\{ q \in H^1(\mathbb{R}^d);\ q \geq 0 \text{ a.e. in } \mathbb{R}^d \text{ and } q(x)(1 - \rho(x)) = 0 \text{ a.e.} \right\}$$

(we can also write $H^1_\rho(\mathbb{R}^d) = \{q \in H^1(\mathbb{R}^d);\ q(x) \in \partial f_\infty(\rho(x)) \text{ a.e.}\}$, but the condition $q(x)(1 - \rho(x)) = 0$ is a common way of writing the condition $q(x) = 0$ a.e. in $\{\rho < 1\}$).

Formally at least, this projection operator guarantees that the density satisfies the incompressibility constraint $\rho \leq 1$ since it implies $\operatorname{div} U \geq 0$ in (the interior of) the set $\{\rho = 1\}$. In order to develop a well-posedness theory for this model, we first show that it can be rewritten without the projection operator via the introduction of a pressure term (as is commonly done in incompressible fluid dynamic—although in our case the constraint is $\rho \leq 1$ and not $\rho = 1$).

To derive this formulation, we note that the projection $U := \mathbb{P}_{C_\rho}(\nabla\phi)$ is characterized by the following inequality:

$$\int_{\mathbb{R}^d} (U - \nabla\phi) \cdot (U - Q)\, dx \leq 0, \qquad \forall Q \in C_\rho. \tag{3.23}$$

This projection gives rise to a Lagrange multiplier in the form of a pressure term: One can show[6] that there must exist $p \in H^1_\rho(\mathbb{R}^d)$ such that

$$U = \nabla\phi - \nabla p.$$

The velocity $U = \mathbb{P}_{C_\rho}(\nabla\phi)$ is in C_ρ and thus satisfies $\int_{\mathbb{R}^d} U \cdot \nabla q\, dx \le 0$ for all $q \in H^1_\rho(\Omega)$ and the pressure satisfies the orthogonality condition (which follows from (3.23) by taking $Q = 0$ and $Q = 2U$): $\int_{\mathbb{R}^d} U \cdot \nabla p\, dx = 0$. We deduce that $-\int_{\mathbb{R}^d} U \cdot \nabla(p - q)\, dx \le 0$ for all $q \in H^1_\rho$, which implies that p solves the variational inequality:

$$\begin{cases} p \in H^1_\rho \\ \displaystyle\int_{\mathbb{R}^d} [\nabla p - \nabla\phi] \cdot \nabla(p - q) \le 0, \qquad \forall q \in H^1_\rho. \end{cases} \tag{3.24}$$

This variational inequality is in fact an obstacle problem (because of the constraint $p \ge 0$). We can summarize this discussion with the following proposition (see [85, Appendix B]):

Proposition 3.1 *Assume $\phi \in H^1$ and let $p(x)$ be the (unique) solution of the variational inequality* (3.24). *Then $\mathbb{P}_{C_\rho}(\nabla\phi) = \nabla\phi - \nabla p$.*

The macroscopic hard-sphere model can thus be written as

$$\partial_t \rho + \operatorname{div}(\rho\nabla\phi) = \operatorname{div}(\rho\nabla p), \qquad p \in \partial f_\infty(\rho). \tag{3.25}$$

3.2.3 Well-Posedness of the PKS Hard-Sphere Model

In this section, we discuss the well-posedness of the model (3.25) when the velocity field V is the attractive chemotaxis gradient from (1.10). We can also write this as

$$\partial_t \rho + \operatorname{div}(\rho\nabla G * \rho) = \operatorname{div}(\rho\nabla p), \qquad p \in \partial f_\infty(\rho). \tag{3.26}$$

A solution of (3.26) can be constructed either as a gradient flow or as the singular limit of the nonlinear diffusion model.

The gradient flow approach was first developed in the context of congested crowd motion in [101]. In that paper, the velocity field was given by $V = -\nabla\phi$ for a continuous and λ-convex potential ϕ. A similar approach was later carried out with the chemotaxis potential in [85]: Eq. (3.26) is the 2-Wasserstein gradient flow of the

---

[6] This follows from Moreau's decomposition theorem since the set $\mathcal{K}_\rho = \{\nabla p\,;\ p \in H^1_\rho(\mathbb{R}^d)\}$ is a closed convex cone in $L^2(\mathbb{R}^d;\ \mathbb{R}^d)$ and its polar cone is exactly $\mathcal{K}^\circ_\rho = C_\rho$. It follows that all $\nabla\phi \in L^2(\mathbb{R}^d;\ \mathbb{R}^d)$ can be written in a unique way as $\nabla\phi = \mathbb{P}_{C_\rho}(\nabla\phi) + \mathbb{P}_{\mathcal{K}_\rho}(\nabla\phi)$.

energy $\mathscr{E}_\infty$ defined by (2.3) and existence of a solution can be proved via a discrete-time approximation based on a classical JKO minimization scheme: For a given initial data ρ_{in} such that $\mathscr{E}_\infty[\rho_{in}] < \infty$ (where $\mathscr{E}_\infty$ is defined by (2.3)), we fix a time step $\tau > 0$ and define the sequence ρ^n by

$$\rho^0 = \rho_{in}, \qquad \rho^n \in \operatorname{argmin}\left\{\frac{1}{2\tau}W_2^2(\rho, \rho^{n-1}) + \mathscr{E}_\infty[\rho]\right\} \qquad \forall n \geq 1. \tag{3.27}$$

The piecewise constant function defined by $\rho^\tau(t) := \rho^{n+1}$ for $t \in [n\tau, (n+1)\tau)$ can then be proved to converge (as $\tau \to 0$) to a solution of (3.26) which gives the following theorem (see [85]):

Theorem 3.4 *Let G be given by* (1.4). *Given $\rho_{in} \in \mathcal{P}_2(\mathbb{R}^d)$ such that $\rho_{in} \leq 1$ and* $\operatorname{Supp}\rho_{in} \in B_R$. *Then Eq.* (1.10) *(or* (3.26)*) has a unique global weak solution $\rho(t,x)$ with initial data $\rho_{in}(x)$. This solution is such that $\rho \in L^1(0,\infty; L^\infty(\Omega)) \cap C^{1/2}(0,\infty; \mathcal{P}(\Omega))$, $p \in L^2(0,\infty; H^1(\Omega))$ with*

$$0 \leq \rho \leq 1, \quad p \geq 0, \quad p(1-\rho) = 0 \quad \text{a.e. in } [0,\infty) \times \mathbb{R}^d$$

(note that the conditions $\rho \in C^{1/2}([0,\infty); \mathcal{P}_2(\mathbb{R}^d))$ and $p \in L^2(0,\infty; H^1(\mathbb{R}^d))$ are enough to make sense of the product $p(1-\rho) = 0$). Furthermore, there exists $R(t)$ such that

$$\operatorname{Supp}\rho(t) \subset B_{R(T)} \quad \forall t \in [0,T].$$

When the drift term $\nabla\phi$ is fixed, uniqueness with density constraint is proved in [48] assuming some monotonicity for $\nabla\phi$ and in [75] in the spirit of the DiPerna-Lions theory for transport equations, assuming some Sobolev regularity for ϕ. Uniqueness for a related model (without the drift term) was also proved in [120] using a duality method that turns out to be very adaptable and was used in [85] and [68] to prove uniqueness for (3.26).

Theorem 3.4 (existence and uniqueness) also follows from [68], where a solution is constructed by passing to the limit $m \to \infty$ in the soft-sphere model (1.9). Note that the incompressible limit $m \to \infty$ for (1.9) is a classical problem that goes back to the 1980s: the "stiff (or incompressible) limit" of the porous media equation. We can understand this limit by noticing that (1.9) has the same form as (3.26) but with $p = f_m'(\rho)$. We can also make sense of this limit at the level of the energy. Indeed, Eq. (1.9) is the 2-Wasserstein gradient flow for the energy $\mathscr{E}_m$ defined by (2.2) which Γ-converges to $\mathscr{E}_\infty[\rho]$. The Γ-convergence of the energy does not imply the convergence of the corresponding gradient flow, but it is a good indication that such a convergence might take place.

There are numerous references for this limit $m \to \infty$ in various frameworks: We refer to [52, 63, 135] for the classical porous media equation $\partial_t\rho = \operatorname{div}(\rho\nabla f_m'(\rho))$ and to [41, 42, 87, 89, 90, 105, 120] for more recent results in the context of tumor growth modeling, which includes a nonnegative source term $\partial_t\rho =$

$\operatorname{div}(\rho\nabla f_m'(\rho)) + \rho F(f_m'(\rho))$ (more recently [67], these results were extended to the case of the porous media with an absorption term, which leads to interesting behavior). For the equation with a fixed drift term, the limit was first investigated with a fixed potential $\partial_t \rho + \operatorname{div}(\rho\nabla\phi) = \operatorname{div}(\rho\nabla f_m'(\rho))$ which is an important pedestrian model for congested crowd motion (see [2, 38, 40, 88]). Finally, the limit was established for the attractive chemotaxis problem (1.4) with pure Newtonian attraction $\sigma = 0$ (the analysis can be extended to the case $\sigma > 0$, which does not affect the singularity of the kernel at 0): This was first done in [38] for particular solutions (patch solutions) using variational methods and the viscosity solution approach. The convergence of the gradient flow of $\mathscr{E}_m$ for general initial conditions was then proved in [39], and finally, the convergence of weak solutions was proved in [68] by a more direct approach (which can be generalized to include source terms, for example; see [79]).

The key feature of this limit is the fact that $p_m(\rho) = f_m'(\rho) \to \partial f_\infty(\rho)$ when $m \to \infty$. There are other natural pressure terms that lead to a similar result. For example, one can replace the pressure $p_m(\rho)$ with a singular pressure such as

$$p_\alpha(\rho) = \begin{cases} \alpha\frac{\rho}{1-\rho} & \text{if } \rho < 1, \\ \infty & \text{if } \rho \geq 1, \end{cases} \tag{3.28}$$

or

$$p_\alpha(\rho) = \begin{cases} -\alpha\log(1-\rho) & \text{if } \rho < 1, \\ \infty & \text{if } \rho \geq 1. \end{cases} \tag{3.29}$$

Such pressure terms lead to solutions satisfying $\rho < 1$ (for $\alpha > 0$), but in the limit $\alpha \to 0$, we find $p_\alpha(\rho) \to \partial f_\infty(\rho)$. The corresponding limits were made rigorous in [40, 43, 69] for (3.28) (with a fixed potential ϕ and with convergence rates) and in [82] for (3.29) (with the attractive chemotaxis potential with $\sigma = 0$).

3.2.4 The Projected Velocity

In the approach discussed above, solutions of (3.26) are constructed by imposing the constraint $\rho < 1$, rather than by using the projection $\mathbb{P}_{C_\rho}(\nabla\phi)$. However, it is possible to show that the effective velocity in (3.26), $\nabla\phi - \nabla p$, is indeed the expected projection. Following [101], one can show that

$$\int_{\mathbb{R}^d} \nabla q(x) \cdot (\nabla p(t_0, x) - \nabla\phi(t_0, x))\, dx = 0 \qquad \forall q \in H^1_{\rho(t_0)}, \text{ a.e. } t_0 > 0.$$

Since $p \in H^1_{\rho(t_0)}$, this implies that (3.24) holds for a.e. $t_0 > 0$. In fact it implies the stronger statement:

$$\int_\Omega [\nabla p - \nabla\phi]\cdot\nabla(p-q) = 0 \qquad \forall q \in H^1_\rho.$$

By taking $q(x) = p(t_0, x)\varphi(t_0, x))$, we deduce in particular

$$\int_{\mathbb{R}^d} \nabla(p(t_0, x)\varphi(t_0, x))\cdot(\nabla p(t_0, x) - \nabla\phi(t_0, x))\,dx = 0 \qquad \text{a.e. } t_0 > 0,$$

which implies the so-called complementarity condition :

$$p(\Delta p - \Delta\phi) = 0, \text{ in } \mathcal{D}'((0, T)\times\mathbb{R}^d), \quad p(t) \in H^1_{\rho(t)}. \tag{3.30}$$

It is also possible to derive these properties as part of the construction process. Indeed, both the solution of the nonlinear diffusion equation and the discrete-time approximation given by the JKO scheme satisfy the approximation of (3.30). Intuitively, passing to the limit in these approximations require the strong convergence of the pressure in $H^1(\Omega)$. This is done in [120] in the time-monotone model for tumor growth. The case of the more complex model with a term representing nutrients (which is no longer monotone) was then treated in [41] by generalizing the so-called Aronson-Bénilan estimates to get some strong convergence on the gradient of the pressure.

This derivation was also performed in [67] by showing directly that for all $t > 0$, the pressure $p(t)$ is the unique solution of the variational inequality (3.24). This approach has two advantages: This formulation of the complementarity equation can be derived without requiring any strong convergence on ∇p, and unlike (3.30), it identifies the pressure for all time (rather than a.e.). In problems without monotonicity in time, the support of the pressure can indeed be smaller than the set $\{\rho = 1\}$, an important phenomenon which is clearly identified with this obstacle problem formulation. This obstacle problem (3.24) was derived for the limit of the JKO solution in [85] as well.

3.2.5 Relation Between the Microscopic and Macroscopic Models

Since both the microscopic particles model (3.19) and macroscopic model (3.25) have solutions, it is natural to ask whether these solutions converge to one another when $\delta \to 0$ (for the purpose of this discussion, we focus on the case of a fixed velocity field $V(x) = \nabla\phi(x)$). However, unlike the soft-sphere model discussed in Sect. 3.1, the relation between the microscopic and macroscopic models is far from clear in the hard-sphere case. We recall that in the soft-sphere case, this relationship was addressed in two steps: First, there was a one-to-one correspondence between the solutions of the ODE system (3.9) and the empirical measures solutions of the PDE (1.6). Second, solutions of (1.6) (for absolutely continuous initial data) were proved to converge to solutions of (1.9) as $\delta \to 0$. We will show below that the first step still holds in the hard-sphere models: Solutions of the system of ODE with hard-sphere constraint (3.17) are associated to empirical measures solutions of (1.8). But the second step in the limiting process ($\delta \to 0$) is delicate at best. In fact, it is not obvious that Eq. (1.8) is well-posed outside of the particular case of empirical measures associated to (3.9). Without a well-posedness theory, a convergence result (the equivalent of Theorem 3.3) seems out of reach.

Beyond the technical difficulties, we should also point out that the dynamics of the microscopic and macroscopic models are quite different. For example, when the particle i is in contact with several other particles, which is expected in congested configurations, the microscopic condition requires control on the velocity v_i in every direction $x_i - x_j$ for which $D_{i,j} = 0$. By contrast, the macroscopic constraint is a unique scalar constraint on the divergence of the velocity field. The dynamic of the microscopic model is thus much richer than that of the macroscopic model. Simple examples (see [100]) show that slightly different arrangements of the particles at the microscopic level (corresponding to the same macroscopic distribution) can lead to completely different behavior.

We now show that the nonlocal PDE (1.8) corresponds to the microscopic hard-sphere model when $K_\delta = \frac{1}{\omega_d \delta^d} \chi_{B_\delta}$. We already noted (see (3.18)) that the nonoverlapping constraint can be written as $\mu_N = K_\delta * \rho_N \leq 1$. Next, we describe the set of admissible velocity: If $\partial_t \rho_N + \operatorname{div}(\rho_N U) = 0$, and $p \geq 0$ is a measure supported on the set $\{\mu_N = 1\}$, then the condition "$\partial_t \mu_N \leq 0$ when $\mu_N = 1$" implies

$$\frac{d}{dt} \int K_\delta * \rho_N \, p \, dx = \int \rho_N \, U \cdot \nabla K_\delta * p \, dx \leq 0.$$

We can thus define the set of admissible velocities as

$$\widetilde{C}_\delta(\rho_N) = \Big\{ U : \mathbb{R}^d \to \mathbb{R}^d \, ; \, \int \rho_N \, U \cdot \nabla K_\delta * p \, dx \leq 0, \text{ for all } p \geq 0,$$
$$p(1 - K_\delta * \rho_N) = 0 \Big\}. \tag{3.31}$$

Importantly, this description at the level of empirical measures is consistent with (3.20): If $U(x_i) = u_i(\mathbf{x})$, then the condition $\mathbf{u} \in C_\delta(\mathbf{x})$ from (3.20) is equivalent to $U \in \widetilde{C}_\delta(\rho_N)$. For example, if the two particles i and j are touching (up to a translation, we can take $x_i = -\delta e$ and $x_j = +\delta e$), then taking $p(x) = (\delta - |x \cdot e|)_+ \mathcal{H}^1|_{\mathbb{R}e}$ in (3.31) will lead to the condition in (3.20). The simple illustration in Fig. 3 sheds some light on another discrepancy between the microscopic model (3.19) and the macroscopic model (3.25) (already discussed in [101, 102]): The set of feasible velocities C_δ depends not only on the local density but also on the orientation of particles. In the macroscopic model (3.25), the feasible velocity only depends on the local density.

The set $\widetilde{C}_\delta(\rho_N)$ is the polar cone of

$$\widetilde{\mathcal{K}}_\delta(\rho_N) = \{ \nabla K_\delta * p \, ; \, p \geq 0, \quad p(1 - K_\delta * \rho_N) = 0 \},$$

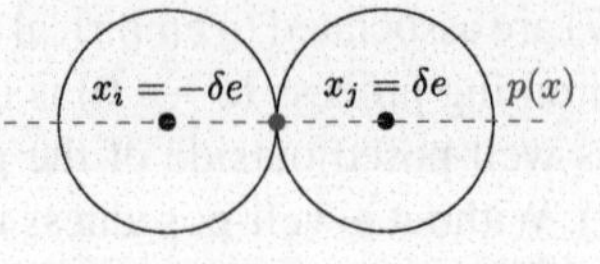

Fig. 3 Schematic of two particles with radius δ in contact along the axis e. The pressure $p(x) = (\delta - |x \cdot e|)_+ \mathcal{H}^1|_{\mathbb{R}e}$ can be highly singular since it only needs to activate along the axis e

in the Hilbert space $L^2(\mathbb{R}^d, d\rho_N)^d$. In keeping with Moreau's decomposition theorem, we expect to be able to write

$$U = \mathbb{P}_{\widetilde{C}_\delta(\rho_N)}(\nabla\phi) = \nabla\phi - \nabla K_\delta * p,$$

for some pressure $p \geq 0$, satisfying $p(1 - \mu_N) = 0$. This is in fact correct when ρ_N is an empirical distribution: This decomposition is equivalent to the (finite dimensional) projection of $\mathbf{v}$ given by (3.21). The existence of a solution to the microscopic model (see Sect. 3.2.1) thus implies the existence of a solution (in the form of the associated empirical measure) of

$$\partial_t \rho + \operatorname{div}(\rho\nabla\phi) = \operatorname{div}(\rho\nabla K_\delta * p), \qquad p \in \partial f_\infty(K_\delta * \rho), \tag{3.32}$$

when $\rho_{in} = \frac{1}{N}\sum_{i=1}^N \delta(x - x_i(0))$ (where we recall that $p \in \partial f_\infty(K_\delta * \rho)$ is equivalent to $K_\delta * \rho \leq 1,\ p \geq 0,\ p(1 - K_\delta * \rho) = 0$).

It is now natural to try to mirror the analysis described in Sect. 3.1 with this model for general density, namely, to prove a mean field limit, as in Theorem 3.1 (the constant λ_0 given by (3.12) depends on m, so the extension to the case $m = \infty$ is not obvious) and the limit $\delta \to 0$. There is however a serious obstruction: Except for the particular case of empirical measures (for which the existence of a solution follows from the existence of a solution for the microscopic model), the existence of solution for (3.32) is far from clear. A discrete-time approximation can be constructed via a JKO scheme with the energy

$$-\int_\Omega \phi(x)\rho(x)\,dx + \mathcal{F}_\infty[K_\delta * \rho],$$

but proving that the resulting velocity is indeed the projection of $\nabla\phi$ onto $\widetilde{C}_\delta(\rho)$ is challenging. Even then, it should also be noted that for general density ρ, it is not clear that the cone $\widetilde{\mathcal{K}}_\delta(\rho)$ is closed and thus Moreau's decomposition theorem does not apply. Similar difficulties arise when trying to obtain the existence of a solution of (1.8) by passing to the limit $m \to \infty$ in the soft-sphere model (1.6). This limit has been studied for the related Brinkman law model (3.7) [89, 121], but this case is somewhat easier since the condition $\rho \leq 1$ (as opposed to $\rho * K \leq 1$) is being enforced.

3.3 Hard-Sphere Models and Hele-Shaw Free Boundary Problems

In some settings, the hard-sphere model (1.10) might lead to the formation and propagation of sharp edges. In fact the connection between the hard-sphere model (1.10) and Hele-Shaw free boundary problem (with active potential, but without surface tension) is well documented: If we assume that $\rho(t,x) = \chi_{E(t)}$ and denote $\phi(t,x) = G * \chi_{E(t)}(x)$, then (1.10) implies (formally at least):

$$\Delta(p-\phi) = 0 \quad \text{in } E(t), \qquad V = -\nabla(p-\phi)\cdot\nu \quad \text{on } \partial E(t), \tag{3.33}$$

where V denotes the normal velocity of the interface $\partial E(t)$ and ν is the outward unit normal vector along $\partial E(t)$, with the condition

$$p(t,x) = 0 \quad \text{on } \partial E(t). \tag{3.34}$$

Equations (3.33)–(3.34) (which one can write with the effective pressure $q = p-\phi$) are reminiscent of a Hele-Shaw free boundary problem (without surface tension) or one-phase Muskat problem.

However, the assumption $\rho(t,x) = \chi_{E(t)}$ is not typically satisfied by all solutions of (1.10). Since the equation only imposes the constraint $\rho(t,x) \le 1$, there are no obvious reasons why ρ should be a characteristic function. But it can sometimes be proved that phase separation is propagated by the equation: If $\rho_{in} = \chi_{E_{in}}$, then $\rho(t) = \chi_{E(t)}$ for all $t > 0$. Such a result was proved in [105] and [87] for a different model, involving a nonnegative source term but no drift. The crucial ingredient in these papers was the monotonicity of ρ with respect to t, and it was proved that $\rho(t) = \chi_{E(t)}$ a.e. $t > 0$ for some open set $E(t)$ with finite perimeter. For the aggregation equation, a similar result was proved in [38, Theorem 1.1 (b)] when G is the purely Newtonian attraction kernel using the viscosity solution approach. Similar results were proved in [2] when ϕ is a fixed potential satisfying $-\Delta\phi \ge 0$ (see also [88] where the assumption is on the source term). This assumption, which implies some monotonicity of ρ along characteristic curves, is essential in these works. When G is given by (1.4), the condition $-\Delta\phi \ge 0$ is no longer satisfied but the fact that this potential is always attractive suggests that there is still some intrinsic monotonicity in the model. Indeed, the propagation of phase separation was proved in [85] for the hard-sphere PKS model (that is, (1.10) with G given by (1.4)). The proof is very different from those mentioned above and the result is somewhat weaker: It is proved that for appropriate initial data and for a.e. $t > 0$ there is a set $E(t)$ such that $\rho(t) = 1$ a.e. in $E(t)$ and $\rho(t) = 0$ a.e. in $\mathbb{R}^d \setminus \overline{E(t)}$. The condition $\rho = \chi_{E(t)}$ would follow if one could show that $|\partial E(t)| = 0$, a property that appears difficult to get.

Making rigorous the connection between (1.10) and the common notions of weak solutions for the Hele-Shaw free boundary problem (e.g., viscosity solutions) is a delicate issue. Besides the question of phase separation, there is the issue of giving a meaning to (3.33) when the set $E(t)$ is not smooth. As discussed in Sect. 3.2.4, the first condition in (3.33) can be derived rigorously (it is the complementarity condition). The fact that the Hele-Shaw velocity law is satisfied in some viscosity solutions' sense was proved in [38, Theorem 1.1] for the Newtonian kernel.

4 Part 2: Sharp Interface Limits and Free Boundary Problems

In the second part of this chapter, we review and extend recent results showing that when $m > 2$ and at appropriate scales, the competition between local repulsion and nonlocal attraction as described by Eq. (1.9) leads to phase separation and to the formation of an interface separating regions of high and low (or 0) density, interface whose evolution can be described by models of geometric nature. As noted in the introduction, phase separation plays a critical role in many biological processes and it has been shown that it can be explained by the simple attraction-repulsion mechanisms modeled by Eq. (1.1). We refer, for instance, to [11, 134] (for biological aggregations such as insect swarms) and to [57] (where phase separation is discussed for a continuum model of cell-cell adhesion). From a mathematical viewpoint, the main idea behind this phenomena is the close relationship between the aggregation-diffusion equation (1.1) and a degenerate Cahn-Hilliard equation. In this part of the chapter, we describe an approach, based on the recent work [86, 104], to rigorously investigate phase separation phenomena and derive effective free boundary problems by considering appropriate singular limits of (1.1).

4.1 A Nonlocal Cahn-Hilliard Equation

Our goal is to derive effective models for aggregation-diffusion phenomena when the total mass $\int \rho(t,x)\,dx$ is very large and the phenomena are observed from far away (at an appropriate timescale): We introduce $\varepsilon \ll 1$ such that $\int_{\mathbb{R}^d} \rho(t,x)\,dx = \varepsilon^{-d}$ and rescale the space and time variables as follows:

$$x = \varepsilon \bar{x}, \qquad t = \varepsilon^2 \bar{t}.$$

The function $\rho^\varepsilon(t,x) := \rho(t/\varepsilon^2, x/\varepsilon)$ now satisfies $\int_{\mathbb{R}^d} \rho^\varepsilon(t,x)\,dx = 1$ and solves (1.11) which we recall here for the reader's convenience (note in particular that $G_\varepsilon * \rho^\varepsilon(x) = G * \rho(x/\varepsilon)$):

$$\partial_t \rho + \operatorname{div}(\rho \nabla (G_\varepsilon * \rho)) = \operatorname{div}(\rho \nabla f'(\rho)), \qquad G_\varepsilon(x) = \varepsilon^{-d} G(\varepsilon^{-1} x)$$

(alternatively, this amounts to studying (1.9) when the range of the nonlocal attraction is small compared to typical macroscopic length). Throughout this section, we will assume that f satisfies

$$f(\rho) = f_m(\rho), \quad m > 2, \quad \text{or} \quad f(\rho) = f_\infty(\rho). \tag{4.1}$$

This assumption ensures that the function $h(\rho)$ defined by (1.13) is a double-well potential (see (1.14) and (1.15)), which is all we really need for the results presented here to hold.

Most of the rigorous results in this direction have only been proved when G is the chemotaxis potential (1.4) or some other very particular kernel (such as the heat kernel). Extending the theory (in particular Theorem 4.7) to general interaction potentials G describing local attraction is an interesting and important open problem. For the formal discussion below, it is enough to assume that

$$G(x) = G(|x|) \geq 0, \quad \int G(x)\,dx = \sigma^{-1} \in (0, \infty) \text{ and } \int_{\mathbb{R}^d} |x|^2 G(x)\,dx < \infty.$$

A crucial observation mentioned in the introduction is the fact that (1.11) can be written as a nonlocal Cahn-Hilliard equation (with a singular double-well potential). Indeed, since $G_\varepsilon * \rho \to \sigma^{-1}\rho$ when $\varepsilon \to 0$, it makes sense to rewrite (1.11) as

$$\partial_t \rho = \operatorname{div}(\rho \nabla [f'(\rho) - \sigma^{-1}\rho]) - \operatorname{div}(\rho \nabla [G_\varepsilon * \rho - \sigma^{-1}\rho]). \tag{4.2}$$

Introducing the symmetric nonlocal operator

$$B_\varepsilon[\rho](x) := (G_\varepsilon * \rho - \sigma^{-1}\rho)(x) = \int_{\mathbb{R}^d} G_\varepsilon(x-y)[\rho(y) - \rho(x)]\,dy,$$

and using the function h defined by (1.13), we can write (4.2) as

$$\partial_t \rho + \operatorname{div}(\rho v) = 0, \qquad v = -\nabla h'(\rho) + \nabla B_\varepsilon[\rho]. \tag{4.3}$$

This equation is the gradient flow for the energy $\mathscr{J}_\varepsilon$ defined in (2.6) (which can be obtained from the energy (2.5) by a similar computation) with respect to the 2-Wasserstein distance. Equation (4.3) is closely related to a model derived in [61] by Giacomin and Lebowitz as a macroscopic model for phase separation in particle systems with long-range interactions. This equation can be seen as a nonlocal approximation of the degenerate Cahn-Hilliard equation [20, 50]: When ρ is a smooth function, we have (using the symmetry of G)

$$B_\varepsilon[\rho] = \varepsilon^2 \beta \Delta \rho + \mathcal{O}(\varepsilon^4), \qquad \beta := \int_{\mathbb{R}^d} (z \cdot e)^2 G(z)\,dz$$

(for any $e \in \mathbb{S}^{d-1}$) and so (4.3) is close to

$$\partial_t \rho + \operatorname{div}(\rho v) = 0, \qquad v = -\nabla h'(\rho) + \varepsilon^2 \beta \nabla \Delta \rho. \tag{4.4}$$

This connection between aggregation-diffusion equations and degenerate Cahn-Hilliard equations is discussed and used, for instance, in [11, 24, 57, 86, 104]. When h is a double-well potential (as is the case when (4.1) holds), it is well known that (4.4) leads to phase separation and can be approximated—at appropriate timescales—by Stefan or Hele-Shaw free boundary problems. The sharp interface limit $\varepsilon \to 0$ was first investigated via the formal expansion method by Pego [119] (for the nondegenerate equation) and by Glasner [64] (for the degenerate equation). Rigorous justification of these formal asymptotic was established by Alikakos, Bates, and Chen [4, 34] for the nondegenerate equation. Formal results were also obtained for the nonlocal equation (4.3) by Giacomin and Lebowitz in [62]. In this part, we present some rigorous results for such limits for the aggregation diffusion equation (4.3).

As a final comment, we note that the convergence of the rescaled version of (4.3)

$$\partial_t \rho + \operatorname{div}(\rho v) = 0\,, \qquad v = -\nabla h'(\rho) + \varepsilon^{-2} \nabla B_\varepsilon[\rho],$$

to the degenerate Cahn-Hilliard equation,

$$\partial_t \rho + \operatorname{div}(\rho v) = 0\,, \qquad v = -\nabla h'(\rho) + \beta \Delta \rho,$$

was recently proved in [50] on the torus and when G_ε is a compactly supported approximation of unity (see also [26] for a similar result for systems of Cahn-Hilliard equations).

4.2 *Phase Separation: Stefan Free Boundary Problems*

Going back to (4.3), we now investigate the asymptotic behavior of the solutions as $\varepsilon \to 0$. Since $B_\varepsilon[\rho]$ vanishes as ε vanishes, we expect the asymptotic dynamics to be close to that of the forward-backward diffusion equation:

$$\partial_t \rho = \operatorname{div}(\rho \nabla h'(\rho)), \tag{4.5}$$

(recall that under (4.1), $\rho \mapsto h(\rho)$ is not convex). Such equations are commonly associated with the modeling of phase separation mechanisms, but are ill-posed from a mathematical point of view (see [73]). One would expect the limits of solutions of (4.3) to inherit some additional properties, but characterizing these limits is quite challenging, even for the classical Cahn-Hilliard equation. The simplest setting is when the initial data does not take values in the mushy region $(0, \theta)$: Such initial data are usually referred to as well prepared. In that case, the limit $\varepsilon \to 0$ of (4.3) leads to the equation

$$\partial_t \rho = \operatorname{div}(\rho \nabla h^{**\prime}(\rho)), \tag{4.6}$$

where h^{**} denotes the convex hull of h. When $h = h_m$ in (1.14), we have

$$h_m^{**}(\rho) = \begin{cases} 0 & \text{for } \rho \in [0, \theta_m]; \\ h_m(\rho) & \text{for } \rho \geq \theta_m. \end{cases}$$

With this choice of h (4.6) corresponds to a generalized Stefan free boundary problem, where ρ is the enthalpy variable. This equation propagates phase separation in the sense that the mushy region $\{\rho(t,x) \in (0,\theta)\}$ is nonincreasing with respect to t (see, for example, [66]). So when $|\{\rho(x,0) \in (0,\theta)\}| = 0$, this property remains true for all $t > 0$, and there exists a set $E(t)$ such that

$$\rho > \theta \text{ in } E(t), \qquad \rho = 0 \text{ in } \mathbb{R}^d \setminus E(t).$$

The evolution of the sharp interface $\partial E(t)$ is then described by the Stefan problem (4.6) which is a weak formulation of (1.17).

When the initial data is not well prepared, further characterization of the limits for solutions of (4.3) is required to understand how phase separation arises. This has been studied for the nondegenerate Cahn-Hilliard equation. First, we note (see Fig. 2) that there exists $\underline{\theta} \in (0,\theta)$ such that $h'' > 0$ in $(\underline{\theta}, \infty)$ and $h'' < 0$ in $(0, \underline{\theta})$. When ρ_{in} does not take values in the unstable region $(0, \underline{\theta})$, Plotnikov [123] characterized the limits in dimension 1 and showed that the limiting problem may exhibit hysteresis phenomena. In general the problem is very unstable and numerical computations show that oscillations appear in the region where $\rho_{in}(x) \in (0, \underline{\theta})$ (see [10, 51]). Passing to the limit $\varepsilon \to 0$ requires a weak formulation of (4.5) involving Young's measures (see, for example, Plotnikov [124] and Slemrod [130]). We will not review these results here, but we stress the fact that the dynamics of this phase separation process in the set $\{\rho_{in} \in (0,\theta)\}$ is very rich, even in dimension 1 (see [10]). We are not aware of any similar analysis for the degenerate Cahn-Hilliard equation or for its nonlocal approximation (4.3).

Even with well-prepared initial data, justifying the convergence to the Stefan problem (4.6) is delicate. A similar limit has been studied for the degenerate Cahn-Hilliard equation (4.4). The convergence of (4.4) to (4.6), studied formally in [119], was established rigorously only in dimension 1 by Delgadino in [45], for well-prepared initial data. Delgadino relied on the approach developed by Sandier and Serfaty [129] to prove the convergence of gradient flows using the fact that the Allen-Cahn energy (2.7) Γ-converges to $\int_{\mathbb{R}^d} h^{**}(\rho(x))\,dx$. Convergence to a Stefan free boundary problem was also proved by a similar approach for the 1-dimensional nondegenerate Cahn-Hilliard equation in [9]. But as far as we know no such results are known for our nonlocal equation (4.3) except for some formal results for a nonlocal equation similar to (4.3) in [62] (via formal expansion and matched asymptotic analysis). Extending the rigorous analysis of [45] to this equation is an interesting open problem. We note that such a result appears plausible at the level of the energy: Indeed a corresponding Γ-convergence result can be obtained for the energy $\mathcal{J}_\varepsilon$ given by (2.6).

Proposition 4.2 *Assume that* $G \in L^1(\mathbb{R}^d)$ *and that* $\int_{\mathbb{R}^d} |z|^2 G(z)\, dz < \infty$. *Then* $\mathcal{J}_\varepsilon$ *(defined by* (2.6)*)* Γ*-converges to* $\mathcal{J}^*$ *(defined by* (2.8)*). More precisely, we have:*

(i) For any sequence $\rho^\varepsilon \in \mathcal{P}_2(\mathbb{R}^d)$ *that converges weakly to* $\rho \in \mathcal{P}_2(\mathbb{R}^d)$, *we have*

$$\liminf_{\varepsilon \to 0} \mathcal{J}_\varepsilon(\rho^\varepsilon) \geq \mathcal{J}^*(\rho).$$

(ii) For any $\rho \in \mathcal{P}_2(\mathbb{R}^d)$, *there exists a sequence* $\rho^\varepsilon \in \mathcal{P}_2(\mathbb{R}^d)$ *such that* ρ_ε *converges weakly to* ρ *and*

$$\limsup_{\varepsilon \to 0} \mathcal{J}_\varepsilon(\rho^\varepsilon) \leq \mathcal{J}^*(\rho).$$

We provide a short proof of this result in Appendix section "Appendix B Convergence Results for the Stefan Problem".

While we will not attempt to extend the analysis of [45] to our nonlocal equation (4.3), we can prove the following conditional result:

Theorem 4.5 *Assume that* (4.1) *holds and that* G *is given by* (1.4). *Let* ρ_{in}^ε *be a sequence of initial data in* $\mathcal{P}_2(\mathbb{R}^d)$ *satisfying* $\mathcal{J}_\varepsilon(\rho_{in}^\varepsilon) \leq C$ *and let* $\rho^\varepsilon(t,x)$ *be the corresponding solution of* (1.11) *up to time* $T > 0$. *There exists a subsequence* ε_n *such that* $\rho^{\varepsilon_n}(t)$ *converges narrowly (locally uniformly in* t*) to* $\rho \in AC([0,T]; \mathcal{P}_2(\mathbb{R}^d))$. *Furthermore, if we have*

$$\lim \int_0^T \mathcal{J}_{\varepsilon_n}(\rho^{\varepsilon_n}(t))\, dt \to \int_0^T \mathcal{J}^*(\rho(t))\, dt, \tag{4.7}$$

then ρ *solves* (4.6) *on* $[0,T] \times \mathbb{R}^d$ *and* ρ^{ε_n} *converges strongly (in* L^2*) to* ρ *in* $\{h^{**}(\rho) > 0\}$.

As mentioned in the introduction, the convergence assumption (4.7) offers a classical approach to (conditional) convergence results for free boundary problems (see also the next section). We recall that the Γ-convergence of $\mathcal{J}_\varepsilon$ implies the liminf inequality, so (4.7) says that there is no loss of energy in the limit. Theorem 4.5 appears to be new and we provide a proof in Appendix section "Appendix B Convergence Results for the Stefan Problem". While we do not explicitly require the initial data to be well prepared in this result, assumption (4.7) is in fact likely stronger: In [45] it is proved (in dimension 1 and for the Cahn-Hilliard equation) that condition (4.7) is satisfied when the convergence of the energy holds for the initial data, that is, $\mathcal{J}_\varepsilon(\rho_{in}^\varepsilon) \to \mathcal{J}^*(\rho_{in})$.

4.3 Surface Tension Phenomena: Hele-Shaw Flows

When $t \to \infty$, solutions of the Stefan problem (4.6) converge to functions of the form $\theta\chi_{E_\infty}$, which are steady states of the Stefan problem (for any set E_∞). Going back to our aggregation-diffusion equation (1.11) (or (4.3)), this means that solutions experience phase separation. However, these simple characteristic functions are not steady states of (1.11) and will thus continue to evolve at a much slower timescale. In order to characterize this evolution, we now take an initial condition of the form

$$\rho_{in} = \theta\chi_{E_{in}}$$

in (4.3) and rescale the time variable $t \mapsto \varepsilon^{-1}t$, leading to Eq. (1.18), which we rewrite as

$$\varepsilon\partial_t\rho = \operatorname{div}(\rho\nabla h'(\rho)) - \operatorname{div}(\rho\nabla B_\varepsilon[\rho]). \tag{4.8}$$

In this section, we describe how the limit $\varepsilon \to 0$ gives rise to surface tension (mean-curvature) effects. In the next Sect. 4.4, we will show that when the equation is set on a bounded domain this limit also yields contact angle conditions. This is not surprising, since for the corresponding rescaled Cahn-Hilliard equation,

$$\varepsilon\partial_t\rho = \operatorname{div}(\rho\nabla[h'(\rho) - \varepsilon^2\Delta\rho]),$$

it was formally shown by Glasner [64] (using formal expansion and matched asymptotics) that the limit $\varepsilon \to 0$ leads to a one-phase Hele-Shaw free boundary problem with surface tension. We can also make sense of this limit by looking at the associated energy functional: Since $h(\rho_{in}) = 0$, (2.6) reduces to

$$\begin{aligned}
\mathscr{J}_\varepsilon[\rho_{in}] &= \frac{\theta^2}{4}\iint_{\mathbb{R}^d\times\mathbb{R}^d} G_\varepsilon(x-y)[\chi_{E_{in}}(x) - \chi_{E_{in}}(y)]^2\,dx\,dy \\
&= \frac{\theta^2}{4}\iint_{\mathbb{R}^d\times\mathbb{R}^d} G_\varepsilon(x-y)|\chi_{E_{in}}(x) - \chi_{E_{in}}(y)|\,dx\,dy \\
&= \varepsilon\gamma_0 P(E_{in}) + o(1)
\end{aligned}$$

for some $\gamma_0 > 0$ (see, for instance, [15, 106]) where $P(E)$ denotes the perimeter of E, defined by (2.9). We thus introduce the rescaled energy $\mathscr{G}_\varepsilon[\rho] := \varepsilon^{-1}\mathscr{J}_\varepsilon[\rho]$, that is,

$$\mathscr{G}_\varepsilon[\rho] = \begin{cases} \dfrac{1}{\varepsilon}\displaystyle\int_{\mathbb{R}^d} h(\rho)\,dx + \dfrac{1}{4\varepsilon}\iint_{\mathbb{R}^d\times\mathbb{R}^d} G_\varepsilon(x-y)\;\;[\rho(x) - \rho(y)]^2\,dx\,dy \\ \qquad\qquad\qquad\qquad\qquad\qquad \text{if } \rho \in \mathcal{P}_2(\mathbb{R}^d),\ d\rho \ll dx, \\ \infty \qquad\qquad\qquad\qquad\qquad\qquad \text{otherwise,} \end{cases} \tag{4.9}$$

which is a nonlocal approximation of the rescaled Allen-Cahn functional:

$$\frac{1}{\varepsilon}\int_{\mathbb{R}^d} h(\rho)\,dx + \frac{\varepsilon}{2}\int_{\mathbb{R}^d} |\nabla\rho|^2\,dx,$$

whose Γ-convergence toward the perimeter functional is a classical result of Modica-Mortola [109] (see also Modica [108] and Sternberg [131]). A corresponding result for $\mathcal{G}_\varepsilon$ was proved for rather general kernels G when h is a smooth double-well functional in [1]. We recall that in our framework h is not a smooth potential at 0 (since $h(\rho) = -\infty$ when $\rho < 0$ and $h(\rho) = +\infty$ when $\rho > 1$ in the hard-sphere case $m = \infty$). Nevertheless, the proof can be adapted to this case. When G is given by (1.4), we have the following result (see [106] for the case $m = +\infty$ and [104] for the case $m \in (2, \infty)$):

Theorem 4.6 *Assume that G is given by* (1.4) *and that f satisfies* (4.1). *Then the energy functional $\mathcal{G}_\varepsilon$ defined by* (4.9) *Γ-converges to $\mathcal{G}_0$ defined by*

$$\mathcal{G}_0(\rho) = \begin{cases} \gamma\displaystyle\int_{\mathbb{R}^d} |\nabla\rho| & \text{if } \rho \in \mathcal{P}_2(\mathbb{R}^d), \quad \rho \in \mathrm{BV}(\mathbb{R}^d; \{0, \theta\}), \\ \infty & \text{otherwise}, \end{cases}$$

for some constant γ defined by (4.10) *below. More precisely, we have:*

(i) ***liminf property***: *For all sequences $\rho^\varepsilon \in L^1(\Omega)$ such that $\rho^\varepsilon \to \rho$ in $L^1(\Omega)$, we have*

$$\liminf_{\varepsilon\to 0} \mathcal{G}_\varepsilon(\rho^\varepsilon) \geq \mathcal{G}_0(\rho).$$

(ii) ***limsup property***: *For all $\rho \in L^1(\Omega)$, there exists a sequence $\rho^\varepsilon \in L^1(\Omega)$ such that $\rho^\varepsilon \to \rho$ in $L^1(\Omega)$ and*

$$\limsup_{\varepsilon\to 0} \mathcal{G}_\varepsilon(\rho^\varepsilon) \leq \mathcal{G}_0(\rho).$$

The definition of the constant γ appearing in the definition of $\mathcal{G}_0$ requires the introduction of the function

$$g(s) := \inf_{\rho\geq 0}\left[h(\rho) + \frac{1}{2\sigma}(\rho - \sigma s)^2\right].$$

We note that $g(s) \geq 0$ for all $s \geq 0$ and $g(0) = g(\theta/\sigma) = 0$ (in fact g is a double-well potential). The constant γ is then given by[7]

[7] When $h = h_\infty$, we have $\theta = 1$ and $g(s) = \frac{\sigma}{2}\min\{s^2, (s - \sigma^{-1})^2\}$ which leads to $\gamma = \frac{1}{4\sigma^{3/2}}$.

$$\gamma := \frac{1}{\theta} \int_0^{\theta/\sigma} \sqrt{2g(s)}\, ds. \tag{4.10}$$

A key tool in the proof of this theorem in [104, 106] is the following alternative formula for $\mathscr{G}_\varepsilon$ (with $\phi_\varepsilon = G_\varepsilon * \rho$):

$$\mathscr{G}_\varepsilon(\rho) = \frac{1}{\varepsilon} \int_{\mathbb{R}^d} h(\rho) + \frac{1}{2\sigma}(\rho - \sigma\phi^\varepsilon)^2 \, dx + \varepsilon \int_{\mathbb{R}^d} \frac{1}{2} |\nabla \phi_\varepsilon|^2 \, dx \tag{4.11}$$

$$\geq \frac{1}{\varepsilon} \int_{\mathbb{R}^d} g(\phi_\varepsilon) dx + \varepsilon \int_{\mathbb{R}^d} \frac{1}{2} |\nabla \phi_\varepsilon|^2 \, dx. \tag{4.12}$$

Inequality (4.12) shows a stronger connection between $\mathscr{G}_\varepsilon$ and the Modica-Mortola functional with double-well potential g (which explains the role of the function g in determining the constant γ in $\mathscr{G}_0$). Formula (4.11) is the key to proving another important property of this energy: The convergence of the energy implies the convergence of the corresponding first variation. Such results have a long history: Reshetnyak [125] proved that if χ_{E_ε} converges to χ_E strongly in L^1, then the convergence of the perimeter $P(E_\varepsilon) \to P(E)$ implies the convergence of its first variation:

$$\delta P(E_\varepsilon)[\xi] := \int (\operatorname{div} \xi - \nu_\varepsilon \otimes \nu_\varepsilon : D\xi) \, |D\chi_{E_\varepsilon}|$$

$$\longrightarrow \int (\operatorname{div} \xi - \nu \otimes \nu : D\xi) |D\chi_E| \qquad \forall \xi \in C_c^1(\mathbb{R}^d; \mathbb{R}^d),$$

where ν is the density $\frac{D\chi_E}{|D\chi_E|}$ (which exists by Radon-Nikodym's differentiation theorem) and can be interpreted as the normal vector to ∂E. We recall that the first variation of P is the mean curvature. Indeed, for a smooth interface ∂E we have

$$\int (\operatorname{div} \xi - \nu \otimes \nu : D\xi) |D\chi_E| = \int_{\partial E} \operatorname{div} \xi - \nu \otimes \nu : D\xi \, d\mathcal{H}^{d-1} = \int_{\partial E} \kappa \, \xi \cdot \nu \, d\mathcal{H}^{d-1} \tag{4.13}$$

where κ denotes the mean curvature of ∂E. A similar result was proved by Luckhaus and Modica [97] for the Ginzburg-Landau functional:

$$\int_\Omega \frac{1}{\varepsilon}(1 - \rho^2)^2 + \varepsilon |\nabla \rho|^2 \, dx.$$

But as far as we know, corresponding results for the nonlocal functional $\mathscr{G}_\varepsilon$ are only known for particular choices of the kernel G: Laux and Otto [94] proved it when G is the Gaussian and used it in the study of the thresholding scheme for multiphase mean-curvature flow (see also [53–55]). Jacobs, Kim, and Mészáros [78] proved a similar result and used it to derive the Muskat problem. It should be noted however that in these references, the convergence is only proved when $\mathscr{G}_\varepsilon$ is restricted to characteristic functions. In [86], we proved the corresponding result when G is given by (1.4) in the hard-sphere case $m = +\infty$ and then in [104] for the same kernel in the soft-sphere case $m \in (2, \infty)$ without restriction to characteristic functions. We state here the corresponding result in our framework (see [104]—see also [86] for the case $m = \infty$):

Proposition 4.3 *Assume that G is given by* (1.4) *and that f satisfies* (4.1). *Given any sequence $\rho^\varepsilon \in L^1$ which strongly converges to ρ in $L^1(\mathbb{R}^d)$, if*

$$\lim_{\varepsilon \to 0} \mathcal{G}_\varepsilon(\rho^\varepsilon) = \mathcal{G}_0(\rho) < \infty,$$

then for all $\xi \in C_c^1(\mathbb{R}^d, \mathbb{R}^d)$ we have

$$\lim_{\varepsilon \to 0} -\varepsilon^{-1} \int_{\mathbb{R}^d} (h'(\rho^\varepsilon) - B_\varepsilon[\rho^\varepsilon]) \nabla \rho^\varepsilon \cdot \xi \, dx = \gamma \int_{\mathbb{R}^d} [\operatorname{div} \xi - \nu \otimes \nu : D\xi] \, |\nabla \rho|,$$

where $\nu = \frac{\nabla \rho}{|\nabla \rho|}$.

This proposition is the cornerstone of the proof of the following theorem which makes clear the connection between the rescaled aggregation-diffusion equation (4.8) and the Hele-Shaw free boundary problem with surface tension (1.19), where the latter is the gradient flow of the perimeter functional $\mathcal{G}_0$ with respect to the 2-Wasserstein distance.

Theorem 4.7 *Given a sequence of initial data $\rho_{in}^{\varepsilon_n}$ such that $\mathcal{G}_{\varepsilon_n}(\rho_{in}^{\varepsilon_n}) \leq M$ and $\rho_{in}^{\varepsilon_n} \to \rho_{in} = \theta \chi_{E_{in}} \in BV(\mathbb{R}^d; \{0, \theta\})$, let $\rho^{\varepsilon_n}(t, x)$ be the weak solution of* (1.18) *with initial data $\rho_{in}^{\varepsilon_n}$. Then the following hold:*

(i) Along a subsequence, the density $\rho^{\varepsilon_n}(t, x)$ converges strongly in $L^\infty(0, T; L^1(\mathbb{R}^d))$ to

$$\rho(t, x) = \theta \chi_{E(t)} \in L^\infty(0, \infty; \mathrm{BV}(\mathbb{R}^d; \{0, \theta\}))$$

and there exists a velocity function $v(t, x)$ such that

$$\begin{cases} \partial_t \rho + \operatorname{div}(\rho v) = 0, \\ \rho(x, 0) = \theta \chi_{E_{in}}. \end{cases} \tag{4.14}$$

(ii) If the following energy convergence assumption holds:

$$\lim_{n \to \infty} \int_0^T \mathcal{G}_{\varepsilon_n}(\rho^{\varepsilon_n}(t)) \, dt = \int_0^T \mathcal{G}_0(\rho(t)) \, dt,$$

then there exists $p \in L^2(0, T; (C^s(\mathbb{R}^d))^*)$ (for any $s > 0$) such that

$$\int_0^\infty \int_{\mathbb{R}^d} (\rho v \cdot \xi - \theta p \operatorname{div} \xi)\, dx\, dt = -\gamma \int_0^\infty \int_{\mathbb{R}^d} [\operatorname{div} \xi - \nu \otimes \nu : D\xi] |\nabla \rho|\, dt \tag{4.15}$$

for any vector field $\xi \in C_c^\infty((0,\infty)\times\mathbb{R}^d; \mathbb{R}^d)$.[8] *Together, Eqs.* (4.14) *and* (4.15) *say that* $E(t)$ *is a weak solution of the Hele-Shaw free boundary problem with surface tension* (1.19) *with initial condition* E_{in}.

The notion of weak solution of (1.19) that we recover with this theorem is similar to the definition considered, for example, in [78, 86, 92]: The continuity equation (4.14) encodes the incompressibility condition $\operatorname{div} v = 0$ in $E(t)$ and the free boundary condition $V = -v \cdot \nu$. By taking test functions ξ supported in either $E(t)$ or $\mathbb{R}^d \setminus E(t)$, we see that (4.15) implies $\nabla p = 0$ outside of $E(t)$ and $v = -\nabla p$ in $E(t)$. For general test functions ξ and taking into account the right-hand side of (4.15), we further get the surface tension condition $[p] = \gamma\kappa$ on $\partial E(t)$ as a consequence of (4.13). In the radial setting, (4.13) implies that p equals the constant curvature $\partial E(t)$ in a strong sense, thus yielding discontinuity of the pressure across the interface.

Proof This result is proved in [86] when $m = \infty$ and [104] when $m \in (2,\infty)$ when the equation is set in a bounded domain. The proof can be extended to $\mathbb{R}^d$ with the use of the estimate on the second moment $\int_{\mathbb{R}^d} |x|^2 \rho\, dx$. Indeed, the only required adjustment is in the proof of [104, Proposition 4.2], where the argument only yields the strong convergence of ρ^{ε_n} to ρ in $L^\infty(0,T; L^1_{loc}(\mathbb{R}^d))$. But we can write

$$\begin{aligned}
&\limsup \int_{\mathbb{R}^d} |\rho^{\varepsilon_n}(t) - \rho(t)|\, dx \\
&\quad \le \limsup \int_{B_R} |\rho^{\varepsilon_n}(t) - \rho(t)|\, dx + \limsup \int_{\mathbb{R}^d \setminus B_R} |\rho^{\varepsilon_n}(t) - \rho(t)|\, dx \\
&\quad \le 0 + \frac{1}{R^2} \int_{\mathbb{R}^d} |x|^2 (\rho^{\varepsilon_n}(t) + \rho(t))\, dx
\end{aligned}$$

and the strong convergence in $L^\infty(0,T; L^1(\mathbb{R}^d))$ follows by taking $R \to \infty$.

More General Nonlinearity f and Interaction Kernel G Theorem 4.7 holds for general convex nonlinearities $f(\rho)$, provided the function $h(\rho)$ is a double-well potential. For technical reasons, the proof also requires that $f(\rho) \ge \left(\frac{1}{2\sigma} + \eta\right)\rho^2$ for large ρ. In particular, adding a small linear diffusion term to the equation, which adds the term $\nu\rho \log \rho$ to the function $h(\rho)$, will displace the well at $\rho = 0$ to some (small) positive value. In that case we expect the limit $\varepsilon \to 0$ to lead to a two-phase free boundary problem (Mullins-Sekerka).

[8] The integral $\int_{\mathbb{R}^d} p \operatorname{div} \xi\, dx$ above should be understood as the duality bracket $\langle q, \operatorname{div} \xi \rangle_{(C^s(\Omega))^*, C^s(\Omega)}$.

While it is not difficult to extend Theorem 4.7 to general nonlinearities, the proofs provided in [86, 104] are strongly dependent on the particular form of the interaction kernel G. In fact the proofs make use of the elliptic equation (1.4). One could extend the proof to slightly more general equations, for example, by replacing the Laplace operator by a general elliptic operator (which would introduce both space and direction dependence in the free boundary condition in (1.19)), but treating the case of general interaction kernels seems to require new ideas. As noted earlier, the Γ-convergence of the energy (Theorem 4.6) is known to hold for a large class of such kernel, and the main obstacle is rather to prove Proposition 4.3 in a more general framework.

4.4 Bounded Domains and Contact Angle Conditions

We have so far only considered problems set in the whole space. But in many experimental settings, the particles are confined to a bounded domain $\Omega \subset \mathbb{R}^d$. The equation for ρ is then set on Ω and supplemented with no-flux boundary conditions on $\partial\Omega$. But the domain can also have a nontrivial influence on the interaction kernel.

In some settings, it might make sense to stick to interactions that depend only on the distance between the particles (mathematically, this is done by extending ρ by zero outside the domain and keeping the convolution $G_\varepsilon * \rho$ in the equation). But in the context of chemotaxis (1.5), the potential $\phi = \phi_\varepsilon = G_\varepsilon * \rho$ represents the concentration of a chemoattractant chemical which is now diffusing in the same bounded domain Ω so that this convolution should be replaced with the solution of some elliptic boundary value problem.

In the first part of this section, we present results that have been proved in this later setting. We will then discuss the case of interactions that depend only on the distance between the particles.

4.4.1 Chemotaxis Potential in Bounded Domains

In the context of chemotaxis (1.5), we replace the convolution $G_\varepsilon * \rho$ by the solution ϕ_ε of the following elliptic boundary value problem:

$$\begin{cases} \sigma\phi_\varepsilon - \varepsilon^2\Delta\phi_\varepsilon = \rho & \text{in } \Omega, \\ a\phi_\varepsilon + \varepsilon b\nabla\phi_\varepsilon \cdot n = 0 & \text{on } \partial\Omega. \end{cases} \tag{4.16}$$

We are taking general Robin-type boundary conditions at the boundary $\partial\Omega$ for further generality, even though Neumann conditions ($a = 0$) are probably the most natural type of conditions. When $a > 0$ (4.16) takes into account the possible leakage (or destruction) of the chemical at the boundary. We can take a and b to be constant or functions of $x \in \partial\Omega$ and we will always assume that $a \geq 0$, $b \geq 0$

and $a+b>0$. The solution of (4.16) can be written as $\phi_\varepsilon(x)=\int_\Omega G_\varepsilon(x,y)\rho(y)\,dy$ with $G_\varepsilon(x,y)\geq 0$ a symmetric kernel and we set

$$B_\varepsilon[\rho]:=\phi_\varepsilon-\sigma^{-1}\rho,$$

so that (4.8) corresponds to

$$\begin{cases}\varepsilon\partial_t\rho=\operatorname{div}(\rho\nabla h'(\rho))-\operatorname{div}(\rho\nabla B_\varepsilon[\rho]) & \text{in } (0,T)\times\Omega;\\ \rho\nabla(h'(\rho)-B_\varepsilon[\rho])\cdot n=0 & \text{on } (0,T)\times\partial\Omega;\\ \rho(x,0)=\rho_{in} & \text{in } \Omega.\end{cases} \tag{4.17}$$

However, we do not have $\int_\Omega G_\varepsilon(x,y)\,dy=\sigma^{-1}$ so we cannot write $B_\varepsilon[\rho]$ as $\int_\Omega G_\varepsilon(x,y)[\rho(y)-\rho(x)]\,dy$. We can still write $\int_\Omega G_\varepsilon(x,y)\,dy=\sigma^{-1}-\tau_\varepsilon(x)$ where $\tau_\varepsilon\geq 0$ solves

$$\begin{cases}\sigma\tau_\varepsilon-\varepsilon^2\Delta\tau_\varepsilon=0 & \text{in } \Omega,\\ a\tau_\varepsilon+\varepsilon b\nabla\tau_\varepsilon\cdot n=\frac{a}{\sigma} & \text{on } \partial\Omega.\end{cases}$$

We thus have

$$B_\varepsilon[\rho](x)=\int_\Omega G_\varepsilon(x,y)[\rho(y)-\rho(x)]\,dy-\tau_\varepsilon(x)\rho(x)$$

and the corresponding energy functional can be written as

$$\begin{aligned}\mathscr{G}_\varepsilon[\rho]&=\frac{1}{\varepsilon}\int_\Omega h(\rho)\,dx-\frac{1}{2\varepsilon}\int_\Omega\rho B_\varepsilon[\rho]\,dx\\&=\frac{1}{\varepsilon}\int_\Omega h(\rho)\,dx+\frac{1}{4\varepsilon}\iint_{\Omega\times\Omega}G_\varepsilon(x,y)[\rho(x)-\rho(y)]^2\,dx\,dy\\&\quad+\frac{1}{2\varepsilon}\int_\Omega\tau_\varepsilon(x)|\rho(x)|^2\,dx.\end{aligned}$$

The last term in this energy accounts for the boundary conditions on ϕ_ε. In the case of Neumann boundary condition $a=0$, we have $\tau_\varepsilon=0$, and this new term vanishes. But in general this new term will contribute to the limiting energy. Indeed, when $\partial\Omega$ is regular enough, we can show[9] that $\frac{1}{\varepsilon}\tau_\varepsilon$ converges as a distribution to $\frac{1}{\sigma^{3/2}}\frac{a}{a+b\sqrt{\sigma}}d\mathcal{H}^{n-1}|_{\partial\Omega}$. This suggests that if E is a subset of Ω with C^1 boundary, then

[9] To see this, notice that when Ω is the half-space $\{x_d>0\}$, we have $\tau_\varepsilon(x)=\frac{1}{\sigma}\frac{a}{a+b\sqrt{\sigma}}e^{-\frac{\sqrt{\sigma}}{\varepsilon}x_d}$ and so $\varepsilon^{-1}\tau_\varepsilon\to\frac{1}{\sigma^{3/2}}\frac{a}{a+b\sqrt{\sigma}}\delta(x_d)$.

$$\mathscr{G}_\varepsilon[\theta\chi_E] \to \gamma\theta P(E,\Omega) + \frac{1}{2\sigma^{3/2}}\frac{a\theta}{a+b\sqrt{\sigma}}\mathcal{H}^{n-1}(\partial\Omega\cap E),$$

where the local perimeter $P(E,\Omega)$ is defined by

$$P(E,\Omega) := \sup\left\{\int_E \operatorname{div} g\,dx\,;\ g\in (C_0^1(\Omega))^d,\quad |g(x)|\le 1\ \forall x\in\Omega\right\} = \int_\Omega |\nabla\chi_E|.$$

This limit was proved rigorously in [106] (see also [86, Appendix C]) for the hard-sphere case $f = f_\infty$. In that case $\theta = 1$ and $\gamma = \frac{1}{4\sigma^{3/2}}$, so the formula for the limit becomes

$$\frac{1}{4\sigma^{3/2}}\left(P(E,\Omega) + \frac{2a}{a+b\sqrt{\sigma}}\mathcal{H}^{n-1}(\partial\Omega\cap E)\right). \tag{4.18}$$

Here we make a couple of remarks: First, functionals like (4.18), in which the area of the free surface $\partial E\cap\Omega$ and the contact area $\partial\Omega\cap E$ are weighted differently, appear naturally in various applications, such as the capillary energy for the modeling of sessile liquid drop in contact with a solid support. It is well known that the minimization of such energies leads to contact angle conditions. Second, we note that when $\frac{2a}{a+b\sqrt{\sigma}} > 1$, the energy (4.18) is not lower semicontinuous and can therefore not be the Γ-limit of $\mathscr{G}_\varepsilon$. In fact, the following result was proved in [86, 106] when $f = f_\infty$:

Theorem 4.8 *Assume that Ω is a bounded open set with $C^{1,\alpha}$ boundary and take $f = f_\infty$ and G given by (1.4). Assume further that a and b are nonnegative constants such that $a+b>0$. The functional $\mathscr{G}_\varepsilon$ Γ-converges, when $\varepsilon\to 0$ to*

$$\mathscr{G}_0(\rho) := \begin{cases} \dfrac{1}{4\sigma^{3/2}}\left[\displaystyle\int_\Omega |\nabla\rho| + \int_{\partial\Omega}\min\left(1, \frac{2a}{a+\sqrt{\sigma}b}\right)\rho\,d\mathcal{H}^{n-1}(x)\right] & \\ \qquad\qquad\qquad\qquad\qquad\qquad \text{if } \rho\in \mathrm{BV}(\Omega;\{0,1\}), \\ \infty \qquad\qquad\qquad\qquad\qquad\qquad \text{otherwise.} \end{cases}$$

The result also holds when $a(x)$, $b(x)$ are bounded, Lipschitz nonnegative functions defined on $\partial\Omega$ such that $\alpha(x)+\beta(x)\ge\eta>0$ and that $\mathcal{H}^{n-2}\left(\partial\left\{\frac{2a}{a+\sqrt{\sigma}b} > 1\right\}\right) < \infty$ (this technical condition ensures that the function $x\mapsto \frac{2a(x)}{a(x)+\sqrt{\sigma}b(x)}$ does not oscillate too much around the value 1). A crucial tool for the proof of this result is the following alternative formula for $\mathscr{G}_\varepsilon$:

$$\begin{aligned}\mathscr{G}_\varepsilon(\rho) = {}& \frac{1}{2\varepsilon}\int_\Omega h_\infty(\rho)\,dx + \frac{1}{2\sigma\varepsilon}\int_\Omega(\rho-\sigma\phi_\varepsilon)^2\,dx + \frac{\varepsilon}{2}\int_\Omega|\nabla\phi_\varepsilon|^2\,dx \\ & + \frac{\varepsilon^2}{2}\int_{\partial\Omega}\frac{b}{a+b}|\nabla\phi_\varepsilon\cdot n|^2\,d\mathcal{H}^{n-1}(x) + \frac{1}{2}\int_{\partial\Omega}\frac{a}{a+b}|\phi_\varepsilon|^2\,d\mathcal{H}^{n-1}(x),\end{aligned}$$

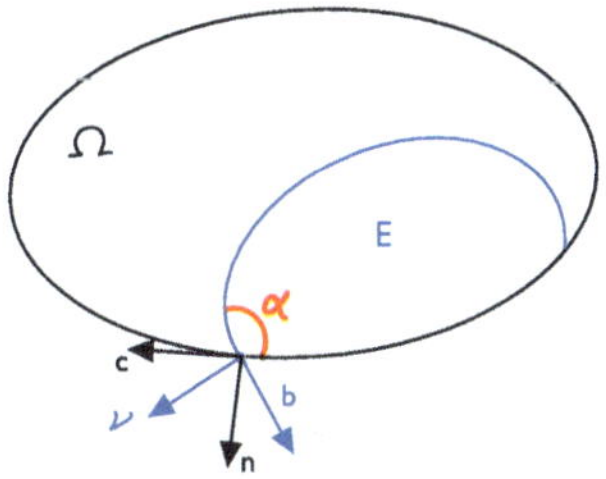

Fig. 4 The contact angle condition

where $\phi_\varepsilon(x) = \int_\Omega G_\varepsilon(x, y)\rho(y)\,dy$ solves (4.16).

Theorem 4.8 was also proved in [104] for the soft-sphere case $f = f_m$ with $m > 2$ when $a = 0$ (Neumann boundary condition). The approach of [106] and [104] can be combined to get the result when $f = f_m$ and for general coefficients a and b (although the formula for the coefficient in the boundary integral is not straightforward).

When $a = 0$, then $\mathscr{G}_0(\chi_E)$ is proportional to the local perimeter $P(E, \Omega)$ and thus only measures the area of $\partial E \cap \Omega$, while when $b = 0$ (Dirichlet boundary conditions), then $\mathscr{G}_0$ is proportional to the full perimeter $P(E)$. When $\frac{2a}{a+\sqrt{\sigma}b} \in (0, 1)$, the areas of $\partial E \cap \Omega$ and $E \cap \partial\Omega$ are weighted differently.

As mentioned above, the boundary term in $\mathscr{G}_0$ translates into a contact angle condition in the limiting Hele-Shaw free boundary problem: It was proved in [85] that the limit $\varepsilon \to 0$ of (4.16)–(4.17) leads to the same Hele-Shaw free boundary problem (1.19) in Ω, supplemented with the Neumann boundary condition:

$$\nabla p \cdot n = 0 \qquad \text{on } \partial\Omega$$

and the condition (see Fig. 4)

$$\cos(\alpha) := -\min\left(1, \frac{2a}{a + \sqrt{\sigma}b}\right) \quad \text{on } \partial E(t) \cap \partial\Omega, \tag{4.19}$$

where α denotes the angle between the free surface ∂E and the fixed boundary $\partial\Omega$ (see Fig. 4). In particular, for Neuman condition $a = 0$ (no absorption of the chemical on $\partial\Omega$), we get $\alpha = \frac{\pi}{2}$ so the contact must be orthogonal, while for Dirichlet condition $b = 0$ (and whenever the absorption rate satisfies $a/b \geq \sqrt{\sigma}$), the contact must be tangential ($\alpha = \pi$). Let us note that a similar contact angle condition was derived recently in [70] for the geometric mean-curvature flow.

The precise result in this case, which is proved by establishing the equivalent of Proposition 4.3 with the boundary term (see [86, Proposition 5.2]), is as follows:

Theorem 4.9 *[86, Theorem 1.6] Under the assumptions of Theorem 4.8 and given a sequence ε_n and initial data $\rho_{in}^{\varepsilon_n}$ such that $\mathscr{G}_{\varepsilon_n}(\rho_{in}^{\varepsilon_n}) \leq M$ and $\rho_{in}^{\varepsilon_n} \to \rho_{in} = \chi_{E_{in}} \in BV(\Omega)$ as $\varepsilon_n \to 0$, let $\rho^{\varepsilon_n}(t, x)$ be the weak solution of* (4.17).

Then the conclusion of Theorem 4.7 holds, with (4.15) *replaced by*

$$\int_0^\infty \int_\Omega \rho v \cdot \xi - \theta p \operatorname{div} \xi(t)\, dx\, dt = -\gamma \int_0^\infty \int_\Omega [\operatorname{div} \xi - \nu \otimes \nu : D\xi]\, |\nabla \rho|\, dt \tag{4.20}$$

$$- \gamma \min\left(1, \frac{2a}{a + \sqrt{\sigma} b}\right) \int_0^\infty \int_{\partial\Omega} [\operatorname{div} \xi - n \otimes n : D\xi]\, \rho\, d\mathcal{H}^{n-1}(x)\, dt,$$

for any vector field $\xi \in C_c^\infty((0,\infty) \times \overline{\Omega}; \mathbb{R}^d)$ *such that* $\xi \cdot n = 0$ *on* $\partial\Omega$.

The fact that (4.20) implies the contact angle condition (4.19) can be seen by using the classical formula (for a smooth interface Σ):

$$\int_\Sigma \operatorname{div} \xi - \nu \otimes \nu : D\xi = \int_\Sigma \kappa\, \xi \cdot \nu + \int_\Gamma b \cdot \xi,$$

where ν is the normal vector to Σ, κ denotes the mean curvature of Σ, and b is the conormal vector along $\Gamma = \partial \Sigma$. Indeed, formally at least, the right-hand side of (4.20) is (using the fact that $\xi \cdot n = 0$ on $\partial\Omega$)

$$-\gamma \left[\int_\Sigma \kappa \xi \cdot \nu + \int_\Gamma \vec{b} \cdot \xi\right] + \gamma \min\left(1, \frac{2a}{a + \sqrt{\sigma} b}\right) \left[\int_{\partial\Omega \cap \Omega_s} \kappa \xi \cdot n + \int_\Gamma \vec{c} \cdot \xi\right]$$

$$= \gamma \left[-\int_\Sigma \kappa \xi \cdot \nu + \int_\Gamma \min\left(1, \frac{2a}{a + \sqrt{\sigma} b}\right) \vec{c} \cdot \xi - \vec{b} \cdot \xi\right],$$

where $\vec{b}$ and $\vec{c}$ are unit conormal vectors along $\Gamma = \partial\Sigma \cap \partial\Omega$: $\vec{b}$ is tangent to Σ while $\vec{c}$ is tangent to $\partial\Omega$ (see Fig. 4). When $\xi = \vec{c}$ on $\partial\Omega$, we deduce $\cos\alpha = \vec{b} \cdot \vec{c} = \min\left(1, \frac{2a}{a + \sqrt{\sigma} b}\right)$.

4.4.2 Interactions Dependent Only on the Distance in a Bounded Domain

While the framework discussed above makes sense in the context of chemical interactions, there are other settings in which interactions depend on the distance between particles, even in the presence of fixed boundaries (or obstacles). This is often the case of social interactions which are facilitated by sight [11]. Another example is the modeling of cell-cell adhesion [28, 57] by an attractive force between cells that are within a given distance from each other and the formation of membraneless organelles in eukaryotic cells [72, 107] (in that case, the bounded domain is the cell). In that case, we have

$$\phi_\varepsilon(x) = \int_\Omega G_\varepsilon(x - y) \rho(y)\, dy = \int_{\mathbb{R}^d} G_\varepsilon(x - y) \bar{\rho}(y)\, dy,$$

where $\bar{\rho}$ is the extension of ρ to $\mathbb{R}^d$ by 0. For simplicity, we focus on the hard-sphere case $f = f_\infty$, although the soft-sphere case could be treated similarly. The energy functional then reads

$$\varepsilon^{-1}\int_\Omega h_\infty(\rho(x)) - \frac{1}{2}\rho(x)\phi_\varepsilon(x)\,dx = \varepsilon^{-1}\int_{\mathbb{R}^d} h_\infty(\bar{\rho}(x)) - \frac{1}{2}\bar{\rho}(x)G_\varepsilon * \bar{\rho}(x)\,dx,$$

which is the same energy as studied in Sect. 4.3 applied to $\bar{\rho}$ instead of ρ. Its Γ-limit is proportional to the (full) perimeter functional $P(E)$. The limit $\varepsilon \to 0$ of the aggregation-diffusion equation (1.20) then leads to the Hele-Shaw free boundary problem (1.19) supplemented with the contact angle condition:

$$\cos\alpha = -1$$

that is, $\alpha = \pi$: The boundary of the domain $\partial\Omega$ acts like a hydrophobic surface and any contact of the free surface with the fixed boundary will be tangential.

Other contact angle conditions may be derived by taking into account the interactions of the particles with the boundary of the domain (in the modeling of cell-cell adhesion, the fixed boundary may represent a different species with a different interspecies adhesion strength). To illustrate this effect, we introduce the potential $\tau_\varepsilon(x) = G_\varepsilon * \chi_{\mathbb{R}^d\setminus\Omega}$ and consider the equation

$$\begin{cases} \varepsilon\partial_t\rho + \operatorname{div}(\rho\nabla(G_\varepsilon * \rho)) + \eta\operatorname{div}(\rho\nabla\tau_\varepsilon) = \operatorname{div}(\rho\nabla p), \quad p \in \partial f_\infty(\rho) & \text{in } [0,\infty)\times\Omega, \\ \rho\nabla[G_\varepsilon * \rho + \eta\tau_\varepsilon - p]\cdot n = 0 & \text{on } [0,\infty)\times\partial\Omega, \\ \rho(x,0) = \rho_{in}^\varepsilon(x) & \text{in } \Omega. \end{cases} \tag{4.21}$$

The new drift term appearing in this equation models the attraction of the particles toward $\mathbb{R}^d \setminus \Omega$ (or repulsion if $\eta < 0$). It describes the interactions of our particles with (different) particles located outside of Ω. These interactions are assumed to be of similar nature but possibly different strength (another classical example is that of a fluid resting on a solid surface: The fluid molecules interact with each other as well as with the solid).

The respective strength of the interactions—the coefficient η—determines the contact angle condition. This can be seen by considering the energy associated to (4.21), which reads as follows:

$$\mathcal{G}_{\eta,\varepsilon}(\rho) = \varepsilon^{-1}\int_\Omega f_\infty(\rho) + \frac{1}{2\sigma}\rho - \frac{1}{2}\rho\phi_\varepsilon - \eta\rho\tau_\varepsilon\,dx.$$

Using the fact that $\tau_\varepsilon(x) = \frac{1}{\sigma} - \int_\Omega G_\varepsilon(x-y)\,dy$, we can write

$$\mathcal{G}_{\eta,\varepsilon}(\rho) = \varepsilon^{-1}\int_\Omega f_\infty(\rho) + \frac{1}{2}\phi_\varepsilon(1-\rho) - \eta\rho\tau_\varepsilon + \frac{1}{2}\left(\frac{1}{\sigma}\rho - \phi_\varepsilon\right)dx$$

$$= \varepsilon^{-1} \int_\Omega f_\infty(\rho) + \frac{1}{2}\phi_\varepsilon(1-\rho)\,dx + \left(\frac{1}{2} - \eta\right)\varepsilon^{-1}\int_\Omega \rho\tau_\varepsilon\,dx.$$

This energy has two terms: The first term was studied, for example, in [78] where it was shown (for a different kernel, but the proof extends to (1.4) using the arguments of [86]) that it Γ-converges to the local perimeter:

$$\frac{1}{4\sigma^{3/2}} P(E, \Omega).$$

The second term yields a contribution from the boundary. When G is the Newtonian attractive kernel (1.4), then τ_ε solves

$$\tau_\varepsilon - \varepsilon^2 \Delta \tau_\varepsilon = \chi_{\mathbb{R}^d \setminus \Omega}. \tag{4.22}$$

When Ω is the half-space $\{x \in \mathbb{R}^d\,;\, x_d > 0\}$, we can solve (4.22) explicitly and find $\tau_\varepsilon(x) = \frac{1}{2\sigma} e^{-\frac{\sqrt{\sigma}x_d}{\varepsilon}}$ for $x_d \geq 0$ so that $\frac{1}{\varepsilon}\tau_\varepsilon(x)\chi_{x_d>0}$ converges to $\frac{1}{2\sigma^{3/2}}\delta(x_d)$ as $\varepsilon \to 0$ and

$$\varepsilon^{-1} \int_\Omega \rho\tau_\varepsilon\,dx \to \frac{1}{2\sigma^{3/2}} \int_{\partial\Omega} \rho\,d\mathcal{H}^{n-1}(x).$$

This convergence can be justified for general domains Ω with smooth boundary. In fact, we can show:

Proposition 4.4 *Assume that Ω is a bounded open set with $C^{1,\alpha}$ boundary and take $f = f_\infty$ and G given by* (1.4). *For all $\eta \in [0, 1/2]$, the energy functional $\mathcal{G}_{\eta,\varepsilon}$ Γ-converges to*

$$\frac{1}{4\sigma^{3/2}} P(E, \Omega) + \frac{1}{2\sigma^{3/2}}\left(\frac{1}{2} - \eta\right) H^{n-1}(\partial\Omega \cap E). \tag{4.23}$$

Note that the result holds for $\eta \in [0, 1)$ but the proof is more delicate than the proof given below when $\eta > 1/2$. This proposition suggests that a result similar to Theorem 4.9 holds for (4.21): Under some assumptions on the convergence of the energy, the limit $\varepsilon \to 0$ in (4.21) leads to the Hele-Shaw free boundary problem (with surface tension) (1.19), supplemented with the contact angle condition:

$$\cos\alpha = 2\eta - 1$$

at the triple junction. We can thus have contact angles between π and 0 when η takes values between 0 and 1. When $\eta \geq 1$, the attraction toward the boundary dominates the dynamics and the particles will spread over the whole boundary to maximize the contact area. A rigorous justification of this result would require proving the convergence of the first variation of the energy (the equivalent of Proposition 4.3).

Idea of the Proof of Proposition 4.4 We note that

$$\int_\Omega \rho\tau_\varepsilon\,dx = \int_\Omega \rho * G_\varepsilon \chi_{\mathbb{R}^d\setminus\Omega}\,dx = \int_\Omega \phi_\varepsilon \chi_{\mathbb{R}^d\setminus\Omega}\,dx = \int_\Omega \frac{1}{\sigma}\rho - \phi_\varepsilon\,dx$$

and we write

$$\begin{aligned}\mathscr{G}_{\eta,\varepsilon}(\rho) &= \varepsilon^{-1}(1-2\eta)\int_\Omega f_\infty(\rho) + \frac{1}{2\sigma}\rho - \frac{1}{2}\rho\phi_\varepsilon\,dx + \varepsilon^{-1}2\eta\int_\Omega f_\infty(\rho)\\ &\quad + \frac{1}{2\sigma}\rho - \frac{1}{2}\rho\phi_\varepsilon - \frac{1}{2}\rho\tau_\varepsilon\,dx\\ &= \varepsilon^{-1}(1-2\eta)\int_{\mathbb{R}^d} h_\infty(\bar\rho) + \frac{1}{2}\bar\rho(1-\phi_\varepsilon)\,dx + \varepsilon^{-1}2\eta\int_\Omega f_\infty(\rho)\\ &\qquad + \frac{1}{2}\phi_\varepsilon(1-\rho)\,dx\end{aligned}$$

where every term is nonnegative. The first term is (up to a multiplicative constant) the energy $\mathscr{G}_\varepsilon$ and thus Γ-converges to $\frac{1}{4\sigma^{3/2}}(1-2\eta)P(E)$ (when $\eta \le 1/2$). The second term can be written as

$$\varepsilon^{-1}2\eta\int_\Omega f_\infty(\rho)+\frac{1}{2}\tau_\varepsilon(x)\rho(x)(1-\rho(x))\,dx+\frac{1}{4}\iint_{\Omega\times\Omega} G(x-y)[\rho(y)-\rho(x)]^2\,dx\,dy,$$

where the first integral is the double-well potential and the second integral is similar to the energy studied, for example, in [1]. This energy Γ-converges to $2\eta\frac{1}{4\sigma^{3/2}}P(E,\Omega)$. The limiting energy is thus

$$\frac{1}{4\sigma^{3/2}}(1-2\eta)P(E) + 2\eta\frac{1}{4\sigma^{3/2}}P(E,\Omega),$$

which is equivalent to (4.23).

5 Conclusion

In this chapter, we presented a road map to investigate phase separation phenomena in biological (and other) systems. We start from an agent-based microscopic description of attractive behavior limited by a volume exclusion principle, accounting for the natural incompressibility of the agents. The key takeaway is that phase separation is natural when repulsion is modeled by nonlinear diffusion with $m > 2$ (soft-sphere models) or with density constraint (hard-sphere models) and when the long-range attractive kernel is integrable. We note that the restriction $m > 2$, which is required for the function $h(\rho)$ defined by (1.13) to be a double-well potential, is also consistent with known results on the existence of stationary states and energy minimizers for (1.5) when $\sigma > 0$ (see [6, 18, 81])—the existence of such stationary states seems necessary for our analysis which concerns the behavior of solutions of (1.5) over large timescales.

Based on recent works, we proposed a rigorous mathematical approach to derive effective models of geometric nature which describe the evolution of the sharp interface over appropriate timescales after aggregation occurs. In particular, the results presented here show that surface tension and contact angle dynamics, classically associated with fluid interfaces, are natural consequences of volume-limited aggregation. This is in agreement with many real-life observations: For instance, the liquid-like behavior of membraneless organelles in eukaryotic cells has been well documented [16, 72].

This review leads to several important open questions. Among them is the well-posedness of the hard-sphere microscopic model (1.8) for general initial conditions and its convergence as $\delta \to 0$ to the macroscopic model (1.10). Solutions to (3.32) can be formally obtained in the limit $m \to \infty$ in the soft-sphere model (1.6) or via a JKO-type discrete scheme, but neither approaches has been made rigorous.

Concerning the derivation of the free boundary problems, we emphasize once more that all stated rigorous results are conditional, which require an assumption on the energy convergence. Obtaining nonconditional convergence results remains out of reach, not only in the context of the problems discussed in this review, but for related free boundary problems as well. Last but not least, these conditional results have been proved only for the particular kernel G, solution of (1.4). On the other hand, the Γ-convergence of the energies $\mathscr{J}_\varepsilon$ and $\mathscr{G}_\varepsilon$ has been proved for very general G. Thus, we may conjecture that these conditional results can hold for a large class of nonnegative integrable kernels. Further potential generalizations of our analysis also include the derivation of two-phase free boundary problems, either by considering pressure function f for which the two wells of the function h are both positive (leading to a Mullins-Sekerka free boundary problem) or for multispecies models such as those proposed in [57].

Appendix A From Micro to Macro: Convergence $\delta \to 0$

Proof of Theorem 3.3 The continuity equation and (3.13) imply that ρ^δ is bounded in $C^{1/2}([0,T]; \mathcal{P}_2(\mathbb{R}^d))$. It follows from a refined version of Ascoli-Arzelà theorem, as in [31, Proposition 4.1], that there exists a subsequence such that $\rho^\delta(t)$ narrowly converge to some function $\bar{\rho}(t)$ (uniformly in t).

Since K_δ is supported in B_δ, we have $K_\delta * \varphi \to \varphi$ uniformly for any uniformly continuous function φ. Using the bound on the second moment (3.14), we can then show that $K_\delta * \rho^\delta(t)$ converges narrowly to $\bar{\rho}(t)$.

We also note that (3.13) implies that $\rho^\delta * K_\delta$ is bounded in $L^\infty(0,T; L^m(\mathbb{R}^d))$ and thus also in $L^\infty(0,T; L^2(\mathbb{R}^d))$. Furthermore, using the fact that $\phi^\delta = G * \rho^\delta$ solves $\sigma\phi^\delta - \eta\Delta\phi^\delta = \rho^\delta$, we see that ϕ^δ and $K_\delta * \phi^\delta$ are bounded in $L^\infty(0,T; H^1(\mathbb{R}^d))$.

However, this convergence is not enough to pass to the limit in the equation and the main idea in [31] is to rewrite the limiting equation (1.9) (when f is given by (1.2)) as

$$\partial_t \rho = -\operatorname{div}(\rho \nabla G * \rho) + 2\nabla \cdot (\rho^{\frac{m}{2}} \nabla \rho^{\frac{m}{2}}) \tag{A.1}$$

and to adjust, in a similar way, the weak formulation of the regularized equation (1.6). Indeed, given a test function $\psi \in C_c^\infty([0,T] \times \mathbb{R}^d)$, a straightforward computation with integration by parts yields

$$\begin{aligned}
&\int \psi(T,x)\, d\rho^\delta(T,x) - \int \psi(0,x)\, d\rho_{\text{in}}(x) - \int_0^T \int \partial_t \psi\, d\rho^\delta(t,x)\, dt \\
&= \int_0^T \int \nabla(K_\delta * K_\delta * \phi^\delta) \cdot \nabla \psi\, d\rho^\delta(t,x)\, dt - \frac{m}{m-1} \\
&\quad \times \int_0^T \int \nabla K_\delta * (K_\delta * \rho^\delta)^{m-1} \cdot \nabla \psi\, d\rho^\delta(t,x)\, dt \\
&= \int_0^T \int \nabla(K_\delta * \phi^\delta) \cdot [K_\delta * (\rho^\delta \nabla \psi)] \\
&\quad - m \int_0^T \int (K_\delta * \rho^\delta)^{m-2} \nabla (K_\delta * \rho^\delta) \cdot [K_\delta * (\rho^\delta \nabla \psi)] \\
&= \int_0^T \int (K_\delta * \rho^\delta) \nabla (K_\delta * \phi^\delta) \cdot \nabla \psi - 2 \int_0^T \int (K_\delta * \rho^\delta)^{\frac{m}{2}} \nabla (K_\delta * \rho^\delta)^{\frac{m}{2}} \cdot \nabla \psi
\end{aligned} \tag{A.2}$$

$$+ \int_0^T \int \nabla(K_\delta * \phi^\delta) \cdot E^\delta - 2 \int_0^T \int (K_\delta * \rho^\delta)^{\frac{m}{2}-1} \nabla (K_\delta * \rho^\delta)^{\frac{m}{2}} \cdot E^\delta. \tag{A.3}$$

The error term in (A.3) comes from the commutation between convolution and multiplication, i.e., $E^\delta(x) := [K_\delta * (\rho^\delta \nabla \psi)](x) - (K_\delta * \rho^\delta) \nabla \psi(x)$. We will see that this error term vanishes in the limit. Passing to the limit in the second term of (A.2) (which has the same structure as (A.1)) requires the following convergences:

$$(K_\delta * \rho^\delta)^{\frac{m}{2}} \to \rho^{\frac{m}{2}}, \text{ (strong) in } L^2_{t,x}, \text{ and } \nabla (K_\delta * \rho^\delta)^{\frac{m}{2}} \rightharpoonup \nabla \rho^{\frac{m}{2}}, \quad \text{(weak) in } L^2_{t,x}.$$

This weak-strong convergence pair is proved by considering the evolution of $\int \rho^\delta \log \rho^\delta$. Indeed, since $\mathscr{E}[\rho_{\text{in}}] < +\infty$ and $\rho \log \rho \lesssim \rho^m$ for $m > 1$, we see that $\int \rho_{\text{in}} \log \rho_{\text{in}} < +\infty$. Formally (these steps can be made rigorous by using the piecewise constant interpolation given by the JKO scheme as in [31]), by differentiating $\int \rho^\delta \log \rho^\delta$ in time we get

$$\int \rho^\delta(t,x)\log\rho^\delta(t,x) - \int \rho_{\rm in}\log\rho_{\rm in}$$
$$= -\frac{2}{m}\int_0^t\int |\nabla (K_\delta*\rho^\delta)^{\frac{m}{2}}|^2 + \int_0^t\int \nabla(K_\delta*\phi^\delta)\cdot\nabla K_\delta*\rho^\delta$$

where

$$\int_0^t\!\!\int \nabla(K_\delta*\phi^\delta)\cdot\nabla K_\delta*\rho^\delta = -\int_0^t\!\!\int \Delta(K_\delta*\phi^\delta)\,K_\delta*\rho^\delta$$
$$= -\frac{\sigma}{\eta}\int_0^t\!\!\int (K_\delta*\phi^\delta)\,K_\delta*\rho^\delta + \frac{1}{\eta}\int_0^t\!\!\int |K_\delta*\rho^\delta|^2$$

and so

$$\int \rho^\delta(t,x)\log\rho^\delta(t,x) + \frac{2}{m}\int_0^t\!\!\int |\nabla (K_\delta*\rho^\delta)^{\frac{m}{2}}|^2 + \frac{\sigma}{\eta}\int_0^t\!\!\int (K_\delta*\phi^\delta)\,K_\delta*\rho^\delta$$
$$= \int \rho_{\rm in}\log\rho_{\rm in} + \frac{1}{\eta}\int_0^t\!\!\int |K_\delta*\rho^\delta|^2.$$

We recall (see [31, Remark 4.2]) that the entropy is bounded below by the second moment as follows:

$$\int \rho^\delta(t,x)\log\rho^\delta(t,x) \geq -d\log(2\pi) - \frac{1}{2}\int |x|^2\rho^\delta,$$

and since $m > 2$, we have

$$\int_0^t\int |K_\delta*\rho^\delta|^2 \leq C\int_0^t\int K_\delta*\rho^\delta + |K_\delta*\rho^\delta|^m\,dx$$
$$\leq C\,(1+\mathscr{E}_\delta[\rho_{\rm in}])\,t.$$

This implies that the sequence $(K_\delta*\rho^\delta)^{\frac{m}{2}}$ is bounded in $L^2(0,T;H^1(\mathbb{R}^d))$ uniformly in $\delta>0$. Proceeding as in [31, Proposition 4.3] (using a refined version of the Aubin-Lions Lemma [127, Theorem 2]) we can then show that up to another subsequence we have

$$K_\delta*\rho^\delta \to \rho \qquad \text{in } L^m([0,T]\times\mathbb{R}^d), \qquad \nabla(K_\delta*\rho^\delta)^{\frac{m}{2}} \rightharpoonup \nabla\rho^{\frac{m}{2}} \text{ in } L^2([0,T]\times\mathbb{R}^d).$$

We already mentioned that $K_\delta*\phi^\delta$ was bounded in $L^\infty(0,T;H^1(\mathbb{R}^d))$, which implies

$$\nabla(K_\delta*\phi^\delta) \rightharpoonup \nabla\phi \text{ in } L^2([0,T]\times\mathbb{R}^d).$$

Finally (see [31, Lemma 5.1]), we have that $E^{\delta}(t,x)$ converges to zero in $L^m([0,T]\times\mathbb{R}^d)$. These convergences are enough to show that (A.3) vanishes as $\delta\to 0$ and to pass to the limit in (A.2).

Appendix B Convergence Results for the Stefan Problem

Proof of Proposition 4.2 We have $\mathscr{J}_\varepsilon[\rho^\varepsilon]\geq\mathscr{J}^*[\rho^\varepsilon]$ and the functional $\mathscr{J}^*$ is convex and thus lower semicontinuous with respect to the weak topology. The liminf property follows.

To prove the limsup property, we will simply use the fact that $\mathscr{J}_\varepsilon$ can be bounded by some Allen-Cahn functional for which the Γ convergence has been proved elsewhere. Indeed, we can write

$$\iint_{\mathbb{R}^d\times\mathbb{R}^d} G_\varepsilon(x-y)[\rho(x)-\rho(y)]^2\,dx\,dy=\iint_{\mathbb{R}^d\times\mathbb{R}^d} G_\varepsilon(z)[\rho(y+z)-\rho(y)]^2\,dz\,dy$$

where

$$\begin{aligned}|\rho(y+z)-\rho(y)|^2&=\left|\int_0^1\nabla\rho(y+tz)\cdot z\,dt\right|^2\\&\leq\int_0^1|\nabla\rho(y+tz)|^2\,dt\,|z|^2.\end{aligned}$$

We deduce

$$\begin{aligned}&\iint_{\mathbb{R}^d\times\mathbb{R}^d} G_\varepsilon(x-y)[\rho(x)-\rho(y)]^2\,dx\,dy\\&\leq\int_0^1\iint_{\mathbb{R}^d\times\mathbb{R}^d}|z|^2G_\varepsilon(z)\,|\nabla\rho(y+tz)|^2\,dz\,dy\,dt\\&\leq\int_0^1\int_{\mathbb{R}^d}|z|^2G_\varepsilon(z)\int_{\mathbb{R}^d}|\nabla\rho(y)|^2\,dy\,dz\,dt\\&\leq\varepsilon^2\int_{\mathbb{R}^d}|z|^2G(z)\,dz\int_{\mathbb{R}^d}|\nabla\rho(y)|^2\,dy.\end{aligned}$$

We thus have

$$\mathscr{J}_\varepsilon[\rho^\varepsilon]\leq\int_{\mathbb{R}^d}h(\rho^\varepsilon)\,dx+\varepsilon^2D\int_{\mathbb{R}^d}\left|\nabla\rho^\varepsilon(x)\right|^2\,dx$$

for some constant D. The functional in the right-hand side is an Allen-Cahn functional for which the Γ-convergence is proved, for example, in [9]. Any recovery sequence for this functional will be a recovery sequence for $\mathcal{J}_\varepsilon$.

Proof of Theorem 4.5 The first part of the proof is classical (see, for instance, [104, Proposition 4.2] we recall the main steps here: We note that ρ^ε solves (4.3) with

$$\mathcal{J}_\varepsilon[\rho^\varepsilon(t)] + \int_0^t \int_{\mathbb{R}^d} \rho^\varepsilon |v^\varepsilon|^2 \, dx \, dt \le \mathcal{J}_\varepsilon[\rho^\varepsilon_{in}], \qquad \sup_{t\in[0,T]} \int_{\mathbb{R}^d} |x|^2 \rho^\varepsilon(t) \, dx \le C(T).$$

We have in particular (see [5])

$$d_W(\rho^\varepsilon(t), \rho^\varepsilon(s)) \le \left(\int_0^T \int_{\mathbb{R}^d} \rho^\varepsilon |v^\varepsilon|^2 \, dx \, dt \right)^{1/2} (t-s)^{1/2}$$

so ρ^ε is bounded in $C^{1/2}([0,T]; \mathcal{P}_2(\mathbb{R}^d))$. Together with the bound on the second moment, this implies that there exists a subsequence ρ^{ε_n} such that $\rho^{\varepsilon_n}(t)$ narrowly converges to $\rho(t)$ locally uniformly in t (see, for example, [19, Proposition 4.1]). In addition, the fact that $m > 2$ together with the bound on the energy implies that ρ^ε is bounded in $L^\infty(0,\infty; L^2(\mathbb{R}^d))$ and so $j^\varepsilon = \rho^\varepsilon v^\varepsilon$ is bounded in $L^2(0,\infty; L^{4/3}(\mathbb{R}^d))$. Up to another subsequence, we can thus assume that j^ε converges to j (weakly) and a classical argument implies that $j = \rho v$ for some $v \in L^2(d\rho(t))$ (see, for example, [86, Proposition 4.1]). We can pass to the limit in the continuity equation to show that ρ, v satisfy the continuity equation:

$$\partial_t \rho + \operatorname{div}(\rho v) = 0.$$

To complete the proof, we will use the assumption of convergence of the energy (4.7) to pass to the limit in the momentum equation. First, since $h(\rho^\varepsilon) \ge h^{**}(\rho^\varepsilon) \ge 0$ and

$$\liminf_{\varepsilon\to 0} \int_0^T \int_{\mathbb{R}^d} h^{**}(\rho^\varepsilon) \, dx \, dt \ge \int_0^T \int_{\mathbb{R}^d} h^{**}(\rho) \, dx \, dt$$

(by the convexity of h^{**}) and since $\frac{1}{4} \iint_{\mathbb{R}^d\times\mathbb{R}^d} G_\varepsilon(x-y)[\rho(x)-\rho(y)]^2 \, dx \, dy \ge 0$, we see that (4.7) implies that

$$\int_0^T \int_{\mathbb{R}^d} h(\rho^{\varepsilon_n})\varphi \, dx \, dt \to \int_0^T \int_{\mathbb{R}^d} h^{**}(\rho)\varphi \, dx \, dt \text{ for any } \varphi \in L^\infty, \tag{B.1}$$

$$h(\rho^{\varepsilon_n}) - h^{**}(\rho^{\varepsilon_n}) \to 0 \quad \text{in } L^1((0,T)\times\mathbb{R}^d), \tag{B.2}$$

$$\int_0^T \iint_{\mathbb{R}^d\times\mathbb{R}^d} G_{\varepsilon_n}(x-y)[\rho^{\varepsilon_n}(x) - \rho^{\varepsilon_n}(y)]^2 \, dx \, dy \, dt \to 0. \tag{B.3}$$

We now write the weak formulation of the momentum equation, using the fact that $v^{\varepsilon} = -\nabla h'(\rho^{\varepsilon}) + \nabla B_{\varepsilon}(\rho^{\varepsilon})$ (see (4.3)):

$$\int_0^T \int_{\mathbb{R}^d} \rho^{\varepsilon_n} v^{\varepsilon_n} \cdot \xi \, dx \, dt = \int_0^T \int_{\mathbb{R}^d} \rho^{\varepsilon_n} \nabla[-h'(\rho^{\varepsilon_n}) + B_{\varepsilon}[\rho^{\varepsilon_n}]] \cdot \xi \, dx \, dt$$
$$= \int_0^T \int_{\mathbb{R}^d} [\rho^{\varepsilon_n} h'(\rho^{\varepsilon_n}) - h(\rho^{\varepsilon_n})] \mathrm{div}\, \xi \, dx \, dt$$
$$- \int_0^T \int_{\mathbb{R}^d} \rho^{\varepsilon_n} \nabla B_{\varepsilon_n}[\rho^{\varepsilon_n}] \cdot \xi \, dx \, dt. \quad \text{(B.4)}$$

We will complete the proof of the theorem by passing to the limit in (B.4) thanks to the following Lemma:

Lemma B.1 *Assume* (B.1), (B.2), *and* (B.3). *Then:*
For all $\xi \in C_c^{\infty}((0, \infty) \times \mathbb{R}^d; \mathbb{R}^d)$, *there holds*

$$\lim_{n\to\infty} \int_0^T \int_{\mathbb{R}^d} \rho^{\varepsilon_n} \nabla B_{\varepsilon_n}[\rho^{\varepsilon_n}] \cdot \xi \, dx \, dt = 0. \quad \text{(B.5)}$$

For all $\xi \in C_c^{\infty}((0, \infty) \times \mathbb{R}^d; \mathbb{R}^d)$, *there exists a subsequence* ε_n' *such that*

$$\lim_{n\to\infty} \int_0^T \int_{\mathbb{R}^d} \left[\rho^{\varepsilon_n'} h'(\rho^{\varepsilon_n'}) - h(\rho^{\varepsilon_n'})\right] \mathrm{div}\, \xi \, dx \, dt$$
$$= \int_0^T \int_{\mathbb{R}^d} \left[\rho h^{**\prime}(\rho) - h^{**}(\rho)\right] \mathrm{div}\, \xi \, dx \, dt. \quad \text{(B.6)}$$

This lemma implies (for a given $\xi \in C_c^{\infty}((0, \infty) \times \mathbb{R}^d; \mathbb{R}^d)$) and with the corresponding subsequence such that (B.6) holds)

$$\int_0^T \int_{\mathbb{R}^d} \rho v \cdot \xi \, dx \, dt = \lim_{n\to\infty} \int_0^T \int_{\mathbb{R}^d} \rho^{\varepsilon_n'} v^{\varepsilon_n'} \cdot \xi \, dx \, dt$$
$$= \int_0^T \int_{\mathbb{R}^d} \left[\rho h^{**\prime}(\rho) - h^{**}(\rho)\right] \mathrm{div}\, \xi \, dx \, dt$$

and since this holds for all $\xi \in C_c^{\infty}((0, \infty) \times \mathbb{R}^d; \mathbb{R}^d)$, the theorem follows.

It remains now to prove the above lemma.

Proof of Lemma B.1 In order to prove (B.5), we use the same computation used in [86, 106]: Recalling that $\phi^{\varepsilon} = G_{\varepsilon} * \rho$ solves $\sigma\phi^{\varepsilon} - \varepsilon^2 \Delta\phi^{\varepsilon} = \rho$, we can write

$$\frac{1}{2} \int_0^T \iint_{\mathbb{R}^d \times \mathbb{R}^d} G_{\varepsilon}(x-y)[\rho(x) - \rho(y)]^2 \, dx \, dy \quad \text{(B.7)}$$

$$= \int_0^T \iint_{\mathbb{R}^d \times \mathbb{R}^d} \rho \left[\frac{1}{\sigma}\rho - \phi_\varepsilon \right] dx\, dy$$

$$= \sigma \int_0^T \iint_{\mathbb{R}^d \times \mathbb{R}^d} \left[\frac{1}{\sigma}\rho - \phi_\varepsilon \right]^2 dx\, dy + \sigma \int_0^T \iint_{\mathbb{R}^d \times \mathbb{R}^d} \phi^\varepsilon \left[\frac{1}{\sigma}\rho - \phi_\varepsilon \right] dx\, dy$$

$$= \sigma \int_0^T \iint_{\mathbb{R}^d \times \mathbb{R}^d} \left[\frac{1}{\sigma}\rho - \phi_\varepsilon \right]^2 dx\, dy + \varepsilon^2 \int_0^T \iint_{\mathbb{R}^d \times \mathbb{R}^d} |\nabla \phi_\varepsilon|^2\, dx\, dy$$

and

$$\int_0^T \int_{\mathbb{R}^d} \rho^\varepsilon \nabla B_\varepsilon[\rho^\varepsilon] \cdot \xi\, dx\, dt = \int_0^T \int_{\mathbb{R}^d} \rho^\varepsilon \nabla [\phi_\varepsilon - \sigma^{-1}\rho^\varepsilon] \cdot \xi\, dx\, dt$$

$$= \int_0^T \int_{\mathbb{R}^d} [\rho^\varepsilon - \sigma\phi_\varepsilon] \nabla [\phi_\varepsilon - \sigma^{-1}\rho^\varepsilon] \cdot \xi\, dx\, dt$$

$$+ \int_0^T \int_{\mathbb{R}^d} -\sigma\phi_\varepsilon \nabla [\phi_\varepsilon - \sigma^{-1}\rho^\varepsilon] \cdot \xi\, dx\, dt$$

$$= \frac{\sigma}{2} \int_0^T \int_{\mathbb{R}^d} [\phi_\varepsilon - \sigma^{-1}\rho^\varepsilon]^2 \operatorname{div} \xi\, dx\, dt + \varepsilon^2 \int_0^T \int_{\mathbb{R}^d} \phi_\varepsilon \nabla \Delta \phi_\varepsilon \cdot \xi\, dx\, dt$$

$$= \frac{\sigma}{2} \int_0^T \int_{\mathbb{R}^d} [\phi_\varepsilon - \sigma^{-1}\rho^\varepsilon]^2 \operatorname{div} \xi\, dx\, dt$$

$$+ \varepsilon^2 \int_0^T \int_{\mathbb{R}^d} \frac{|\nabla \phi_\varepsilon|^2}{2} \operatorname{div} \xi + D\xi : \nabla \phi_\varepsilon \otimes \nabla \phi_\varepsilon + \phi_\varepsilon \nabla \phi_\varepsilon \cdot \Delta \xi\, dx\, dt,$$

which together imply

$$\left| \int_0^T \int_{\mathbb{R}^d} \rho^{\varepsilon_n} \nabla B_{\varepsilon_n}[\rho^{\varepsilon_n}] \cdot \xi\, dx\, dt \right|$$

$$\le \|\operatorname{div} \xi\|_{L^\infty} \int_0^T \int_{\mathbb{R}^d} [\phi_\varepsilon - \sigma^{-1}\rho^\varepsilon]^2\, dx\, dt$$

$$+ C(1 + \|D\xi\|_{L^\infty})\varepsilon^2 \int_0^T \int_{\mathbb{R}^d} |\nabla \phi_\varepsilon|^2\, dx\, dt + C\varepsilon^2 \|\Delta \xi\|_{L^2}$$

$$\le C(\xi) \left(\int_0^T \iint_{\mathbb{R}^d \times \mathbb{R}^d} G_{\varepsilon_n}(x-y)[\rho^{\varepsilon_n}(x) - \rho^{\varepsilon_n}(y)]^2\, dx\, dy + \varepsilon^2 \right).$$

Thus, (B.3) yields (B.5).

We now turn to the proof of (B.6). Given $\xi \in C_c^\infty((0,\infty)\times\mathbb{R}^d;\mathbb{R}^d)$, we fix $Q=(0,T)\times B_R$ such that $\operatorname{Supp}\xi \in Q$ and we introduce the measurable sets:

$$E=\{h^{**}(\rho)=0\}\cap Q, \qquad F=\{h^{**}(\rho)>0\}\cap Q.$$

First, we claim that there exists a subsequence ε_n' such that

$$\lim_{n\to\infty}\iint_E\left[\rho^{\varepsilon_n'}h'(\rho^{\varepsilon_n'})-h(\rho^{\varepsilon_n'})\right]\operatorname{div}\xi\,dx\,dt\to 0. \tag{B.8}$$

Indeed, using (B.1), we get $h(\rho^{\varepsilon_n'})\to 0$ a.e. in E (for some subsequence). Furthermore, for any $\delta>0$, there exists C_δ such that $|\rho h'(\rho)-h(\rho)|\le\delta+C_\delta h(\rho)$ (this is because the function in the left-hand side vanishes when $\rho=0$ or θ which are the only zeroes of h and the two functions have the same growth at ∞). We deduce

$$\begin{aligned}&\limsup_{n\to\infty}\iint_E\left|\rho^{\varepsilon_n'}h'(\rho^{\varepsilon_n'})-h(\rho^{\varepsilon_n'})\right||\operatorname{div}\xi|\,dx\,dt\\&\quad\le\|\operatorname{div}\xi\|_{L^\infty}\limsup_{n\to\infty}\iint_E\delta+C_\delta h(\rho^{\varepsilon_n'})\,dx\,dt\le\|\operatorname{div}\xi\|_{L^\infty}|Q|\delta\end{aligned}$$

and (B.8) follows by taking $\delta\to 0$.

Next, we will prove that the strict convexity if h^{**} in $\{h^{**}(s)>0\}$ implies the strong convergence of ρ^ε to ρ in F. We define $w^\varepsilon=\max\{\rho^\varepsilon,\theta\}$, which only takes values in the set where $\rho\mapsto h^{**}(\rho)$ is strictly convex. Since w^{ε_n} and ρ^{ε_n} are bounded in $L^2(Q)$, they both have some weak limit w and ρ (up to some subsequence). And since $w^{\varepsilon_n}\ge\rho^{\varepsilon_n}$, we have $w\ge\rho$. Furthermore, we have

$$h^{**}(w^{\varepsilon_n})=h^{**}(\rho^{\varepsilon_n})\le h(\rho^{\varepsilon_n})$$

(the first equality holds since $h^{**}(\rho)=0$ when $\rho<\theta$) and the convexity of h^{**} implies

$$\begin{aligned}\int_Q h^{**}(w)&\le\liminf\int_Q h^{**}(w^{\varepsilon_n'})\le\limsup\int_Q h^{**}(w^{\varepsilon_n'})\le\limsup\int_Q h(\rho^{\varepsilon_n'})\\&=\int_Q h^{**}(\rho)\le\int_Q h^{**}(w)\end{aligned}$$

where we used the assumption of convergence of the energy (4.7). We deduce that we have equality in all these inequalities. In particular

$$\lim\int_Q h^{**}(w^{\varepsilon_n'})=\int_Q h^{**}(w)$$

and

$$\int_Q h^{**}(\rho) = \int_Q h^{**}(w).$$

The first equality and strict convexity of h^{**} in the set of values of $w^{\varepsilon'_n}(t,x)$ implies that $w^{\varepsilon'_n}$ converges strongly (and a.e.) to w. The second equality (together with the fact that $w \geq \rho$ and the strict monotonicity of h^{**} in (θ, ∞)) implies that $\rho = w$ in F. Finally, since $w^{\varepsilon'_n} \geq \rho^{\varepsilon'_n}$, we have

$$\int_F |w^{\varepsilon'_n} - \rho^{\varepsilon'_n}| = \int_F w^{\varepsilon'_n} - \rho^{\varepsilon'_n} \to \int_F w - \rho = 0$$

and so $\rho^{\varepsilon'_n}$ converges to $w = \rho$ in $L^1(F)$ and a.e. (up to yet another subsequence). Together with (B.8), this implies (B.6).

References

1. G. Alberti, G. Bellettini, A non-local anisotropic model for phase transitions: asymptotic behaviour of rescaled energies. European J. Appl. Math. **9**(3), 261–284 (1998)
2. D. Alexander, I. Kim, Y. Yao, Quasi-static evolution and congested crowd transport. Nonlinearity **27**(4), 823–858 (2014)
3. V. Alexandria, S. Björn, Modelling stripe formation in zebrafish: an agent-based approach. J. R. Soc. Interface **12**, 20150812 (2015)
4. N.D. Alikakos, P.W. Bates, X. Chen, Convergence of the Cahn-Hilliard equation to the Hele-Shaw model. Arch. Rational Mech. Anal. **128**(2), 165–205 (1994)
5. L. Ambrosio, N. Gigli, G. Savaré, *Gradient Flows in Metric Spaces and in the Space of Probability Measures*. Lectures in Mathematics ETH Zürich, 2nd edn. (Birkhäuser Verlag, Basel, 2008)
6. J. Bedrossian, Global minimizers for free energies of subcritical aggregation equations with degenerate diffusion. Appl. Math. Lett. **24**(11), 1927–1932 (2011)
7. J. Bedrossian, N. Rodríguez, Inhomogeneous Patlak-Keller-Segel models and aggregation equations with nonlinear diffusion in $\mathbb{R}^d$. Discrete Contin. Dyn. Syst. Ser. B **19**(5), 1279–1309 (2014)
8. J. Bedrossian, N. Rodríguez, A.L. Bertozzi, Local and global well-posedness for aggregation equations and Patlak–Keller–Segel models with degenerate diffusion. Nonlinearity **24**(6), 1683 (2011)
9. G. Bellettini, L. Bertini, M. Mariani, M. Novaga, Convergence of the one-dimensional Cahn-Hilliard equation. SIAM J. Math. Anal. **44**(5), 3458–3480 (2012)
10. G. Bellettini, G. Fusco, N. Guglielmi, A concept of solution and numerical experiments for forward-backward diffusion equations. Discrete Contin. Dyn. Syst. **16**(4), 783 (2006)
11. A.J. Bernoff, C.M. Topaz, Biological aggregation driven by social and environmental factors: a nonlocal model and its degenerate Cahn-Hilliard approximation. SIAM J. Appl. Dyn. Syst. **15**(3), 1528–1562 (2016)
12. A. Blanchet, P. Degond, Topological interactions in a Boltzmann-type framework. J. Stat. Phys. **163**(1), 41–60 (2016)
13. A. Blanchet, P. Degond, Kinetic models for topological nearest-neighbor interactions. J. Stat. Phys. **169**(5), 929–950 (2017)

14. G.A. Bonaschi, J.A. Carrillo, M. Di Francesco, M.A. Peletier, Equivalence of gradient flows and entropy solutions for singular nonlocal interaction equations in 1D. ESAIM Control Optim. Calc. Var. **21**(2), 414–441 (2015)
15. J. Bourgain, H. Brezis, P. Mironescu, Another look at Sobolev spaces, in *Optimal Control and Partial Differential Equations* (IOS, Amsterdam, 2001), pp. 439–455
16. C.P. Brangwynne, et al., Germline p granules are liquid droplets that localize by controlled dissolution/condensation. Science **324**, 1729–1732 (2009)
17. D. Bresch, P.E. Jabin, Z. Wang, On mean-field limits and quantitative estimates with a large class of singular kernels: application to the Patlak-Keller-Segel model. C. R. Math. Acad. Sci. Paris **357**(9), 708–720 (2019)
18. M. Burger, M. di Francesco, M. Franek, Stationary states of quadratic diffusion equations with long-range attraction. Commun. Math. Sci. **11**(3), 709–738 (2013)
19. M. Burger, A. Esposito, Porous medium equation and cross-diffusion systems as limit of nonlocal interaction. Nonlinear Anal. **235**(Paper No. 113347), 30 (2023)
20. J.W. Cahn, J.E. Hilliard, Free energy of a nonuniform system. I. Interfacial free energy. J. Chem. Phys. **28**, 258–267 (1958)
21. V. Calvez, J.A. Carrillo, Volume effects in the Keller-Segel model: energy estimates preventing blow-up. J. Math. Pures Appl. (9) **86**(2), 155–175 (2006)
22. J.A. Carrillo, M. DiFrancesco, A. Figalli, T. Laurent, D. Slepčev, Global-in-time weak measure solutions and finite-time aggregation for nonlocal interaction equations. Duke Math. J. **156**(2), 229–271 (2011)
23. J.A. Carrillo, S. Hittmeir, B. Volzone, Y. Yao, Nonlinear aggregation-diffusion equations: Radial symmetry and long time asymptotics. Invent. Math. **218**, 889–977 (2019)
24. J.A. Carrillo, K. Craig, Y. Yao, Aggregation-diffusion equations: dynamics, asymptotics, and singular limits, in *Active Particles. Vol. 2. Advances in Theory, Models, and Applications*. Modeling and Simulation in Science, Engineering, & Technology. (Birkhäuser/Springer, Cham, 2019), pp. 65–108
25. J.A. Carrillo, M.G. Delgadino, J.S.H. Wu, Convergence of a particle method for a regularized spatially homogeneous Landau equation. Math. Models Methods Appl. Sci. **33**(5), 971–1008 (2023)
26. J.A. Carrillo, C. Elbar, J. Skrzeczkowski, Degenerate Cahn-Hilliard systems: From nonlocal to local. arXiv preprint arXiv:2303.11929 (2023)
27. J.A. Carrillo, R.J. McCann, C. Villani, Contractions in the 2-Wasserstein length space and thermalization of granular media. Arch. Ration. Mech. Anal. **179**(2), 217–263 (2006)
28. J.A. Carrillo, H. Murakawa, M. Sato, H. Togashi, O. Trush, A population dynamics model of cell-cell adhesion incorporating population pressure and density saturation. J. Theor. Biol. **474**, 14–24 (2019)
29. J.A. Carrillo, Y.-P. Choi, M. Hauray, The derivation of swarming models: mean-field limit and Wasserstein distances, in *Collective Dynamics from Bacteria to Crowds*. CISM Courses and Lectures, vol. 553 (Springer, Vienna, 2014), pp. 1–46
30. J.A. Carrillo, K. Craig, F.S. Patacchini, A blob method for diffusion. Calc. Var. Partial Differential Equations **58**(2), Paper No. 53, 53 (2019)
31. J.A. Carrillo, A. Esposito, J.S.-H. Wu, Nonlocal approximation of nonlinear diffusion equations. Calc. Var. Partial Differ. Equations **63**(4), 100 (2024)
32. L.-P. Chaintron, A. Diez, Propagation of chaos: a review of models, methods and applications. I. Models and methods. Kinet. Relat. Models **15**(6), 895–1015 (2022)
33. L.-P. Chaintron, A. Diez, Propagation of chaos: a review of models, methods and applications. II. Applications. Kinet. Relat. Models **15**(6), 1017–1173 (2022)
34. X. Chen, Global asymptotic limit of solutions of the Cahn-Hilliard equation. J. Differential Geom. **44**(2), 262–311 (1996)
35. R.M. Colombo, M. Garavello, M. Lécureux-Mercier, A class of nonlocal models for pedestrian traffic. Math. Models Methods Appl. Sci. **22**(04), 1150023 (2012)
36. K. Craig, A.L. Bertozzi, A blob method for the aggregation equation. Math. Comp. **85**(300), 1681–1717 (2016)

37. K. Craig, K. Elamvazhuthi, M. Haberland, O. Turanova, A blob method for inhomogeneous diffusion with applications to multi-agent control and sampling. Math. Comp. **92**(344), 2575–2654 (2023)
38. K. Craig, I. Kim, Y. Yao, Congested aggregation via Newtonian interaction. Arch. Ration. Mech. Anal. **227**(1), 1–67 (2018)
39. K. Craig, I. Topaloglu, Aggregation-diffusion to constrained interaction: minimizers & gradient flows in the slow diffusion limit. Ann. Inst. H. Poincaré C Anal. Non Linéaire **37**(2), 239–279 (2020)
40. N. David, T. Dębiec, B. Perthame, Convergence rate for the incompressible limit of nonlinear diffusion-advection equations. Ann. Inst. H. Poincaré C Anal. Non Linéaire **40**(3), 511–529 (2023)
41. N. David, B. Perthame, Free boundary limit of a tumor growth model with nutrient. J. Math. Pures Appl. **155**, 62–82 (2021)
42. N. David, M. Schmidtchen, On the incompressible limit for a tumour growth model incorporating convective effects. Commun. Pure Appl. Math. **77**(5), 2613–2650 (2024)
43. P. Degond, S. Hecht, N. Vauchelet, Incompressible limit of a continuum model of tissue growth for two cell populations. Netw. Heterog. Media **15**(1), 57–85 (2020)
44. P. Degond, F.-J. Mustieles, A deterministic approximation of diffusion equations using particles. SIAM J. Sci. Statist. Comput. **11**(2), 293–310 (1990)
45. M.G. Delgadino, Convergence of a one-dimensional Cahn-Hilliard equation with degenerate mobility. SIAM J. Math. Anal. **50**(4), 4457–4482 (2018)
46. M.G. Delgadino, X. Yan, Y. Yao, Uniqueness and nonuniqueness of steady states of aggregation-diffusion equations. Comm. Pure Appl. Math. **75**(1), 3–59 (2022)
47. M. Di Francesco, M.D. Rosini, Rigorous derivation of nonlinear scalar conservation laws from follow-the-leader type models via many particle limit. Arch. Ration. Mech. Anal. **217**(3), 831–871 (2015)
48. S. Di Marino, A.R. Mészáros, Uniqueness issues for evolution equations with density constraints. Math. Models Methods Appl. Sci. **26**(9), 1761–1783 (2016)
49. R.L. Dobrušin, Vlasov equations. Funktsional. Anal. i Prilozhen. **13**(2), 48–58, 96 (1979)
50. C. Elbar, J. Skrzeczkowski, Degenerate Cahn-Hilliard equation: From nonlocal to local. J. Differ. Equations **364**, 576–611 (2023)
51. C.M. Elliott, The Stefan problem with a nonmonotone constitutive relation. IMA J. Appl. Math. **35**(2), 257–264 (1985). Special issue: IMA conference on crystal growth (Oxford, 1985)
52. C.M. Elliott, M.A. Herrero, J.R. King, J.R. Ockendon, The mesa problem: Diffusion patterns for $u_t = \Delta u^m$ as $m \to \infty$. IMA J. Appl. Math. **37**(2), 147–154 (1986)
53. M. Elsey, S. Esedoḡlu, Threshold dynamics for anisotropic surface energies. Math. Comp. **87**(312), 1721–1756 (2018)
54. S. Esedoḡlu, M. Jacobs, Convolution kernels and stability of threshold dynamics methods. SIAM J. Numer. Anal. **55**(5), 2123–2150 (2017)
55. S. Esedoḡlu, F. Otto, Threshold dynamics for networks with arbitrary surface tensions. Comm. Pure Appl. Math. **68**(5), 808–864 (2015)
56. S. Fagioli, O. Tse, On gradient flow and entropy solutions for nonlocal transport equations with nonlinear mobility. Nonlinear Anal. **221**(Paper No. 112904), 35 (2022)
57. C. Falcó, R.E. Baker, J.A. Carrillo, A local continuum model of cell-cell adhesion. SIAM J. Appl. Math. **84**, S17–S42 (2023)
58. S. Faure, B. Maury. Crowd motion from the granular standpoint. Math. Models Methods Appl. Sci. **25**(3), 463–493 (2015)
59. A. Figalli, R. Philipowski, Convergence to the viscous porous medium equation and propagation of chaos. ALEA Lat. Am. J. Probab. Math. Stat. **4**, 185–203 (2008)
60. M. Fischer, L. Kanzler, C. Schmeiser, One-dimensional short-range nearest-neighbor interaction and its nonlinear diffusion limit. SIAM J. Appl. Math. **84**(1), 1–18 (2024)
61. G. Giacomin, J.L. Lebowitz, Phase segregation dynamics in particle systems with long range interactions. I. Macroscopic limits. J. Statist. Phys. **87**(1-2), 37–61 (1997)

62. G. Giacomin, J.L. Lebowitz, Phase segregation dynamics in particle systems with long range interactions. II. Interface motion. SIAM J. Appl. Math. **58**(6), 1707–1729 (1998)
63. O. Gil, F. Quirós, Boundary layer formation in the transition from the porous media equation to a Hele-Shaw flow. Annales de l'IHP Analyse Non Linéaire **20**(1), 13–36 (2003)
64. K. Glasner, A diffuse interface approach to Hele-Shaw flow. Nonlinearity **16**(1), 49–66 (2003)
65. F. Golse, On the dynamics of large particle systems in the mean field limit, in *Macroscopic and Large Scale Phenomena: Coarse Graining, Mean Field Limits and Ergodicity*. Lecture Notes in Applied Mathematics and Mechanics, vol. 3 (Springer, Cham, 2016), pp. 1–144
66. I.G. Götz, B.B. Zaltzman, Nonincrease of mushy region in a nonhomogeneous Stefan problem. Quart. Appl. Math. **49**(4), 741–746 (1991)
67. N. Guillen, I. Kim, A. Mellet, A Hele-Shaw limit without monotonicity. Arch. Ration. Mech. Anal. **243**(2), 829–868 (2022)
68. Q. He, H.-L. Li, B. Perthame, Incompressible limits of Patlak-Keller-Segel model and its stationary state. Acta Appl. Math. **188**(1), 11 (2023)
69. S. Hecht, N. Vauchelet, Incompressible limit of a mechanical model for tissue growth with non-overlapping constraint. Commun. Math. Sci. **15**(7), 1913–1932 (2017)
70. S. Hensel, T. Laux, BV solutions for mean curvature flow with constant contact angle: Allen-Cahn approximation and weak-strong uniqueness. arXiv preprint arXiv: 2112.11150 (2021)
71. T. Hillen, K.J. Painter, A user's guide to PDE models for chemotaxis. J. Math. Biol. **58**(1), 183–217 (2009)
72. T. Hirose, K. Ninomiya, S. Nakagawa, et al., A guide to membraneless organelles and their various roles in gene regulation. Nat Rev Mol Cell Biol **24**, 288–304 (2023)
73. K. Höllig, Existence of infinitely many solutions for a forward backward heat equation. Trans. Am. Math. Soc. **278**(1), 299–316 (1983)
74. R.L. Hughes, A continuum theory for the flow of pedestrians. Transp. Res. B Methodol. **36**(6), 507–535 (2002)
75. N. Igbida, L^1-theory for Hele-Shaw flow with linear drift. Math. Models Methods Appl. Sci. **33**(7), 1545–1576 (2023)
76. P.-E. Jabin, Z. Wang, Mean field limit for stochastic particle systems, in *Active Particles, Volume 1 : Advances in Theory, Models, and Applications* (2017), pp. 379–402
77. P.-E. Jabin, Z. Wang, Quantitative estimates of propagation of chaos for stochastic systems with $W^{-1,\infty}$ kernels. Invent. Math. **214**(1), 523–591 (2018)
78. M. Jacobs, I. Kim, A.R. Mészáros, Weak solutions to the Muskat problem with surface tension via optimal transport. Arch. Ration. Mech. Anal. **239**(1), 389–430 (2021)
79. M. Jacobs, I. Kim, J. Tong, Darcy's law with a source term. Arch. Ration. Mech. Anal. **239**(3), 1349–1393 (2021)
80. R. Jordan, D. Kinderlehrer, F. Otto, The variational formulation of the Fokker-Planck equation. SIAM J. Math. Anal. **29**(1), 1–17 (1998)
81. G. Kaib, Stationary states of an aggregation equation with degenerate diffusion and bounded attractive potential. SIAM J. Math. Anal. **49**(1), 272–296 (2017)
82. T.K. Karper, E. Lindgren, E. Tadmor, A new model for chemotaxis with saturation effects and its incompressible limit. Preprint.
83. E.F. Keller, L.A. Segel, Model for chemotaxis. J. Theor. Biol. **30**(2), 225–234 (1971)
84. I. Kim, A. Mellet, J.S.-H. Wu, Mean field limit for congestion dynamics. Communications in Partial Differential Equations. To appear.
85. I. Kim, A. Mellet, Y. Wu, A density-constrained model for chemotaxis. Nonlinearity **36**(2), 1082–1119 (2023)
86. I. Kim, A. Mellet, Y. Wu, Density-constrained chemotaxis and Hele-Shaw flow. Trans. Am. Math. Soc. **377**(01), 395–429 (2024)
87. I. Kim, N. Požár, Porous medium equation to Hele-Shaw flow with general initial density. Trans. Am. Math. Soc. **370**(2), 873–909 (2018)
88. I. Kim, N. Požár, B. Woodhouse, Singular limit of the porous medium equation with a drift. Adv. Math. **349**, 682–732 (2019)

89. I. Kim, O. Turanova, Uniform convergence for the incompressible limit of a tumor growth model. Ann. Inst. H. Poincaré Anal. Non Linéaire **35**(5), 1321–1354 (2018)
90. I.C. Kim, B. Perthame, P.E. Souganidis, Free boundary problems for tumor growth: a viscosity solutions approach. Nonlinear Anal. **138**, 207–228 (2016)
91. R. Kowalczyk, Preventing blow-up in a chemotaxis model. J. Math. Anal. Appl. **305**(2), 566–588 (2005)
92. M. Kroemer, T. Laux, The Hele-Shaw flow as the sharp interface limit of the Cahn-Hilliard equation with disparate mobilities. Commun. Partial Differ. Equations **47**(12), 2444–2486 (2022)
93. A. Lachapelle, M.-T. Wolfram, On a mean field game approach modeling congestion and aversion in pedestrian crowds. Transp. Res. B Methodol. **45**(10), 1572–1589 (2011)
94. T. Laux, F. Otto, Convergence of the thresholding scheme for multi-phase mean-curvature flow. Calc. Var. Partial Differ. Equations **55**(5), 1–74 (2016)
95. P.-L. Lions, S. Mas-Gallic, Une méthode particulaire déterministe pour des équations diffusives non linéaires. C. R. Acad. Sci. Paris Sér. I Math. **332**(4), 369–376 (2001)
96. Q.X. Liu, A. Doelman, V. Rottschäfer, et al., Phase separation explains a new class of self-organized spatial patterns in ecological systems. Proc. Natl. Acad. Sci. U.S.A. **110**, 11905–11910 (2013)
97. S. Luckhaus, L. Modica, The Gibbs-Thompson relation within the gradient theory of phase transitions. Arch. Ration. Mech. Anal. **107**(1), 71–83 (1989)
98. P.M. Lushnikov, N. Chen, M. Alber, Macroscopic dynamics of biological cells interacting via chemotaxis and direct contact. Phys. Rev. E. **78**, 061904 (2008)
99. S. Mas-Gallic, A deterministic particle method for the linearized Boltzmann equation. Transport Theory Statist. Phys. **16**(4–6), 855–887 (1987)
100. B. Maury, Congested transport and microscopic and macroscopic scales, in *European Congress of Mathematics* (European Mathematical Society, Zürich, 2018), pp. 427–442
101. B. Maury, A. Roudneff-Chupin, F. Santambrogio, A macroscopic crowd motion model of gradient flow type. Math. Models Methods Appl. Sci. **20**(10), 1787–1821 (2010)
102. B. Maury, A. Roudneff-Chupin, F. Santambrogio, J. Venel, Handling congestion in crowd motion modeling. Netw. Heterog. Media **6**(3), 485–519 (2011)
103. B. Maury, J. Venel, A discrete contact model for crowd motion. ESAIM Math. Model. Numer. Anal. **45**(1), 145–168 (2011)
104. A. Mellet, Hele-shaw flow as a singular limit of a keller-segel system with nonlinear diffusion. Calc. Var. Partial Differential Equations **63**(8), (2024)
105. A. Mellet, B. Perthame, F. Quiros, A Hele-Shaw problem for tumor growth. J. F. Analysis **273**(10), 3061–3093 (2017)
106. A. Mellet, Y. Wu, Γ-convergence of some nonlocal perimeters in bounded domains with general boundary conditions. SIAM J. Math. Anal. **55**(4), 3226–3261 (2023)
107. D.M. Mitrea, R.W. Kriwacki, Phase separation in biology; functional organization of a higher order. Cell Commun. Signal **14**, 1–20 (2016)
108. L. Modica, The gradient theory of phase transitions and the minimal interface criterion. Arch. Rational Mech. Anal. **98**(2), 123–142 (1987)
109. L. Modica, S. Mortola, Un esempio di Γ^--convergenza. Boll. Un. Mat. Ital. B (5) **14**(1), 285–299 (1977)
110. D. Morale, V. Capasso, K. Oelschläger, An interacting particle system modelling aggregation behavior: from individuals to populations. J. Math. Biol. **50**(1), 49–66 (2005)
111. S. Motsch, D. Peurichard, From short-range repulsion to Hele-Shaw problem in a model of tumor growth. J. Math. Biol. **76**(1–2), 205–234 (2018)
112. K. Oelschläger, A sequence of integro-differential equations approximating a viscous porous medium equation. Z. Anal. Anwendungen **20**(1), 55–91 (2001)
113. K. Oelschläger, Large systems of interacting particles and the porous medium equation. J. Differ. Equations **88**(2), 294–346 (1990)
114. F. Otto, Dynamics of labyrinthine pattern formation in magnetic fluids: a mean-field theory. Arch. Rational Mech. Anal. **141**(1), 63–103 (1998)

115. F. Otto, The geometry of dissipative evolution equations: the porous medium equation. Comm. Partial Differential Equations **26**(1-2), 101–174 (2001)
116. K.J. Painter, T. Hillen, Volume-filling and quorum-sensing in models for chemosensitive movement. Can. Appl. Math. Q. **10**(4), 501–543 (2002)
117. J.K. Parrish, L. Edelstein-Keshet, Complexity, pattern, and evolutionary trade-offs in animal aggregation. Science **284**, 99–101 (1999)
118. C.S. Patlak, Random walk with persistence and external bias. Bull. Math. Biophys. **15**(3), 311–338 (1953)
119. R.L. Pego, Front migration in the nonlinear Cahn-Hilliard equation. Proc. Roy. Soc. London Ser. A **422**(1863), 261–278 (1989)
120. B. Perthame, F. Quirós, J.L. Vázquez, The Hele-Shaw asymptotics for mechanical models of tumor growth. Arch. Ration. Mech. Anal. **212**(1), 93–127 (2014)
121. B. Perthame, N. Vauchelet, Incompressible limit of a mechanical model of tumour growth with viscosity. Philos. Trans. Roy. Soc. A **373**(2050), 20140283, 16 (2015)
122. R. Philipowski, Interacting diffusions approximating the porous medium equation and propagation of chaos. Stochastic Process. Appl. **117**(4), 526–538 (2007)
123. P.I. Plotnikov, Forward-backward parabolic equations and hysteresis. Zap. Nauchn. Sem. S.-Peterburg. Otdel. Mat. Inst. Steklov. (POMI) **233**, 183–209, 257–258 (1996)
124. P.I. Plotnikov, Passage to the limit with respect to a small parameter in Cahn-Hilliard equations. Sibirsk. Mat. Zh. **38**(3), 638–656, iii (1997)
125. J.G. Rešetnjak, The weak convergence of completely additive vector-valued set functions. Sibirsk. Mat. **9**, 1386–1394 (1968)
126. M. Rosenzweig, S. Serfaty, Global-in-time mean-field convergence for singular Riesz-type diffusive flows. Ann. Appl. Probab. **33**(2), 954–998 (2023)
127. R. Rossi, G. Savaré, Tightness, integral equicontinuity and compactness for evolution problems in Banach spaces. Ann. Sc. Norm. Super. Pisa Cl. Sci. (5) **2**(2), 395–431 (2003)
128. G. Russo, Deterministic diffusion of particles. Comm. Pure Appl. Math. **43**(6), 697–733 (1990)
129. E. Sandier, S. Serfaty, Gamma-convergence of gradient flows with applications to Ginzburg-Landau. Comm. Pure Appl. Math. **57**(12), 1627–1672 (2004)
130. M. Slemrod, Dynamics of measure valued solutions to a backward-forward heat equation. J. Dynam. Differential Equations **3**(1), 1–28 (1991)
131. P. Sternberg, The effect of a singular perturbation on nonconvex variational problems. Arch. Rational Mech. Anal. **101**(3), 209–260 (1988)
132. Y. Sugiyama, Time global existence and asymptotic behavior of solutions to degenerate quasi-linear parabolic systems of chemotaxis. Differ. Integr. Equations **20**(2), 133–180 (2007)
133. A.-S. Sznitman, Topics in propagation of chaos, in *École d'Été de Probabilités de Saint-Flour XIX—1989*. Lecture Notes in Mathematics, vol. 1464 (Springer, Berlin, 1991), pp. 165–251
134. C.M. Topaz, A.L. Bertozzi, M.A. Lewis, A nonlocal continuum model for biological aggregation. Bull. Math. Biol. **68**(7), 1601–1623 (2006)
135. J.L. Vázquez, *The Porous Medium Equation*. Oxford Mathematical Monographs (The Clarendon Press, Oxford University Press, Oxford, 2007). Mathematical theory

Conservative Semi-Lagrangian Methods for Kinetic Equations

Seung-Yeon Cho, Maria Groppi, Jing-Mei Qiu, Giovanni Russo, and Seok-Bae Yun

Abstract The chapter is a review on semi-Lagrangian methods for kinetic equations. In semi-Lagrangian schemes for Boltzmann and BGK models, the equation for the distribution function is integrated along the characteristics, and the value of the unknown is computed on grid nodes in phase space. The value at the foot of the characteristic is computed by suitable interpolation. This approach allows the use of much larger time steps, compared to a Eulerian treatment of the transport term, since the latter requires CFL-type limitations on the time step, thus resulting in more efficient schemes. The use of standard non-oscillatory reconstructions may cause lack of conservation of semi-Lagrangian schemes. Conservation is restored by adopting a suitable conservative reconstruction, which allows at the same time high-order approximation and lack of spurious oscillations. Constrained L^2-minimization is adopted to impose conservation also at the level of the discrete collisional term. The technique is general and can be applied to a variety of kinetic equations, including BGK model for single gas and for mixtures, Boltzmann equation for rarefied gas dynamics, and Vlasov-Poisson systems. In the latter case

Research is supported by NSF grant NSF-DMS-2111253, Air Force Office of Scientific Research FA9550-22-1-0390 and Department of Energy DE-SC0023164.

S.-Y. Cho
Department of Mathematics, Gyeongsang National University, Jinju-si, Republic of Korea

M. Groppi
Department of Mathematical, Physical and Computer Sciences, University of Parma, Parma, Italy

J.-M. Qiu
Department of Mathematical Sciences, University of Delaware, Newark, DE, USA

G. Russo (✉)
Department of Mathematics and Computer Science, University of Catania, Catania, Italy
e-mail: giovanni.russo1@unict.it

S.-B. Yun
Department of Mathematics, Sungkyunkwan University, Seoul, Republic of Korea

J. A. Carrillo, E. Tadmor (eds.), *Active Particles, Volume 4*, Modeling and Simulation in Science, Engineering and Technology,
https://doi.org/10.1007/978-3-031-73423-6_7

the conservative reconstruction can be directly applied by using splitting methods, while non-splitting methods require a slightly different treatment. The use of high-order reconstruction and high-order time integrators, together with spectral methods in velocity, allows very accurate schemes for the numerical solution of collisional kinetic equations. Finally, an account of the theoretical results on the convergence properties of the method is provided at the end of the chapter.

1 Introduction

In this chapter, we focus on kinetic models for rarefied gases and plasmas. Kinetic equations provide a mesoscopic description of a large collection of particles, with applications ranging from gas dynamics [36] to plasma physics [86] and from traffic flow models [76] to swarming dynamics [33]. A monoatomic gas such as a noble gas is made up of a collection of atoms which interact by short-range interactions when the atoms are sufficiently close. If the gas is rarefied, i.e., if the average distance between molecules is much larger than the interaction distance, then multiple collisions can be neglected, and the interaction among atoms is essentially described in terms of binary collisions. The dynamics of the system could in principle be governed by Newton's equation of motion for each individual molecule. This is what is typically done in molecular dynamics if the number of interacting particles is not too large. However, when such a number in the region of interest is large, this description is not practical since it would require the solution of an enormous number of equations. Furthermore, this amount of information is unnecessary to grasp the collective behavior of the gas. With a large number of particles a kinetic description is sufficiently accurate because it relies on relatively small statistical fluctuations. Molecular dynamics is still adopted when the particle interaction is so complex that a kinetic model has not yet been derived. This is typically the case in many biological systems [70].

In the kinetic description of a monoatomic gas, the state of the gas is completely described in terms of a distribution function in phase space, say $\tilde{f}(x, p, t)$, which, suitably normalized, provides the density of particles per unit volume in phase space at a given time t, i.e., $\tilde{f}(x, p, t)d^3x\, d^3p$ represents the number of molecules in the infinitesimal volume $d^3x\, d^3p$ at time t. Here $x \in \mathbb{R}^3$ denotes the particle position in space and $p \in \mathbb{R}^3$ the particle momentum. In a monoatomic nonrelativistic gas, momentum and velocity are proportional, since $p = mv$, so one can equivalently describe the gas using a function $f(x, v, t)$, related to $\tilde{f}$ by $\tilde{f}(x, mv, t) = m^3 f(x, v, t)$.

The Boltzmann transport equation (BTE) provides a kinetic description of a rarefied gas undergoing binary elastic collisions. The interaction between the molecules is modeled by the collision operator, which is a bilinear integral operator, whose action requires, for each point in phase space, to perform an integration over a five-dimensional space. The collisional operator preserves the conservation of mass, momentum, and energy satisfied by the physical system, while the irreversibility of the gas evolution is quantitatively described by the H-theorem [36].

The parameter that determines the degree of rarefaction of the flow is the so-called Knudsen number, Kn, which is the ratio between the mean free path of the molecules and a characteristic macroscopic length. As Kn gets smaller, the effect of the collision increases, and for very small values of Kn (fluid dynamic limit), the gas is well described by macroscopic equations such as Euler or Navier-Stokes equations.

The numerical solution of kinetic equations presents several challenges. The first one concerns the dimensionality of the problem: in the fully 3D case in space and velocity, f has seven independent variables, which reduce to six for stationary problems. In the case of the BTE the computation of the collisional operator itself requires a great effort because of its integral nature and because it is desirable to maintain, at discrete level, several qualitative properties of the BTE, such as preservation of the collision invariants, and validity of a discrete version of the H-theorem.

The most commonly adopted methods for the numerical solution of BTE and other kinetic equations may be grouped into two main classes: stochastic and deterministic methods. Stochastic methods are based on a simulation of binary collisions among particles: a pair of particles which is sufficiently close to undergo a collision is randomly selected, the outcome of the collision is randomly sampled according to the differential scattering cross section, and the new particle velocities are then computed, automatically satisfying conservation of momentum and energy of the pair. The most widely adopted stochastic method, direct simulation Monte Carlo (DSMC), was proposed by G. Bird more than 60 years ago [10] and has several advantages:

- It is memory effective, since particles automatically occupy the region of phase space they need, and no memory is wasted with (almost) empty grid points in phase space.
- It is cost-effective, since the simulation of each collision can be sampled very quickly without computing any integral.
- It automatically satisfies at the microscopic level conservation of mass, momentum, and energy.
- Between collisions, particles are freely advected in space according to their velocity, with no stability restriction on the time step.

The main drawback of DSMC and related stochastic methods is that they suffer from statistical fluctuations, whose reduction (use of a large number of particles, repeat the simulation for time-dependent problems, or accumulate statistics by averaging over time for stationary problems) often needs a considerable amount of computational effort. If one is interested in low Mach flows or in accurate solutions required to formulate or corroborate conjectures, DSMC is not the preferred method. On the other hand, for situations very far from local equilibrium, then DMSC is, at the moment, the only valid option.

Deterministic methods, on the other hand, are based on some discretization of the equation in phase space and on the computation of the discrete solution by more classical methods. The most accurate deterministic schemes for the BTE are based on Fourier spectral methods, as illustrated in Sect. 5, which allow to reduce the computational cost of the evaluation of the collisional operator to $N_V \log N_V$, where N_V is the total number of grid points in velocity for each grid point in space. These methods provide also spectral accuracy in velocity. High accuracy in space and time is obtained by using high-order discretization of the transport term, obtained by high-order reconstruction in space and high-order integration in time, obtained, for example, by Runge-Kutta or multistep methods. Deterministic schemes are able to provide very accurate solutions for special geometries, especially if the system is not very far from equilibrium, for example, for low or moderate Mach numbers. In many cases, simpler kinetic models are used in place of the full Boltzmann equation. The most widely adopted is probably the BGK model, in which the collisional operator is replaced by a relaxation approximation. For such model it is easy to construct schemes which are able to capture the fluid dynamic limit as the Knudsen number vanishes. Such schemes are said *asymptotic preserving* (AP), since in the fluid dynamic limit they become kinetic schemes for the numerical solution of the underlying Euler or Navier-Stokes equations.

An account of most common numerical methods for BTE and other kinetic equation is provided in the review paper [48]. Most schemes in the literature are based on Eulerian discretization of the transport term. This approach has been thoroughly tested for advection-type equations and for hyperbolic systems. It allows the construction of high-order schemes which preserve several qualitative properties of the solution, including positivity [71]. All explicit schemes based on Eulerian discretization of the convection term, however, suffer from the classical CFL restriction on the time step, which can be written as $V_{\max}\Delta t \leq c\Delta x$, where c is a constant close to 1 that the depends on the particular scheme, and $V_{\max}$ is the largest speed adopted in the computational domain.

An alternative to deterministic schemes based on Eulerian discretization of the advection equation is offered by (backward) semi-Lagrangian (SL) schemes: the distribution function is still discretized on a fixed grid in phase space, and the advection term is treated by writing the equation in characteristic form (see Sect. 2). High order in space is achieved by high-order reconstruction at the foot of the characteristic at previous time step(s), while high order in time can be obtained by Runge-Kutta or multistep methods. The advantage of this approach is that stability restriction on the time step is much milder than in the case of Eulerian-based methods (see, for example, [64]), resulting in more efficient schemes. As we shall see, straightforward implementation of high-order SL schemes will not preserve the conservation properties of the original kinetic equation.

The goal of the current chapter is to review SL schemes for kinetic equations, analyzing the cause of loss of conservation and illustrating the techniques that allow the construction of conservative SL schemes, which provide high-order accuracy in all the discretization parameters and preservation of global conservation of mass, momentum, and energy (for periodic boundary conditions), with mild stability restriction on the time step.

The plan of the chapter is the following. After this introduction, the second section introduces first- and high-order SL schemes for the BGK equations. The use of implicit time discretization will provide AP schemes for the BGK model. However, it is shown that high-order schemes do not preserve global momentum and energy. In the third section the cause of loss of conservation is analyzed, and it is shown how to restore conservation by two main ingredients, namely, L^2-minimization and conservative reconstruction. Several tests are presented illustrating the properties of high-order accuracy and conservation of the proposed schemes. An extension of the method to nonuniform grid in velocity is discussed in the last part of the section. Section 4 deals with the application of high-order conservative SL schemes to systems of BGK equations for inert gas mixtures. The next section is devoted to the construction of conservative SL schemes to the Boltzmann equation of gas dynamics. Section 6 illustrates applications of the method to Vlasov-type equations. In Sect. 7 we summarize some of the main theoretical results concerning the convergence properties of conservative SL schemes for BGK-type models. Finally, in the last section we draw some conclusions.

2 SL Methods for the BGK Model

In this section, we describe how to construct first- and higher-order SL schemes for the BGK model of the Boltzmann equation.

2.1 BGK Equation

In the field of kinetic theory of rarefied gas flows, the celebrated Boltzmann equation is well known for its good predictability of rarefied gas dynamics. However, the computational cost for computing the fivefold integral of the Boltzmann collision operator makes numerical solution hard to compute.

As an alternative to the Boltzmann equation, the BGK equation [9] has been widely adopted as an approximate model. In the BGK model the Boltzmann collision operator is replaced with a relaxation operator towards the local thermodynamic equilibrium. The BGK model reads

$$\frac{\partial f}{\partial t} + v \cdot \nabla_x f = \frac{1}{\varepsilon} \left(\mathcal{M}(f) - f \right), \tag{1}$$

where $f(x, v, t)$ denotes the number density of nonionized monoatomic molecules on the phase space $\mathbb{R}^d \times \mathbb{R}^d$ at time $t \in \mathbb{R}_+$. Here $d \in \{1, 2, 3\}$ is the dimension for space and velocity variables. In the model, ε represents the relaxation time. Even if gas molecules move in three-dimensional space, one can consider models in which the gas moves in a smaller number of dimensions. For example, one could imagine a two-dimensional gas, in which both space (and consequently velocity) is two-dimensional, just as a simplified mathematical model of the gas. In such a case

$d = 2$. Mathematically, the BGK model can be defined even with $d = 1$, again as a mathematical model. A different simplification is obtained when one looks for solutions in which the distribution function does not depend on some of the space coordinates. For example, in a slab geometry one could look for solutions in which f depends only on one coordinate, say $x \in \mathbb{R}$, while it still depends on three velocity coordinates, say v_x, v_y, v_z. In such cases it would be more appropriate to write Eq. (1) as

$$\frac{\partial f}{\partial t} + v_x \frac{\partial f}{\partial x} = \frac{1}{\varepsilon} (\mathcal{M}(f) - f),$$

with $f = f(x, v_x, v_y, v_z, t)$. Therefore, we may consider that $x \in \mathbb{R}$, while $v \in \mathbb{R}^3$, i.e., we distinguish between number of dimensions in space, d_x, and in velocity, d_v. In this case, we have $d_x = 1$ and $d_v = 3$. With a slight abuse of notation, we shall write $x \in \mathbb{R}^{d_x}$ and $v \in \mathbb{R}^{d_v}$, with $1 \leq d_x \leq d_v \leq 3$. For simplicity of notation, we keep writing the BGK equation in the form (1), even if f depends only on the first d_x component of the space vector.

Equation (1) has to be supplemented by initial conditions in $\mathbb{R}^{d_x} \times \mathbb{R}^{d_v}$. In case the space variable x is confined in a domain $\Omega \subset \mathbb{R}^{d_x}$, one has to provide appropriate boundary conditions on $\partial\Omega$. For simplicity, unless otherwise stated, throughout the chapter we shall assume periodic boundary conditions in space.

In (1), the collision integral of the Boltzmann equation has been replaced with the relaxation process towards the local thermodynamical equilibrium, the so-called local Maxwellian $\mathcal{M}(f)$:

$$\mathcal{M}(f)(x, v, t) := \frac{\rho(x,t)}{\sqrt{(2\pi T(x,t))^{d_v}}} \exp\left(-\frac{|v - U(x,t)|^2}{2T}\right). \tag{2}$$

Here macroscopic densities for mass ρ, momentum ρU, energy E, and temperature T are defined by

$$(\rho, \rho U, 2E, d_v \rho T) = \int_{\mathbb{R}^{d_v}} \left(1, v, |v|^2, |v - U|^2\right) f dv.$$

Note that the relaxation operator still maintains some qualitative features of the Boltzmann collision operator:

- Collision invariants $1, v, |v|^2$:

$$\int_{\mathbb{R}^{d_v}} (\mathcal{M}(f) - f) \begin{pmatrix} 1 \\ v \\ |v|^2 \end{pmatrix} dv = 0.$$

- Conservation laws for mass, momentum, and energy:

$$\frac{d}{dt}\int_{\mathbb{R}^{d_x}\times\mathbb{R}^{d_v}} f\phi(v)\,dx\,dv = 0,$$

where $\phi(v) := \left(1, v, \frac{|v|^2}{2}\right)^\top$.

- The H-theorem:

$$\frac{d}{dt}\int_{\mathbb{R}^{d_x}\times\mathbb{R}^{d_v}} f\log\frac{1}{f}\,dx\,dv \geq 0.$$

- In the fluid regime $\varepsilon \to 0$, the solution f tends to $\mathcal{M}$. Then, the macroscopic moments of f satisfy the compressible Euler system:

$$\begin{aligned}
&\rho_t + \nabla_x \cdot (\rho u) = 0,\\
&(\rho u)_t + \nabla_x \cdot (\rho u \otimes u + \rho T I_d) = 0,\\
&E_t + \nabla_x \cdot [(E + \rho T)\,u] = 0,
\end{aligned}$$

where I_d denotes the identity matrix in d_x dimension.

2.2 *SL Methods for the BGK Model*

Here, we review a class of non-splitting implicit SL methods for the BGK model introduced in [64, 98]. For description, we introduce the notation which will be used in the rest of this chapter:

$$\begin{aligned}
x_i &= x_{\min} + i\Delta x, \quad i \in \mathcal{I},\\
v_j &= v_{\min} + j\Delta v, \quad j \in \mathcal{J},\\
t^n &= n\Delta t.
\end{aligned}$$

Here the set $\mathcal{I}$ and $\mathcal{J}$ denote the index sets for spatial and velocity grids, respectively. Thus, $f_{i,j}^n$ denotes a numerical approximation of $f(x_i, v_j, t^n)$.

2.2.1 Lagrangian Formulation and Derivation of First-Order Method

A first-order SL method for the BGK model (1) is obtained by applying the implicit Euler method to its characteristic form (see Fig. 1):

$$f_{i,j}^{n+1} = \tilde{f}_{i,j}^n + \frac{\Delta t}{\varepsilon}\left(\mathcal{M}[f_i^{n+1}]_j - f_{i,j}^{n+1}\right), \tag{3}$$

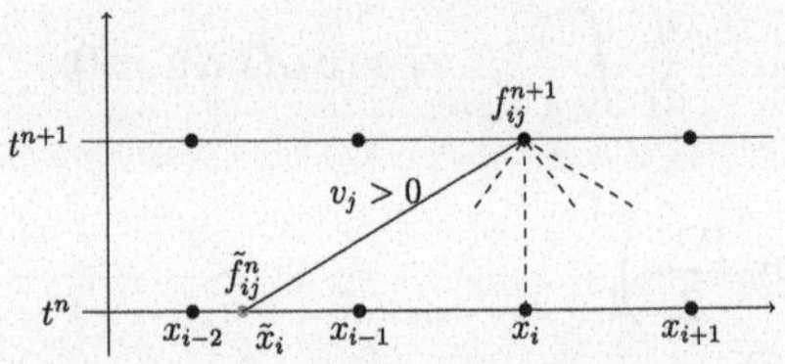

Fig. 1 First-order implicit scheme. The foot of the characteristic $\tilde{x}_i = x_i - v_j \Delta t$ does not lie on the grid. The value of $\tilde{f}_{ij}$ can be reconstructed by linear interpolation using the values of $f_{\cdot j}$ at the boundary of the interval containing $\tilde{x}_i$

where $\tilde{f}^n_{i,j}$ is an approximation of $f(x_i - v_j \Delta t, v_j, t^n)$ which can be reconstructed from $\{f^n_{i,j}\}_{i \in \mathcal{I}}$ by linear interpolation. The quantity $\mathcal{M}[f_i^{n+1}]_j$ is the local Maxwellian computed by

$$\mathcal{M}[f_i^{n+1}]_j = \frac{\rho_i^{n+1}}{\sqrt{\left(2\pi T_i^{n+1}\right)^{d_v}}} \exp\left(-\frac{|v_j - U_i^{n+1}|^2}{2T_i^{n+1}}\right), \tag{4}$$

with discrete macroscopic quantities computed as

$$\left(\rho_i^{n+1},\ \rho_i^{n+1} U_i^{n+1},\ d_v \rho_i^{n+1} T_i^{n+1}\right) := \sum_{j \in \mathcal{J}} f_{i,j}^{n+1} \left(1, v_j, |v_j - U_i^{n+1}|^2\right) (\Delta v)^{d_v}. \tag{5}$$

In order to compute the moments we proceed as follows: using the structural feature of the relaxation term of (1), one can compute (3) explicitly. We multiply both sides of (3) by the collision invariants $\phi_j := (1, v_j, \frac{|v_j|^2}{2})$, and sum over $j \in \mathcal{J}$:

$$\sum_{j \in \mathcal{J}} (f_{i,j}^{n+1} - \tilde{f}_{i,j}^n) \phi_j (\Delta v)^{d_v} = \frac{\Delta t}{\varepsilon} \sum_{j \in \mathcal{J}} \left(\mathcal{M}[f_i^{n+1}]_j - f_{i,j}^{n+1}\right) \phi_j (\Delta v)^{d_v}.$$

Here the right-hand side can be negligible if the discrete summation is computed with a sufficiently refined grid on the appropriate velocity domain, because midpoint rule is spectrally accurate when applied to a Maxwellian. Then, the discrete macroscopic quantities ρ_i^n, U_i^n, and T_i^n can be computed as follows:

$$\begin{aligned}\left(\rho_i^{n+1},\ \rho_i^{n+1} U_i^{n+1},\ E_i^{n+1}\right) &= \sum_{j \in \mathcal{J}} f_{i,j}^{n+1} \phi_j (\Delta v)^{d_v} \\ &= \sum_{j \in \mathcal{J}} \tilde{f}_{i,j}^n \phi_j (\Delta v)^{d_v} =: \left(\tilde{\rho}_i^n,\ \tilde{\rho}_i^n \tilde{U}_i^n,\ \tilde{E}_i^n\right);\end{aligned}$$

this further gives

$$T_i^{n+1} = \tilde{T}_i^n := \frac{1}{d_v \tilde{\rho}_i^n} \sum_{j \in \mathcal{J}} \tilde{f}_{i,j}^n |v_j - \tilde{U}_i^n|^2.$$

Consequently, one can update the solution as follows:

$$f_{i,j}^{n+1} = \tilde{f}_{i,j}^n + \frac{\Delta t}{\varepsilon}\left(\mathcal{M}[\tilde{f}_i^n]_j - f_{i,j}^{n+1}\right), \tag{6}$$

with

$$\mathcal{M}[\tilde{f}_i^n]_j = \frac{\tilde{\rho}_i^n}{\sqrt{\left(2\pi \tilde{T}_i^n\right)^{d_v}}} \exp\left(-\frac{|v_j - \tilde{U}_i^n|^2}{2\tilde{T}_i^n}\right).$$

Equation (6) can be explicitly solved giving

$$f_{i,j}^{n+1} = \frac{\varepsilon \tilde{f}_{i,j} + \Delta t \mathcal{M}[\tilde{f}_i^n]_j}{\varepsilon + \Delta t}.$$

Note that as $\varepsilon \to 0$, the method projects the numerical solution onto the corresponding local Maxwellian, suggesting that it becomes a *kinetic scheme* for the underlying Euler limit (see, for example, [90]).

We shall discuss this issue further in Sect. 2.2.4.

2.2.2 Runge-Kutta Methods

The method illustrated in the previous section is first-order accurate in space and time. More accurate schemes can be constructed by adopting higher-order reconstruction in space and higher-order time integrators.

If we want to avoid time step restrictions for small relaxation time, it is advisable to adopt L-stable implicit schemes [68]. In [64, 98], considering the stiffness arising in the fluid limit $\varepsilon \to 0$ of the BGK model (1), high-order SL methods are obtained by combining high-order L-stable diagonally implicit Runge-Kutta methods (DIRK) or backward-difference formulas (BDFs) with high-order spatial reconstructions. We first review the methods based on s-stage, L-stable diagonally implicit Runge-Kutta (DIRK) schemes [69], represented by the Butcher's table:

$$\begin{array}{c|c} c & A \\ \hline & b^\top \end{array}$$

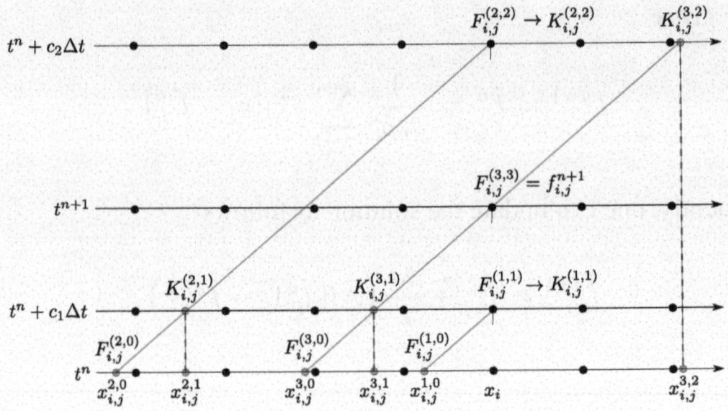

Fig. 2 Representation of DIRK3 scheme [41]. Black circles, grid nodes; grey circles, points where interpolation is needed

where $A = [a_{ij}]$ is a $s \times s$ matrix such that $a_{ij} = 0$ for $i < j$, $c = (c_1, ..., c_s)^\top$ and $b = (b_1, ..., b_s)^\top$ are coefficients vectors. In particular, we shall adopt *stiffly accurate* schemes, i.e., Runge-Kutta schemes in which the weight vector $b^\top$ is equal to the last row of matrix A. As a consequence, the numerical solution is given by the last stage value. We remind that an A-stable Runge-Kutta method which is stiffly accurate is also L-stable, i.e., its absolute stability function vanishes at infinity.

In the description of an s-stage DIRK method, we define the following quantities (see Fig. 2):

- The location of the characteristic foot on the ℓth stage along the kth backward characteristic starting from x_i with velocity v_j at time $t^n + c_k\Delta t$:

$$x_{i,j}^{k,\ell} := x_i - (c_k - c_\ell)v_j\Delta t.$$

- The interpolation of the distribution function f on $x_{i,j}^{k,\ell}$:

$$F_{i,j}^{(k,\ell)} := f(x_{i,j}^{k,\ell}, v_j, t^n + c_\ell\Delta t).$$

  Setting $c_0 = 0$, we define

$$F_{i,j}^{(k,0)} := f(x_i - v_j c_k \Delta t, v_j, t^n).$$

- RK fluxes:

$$K(x, v, t) = \frac{1}{\varepsilon}\left(\mathcal{M}(x, v, t) - f(x, v, t)\right),$$

  and its discrete values on $x_{i,j}^{k,\ell}$

$$K_{i,j}^{(k,\ell)} = K(x_{i,j}^{k,\ell}, v_j, t^n + c_\ell \Delta t).$$

Note that one has to use a suitable interpolation if the location of the characteristic foot $x_{i,j}^{k,\ell}$ is located on off-grid points. In [64] the authors consider G-WENO reconstruction, i.e., the classical point-wise WENO methods (in particular third-order WENO23 and fifth-order WENO35), suitably modified to reconstruct the function on an arbitrary point of the interval, rather than just at cell edges (see [32]).

We summarize the scheme as follows:

Algorithm for s-Order DIRK-Based Methods
For $k = 1, \ldots, s$:

1. Interpolate $F_{i,j}^{(k,0)}$ on $x_{i,j}^{k,0}$ from $\{F_{i,j}^{(k,0)}\}_{i \in \mathcal{I}}$.
2. For $\ell = 1 \ldots, k-1$, interpolate $K_{i,j}^{(k,\ell)}$ on $x_{i,j}^{k,\ell}$ from $\{K_{i,j}^{(\ell,\ell)}\}_{i \in \mathcal{I}}$. (Skip for $k = 1$.)
3. Compute:

$$F_{i,j}^{(k,k)} = F_{i,j}^{(k,0)} + \Delta t \sum_{\ell=1}^{k-1} a_{k\ell} K_{i,j}^{(k,\ell)} + \frac{\Delta t}{\varepsilon} a_{kk} \left(\mathcal{M}\left[F_i^{(k,k)}\right]_j - F_{i,j}^{(k,k)} \right), \quad (7)$$

where we compute the macroscopic moments of $\mathcal{M}\left[F_i^{(k,k)}\right]_j$ from $\left\{F_{i,j}^{(k,0)} + \Delta t \sum_{\ell=1}^{k-1} a_{k\ell} K_{i,j}^{(k,\ell)}\right\}_{j \in \mathcal{J}}$.

Since the adopted DIRK methods are stiffly accurate (SA), one can update solutions by setting $f_{i,j}^{n+1} = F_{i,j}^{(s,s)}$.

In our numerical tests, we use the third-order L-stable DIRK (DIRK3) method introduced in [20]. The corresponding Butcher table is

$$\begin{array}{c|ccc} \gamma & \gamma & 0 & 0 \\ c_2 & c_2 - \gamma_2 & \gamma_2 & 0 \\ 1 & 1 - b_2 - \gamma & b_2 & \gamma \\ \hline & 1 - b_2 - \gamma & b_2 & \gamma \end{array}$$

where $\gamma = 0.3$, $\gamma_2 = 13/3$, $b_2 = -3/710$, and $c_2 = 8/3$.

Schematic illustration of DIRK3-based scheme is represented in Fig. 2.

2.2.3 BDF Methods

The backward differentiation formula (BDF) is a linear multistep method (see [68]). The s-order BDF method for $y' = g(t, y)$ is represented by

$$BDF: y^{n+1} = \sum_{k=1}^{s} \alpha_k y^{n+1-k} + \beta_s \Delta t g(t_{n+1}, y^{n+1}),$$

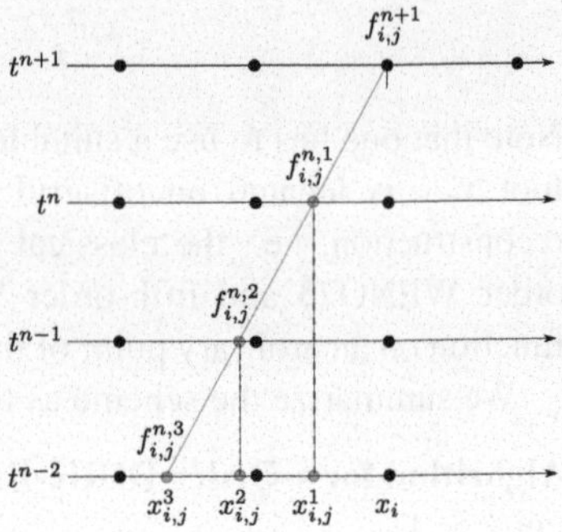

Fig. 3 Representation of BDF3 scheme [41]. Black circles, grid nodes; grey circles, points where interpolation is needed

where α_k and β_s are constants depending on s. Here and throughout the chapter we assume that the time step Δt is constant during the computation. The main advantage of the s-order BDF method against an s-stage DIRK method is that it enables to design SL schemes with fewer interpolations and computations of the local Maxwellian. The BDF-based SL scheme works as follows:

Algorithm for s-Order BDF-Based Methods

1. For $k = 1, \ldots, s$, interpolate $f_{i,j}^{n,k} = f(x_i - kv_j\Delta t, v_j, t^{n+1-k})$ from $\left\{f_{i,j}^{n+1-k}\right\}_{i\in\mathcal{I}}$.
2. Compute:

$$f_{i,j}^{n+1} = \sum_{k=1}^{s} \alpha_k f_{i,j}^{n,k} + \beta_s \frac{\Delta t}{\varepsilon}\left(\mathcal{M}\left[f_i^{n+1}\right]_j - f_{i,j}^{n+1}\right),$$

where we compute the macroscopic moments of $\mathcal{M}\left[f_i^{n+1}\right]_j$ from $\left\{\sum_{k=1}^{s} \alpha_k f_{i,j}^{n,k}\right\}_{j\in\mathcal{J}}$.

For numerical tests, we will consider the BDF3 ($s = 3$) method, which, when applied to the equation $y' = g(t, y)$, reads

$$y^{n+1} = \frac{18}{11}y^n - \frac{9}{11}y^{n-1} + \frac{2}{11}y^{n-2} + \frac{6}{11}\Delta t\, g(t^{n+1}, y^{n+1}).$$

The scheme is illustrated in Fig. 3. We note that for each grid node in space and velocity and for each time step, the BDF3 scheme involves three interpolations and one evaluation of the Maxwellian, while the DIRK3-based scheme performs six interpolations and three evaluations of the Maxwellian.

2.2.4 Asymptotic-Preserving Property of SL Schemes for the BGK Model

In [41] the authors we have shown that BDF-based SL methods update solutions as

$$f_{i,j}^{n+1} = \frac{\varepsilon f_{i,j}^* + \beta_s \Delta t \mathcal{M}[f_i^*]_j}{\varepsilon + \beta_s \Delta t} \tag{8}$$

where $f_{i,j}^* = \sum_{k=1}^s \alpha_k f_{i,j}^{n,k}$ and $\mathcal{M}[f_i^*]_j$ is the Maxwellian constructed from $\{f_{i,j}^*\}_{j\in\mathcal{J}}$. Then, we can rewrite (8) as

$$\mathcal{M}[f_i^*]_j - f_{i,j}^{n+1} = \frac{\varepsilon}{\varepsilon + \beta_s \Delta t}\left(\mathcal{M}[f_i^*]_j - f_{i,j}^*\right).$$

Under the assumption that $f_{i,j}^*$, $\mathcal{M}[f_i^*]_j = \mathcal{O}(1)$ and $\beta_s \neq 0$, for a fixed time step $\Delta t \gg \varepsilon$, we have

$$f_{i,j}^{n+1} = \mathcal{M}[f_i^*]_j + \mathcal{O}(\varepsilon), \tag{9}$$

for any $n \geq 0$ regardless of initial data. We furthermore note that the following identity holds: $\mathcal{M}[f_i^{n+1}]_j = \mathcal{M}[f_i^*]_j$ which comes from the fact that $\{f_{i,j}^{n+1}\}_{j\in\mathcal{J}}$ and $\{f_{i,j}^*\}_{j\in\mathcal{J}}$ have the same moments. Consequently, this together with (9) gives

$$f_{i,j}^{n+1} = \mathcal{M}[f_i^{n+1}]_j + \mathcal{O}(\varepsilon). \tag{10}$$

In a similar manner, DIRK-based high-order SL methods also satisfy relation (10). In order to show that, we rewrite (7) in the following form:

$$\mathcal{M}[F_i^{(k,k)}]_j - F_{i,j}^{(k,k)} = \frac{\varepsilon}{\varepsilon + a_{kk}\Delta t}\left(\mathcal{M}\left[\hat{F}_i^{(k,\ell)}\right]_j - \hat{F}_{i,j}^{(k,\ell)}\right),$$

where $\hat{F}_{i,j}^{(k,\ell)} = F_{i,j}^{(k,0)} + \Delta t \sum_{\ell=1}^{k-1} a_{k\ell} K_{i,j}^{(k,\ell)}$ and $\mathcal{M}\left[\hat{F}_i^{(k,\ell)}\right]_j$ is computed from $\left\{\hat{F}_{i,j}^{(k,\ell)}\right\}_{j\in\mathcal{J}}$. Then, this formally gives

$$\mathcal{M}[F_i^{(k,k)}]_j - F_{i,j}^{(k,k)} = \mathcal{O}(\varepsilon)$$

for the internal stages. Now, assuming that the DIRK scheme is SA, i.e., $f_{i,j}^{n+1} = F_{i,j}^{(s,s)}$, for the numerical solution we get

$$f_{i,j}^{n+1} = \mathcal{M}[f_i^{n+1}]_j + \mathcal{O}(\varepsilon).$$

Here we only showed that our solution converges to Maxwellian as $\varepsilon \to 0$. In order to show that our methods are *asymptotic preserving* (AP), we present related numerical tests in Sects. 3.3.4 and 3.3.5. Note that another important property for a numerical scheme is the *asymptotic accuracy* (AA) property, i.e., the scheme is AP, and in addition, it maintains its order of temporal accuracy when $\varepsilon \to 0$. Numerical tests in Sects. 3.3.2 and 3.3.3 show that BDF schemes are AA; on the

Table 1 Accuracy test in space and time for the 1D BGK equation. Initial data is given in (33). Here we used CFL= 4 and $N_v = 32$. Notice that the scheme based on BDF provides uniformly accurate solutions for all values of ε, while RK-based scheme suffers from order reduction

Relative L^1 error and order of density									
		$\varepsilon = 10^{-6}$		$\varepsilon = 10^{-4}$		$\varepsilon = 10^{-2}$		$\varepsilon = 10^{-0}$	
Scheme	$(N_x, 2N_x)$	Error	Rate	Error	Rate	Error	Rate	Error	Rate
SL-RK3-	(320, 640)	1.78e-05	2.33	1.52e-05	2.46	1.12e-06	2.99	4.29e-06	2.99
WENO23	(640, 1280)	3.55e-06	2.15	2.75e-06	2.37	1.41e-07	2.99	5.41e-07	3.00
	(1280, 2560)	8.00e-07		5.33e-07		1.77e-08		6.78e-08	
SL-BDF3-	(320, 640)	2.74e-05	3.02	2.46e-05	3.06	1.39e-06	3.23	1.05e-05	2.97
WENO23	(640, 1280)	3.37e-06	3.10	2.95e-06	3.10	1.47e-07	3.13	1.34e-06	2.99
	(1280, 2560)	3.93e-07		3.45e-07		1.69e-08		1.68e-07s	

other hand, DIRK schemes are AP, but not AA. This means that such schemes exhibit a loss of accuracy in the limit $\varepsilon \to 0$. In the literature, this is known as *order reduction phenomenon*. In [49], it was demonstrated that additional order conditions are needed to satisfy the AA property and new RK schemes were developed with both AP and AA properties.

We also remark that the methods are able to capture the Euler limit of the BGK model, while there is no guarantee that they automatically capture the Navier-Stokes limit, i.e., the equations that arise from the Chapman-Enskog expansion of the BGK model to first order in ε. Both BGK and Boltzmann equations admit a Navier-Stokes limit; however, since the BGK model has only one adjustable parameter, it cannot match both viscosity and thermal conductivity of the Navier-Stokes equations obtained from the BTE. To this purpose one has to adopt a more sophisticated BGK model such as, for example, ES-BGK [4, 100], which is able to match both dissipation parameters.

Numerical tests confirm that both approaches, i.e., SL-RK-WENO and SL-BDF-WENO, achieve high-order accuracy for smooth solution and are able to capture shocks without spurious oscillations (see Table 1 and Fig. 10 shown later in Sect. 3.3.4). However, as we shall show in the next section, these schemes present nonnegligible conservation error for non-smooth solutions.

2.2.5 Single Shock Test

We apply the scheme to the propagation of a single shock, where we can compare the numerical solution to the exact one and therefore accurately check the conservation properties of the scheme.

As initial data, we use the Maxwellian with moments corresponding to an exact shock for a polytropic gas, parametrized by the Mach number M: [112]

Table 2 Single shock test [20]. CFL = 4, $\varepsilon = 10^{-6}$. Conservation errors of discrete moments in the relative L_1 norm for first-order scheme applied to single shock with velocity domain $[-20, 20]$

	IE-SL-linear-CM, $N_x = 100$			IE-SL-linear-CM, $N_x = 200$		
N_v	Mass	Momentum	Energy	Mass	Momentum	Energy
30	3.63e-04	0.0012	0.0021	9.10e-04	0.0030	0.0051
40	5.54e-08	3.26e-07	6.03e-07	1.15e-07	6.43e-07	1.25e-06
50	8.55e-13	7.81e-12	1.43e-11	1.78e-12	1.54e-11	2.97e-11
60	3.55e-14	4.96e-14	3.89e-14	7.45e-14	8.24e-14	7.23e-14
90	3.24e-14	4.82e-14	3.77e-14	7.16e-14	7.32e-14	7.45e-14

$$(\rho_0, u_0, p_0) = \begin{cases} \left(\frac{(\gamma+1)M^2}{(\gamma-1)M^2+2}, \frac{2\sqrt{\gamma}(M^2-1)}{(\gamma+1)M}, 1 + \frac{2\gamma(M^2-1)}{(\gamma+1)}\right), & \text{for } x \leq 0.5 \\ (1, 0, 1), & \text{for } x > 0.5. \end{cases}$$

We take the relaxation time $\varepsilon = 10^{-6}$, the polytropic constant $\gamma = 3$ (corresponding to a polytropic gas with one degree of freedom per gas molecule), and Mach number $M = 2$. To prevent the solution from reaching the boundary, the final time is taken $t_f = 0.4$. We used free-flow boundary conditions i.e., zero Neumann conditions on the distribution function at the space boundary, and performed the computation on $(x, v) \in [0, 5] \times [-20, 20]$. The results are summarized in Table 2, where the conservation errors are reported for various values of N_x and N_v. Here CM stands for continuous Maxwellian (see Sect. 3.1).

From the results we can make the following observations:

1. Table 1 shows that the first-order implicit-explicit Euler semi-Lagrangian scheme, denoted IE-SL, with a sufficiently large number of grid points in velocity space maintains conservation within machine precision, independently of the number of grid point in space.
2. The same scheme with smaller number of points in velocity produces nonnegligible conservation errors.

This numerical evidence suggests that the convection part is conservative, while errors in conservation are a consequence of the numerical approximation of the relaxation term. The lack of conservation is indeed due to the use of a continuous Maxwellian on a discrete scheme in velocity: the parameters of the continuous Maxwellian do not coincide with the discrete moments; they are just approximated by them with spectral accuracy when the integrals are replaced by a summation. The spectral accuracy of the quadrature explains, for example, the dramatic drop of the conservation error when the number of points in velocity is increased from 40 to 50.

We repeat the same moving shock test using third-order schemes, and the results are summarized in Table 3 for SL schemes using Runge-Kutta time advancement (RK3-W35). Fully resolved high-order schemes both in space and velocity produce finite conservation error, which is much larger than the conservation error of the

Table 3 Single shock test [20]. CFL $= 2$, $\varepsilon = 10^{-6}$. Conservation errors of discrete moments in the relative L_1 norm for high-order schemes applied to single shock problem with velocity domain $[-20, 20]$

	Classical RK3-W35-CM		
(N_x, N_v)	Mass	Momentum	Energy
(200, 42)	1.54e-03	1.30e-02	1.45e-02
(200, 50)	1.30e-03	1.35e-02	1.51e-02
(200, 60)	1.68e-03	1.30e-02	1.53e-02
(200, 90)	1.60e-03	1.32e-02	1.53e-02

first-order scheme, shown in Table 2. This indicates that there are cases where high-order schemes may show even bigger conservation errors compared to those obtained by the first-order scheme. Similar results are obtained by using third-order BDF scheme with GWENO reconstruction[1] [20].

The main qualitative difference between first-order and high-order methods is that the former uses a fixed stencil for the linear interpolation at the feet of the characteristics, while high-order non-oscillatory reconstructions such as GWENO use a weighted sum of reconstructions on different stencils, the weight depending on the local regularity properties of the function to be reconstructed. As a result, in the first-order SL scheme the interpolation weights are the same for all intervals, whereas in high-order SL schemes, due to the nonlinearity of the non-oscillatory reconstruction, the interpolation weights are not the same for all intervals, thus destroying the translation invariance which is at the basis of the conservation property of the schemes.

3 Conservative SL Methods

3.1 Conservative Treatment of Maxwellian

The treatment of the collision term as described in the previous section does not produce a strictly conservative method. Besides unavoidable roundoff errors, there are two reasons for lack of strict conservation. The first one is the fact that the unbounded domain in velocity is truncated by a finite computational domain. The second one is that the moments $(\rho, \rho u, E)$ of the continuous Maxwellian (2) are exact only if computed by integration over v. If we compute the moments of the Maxwellian (4) by the discrete summation (5), there is a small discrepancy which is essentially due to the approximation of an integral by a discrete sum. The first source of error can be reduced by using a large enough computational domain in velocity,

[1] GWENO is a generalization of WENO introduced in [32], which allows high-order non-oscillatory reconstruction in the whole cell rather than just at cell boundaries.

which essentially includes the numerical support of the distribution function. The second source of error can be reduced by using a relatively large number of grid points in velocity domain. Because the classical midpoint rule is spectrally accurate when applied to a rapidly decaying smooth function such as a Maxwellian, even a moderate number of grid points in velocity is usually sufficient to make conservation errors very small.

However, for small Knudsen number the behavior of the gas should be adequately described by a relatively small number of velocities. In such a case, the conservation error introduced by the use of a continuous Maxwellian may be relevant. Here we adopt two techniques which restore full conservation in the approximation of the moments, namely, constrained L^2-minimization and discrete Maxwellian.

3.1.1 Constrained L^2-Minimization (dM_1)

This technique has been introduced by Irene Gamba and coworkers in [63].

Let us define the discrete moments obtained by function $\tilde{f}$ by $\mathcal{U} := (\rho, \rho U, E)^\top \in \mathbb{R}^{d_v+2}$. The continuous Maxwellian with continuous moments equal to $\mathcal{U}$ will not have the same discrete moments of $\tilde{f}$. Therefore, we define a new discrete Maxwellian $g = \{g_j\}_{j\in\mathcal{J}}$ by imposing that it is as close as possible to $\mathcal{M}[\mathcal{U}](v_j)$ and has exactly the same discrete moments $\mathcal{U}$.

Let us consider d_v-dimensional velocity domain with grid points v_j for $j \in \mathcal{J}$. Then we solve the following minimization problem:

$$\min \left\|\mathcal{M} - g\right\|_2^2 \quad \text{s.t} \quad Cg = \mathcal{U} \tag{11}$$

where

$$\mathcal{M} := (\mathcal{M}_1, \mathcal{M}_2, \cdots, \mathcal{M}_{(N_v+1)^{d_v}})^\top \in \mathbb{R}^{(N_v+1)^{d_v}},$$

$$g := (g_1, g_2, \cdots, g_{(N_v+1)^{d_v}})^\top \in \mathbb{R}^{(N_v+1)^{d_v}},$$

$$C := \begin{pmatrix} (\Delta v)^{d_v} \\ v_j(\Delta v)^{d_v} \\ \frac{|v_j|^2}{2}(\Delta v)^{d_v} \end{pmatrix} \in \mathbb{R}^{(d_v+2)\times(N_v+1)^{d_v}}, \quad \mathcal{U} = (\rho, \rho U, E)^\top \in \mathbb{R}^{(d_v+2)\times 1},$$

$$\mathcal{M}_j := \frac{\rho}{\sqrt{(2\pi T)^{d_v}}} \exp\left(-\frac{\left|v_j - U\right|^2}{2T}\right), \quad j \in \mathcal{J},$$

where $d_v \rho T = 2E - \rho|U|^2$.

To solve the minimization problem (11), we use the method of Lagrange multiplier with the following Lagrangian $\mathcal{L}(g, \lambda)$:

$$\mathcal{L}(g, \lambda) = \left\|\mathcal{M} - g\right\|_2^2 + \lambda^\top (Cg - \mathcal{U}).$$

We first compute the gradient of $\mathcal{L}$ with respect to g and λ:

$$\nabla_g \mathcal{L} = 0 \Leftrightarrow \quad g = \mathcal{M} + \tfrac{1}{2} C^\top \lambda$$
$$\nabla_\lambda \mathcal{L} = 0 \Leftrightarrow Cg = \mathcal{U}.$$

Then, we use these to derive

$$\begin{aligned} Cg = C\mathcal{M} + \frac{1}{2} CC^\top \lambda \Leftrightarrow \mathcal{U} &= C\mathcal{M} + \frac{1}{2} CC^\top \lambda \\ &\Leftrightarrow \lambda = 2(CC^\top)^{-1}(\mathcal{U} - C\mathcal{M}). \end{aligned}$$

Note that the matrix $CC^\top$ is invertible because it is symmetric and positive definite. Consequently,

$$g = \mathcal{M} + C^\top (CC^\top)^{-1}(\mathcal{U} - C\mathcal{M}). \tag{12}$$

Once the matrix $C^\top (CC^\top)^{-1}$ is computed, it can be stored and reused for other Maxwellian computations. In view of this, the minimization approach (11) only requires additional matrix computations in determining g.

Note that this technique does not necessarily preserve positivity of the solution. A weighted version of the constrained L^2-minimization has been proposed in [18], which helps in reducing positivity errors near the tail of the distribution and is recalled in the next subsection.

3.1.2 Weighted L^2-Minimization for Moment Correction

In [18] the L^2-minimization technique [63] has been extended to a weighted version, which helps in preventing positivity errors near the tails of the distribution function.

Given an initial guess f, here we consider a weight function $1/h$ and look for a solution g of the following weighted L^2-minimization problem:

$$\min_g \left\| f \circ \frac{1}{h} - g \right\|_2^2 \quad \text{s.t} \quad C_h g = \mathcal{U}$$

where $\circ$ denotes the component-wise multiplication and

$$f := (f_1, f_2, \cdots, f_{(N_v+1)^{d_v}})^\top \in \mathbb{R}^{(N_v+1)^{d_v}},$$

$$g = (g_1, g_2, \cdots, g_{(N_v+1)^{d_v}})^\top \in \mathbb{R}^{(N_v+1)^{d_v}},$$

$$h := (h_1, h_2, \cdots, h_{(N_v+1)^{d_v}})^\top \in \mathbb{R}^{(N_v+1)^{d_v}},$$

$$C_h := \begin{pmatrix} h_j(\Delta v)^{d_v} \\ h_j v_j (\Delta v)^{d_v} \\ h_j \frac{|v_j|^2}{2}(\Delta v)^{d_v} \end{pmatrix} \in \mathbb{R}^{(d_v+2)\times(N_v+1)^{d_v}}, \quad \mathcal{U} = (\rho, \rho U, E)^\top \in \mathbb{R}^{(d_v+2)\times 1}.$$

Here g is constructed as close as possible to the ratio of f and h which corresponds to the macroscopic quantities $\mathcal{U}$, while $g \circ h$ gives the approximation of f reproducing exactly the same discrete moments $\mathcal{U}$. Note that the use of weight $h_j = 1$ for all j leads to the classical L^2-minimization. In order to emphasize the effect of the tails, the function h has to be a decreasing function of $(v - u)^2$, such as, for example, the local Maxwellian corresponding to f.

As in the L^2-minimization without weight, the method of Lagrange multiplier gives

$$g \circ h = f + C^\top (CC^\top)^{-1} \left(\mathcal{U} - C \left(f \circ \frac{1}{h} \right) \right) \circ h, \tag{13}$$

where we use the fact that the matrix $CC^\top$ is invertible because it is symmetric and positive definite. The additional computational cost for this procedure with respect to the classical approach [63] is just a few component-wise division and multiplication of vectors of size $(N_v + 1)^{d_v}$. A simple example that shows the advantage of this method can be found in [18].

3.1.3 Discrete Maxwellian

Discrete Maxwellian (Entropy Minimization, $dM2$)

The second approach, introduced by Mieussens [83], determines the value of g by solving the following entropy minimization problem:

$$\min \left\{ \sum_{j \in \mathcal{J}} g_j \log g_j,\ g_j \geq 0 \quad \text{s.t.} \quad \sum_{j \in \mathcal{J}} g_j \phi_j (\Delta v)^{d_v} = \mathcal{U} \right\}, \tag{14}$$

where, as usual, ϕ_j denote the collision invariants. This is equivalent to solving the following equation:

$$\sum_{j \in \mathcal{J}} \left(\exp(a \cdot \phi_j) \right) \phi_j (\Delta v)^{d_v} = \mathcal{U} \quad \text{where} \quad a \in \mathbb{R}^{d_v+2}. \tag{15}$$

Then, the solution $a \in \mathbb{R}^{d_v+2}$ of (15) gives the solution to (14):

$$g_j = \exp(a \cdot \phi_j) \text{ for } j \in \mathcal{J}. \tag{16}$$

Compared to the methods (12) or (13), this one may be more expensive in that it requires an iterative solution of the nonlinear system (15), say by Newton's methods. However, this approach always gives nonnegative Maxwellian; therefore, it is more helpful in preserving positivity of the numerical solution of the BGK model (1).

Alternative techniques to preserve the moments are possible. For example, in [58] the authors use a different correction previously introduced by Aristov and Tcheremissine [5].

3.2 Conservative Reconstruction

In the previous section, we confirmed that the nonlinear weights used for GWENO methods may lead to loss of conservation. To avoid this problem, we adopt a conservation reconstruction technique proposed in [40]. In this section, we review the basic idea of the conservative reconstruction.

3.2.1 Conservative Reconstruction in 1D

Let $u : \mathbb{R} \to \mathbb{R}$ be a smooth function and $\bar{u}_h : \mathbb{R} \to \mathbb{R}$ be a corresponding sliding average function:

$$\frac{1}{h}\int_{x-h/2}^{x+h/2} u(y)\,dy.$$

Assume that we are given the values of $\bar{u}_h$ on uniform grids $x_i = ih$:

$$\frac{1}{h}\int_{I_i} u(x)\,dx = \bar{u}_h(x_i), \quad I_i = [x_{i-\frac{1}{2}}, x_{i+\frac{1}{2}}] = [x_i - h/2, x_i + h/2],$$

where $i \in \mathcal{I}$. Our goal is to construct an approximation $Q(x)$ of the sliding average $\bar{u}_h(x)$, which is conservative in the sense that for any periodic function $u(x)$ with period $L = Nh$, $N \in \mathbb{N}$, we have

$$\sum_{i=1}^{N} Q(x_i + \theta h) = \sum_{i=1}^{N} \bar{u}(x_i), \quad \theta \in [0, 1).$$

Assume we have a piecewise smooth reconstruction $R(x) = \sum_i R_i(x)\chi_i(x)$, for $i \in \mathcal{I}$, where $\chi_i(x)$ denotes the characteristic function of interval I_i and each $R_i(x)$ denotes a polynomial of degree k and has the following properties:

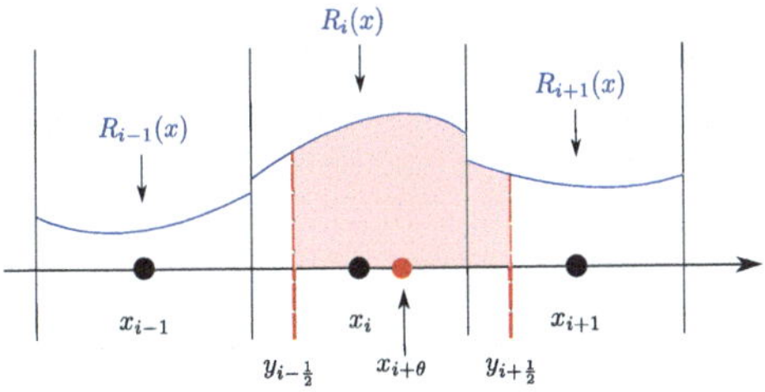

Fig. 4 Description of one-dimensional conservative reconstruction [40]

1. High-order accuracy in the approximation of $u(x)$:

$$u(x) = R_i(x) + \mathcal{O}\left(h^{k+1}\right), \quad x \in I_i. \tag{17}$$

2. Conservation in the sense of cell averages:

$$\frac{1}{h}\int_{x_{i-\frac{1}{2}}}^{x_{i+\frac{1}{2}}} R_i(x)\,dx = \bar{u}_i.$$

Consider a shifted interval $[y_{i-\frac{1}{2}}, y_{i+\frac{1}{2}}]$ whose center is $x_{i+\theta} := x_i + \theta h, \theta \in [0, 1)$, and denote by $\bar{u}_h(x_{i+\theta})$ the sliding average of u at $x_{i+\theta}$ (see Fig. 4). We see that

$$x_{i-\frac{1}{2}} \le y_{i-\frac{1}{2}} < x_{i+\frac{1}{2}} \le y_{i+\frac{1}{2}} < x_{i+\frac{3}{2}}.$$

Our strategy is to approximate $\bar{u}(x_{i+\theta})$ by $Q_{i+\theta} := Q(x_{i+\theta})$, where

$$Q_{i+\theta} = \frac{1}{h}\int_{y_{i-\frac{1}{2}}}^{y_{i+\frac{1}{2}}} R(x)\,dx = \frac{1}{h}\int_{x_{i-\frac{1}{2}+\theta}}^{x_{i+\frac{1}{2}+\theta}} R(x)\,dx, \tag{18}$$

which is equivalent to

$$Q_{i+\theta} = \frac{1}{h}\int_{x_{i-\frac{1}{2}+\theta}}^{x_{i+\frac{1}{2}}} R_i(x)\,dx + \frac{1}{h}\int_{x_{i+\frac{1}{2}}}^{x_{i+\frac{1}{2}+\theta}} R_{i+1}(x)\,dx.$$

From now on, we consider $R_i(x)$ to be piecewise polynomials of degree k of the form

$$R_i(x) = \sum_{\ell=0}^{k} \frac{R_i^{(\ell)}}{\ell!}(x - x_i)^{\ell}. \tag{19}$$

Making use of (19) in the first term, we obtain

$$\frac{1}{h}\int_{x_{i-\frac{1}{2}+\theta}}^{x_{i+\frac{1}{2}}} R_i(x)\,dx = \frac{1}{h}\sum_{\ell=0}^{k} R_i^{(\ell)} \int_{x_{i-\frac{1}{2}+\theta}}^{x_{i+\frac{1}{2}}} \frac{1}{\ell!}(x-x_i)^\ell\,dx = \sum_{\ell=0}^{k} h^\ell R_i^{(\ell)} \alpha_\ell(\theta)$$

where

$$\alpha_\ell(\theta) = \frac{1-(2\theta-1)^{\ell+1}}{2^{\ell+1}(\ell+1)!}.$$

Similarly, we can write

$$\frac{1}{h}\int_{x_{i+\frac{1}{2}}}^{x_{i+\frac{1}{2}+\theta}} R_{i+1}(x)\,dx := \sum_{\ell=0}^{k} h^\ell R_{i+1}^{(\ell)} \beta_\ell(\theta),$$

with

$$\beta_\ell(\theta) = \frac{(2\theta-1)^{\ell+1}-(-1)^{\ell+1}}{2^{\ell+1}(\ell+1)!}.$$

Letting $Q_{i+\theta}$ denote the approximation of $\bar{u}_h(x_{i+\theta})$, we obtain

$$Q_{i+\theta} := \sum_{\ell=0}^{k} h^\ell \left(\alpha_\ell(\theta) R_i^{(\ell)} + \beta_\ell(\theta) R_{i+1}^{(\ell)}\right). \tag{20}$$

Here, we note that $\alpha_\ell(\theta)$ and $\beta_\ell(\theta)$ satisfy the following relations:

- If $\ell = 2n$, $0 \le n$, is an even number,

$$\alpha_\ell(\theta) + \beta_\ell(\theta) = \frac{1}{2^{2n}(2n+1)!}$$

- If $\ell = 2n+1$, $0 \le n$, is an odd number,

$$\alpha_\ell(\theta) + \beta_\ell(\theta) = 0.$$

We list the explicit form of $\alpha_\ell(\theta)$ and $\beta_\ell(\theta)$ for $\ell = 0, 1, 2$:

$$\alpha_0(\theta) = 1-\theta, \qquad \alpha_1(\theta) = \frac{\theta(1-\theta)}{2}, \qquad \alpha_2(\theta) = \frac{1-q(\theta)}{24}$$

$$\beta_0(\theta) = \theta, \qquad \beta_1(\theta) = -\frac{\theta(1-\theta)}{2}, \qquad \beta_2(\theta) = \frac{q(\theta)}{24},$$

where $q(\theta) = 3\theta - 6\theta^2 + 4\theta^3$, $\theta \in [0, 1)$.

Remark 3.1 (Finite Difference Framework) Given a smooth function u, there exists a function $\hat{u}$ whose sliding average recovers the value of u:

$$u(x) = \frac{1}{h} \int_{x-h/2}^{x+h/2} \hat{u}(y)\, dy.$$

Since we can look for a basic reconstruction $R = \sum_i R_i(x)\chi_i(x)$ with R_i satisfying

$$\frac{1}{h} \int_{x_i-h/2}^{x_i+h/2} R(y)\, dy = u(x_i),$$

our reconstruction can be used also in the point-wise framework (see [103]). In view of this, we will apply it to the construction of SL schemes in the finite difference framework.

3.2.2 General Properties

In this section, we provide several properties of the reconstruction (20) such as accuracy, conservation, and consistency with the classical interpolation by making a suitable choice of $R_i^{(\ell)}$ in the reconstruction.

Recalling the assumption (17), we have a function $R(x)$ which approximates point values of u and our goal is to approximate the sliding average function $\bar{u}$ with our reconstruction (20). Before checking the accuracy order, we note that the cell average function $\bar{u}(x)$ can be expressed in terms of derivatives of function $u(x)$:

$$\begin{aligned} \bar{u}(x) &= \frac{1}{h} \int_{x-h/2}^{x+h/2} u(y)\, dy \\ &= \frac{1}{h} \int_{x-h/2}^{x+h/2} \sum_{\ell=0}^{\infty} \frac{u^{(\ell)}(x)}{\ell!}(y-x)^\ell\, dy \\ &= \sum_{\ell=\text{even}}^{\infty} h^\ell u^{(\ell)}(x) \frac{1}{2^\ell(\ell+1)!}. \end{aligned} \tag{21}$$

Inserting $x = x_{i+\theta}$ into (21), we obtain

$$\begin{aligned} \bar{u}(x_{i+\theta}) &= \sum_{\ell=\text{even}}^{\infty} h^\ell u^{(\ell)}(x_{i+\theta}) \frac{1}{2^{2\ell}(\ell+1)!} \\ &= u^{(0)}(x_{i+\theta}) + \frac{h^2}{24} u^{(2)}(x_{i+\theta}) + \frac{h^4}{1920} u^{(4)}(x_{i+\theta}) + \cdots . \end{aligned}$$

With this formula, in the following proposition, we provide a sufficient condition for a polynomial reconstruction $Q_{i+\theta}$ to be a $(k+2)$th-order accurate approximation of $\bar{u}_h(x+\theta)$ for $\theta \in [0, 1)$.

Proposition 3.1 *Let $k \geq 0$ be an even integer, $R_i \in \mathbb{P}^k$ be given by (19), and u be a smooth function $u : \mathbb{R} \to \mathbb{R}$. Suppose we have a piecewise polynomial $R(x) = \sum_i R_i(x)\chi_i(x)$, which satisfies*

$$\begin{aligned} u_i^{(\ell)} &= R_i^{(\ell)} + \mathcal{O}(h^{k+2-\ell}), \quad 0 \leq \ell \leq k, \text{ whenever } \ell \text{ is an even integer} \\ u_i^{(\ell)} - u_{i+1}^{(\ell)} &= R_i^{(\ell)} - R_{i+1}^{(\ell)} + \mathcal{O}(h^{k+2-\ell}), \; 0 \leq \ell < k, \text{ whenever } \ell \text{ is an odd integer.} \end{aligned} \tag{22}$$

Then, the reconstruction $Q_{i+\theta}$ gives a $(k+2)$th-order approximation of the sliding average $\bar{u}(x_{i+\theta})$ for any $\theta \in [0, 1)$.

Proof For detailed proof, we refer to [40].

Remark 3.2 Condition consistency can be satisfied with mild regularity assumptions on the function $u(x)$.

Let $k \geq 0$ be an even integer, and consider a function $u(x) \in C^{k+2}(\mathbb{R})$ and its primitive function $U(x) := \int_{-\infty}^{x} u(y)\,dy \in C^{k+3}(\mathbb{R})$. Then, by looking for a polynomial $P_i(x) \in \mathbb{P}^{k+1}$ such that

$$P_i(x_{i-\frac{1}{2}+j}) = U(x_{i-\frac{1}{2}+j}), \quad j = -r, \cdots, s+1, \quad r+s = k,$$

one can show that the second condition in (22) is satisfied. Furthermore, if $R_i^{(\ell)}$ can be represented with a Lipschitz function F_ℓ

$$R_i^{(\ell)} = F_\ell\,(\bar{u}_{i-r}, \cdots, \bar{u}_{i+s})$$

which satisfies

$$F_\ell\,(\bar{u}_{i-r}, \cdots, \bar{u}_{i+s}) - u^{(\ell)}(x_i) = \mathcal{O}\left((\Delta x)^{k+1-\ell}\right),$$

the condition (22) is also satisfied. For more details, we refer to [40].

In Proposition 3.1, we see that the choice of an even integer $k \geq 0$ leads to the improvement of accuracy. In such a case, one can show that the reconstruction $Q_{i+\theta}$ based on linear weights coincides with the classical polynomial interpolation.

Proposition 3.2 *Let $k \geq 0$ be an even integer with $k = 2r$. For each $i \in \mathcal{I}$, assume that we have a basic reconstruction $R_i(x) \in \mathbb{P}^k$, which is a polynomial of degree k in (19) and interpolates the function u in the sense of cell averages:*

$$\frac{1}{h}\int_{x_{i+j-\frac{1}{2}}}^{x_{i+j+\frac{1}{2}}} R_i(x)\,dx = \bar{u}_{i+j}, \quad -r \le j \le r,$$

with a symmetric stencil $S_i := \{i-r, i-r+1, \cdots, i+r\}$. *Then, the reconstruction* $Q_{i+\theta}$ *in* (20) *based on* R_i *and* R_{i+1} *is the Lagrange polynomial* $L(x)$ *that interpolates* $\bar{u}_{i+j}$, *for* $-r \le j \le r+1$, *where* $x = x_i + \theta h$ *and* $\theta \in [0, 1)$.

The proof is based on the observation that interpolation in the sense of the cell averages is equivalent to the point-wise interpolation of sliding averages at cell center, which in turn, is equivalent to the point-wise interpolation of primitive function at cell edges (see [51]).

Remark 3.3 For $k = 0$, the only possible choice is to set $R_i(x) = \bar{u}_i$ and the resulting reconstruction $Q_{i+\theta}$ reduces to the linear interpolation constructed from two points $\bar{u}_i$ and $\bar{u}_{i+1}$.

In the following proposition, we show that total mass is preserved for any θ-shifted summation, $\theta \in [0, 1)$.

Proposition 3.3 *Assume that* $R_i(x)$ *of the form* (19) *are polynomials satisfying*

$$\frac{1}{h}\int_{x_{i-\frac{1}{2}}}^{x_{i+\frac{1}{2}}} R_i(x)\,dx = \bar{u}_i, \quad i \in \mathcal{I} \tag{23}$$

and $Q_{i+\theta}$ *is constructed according to* (20). *Then, for periodic functions* $\bar{u}(x)$ *with period* $L = N\Delta x$, $N \in \mathbb{N}$,

$$\sum_{i=1}^{N} Q_{i+\theta} = \sum_{i=1}^{N} \bar{u}_i, \tag{24}$$

for any $\theta \in [0, 1)$.

The proof is left to the reader and can be found in [40].

Remark 3.4 We remark that this summation preserving property can be useful when our reconstruction is applied to the SL treatment of a constant convection term, where characteristic curves are given by parallel lines for each grid point. In such cases, the proposed reconstruction attains conservation at a discrete level; hence, it can be applied to the simulation of physical models satisfying this conservation property. Considerable examples are the BGK-type models of the Boltzmann equation of rarefied gas dynamics. We can also apply this to the splitting method for the Vlasov-Poisson system in plasma physics. The latter problems will be considered in Sect. 6.

In the following section, we will show that our reconstruction (20) inherits some properties of the basic reconstruction $R_i(x)$ such as non-oscillatory property and positivity.

Table 4 Relative conservation errors (26) of the reconstruction for $\bar{u}_1$ and $\bar{u}_2$ in (25). Numerical solutions are obtained with $N_x = 40$

Reconstruction	$\bar{u}_1$	$\bar{u}_2$
Q-CWENO23	4.4409e-16	7.9232e-16
Q-P2	4.4409e-16	7.9232e-16
GWENO23	1.1102e-15	2.0015e-03

3.2.3 Non-Oscillatory Property

The most common reconstructions adopted in the derivation of high-resolution shock-capturing schemes are at the same time conservative and (essentially) non-oscillatory. The WENO methodology [103], for example, satisfies such properties and is widely adopted. Using it as a basic reconstruction R, the resulting sliding average Q based on (20) will be smoother than R and will inherit its non-oscillatory property.

Unless otherwise indicated, throughout this chapter we choose CWENO [31, 78] as our basic reconstruction. We denote by QCWENO the conservative reconstruction obtained using CWENO as basic reconstruction.

We compare QCWENO23, based on CWENO23 [79], with GWENO34 introduced in [32] in the context of SL methods. The result is summarized in Table 4 for the two periodic test functions u_1 and u_2 in $[-1, 1]$:

$$\bar{u}_1(x) = 1 + \frac{1}{2}\sin(4\pi x), \quad \bar{u}_2(x) = \frac{1}{2}(1 + 0.4\cos(1.5\pi x)\mathrm{sign}(x))\exp(-10x^2), \tag{25}$$

where the conservation errors are computed as

$$\mathrm{Err}_\ell = \frac{\max_\theta \left|\sum_i \mathrm{Recon}(x_i + h\theta) - \sum_i \bar{u}_\ell(x_i)\right|}{\sum_i \bar{u}_\ell(x_i)}, \quad \ell = 1, 2. \tag{26}$$

Both reconstructions produce non-oscillatory solutions. If a function $\bar{u}$ is sufficiently smooth, then conservation errors in the GWENO reconstruction will be negligible; however, if the function changes abruptly, then GWENO may suffer from noticeable conservation errors. (See Table 4.) In Fig. 5, we show that the solutions obtained by GWENO34 and Q-CWENO23 gives non-oscillatory solutions for non-smooth function. In contrast, the conservative reconstruction based on parabola (Q-P2) produces spurious oscillation near discontinuity. On the other hand, the conservative reconstructions maintain conservation up to machine precision, while GWENO34 shows nonnegligible conservation error.

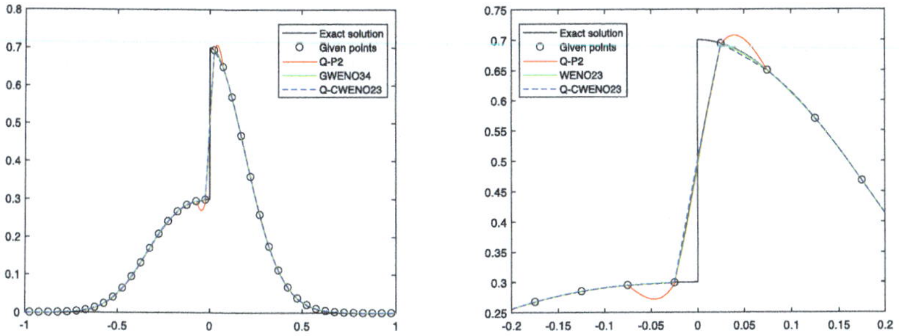

(a) (b)

Fig. 5 Comparison of reconstructions between GWENO34, Q-CWENO23, and Q-P2 [40]. Black circles are given values on grid points of $\bar{u}_2(x)$ given in (25). (**a**) Comparison of reconstructions. (**b**) Solutions on $x \in [-0.2, -2]$

3.2.4 Positivity-Preserving Property

In several circumstances the solution one is looking for is a nonnegative function. This is the case, for example, of distribution function in kinetic equations. In such cases, it may be important to preserve at a discrete level the positivity of the solution. Standard piecewise polynomial reconstructions (linear reconstructions) do not preserve positivity; however, several techniques exist in the literature that can be adopted to ensure positivity in the reconstruction [30, 102]. Here we remark that if the basic reconstruction R is positive preserving, then the sliding average of R will provide a conservative and positivity-preserving reconstruction. Given nonnegative basic reconstructions $R_i(x) \geq 0$, obtained from positive cell averages $\bar{u}_i > 0\, \forall i$, the positivity of the reconstruction (20) directly follows from (18). Here we verify this with a numerical example. Let us consider a basic reconstruction R_i, obtained by the *positive flux conservative* (PFC) technique in [46].

The technique has been proposed to bound the numerical solution between zero and a prescribed maximum value. This property is important when dealing with collisionless Vlasov-like equations, whose solutions are just transported in phase space, therefore preserving their maximum and minimum values. Hereafter we denote by Q-parabola the reconstruction (20) based on this. In Fig. 6, we compare Q-parabola with Q-CWENO23 reconstructions. For this, we consider a function on the periodic domain $[-1, 1]$:

$$g(x) = \begin{cases} 10^{-5} + 0.1\,(1 + \sin(\pi x)), & -0.5 \leq x \leq 0.4 \\ 10^{-5}, & \text{otherwise} \end{cases} \quad . \tag{27}$$

In Fig. 6a and b, the difference between two reconstructions appears near $[-0.65, -0.55]$ and $[0.45, 0.55]$. In case of Q-parabola, the use of positive limiter always guarantees positive reconstructions for any $x \in [-1, 1]$, while very small

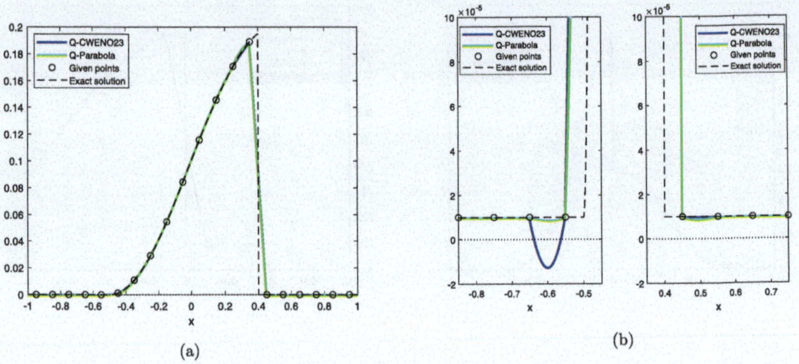

Fig. 6 Comparison of reconstructions between Q-parabola and Q-CWENO23 [40]. Dashed lines are exact solutions $g(x)$ and black circles are given values on grid points of $g(x)$ given in (27). **(a)** Comparison of g given by (27) and its reconstructions. **(b)** Comparison of g given by (27) and its reconstructions on $[-0.85, -0.45]$ and $[0.35, 0.75]$

oscillations appear near discontinuities. On the other hand, although Q-CWENO23 always prevents spurious oscillation, negative solutions may occur depending on the choice of ϵ used for nonlinear weights (121). In this case, we took $\epsilon = 10^{-6}$, and Eq. (121) of CWENO23 returns weights very close to the linear ones on the cell $[-0.6, -0.5]$, which gives negative values on the interval $[-0.65, -0.55]$. We remark that if CWENO23 reconstructions give linear polynomials on two consecutive cells, the corresponding reconstruction (20) is to be positive between the two cell centers. Consequently, the suitable choice of ϵ can enable Q-CWENO23 to avoid both negative reconstructions and spurious oscillations. Other possible ways to guarantee the positivity of basic reconstructions are to adopt a linear scaling approach [59, 120, 121] or use positive limiters [46, 56].

Remark 3.5 We remark that both WENO and CWENO reconstructions and PFC satisfy condition (23) required in Proposition 3.3. Therefore, the Q reconstruction (20) based on such reconstructions satisfy the conservation property (24).

For completeness, CWENO23 reconstruction is summarized in Appendix 8.

3.2.5 Maximum Preserving Property

Here, we introduce a way to impose the maximum principle preserving (MPP) property on the conservative reconstruction by using a linear scaling technique [59, 120, 121].

The procedure works as follows: we first compute the linear scaling parameter in cell i:

$$\xi_i = \max\left\{1, \frac{|M - u_i|}{|M_i' - u_i|}, \frac{|m - u_i|}{|m_i' - u_i|}\right\}, \tag{28}$$

where M and m are maximum and minimum values of a known function u on the whole computational domain, respectively, and M_i' and m_i' are maximum and minimum values of local reconstruction $R_i(x)$ on the cell I_i, respectively. Next, we use ξ_i to basic reconstructions $R_i(x)$ as follows:

$$\tilde{R}_i(x) = u_i + \xi_i(R_i(x) - u_i),$$

where $\tilde{R}_i(x)$ is the rescaled polynomial defined on I_i:

$$\tilde{R}_i(x) = \sum_{\ell=0}^{k} \frac{\tilde{R}_i^{(\ell)}}{\ell!}(x - x_i)^\ell.$$

Consequently,

$$\tilde{R}_i^{(0)} = u_i + \xi_i(R_i^{(0)} - u_i), \quad \tilde{R}_i^{(\ell)} = \xi_i R_i^{(\ell)}, \quad 1 \le \ell \le k,$$

from which we obtain a positive reconstruction $Q(x_i + \theta\Delta x)$:

$$Q_{i+\theta} = \sum_{\ell=0}^{k} (\Delta x)^\ell \left(\alpha_\ell(\theta)\tilde{R}_i^{(\ell)} + \beta_\ell(\theta)\tilde{R}_{i+1}^{(\ell)}\right).$$

We remark that the maximum preserving property still holds for $Q(x)$, $\forall x$, because $Q(x)$ is obtained by taking the sliding average of a piecewise reconstruction satisfying the maximum preserving property:

$$m \le Q(x) = \frac{1}{\Delta x}\int_{x-\Delta x/2}^{x+\Delta x/2} R(y)\,dy \le M.$$

In some applications like the Vlasov-Poisson model, preliminary information about the maximum M and minimum m values of the exact solution are given. In such case, the limiter can be useful. Even in the case of the advection equation or rigid body rotation, the range of the solution is determined by initial data.

3.3 *Numerical Test*

Here we present numerical results for the BGK model. In 1D problem $(d_x, d_v) = (1, 1)$, we consider uniform mesh Δx and Δv for the space and velocity domain. For 1D problem, we use a CFL number defined by

$$\mathrm{CFL} := \max_j |v_j| \frac{\Delta t}{\Delta x}. \tag{29}$$

Similarly, in 2D problem $(d_x, d_v) = (2, 2)$, we take uniform meshes both in space $(\Delta x = \Delta y)$ and in velocity $(\Delta v_x = \Delta v_y = \Delta v)$. Then, we use the CFL number defined by

$$\mathrm{CFL} := \max_{1 \le \ell \le d_v} \max_j \left| v_j^{(\ell)} \right| \frac{\Delta t}{\Delta x}. \tag{30}$$

For accuracy tests, we measure numerical errors and the corresponding convergence rate using a relative L^1-norm of the numerical solutions, computed by standard Richardson extrapolation. For example, in case of 1D a space-time-dependent solution $u(x,t)$, let $u_i^n(h)$ be the solution obtained by using $h = \Delta x$ with N_x grid points. Similarly, we denote by $u_i^n(h/2)$ the solution obtained with $h/2$ with $2N_x$ grid points.

- Relative L^1-norm of u^n using N_x

$$\mathrm{error}(N_x, 2N_x) = \frac{\sum_{i=1}^{N_x^{d_x}} |u_i^n(h) - u_{2i-1}^n(h/2)|}{\sum_{i=1}^{N_x^{d_x}} |u_{2i-1}^n(h/2)|}. \tag{31}$$

  We measure two-dimensional relative L^1-norm in the same way.
- Convergence rate

$$\mathrm{Rate}(N_x) = \log_2 \left(\frac{\mathrm{error}(N_x, 2N_x)}{\mathrm{error}(2N_x, 4N_x)} \right). \tag{32}$$

For two-dimensional problems, we assume uniform square meshes both in space and velocity domains, i.e., $\Delta x = \Delta y = h$ for space and $\Delta v^1 = \Delta v^2 = \Delta v$ for velocity. In two-dimensional accuracy tests, we will compute errors and convergence rates using relative L^1-norm as in (31) and (32).

3.3.1 1D Conservation Test

To demonstrate the necessity of both conservative reconstruction and conservative treatment of Maxwellian in the construction of conservative methods, we repeat the single shock test in Sect. 2.2.5. Here we denote by $dM1$ and $dM2$ the methods in Sects. 3.1.1 and 3.1.3, respectively. In Table 5, we first show that nonconservative reconstruction could introduce finite errors even with the conservative treatment of Maxwellian. On the other hand, in Table 6 the three columns for RK3-QCWENO35-CM method imply that the conservative reconstruction without any conservative treatment of Maxwellian may not guarantee the exact conservation if velocity

Table 5 Single shock test. CFL $= 2$, $\varepsilon = 10^{-6}$. Conservation errors of discrete moments in the relative L_1 norm for high-order schemes applied to single shock problem with velocity domain $[-20, 20]$

	RK3-WENO35-$dM1$			RK3-WENO35-$dM2$		
(N_x, N_v)	Mass	Momentum	Energy	Mass	Momentum	Energy
(100, 42)	1.22e-03	1.29e-02	1.47e-02	1.22e-03	1.29e-02	1.47e-02
(100, 50)	1.06e-03	1.31e-02	1.47e-02	1.06e-03	1.31e-02	1.47e-02
(100, 60)	1.43e-03	1.26e-02	1.49e-02	1.43e-03	1.26e-02	1.49e-02
(100, 90)	1.35e-03	1.28e-02	1.48e-02	1.35e-03	1.28e-02	1.48e-02

grid points are not enough. In contrast, the other results obtained by conservative reconstruction combined with $dM1$ or $dM2$ guarantee the conservation of mass/momentum/energy up to machine precision regardless of number of velocity grid points.

3.3.2 1D Accuracy Test

To check accuracy, we consider a test proposed in [91]. The initial distribution is given by the Maxwellian

$$f_0(x, v) = \frac{\rho_0}{\sqrt{2\pi T_0}} \exp\left(-\frac{|v - u_0(x)|^2}{2T_0}\right), \tag{33}$$

where initial density and temperature are assumed to be uniform, with constant value $\rho_0(x) = 1$ and $T_0(x) = 1$. The initial velocity profile is given by

$$u_0(x) = 0.1 \exp\left(-(10x - 1)^2\right) - 2 \exp\left(-(10x + 3)^2\right).$$

For space, we assume periodic boundary conditions in the interval $[-1, 1]$ with $N_x = 320, 640, 1280$, and 2560. For velocity domain, we consider the interval $[-10, 10]$ with $N_v = 20$, which is enough when we use correction (12), as described in Sect. 3.1. We set time steps using (29) with CFL$=4$ and compute numerical solutions up to $t_f = 0.32$ because for a small Knudsen number shock appears at $t = 0.35$. For RK3- and BDF3-based methods, denoted by RK3-QCWENO23 and BDF3-QCWENO23, respectively, we use QCWENO23 reconstruction, and hence we expected a convergence rate of approximately 3. In Table 7, we report errors and convergence rate computed with the relative L^1-norm for density ρ, as in (31) and (32). The result shows that order reductions appear for RK3-based method as $\varepsilon \to 0$, while it attains the expected order 3 of reconstruction for $\varepsilon = 1$. On the contrary, BDF3 does not suffer from order reduction and gives the desired convergence rate of 3 for all values of ε. We remark that similar results are obtained

Table 6 Single shock test. CFL = 2, $\varepsilon = 10^{-6}$. Conservation errors of discrete moments in the relative L_1 norm for high-order schemes applied to single shock problem with velocity domain $[-20, 20]$

	RK3-QCWENO35-CM			RK3-QCWENO35-$dM1$			RK3-QCWENO35-$dM2$		
(N_x, N_v)	Mass	Momentum	Energy	Mass	Momentum	Energy	Mass	Momentum	Energy
(100, 42)	2.90e-07	1.96e-06	3.49e-06	2.68e-14	5.81e-14	3.03e-14	1.59e-13	1.95e-13	8.84e-14
(100, 50)	3.46e-11	3.21e-10	5.69e-10	6.15e-14	9.06e-14	1.01e-13	6.08e-13	5.05e-13	9.64e-14
(100, 60)	1.03e-12	3.13e-13	6.21e-14	9.51e-14	6.26e-14	1.66e-13	1.03e-12	2.20e-13	4.94e-15
(100, 90)	7.38e-13	2.50e-13	3.71e-15	1.91e-14	2.43e-14	1.54e-14	7.44e-13	2.63e-13	2.78e-14

Table 7 Accuracy test for the 1D BGK equation. Initial data is given in (33). Here we used CFL= 4

Relative L^1 error and order of density									
		$\varepsilon = 10^{-6}$		$\varepsilon = 10^{-4}$		$\varepsilon = 10^{-2}$		$\varepsilon = 10^{-0}$	
Scheme	$(N_x, 2N_x)$	Error	Rate	Error	Rate	Error	Rate	Error	Rate
RK3-QCWENO23	(320, 640)	2.010e-05	2.50	1.779e-05	2.67	1.440e-06	3.22	1.242e-05	3.96
	(640, 1280)	3.537e-06	2.13	2.792e-06	2.38	1.550e-07	3.09	7.979e-07	3.43
	(1280, 2560)	8.051e-07		5.378e-07		1.822e-08		7.416e-08	
BDF3-QCWENO23	(320, 640)	2.740e-05	3.02	2.461e-05	3.06	1.385e-06	3.23	1.050e-05	2.97
	(640, 1280)	3.367e-06	3.10	2.953e-06	3.04	1.475e-07	3.13	1.336e-06	2.99
	(1280, 2560)	3.933e-07		3.450e-07		1.687e-08		1.681e-07	

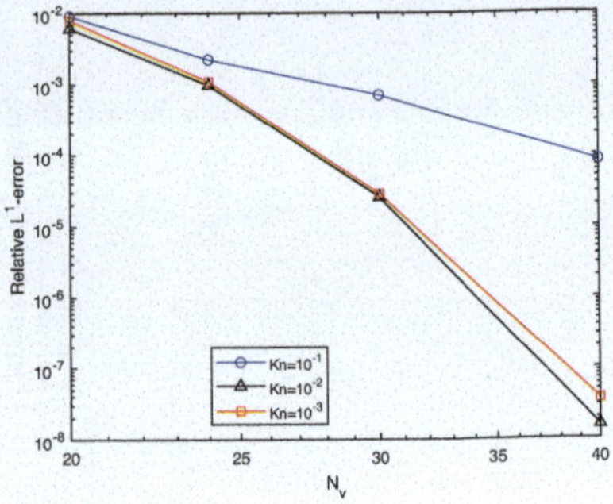

Fig. 7 Error versus N_v. Initial data is given in (33)

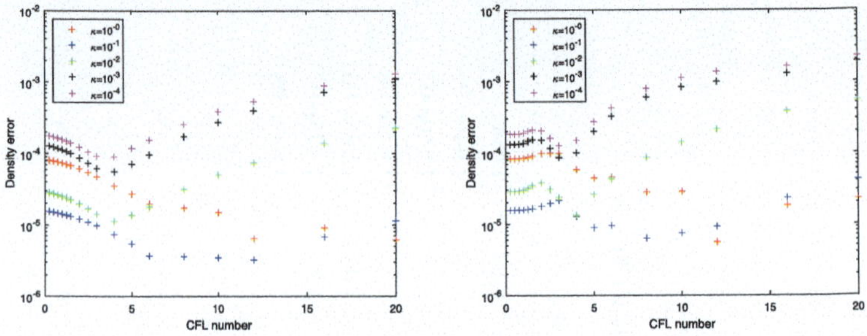

(a) (b)

Fig. 8 Error versus CFL number [41]. Initial data is given in (33). (**a**) RK3-QCWENO23. (**b**) BDF3-QCWENO23

for other macroscopic quantities such as momentum and energy, but we omit to show them here.

To further investigate the errors from velocity discretization, we fix Δx and Δt determined by $N_x = 640$ and CFL$= 2$. Then, we compare the numerical solutions obtained by $N_v = 20, 24, 30, 40, 120$. As reference solution, we use the one obtained by $N_v = 120$. In this test, we do not use any conservative correction for the Maxwellian. We report the result of the accuracy test in Fig. 7. Note that errors decrease dramatically as we increase the number of grid points in velocity direction, which implies the spectral accuracy of velocity discretization. It is also noticeable that errors decay slowly for larger Knudsen number, since in such cases we need more velocity grid points to resolve the shape of distribution functions which are far from local Maxwellian.

Since SL schemes allow large CFL numbers, the choice of CFL number is important to secure both accuracy and efficiency. For such a reason we compute the errors obtained with different values of CFL numbers for various Knudsen numbers. In Fig. 8, we report the relative L^1 errors obtained by $N_x = 160, 320$ for each scheme. From Fig. 8, we observe that the error has a non-monotone behavior with respect to the CFL number. This is due to the fact that for smaller time steps Δt the interpolation error accumulates more frequently, making it dominant for small

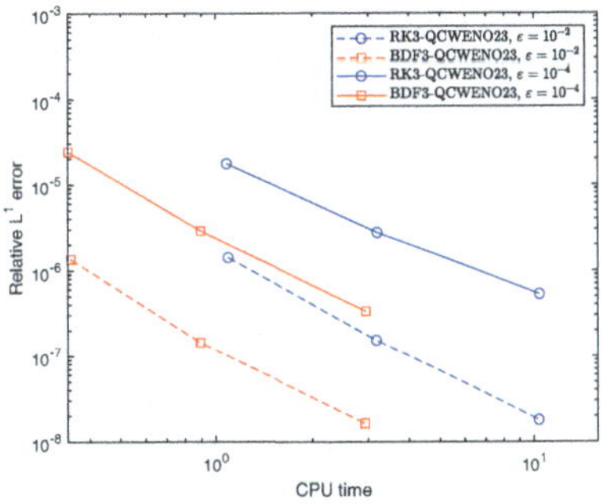

Fig. 9 Error versus CPU time. One-dimensional accuracy test for the BGK model. BDF schemes (red lines) appear to be more efficient than RK schemes (blue lines). Initial data is given in (33)

enough CFL numbers; the behavior is in agreement with consistency analysis of SL schemes, as illustrated, for example, in the book by Falcone and Ferretti [50, Chapter 6]. As a consequence there is an optimal CFL number that depends on the Knudsen number and is in general larger than one. For large enough Courant number the error increases with CFL, as expected. Similar results are obtained using a different DIRK3 scheme, such as the one adopted in [64].

Now, we compare the efficiency of RK and BDF methods. In Fig. 9, we report the error versus CPU time for third-order schemes. For both schemes, we used $\varepsilon = 10^{-4}, 10^{-2}$, and CFL$=4$. From the figure, it appears that the BDF-based scheme is more efficient, in agreement with what is expected from the operation count reported at the end of Sect. 2.2.3, given that the two schemes adopt the same CFL number (and therefore the same time step).

3.3.3 2D Accuracy Test

Here, we check the accuracy of SL methods for the 2D BGK model. Initial conditions are taken by the local Maxwellian with the macroscopic quantities:

$$\rho_0(x, y) = 1 + (\sin \pi x)^2 (\sin \pi y)^2, \quad u_0(x, y) = (0, 0), \quad T_0(x, y) = 1, \tag{34}$$

where periodic boundary conditions are imposed on $[0, 1]^2$. We take $N_v = 32$ for the velocity domain $[-10, 10]^2$, which means that we use $(32+1)^2$ grid points. The final time is $t_f = 0.6$ up to which the solution remains smooth. We use a fixed time step Δt with CFL$=4$ using (30). In this test, the positivity of Maxwellian is not strictly demanded; hence, we enforce conservation by adopting the L^2-projection method of Sect. 3.1.1, expressed by Eq. (12). Here we consider RK3- and BDF3-based SL schemes combined with the conservative reconstruction based on the 2D CWENO23 method, so the expected convergence rates are between 2 and 3.

In Table 8, we report the numerical errors and the corresponding accuracy orders for density. The result of the BDF3 scheme shows the desired convergence rate 3 for all values of Knudsen number. On the other hand, in case of the RK3 method,

Table 8 Accuracy test for the 2D BGK equation [41]. Initial data is given in (34)

Relative L^1 error and order of density

Scheme	$(N_x^2, (2N_x)^2)$	$\varepsilon = 10^{-6}$ Error	Rate	$\varepsilon = 10^{-4}$ Error	Rate	$\varepsilon = 10^{-2}$ Error	Rate	$\varepsilon = 10^{-0}$ Error	Rate
RK3	$(40^2, 80^2)$	5.39e-04	2.53	5.13e-04	2.58	8.71e-05	2.70	1.74e-05	3.03
	$(80^2, 160^2)$	9.31e-05	2.63	8.57e-05	2.70	1.34e-05	2.66	2.14e-06	3.02
	$(160^2, 320^2)$	1.50e-05		1.32e-05		2.12e-06		2.64e-07	
BDF3	$(40^2, 80^2)$	7.18e-04	2.34	6.94e-04	2.37	1.35e-04	2.68	2.74e-05	3.03
	$(80^2, 160^2)$	1.42e-04	2.92	1.35e-04	2.94	2.11e-05	2.89	3.36e-06	3.02
	$(160^2, 320^2)$	1.87e-05		1.75e-05		2.84e-06		4.15e-07	

we observe order reduction in the limit $\varepsilon = 10^{-6}$ as in the 1D accuracy test (see Table 7). A similar result can be found, for example, in [64].

3.3.4 1D Riemann Problem

Here we consider the classical Riemann problem to check the shock-capturing capability of high-order conservative SL schemes in the fluid limit $\varepsilon \to 0$. We compare our scheme with nonconservative SL schemes [64] and with conservative correction-based SL methods [20]. As initial data we consider local Maxwellians with piecewise constant macroscopic quantities used in [64]:

$$(\rho_0, u_0, p_0) = \begin{Bmatrix} (2.25, 0, 1.125), & \text{for } x \leq 0.5 \\ (3/7, 0, 1/6), & \text{for } x > 0.5 \end{Bmatrix}. \tag{35}$$

We impose free-flow boundary condition on the interval $x \in [0, 1]$ and velocity domain $v \in [-10, 10]$ up to final time $t_f = 0.16$. We take uniform grids with $N_x = 200$, $N_v = 30$, and CFL$= 2$, which means a time step by $\Delta t = \Delta x/5$. For this test, we enforce conservation by adopting a discrete Maxwellian, as described in Sect. 3.1, (16), because it guarantees the positivity of local Maxwellian with a relatively small number of velocity grid points $N_v = 30$, while the approach based on L^2-minimization may produce negative values, thus leading to instability. In Fig. 10, we show the results obtained with the BDF3 scheme, with two different reconstructions: GWENO23 and QCWENO23. The left panel shows the density profiles obtained by the schemes. It is evident that far from the shock the results are very similar; however, near the shock only the conservative reconstruction guarantees the correct shock speed. In Fig. 10b, we report a zoom of the pressure profile near the shock: it appears that conservative solutions converge to the exact solution as we increase the number grid points.

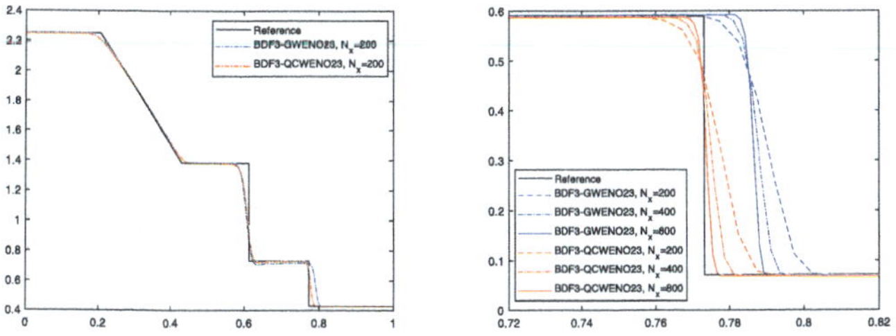

(a) (b)

Fig. 10 Riemann problem for the 1D BGK model with $\varepsilon = 10^{-6}$ [41]. BDF3-based method with QCWENO23. Initial data is given in (35). (**a**) Density. (**b**) Pressure

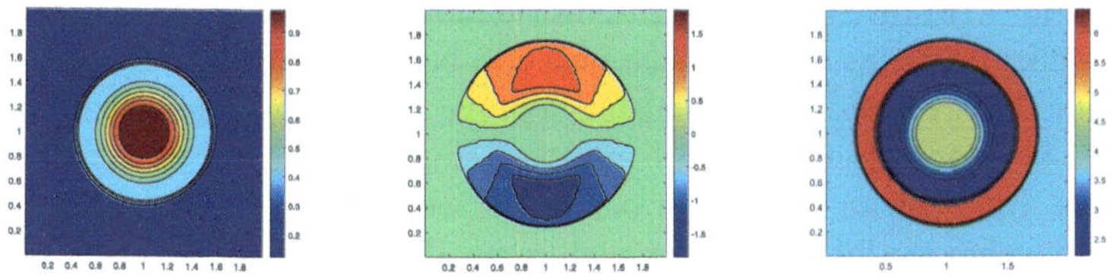

(a) (b) (c)

Fig. 11 2D Riemann problem [41] with $\varepsilon = 10^{-4}$. Macroscopic moments. (**a**) Density. (**b**) Velocity in y-direction. (**c**) Temperature

3.3.5 2D Riemann Problem

Now, we perform a test on the 2D Riemann problem presented in [47]. We use $\varepsilon = 10^{-4}$ so that the solution is rather close to a local Maxwellian. The initial distribution function is a local Maxwellian with macroscopic moments:

$$(\rho_0, u_{10}, u_{20}, T_0) = \left\{ \begin{array}{ll} (1, 0, 0, 5), & \text{for } (x-1)^2 + (y-1)^2 \leq 0.2 \\ (0.125, 0, 0, 4), & \text{for } (x-1)^2 + (y-1)^2 > 0.2 \end{array} \right\}.$$

The computation is performed on the space domain $[0, 2]^2$ with free-flow boundary conditions; the velocity domain is $[-15, 15]^2$, and the final time $t_f = 0.07$. We take uniform grids with $N_x = 200$, $N_y = 200$, and $N_v = 20$ and fix a time step Δt with CFL$=2.1$ using (30). Here we use the correction to the Maxwellian according to the L^2-projection method, (12). The results of the computation (see [41] for more details) are reported in Figs. 11–12 and are in agreement with results from the literature [47].

3.4 Velocity Adaptation

Most numerical schemes proposed for solving BGK models for rarefied gas dynamics are based on the discrete velocity approximation. Since such approaches

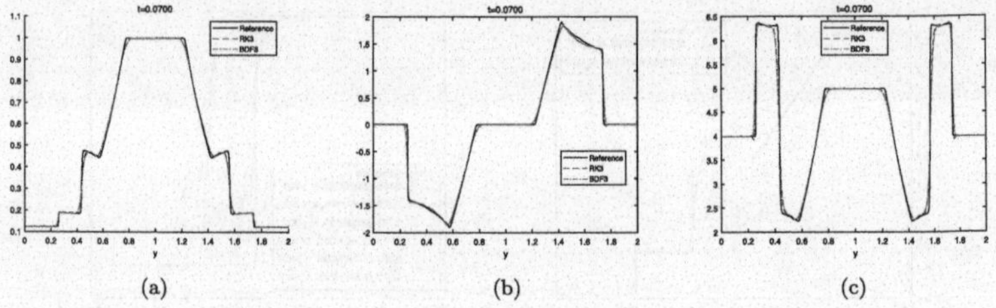

Fig. 12 Riemann problem for the 2D BGK model [41] with $\varepsilon = 10^{-4}$. Macroscopic moments at $x = x_{100}$. (a) Density. (b) Velocity in y-direction. (c) Temperature

use fixed velocity grids, and one must secure a sufficiently large domain with fine velocity grids to resolve the structure of distribution functions. When one treats high Mach number problems, the computational cost becomes prohibitively expensive.

A methodology based on the adjustment of the velocity grids locally in time and space has been applied to the BGK model of single gas [7, 25] and gas mixtures [26] in the Eulerian framework. In our recent work [18], we extend the velocity adaptation technique for the BGK model to the SL framework. Here we provide a brief description of the technique.

3.4.1 First-Order SL Method for the BGK Model

Before describing the method, we introduce the notation we will use throughout this section. We consider a fixed time step Δt and a uniform mesh size Δx. The nth time step and ith node is denoted by t^n and $x_i = x_{min} + \left(i - \frac{1}{2}\right)\Delta x$. For each i and n, let us denote the set of velocity grid points by $\mathcal{G}_i^n$ with uniform mesh size Δv_i^n. The set of indices corresponding to space and to each $\mathcal{G}_i^n$ will be denoted by $\mathcal{I}$ and $\mathcal{J}_i^n$, respectively. A cell assigned to each $(i, j) \in \mathcal{I} \times \mathcal{J}_i^n$ will be defined by $I_{ij}^n := [x_i - \Delta x/2, x_i + \Delta x/2] \times \left[v_j - \Delta v_i^n/2, v_j + \Delta v_i^n/2\right]$. The notation $\tilde{I}_{ij}^n$ is the region such that

$$\tilde{I}_{ij}^n := \{(x, v) \in \mathbb{R}^2 | \, x_{i-\frac{1}{2}} - v\Delta t \leq x \leq x_{i+\frac{1}{2}} - v\Delta t, \; v_{i,j}^n - \Delta v_i^n/2 \leq v \leq v_{i,j}^n + \Delta v_i^n/2.\}.$$

Now, we describe how to apply local velocity adaptation approach to a first-order SL scheme for a 1D BGK model.

We start from the characteristic of (1):

$$\begin{aligned} \frac{df}{ds} &= \frac{1}{\varepsilon}\left(\mathcal{M}(f) - f\right), \quad t > 0, \quad x, v \in \mathbb{R} \\ \frac{dX}{ds} &= v, \quad X(t) = x, \quad X(s) = x - v(t - s). \end{aligned} \tag{36}$$

To treat the stiffness coming from ε, we begin by applying the implicit Euler scheme to (36). Then, we advance the solution from t^n to t^{n+1} using the following discretization:

$$f^{n+1}(x,v) = f^n(x - v\Delta t, v) + \frac{\Delta t}{\varepsilon}\left(\mathcal{M}^{n+1}(x,v) - f^{n+1}(x,v)\right), \tag{37}$$

where f^n denotes the solution at time t^n.

Now, we explain our method with five steps. **Step 1**, we first predict ρ_i^*, U_i^*, E_i^*, and T_i^* for each cell using the grid points $\mathcal{G}_i^n$ previously defined. **Step 2**, we use the mean velocity U_i^* and temperature T_i^* to define a set of new grid points $\mathcal{G}_i^{n+1}$ for each i. **Step 3**, we improve the computation of ρ_i^{n+1}, U_i^{n+1}, E_i^{n+1}, and T_i^{n+1} for each cell i. **Step 4**, we correct the Maxwellians and the numerical solution using the weighted L^2-minimization technique to impose the correct conservation. **Step 5**, we update the numerical solution.

Step 1: Prediction of macroscopic quantities at t^{n+1} using $\mathcal{G}_i^n$. First, we aim to define the new velocity nodes for time t^{n+1}. For this, we integrate (37) over $I_i \times \mathbb{R}$, which gives

$$\begin{aligned}\int_{\mathbb{R}}\int_{I_i} f^{n+1}(x,v)\begin{pmatrix}1\\ v\\ v^2\end{pmatrix} dx\,dv &= \int_{\mathbb{R}}\int_{I_i} f^n(x - v\Delta t, v)\begin{pmatrix}1\\ v\\ v^2\end{pmatrix} dx\,dv\\ &= \int_{\mathbb{R}}\int_{\tilde{I}_i(v,\Delta t)} f^n(x,v)\begin{pmatrix}1\\ v\\ v^2\end{pmatrix} dx\,dv\end{aligned}$$

where

$$I_i = \left[x_{i-\frac{1}{2}}, x_{i+\frac{1}{2}}\right], \quad \tilde{I}_i(v,\Delta t) = \left[x_{i-\frac{1}{2}} - v\Delta t, x_{i+\frac{1}{2}} - v\Delta t\right].$$

Based on this, we precompute ρ_i^*, u_i^* and E_i^*, and temperature T_i^*:

$$\begin{aligned}\begin{pmatrix}\rho_i^*\\ \rho_i^* U_i^*\\ 2E_i^*\end{pmatrix} &:= \frac{1}{\Delta x}\int_{\mathbb{R}}\int_{\tilde{I}_i(v,\Delta t)} f^n(x,v)\begin{pmatrix}1\\ v\\ v^2\end{pmatrix} dx\,dv\\ &\approx \sum_{k\in\mathcal{J}_i^n}\frac{1}{\Delta x}\int_{\tilde{I}_{i,k}^n}\begin{pmatrix}P^n(x,v)\\ Q^n(x,v)\\ R^n(x,v)\end{pmatrix} dx\,dv,\\ d_v R T_i^* &:= (2E_i^* - \rho_i^*|U_i^*|^2)/\rho_i^*.\end{aligned} \tag{38}$$

This relation means that we compute macroscopic quantities based on the projected area of the grey region along characteristics. (In Fig. 13, the parallelogram-shaped

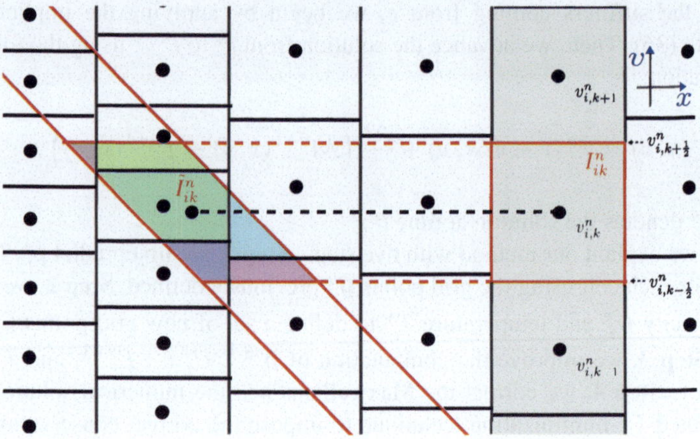

Fig. 13 Phase space discretization at time $t = t^n$ and computation of moment prediction by integration over the cell $\tilde{I}^n_{i,k}$ (**Step 1**) [18]. In each colored patch, we use three different polynomials P^n, Q^n, R^n for the computation of the contribution to ρ_i^*, $\rho_i^* U_i^*$, E_i^*. The grey region means the distribution function on $\mathcal{G}_i^n$

domain $\tilde{I}^n_{i,k}$ is the projected area of I^n_{ik} along the characteristics.) To construct a first-order scheme, it is enough to take

$$P^n(x,v) := \sum_{i,k \in \mathcal{J}_i^n} f^n_{ik}\chi^n_{ik}(x,v), \quad Q^n(x,v) := \sum_{i,k \in \mathcal{J}_i^n} v_k f^n_{ik}\chi^n_{ik}(x,v),$$

$$R^n(x,v) := \sum_{i,k \in \mathcal{J}_i^n} v_k^2 f^n_{ik}\chi^n_{ik}(x,v), \tag{39}$$

where $\chi^n_{ik}(x,v)$ is the characteristic function on I^n_{ik}. The integration in (38) will be obtained by summation of the integrals for trapezoidal patches. An explicit formula of such integral is illustrated in [18]. Concerning the high-order reconstruction of the polynomials $P^n(x,v)$, $Q^n(x,v)$, and $R^n(x,v)$, we also refer to [18].

Step 2: Choice of local velocity grids $\mathcal{G}_i^{n+1}$. Since our interests lie on the problems for small Knudsen number ε, we consider the situation when the shape of the distribution function is close to a local Maxwellian. Since it takes the form of the normal distribution whose mean is U and standard deviation is $\sqrt{RT}$, most of the distribution function is concentrated in the interval

$$U_i^* - \alpha\sqrt{RT_i^*} \le v \le U_i^* + \alpha\sqrt{RT_i^*}.$$

The α will be chosen large enough so that distributions outside this interval are acceptably small and can be negligibly small, which helps us to guarantee the approximate conservation of mass/momentum/energy. To define velocity grids $\mathcal{G}_i^{n+1}$, it remains to set a size of mesh Δv_i^{n+1}.

To resolve the shape of the Maxwellian, we impose that between the two inflection points of the Maxwellian there should be at least three grid points:

$$\Delta v_i^{n+1} \leq \sqrt{RT_i^*}. \tag{40}$$

Now, based on this, we calculate $N_v = 2\left\lceil \alpha\sqrt{RT_i^*}/\Delta v_i^{n+1} \right\rceil$:

$$v_{min,i}^{n+1} := U_i^* - \frac{N_v}{2}\Delta v_i^{n+1}, \quad v_{max,i}^{n+1} = U_i^* + \frac{N_v}{2}\Delta v_i^{n+1}.$$

Here, for any $\xi \in \mathbb{R}$, $\lceil \xi \rceil$ means the smallest integer equal or greater than ξ. Finally, we newly define a set of grid points $\mathcal{G}_i^{n+1}$ by

$$\mathcal{G}_i^{n+1} := \{v_{i,j}^{n+1} | v_{i,j}^{n+1} := v_{min,i}^{n+1} + j\Delta v_i^{n+1}, 0 \leq j \leq N_v\}.$$

This, however, may be insufficient to resolve solutions where abrupt changes in velocity and temperature appear. Moreover, since we are considering a SL framework, the information of nearby cells should be involved considering the CFL number. For this reason, we use

$$T_{i,max}^* = \max_{-\delta \leq j \leq \delta} T_{i+j}^*, \quad T_{i,min}^* = \min_{-\delta \leq j \leq \delta} T_{i+j}^*, \quad \Delta v_i^{n+1} \leq \sqrt{RT_{i,min}^*},$$

where $\delta = \lceil \text{CFL} \rceil + 1$. The additional number 1 is due to the use of the integral based on the parallelogram.

Note that we determine Δv_i^{n+1} as follows:

$$\Delta v_i^{n+1} = \beta\sqrt{RT_{i,min}^*}, \quad 0 < \beta \leq 1.$$

Regarding the choice of α and β, we refer to the numerical test in Sect. 3.4.3.

Remark 3.6 When the shape of the distribution function is far from a local Maxwellian, its numerical support may be relatively large, that is, the bounds could be far from the bounds computed from the Maxwellian. For treating this problem, in [7, 25] the authors did as follows: once a transport step is performed for the prediction of the distribution function on the boundary of newly defined velocity domain, its relative scale is compared with a given tolerance. If the value is larger than the tolerance, the new velocity node is considered and the procedure is repeated until the values on end points become small enough. In our approach, we could treat the problem similarly. In the following step, we predict $f_{i,j}^{n+1}$ for each newly defined velocity node $v_{i,j}^{n+1}$, $j = 0, ..., N_v$. Then, additionally, we compute $f_{i,-1}^{n+1}$ (or f_{i,N_v+1}^{n+1}). If the value of $f_{i,-1}^{n+1}$ (or f_{i,N_v+1}^{n+1}) is larger than a tolerance tol, i.e.,

$$\frac{f_{i,-1}^{n+1}}{\max_{j\in\mathcal{J}_i^{n+1}} f_{i,j}^{n+1}} > tol \quad \left(\text{or } \frac{f_{i,N_v+1}^{n+1}}{\max_{j\in\mathcal{J}_i^{n+1}} f_{i,j}^{n+1}} > tol\right),$$

we include the velocity node $j = 0$ (or $j = N_v + 1$) in G_i^{n+1} and repeat the procedure until the value of distribution becomes small near boundary. As an alternative approach, we also refer to the work [7].

Step 3: Correction of macroscopic quantities at t^{n+1} and computation of $\tilde{f}_{ij}^n$ and $\mathcal{M}_{ij}^{n+1}$. In this step, we improve the prediction of the moments at time t^{n+1}. For this, we first compute $\tilde{f}_{i,j}^n$ as an approximation of $f(x_i - v_j\Delta t, v_j, t^n)$:

$$\tilde{f}_{i,j}^n = \frac{1}{\Delta x \Delta v_i^{n+1}} \int_{\tilde{I}_{i,j}^{n+1}} P^n(x, v)\, dx\, dv. \tag{41}$$

That is, we compute $\tilde{f}_{i,j}^n$ as the average of P^n on $\tilde{I}_{i,j}^{n+1}$. Now, we use this to compute ρ_i^{n+1}:

$$\rho_i^{n+1} := \sum_{j\in\mathcal{J}_i^{n+1}} \tilde{f}_{i,j}^n \Delta v_i^{n+1}. \tag{42}$$

The other quantities U_i^{n+1}, E_i^{n+1}, and temperature T_i^{n+1} are obtained by

$$\begin{aligned} \begin{pmatrix} \rho_i^{n+1} U_i^{n+1} \\ 2E_i^{n+1} \end{pmatrix} &:= \sum_{j\in\mathcal{J}_i^{n+1}} \frac{1}{\Delta x} \int_{\tilde{I}_{i,j}^{n+1}} \begin{pmatrix} Q^n(x, v) \\ R^n(x, v) \end{pmatrix} dx\, dv, \\ d_v R T_i^{n+1} &:= (2E_i^{n+1} - \rho_i^{n+1}|U_i^{n+1}|^2)/\rho_i^{n+1}. \end{aligned} \tag{43}$$

This step is needed because the new velocity domain could be quite different from that of time t^n. Notice that this step is not expensive because the polynomials P^n, Q^n, and R^n are already known from **Step 1**. Then, the local Maxwellian $\mathcal{M}_{ij}^{n+1}$ is computed by

$$\mathcal{M}_{ij}^{n+1} = \frac{\rho_i^{n+1}}{\sqrt{\left(2\pi T_i^{n+1}\right)^2}} \exp\left(-\frac{|v_j - U_i^{n+1}|^2}{2T_i^{n+1}}\right).$$

Step 4: Correction of $\tilde{f}_{ij}^n$ and $\mathcal{M}_{ij}^{n+1}$ using a weighted L^2-minimization. Now, for each newly defined $v_j \in \mathcal{G}_i^{n+1}$ we attempt to compute the solution as

$$f_{i,j}^{n+1} = \tilde{f}_{i,j}^n + \frac{\Delta t}{\varepsilon}\left(\mathcal{M}_{ij}^{n+1} - f_{i,j}^{n+1}\right). \tag{44}$$

Relation (44) implies that the discrete summation of f_{ij}^{n+1} and $\tilde{f}_{ij}^{n}$ over $1, v_j, v_j^2$ should reproduce the same macroscopic quantities. However, this may not be a good approximation if there are not enough grid points in velocity space.

To enforce conservation, we use the weighted L^2-minimization technique described in Sect. 3.1.2. First, we solve

$$\min_p \left\| \mathbf{1} - p_i^{n+1} \right\|_2^2 \quad \text{s.t} \quad C_i^{n+1} p_i^{n+1} = \begin{pmatrix} \rho_i^{n+1} \\ \rho_i^{n+1} U_i^{n+1} \\ E_i^{n+1} \end{pmatrix} \tag{45}$$

where

$$\mathbf{1} := (1, 1, \cdots, 1)^\top \in \mathbb{R}^{(N_v+1)^{d_v}}.$$

$$p_i^{n+1} := (p_{i,1}^{n+1}, p_{i,2}^{n+1}, \cdots, p_{i,N_v+1}^{n+1})^\top \in \mathbb{R}^{(N_v+1)^{d_v}}$$

$$C_i^{n+1} := \begin{pmatrix} \mathcal{M}_{i,j}^{n+1} (\Delta v)^{d_v} \\ \mathcal{M}_{i,j}^{n+1} v_j (\Delta v)^{d_v} \\ \mathcal{M}_{i,j}^{n+1} \frac{|v_j|^2}{2} (\Delta v)^{d_v} \end{pmatrix} \in \mathbb{R}^{(d_v+2)\times(N_v+1)^{d_v}}$$

where $v_j \in \mathcal{G}_i^{n+1}$ and $|\mathcal{G}_i^{n+1}| = N_v + 1$. Then, we replace $\{\mathcal{M}_{i,j}^{n+1}\}_{j\in\mathcal{J}_i^{n+1}}$ with $M_i^{n+1} := p_i^{n+1} \circ \mathcal{M}_i^{n+1}$.

Furthermore, notice that $\tilde{f}_{i,j}^{n}$ can be used only for computing the density ρ_i^{n+1} by (41). To guarantee conservation, we use again the weighted L^2-minimization and replace $\tilde{f}_{i,j}^{n}$ by a new distribution function $g_{i,j}^{n}$ which is the closest one to $\tilde{f}_{i,j}^{n}$ that provides correct moments ρ_i^{n+1}, $\rho_i^{n+1} U_i^{n+1}$ and E_i^{n+1} previously computed in (42) and (43). Then, we find $g_i^n := q_i^{n+1} \circ \tilde{f}_i^n$ by solving

$$\min_q \left\| \tilde{f}_i^n \circ \frac{1}{M_i^{n+1}} - q_i^{n+1} \right\|_2^2 \quad \text{s.t} \quad D_i^{n+1} q_i^{n+1} = \begin{pmatrix} \rho_i^{n+1} \\ \rho_i^{n+1} U_i^{n+1} \\ E_i^{n+1} \end{pmatrix}$$

where

$$q_i^{n+1} := (q_{i,1}^{n+1}, q_{i,2}^{n+1}, \cdots, q_{i,N_v+1}^{n+1})^\top \in \mathbb{R}^{(N_v+1)^{d_v}}$$

$$D_i^{n+1} := \begin{pmatrix} M_i^{n+1} (\Delta v)^{d_v} \\ M_i^{n+1} v_j (\Delta v)^{d_v} \\ M_i^{n+1} \frac{|v_j|^2}{2} (\Delta v)^{d_v} \end{pmatrix} \in \mathbb{R}^{(d_v+2)\times(N_v+1)^{d_v}}$$

where M_i^{n+1} is the solution obtained from (45).

Step 5: Update of solution f_{ij}^{n+1}. The final step is to update the solution using

$$f_{i,j}^{n+1} = g_{i,j}^{n} + \frac{\Delta t}{\varepsilon}\left(M_{ij}^{n+1} - f_{i,j}^{n+1}\right).$$

3.4.2 Approximation Based on a Parallelogram

To ensure the conservation at the discrete level, we compute macroscopic quantities with (42) and (43). Here we consider a more general form of (39):

$$\begin{pmatrix} P^n(x,v) \\ Q^n(x,v) \\ R^n(x,v) \end{pmatrix} = \sum_{ik} \begin{pmatrix} P_{ik}^n(x,v) \\ Q_{ik}^n(x,v) \\ R_{ik}^n(x,v) \end{pmatrix} \chi_{ik}(x,v)$$

with polynomials $P_{ik}^n(x,v)$, $Q_{ik}^n(x,v)$, $R_{ik}^n(x,v)$ satisfying

$$\begin{pmatrix} f_{ik}^n \\ v_k f_{ik}^n \\ v_k^2 f_{ik}^n \end{pmatrix} = \frac{1}{\Delta x \Delta v_i^n} \int_{I_{ik}^n} \begin{pmatrix} P_{ik}^n(x,v) \\ Q_{ik}^n(x,v) \\ R_{ik}^n(x,v) \end{pmatrix} dx\,dv, \qquad v_k \in \mathcal{G}_i^n$$

Then, as in (42) and (43), we compute

$$\begin{pmatrix} \rho_i^{n+1} \\ \rho_i^{n+1} U_i^{n+1} \\ 2E_i^{n+1} \end{pmatrix} := \sum_{j \in \mathcal{J}_i^{n+1}} \frac{1}{\Delta x} \int_{\tilde{I}_{i,j}^{n+1}} \begin{pmatrix} P^n(x,v) \\ Q^n(x,v) \\ R^n(x,v) \end{pmatrix} dx\,dv,$$

$$d_v R T_i^{n+1} := (2E_i^{n+1} - \rho_i^{n+1} U_i^{n+1})/\rho_i^{n+1},$$

where

$$\tilde{I}_{ij}^{n+1} := \{(x,v) \in \mathbb{R}^2 \,|\, x_{i-\frac{1}{2}} - v\Delta t \le x \le x_{i+\frac{1}{2}} - v\Delta t,$$
$$v_j - \Delta v_i^{n+1}/2 \le v \le v_j - \Delta v_i^{n+1}/2\}.$$

Here $\tilde{I}_{ij}^{n+1}$ denotes the set of the characteristic feet which come from the cell I_{ij}^{n+1}. Then, we obtain

$$\sum_i \begin{pmatrix} \rho_i^{n+1} \\ \rho_i^{n+1} U_i^{n+1} \\ 2E_i^{n+1} \end{pmatrix} \Delta x = \sum_i \sum_{j \in \mathcal{J}_i^{n+1}} \int_{\tilde{I}_{i,j}^{n+1}} \begin{pmatrix} P^n(x,v) \\ Q^n(x,v) \\ R^n(x,v) \end{pmatrix} dx\,dv$$

$$= \sum_i \sum_{k \in \mathcal{J}_i^n} \int_{I_{i,k}^n} \begin{pmatrix} P^n(x,v) \\ Q^n(x,v) \\ R^n(x,v) \end{pmatrix} dx\,dv = \sum_{ik} \begin{pmatrix} f_{ik}^n \\ v_k f_{ik}^n \\ v_k^2 f_{ik}^n \end{pmatrix} \Delta x \Delta v_i^n.$$

Note that the second equation holds only when the new phase space $\cup_i \mathcal{G}_i^{n+1}$ includes the old one $\cup_i \mathcal{G}_i^n$. In practice, here we may introduce small conservation errors due to the truncation of the velocity domain.

Remark 3.7 The proposed technique has been extended to attain second-order accuracy by adopting the DIRK2 or BDF2 method and linear reconstruction based on the minmod approach. The details can be found in [18].

3.4.3 Numerical Tests

In this section, we perform several tests checking the accuracy, efficiency, and robustness of our scheme. In all tests, the time step is obtained by fixing the CFL number. For simplicity, in each run we use the same value of time step throughout the whole calculation. The time step Δt is determined by

$$\mathrm{CFL} = \max_i \max_{v_j \in \mathcal{G}_i^0} |v_j| \frac{\Delta t}{\Delta x}.$$

For updating velocity grids in each time step, we use $\alpha = 10$, $\beta = 0.5$ for all numerical tests. To obtain initial steps f^1 for the BDF2-based approach, we have used a RK2-based high-order conservative SL scheme.

Accuracy and Conservation Test

Here we revisit numerical results in [18] that confirms the effect of parameters α and β. For this, we consider the same numerical test in Sect. 3.3.2. Here we use CFL= 4. For velocity, we initially set $N_v = 60$ grid points, and the local velocity approach is applied for time step $n > 1$. Initialization for $n = 1$ is obtained by using an RK2-based conservative SL scheme [41]. In order to show the robustness of our method, we check conservation in the presence of shock. To this purpose, we set $N_x = 80$ and take final time $T_f = 0.4$, which is long enough to observe a shock.

To measure the conservation error, we use

$$\mathrm{err}_\rho = \left| \frac{\sum_i (\rho_i^{N_t} - \rho_i^0)}{\sum_i \rho_i^0} \right|, \quad \mathrm{err}_{\rho U} = \left| \frac{\sum_i (\rho_i^{N_t} U_i^{N_t} - \rho_i^0 U_i^0)}{\sqrt{\sum_i \rho_i^0 \sum_i E_i^0}} \right|,$$

$$\mathrm{err}_E = \left| \frac{\sum_i (E_i^{N_t} - E_i^0)}{\sum_i E_i^0} \right|$$

where the time index N_t corresponds to the final time $T_f = 0.4$. In Fig. 14, it appears that the conservation is attained up to machine precision for mass and energy when $\alpha \geq 8$, while the conservation properties are almost independent on the parameter β that determines the resolution in velocity space. In this test, no positivity error has been observed. Here we note that conservation is attained for all $\beta = 0.75, 0.5, 0.25$.

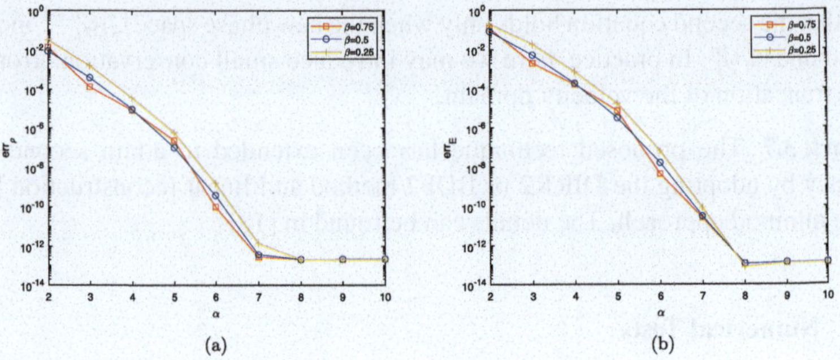

Fig. 14 Test 3.4.3. Conservation errors versus α for different values of β [18]. (**a**) Density. (**b**) Energy

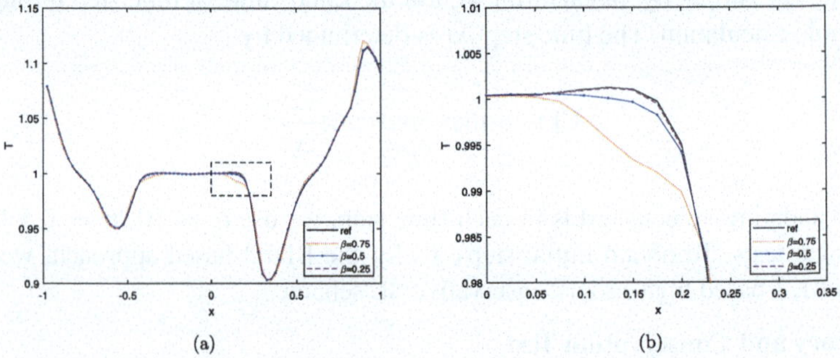

Fig. 15 Temperature profile in test 3.4.3 at final time $T_f = 0.4$. Comparison of numerical solution for $\alpha = 5$ and for various values of β [18]. Panel (**b**) is a zoom of panel (**a**)

In the following Fig. 15, we compare the solutions with a reference solution obtained by the BDF2-based SL scheme without grid adaptation. The result shows that $\beta = 0.75$ leads to noticeable errors in the solution, while $\beta \leq 0.5$ gives more accurate solutions. Although such numbers α and β are not optimized here, to balance the accuracy and efficiency of the method, we choose $\alpha = 10$ and $\beta = 0.5$ in the next test. Numerical results that confirm the second-order accuracy can be found in [18].

Two Interacting Blast Waves

For a high Mach number, the local velocity approach would be more efficient in that it makes use of fewer grid points. Here we consider a test called "the two interacting blast waves" [25]. The initial distribution is given by the Maxwellian with initial density $\rho_0(x) = 1$ and velocity $u_0(x) = 0$. Initial temperature is given by

$$T_0(x) = \begin{cases} 4.8, & \text{for } x \in [0, 0.1] \\ 4.8 \times 10^{-5}, & \text{for } x \in [0.1, 0.9] \\ 4.8 \times 10^{-1}, & \text{for } x \in [0.9, 1] \end{cases}$$

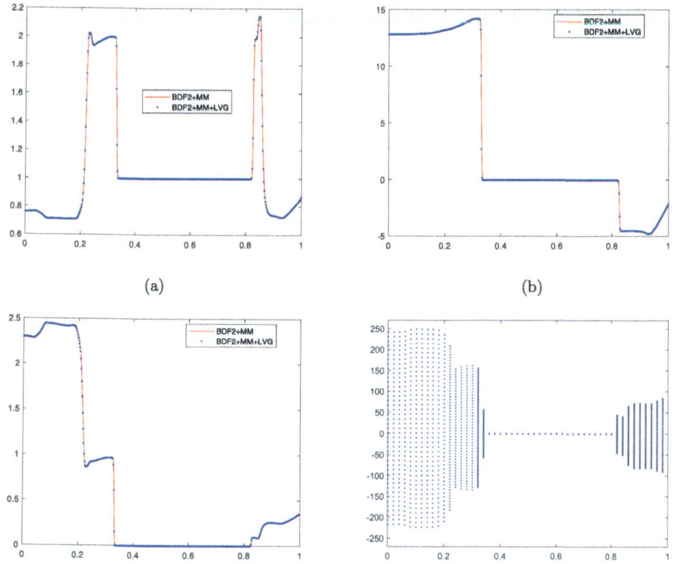

(a) (b)

(c) (d)

Fig. 16 Two interacting blast waves with $C = 1.08 \times 10^{-9}$ [18]. For a classical BDF2+MM scheme (red line) we use $N_v = 3800$ for each spatial node, while for a local velocity grid approaches we take an average of $N_v = 42$. (**a**) Density. (**b**) Velocity. (**c**) Temperature. (**d**) Grid points at final time

with gas constant $R = 208.1$. We impose the free-flow boundary condition on the interval $x \in [0, 1]$ and compute numerical solutions up to final time $T_f = 0.008$. We take uniform grids with $N_x = 500$ and fix a time step based on CFL= 2. We produce the result using $\varepsilon := CT^{\omega}/\rho$ with $\omega = -0.19$ and $C = 1.08 \times 10^{-9}$ as used in [25].

In this problem, the bounds of the global velocity grid of the classical SL schemes are fixed to be $[-190, 190]$ to guarantee the conservation up to machine precision. The size of velocity mesh is fixed by $\Delta v = 0.1$ which is small enough to satisfy the condition (40) and enables to resolve Maxwellian distribution corresponding to the smallest initial temperature.

In Fig. 16, we observe that the solution obtained by the local velocity grid approach is able to capture the correct shock position. Note that the use of coarse velocity grids produces some errors, which becomes negligible as the velocity grid is refined. Finally, in Fig. 16d we depict the local velocity grids used for different space positions. In this figure, along the vertical direction for each spatial cell, the set of local velocity grids is dotted. Note that the size of the mesh in the velocity direction varies depending on the temperature.

Although there are great savings in memory allocation, the method is not optimized in terms of CPU time. In order to make the method competitive also in terms of computational efficiency, one needs to adopt HPC tools in the implementation of the various numerical techniques included in the methodology.

4 BGK Models for Mixtures

In this section we investigate some properties of different consistent BGK models for inert gas mixtures available in literature, using high-order conservative SL schemes.

4.1 *Kinetic Boltzmann and BGK Equations*

Consistent BGK models for inert gas mixtures allow to capture the main properties of the Boltzmann equations which describe multicomponent rarefied gases. Consistency means that conservation laws, H-theorem, and equilibrium solutions which characterize the Boltzmann model are correctly reproduced by the BGK approximations. The generalization to mixtures of the BGK model for a single gas described in Sect. 2.1 is not obvious, and we will consider here three different consistent models available in literature. To briefly describe these models, it is useful to recall the Boltzmann description of a mixture of rarefied gases.

Let us consider a mixture of L monoatomic inert gases. Under the assumption that the sth gas has a mass $m_s > 0$, its dynamics can be described through the distribution functions $f_s(x, v, t)$, $s = 1, \cdots, L$ defined on phase space $(x, v) \in \mathbb{R}^3 \times \mathbb{R}^3$ at time $t > 0$, whose evolution is governed by the Boltzmann-type equations:

$$\frac{\partial f_s}{\partial t} + v \cdot \nabla_x f_s = Q_s, \tag{46}$$

where Q_s is the collision term of the sth species, which collects bi-species collision operators between sth and other kth gases:

$$Q_s = \sum_{k=1}^{L} Q_{sk}(f_s, f_k).$$

The binary collision operator Q_{sk} has the classical form

$$Q_{sk}(f_s, f_k) = \int_{\mathbb{R}^3 \times \mathbb{S}^2} dw\, d\omega\, g_{sk}\left(|y|, \hat{y} \cdot \omega\right) \left[f_s(v') f_k(w') - f_s(v) f_k(w) \right] \tag{47}$$

with a nonnegative scattering kernel g_{sk} depending on intermolecular potentials. Here, $w \in \mathbb{R}^3$, $\omega \in \mathbb{S}^2$ (the unit sphere), $y := v - w$ is the relative velocity, and $\hat{y} := y/|y|$; v' and w' are post-collisional velocities of sth and kth gases whose masses are m_s, m_k and pre-collisional velocities are v, w (for more details, see, for instance, [36] and references therein).

The distribution function f_s, $s = 1, \ldots, L$, can be used to reproduce s-species macroscopic quantities such as number density n_s, average velocity u_s, and absolute temperature T_s:

$$n_s = \langle f_s, 1 \rangle, \quad n_s u_s = \langle f_s, v \rangle, \quad 3 n_s K_B T_s = m_s \langle f_s, |v - u_s|^2 \rangle$$

where $\langle f, h \rangle := \int_{\mathbb{R}^3} dv f(v) h(v)$. Similarly, global macroscopic variables such as number density n, mass density ρ, velocity u, and temperature T of the mixture can be obtained as follows:

$$n = \sum_{s=1}^{L} n_s, \quad \rho = \sum_{s=1}^{L} \rho_s, \quad \rho_s = m_s n_s, \quad s = 1, \cdots, L$$

$$u = \frac{1}{\rho} \sum_{s=1}^{L} \rho_s u_s, \quad 3 n K_B T = 3 \sum_{s=1}^{L} n_s K_B T_s + \sum_{s=1}^{L} \rho_s |u_s - u|^2.$$

The equilibrium solution to (46) is given by Maxwellians which share a common velocity u and temperature T:

$$f_s^{eq} = n_s M\left(v; u, \frac{K_B T}{m_s}\right)$$

where

$$M(v; a, b) := \left(\frac{1}{2\pi b}\right)^{3/2} \exp\left(-\frac{1}{2b}|v - a|^2\right), \quad a \in \mathbb{R}^3,\ b > 0. \tag{48}$$

The structure of the collision operators Q_s allows to guarantee fundamental properties of the Boltzmann equations for inert gas mixtures: exact conservation laws, uniqueness of equilibrium solutions, and H-theorem for the relaxation to equilibrium. These properties will be reproduced by the BGK models we consider in this section.

4.1.1 The BGK Model of Andries, Aoki, and Perthame (AAP Model)

In [2], the first consistent BGK model for mixtures was proposed, which allows to reproduce the same exchange rate in momentum and energy of the Boltzmann equations (46). In this model the collision term Q_s of each Boltzmann equation (46) is replaced by a relaxation operator which drives the evolution towards an attracting auxiliary Maxwellian M^s, depending on fictitious parameters. In spite of such structural change, this BGK model still satisfies the main properties of the Boltzmann equation such as conservation laws, H-theorem, and indifferentiability principle. The scaled model equations are described by

$$\frac{\partial f_s}{\partial t} + v \cdot \nabla_x f_s = \tilde{Q}_s = \frac{\nu_s}{\varepsilon}\left(n_s M^s - f_s\right), \quad s = 1, \cdots, L, \tag{49}$$

where ε is the Knudsen number, ν_s is the collision frequency for s-species gas, and M^s is the attracting Maxwellian:

$$M^s = M\left(v; u^s, \frac{K_B T^s}{m_s}\right),$$

with M defined in (48). Notice that M^s is defined in terms of fictitious parameters u^s, T^s (different from the actual fields u_s, T_s), which, under the assumption of Maxwell molecules' interaction potential, are explicit functions of the actual moments of the distribution function; they are obtained by imposing that

$$m_s\langle \tilde{Q}_s, v\rangle = \frac{\nu_s}{\varepsilon} m_s n_s (u^s - u_s) = m_s \langle Q_s, v\rangle$$

$$\frac{m_s}{2}\langle |v|^2, \tilde{Q}_s\rangle = \frac{\nu_s}{\varepsilon}\left[\frac{3}{2} n_s K_B (T^s - T_s) + \frac{1}{2} m_s n_s (|u^s|^2 - |u_s|^2)\right] = \frac{m_s}{2}\langle |v|^2, Q_s\rangle.$$

The explicit expressions resulting from the previous equalities can be found in [2].

4.1.2 The BGK Model Preserving Global Conservation (GS Model)

Another BGK-type model with one attracting Maxwellian for each species has been proposed in [14, 65]. The fictitious parameters are determined by imposing the same conservation laws of the Boltzmann equation (46), namely, species number densities, global momentum, and total kinetic energy. In [14, 65], the model has been originally designed to describe a bimolecular reversible chemical reaction in a four-species mixture; the model has been then adapted to a general L species inert gas mixture in [65]. The scaled model reads

$$\frac{\partial f_s}{\partial t} + v \cdot \nabla_x f_s = \frac{\nu_s}{\varepsilon}\left(n_s M^s_{GS} - f_s\right). \tag{50}$$

where ν_s is the collision frequency for s-species gas and M^s_{GS} is the attracting Maxwellian:

$$M_{GS}^{s} = M\left(v; \bar{u}, \frac{K_B \bar{T}}{m_s}\right), \tag{51}$$

where M is defined in (48). Note that it depends only on the auxiliary parameters $\bar{u}$ and $\bar{T}$, which are determined by imposing the conservation of total momentum and energy:

$$\sum_{s=1}^{L} \int_{\mathbb{R}^3} m_s \begin{pmatrix} v \\ |v|^2/2 \end{pmatrix} \left(n_s M_{GS}^{s} - f_s\right) dv = 0.$$

Consequently, easy computation allows to obtain the expressions of $\bar{u}$ and $\bar{T}$ in terms of the actual macroscopic fields u_s and T_s, and they can be found in [65]; in that paper, it is also proved that the positivity of auxiliary temperature $\bar{T}$ is guaranteed and the H-theorem holds for the space homogeneous case.

4.1.3 A General Consistent BGK Model for Inert Gas Mixtures (BBGSP Model)

In [16], authors introduce a different BGK-type model whose BGK operators Q_s mimic the structure of the Boltzmann ones, namely, are sums of bi-species operators Q_{sk}, each of them prescribing the same exchange rates of the corresponding term of the Boltzmann equations. These exchange rates can be computed exactly and made explicit in case of Maxwell molecules' interaction potentials. This BBGSP model also satisfies the main qualitative properties of Boltzmann equation such as conservation laws, H-theorem, and indifferentiability principle. The scaled equations read

$$\frac{\partial f_s}{\partial t} + v \cdot \nabla_x f_s = \frac{1}{\varepsilon} \sum_{k=1}^{L} \nu_{sk} \left(n_s M_{sk} - f_s\right), \quad s = 1, \cdots, L, \tag{52}$$

with

$$M_{sk} = M\left(v; u_{sk}, \frac{K_B T_{sk}}{m_s}\right),$$

where M is defined in (48). The auxiliary parameters u_{sk} and T_{sk} are defined by

$$\begin{aligned} u_{sk} &= (1 - a_{sk}) u_s + a_{sk} u_k \\ T_{sk} &= (1 - b_{sk}) T_s + b_{sk} T_k + \frac{\gamma_{sk}}{K_B} |u_s - u_k|^2 \end{aligned} \tag{53}$$

with

$$a_{sk} = \frac{\lambda_{sk} n_k m_k}{\nu_{sk}(m_s + m_k)}, \quad b_{sk} = \frac{2a_{sk} m_s}{m_s + m_k}, \quad \gamma_{sk} = \frac{m_s a_{sk}}{3}\left(\frac{2m_k}{m_s + m_k} - a_{sk}\right), \tag{54}$$

where λ_{sk} are constants in case of Maxwell molecules (for other interaction potentials we refer to Section 3 Equations (28)–(29) in [16]), and ν_{sk} are collision frequencies that have to be properly chosen in order to fulfill consistency properties. In [16], authors proved that the positivity of T_{sk} is guaranteed by

$$T_s > 0, \quad T_k > 0, \quad \nu_{sk} \geq \frac{1}{2}\lambda_{sk} n_k.$$

Considering this, throughout this section, we set $\nu_{sk} = \nu_0^{sk} n_k = \lambda_{sk} n_k$. This implies

$$a_{sk} = \frac{m_k}{m_s + m_k}, \quad b_{sk} = \frac{2a_{sk} m_s}{m_s + m_k}, \quad \gamma_{sk} = \frac{m_s a_{sk}}{3}\left(\frac{2m_k}{m_s + m_k} - a_{sk}\right).$$

The discrepancy between the above BGK models for inert mixtures has been analytically investigated in [17]. The analysis shows that the AAP model (49) and the BBGSP model (52) have little discrepancy, visible only when there is a large variance in the macroscopic species velocities u_s. On the contrary, a more pronounced discrepancy appears between AAP and GS models (details can be found in [17], Section 4 and Proposition 1); this will be confirmed and discussed later in the numerical tests.

4.2 Hydrodynamic Limits at Navier-Stokes Level

An interesting point of (52) is the possibility of allowing different hydrodynamic limits, thanks to the structure of the BBGSP collision operators as a sum of bispecies relaxation terms.

In [11], the hydrodynamic limit of the BBGSP model (52)–(54) at the Navier-Stokes level is derived using the Chapman-Enskog expansion in the regime in which all collisions are dominant. The equations for macroscopic (hydrodynamic) variables n_s, u, and T, obtained as ε-order closure of the macroscopic equations (moments of the BGK ones), are given by

$$
\begin{aligned}
&\frac{\partial n_s}{\partial t} + \nabla \cdot (n_s u) + \varepsilon \nabla \cdot (n_s u_s^{(1)}) = 0, \qquad s = 1, \cdots, L \\
&\frac{\partial}{\partial t}(\rho u) + \nabla \cdot (\rho u \otimes u) + \nabla (n K_B T) + \varepsilon \nabla \cdot (P^{(1)}) = 0, \\
&\frac{\partial}{\partial t}\left(\frac{1}{2}\rho |u|^2 + \frac{3}{2} n K_B T\right) + \nabla \cdot \left[\left(\frac{1}{2}\rho |u|^2 + \frac{5}{2} n K_B T\right) u\right] \\
&\qquad\qquad + \varepsilon \nabla \cdot (P^{(1)} \cdot u) + \varepsilon \nabla \cdot q^{(1)} = 0,
\end{aligned}
\tag{55}
$$

where $u_s^{(1)}$, $P^{(1)}$, $q^{(1)}$ are first-order corrections with respect to ε. The diffusion velocity $u_s^{(1)}$ is given by Fick's law:

$$
u_s^{(1)} = \sum_{k=1}^{L} \frac{L_{sk}}{\rho_s \rho_k} \nabla (n_k K_B T) \tag{56}
$$

with L symmetric matrix (see [11, 17] for details). The first-order correction for pressure tensor $P^{(1)}$ is of the following form:

$$
P_{\alpha\beta}^{(1)} = -\mu \left(\frac{\partial u_\alpha}{\partial x_\beta} + \frac{\partial u_\beta}{\partial x_\alpha} - \frac{2}{3} \nabla \cdot u \delta_{\alpha\beta} \right), \qquad 1 \le \alpha, \beta \le 3, \tag{57}
$$

where the viscosity coefficient μ is given by

$$
\mu := \sum_{s=1}^{L} \frac{n_s K_B T}{\sum_{s=1}^{L} \nu_{sk}^{(0)}},
$$

with $\nu_{sk}^{(0)}$ the leading order term in the expansion of the collision frequency ν_{sk}. Finally, the heat flux $q^{(1)}$ is given by

$$
q^{(1)} = \frac{5}{2} K_B T \sum_{s=1}^{L} n_s u_s^{(1)} - \lambda \nabla T, \tag{58}
$$

where λ is the thermal conductivity coefficient:

$$
\lambda = \frac{5}{2} K_B^2 T \sum_{s=1}^{L} \frac{n_s}{m_s \sum_{k=1}^{L} \nu_{sk}^{(0)}}.
$$

For a detailed description of these results, we refer to [11]. It is remarkable that they are in complete agreement with those obtained from the AAP model (49). In the case of the Boltzmann equation with Maxwellian particles, one obtains the same system, but with different viscosity and thermal conductivity (see [2], Remark 4.3).

Classical Euler equations for a gas mixture can be obtained from (55) by putting $\varepsilon = 0$.

The same structure of the Navier-Stokes equations (55), with first-order corrections (56), (57), and (58), is reproduced only qualitatively by the ε-order asymptotics of the GS model (50)–(51). In fact, the matrix involved in Fick's law (56) and consequently the diffusion velocities are different from the ones deduced from both AAP and BBGSP models (see [15]), while viscosity η and thermal conductivity λ coincide for the three BGK models.

In [13], the authors consider the case in which intraspecies collisions are the dominant process in the evolution of the mixture. This occurs, for instance, in the so-called ε-mixtures of heavy and light gases [61], where molecules with very disparate masses exchange energy more slowly than molecules of the same species, and also in some applications to plasmas and astrophysics [110]. In this case, it is possible to define a proper Knudsen number and obtain the adimensional scaled equations:

$$\frac{\partial f_s}{\partial t} + \mathbf{v} \cdot \nabla_{\mathbf{x}} f_s = \frac{1}{\varepsilon} \nu_{ss} \left(n_s M_{ss} - f_s\right) + \frac{1}{\kappa} \sum_{k \neq s}^{L} \nu_{sk} \left(n_s M_{sk} - f_s\right). \tag{59}$$

This scaling leads to a Navier-Stokes model of multi-velocity and multi-temperature for s-species gases. In this case, the equations for macroscopic (hydrodynamic) variables n_s, u_s, and T_s are given by

$$\begin{aligned}
&\frac{\partial n_s}{\partial t} + \nabla \cdot (n_s u_s) = 0, \\
&\frac{\partial}{\partial t}(\rho_s u_s) + \nabla \cdot (\rho_s u_s \otimes u_s) + \nabla(n_s K_B T_s) + \varepsilon \nabla \cdot (P_s^{(1)}) = \frac{1}{\kappa} \sum_{k \neq s}^{L} \mathcal{R}_{sk}, \\
&\frac{\partial}{\partial t}\left(\frac{1}{2}\rho_s |u_s|^2 + \frac{3}{2} n_s K_B T_s\right) + \nabla \cdot \left[\left(\frac{1}{2}\rho_s |u_s|^2 + \frac{5}{2} n_s K_B T_s\right) u_s\right] \\
&\qquad + \varepsilon \nabla \cdot (P_s^{(1)} \cdot u_s) + \varepsilon \nabla \cdot q_s^{(1)} = \frac{1}{\kappa} \sum_{k \neq s}^{L} \mathcal{S}_{sk},
\end{aligned} \tag{60}$$

for $s = 1, \cdots, L$, with source terms

$$\begin{aligned}
\mathcal{R}_{sk} &= \lambda_{sk} m_{sk} n_s n_k (u_k - u_s), \\
\mathcal{S}_{sk} &= \lambda_{sk} \frac{m_{sk}}{m_s + m_k} n_s n_k \left[(m_s u_s + m_k u_k) \cdot (u_k - u_s) + 3 K_B (T_k - T_s)\right] \\
m_{sk} &= \frac{m_s m_k}{m_s + m_k}.
\end{aligned}$$

The expressions of the pressure tensors and of the heat fluxes can be found in [13] (see also Sect. 5.2 in [17]).

Correspondingly, multi-temperature Euler equations can be obtained by putting $\varepsilon = 0$ in (60). Multi-temperature descriptions at Euler level can be obtained also in the framework of extended thermodynamics [97, 104]. It is worth noticing that, according to [81], shock structure in helium-argon mixtures can be better reproduced by the multi-temperature description. In numerical tests, we will also numerically deal with the helium-argon mixtures.

4.3 Numerical Test

In this section, we present several numerical examples. First, we numerically check the discrepancies between the three BGK-type models for inert gas mixtures given by (49), (50), and (52). Second, concerning the scaled model (59) we approximate the two corresponding systems of NS equations (55) and (60). Third, we consider a binary mixture of noble gases with large mass ratio, in which the multi-velocity and multi-temperature description (59) and (60), respectively, may explain better the behavior of gases.

To compute numerical solutions to (49) and (50), we consider a SL method introduced in [65]. Since the method has been introduced with a nonconservative reconstruction, to make the method conservative we adopt a technique introduced in [40, 41]. In particular, we make use of Q-CWENO23 and Q-CWENO35 reconstructions, which are based on CWENO23 [77] and CWENO35 [44], respectively. For details, we refer to [40, 41]. For (52), we use a conservative SL method introduced in [39]. For the time discretization, we consider an implicit Runge-Kutta method (DIRK) and a backward difference formula (BDF). In particular, we here consider a second-order DIRK method and a third-order BDF3 method in [39].

Note that we perform numerical simulations based on Chu reduction [43] as in [39, 65], which allows to reduce the problem from 3D to 1D in velocity and space, under suitable symmetry assumptions. Below we list the name of schemes, which will be used in this section:

1. RK2-QCWENO23: DIRK2 with Q-CWENO23
2. BDF3-QCWENO35: BDF3 with Q-CWENO35

For discretization of the space and velocity (1D) domain, we use N_x and $N_v + 1$ grid points with uniform mesh sizes Δx and Δv, respectively. Based on this, we will use grid points x_i and v_j over computation domain $[x_{min}, x_{max}] \times [v_{min}, v_{max}]$. To fix a time step Δt, we use a CFL number defined by

$$\mathrm{CFL} = \max\{|v_{min}|, |v_{max}|\} \frac{\Delta t}{\Delta x}.$$

To compute the solutions of NS equations (55) and (60), we use spectral methods, which make the treatment of the many derivatives appearing there easier. Let us consider a general representation of the two NS equations (in equivalent form, see [17] for details):

$$U_t = F(U, U_x, U_{xx}), \tag{61}$$

where $U := U(x,t)$ is assumed to be smooth and periodic on the spatial domain. Given values of $\{U_i^n\}$, we first compute Fourier coefficients $\hat{U}_k$, $k = -N_x/2, \cdots, N_x/2 - 1$ using the fast Fourier transform (FFT). Then, we compute the nth spatial derivatives of functions involved in equations of interest using the inverse fast Fourier transform of $\{(jk)^n \hat{U}_k\}$ where j denotes the imaginary number. For the time integration in (61), we adopt an explicit RK4 scheme.

4.3.1 Comparison Among the BGK Models

Here we numerically investigate the discrepancy between the three BGK-type models for gas mixture. For this, as in [1, 65], we consider a mixture of four monoatomic gases whose molecular masses are given by

$$m_1 = 58.5, \quad m_2 = 18, \quad m_3 = 40, \quad m_4 = 36.5. \tag{62}$$

For the choice of the collision frequencies we refer to the values in [17].

To show the discrepancy between the AAP model (49) and the BBGSP model (52), which as already mentioned in Sect. 4.1 becomes apparent when there is a large variance in the macroscopic velocities of gases, we set as initial data Maxwellians whose macroscopic fields are given by

$$\begin{gathered} n_0^s(x) = \frac{1}{m_s}, \quad T_0^s(x) = \frac{4}{\sum_{s=1}^4 n_0^s}, \\ u_0^s(x) = \frac{\eta_s}{\sigma_s}\left[\exp\left(-\left(\sigma_s x - 1 + \frac{s}{3}\right)^2\right) + \exp\left(-\left(\sigma_s x + 3 - \frac{s}{10}\right)^2\right)\right] \end{gathered} \tag{63}$$

where $\sigma_s = (10, 13, 16, 9)$ and $\eta_s = (-30, -10, 10, 30)$, for $s = 1, \cdots, 4$. We impose periodic boundary conditions on the space domain $[-1, 1]$ and truncate velocity domain by $[-15, 15]$. To compute numerical solutions, we use RK2-QCWENO23 for $N_x = 200$ and $N_v = 60$. We first check the discrepancy of the three models at final time $t_f = 0.04$. For this, we use CFL $= 0.2$ up to $t = 0.004$ and CFL$= 2$ for $t \in [0.004, 0.04]$ in order to be able to resolve the relaxation towards local equilibrium.

In Figs. 17 and 18, we plotted the numerical solutions of the four-gas mixture obtained with the three BGK methods at $t_f = 0.2$, using CFL$= 0.2$ up to $t = 0.02$ and CFL$= 2$ for $t \in [0.02, 0.2]$. Since the heaviest ($s = 1$) and the lightest ($s = 2$) gases show the main largest differences in the behavior compared to the

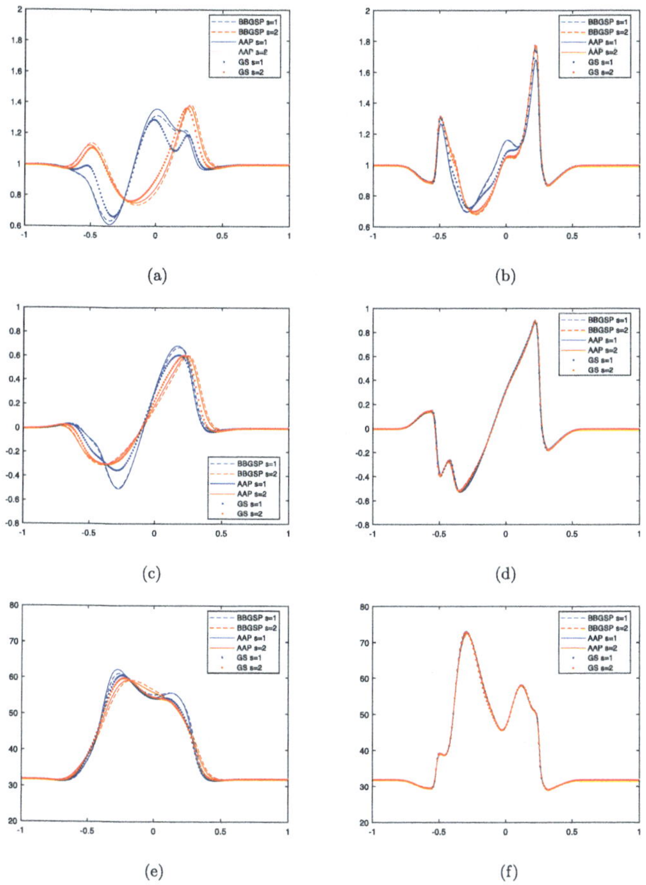

Fig. 17 Comparison of the three BGK models for $\varepsilon = 10^{-2}$ (left) and $\varepsilon = 10^{-3}$ (right) with initial data in (63). (**a**) Density ρ_s, $s = 1, 2$. (**b**) Density ρ_s, $s = 1, 2$. (**c**) Velocity u_s, $s = 1, 2$. (**d**) Velocity u_s, $s = 1, 2$. (**e**) Temperature T_s, $s = 1, 2$. (**f**) Temperature T_s, $s = 1, 2$

other two gases, we only report these two components in Figs. 17 and 18. The results show that some difference appears for $\varepsilon = 10^{-2}$, especially between GS models and the other ones, which instead remain closer to each other. This is in agreement with the results mentioned in Sect. 4.2, and it is due to the fact that the GS model (50) has a quantitatively different hydrodynamic Navier-Stokes limit with respect to AAP and BGK models, which instead share the same hydrodynamics up to Navier-Stokes level. Anyway, all species velocities and temperatures obtained from the three models are close to global velocity and temperature already from $\varepsilon = 10^{-3}$, but densities equalize slower. Although we took smooth initial data, it is noticeable that shocks appear as time flows around $x = 0.27$. Figure 18a shows even that densities are almost overlapped for $\varepsilon = 10^{-4}$. Here we omit the profiles of temperature since they show similar trends of Fig. 17f.

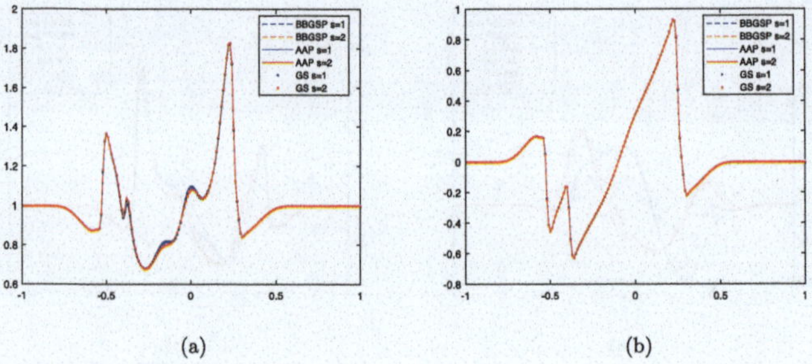

Fig. 18 Comparison of the three BGK models for $\varepsilon = 10^{-4}$ with initial data in (63). (**a**) Density ρ_s, $s = 1, 2$. (**b**) Velocity u_s, $s = 1, 2$

4.4 Comparison with NS Equations

Here we focus on the different hydrodynamic limits that can be obtained from the BGK model (52) to highlight their peculiar behaviors.

4.4.1 Case 1: Global Velocity and Temperature

In this test, we check the capability to capture the hydrodynamic limit (55) by solving model (52). For this, we consider the same test proposed in [1, 65]. Here we consider the mixture of four monoatomic gases with molecular masses in (62).

We set initial data as Maxwellians whose macroscopic fields are

$$\begin{aligned} n_0^s(x) &= \frac{1}{m_s}, \quad T_0^s(x) = \frac{4}{\sum_{s=1}^4 n_0^s}, \\ u_0^s(x) &= \frac{s}{\sigma_s}\left[\exp\left(-\left(\sigma_s x - 1 + \frac{s}{3}\right)^2\right) + \exp\left(-\left(\sigma_s x + 3 - \frac{s}{10}\right)^2\right)\right] \end{aligned} \tag{64}$$

for $s = 1, \cdots, 4$, where $\sigma_s = (10, 13, 16, 19)$. To confirm the agreement between the limiting solutions to (52) and the solutions to NS equations (55), we consider Knudsen numbers $\varepsilon = 10^{-k}$, $k = 2, 4$. Here we compute numerical solutions to (52) using $N_x = 500$ and $N_v = 60$. We take CFL$= 0.2$ up to $t = 0.02$ and CFL$= 2$ in $t \in [0.02, 0.2]$.

In Figs. 19 and 20, we compare the behaviors of densities, global velocity, and temperature. In case of density, we focus on the comparison of the behaviors of the heaviest ($s = 1$) and the lightest ($s = 2$) gases which show the largest difference. Figure 19 shows that for these parameters and initial data BGK solutions (52) and NS solutions (55) are quite different for $\varepsilon = 10^{-2}$. However, the results in Fig. 20 show that both solutions give similar values of macroscopic quantities as we take

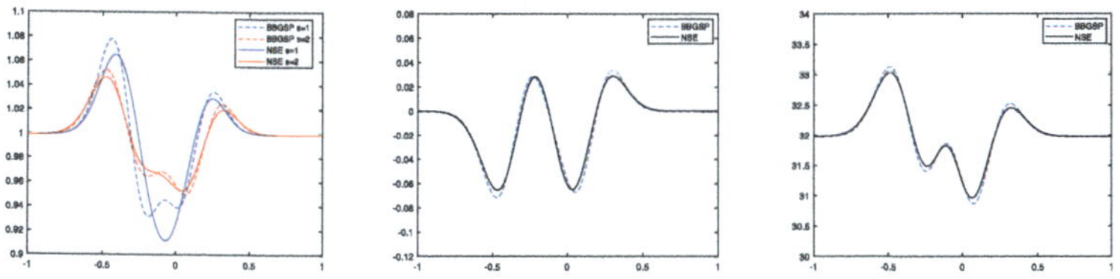

(a) (b) (c)

Fig. 19 Comparison of BGK model (52) and NS equations (55) for $\varepsilon = 10^{-2}$ with initial data in (64). (**a**) Density ρ_s, $s = 1, 2$. (**b**) Velocity u. (**c**) Temperature T

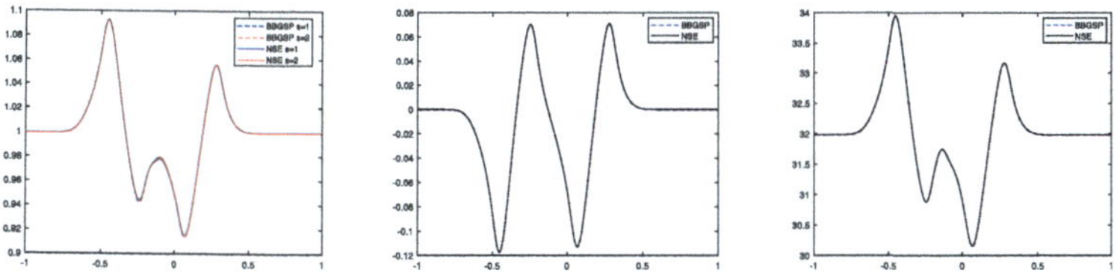

(a) (b) (c)

Fig. 20 Comparison of BGK model (52) and NS equations (55) for $\varepsilon = 10^{-4}$ with initial data in (64). (**a**) Density ρ_s, $s = 1, 2$. (**b**) Velocity u. (**c**) Temperature T

smaller Knudsen numbers, and all fields overlap for a sufficiently small Knudsen number $\varepsilon = 10^{-4}$.

4.4.2 Case 2: Multi-Velocity and Multi-Temperature

Here we consider the case in which intraspecies collisions are dominant, and hence, we impose $\kappa \neq \varepsilon$ in (59) so that the scaled version of the BBGSP model (52) leads to the multi-velocity and multi-temperature NS equations (60) for small values of ε. Here, we take Knudsen number $\varepsilon = 10^{-k}$, $k = 2, 4$ and set $\kappa = 1$. We compute numerical solutions with the same numerical setting of Sect. 4.4.1.

In Figs. 21 and 22, we show species velocity and species temperature as well as densities. As in the previous test, we focus on the behaviors of the heaviest gas ($s = 1$) and the lightest one ($s = 2$). In Fig. 21, we can observe, contrary to the previous case (which can be reproduced by taking $\kappa = \varepsilon$), a clear separation between macroscopic quantities of the different species; moreover, the behavior of the BGK solutions of (59) is similar to those of multi-velocity and multi-temperature NS description (60). We note that in Fig. 22 as ε becomes smaller, the kinetic and macroscopic solutions show very good agreement in each species velocity and temperature.

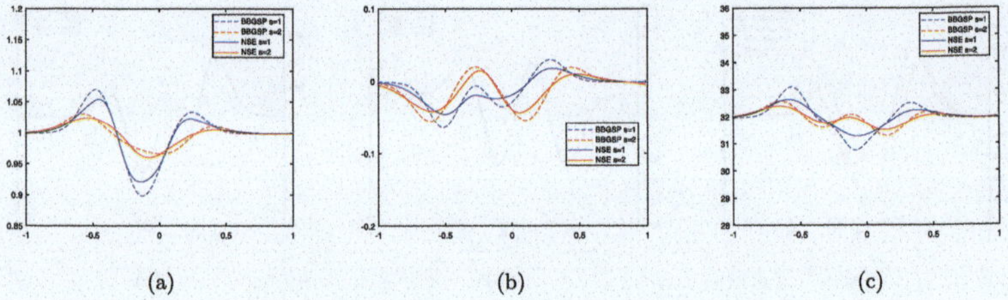

Fig. 21 Comparison of the scaled BBGSP model (59) and Navier-Stokes equations (60) with $\kappa = 1$ for $\varepsilon = 10^{-2}$ with initial data in (64). (**a**) Density ρ_s, $s = 1, 2$. (**b**) Velocity u_s, $s = 1, 2$. (**c**) Temperature T_s, $s = 1, 2$

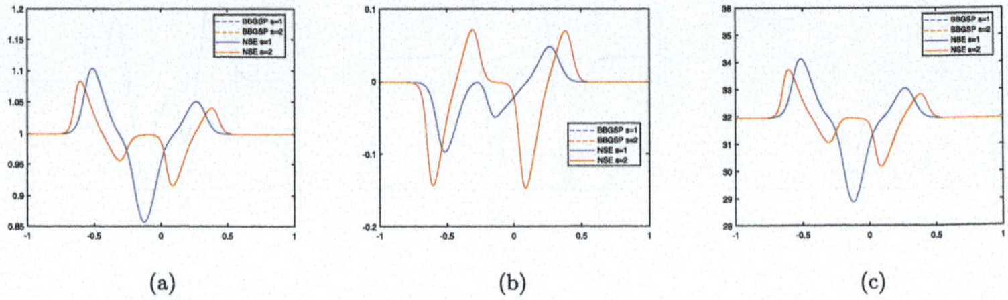

Fig. 22 Comparison of the scaled BBGSP model (59) and Navier-Stokes equations (60) with $\kappa = 1$ for $\varepsilon = 10^{-3}$ (left) and $\varepsilon = 10^{-4}$ (right) with initial data in (64). (**a**) Density ρ_s, $s = 1, 2$. (**b**) Velocity u_s, $s = 1, 2$. (**c**) Temperature T_s, $s = 1, 2$

4.5 Shock Problem for Binary Gas Mixture

In this test, we consider a shock problem for the binary mixture of two noble gases, helium (He) and argon (Ar), for which mass ratio becomes relatively large. We aim at checking in this real case which Euler description (for global or species macroscopic fields) is better capable of capturing the behavior of the mixture.

To perform the realistic simulation, we consider the molecular mass of He and Ar as follows:

$$m_1^m \approx 4 \times 10^{-3} Kg/mol \, (He), \quad m_2^m \approx 40 \times 10^{-3} Kg/mol \, (Ar).$$

In view of this, we rewrite the BBGSP model (52) in terms of mole. We divide f_s by Avogadro's number $N_a \approx 6.02 \times 10^{23}$. Now, we denote the molecular density by

$$f_s^m(\mathbf{x}, \mathbf{v}, t) := \frac{f(\mathbf{x}, \mathbf{v}, t)}{N_a}$$

for $s = 1, 2$. Then, we can rewrite (52) as

$$\frac{\partial f_s^m}{\partial t} + \mathbf{v} \cdot \nabla_{\mathbf{x}} f_s^m = \frac{1}{\varepsilon} \nu_{ss} \left(n_s M_{ss} - f_s^m \right) + \frac{1}{\kappa} \sum_{k \neq s}^{L} \nu_{sk} \left(n_s M_{sk} - f_s^m \right),$$

$$M_{sk} = M \left(\mathbf{v}; u_{sk}^m, \frac{RT_{sk}^m}{m_s^m} \right),$$

with macroscopic quantities at the level of mole:

$$n_s^m = \langle f_s^m, 1 \rangle, \quad n_s^m u_s^m = \langle f_s^m, \mathbf{v} \rangle, \quad 3n_s^m RT_s^m = m_s^m \langle f_s^m, |\mathbf{v} - u_s|^2 \rangle.$$

Here we define the universal gas constant R as $R := K_B N_a \approx 8.3145$ J/mol and the molecular mass m_s^m as $m_s^m = m_s N_a$. Note that the values of u_{sk}^m and T_{sk}^m are samely defined as in (53) by using m_s^m, n_s^m, u_s^m, T_s^m. For global macroscopic variables, we use the following expressions:

$$n^m = \sum_{s=1}^{L} n_s^m, \quad \rho^m = \sum_{s=1}^{L} \rho_s^m, \quad \rho_s^m = m_s^m n_s^m, \quad s = 1, \cdots, L$$

$$u^m = \frac{1}{\rho^m} \sum_{s=1}^{L} \rho_s^m u_s^m, \quad 3n^m RT^m = 3 \sum_{s=1}^{L} n_s^m RT_s^m + \sum_{s=1}^{L} \rho_s^m |u_s^m - u^m|^2.$$

Let us consider room temperature $T_s^m = 300$ K. As in [12], here we use the collision frequencies corresponding to $T_s^m = 300$ K based on the following formula:

$$\nu_0^{ss} = \frac{4}{3} \frac{T}{\mu_s(T)}, \qquad s = 1, 2$$

$$\nu_0^{12} = \frac{2\sqrt{2}}{3} \frac{(m_1 + m_2)^{\frac{1}{4}}}{(m_1 m_2)^{\frac{1}{2}}} \frac{T}{(\mu_1(T) \mu_2(T))^{\frac{1}{2}}},$$

where viscosity coefficients μ_s for noble gases are provided in [75]. In case of helium and argon, we have

$$\nu_0^{11} = 19.96, \quad \nu_0^{12} = 1.153$$
$$\nu_0^{21} = 1.153, \quad \nu_0^{22} = 17.52.$$

Note that we set $\nu_{sk} = \nu_0^{sk} n_k$. Now, we consider a shock problem by taking the Maxwellian as initial data which reproduces the following macroscopic variables:

$$\left(\rho_0^m, u_0^m, T_0^m \right) = \begin{cases} (1.7628, \, 0, \, 300), & x < 0.5 \\ (0.8814, \, 0, \, 300), & x > 0.5. \end{cases}$$

The units of ρ_0^m, u_0^m, T_0^m are kg/m^3, m/s, K, respectively. Now, we set

$$(\rho_{01}^m, \rho_{02}^m) = \begin{cases} (0.1598,\ 1.6030), & x < 0.5 \\ (0.0799,\ 0.8015), & x > 0.5 \end{cases}.$$

Note that the density of helium and argon gases for $T_0^m = 300K$ for $p^m = 1\,bar$ are given by

$$\rho_1^m = 0.1598\,kg/m^3, \quad \rho_2^m = 1.603\,kg/m^3.$$

For numerical simulation, we assume the free-flow boundary condition on the spatial domain $x \in [-6, 6]$ with velocity domain $v \in [-160, 160]$. We compute numerical solutions with $N_x = 600$ and $N_v = 320$ up to $t_f = 0.06$. Here we use CFL=1.5 and take different values of $\varepsilon = 10^{-q}$, $q = 3, 4, 5$. Numerical solutions to the BBGSP model is computed with a conservative SL scheme based on BDF3-QCWENO35, while the solutions to global velocity and temperature Euler system and multi-velocity and multi-temperature Euler system are obtained by a high-order finite difference WENO scheme with the Runge-Kutta method.

In Fig. 23, for relatively large values of $\kappa = 10^{-3}$, the panels on the left column show that the global velocity and temperature is closely related to the dynamics of argon gas. This is because its density is ten times bigger than that of helium gas. Thus, it is difficult to describe the dynamics of mixtures involving helium gas with global velocity and temperature description. On the other hand, the panels on the right column enable us to capture the behaviors of helium gas and this is the case where multi-velocity and multi-temperature description can be a suitable model for a better description of this dynamics. In the following Fig. 24, relevant to a smaller $\kappa = 10^{-5}$, both species velocities and temperatures are close to global velocity and temperature. Here the global velocity and temperature Euler system can be already a suitable choice for describing the dynamics of binary mixtures.

Remark 4.1 When the mass ratio of two gases becomes much larger, the species velocities and temperatures could be quite different from each other. This may create a numerical difficulty because the computational cost and memory allocation requirements become prohibitively large due to the use of a fixed velocity grid. As a remedy, one can adopt the local velocity grid approach. As related works, we refer to [18, 25].

5 Boltzmann Equation

The Boltzmann transport equation (BTE) has been well known for its good predictability of dynamics of rarefied gas, when the mean free path of the molecules is not negligible compared to macroscopic length scales [36]. The BTE reads

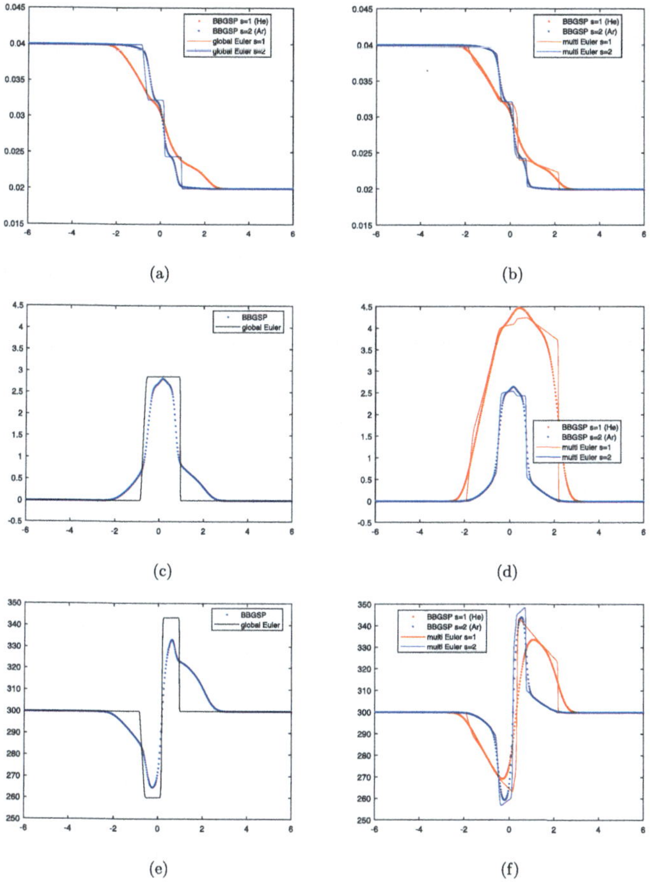

Fig. 23 Comparison of the numerical solution of the scaled BBGSP model (59) for $\varepsilon = \kappa = 10^{-3}$ with (left) global velocity and temperature Euler system (55) for $\varepsilon = 0$ and (right) multi-velocity and multi-temperature Euler system (60) for $\varepsilon = 0$, $\kappa = 10^{-3}$. We use the initial data in (4.5). (**a**) Number density n_s, $s = 1, 2$. (**b**) Number density n_s, $s = 1, 2$. (**c**) Velocity u. (**d**) Velocity u_s, $s = 1, 2$. (**e**) Temperature T. (**f**) Temperature T_s, $s = 1, 2$

$$\frac{\partial f}{\partial t} + v \cdot \nabla_x f = \frac{1}{\kappa} Q(f),$$

where f is the velocity distribution defined on the phase space $(x, v, t) \in \mathbb{R}^{d_x} \times \mathbb{R}^{d_v} \times \mathbb{R}$ and $\nabla_x = (\partial_{x_1}, \ldots, \partial_{x_{d_x}})$.

Here we use the same abuse of notation we adopted for the BGK equation (see Sect. 2.1), i.e., we assume that the distribution function f depends only on the first d_x space coordinates, with $d_x \in \{1, 2, 3\}$, while it depends on $d_v \in \{2, 3\}$ velocity coordinates.

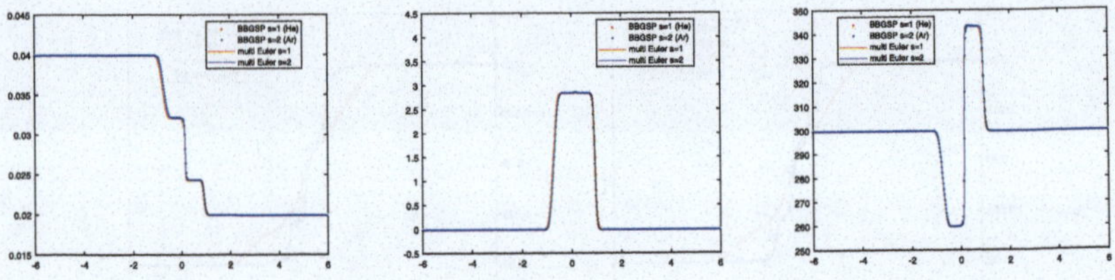

(a) (b) (c)

Fig. 24 Comparison of the scaled BBGSP model (59) for $\varepsilon = \kappa = 10^{-5}$ with multi-velocity and multi-temperature Euler system (60) for $\varepsilon = 0$, $\kappa = 10^{-5}$. (**a**) Number density n_s, $s = 1, 2$. (**b**) Velocity u_s, $s = 1, 2$. (**c**) Temperature T_s, $s = 1, 2$

The Boltzmann collision operator $Q(f)$ describes the binary collision of gas molecules:

$$Q(f) = \int_{\mathbb{R}^{d_v}} \int_{\mathbb{S}^{d_v-1}} B(|v - v_*|, \omega) \left(f'_* f' - f_* f \right) d\omega dv_* \tag{65}$$

where B is the collision kernel, which depends on the relative velocity of the colliding particles and on the deflection angle after the collision, and $\omega \in \mathcal{S}^{d_v-1}$ denotes a unit vector on a sphere. In (65), the pre-collision velocities v, v_* and post-collision velocities v' and v'_* are related to preserve momentum and energy:

$$v + v_* = v' + v'_*, \quad |v|^2 + |v_*|^2 = |v'|^2 + |v'_*|^2$$

i.e.,

$$v' = \frac{1}{2}\left(v + v_* + |v - v_*|\omega\right), \quad v'_* = \frac{1}{2}\left(v + v_* - |v - v_*|\omega\right).$$

We recall that the Knudsen number κ, i.e., the ratio between a typical mean free path of the molecules and a typical macroscopic length, determines the degree of rarefaction of the gas. The operator $Q(f)$ has the following collision invariance: $1, v, |v|^2$,

$$\int_{\mathbb{R}^{d_v}} Q(f) \begin{pmatrix} 1 \\ v \\ |v|^2 \end{pmatrix} dv = \mathbf{0},$$

which leads to the conservation of mass, momentum, and energy. Moreover, BTE satisfies the H-theorem which states that the entropy function, defined by

$$H(f) = \int_{\mathbb{R}^{d_x} \mathbb{R}^{d_v}} f \ln f \, dv \, dx,$$

is a nonincreasing function of time, i.e., $\frac{dH}{dt} \leq 0$, and the equality holds if and only if $Q(f) = 0$ (see [109]). It is possible to show that this relation holds if and only if the distribution function f has the form of

$$\ln f = a|v|^2 + b \cdot v + c.$$

Here the coefficients $a, c \in \mathbb{R}$, $b \in \mathbb{R}^{d_v}$ are related to macroscopic variables, density ρ, mean velocity U, and temperature T by

$$\rho(x,t) = \int_{\mathbb{R}^{d_v}} f dv, \quad \rho(x,t)U(x,t) = \int_{\mathbb{R}^{d_v}} v f dv,$$
$$d_v \rho(x,t) RT(x,t) = \int_{\mathbb{R}^{d_v}} |v - U(x,t)|^2 f dv,$$

where R denotes the gas constant. Therefore, any equilibrium function has the form of local Maxwellian:

$$\mathcal{M}(f)(x,v,t) := \frac{\rho(x,t)}{(2\pi RT(x,t))^{d_v/2}} \exp\left(-\frac{|v - U(x,t)|^2}{2RT(x,t)}\right).$$

Continuum models such as Euler and Navier-Stokes can be derived from the BTE as an asymptotic limit for small values of the Knudsen number through the Chapman-Enskog expansion [36].

Computing the numerical solution to the Boltzmann transport equation has been a challenging problem due to the high dimension of phase space and the complicated structure of the Boltzmann collision operator defined by a fivefold integral (for the physical case $d_v = 3$). In recent years, several numerical techniques have been proposed for efficient computation of the collision operator. Most of them were based on a discretization of the equation on a fixed grid in phase space and adopted Fourier spectral methods, which enables one to accurately compute the Boltzmann collision operator with a relatively small computational effort [85, 89]. On the other hand, together with the development of such a spectral method, there has been progress in the construction of high-order methods (in space and time) for the BTE [52, 55, 80]. Depending on how the convection term is treated, we distinguish between three classes of schemes: fully Lagrangian (particle methods), Eulerian, and Semi-Lagrangian (SL) methods. Particle methods have the advantage of making an efficient use of phase space, providing a simple treatment of convection and allowing computation of high Mach number flows. Their drawback is that they are usually low-order accurate. For this reason, we restrict ourselves to space discretization based on a regular Cartesian mesh. In case of Eulerian approach, splitting methods [55], exponential time integrators [80], and implicit-explicit methods based on Runge-Kutta [22, 52] or multistep [23] have been applied. For large and moderate values of the Knudsen number, the main time step constraint is related to the CFL restriction due to the convection term, which may affect the efficiency of Eulerian methods. This issue could be overcome by adopting the SL approach, which computes the solutions on a fixed grid by integrating the equation along its characteristic curves. In [60] the SL method has been coupled with a splitting technique, obtaining second-order accuracy in time.

Here we review the construction and application of high-order conservative SL methods for the numerical solution of the BTE [19].

Fourth order in time is obtained by the classical fourth-order Runge-Kutta method for the time discretization, while the conservative non-oscillatory reconstruction illustrated in Sect. 3.2 (see [40]) is adopted, achieving fifth-order accuracy in space for smooth solutions. In order to prevent the loss of conservation in momentum and energy that comes from the computation of the collision operator, we have used a weighted L^2-minimization technique [18], which differs from the classical one [63]. An alternative to preserve the moments has been recently developed in the context of the spectral method [88].

5.1 Derivation of the First-Order SL Method in One Space Dimension

To derive the first-order SL method, we begin by writing the BTE in characteristic form:

$$\begin{aligned} &\frac{df}{dt} = \frac{1}{\kappa} Q(f) =: K, \\ &\frac{dx}{dt} = v_x, \\ &x(0) = \tilde{x}, \quad f(0, x, v) = f_0(x, v), \quad t \geq 0, \quad x \in \mathbb{R}, \ v \in \mathbb{R}^{d_v}. \end{aligned} \tag{66}$$

Applying the explicit Euler method to (66), we obtain

$$f^{n+1}(x, v) = f^n(x - v_x \Delta t, v) + \Delta t K^n(x - v\Delta t, v),$$

where $f^n(x, v) = f(x, v, t_n)$ and $K^n(x, v) = \frac{1}{\kappa} Q\left(f(x, v, t_n)\right)$. Discretization in space and velocity further gives

$$f_{i,j}^{n+1} = \tilde{f}_{i,j}^n + \Delta t \tilde{K}_{i,j}^n, \tag{67}$$

where

$$\tilde{f}_{i,j}^n \approx f(x_i - v_{xj}\Delta t, v_j, t_n), \quad \tilde{K}_{i,j}^n \approx K\left(x_i - v_{xj}\Delta t, v_j, t_n\right).$$

Since the characteristic feet $x_i - v_{xj}\Delta t$ could be located on an off-grid point (see Fig. 1), we need to interpolate the values of $\tilde{f}_{i,j}^n$ from $f_{\mathcal{I},j}^n$ and $\tilde{K}_{i,j}^n$ from $K_{\mathcal{I},j}^n$. For this, we consider linear interpolation which is sufficient for the construction of the first-order method. For the computation of $Q^n_{i,j}$, we use the fast spectral method [85] described in Sect. 5.4. Since the macroscopic quantities except for mass are not exactly preserved, we apply the weighted L^2-minimization [18] to $Q^n_{i,\mathcal{J}}$ by imposing

$$(\Delta v)^{d_v} \sum_j Q^n_{i,j} \phi_j = 0, \quad \phi_j = 1,\, v_j,\, v_j^2.$$

Finally, we use the corrected values of $Q^n_{i,\mathcal{J}}$ to define $K^n_{i,\mathcal{J}}$ and then interpolate $\tilde{K}^n_{i,j}$ and update solutions with (67). Note that the linear interpolation prevents the loss of conservation in macroscopic quantities [20].

5.2 *Derivation of the High-Order SL Methods Based on the RK Method*

In this section, we describe our high-order SL methods for the BTE. First, we define a backward characteristic curve to (66) which passes through the location x_i at time $t_n + c_k\Delta t$. We call this the kth characteristic curve and denote by $x_{ij}^{(k,\ell)} := x_i - (c_k - c_\ell)v_j\Delta t$ the location of its intersection with the line $t_n + c_\ell\Delta t$ (see Figs. 25 and 26). The ℓth stage value of the collision term $K = \frac{1}{\kappa}Q(f)$ computed along the kth characteristic corresponding to x_i and v_j will be denoted by $K_{i,j}^{(k,\ell)}$.

Applying s-stage explicit Runge-Kutta methods to (66), we obtain

$$k\text{th-stage value:} \qquad F_{ij}^{(k,k)} = F_{i,j}^{(k,0)} + \Delta t \sum_{\ell=1}^{k-1} a_{k\ell} K_{i,j}^{(k,\ell)}, \quad k = 1, \ldots, s$$

$$\text{numerical solution:} \qquad f_{i,j}^{n+1} = \tilde{f}_{i,j}^n + \Delta t \sum_{k=1}^{s} b_k \tilde{K}_{i,j}^{(k)},$$

where $\tilde{K}_{i,j}^{(k)} \approx K\left(x_i - v_j(1 - c_k)\Delta t, v_j, t_n + c_k\Delta t\right)$ denotes the kth Runge-Kutta flux. For each k-stage, for each i, we compute $Q_{i,\mathcal{J}}^{(\ell,\ell)}$ from $F_{i,\mathcal{J}}^{(\ell,\ell)}$ by the spectral method, and then as in the first-order scheme, we implement the weighted L^2-minimization so that $Q_{i,\mathcal{J}}^{(k,k)}$ are modified to satisfy

$$(\Delta v)^{d_v} \sum_j Q_{i,j}^{(k,k)} \phi_j = 0.$$

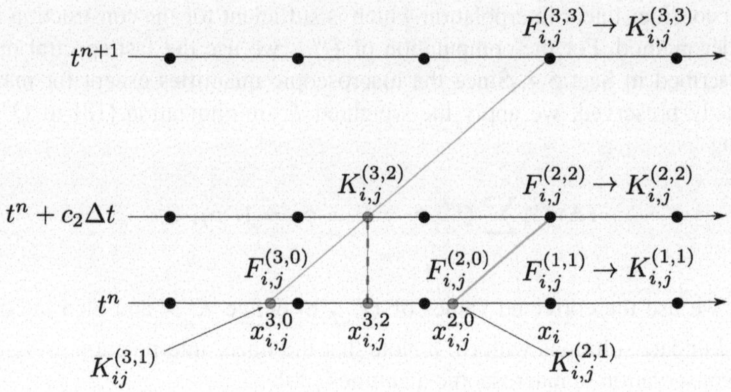

Fig. 25 Illustration of RK3 method [19]. The quantities on grey points need interpolation

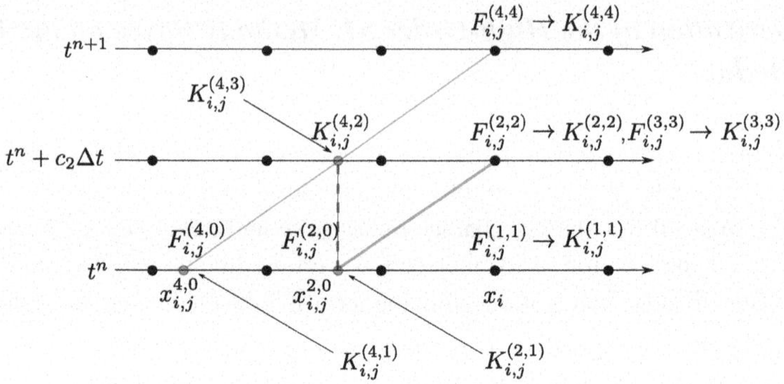

Fig. 26 Illustration of RK4 [19]. The quantities on grey points need interpolation

After we set $K_{i,j}^{(k,k)} = \frac{1}{\kappa} Q_{i,j}^{(k,k)}$, the off-diagonal elements $K_{i,j}^{(k,\ell)}$ are computed from $K_{\mathcal{I},j}^{(\ell,\ell)}$ by high-order conservative reconstruction [40].

The general description of SL-based RK methods can be summarized as follows:

Algorithm for ERK-Based SL Schemes
For $k = 1, ..., s$:

1. Using a conservative reconstruction [40], interpolate $F_{i,j}^{(k,0)}$ on $x_{i,j}^{k,0}$ from $f_{\mathcal{I},j}^n$.
2. (Skip this for $k = 1$) For $\ell = 1, ..., k-1$, we interpolate $K_{i,j}^{(k,\ell)}$ on $x_{i,j}^{k,\ell}$ from $\{K_{i,j}^{(\ell,\ell)}\}_{i\in\mathcal{I}}$.
3. Compute:

$$F_{i,j}^{(k,k)} = F_{i,j}^{(k,0)} + \Delta t \sum_{\ell=1}^{k-1} a_{k\ell} K_{i,j}^{(k,\ell)}.$$

4. Construct $Q_{i,\cdot}^{(k,k)}$ using $\{F_{i,j}^{(k,k)}\}_{j\in\mathcal{J}}$ using the fast spectral method described in Sect. 5.4.
5. For each i, using the weighted L^2-minimization procedure described in Sect. 3.1.2, we modify $Q_{i,\mathcal{J}}^{(k,k)}$ so that it satisfies

$$(\Delta v)^{d_v} \sum_j Q_{i,j}^{(k,k)} \phi_j = 0, \quad \phi_j = 1, v_j, |v_j|^2/2$$

up to a tolerance (tol).
6. Compute $K_{i,j}^{(k,k)} = \dfrac{Q_{i,j}^{(k,k)}}{\kappa}$ for each i, j.
7. If iterations over, update numerical solution:

$$f_{i,j}^{n+1} = \tilde{f}_{i,j}^n + \Delta t \sum_{\ell=1}^{s} b_\ell \tilde{K}_{i,j}^{(\ell)},$$

where $\tilde{K}_{i,j}^{(\ell)} \approx K(x_i - v_j(1 - c_\ell)\Delta t, v_j, t_n + c_\ell \Delta t)$.

In this work, we consider two Runge-Kutta methods for time discretization: the third-order method (RK3) proposed in [101] and the classical RK4. Here we give Butcher's tables for these methods.

$$\text{RK3} = \begin{array}{c|ccc} 0 & 0 & 0 & 0 \\ \alpha & \alpha & 0 & 0 \\ 1 & 1 + \frac{1-\alpha}{\alpha(3\alpha-2)} & -\frac{1-\alpha}{\alpha(3\alpha-2)} & 0 \\ \hline & \frac{1}{2} - \frac{1}{6\alpha} & \frac{1}{6\alpha(1-\alpha)} & \frac{2-3\alpha}{6(1-\alpha)} \end{array}, \qquad \text{RK4} = \begin{array}{c|cccc} 0 & 0 & 0 & 0 & 0 \\ 1/2 & 1/2 & 0 & 0 & 0 \\ 1/2 & 0 & 1/2 & 0 & 0 \\ 1 & 0 & 0 & 1 & 0 \\ \hline & \frac{1}{6} & \frac{1}{3} & \frac{1}{3} & \frac{1}{6} \end{array}$$

In both cases, $c_s = 1$ gives $\tilde{K}_{i,j}^{(k)} = K_{i,j}^{(s,k)}$, $k = 1, \ldots, s$, and thus, no interpolation is needed for the last Runge-Kutta flux. In the RK3 method, the choice $\alpha = \frac{1}{3}$ gives $b_1 = 0$, which allows us to save memory for the numerical solution. Moreover, on the RK4 method we have $a_{31} = a_{41} = a_{42} = 0$, which simplifies the computation of stage values by avoiding interpolation.

$$\begin{aligned} &\text{1st-stage value:} && F_{i,j}^{(1,1)} = f_{i,j}^n \\ &\text{2nd-stage value:} && F_{i,j}^{(2,2)} = F_{i,j}^{(2,0)} + \frac{\Delta t}{3} K_{i,j}^{(2,1)} \\ &\text{3rd-stage value:} && F_{i,j}^{(3,3)} = F_{i,j}^{(3,0)} - \Delta t K_{i,j}^{(3,1)} + 2\Delta t K_{i,j}^{(3,2)} \\ &\text{numerical solution:} && f_{i,j}^{n+1} = F_{i,j}^{(3,0)} + \frac{3\Delta t}{4} K_{i,j}^{(3,2)} + \frac{\Delta t}{4} K_{i,j}^{(3,3)}. \end{aligned}$$

$$
\begin{aligned}
&\text{1st-stage value:} && F_{i,j}^{(1,1)} = f_{i,j}^n \\
&\text{2nd-stage value:} && F_{i,j}^{(2,2)} = F_{i,j}^{(2,0)} + \frac{\Delta t}{2} K_{i,j}^{(2,1)} \\
&\text{3rd-stage value:} && F_{i,j}^{(3,3)} = F_{i,j}^{(2,0)} + \frac{\Delta t}{2} K_{i,j}^{(2,2)} \\
&\text{4th-stage value:} && F_{i,j}^{(4,4)} = F_{i,j}^{(4,0)} + \Delta t K_{i,j}^{(4,3)} \\
&\text{numerical solution:} && f_{i,j}^{n+1} = F_{i,j}^{(4,0)} + \frac{\Delta t}{6} K_{i,j}^{(4,1)} + \frac{\Delta t}{3} K_{i,j}^{(4,2)} + \frac{\Delta t}{3} K_{i,j}^{(4,3)} \\
& && \qquad + \frac{\Delta t}{6} K_{i,j}^{(4,4)}.
\end{aligned}
$$

Remark 5.1 The RK3-based SL method involves three computations for the collision operator and five interpolations (two for distribution function and three for the collision operator). In the algorithm based on RK4, we can save computational costs by avoiding unnecessary redundant computations. As in the RK3 method, no computation is needed for computing $F_{i,j}^{(1,1)}$ on the first stage. On the third stage, we can take $F_{i,j}^{(3,0)} = F_{i,j}^{(2,0)}$ because $c_2 = c_3$. In addition, no interpolation is needed for computing $Q_{i,j}^{(3,2)}$ because $Q_{i,j}^{(3,2)} = Q_{i,j}^{(2,2)}$. When updating numerical solutions, the value of $c_4 = 1$ enables us to set $\tilde{f}_{i,j}^n = F_{i,j}^{(4,0)}$, $\tilde{Q}_{i,j}^{(1)} = Q_{i,j}^{(3,1)}$, and $Q_{i,j}^{(4,2)} = Q_{i,j}^{(3,2)}$. Totally, four evolutions of collision operator and six interpolations are required (two for distribution function and four for collision operator.)

5.3 *Derivation of the High-Order SL Methods Based on Multistep Methods*

Here we present high-order SL methods for the BTE based on explicit multistep methods. In particular, we focus on the Adams-Bashforth (AB) methods. In the methods, we only need to consider one backward characteristic curve to (66) which passes through the location x_i at time t_{n+1}. The location of the characteristic foot which meets the time level t_{n+1-k} is denoted by $x_{i,j}^{n,k} := x_i - k v_j \Delta t$. We define by $K_{i,j}^{n,k} \approx K(x_{i,j}^{n,k}, v_j, t_{n+1-k})$ the distribution function on $x_{i,j}^{n,k} := x_i - k v_j \Delta t$ (see Fig. 27). Applying s-step explicit Adams-Bashforth methods to (66), we get

$$
f_{i,j}^{n+1} = \tilde{f}_{i,j}^n + \Delta t \sum_{k=1}^{s} a_k K_{i,j}^{n,k},
$$

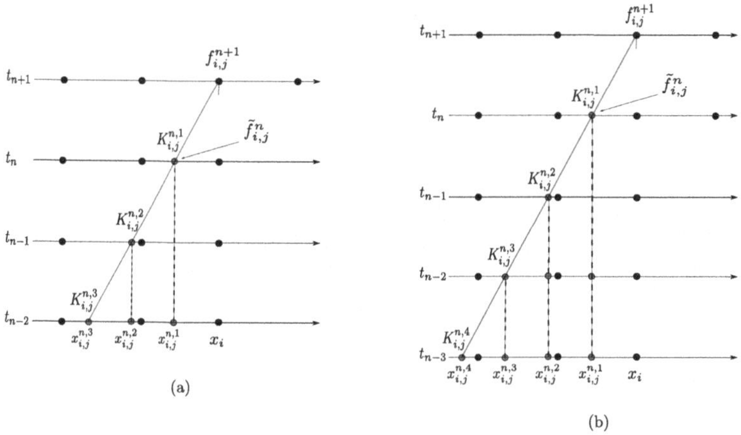

Fig. 27 Illustration of the AB3 and AB4 method [19]. The quantities on grey points need interpolation. (**a**) AB3. (**b**) AB4

where $\tilde{f}^n_{i,j} \approx f\left(x_i - v_j\Delta t, v_j, t_n\right)$. In each time level t_{n+1-k}, we are given $K^{n+1-k}_{i,j}$; therefore, we can interpolate $K^{n,k}_{i,j}$ from $K^{n+1-k}_{\mathcal{I},j}$. For the next step, we compute the collision operator $Q^{n+1}_{i,\mathcal{J}}$ from $f^{n+1}_{i,\mathcal{J}}$ using the fast spectral method and modify it by imposing

$$(\Delta v)^{d_v} \sum_j Q^{n+1}_{i,j}\phi_j = 0.$$

Then, save $K^{n+1}_{i,j} = \frac{1}{\kappa} Q^{n+1}_{i,j}$.

The general description of SL-based AB methods can be summarized as follows:

Algorithm for AB-Based SL Schemes

1. Using a conservative reconstruction [40], interpolate $\tilde{f}^n_{i,j}$ on $x^{n,1}_{i,j}$ from $f^n_{\mathcal{I},j}$.
2. For $k = 1, \ldots, s$, using a conservative reconstruction [40], interpolate $K^{n,k}_{i,j}$ on $x^{n,k}_{i,j}$ from $K^n_{\mathcal{I},j}$.
3. Compute numerical solution:

$$f^{n+1}_{i,j} = \tilde{f}^n_{i,j} + \Delta t \sum_{k=1}^{s} a_k K^{n,k}_{i,j}.$$

4. Construct $Q^{n+1}_{i,\mathcal{J}}$ using $f^{n+1}_{i,\mathcal{J}}$ using the fast spectral method described in Sect. 5.4.
5. For each i, using the weighted L^2-minimization procedure described in Sect. 3.1.2, we modify $Q^{n+1}_{i,\mathcal{J}}$ so that it satisfies

$$(\Delta v)^{d_v} \sum_j Q^{n+1}_{i,j}\phi_j = 0, \quad \phi_j = 1, v_j, |v_j|^2/2$$

up to a tolerance (tol).

6. Compute $K_{i,j}^{n+1} = \dfrac{Q_{i,j}^{n+1}}{\kappa}$ for each i, j for the next step.

For the iteration, we only need the values of $K_{i,j}^{n+1-k}$, $k = 1, ..., s$ and $f_{i,j}^{n}$, and hence, four interpolation procedures and one computation for the collision operator are necessary for each step. In view of this, multistep methods are more efficient compared to the one-step method. However, we note that these multistep-based SL methods have narrow stability regions compared to RK methods.

$$\text{AB3:} \qquad f_{i,j}^{n+1} = \tilde{f}_{i,j}^{n} + \Delta t \left(\frac{23}{12} K_{i,j}^{n,1} - \frac{16}{12} K_{i,j}^{n,2} + \frac{5}{12} K_{i,j}^{n,3} \right)$$

$$\text{AB4:} \qquad f_{i,j}^{n+1} = \tilde{f}_{i,j}^{n} + \Delta t \left(\frac{55}{24} K_{i,j}^{n,1} - \frac{59}{24} K_{i,j}^{n,2} + \frac{37}{24} K_{i,j}^{n,3} + \frac{9}{24} K_{i,j}^{n,4} \right).$$

Remark 5.2 AB-based methods require the computation of the collision operator only one time for each time step. Regarding the number of interpolation procedure, the AB3-based SL method involves four interpolations (one for distribution function and three for collision operator), while the AB4-based SL method involves five interpolations (one for distribution function and four for collision operator). On the other hand, RK-based methods have wider stability region, therefore allowing larger time steps than those allowed by AB-based methods.

5.4 Fast Spectral Method for the Boltzmann Collision Operator

Computing numerical solutions to the Boltzmann equation has been a challenging problem due to the high dimension of its collision integral. For treating this problem, a class of numerical methods based on Fourier spectral approximation has been proposed and studied in numerous papers [22, 52, 55, 60, 80, 85, 87, 89, 113]. In this section, we review the fast spectral method [85], which is employed in this work.

The spectral method (also-called Fourier-Galerkin method) is based on the Fourier series expansion of the velocity distribution function and the Boltzmann collision. Let us consider uniformly distributed frequencies indexed by k, where $k = (k_1, ..., k_{d_v})$ is the multi-index such that $-\frac{N}{2} \leq k_1, ..., k_{d_v} \leq \frac{N}{2} - 1$. The number N will be set N_v in our simulation. We assume that the distribution function is well approximated by a function with the support contained in $D_T := [-T, T]^{d_v}$. We then approximate the solution by the following truncated Fourier series [89]:

$$f_N(v) = \sum_{k=-\frac{N}{2}}^{\frac{N}{2}-1} \hat{f}_k e^{i \frac{\pi}{T} k \cdot v}, \quad i = \sqrt{-1}$$

where the summation with respect to k-means

$$\sum_{k=-\frac{N}{2}}^{\frac{N}{2}-1} = \sum_{k_1,\ldots,k_{d_v}=-\frac{N}{2}}^{\frac{N}{2}-1}$$

and the Fourier mode $\hat{f}_k$ that corresponds to the frequency index k is given by

$$\hat{f}_k = \frac{1}{(2T)^{d_v}} \int_{D_T} f(v) e^{-i\frac{\pi}{T}k\cdot v} dv.$$

More precisely, if a distribution function f is compactly supported on a ball centered at the origin with radius S and the period satisfies the $T \geq \frac{1+3\sqrt{2}}{2} S$, then there is no aliasing error, i.e., the evaluation of the collision operator obtained by the periodized function and the original compactly supported function are the same.

For details, we refer to the paper [85, 89]. In [89], the explicit form of Fourier mode of collision operator is provided as follows:

$$\hat{Q}_k = \sum_{\ell+m=k,\, k=-\frac{N}{2}}^{\frac{N}{2}-1} \hat{\beta}(\ell, m) \hat{f}_\ell \hat{f}_m, \quad -\frac{N}{2} \leq k \leq \frac{N}{2} - 1 \tag{68}$$

where $\ell = (\ell_1, \ldots, \ell_{d_v})$ and $m = (m_1, \ldots, m_{d_v})$. Here $\hat{\beta}(\ell, m) = \beta(\ell, m) - \beta(m, m)$ is the so-called kernel mode . The explicit form of $\beta(\ell, m)$ for the Maxwellian and hard sphere can be found in [89]. If one performs the summation (68) naively, the corresponding computational cost is $O(N^{2d_v})$. As N and d_v become larger, the resulting spectral method becomes expensive. To overcome this, in [85] authors consider the so-called Carleman representation and approximate the coefficients $\beta(\ell, m)$ with $\beta^f(\ell, m)$:

$$\beta(\ell, m) \approx \beta^f(\ell, m) = \frac{\pi}{M} \sum_{p=0}^{M-1} \alpha_p(\ell) \alpha'_p(m),$$

where

$$\alpha_p(\ell) := \phi_R^2(\ell \cdot e_{\theta_p}), \quad \alpha'_p(m) = \phi_R^2(m \cdot e_{\theta_p + \frac{\pi}{2}}).$$

This finally gives us a convolution structure:

$$\hat{Q}_k^f = \frac{\pi}{M} \sum_{p=0}^{M-1} \sum_{m=-\frac{N}{2}}^{\frac{N}{2}-1} \left(\alpha_p(k-m) \hat{f}_{k-m} \alpha'_p(m) \hat{f}_m - \hat{f}_{k-m} \alpha_p(m) \alpha'_p(m) \hat{f}_m \right),$$

$$-\frac{N}{2} \le k \le \frac{N}{2} - 1.$$

Thus, the fast Fourier transform can be implemented for the computation of $\hat{Q}_k^f$ with reduced computational cost $O(MN^{d_v}\log(N^{d_v}))$. Consequently, we can compute the collision operator in velocity space as follows:

$$Q(v) = \sum_{k=-\frac{N}{2}}^{\frac{N}{2}-1} \hat{Q}_k e^{i\frac{\pi}{T}k\cdot v}.$$

For our simulation, we use MATLAB to perform the fast Fourier transform and inverse Fourier transform.

5.5 *Numerical Test*

Throughout this section, we consider two types of numerical tests: (1) one dimension in space and two dimension in velocity with Maxwellian molecules and (2) one dimension in space and three dimensions in velocity with hard-sphere molecules. In each case, the velocity domain is truncated by $[-v_{\max}, v_{\max}]^{d_v}$. When determining the time step, we use the CFL number defined by (29).

The abbreviations of schemes used in numerical tests are reported in Table 9. The performance of other combinations of high-order methods and two-dimensional numerical results can be also found in [19].

5.5.1 Test 1: Accuracy Test and Optimal CFL Number

In this test, we consider smooth initial data to show that the proposed SL scheme gives the expected convergence rate. Then, we look for the optimal CFL numbers for the method.

Convergence Rate

Here we check the accuracy of high-order methods described in Table 9. As in [52, 80], we consider two different initial data. One is the nonequilibrium data:

Table 9 Numerical methods

	Time discretization	Reconstruction	type
RK4+WENO35+Euler	RK4	WENO35	Eulerian
RK4+QCWENO35+SL	RK4	QCWENO35	SL
AB4+QCWENO35+SL	AB4	QCWENO35	SL

Table 10 Test 1: error and convergence rate for $\kappa = 1$. RK4+QCWENO35+SL [19]

Relative L^1 error and convergence rate of density ρ								
	Nonequilibrium				Equilibrium			
	CFL= 4		CFL= 10		CFL= 4		CFL= 10	
$(N_x, 2N_x)$	Error	Rate	Error	Rate	Error	Rate	Error	Rate
(40, 80)	3.442e-06	4.876	1.098e-06	4.899	9.372e-06	4.906	4.181e-06	4.916
(80, 160)	1.172e-07	4.955	3.679e-08	4.964	3.127e-07	4.974	1.386e-07	4.971
(160, 320)	3.780e-09		1.179e-09		9.954e-09		4.417e-09	

Table 11 Test 1: error and convergence rate for $\kappa = 0.1$. RK4+QCWENO35+SL [19]

Relative L^1 error and convergence rate of density ρ								
	Nonequilibrium				Equilibrium			
	CFL= 4		CFL= 10		CFL= 4		CFL= 10	
$(N_x, 2N_x)$	Error	Rate	Error	Rate	Error	Rate	Error	Rate
(40, 80)	2.572e-07	4.892	1.684e-07	4.258	3.534e-06	4.955	1.670e-06	4.812
(80, 160)	8.663e-09	4.958	8.802e-09	4.184	1.139e-07	4.988	5.943e-08	4.867
(160, 320)	2.787e-10		4.844e-10		3.591e-09		2.036e-09	

$$f_0^{neq}(x, v) = \frac{1}{2}\frac{\rho_0(x)}{2\pi T_0(x)}\left(\exp\left(-\frac{(v - U_0(x))^2}{2T_0(x)}\right) + \exp\left(-\frac{(v + U_0(x))^2}{2T_0(x)}\right)\right) \tag{69}$$

which is the sum of two Maxwellians centered on U_0 and $-U_0$, respectively. The other one is the equilibrium initial data:

$$f_0^{eq}(x, v) = \frac{\rho_0(x)}{2\pi T_0(x)}\exp\left(-\frac{|v - U_0|^2}{2T_0(x)}\right). \tag{70}$$

In each case, initial density ρ_0, velocity u_0, and temperature T_0 are given by

$$\rho_0(x) = \frac{2 + \sin 2\pi x}{2}, \quad U_0(x) = (0.75, -0.75), \quad T_0(x) = \frac{5 + 2\cos 2\pi x}{20}.$$

We assume periodic boundary condition in space $x \in [-0.5, 0.5]$ and consider the velocity domain $v \in [-8, 8]$ with $N_v = 32$. The final time is taken by $t = 0.75$ at which solutions remain smooth. We implemented all methods in Table 9 with the weighted L^2-minimization technique using $N_x = 40, 80, 160, 320$ and the fixed time step Δt that corresponds to CFL= 4 or 10. For the comparison of accuracy between SL and Eulerian methods, we implement (RK4+WENO35+Euler) with the time step determined by CFL= 0.5.

In Tables 10, 11, and 12, we report the relative L^1-errors and the corresponding convergence rates for RK-based methods. It appears that spatial errors are dominant for the Knudsen number $\kappa = 0.1$ and 1, which gives order 5, the expected order of

Table 12 Test 1: error and convergence rate for CFL= 0.5. RK4+WENO35+Euler [19]

Relative L^1 error and convergence rate of density ρ								
	Nonequilibrium				Equilibrium			
	$\kappa = 0.1$		$\kappa = 1$		$\kappa = 0.1$		$\kappa = 1$	
$(N_x, 2N_x)$	Error	Rate	Error	Rate	Error	Rate	Error	Rate
(40, 80)	1.593e-06	4.990	2.834e-05	4.672	9.328e-06	4.915	2.216e-05	4.763
(80, 160)	5.013e-08	5.147	1.112e-06	5.186	3.092e-07	5.071	8.160e-07	5.120
(160, 320)	1.415e-09		3.055e-08		9.196e-09		2.347e-08	

Table 13 Test 1: error and convergence rate for $\kappa = 1$. AB4+QCWENO35+SL [19]

Relative L^1 error and convergence rate of density ρ								
	Nonequilibrium				Equilibrium			
	CFL= 4		CFL= 10		CFL= 4		CFL= 10	
$(N_x, 2N_x)$	Error	Rate	Error	Rate	Error	Rate	Error	Rate
(40, 80)	3.424e-06	4.868	8.146e-06	4.033	9.112e-06	4.771	9.647e-06	3.162
(80, 160)	1.173e-07	4.970	4.977e-07	4.068	3.338e-07	4.902	1.078e-06	4.398
(160, 320)	3.741e-09		2.967e-08		1.117e-08		5.116e-08	

Table 14 Test 1: error and convergence rate for $\kappa = 0.1$. AB4+QCWENO35+SL [19]

Relative L^1 error and convergence rate of density ρ								
	Nonequilibrium				Equilibrium			
	CFL= 4		CFL= 10		CFL= 4		CFL= 10	
$(N_x, 2N_x)$	Error	Rate	Error	Rate	Error	Rate	Error	Rate
(40, 80)	1.021e-06	4.188	4.272e-05	4.179	3.739e-06	4.853	5.525e-05	5.001
(80, 160)	5.597e-08	4.137	2.358e-06	4.135	1.294e-07	4.867	1.726e-06	4.294
(160, 320)	3.182e-09		1.342e-07		4.432e-09		8.799e-08	

convergence of the reconstruction QCWENO35. On the other hand, in Table 11 we can also observe order 4, which is the expected order of convergence for the RK4 method. This is because time discretization errors become dominant as we increase the CFL number from 4 to 10. Also, Tables 13 and 14 show that a multistep-based method gives the expected convergence rate.

We observe that SL schemes are more efficient and accurate than the corresponding Eulerian-based schemes with the same discretization of parameter Δx, Δv. The improvement of efficiency is due to the possibility of using much larger CFL numbers, therefore reducing the total number of time steps. SL schemes are also more accurate because of the reduced error in the treatment of the transport term, especially for very rarefied flow, where the transport is dominant over collisions.

Optimal CFL Number

The main advantage of the SL method is the allowance of a large time step. Hence, it is important to find the CFL number which gives the smallest errors. Hereafter we call it the optimal CFL number. For this test, we use the same initial data (69) and (70) to compute errors. In Fig. 28, we report the relative L^1-norm of the error of the density obtained by $N_x = 160, 320$. We first look at the case for equilibrium data in Fig. 28a. For a Knudsen number $\kappa = 1$, it appears that error becomes very small around CFL=16. This is because with the choice of $N_v = 32$ and CFL=16, the feet of the characteristics fall on grid points, so there is no interpolation error, which, for other CFL values, is the dominant source of error because for large Knudsen number convection is dominant. For a relatively smaller Knudsen number $\kappa = 0.5$, the CFL number around 10 gives the smallest error. In Fig. 28b, we can also observe a similar tendency with different optimal CFL numbers. We remark that choosing an optimal CFL number depends on the initial data.

Finally, we compare the efficiency of two methods, RK4+QCWENO35+SL and AB4+QCWENO35+SL. For this, we fix Knudsen number $\kappa = 0.1$. For equilibrium data, RK4 and AB4 methods achieve the smallest errors with CFL = 4 and 12, respectively. In case of nonequilibrium data, RK4 and AB4 methods lead to the smallest errors when CFL = 6 and 10, respectively. Based on these observations, we use the CFL numbers for comparison and the results are reported in Fig. 29. Note that for small CFL numbers, the AB4 method attains smaller errors within a shorter time, while for lager CFL numbers the RK4 method works better than the AB4 method.

5.5.2 Test 2: A Sod Problem

In this test, we show that our scheme is able to capture the correct behavior of the shock profile and it preserves conservative variables. Here we consider initial data as in [80]:

$$(\rho_0, U_{10}, U_{20}, T_0) = \begin{cases} (1, 0, 0, 1), & \text{if} \quad -1 \leq x \leq 0.5 \\ \left(\frac{1}{8}, 0, 0, \frac{1}{4}\right), & \text{otherwise} \end{cases} . \tag{71}$$

In order to check the global conservation, we consider the periodic boundary condition on $x \in [-2, 2]$ with $N_x = 400$. To show the effect of L^2-minimization, we perform the test for various discrete velocities with $N_v = 16, 32, 64$ and $v_{\max} = 8$ or 10. We use CFL= 2, 10 for the SL method while we take CFL= 0.5 for the Eulerian one.

Density, velocity, and temperature at time $t = 0.2$ are plotted in Fig. 30. Even with the use of larger time steps, the SL-based scheme is able to reproduce results which are comparable with those of the Eulerian one. Table 15 shows L^2-minimization can guarantee the conservation of physical quantities up to machine precision even with a fewer number of velocity grids. However, we should not take too few number of velocity grids because it could lead to unstable solutions. Note that even without L^2-minimization, the conservation could be improved if we consider sufficiently large grid numbers in velocity.

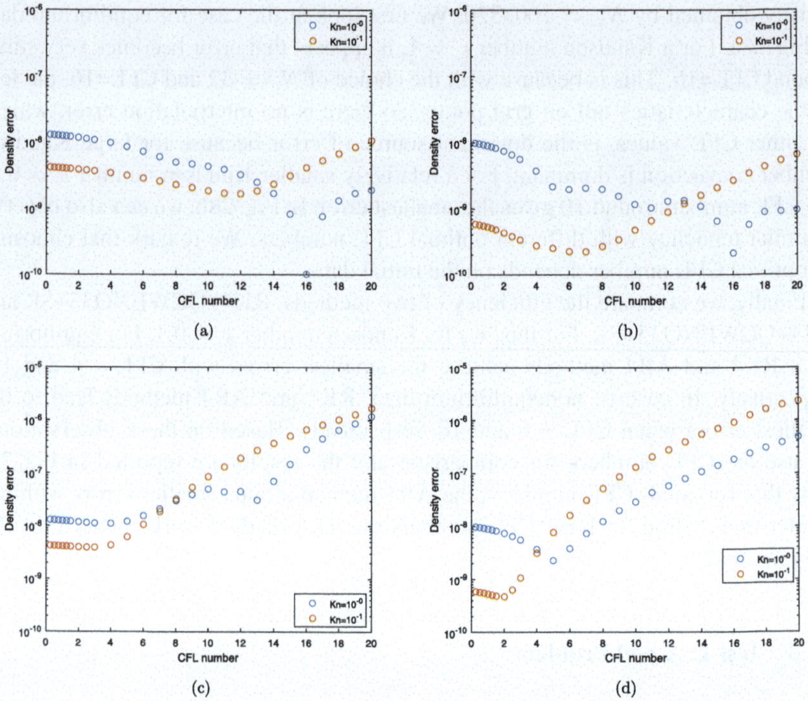

Fig. 28 Test 1: error versus CFL number for different Knudsen numbers. Panels in the first and second rows correspond to results obtained by RK4+QCWENO35+SL [19] and AB4+QCWENO35+SL, respectively. (**a**) Equilibrium data. (**b**) Nonequilibrium data. (**c**) Equilibrium data. (**d**) Nonequilibrium data

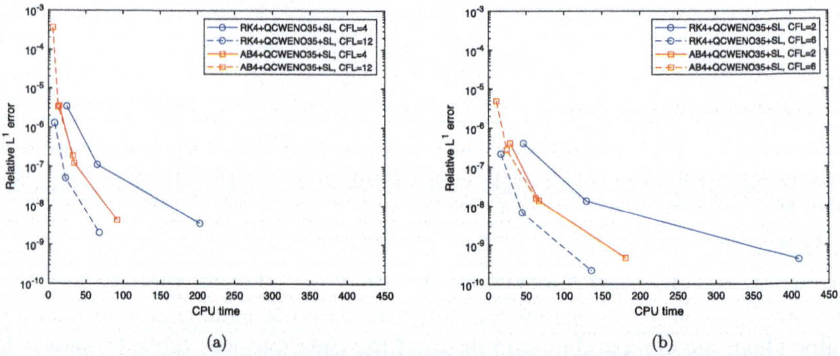

Fig. 29 Test 1: error versus CPU time for the Boltzmann equation. Numerical results are computed for $\kappa = 0.1$. For this case RK-based schemes (blue lines) allow much larger CFL number, thus achieving better efficiency that AB schemes (red lines). (**a**) Equilibrium data. (**b**) Nonequilibrium data

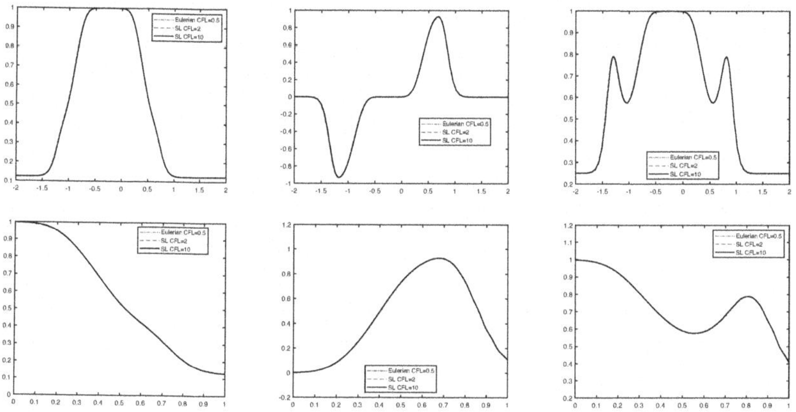

Fig. 30 Test 2: comparison of Eulerian and SL methods for $\kappa = 0.01$ with L^2-minimization [19]. $N_v = 64$, $v_{\max} = 10$

Table 15 Test 2: Conservation error for RK4+QWENO35+SL [19]

		Without L^2-minimization		L^2-minimization	
	N_v	$v_{\max} = 8$	$v_{\max} = 10$	$v_{\max} = 8$	$v_{\max} = 10$
Mass	16	−7.655e-03	−8.710e-02	NaN	NaN
	32	9.493e-08	7.269e-06	−6.096e-15	−9.144e-15
	64	−1.358e-09	−5.628e-09	−7.803e-15	−1.024e-14
Momentum	16	1.152e-03	1.903e-02	NaN	NaN
	32	9.483e-08	2.755e-07	−2.314e-17	6.028e-17
	64	5.533e-08	8.564e-10	2.263e-17	−3.488e-17
Energy	16	−6.587e-02	−1.150e+00	NaN	NaN
	32	−9.281e-06	−1.013e-04	−7.132e-15	−9.090e-15
	64	−5.407e-06	−2.339e-07	−7.132e-15	−1.035e-14

5.5.3 Test 3: Convergence to Equilibrium

In [52], the author investigated the behavior of the quantity $\|\rho(t) - \rho_g\|_1$ to estimate the speed of convergence of distribution function to the global equilibrium and demonstrated that oscillations appear on the difference between global and local equilibria. To reproduce the result, we use the same initial data:

$$
\begin{aligned}
f_0(x, v) &= \frac{1 + A_0 \sin(\pi x)}{2\pi T_0} \left[\exp\left(-\frac{|v - U_0|^2}{2T_0} \right) + \exp\left(-\frac{|v + U_0|^2}{2T_0} \right) \right] \\
A_0 &= \frac{1}{2}, \quad T_0 = 0.125, \quad U_0 = \left(\frac{1}{2}, \frac{1}{2} \right)
\end{aligned}
\tag{72}
$$

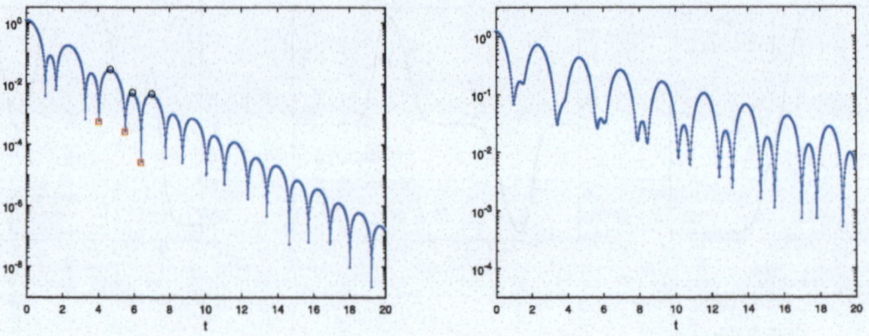

(a) (b)

Fig. 31 Test 3: time evolution of $\|\rho - \rho_g\|_1$ associated to initial data (72) [19]. (**a**) $\kappa = 0.5$. (**b**) $\kappa = 0.1$

with periodic space domain $x \in [-1, 1]$ and velocity domain $v \in [-5, 5]$. Here the corresponding global equilibrium is given by

$$\mathcal{M}_g = \frac{\rho_g}{2\pi T_g} \exp\left(-\frac{|v|^2}{2T_g}\right)$$

where

$$\rho_g = \int_{-1}^{1}\int_{\mathbb{R}^2} f_0(x, v)dv\,dx, \quad T_g = \frac{1}{2\rho_g}\int_{-1}^{1}\int_{\mathbb{R}^2} f_0(x, v)|v|^2 dv\,dx.$$

The Knudsen number is taken by $\kappa = 0.5, 0.1$ and we set the final time by $t = 20$ to observe the long-time behavior of the numerical solution. Our scheme has advantage in this test because it allows us to use a large time step. Here we set Δt that corresponds to CFL$=6$. In Fig. 31 we show the results, and they are consistent with the literature [53]. Furthermore, in Fig. 32 we can observe the fluctuation of densities around 2 which corresponds to the global equilibrium. For a more detailed explanation about the oscillating behavior, we refer to [54].

5.5.4 Test 4: A Sod Problem with Hard-Sphere Molecule in Three Dimensions

Here we solve a Sod problem with three-dimensional hard-sphere molecules. Initial data is the local Maxwellian with similar macroscopic quantities described in (71), namely,

$$(\rho_0, U_{10}, U_{20}, U_{30}, T_0) = \begin{cases} (1, 0, 0, 0, 1), & \text{if} \quad -1 \leq x \leq 0.5 \\ \left(\frac{1}{8}, 0, 0, 0, \frac{1}{4}\right), & \text{otherwise} \end{cases}. \tag{73}$$

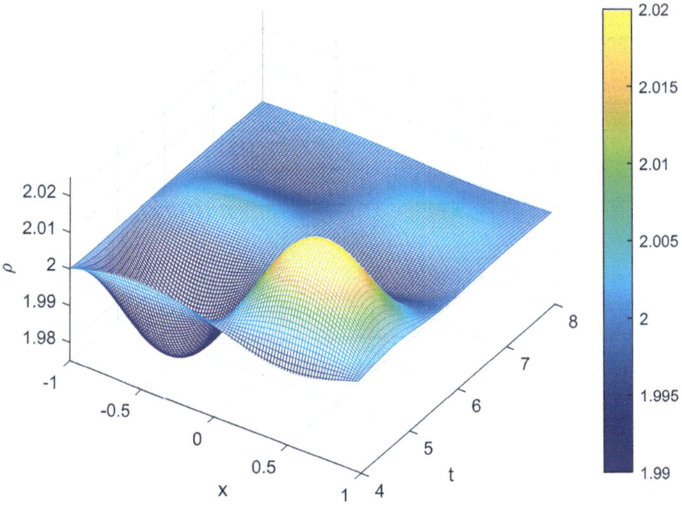

Fig. 32 Test 3: time evolution of $\rho(x, t)$ associated to initial data (72) for $\kappa = 0.5$ [19]. (**a**) $\rho(x, t)$ at $(x, t) \in [-1, 1] \times [4, 8]$ in Fig. 31a

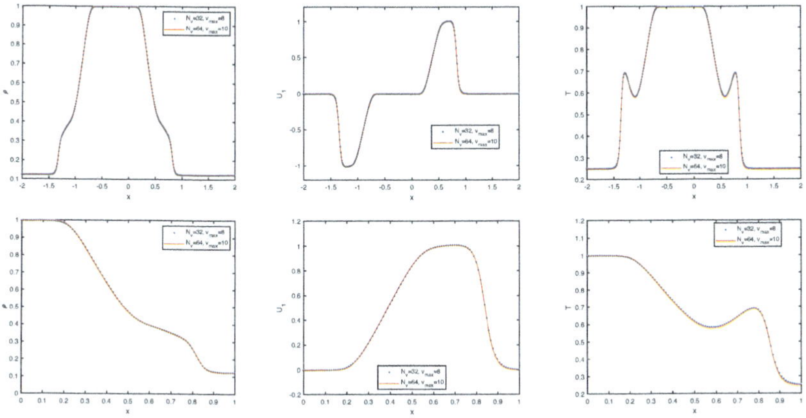

Fig. 33 Test 4: Sod problem with $\kappa = 0.01$ and CFL=4 for three-dimensional hard-sphere molecules [19]

We set the periodic boundary condition on the space domain $x \in [-2, 2]$ with $N_x = 400$. Here we fix $N_v = 32$ for $v_{\max} = 8$ or $N_v = 64$ for $v_{\max} = 10$. The time step is fixed to satisfy CFL= 4. In both cases, we use eight discrete angles in each direction. Density, velocity, and temperature at time $t = 0.5$ are plotted in Fig. 33. We see that the two solutions are very similar. Compared to the two-dimensional case in Fig. 30, for $\kappa = 0.01$, it appears that in the three-dimensional case the width of the shock is narrower than in the two-dimensional case.

Table 16 Test 5: error and convergence rate for $\kappa = 0.1$. RK3+2DQCWENO23+SL

Relative L^1 error and convergence rate								
	ρ		U_1		U_2		T	
$(N_x, 2N_x)$	Error	Rate	Error	Rate	Error	Rate	Error	Rate
(25, 50)	1.041e-04	2.996	2.954e-04	2.887	2.975e-04	2.890	8.713e-05	2.946
(50, 100)	1.305e-05	3.009	3.993e-05	2.969	4.014e-05	2.971	1.131e-05	2.992
(100, 200)	1.622e-06		5.099e-06		5.120e-06		1.422e-06	

5.5.5 Test 5: Two-Dimensional Vortex

Here we present a numerical result for a two-dimensional problem treated in [60]. In the problem, we observe the behavior of a vortex in the space domain $[0, 10]^2$ with periodic boundary conditions. Assuming that the vortex is initially centered at $(5, 5)$ with an ambient flow characterized by a gas with density $\rho_\infty = 1$, mean velocity $U_\infty = (1, 1)$, and temperature $T_\infty = 1$, it returns to the center at time $t = 10$. For the initial distribution function, we take local Maxwellian with macroscopic moments:

$$U(x, 0) = U_\infty + \delta U, \quad T(x, 0) = T_\infty + \delta T$$

where

$$\delta U_1 = -(x_2 - 5)\frac{\beta}{2\pi} \exp\left(\frac{1 - r^2}{2}\right), \quad \delta U_2 = (x_1 - 5)\frac{\beta}{2\pi} \exp\left(\frac{1 - r^2}{2}\right),$$

$$\delta T = -\frac{(\gamma - 1)\beta}{8\gamma\pi^2} \exp\left(1 - r^2\right)$$

with $\gamma = 2$, $\beta = 5$ and $r^2 = (x_1 - 5)^2 + (x_2 - 5)^2$. For density, we use

$$\rho(x, 0) = \rho_\infty \left(\frac{T(x, 0)}{T_\infty}\right)^{\frac{1}{\gamma - 1}}.$$

We consider the periodic boundary condition in space, and the truncated velocity domain is $[-7.5, 7.5]^2$. Numerical solutions are computed up to the final time $t_f = 10$. In Table 16, we report the accuracy of our method using uniform grids with $N_{x_1} = N_{x_2} = 25, 50, 100, 200$ and $N_v = 32$ for each velocity direction. The time step Δt that corresponds to CFL$= 1.05$ is used. We take eight discrete angles $M = 8$ in all cases. From the table, we confirm the third-order accuracy in time and space of the RK3+2DQCWENO23+SL method.

6 Vlasov-Type Equations

The Vlasov-Poisson (VP) system, arising from collisionless plasma applications,

$$\frac{\partial f}{\partial t} + \mathbf{v} \cdot \nabla_{\mathbf{x}} f + \mathbf{E}(\mathbf{x}, t) \cdot \nabla_{\mathbf{v}} f = 0, \tag{74}$$

and

$$\mathbf{E}(\mathbf{x}, t) = -\nabla_{\mathbf{x}} \phi(\mathbf{x}, t), \quad -\Delta_{\mathbf{x}} \phi(\mathbf{x}, t) = \rho(\mathbf{x}, t) - 1, \tag{75}$$

describes the temporal evolution of the particle distribution function in six-dimensional phase space. $f(\mathbf{x}, \mathbf{v}, t)$ is the probability distribution function which describes the probability of finding a particle with velocity $\mathbf{v}$ at position $\mathbf{x}$ at time t, $\mathbf{E}$ is the electric field, and ϕ is the self-consistent electrostatic potential. The probability distribution function couples to the long-range fields via the charge density, $\rho(t, x) = \int_{\mathbb{R}^3} f(x, v, t) dv$, where we take the limit of uniformly distributed infinitely massive ions in the background.

In this section, we review existing SL approaches as two main categories: the dimensional splitting schemes in Sect. 6.1 and the fully two-dimensional SL schemes in Sect. 6.2. For the dimensional splitting schemes, emphasis is placed on achieving high-order dimensional splitting accuracy, while for the nondimensional splitting SL schemes, emphasis is placed on tracking nonlinear characteristics with high-order temporal accuracy.

6.1 Splitting Schemes

The splitting approach has been widely adopted in the context of the Vlasov-Poisson system (74)–(75). The method was originally proposed in [37] and further developed in the finite volume [6, 8, 46, 57, 107], finite difference [34, 92, 94], finite element discontinuous Galerkin framework [95, 96], and a hybrid finite different-finite element framework [67]. The SL approach within the splitting framework allows for simple SL implementation procedures for tracing characteristics with a second-order Strang splitting. On the other hand, if the splitting is not performed properly, numerical instabilities can be observed [72]. In [35, 38, 66], high-order splitting techniques have been applied to the VP simulations. In [42], an integral deferred correction method is proposed for the split SL approach to reduce the splitting error in VP simulations. In this section, we review finite difference high-order splitting methods [41], which adopt a high-order reconstruction technique described in Sect. 3.2.

6.1.1 SL Method for the Transport Equation

A SL method can be obtained by integrating the equation of interest along its characteristic curves. As the simplest case, we consider a one-dimensional transport equation:

$$\partial_t f + a\partial_x f = 0, \quad t \geq 0,\ x \in \mathbb{R},$$

where a is a fixed constant in $\mathbb{R}$. Let us denote the nth time step by $t^n := n\Delta t$, $n \in \mathbb{N}$ for $\Delta t > 0$. Given the distribution function f at time t^n, the exact solution on x_i at t^{n+1} is given by

$$f(x_i, t^{n+1}) = f(X(t^n; x_i, t^{n+1}), t^n),$$

where $X(t; x_i, t^{n+1})$ satisfies

$$\frac{dX}{dt} = a, \quad X(t^{n+1}) = x_i;$$

therefore, $X(t^n; x_i, t^{n+1}) = x_i - a\Delta t$. Considering that the characteristic foot $x_i - a\Delta t$ can be located on off-grid points, one can update solutions after reconstructing $f(x_i - a\Delta t, t^n)$ from given solutions on grid points.

6.1.2 Splitting Methods for the Vlasov-Poisson System

Splitting methods to the VP system (74)–(75) are based on solving separately the two equations:

$$\frac{\partial f}{\partial t} + v \cdot \nabla_x f = 0 \qquad \text{transport in space,} \tag{76}$$

$$\frac{\partial f}{\partial t} + \mathbb{E}(x,t) \cdot \nabla_v f = 0 \qquad \text{drift in velocity.} \tag{77}$$

Each equation can be solved along the characteristics.

1. Transport step over a time step $\tau > 0$:

$$f(x, v, t+\tau) = e^{\tau\mathcal{T}} f(x, v, t) = f(x - v\tau, v, t).$$

2. Drift step over a time step $\tau > 0$:

$$f(x, v, t+\tau) = e^{\tau\mathcal{U}} f(x, v, t) = f(x, v - \mathbb{E}(x,t)\tau, t),$$

where we denoted by $e^{\tau\mathcal{T}}$ and $e^{\tau\mathcal{U}}$, respectively, the operators representing a shift along the x-axis by an amount $-v\tau$ and a shift along the v-axis by an amount $-\mathbb{E}\tau$. Note that both characteristic curves are given by straight lines, which enables one to compute the location of the foot of the characteristic analytically. In each step, we need to compute solutions on off-grid points. For this, we use the conservative reconstruction described in Sect. 3.2. Considering the maximum principle of the Vlasov-Poisson system, it is natural to adopt a basic reconstruction that maintains the maximum and minimum of initial data together with conservation property. For these reasons, we also use the MPP limiter (28) on the basic reconstruction. For details, we refer to [41].

6.1.3 High-Order Splitting Methods for the Vlasov-Poisson System

Using the operators $e^{\tau\mathcal{T}}$ and $e^{\tau\mathcal{U}}$, it is easy to represent splitting methods of various orders. Now, we apply several splitting methods described in [116] to the VP system. Here we illustrate the methods of order 2, 4, and 6 that are used in the present section:

- 2nd-order Strang splitting method:

  1. Shift along x-axis: $f^* = e^{\frac{\Delta t}{2}\mathcal{T}} f^n$.
  2. Shift along v-axis: $f^{**} = e^{\Delta t\mathcal{U}} f^*$.
  3. Shift along x-axis: $f^{n+1} = e^{\frac{\Delta t}{2}\mathcal{T}} f^{**}$.

  We can write this compactly as

$$f^{n+1} = \mathcal{S}_2^{\Delta t} f^n := e^{\frac{\Delta t}{2}\mathcal{T}} e^{\Delta t\mathcal{U}} e^{\frac{\Delta t}{2}\mathcal{T}} f^n. \tag{78}$$

- 4th-order splitting method [116]:

$$f^{n+1} = \mathcal{S}_4^{\Delta t} f^n := \mathcal{S}_2^{x_1\Delta t} \mathcal{S}_2^{x_0\Delta t} \mathcal{S}_2^{x_1\Delta t} f^n, \tag{79}$$

  where $x_0 = -\frac{2^{1/3}}{2-2^{1/3}}$, $x_1 = \frac{1}{2-2^{1/3}}$.

- 6th-order splitting method [35]:

$$\begin{aligned} f^{n+1} = \psi_6^{\Delta t} f^n &:= e^{\Delta t\mathcal{D}_1} e^{a_1\Delta t\mathcal{T}} e^{\Delta t\mathcal{D}_2} e^{a_2\Delta t\mathcal{T}} e^{\Delta t\mathcal{D}_3} e^{a_3\Delta t\mathcal{T}} e^{\Delta t\mathcal{D}_3} e^{a_2\Delta t\mathcal{T}} \\ &\times e^{\Delta t\mathcal{D}_2} e^{a_1\Delta t\mathcal{T}} e^{\Delta t\mathcal{D}_1} f^n, \end{aligned} \tag{80}$$

  where $\mathcal{D}_i = (b_i + 2c_i m(\Delta t)^2 + 4d_i m^2(\Delta t)^4 - 8e_i m^3(\Delta t)^6)\mathcal{U}$, $i = 1, 2, 3$. The value $m = 1$ is used in this work, and the coefficients in D_i are given in Table 17.

When shifting solutions along the v-axis, to obtain electric field $\mathbb{E} = -\nabla_x\phi$, one should find the potential function ϕ in advance. For this, we solve the Poisson

Table 17 Coefficients in (80)

	$i = 1$	$i = 2$	$i = 3$
a_i	0.168735950563437422448196	0.377851589220928303880766	−0.093175079568731452657924
b_i	0.049086460976116245491441	0.264177609888976700200146	0.186735929134907054308413
c_i	−0.000069728715055305084099	−0.000625704827430047189169	−0.002213085124045325561636
d_i	0	$-2.916600457689847816445691 \cdot 10^{-6}$	$3.048480261700038788680723 \cdot 10^{-5}$
e_i	0	0	$4.985549387875068121593988 \cdot 10^{-7}$

equation $-\Delta_x \phi = \rho - 1$ in (74) with fast Fourier transform. Note that we can compute $\{\rho_i^n\}_{i\in\mathcal{I}}$ as follows:

$$\rho_i^n = \sum_j f_{i,j}^n (\Delta v)^{d_v}.$$

In one space dimension, we calculate the discrete electric field $E_i^n = -\partial_x \phi\big|_{(x,t)}$ as follows:

- 4th-order approximation:

$$\partial_x \phi\big|_{(x,t)=(x_i,t^n)} = \frac{8(\phi_{i+1}^n - \phi_{i-1}^n) - (\phi_{i+2}^n - \phi_{i-2}^n)}{12\Delta x} + \mathcal{O}\left((\Delta x)^4\right). \tag{81}$$

- 6th-order approximation:

$$\partial_x \phi\big|_{(x,t)=(x_i,t^n)} = \frac{45(\phi_{i+1}^n - \phi_{i-1}^n) - 9(\phi_{i+2}^n - \phi_{i-2}^n) + (\phi_{i+3}^n - \phi_{i-3}^n)}{60\Delta x} + \mathcal{O}\left((\Delta x)^6\right). \tag{82}$$

- 8th-order approximation:

$$\partial_x \phi\big|_{(x,t)=(x_i,t^n)} = \frac{672(\phi_{i+1}^n - \phi_{i-1}^n) - 168(\phi_{i+2}^n - \phi_{i-2}^n) + 32(\phi_{i+3}^n - \phi_{i-3}^n) - 3(\phi_{i+4}^n - \phi_{i-4}^n)}{840\Delta x} + \mathcal{O}\left((\Delta x)^8\right). \tag{83}$$

The approximation of $\nabla_x \phi$ in multidimension can be done by dimension-wise computation.

Thus, we can obtain high-order approximation to the electric field $\mathbb{E}(x,t)$ from the charge density computed from f. Depending on the desired accuracy in space, we will use one of (81), (82), or (83).

6.1.4 Numerical Results: 1D1V Vlasov-Poisson System

In this section, we present numerical results obtained by conservative SL methods combined with high-order splitting methods. For 1D problems $(d_x, d_v) = (1, 1)$, we adopted a uniform mesh Δx and Δv both in space and velocity. The CFL number is defined as

$$\text{CFL} := \max\left\{\max_j |v_j| \frac{\Delta t}{\Delta x}, \max_i |\mathbb{E}_i| \frac{\Delta t}{\Delta v}\right\}. \tag{84}$$

Table 18 Numerical methods with MPP limiter used for the 1D Vlasov-Poisson model

Scheme	Order of time splitting method	Basic reconstruction
2-P3-MPP	2nd-order Strang splitting method (78)	P3
4-P5-MPP	4th-order Yoshida splitting (79)	P5
6-P7-MPP	6th-order method in (80)	P7

For accuracy tests, we measure numerical errors of density and corresponding convergence rates with (31) and (32). In 2D, we use a similar metric. The numerical methods we will use are described in Table 18. Note that our schemes maintain conservation and MPP properties of analytical solutions for (74).

Accuracy Test
In [41], the accuracy of high-order splitting-based methods is confirmed with smooth initial data [56]:

$$f(x,v,0) = \frac{2}{7\sqrt{2\pi}}(1+5v^2)\left(1+\alpha\left((\cos(2kx)+\cos(3kx))/1.2+\cos(kx)\right)\right) \times \exp\left(-\frac{v^2}{2}\right), \quad \alpha = 0.01, \quad k = 0.5. \tag{85}$$

In the test, the periodic boundary condition is imposed on the interval $[0, 4\pi]$, while zero-boundary condition is imposed on velocity domain $[-8, 8]$. According to the definition of the CFL number (84), the time step is fixed with CFL$=6,\ 30,\ 80$, and we compute numerical solutions up to a final time $t_f = 2$.

From Table 18, we note that each numerical scheme takes a spatial reconstruction whose order is higher than that of time splitting method. This means that for a fixed CFL number and sufficiently refined grid in all independent variables, spatial errors become smaller than time errors, especially for large CFL numbers and/or long integration time. On the contrary, we can expect that convergence rates for reconstructions are easily confirmed for small CFL numbers and/or short integration time.

In Table 19, for CFL$=80$ we obtain the convergence rates of time splitting methods, while for relatively small CFL$=6$ we observe the accuracy of spatial reconstructions. In both cases, the use of the MPP limiter does not lead to any order reduction. In case of CFL$=30$, we observe intermediate results between the two extreme cases. In Fig. 34, we compare the efficiency of schemes applied to the Vlasov-Poisson system with CPU time and numerical errors.

1D Long-Time Simulation
The Vlasov-Poisson system admits several time invariants, such as L^p-norms of the solution total energy and entropy. Therefore, it is important to check if such quantities remain constant in time. In [41], we consider weak and strong Landau damping tests to check the performance of the high-order splitting technique. In these tests, we monitor the behavior of four quantities:

Table 19 Accuracy test for the 1D Vlasov-Poisson system [41]. Initial data is given in (85)

		2-P3-MPP	4-P5-MPP		6-P7-MPP		
	$(N_x, 2N_x)$	Error	Rate	Error	Rate	Error	Rate
CFL= 6	(40, 80)	9.6342e-04	2.83269.2861e-04	4.6418	6.3265e-05	6.5140	
	(80, 160)	1.3524e-04	2.8548	3.7198e-05	4.8619	6.9221e-07	6.8646
	(160, 320)	1.8695e-05	2.7401	1.2792e-06	4.8827	5.9401e-09	6.9705
	(320, 640)	2.7981e-06		4.3361e-08		4.7365e-11	
CFL= 30	(80, 160)	7.9790e-04	2.4358	1.0121e-03	3.4034	7.0842e-07	6.9204
	(160, 320)	1.4747e-04	2.0432	9.5658e-05	3.7276	5.8487e-09	6.9745
	(320, 640)	3.5779e-05	1.8966	7.2211e-06	3.8933	4.6506e-11	6.9745
	(640, 1280)	9.6092e-06		4.8595e-07		3.6982e-13	
CFL= 80	(160, 320)	1.5142e-03	2.4288	1.9771e-03	2.7746	7.0569e-07	8.7136
	(320, 640)	2.8122e-04	2.0837	2.8893e-04	3.6838	1.6810e-09	5.9542
	(640, 1280)	6.6341e-05	2.0194	2.2483e-05	3.9038	2.7113e-11	5.9407
	(1280, 2560)	1.6364e-05		1.5020e-06		4.4141e-13	

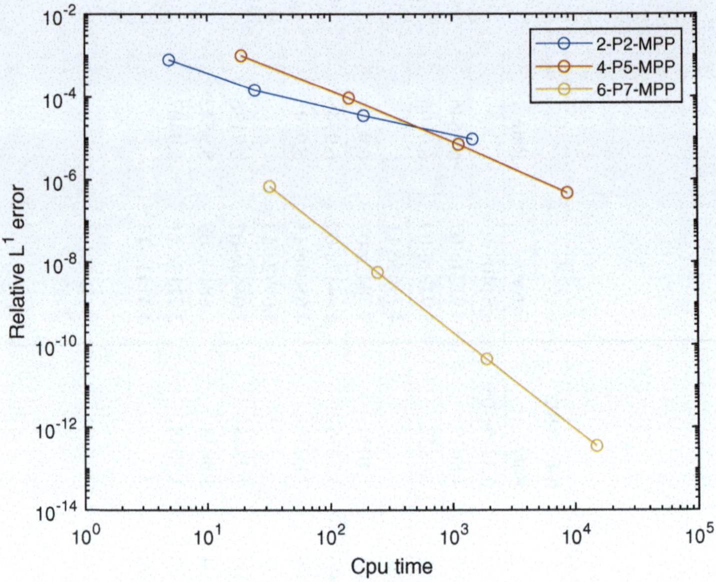

Fig. 34 1D Vlasov-Poisson. Error versus CPU time [41]

- L^p-norm of f^n, $p = 1, 2$

$$\|f^n\|_p = \left(\sum_{i,j} |f^n_{i,j}|^p (\Delta x)^{d_x} (\Delta v)^{d_v} \right)^{1/p}.$$

- Total energy of f^n

$$\text{Energy} = \sum_{i,j} \frac{|v_j|^2}{2} f^n_{i,j} (\Delta x)^{d_x} (\Delta v)^{d_v} + \sum_i \frac{|\mathbb{E}^n_i|^2}{2} (\Delta x)^{d_x}.$$

- Entropy of f^n

$$\text{Entropy} = -\sum_{i,j} f^n_{i,j} \log f^n_{i,j} (\Delta x)^{d_x} (\Delta v)^{d_v}.$$

In addition, we check the behavior of the electric field in time:

- L^2-norm of $\mathbb{E}^n$

$$\|\mathbb{E}^n\|_2 = \left(\sum_i |\mathbb{E}^n_i|^2 (\Delta x)^{d_x} \right)^{1/2}.$$

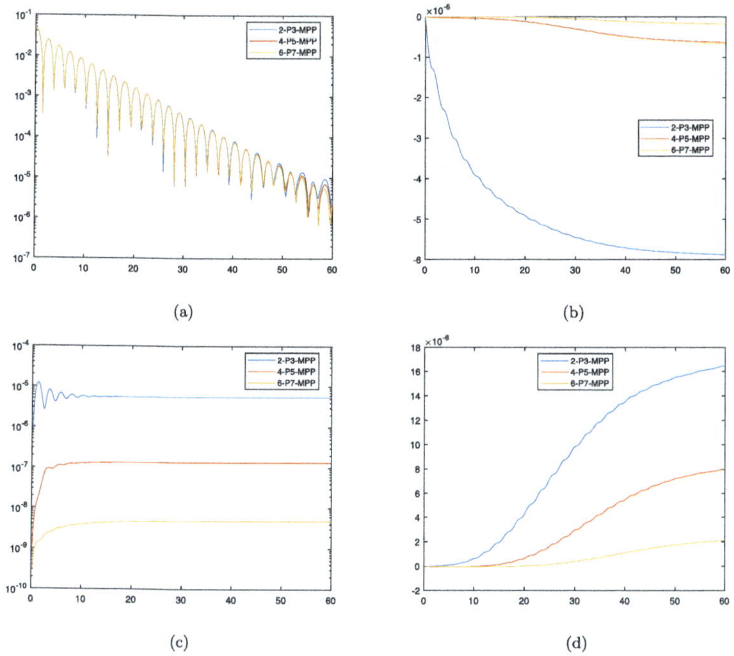

Fig. 35 Weak Landau damping in the 1D Vlasov-Poisson [41]. $N_x = 64$, $N_v = 128$. (**a**) L^2-norm of E. (**b**) $\|f^n\|_2 - \|f^0\|_2$. (**c**) $|\text{Energy}(t) - \text{Energy}(0)|$. (**d**) Entropy$(t)-$Entropy$(0)$

We consider the two benchmark problems in [96] using the initial data:

$$f(x, v, 0) = \frac{1}{\sqrt{2\pi}}\Big(1 + \alpha\cos(kx)\Big)\exp\left(-\frac{v^2}{2}\right),$$

where we use $(\alpha, k) = (0.01, 0.5)$ for weak Landau damping and $(\alpha, k) = (0.5, 0.5)$ for strong Landau damping.

For both problems, we impose periodic boundary condition on the physical domain $[-2\pi, 2\pi]$ and zero-boundary condition on the velocity domain $[-2\pi, 2\pi]$. Numerical solutions are computed up to the final time $t_f = 60$ with a time step determined by CFL$=6$ using (84). The grid size is $N_x = 64$ and $N_v = 128$

In Fig. 35, we plot numerical results obtained by high-order splitting methods listed in Table 18. The time evolution of the L^2-norm of the electric field $\mathbb{E}^n$ shows a theoretical damping rate for all schemes. After $t = 50$, only scheme 6-P7-MPP maintains the correct damping rate.

Regarding the preservation of conserved quantities such as the L^2-norm/energy/entropy of numerical solutions, all schemes produce small conservation errors within 10^{-6}. Among them, the 6-P7-MPP scheme gives the most accurate profiles. These results are comparable with those obtained with

Table 20 Weak Landau damping in the 1D Vlasov-Poisson. $\left| \|f^{N_t}\|_1 - \|f^0\|_1 \right|$

		2-P3-MPP	4-P5-MPP	6-P7-MPP
	(N_x, N_v)	Error	Error	Error
$\max_j \lvert v_j \rvert = 2\pi$	(64, 128)	2.7274e-10	4.1143e-10	2.4873e-10
$\max_j \lvert v_j \rvert = 2.5\pi$	(64, 160)	7.1054e-15	3.6344e-12	1.9540e-14

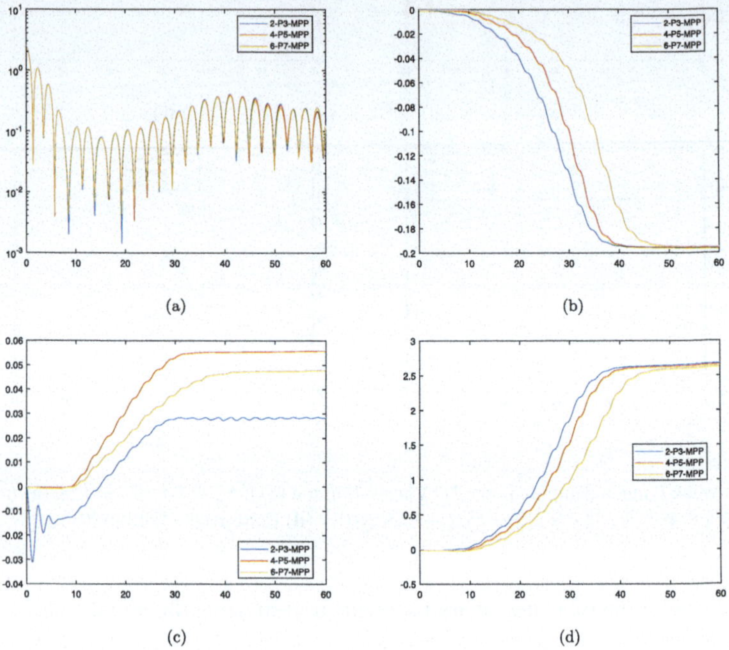

Fig. 36 Strong Landau damping in the 1D Vlasov-Poisson [41]. $N_x = 64$, $N_v = 128$. (**a**) L^2-norm of E. (**b**) $\|f^n\|_2 - \|f^0\|_2$. (**c**) Energy$(t)-$Energy(0). (**d**) Entropy$(t)-$Entropy(0)

other conservative finite difference methods available in the literature, such as, for example, [115].

In Table 20, we also report the L^1-norm of f at a final time for two different velocity domains. When we increase the size of the domain by 25% keeping Δv unchanged, the error decreases by orders of magnitude, confirming that the lack of conservation is only due to finite computational domain.

For strong Landau damping, we first compute numerical solutions taking the same domain and meshes used for the weak Landau damping. In Fig. 36, numerical results are plotted for $N_x = 64$, $N_v = 128$ with velocity domain $[-2\pi, 2\pi]$. Due to the increased perturbation amplitude $\alpha = 0.5$, nonlinear damping begins to appear around $t = 10$. Such nonlinear tendency of the electric field is well captured by each scheme, but all conservation errors are nonnegligible.

In Table 21, the small error in the L^1-norm is due to the loss of conservation caused by zero-boundary condition at the edge of the computational domain in

Table 21 Strong Landau damping in the 1D Vlasov-Poisson. $| \, \|f^{N_t}\|_1 - \|f^0\|_1 \, |$

		2-P3-MPP	4-P5-MPP	6-P7-MPP
	(N_x, N_v)	Error	Error	Error
$\max_j \lvert v_j \rvert = 2\pi$	(64, 128)	4.3616e-08	6.5483e-08	5.0082e-08
$\max_j \lvert v_j \rvert = 2.5\pi$	(64, 160)	4.6185e-13	2.5722e-12	6.9633e-13

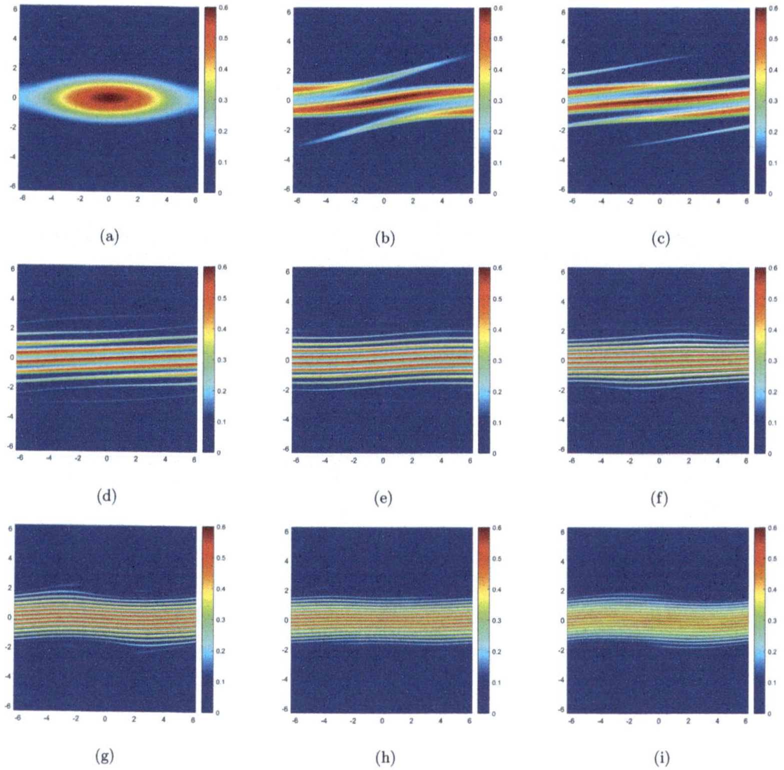

Fig. 37 Phase space portrait of the distribution function in the case of strong Landau damping in the 1D Vlasov-Poisson [41]. Here we use a velocity domain $[-2\pi, 2\pi]$ with $N_x = 128$, $N_v = 256$. **(a)** $t = 0$. **(b)** $t = 6$. **(c)** $t = 9$. **(d)** $t = 18$. **(e)** $t = 27$. **(f)** $t = 36$. **(g)** $t = 45$. **(h)** $t = 54$. **(i)** $t = 60$

velocity. If we slightly enlarge the domain in velocity space, say from $[-2\pi, -2\pi]$ to $[-2.5\pi, -2.5\pi]$, then conservation of the L^1-norm improves by four orders of magnitude. Although all schemes exhibit finite conservation errors in L^2-norm, entropy, and energy, we confirm that scheme 6-P7-MPP has less conservation errors than schemes 2-P3-MPP and 4-P5-MPP. For comparison with the results in [96], we plot the distribution functions in phase space at different times in Fig. 37.

Table 22 Accuracy test for the 2D Vlasov-Poisson system [41]. Initial data is given in (87)

	CFL= 1		CFL= 2		CFL= 3		CFL= 4		CFL= 5	
$(N_x^2, (2N_x)^2)$	error	rate	error	rate	error	rate	error	rate	error	rate
$(20^2, 40^2)$	1.19e-03	2.8278	1.24e-03	2.73	1.43e-03	2.46	2.22e-03	2.47	2.55e-03	2.11
$(40^2, 80^2)$	1.68e-04		1.87e-04		2.61e-04		4.01e-04		5.92e-04	

6.1.5 Numerical Results: 2D2V Vlasov-Poisson System

For two-dimensional problems, we assume uniform square meshes both in space and velocity domains, i.e., $\Delta x = \Delta y = h$ for space and $\Delta v^1 = \Delta v^2 = \Delta v$ for velocity. In two-dimensional accuracy tests, we will compute errors and convergence rates using the relative L^1-norm as in (31) and (32).

In 2D problem $(d_x, d_v) = (2, 2)$, we take the uniform mesh $\Delta x = \Delta y$ for space and Δv for each velocity dimension. We use a CFL number defined by

$$\mathrm{CFL} := \max\left\{\max_j |v_j^1|\frac{\Delta t}{\Delta x} + \max_j |v_j^2|\frac{\Delta t}{\Delta y},\ \max_i |E_i^1|\frac{\Delta t}{\Delta v} + \max_i |E_i^2|\frac{\Delta t}{\Delta v}\right\}, \tag{86}$$

for $E_i := (E_i^1, E_i^2)$.

Accuracy Test

To check the accuracy of splitting methods for the 2D Vlasov-Poisson system, we consider a setup in [57], where the initial data is given by

$$\begin{aligned} f(x, y, v^1, v^2, 0) &= \frac{1}{2\pi}\Big(1 + \alpha\big(\cos(kx) + \cos(ky)\big)\Big)\exp\left(-\frac{|v^1|^2 + |v^2|^2}{2}\right), \\ \alpha &= 0.05, \quad k = 0.5. \end{aligned} \tag{87}$$

We impose the periodic boundary condition on the space domain $[0, 4\pi]^2$ and assume zero-boundary condition on the velocity domain $[-6, 6]^2$. Numerical solutions are computed up to final time $t_f = 2$ taking uniform grids with $N_x = N_y = N_{v^1} = N_{v^2} = 20, 40, 80$ and different time steps based on CFL$= 1, 2, 3, 4, 5$ using (86).

In this test, we consider a SL method based on the 2nd-order Strang splitting method, i.e., in the advection step we solve (76) and in the drift step we solve (77). Note that we can solve each step without dimensional splitting, because during the advection step the velocities are constant in time, and during the drift step the electric field is constant in time. For a basic reconstruction, together with MPP limiter, we again take 2D-P3, which is the two-dimensional optimal polynomial of degree 2 used for CWENO23 reconstruction [78].

In Table 22, we report errors based on relative L^1-norm (31) and convergence rates (32) for two-dimensional solutions ($d_x = d_v = 2$). The results show that, for smaller CFL numbers, numerical errors tend to be smaller, and the desired accuracy of spatial reconstruction appears. On the contrary, as the CFL number gets bigger, the expected accuracy of the method is observed. As in the 1D case, the use of the MPP limiter does not lead to order reduction.

2D Long-Time Simulations

For the 2D Vlasov-Poisson model, we again consider a strong Landau damping problem as in [45]:

$$f(x, y, v^1, v^2, 0) = \frac{1}{2\pi}\Big(1 + \alpha\big(\cos(kx) + \cos(ky)\big)\Big)\exp\left(-\frac{|v^1|^2 + |v^2|^2}{2}\right),$$
$$\alpha = 0.5, \quad k = 0.5. \tag{88}$$

We impose the periodic boundary condition in space domain $[0, 4\pi]^2$ and zero-boundary condition in the velocity domain $[-6, 6]^2$. Using the scheme based on Strang splitting and 2D-P3, numerical solutions are computed up to the final time $t_f = 40$. Here take a fixed time step $\Delta t = \frac{1}{8}$, and we set $N_x = N_y = 32$ and $N_{v^1} = N_{v^2} = 128$.

As in 1D case, Fig. 38 shows that a phase transition appears for the L^2-norm of electric field during time evolution, in agreement with that already obtained in [45]. At final time, due to the truncation of the velocity domain, the L^1-norm of the numerical solution is maintained with a loss equal to 6.67×10^{-6}, which is reduced to 2.02×10^{-10} by using the 25% enlarged velocity domain $[-7.5, 7.5]^2$ and same Δv. The other conservative quantities such as L^2-norm, entropy, and energy are not fully conserved at a numerical level.

6.2 *Non-Splitting Schemes*

In this section, we discuss a high-order truly multidimensional SL finite difference approach for solving the VP system. The "truly multidimensional" means that no operator splitting is involved. We focus on resolving the difficulty in tracing *nonlinear* characteristics with high-order temporal accuracy [93, 114, 115]. A high-order two-stage multi-derivative predictor-corrector algorithm is proposed to build up a high-order characteristic-tracing algorithm based on lower-order ones, with the help of moment equations of the VP system. We use a high-order WENO interpolation to recover information among grid points. Notice that the WENO interpolation in the finite difference framework does not guarantee mass conservation; subsequent work on utilizing the finite element discontinuous Galerkin framework [28, 29] will lead to mass conservative non-splitting SL schemes.

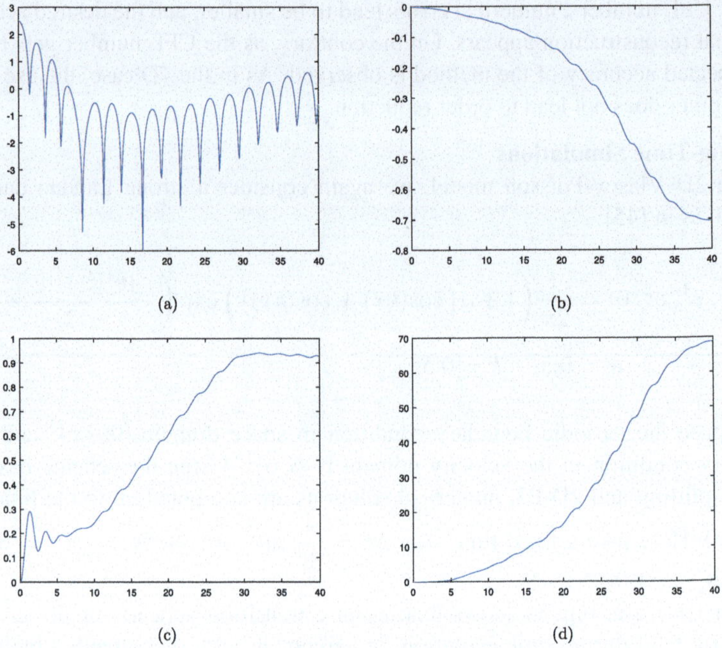

Fig. 38 2D Vlasov-Poisson. Strong Landau damping with initial data (88). $N_x = N_y = 32$, $N_v^1 = N_v^2 = 128$ [41]. (a) Logarithmic scale of L^2-norm of E. (b) $\|f^{N_t}\|_2 - \|f^0\|_2$. (c) Energy($t$)−Energy(0). (d) Entropy(t)−Entropy(0)

6.2.1 Tracking Nonlinear Characteristics for the 1D1V Vlasov-Poisson System

We design a high-order SL finite difference scheme for the VP system without operator splitting. Consider the VP system (74) with 1D in x and 1D in v. The 2D $x - v$ plane is discretized into uniformly spaced rectangular meshes:

$$x_{\frac{1}{2}} < x_{1+\frac{1}{2}} < \cdots < x_{i+\frac{1}{2}} < \cdots < x_{N_x+\frac{1}{2}},$$

$$v_{\frac{1}{2}} < v_{1+\frac{1}{2}} < \cdots < v_{j+\frac{1}{2}} < \cdots < v_{N_v+\frac{1}{2}}.$$

The center of each of the rectangular cell $[x_{i-\frac{1}{2}}, x_{i+\frac{1}{2}}] \times [v_{j-\frac{1}{2}}, v_{j+\frac{1}{2}}]$ is denoted as (x_i, v_j). We consider evolving the numerical solution $f_{i,j}^n$, $i = 1, \cdots N_x$, $j = 1, \cdots, N_v$, where $f_{i,j}^n$ denotes the numerical solution at (x_i, v_j) at the time level t^n. The proposed SL algorithm in updating the solution f_{ij}^{n+1} consists of the following steps:

1. Characteristics are traced backward in time to t^n. Let the foot of the characteristic at the time level t^n emanating from (x_i, v_j) at t^{n+1} be denoted as $(x_i^\star, v_j^\star)$. It is

approximated by numerically solving the following final value problem:

$$\begin{cases} \frac{dx(t)}{dt} = v(t), \\ \frac{dv(t)}{dt} = E(x(t), t), \\ x(t^{n+1}) = x_i, \\ v(t^{n+1}) = v_j. \end{cases} \tag{89}$$

Here, we remark that solving (89) with high-order temporal accuracy is non-trivial. Especially, the electric field E depends on the unknown function f via Poisson's equation (75) in a global rather than local fashion. Moreover, being a final value problem, the electric field E is known initially only at the time step t^n. Below we first discuss the proposed high-order (up to third-order) way of tracing characteristics in time.

2. The solution is updated as

$$f_{i,j}^{n+1} = f(x_i^{n,(l)}, v_j^{n,(l)}, t^n) \approx f(x_i^\star, v_j^\star, t^n).$$

We propose to recover $f(x_i^{n,(l)}, v_j^{n,(l)}, t^n)$ by a high-order (up to sixth-order) WENO interpolation from $f_{i,j}^n$, $i = 1, \cdots N_x$, $j = 1, \cdots, N_v$. Here the superscript (l) denotes the formal order of temporal approximation in locating the feet of characteristics.

The proposed method introduces a predictor-corrector procedure to determine the foot of characteristics. Initially, a first-order scheme is described for tracing characteristics. Following this, a second-order scheme is developed, which builds upon the predictions made by the first-order scheme. Lastly, a third-order scheme is proposed, which is based on the predictions from the second-order scheme. In our notations, the superscript ^{n} denotes the time level; the subscripts i and j denote the location x_i and v_j in x and v directions, respectively; and the superscript $^{(l)}$ denotes the formal order of approximation. For example, in Eq. (90) below, $x_i^{n,(1)}$ (or $v_j^{n,(1)}$) approximates $x_i^\star$ (or $v_j^\star$) with first order, and $E_i^n = E(x_i, t^n)$. $\frac{d}{dt} = \frac{\partial}{\partial t} + \frac{\partial x}{\partial t}\frac{\partial}{\partial x}$ denotes the material derivatives along characteristics. The order of approximation we mentioned in this subsection is for temporal accuracy. We use a spectrally accurate fast Fourier transform (FFT) in solving Poisson's equation (75), whose r.h.s. function $\rho(x, t) = \int f(x, v, t)dv$ is evaluated by a midpoint rule with spectral accuracy [24].

First-Order Scheme We let

$$x_i^{n,(1)} = x_i - v_j\Delta t; \quad v_j^{n,(1)} = v_j - E_i^n \Delta t, \tag{90}$$

which are first-order approximations to $x_i^\star$ and $v_j^\star$. Let

$$f_{i,j}^{n+1,(1)} = f(x_i^{n,(1)}, v_j^{n,(1)}, t^n), \tag{91}$$

which is a first order in time approximation to the exact solution $f(x_i, v_j, t^{n+1})$. Note that the spatial approximation in Eq. (91) is performed via high-order WENO interpolation. Based on $\{f_{i,j}^{n+1,(1)}\}$, we compute

$$\rho_i^{n+1,(1)}, \quad E_i^{n+1,(1)}$$

by using a midpoint rule and FFT based on Poisson's equation (75). Note that $\rho_i^{n+1,(1)}$ and $E_i^{n+1,(1)}$ also approximate $\rho(x_i, t^{n+1})$ and $E(x_i, t^{n+1})$ with first-order temporal accuracy.

Second-Order Scheme We let

$$x_i^{n,(2)} = x_i - \frac{1}{2}(v_j + v_j^{n,(1)})\Delta t, \quad v_j^{n,(2)} = v_j - \frac{1}{2}\left(E(x_i^{n,(1)}, t^n) + E_i^{n+1,(1)}\right)\Delta t, \tag{92}$$

which are second-order approximations to $x_i^\star$ and $v_j^\star$. Note that $E(x_i^{n,(1)}, t^n)$ in Eq. (92) can be approximated by the WENO interpolation from $\{E_i^n\}_{i=1}^{N_x}$. Let $f_{i,j}^{n+1,(2)} = f(x_i^{n,(2)}, v_j^{n,(2)}, t^n)$, approximating $f(x_i, v_j, t^{n+1})$ to second order in time. Based on $\{f_{i,j}^{n+1,(2)}\}$, we compute $\rho_i^{n+1,(2)}, \quad E_i^{n+1,(2)}$ approximating $\rho(x_i, t^{n+1})$ and $E(x_i, t^{n+1})$ with second-order temporal accuracy.

Third-Order Scheme We let

$$x_i^{n,(3)} = x_i - v_j\Delta t + \frac{\Delta t^2}{2}\left(\frac{2}{3}E_i^{n+1,(2)} + \frac{1}{3}E(x_i^{n,(2)}, t^n)\right),$$

$$v_j^{n,(3)} = v_j - E_i^{n+1,(2)}\Delta t + \frac{\Delta t^2}{2}\left(\frac{2}{3}(\frac{d}{dt}E(x_i, t^{n+1}))^{(2)} + \frac{1}{3}\frac{d}{dt}E(x_i^{n,(2)}, t^n)\right),$$

which are third-order approximations to $x_i^\star$ and $v_j^\star$. Note that $\frac{d}{dt}E$ terms will be obtained by using the macro-equations. Let $f_{i,j}^{n+1,(3)} = f(x_i^{n,(3)}, v_j^{n,(3)}, t^n)$, approximating $f(x_i, v_j, t^{n+1})$ with third order in time. Based on $\{f_{i,j}^{n+1,(3)}\}$, we compute $\rho_i^{n+1,(3)}, \quad E_i^{n+1,(3)}$ approximating $\rho(x_i, t^{n+1})$ and $E(x_i, t^{n+1})$ with third-order temporal accuracy. We note that the mechanism to build this third-order scheme is different from Runge-Kutta methods where intermediate-stage solutions are constructed. It has some similarity in spirit to the Lax-Wendroff-type method, where higher-order time derivatives are recursively transformed into spatial derivatives. The difference with the Lax-Wendroff-type time integration is that the Lax-Wendroff method only uses spatial derivatives at one time level, while the proposed method uses the spatial derivatives (or its high-order approximations) at both t^n and t^{n+1} via a predictor-corrector procedure. In a sense, the above method is a two-stage multi-derivative method.

Below, we present our results for the 1D1V Vlasov-Poisson system from [93]. In particular, we consider two-stream instability [56], with an unstable initial distribution function:

$$f(x, v, t = 0) = \frac{2}{7\sqrt{2\pi}}(1 + 5v^2)(1 + \alpha((\cos(2kx) + \cos(3kx))/1.2 + \cos(kx)) \exp(-\frac{v^2}{2})$$

with $\alpha = 0.01$, $k = 0.5$, the length of the domain in the x direction is $L = \frac{2\pi}{k}$, and the background ion distribution function is fixed, uniform, and chosen so that the total net charge density for the system is zero.

We test both spatial and temporal convergence of the proposed truly multidimensional SL method. We first test the spatial convergence by using a sequence of meshes with $N_x = N_v = \{210, 126, 90, 70\}$. The meshes are designed so that the coarse mesh grid coincides with part of the reference fine mesh grid ($N_x = N_v = 630$). The errors are measured by comparing the numerical solution with that on the reference fine grid with 630×630 grid points. We set CFL $= 0.01$ so that the spatial error is the dominant error. Table 23 is the spatial convergence table for the proposed schemes with sixth-order WENO interpolation. The expected fifth-order convergence globally in time is observed. We then test the temporal convergence of the proposed first-, second-, and third-order schemes. Table 24 provides the temporal convergence rate for the scheme with the first- to third-order temporal accuracy. We use the sixth-order WENO interpolation and a spatial mesh of $N_x = N_v = 160$, so that the temporal error is the dominant error. The reference solution is computed by using the same spatial mesh but with relatively small CFL $= 0.1$. In Table 24, the time step size is about 6 to 10 times that from a Eulerian method, yet highly accurate numerical results are achieved. Expected first-, second-, and third-order temporal accuracy are observed. On the rightmost column of Table 24, we show errors and the corresponding orders of convergence with CFL ranging from 1 to 5. One can see that the spatial error starts to dominate as CFL decreases; thus, the theoretical order of convergence in time can no longer be observed.

6.2.2 Tracking Nonlinear Characteristics for the Guiding Center Vlasov System

The other model of interest is the guiding center Vlasov system [107]:

$$\frac{\partial \rho}{\partial t} + \nabla \cdot (\mathbf{E}^{\perp}\rho) = 0, \qquad \mathbf{E}^{\perp} = (-\frac{\partial \psi}{\partial y}, \frac{\partial \psi}{\partial x}), \tag{93}$$

Table 23 Two-stream instability: test order of accuracy in space for the SL WENO schemes by comparing numerical solutions with a reference solution from a simulation with a set of refined mesh. The scheme use sixth-order WENO interpolation and has a third-order temporal accuracy in tracing characteristics. $T = 1$ and $CFL = 0.01$. We set $CFL = 0.01$ so that the spatial error is the dominant error

$N_x \times N_v$	L^1 error	Order
70×70	7.01E-7	–
90×90	2.06E-7	4.88
126×126	3.96E-8	4.89
210×210	3.20E-9	4.95

Table 24 Two-stream instability. $N_x = N_v = 160$ and $T = 5$. Order of accuracy in time for the SL WENO schemes with sixth-order WENO interpolation and various orders of temporal accuracy. The rightmost column is for the third-order scheme with CFL ranging from 1 to 5. One can see that the spatial error starts to dominate as CFL decreases; thus, the theoretical order of convergence in time can no longer be observed

	First order		Second order		Third order		CFL	Third order	
CFL	L^1 error	Order	L^1 error	Order	L^1 error	Order	CFL	L^1 error	Order
6	1.17E-4	–	2.40E-6	–	1.13E-7	3.00	1	1.70E-9	–
7	1.40E-4	1.13	2.80E-6	2.04	1.79E-7	3.02	2	5.00E-9	1.60
8	1.63E-4	1.16	3.69E-6	2.07	2.69E-7	3.02	3	1.44E-8	2.60
9	1.87E-4	1.16	4.69E-6	2.04	3.84E-7	3.03	4	3.35E-8	2.93
10	2.12E-4	1.20	5.84E-6	2.08	5.31E-7	3.065	5	6.52E-8	2.97

where ρ is the charge density, with the electric field $\mathbf{E}$ determined by the 2D Poisson's equation for the potential ψ:

$$-\Delta\psi = \rho, \tag{94}$$

The guiding center model is an approximation to the 2D Vlasov-Poisson system, assuming a strong constant external magnetic field orthogonal to the plane of motion.

Below, we discuss a high-order strategy in tracking nonlinear characteristic curves for the guiding center model (93). We consider a uniformly spaced rectangular mesh:

$$x_{\frac{1}{2}} < x_{1+\frac{1}{2}} < \cdots < x_{i+\frac{1}{2}} < \cdots < x_{N_x+\frac{1}{2}},$$
$$y_{\frac{1}{2}} < y_{1+\frac{1}{2}} < \cdots < y_{j+\frac{1}{2}} < \cdots < y_{N_y+\frac{1}{2}}.$$

The center of each rectangular cell $I_{i,j} = [x_{i-\frac{1}{2}}, x_{i+\frac{1}{2}}] \times [y_{j-\frac{1}{2}}, y_{j+\frac{1}{2}}]$ is denoted as (x_i, y_j) and the mesh sizes by Δx and Δy. Let $\rho^n_{i,j}$ denote the numerical solution at (x_i, y_j) at the time level t^n, and the time stepping size is $\Delta t = t^{n+1} - t^n$.

As in the previous setting, the non-splitting SL scheme updates the solution by tracking characteristic curves and using the fact that solution values stay constant along characteristics with the following two steps:

1. **Characteristic tracking.** Characteristics are traced backward in time from t^{n+1} to t^n. Let $(x^\star_{i,j}, y^\star_{i,j})$ denote the foot of the characteristic, reaching the grid point (x_i, y_j) at t^{n+1}. It is approximated by solving the following final value problem numerically:

$$\begin{cases} \frac{dx(t)}{dt} = \mathbf{E}^{\perp,1}(x(t), y(t), t), \\ \frac{dy(t)}{dt} = \mathbf{E}^{\perp,2}(x(t), y(t), t), \\ x(t^{n+1}) = x_i, \\ y(t^{n+1}) = y_j. \end{cases} \tag{95}$$

   Here $\frac{d}{dt}$ is the material derivative along a characteristic curve. Notice that solving (95) with high-order temporal accuracy is nontrivial, especially since the velocity field $\mathbf{E}^\perp = (\mathbf{E}^{\perp,1}, \mathbf{E}^{\perp,2})$ depends on the unknown function ρ via the 2D Poisson's equation (94) in a global nonlinear fashion. In the following, we use $\mathbf{u} = (u_1, u_2)$ as $\mathbf{E}^\perp$ for simplicity of notation. Below, we will describe our proposed high-order prediction-correction procedure in approximating the foot of the characteristic $(x^\star_{i,j}, y^\star_{i,j})$ at time level t^n via system (95). We denote $(x^{n,(l)}_{i,j}, y^{n,(l)}_{i,j})$ as the numerical approximate values, and the superscript $^{(l)}$ means that $(x^{n,(l)}_{i,j}, y^{n,(l)}_{i,j})$ is an lth-order (in time) approximation to $(x^\star_{ij}, y^\star_{ij})$.
2. **Solution updating.** The solution is updated from

$$\rho^{n+1}_{i,j} = \rho(x^{n,(l)}_{i,j}, y^{n,(l)}_{i,j}, t^n) \approx \rho(x^\star_{i,j}, y^\star_{i,j}, t^n),$$

   since the solution stays constant along characteristics. We propose to recover $\rho(x^{n,(l)}_{i,j}, y^{n,(l)}_{i,j}, t^n)$ by a high-order (up to 6th-order) WENO interpolation from numerical grid point values $\{\rho^n_{i,j}\}_{i,j=1}^{N_x,N_y}$ in its neighborhood. Such a procedure has been described in detail in [93].

A similar predictor-corrector procedure for locating the feet of characteristic curves, but tailored for the nonlinear guiding center system, is proposed in [114]. Below, we adopt the following notations: the superscript ^{n} indicates that the function is at time level t^n. The subscripts i and j are indices associated with the grid point (x_i, y_j). The superscript $^{(l)}$ is for the formal order of accuracy in temporal discretization. For example, in Eq. (96) below, $(x^{n,(1)}_{i,j}, y^{n,(1)}_{i,j})$ approximates the foot of a characteristic curve emanating from the grid point (x_i, y_j) at time level t^n (i.e., $(x^\star_{i,j}, y^\star_{i,j})$) with first-order accuracy; and $u^n_{k,i,j} = u_k(x_i, y_j, t^n)$ is the velocity along x or y direction at the grid point (x_i, y_j) at time level t^n, for $k = 1, 2$ respectively.

First-Order Scheme We let

$$x_{i,j}^{n,(1)} = x_i - u_1(x_i, y_j, t^n)\Delta t; \quad y_{i,j}^{n,(1)} = y_j - u_2(x_i, y_j, t^n)\Delta t, \tag{96}$$

which are first-order approximations to $x_{i,j}^{\star}$ and $y_{i,j}^{\star}$. Let

$$\rho_{i,j}^{n+1,(1)} = \rho(x_{i,j}^{n,(1)}, y_{i,j}^{n,(1)}, t^n), \tag{97}$$

denote a first-order approximation (in time) to $\rho_{i,j}^{n+1}$. Notice that the spatial approximation in Eq. (97) (and in other similar equations in this subsection) is performed via a high-order WENO interpolation. Based on $\{\rho_{i,j}^{n+1,(1)}\}_{i,j=1}^{N_x,N_y}$, we compute $\mathbf{E}_{i,j}^{\perp,n+1,(1)} = \mathbf{u}$ by using FFT solving the 2D Poisson's equation (94). We note that $\mathbf{E}_{i,j}^{\perp,n+1,(1)}$ approximates $\mathbf{E}_{i,j}^{\perp,n+1}$ with first-order temporal accuracy.

Second-Order Scheme We let

$$x_{i,j}^{n,(2)} = x_i - \frac{1}{2}\left(u_{1,i,j}^{n+1,(1)} + u_1^n(x_{i,j}^{n,(1)}, y_{i,j}^{n,(1)}, t^n)\right)\Delta t, \tag{98}$$

$$y_{i,j}^{n,(2)} = y_j - \frac{1}{2}\left(u_{2,i,j}^{n+1,(1)} + u_2^n(x_{i,j}^{n,(1)}, y_{i,j}^{n,(1)}, t^n)\right)\Delta t, \tag{99}$$

in spirit of the trapezoid rule using solutions at two end points of the characteristic curve (one is at the starting grid point at t^{n+1} and the other is at the foot of the characteristic at t^n). In [114], we show that they are second-order approximations to $x_{i,j}^{\star}$ and $y_{i,j}^{\star}$. Note that $\mathbf{u}(x_{i,j}^{n,(1)}, y_{i,j}^{n,(1)}, t^n)$ in Eqs. (98) and (99) can be approximated by the WENO interpolation from $\{\mathbf{u}_{i,j}^n\}_{i,j=1}^{N_x,N_y}$. Let $\rho_{i,j}^{n+1,(2)} = \rho(x_{i,j}^{n,(2)}, y_{i,j}^{n,(2)}, t^n)$; it approximates $\rho_{i,j}^{n+1}$ with second-order accuracy in time. Based on $\{\rho_{i,j}^{n+1,(2)}\}_{i,j=1}^{N_x,N_y}$, we compute $\mathbf{u}_{i,j}^{n+1,(2)} = (u_{1,i,j}^{n+1,(2)}, u_{2,i,j}^{n+1,(2)})$, approximating $\mathbf{u}_{i,j}^{n+1}$ with second-order temporal accuracy.

Third-Order Scheme We let

$$x_{i,j}^{n,(3)} = x_i - u_{1,i,j}^{n+1,(2)}\Delta t + \frac{\Delta t^2}{2}\Bigg(\frac{2}{3}\left(\frac{\partial u_1}{\partial t} + \frac{\partial u_1}{\partial x}u_1 + \frac{\partial u_1}{\partial y}u_2\right)_{i,j}^{n+1,(2)} + \frac{1}{3}\left(\frac{\partial u_1}{\partial t} + \frac{\partial u_1}{\partial x}u_1 + \frac{\partial u_1}{\partial y}u_2\right)(x_{i,j}^{n,(2)}, y_{i,j}^{n,(2)}, t^n)\Bigg), \tag{100}$$

$$y_{i,j}^{n,(3)} = y_j - u_{2,i,j}^{n+1,(2)}\Delta t + \frac{\Delta t^2}{2}\Bigg(\frac{2}{3}\left(\frac{\partial u_2}{\partial t} + \frac{\partial u_2}{\partial x}u_1 + \frac{\partial u_2}{\partial y}u_2\right)_{i,j}^{n+1,(2)}$$

$$+\frac{1}{3}\left(\frac{\partial u_2}{\partial t}+\frac{\partial u_2}{\partial x}u_1+\frac{\partial u_2}{\partial y}u_2\right)(x_{i,j}^{n,(2)},y_{i,j}^{n,(2)},t^n)\Big). \tag{101}$$

Let $\rho_{i,j}^{n+1,(3)}=\rho(x_{i,j}^{n,(3)},y_{i,j}^{n,(3)},t^n)$, it approximates $\rho_{i,j}^{n+1}$ with third-order accuracy in time globally. Based on $\{\rho_{i,j}^{n+1,(3)}\}_{i,j=1}^{N_x,N_y}$, we compute $\mathbf{u}_{i,j}^{n+1,(3)}$ by FFT, approximating $\mathbf{u}_{i,j}^{n+1}$ with third-order temporal accuracy. Note that Eq. (100) is formulated, by mimicking the following expansion:

$$x_{i,j}^{*}=x_i-\frac{dx}{dt}|_{t=t^{n+1}}\Delta t+\frac{\Delta t^2}{2}\left(\frac{2}{3}\frac{d^2x}{dt^2}\Big|_{t=t^{n+1}}+\frac{1}{3}\frac{d^2x}{dt^2}\Big|_{t=t^n}\right)+\mathcal{O}(\Delta t^4),$$

which can be easily verified by Taylor expansion, similarly for Eq. (101). In [114], we show that Eqs. (100)–(101) provide third-order approximations to $x_{i,j}^{\star}$ and $y_{i,j}^{\star}$. The second term on the right-hand side (RHS) of (100) involves a second-order approximation (therefore the superscript $^{(2)}$) to the first-order material derivative (i.e., $\frac{d}{dt}$) of x at t^{n+1}. The third term involves second-order approximations to the second-order material derivatives (i.e., $\frac{d^2}{dt^2}$) of x at two end points of the characteristic curve. For example,

$$\frac{d^2x}{dt^2}=\frac{d}{dt}\left(\frac{dx}{dt}\right)\overset{(95)}{=}\frac{d}{dt}u_1=\frac{\partial u_1}{\partial t}+\frac{\partial u_1}{\partial x}\frac{\partial x}{\partial t}+\frac{\partial u_1}{\partial y}\frac{\partial y}{\partial t}\overset{(95)}{=}\frac{\partial u_1}{\partial t}+\frac{\partial u_1}{\partial x}u_1+\frac{\partial u_1}{\partial y}u_2.$$

The term with coefficient $\frac{2}{3}$ on the RHS of Eq. (100) is at t^{n+1}. Although the solution is not yet available, its second order approximation is available in the prediction step. The term with the coefficient $\frac{1}{3}$ is for the foot of the characteristic at t^n: the solution is known at t^n and the location for the characteristic foot can be approximated with second-order from the previous second-order prediction step too. Similar arguments apply to the y component. Finally, $\frac{\partial \mathbf{u}}{\partial x}$, $\frac{\partial \mathbf{u}}{\partial y}$ are approximated by FFT at grid points, and $\frac{\partial \mathbf{u}}{\partial t}$ is computed by taking the time derivative on the 2D Poisson's equation (94). By utilizing Eqs. (93)–(94), we have a revised 2D Poisson's equation:

$$\Delta\psi_t=-(u_1\rho)_x-(u_2\rho)_y. \tag{102}$$

After obtaining $\mathbf{u}$ by solving Poisson's equation (94), the RHS of (102) can be computed, e.g., by a sixth-order central finite difference scheme. Then we can solve (102) by FFT to get ψ_t and thus obtain $\frac{\partial \mathbf{u}}{\partial t}=(-(\psi_t)_y,(\psi_t)_x)$. Finally, these partial derivatives at the foot of the characteristic are further approximated by WENO interpolations from their values at grid points.

Below, we showcased an example on the 2D guiding center model problem (93) with the initial condition:

$$\rho_0(x,y)=\sin(y)+0.015\cos(kx) \tag{103}$$

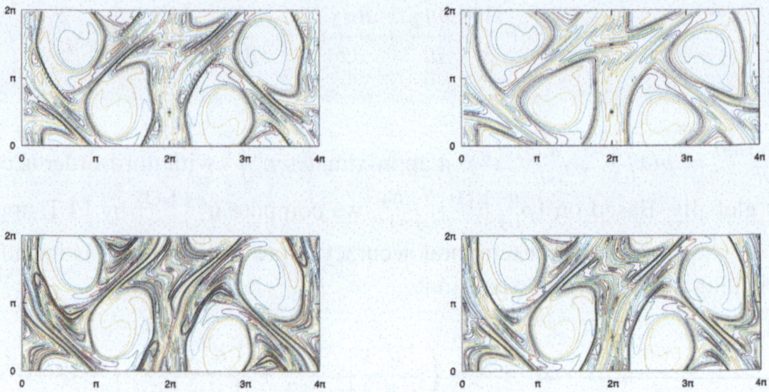

Fig. 39 Contour plots of Kelvin-Helmholtz instability problem. Top two, mesh grid $N_x \times N_y = 128 \times 128$ with CFL $= 10$ (top up), 0.5 (top down); bottom two, mesh grid $N_x \times N_y = 256 \times 256$ with CFL $= 10$ (bottom up), 0.5 (bottom down). Contours: 21 equally spaced lines in the range $[-1.05, 1.05]$

on the domain $[0, 2\pi] \times [0, 4\pi]$ and periodic boundary conditions in both x and y directions. We let $k = 0.5$ to create a Kelvin-Helmholtz (K-H) instability and consider a long-time simulation by evolving the solution up to $T = 50$. We call the 6th-order WENO interpolation in space. We first show the contour plots of the charge density ρ at $T = 40$ in Fig. 39 with the mesh grids $N_x \times N_y$ to be 128×128 and 256×256 and CFL $= 10, 0.5$. For this example with long-time simulation, we can observe larger dissipation due to more time step evolution with a small CFL number 0.5 than CFL $= 10$. In Fig. 40, we compare the cuts of the solution along $y = 3\pi/2$, $y = \pi/2$, and $x = 2\pi$ for three different temporal orders, 1st, 2nd, and 3rd, respectively, on a mesh grid $N_x \times N_y = 128 \times 128$ with CFL $= 10$. The reference solution is obtained by the 3rd-order scheme on a mesh grid $N_x \times N_y = 256 \times 256$ with CFL $= 0.5$. For this example, due to a long-time simulation, we can clearly observe the differences among different temporal order of schemes. We can see that the 3rd-order scheme is much closer to the reference solution than the other two low-order ones, which is a good example to demonstrate the good performance of a higher-order scheme. We also plot the time history of the relative deviations for the analytically conserved quantities of mass, energy, and entropy in Fig. 40. Similarly, due to a nonconservative scheme, for the long-time simulation, the mass is preserved only up to 10^{-6}, and 3rd order is the worst. For energy and entropy, the 1st-order scheme is worse than the 2nd- and 3rd-order ones.

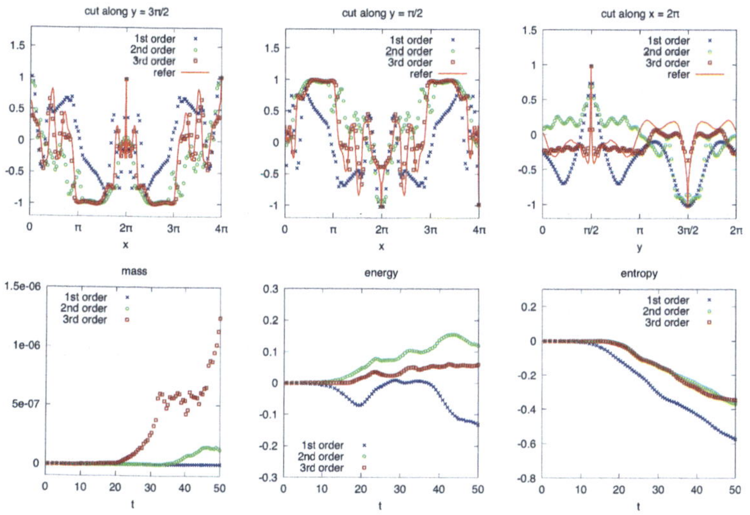

Fig. 40 Kelvin-Helmholtz instability problem. Symbols: mesh grid $N_x \times N_y = 128 \times 128$ with CFL = 10 for three different temporal order, 6th order in space. Lines: reference solutions with mesh grid $N_x \times N_y = 256 \times 256$ and CFL = 0.5, 3rd order in time and 6th order in space. Top left and right: cuts along $y = 3\pi/2$ and $y = \pi/2$ respectively. Middle left: cut along $x = 2\pi$. Middle right, bottom left, and bottom right are the history of relative deviations of mass, energy, and entropy, respectively

7 Convergence Results

7.1 Introduction

The convergence result of a numerical scheme is important in that it guarantees the reliability of the simulation generated using the scheme. Even though the BGK model has been widely used in place of the Boltzmann equation for various flow problems, satisfactory convergence estimates were not established until very recently. The first such result was established by Issautier [73], who proved the convergence of a weighted particle method for the BGK model. Then, Mieussens proved the convergence of a discrete-velocity model for the BGK model to a weak solution of the BGK model. The first convergence result for the fully discretized scheme was obtained for the 1st-order SL scheme for the BGK model in [99]. In [99], the convergence rate is measured in a weighted L^1 norm:

$$\|f(T^f) - f^{N_t}\|_{L^1_2} \leq C\left\{(\Delta x)^2 + \Delta v + \Delta t + \frac{(\Delta x)^2}{\Delta t}\right\}.$$

Here, T^f is the final time, N_t is the number of nodes for temporal variable, and Δx, Δv, Δt denote the spatial, velocity, and temporal node sizes, respectively. Later, an

L^∞-based convergence argument is developed in [100]:

$$\|f(T^f) - f^{N_t}\|_{L^\infty_q} \le C\Big\{(\Delta x)^2 + \Delta v + \Delta t + \frac{(\Delta x)^2}{\Delta t}\Big\},$$

for some constant $C > 0$. The L^∞ theory turns out to have several advantages over the L^1 result. First, such weighted L^∞ norm-based estimate provides a more precise and detailed picture of the convergence of the scheme since the error estimate in L^∞ norm gives a node-wise convergence estimate. Second, and more importantly, the proof is greatly simplified, enabling one to extend the convergence theory to various intricate versions of the BGK model. For example, it was successfully extended to the ellipsoidal BGK model, which is a modification of the BGK model to yield the correct Prandtl number in the Navier-Stokes limit [100]. It is also employed in the study of convergence problems for the SL scheme for the BGK model for a gaseous system consisting of polyatomic molecules [21]. In this section, we briefly review the L^∞-convergence theory developed for the ellipsoidal BGK model in [100].

• **Ellipsoidal BGK model** The BGK model [9, 111] has been widely used as an efficient model for the Boltzmann equation because the BGK model is much more amenable to numerical treatment and still maintains many of the important qualitative properties of the Boltzmann equation. But several shortcomings are also reported where this model fails to reproduce the correct physical data of the Boltzmann equation. One such example is the Prandtl number, which is defined as the ratio between the viscosity and the heat conductivity. The Prandtl number computed via the BGK model does not match with the one derived from the Boltzmann equation, resulting in the incorrect hydrodynamic limit at the Navier-Stokes level. To overcome this, Holway [74] suggested a generalized version of the BGK model by replacing the local Maxwellian with an ellipsoidal Gaussian parametrized by a free parameter $-1/2 \le \nu < 1$. This model is called the ellipsoidal BGK model (ES-BGK model), whose initial value problem reads

$$\begin{aligned} \partial_t f + v \cdot \nabla_x f &= \frac{1}{\kappa} A_\nu(\mathcal{M}_\nu(f) - f), \\ f(x, v, 0) &= f_0(x, v). \end{aligned} \tag{104}$$

The velocity distribution function $f(x, v, t)$ is the number density of the particle system on the phase point $(x, v) \in \mathbb{T}^{d_1} \times \mathbb{R}^{d_2}$ $(d_1 \le d_2)$ at time $t \in \mathbb{R}_+$. The Knudsen number κ is a dimensionless number defined by the ratio between the mean free path and the characteristic length.

The collision frequency A_ν takes various forms depending on modeling assumptions. In this section, we only consider the fixed collision frequency: $A_\nu = (1-\nu)^{-1}$. The ellipsoidal Gaussian $\mathcal{M}_\nu(f)$ reads

$$\mathcal{M}_\nu(f) = \frac{\rho}{\sqrt{\det(2\pi \mathcal{T}_\nu)}} \exp\left(-\frac{1}{2}(v - U)^\top \mathcal{T}_\nu^{-1}(v - U)\right),$$

where the macroscopic density, velocity, temperature, and stress tensor are defined by

$$\rho(x,t) = \int_{\mathbb{R}^{d_2}} f(x,v,t)dv,$$

$$\rho(x,t)U(x,t) = \int_{\mathbb{R}^{d_2}} f(x,v,t)vdv,$$

$$d_2\rho(x,t)T(x,t) = \int_{\mathbb{R}^{d_2}} f(x,v,t)|v-U(x,t)|^2dv,$$

$$\rho(x,t)\Theta(x,t) = \int_{\mathbb{R}^{d_2}} f(x,v,t)(v-U)\otimes(v-U)\,dv.$$

The temperature tensor $\mathcal{T}_\nu$ is given by a convex combination of T and Θ:

$$\mathcal{T}_\nu = (1-\nu)TId + \nu\Theta,$$

where Id is the $d_2 \times d_2$ identity matrix. The ellipsoidal relaxation operator satisfies the following cancellation property:

$$\int_{\mathbb{R}^{d_2}} (\mathcal{M}_\nu(f) - f)\left(1, v, |v|^2\right) dv = 0,$$

which leads to the conservation of mass, momentum, and energy:

$$\frac{d}{dt}\int_{\mathbb{R}^{d_1}\times\mathbb{R}^{d_2}} f dxdv = \frac{d}{dt}\int_{\mathbb{R}^{d_1}\times\mathbb{R}^{d_2}} vf dxdv = \frac{d}{dt}\int_{\mathbb{R}^{d_1}\times\mathbb{R}^{d_2}} f|v|^2dxdv = 0.$$

When the ES-BGK model was first suggested, H-theorem was not verified, which was the main reason why the ES-BGK model has been neglected in the literature until very recently. It was resolved in [4] (see also [27, 119]):

$$\frac{d}{dt}\int_{\mathbb{R}^{d_1}\times\mathbb{R}^{d_2}} f \ln f\, dxdv \le 0,$$

and ignited the interest on this model [3, 27, 53, 62, 82, 84, 108, 117–119, 122].

It can be verified via the Chapman-Enskog expansion that the Prandtl number computed using the ES-BGK model is $1/(1-\nu)$. Therefore, the correct physical Prandtl number can be recovered by choosing appropriate ν, namely, $\nu = 1 - 1/Pr = -1/2$, where Pr denotes the correct Prandtl number. When $\nu = 0$, the ES-BGK model reduces to the original BGK model.

Before we finish this introduction, we summarize the notational convention kept throughout this section:

- We use the following notation for weighted L^∞-Sobolev norm for continuous solution:

$$\|f\|_{L^\infty_q} = \sup_{x,v} |f(x,v)(1+|v|)^q|, \quad \|f\|_{W^{\ell,\infty}_q} = \sum_{|\alpha|+|\beta|\le\ell} \|\partial^\alpha_\beta f\|_{L^\infty_q},$$

where $\alpha, \beta \in \mathbb{Z}_+$, and the differential operator ∂^α_β stands for $\partial^\alpha_\beta = \partial^\alpha_x \partial^\beta_v$.

- For any sequence $a^n_{i,j}$, we use the following notation for the weighted L^∞-norm of the sequence:

$$\|a^n\|_{L^\infty_q} = \sup_{i,j} |a^n_{i,j}(1+|v_j|)^q|.$$

- In view of the above norms, $\|f(t_n) - f^n\|_{L^\infty_q}$ is understood, with a slight abuse of notation, as

$$\|f(t_n) - f^n\|_{L^\infty_q} = \sup_{i,j} \left| \left\{ f(x_i, v_j, t_n) - f^n_{i,j} \right\} (1+|v_j|)^q \right|,$$

for simplicity.

7.2 Description of the Numerical Scheme

We fix $d_1 = 1$ and $d_2 = 3$ case with periodic boundary condition throughout this section. We choose a constant time step Δt with final time T^f. The spatial domain and the velocity domain are divided into uniform grids with mesh size Δx, Δv, respectively:

$$t_n = n\Delta t, \qquad n = 0, 1, \ldots, N_t,$$
$$x_i = i\Delta x, \qquad i = 0, \pm 1, \ldots, \pm N_x, \pm(N_x+1), \cdots$$

where $N_t \Delta t = T^f$, $N_x \Delta x = 1$, and

$$v_j = (v_{j_1}, v_{j_2}, v_{j_3}) = (j_1\Delta v, j_2\Delta v, j_3\Delta v), \quad j = (j_1, j_2, j_3) \in \mathbb{Z}^3.$$

Note that the spatial node is defined on the whole line instead of the unit interval, even though we are considering the periodic problem. Periodicity will be imposed on the initial data f_0, which is defined on the whole line with period 1. This facilitates the proof in several places. We denote the approximate solution of $f(x_i, v_j, t_n)$ by $f^n_{i,j}$. To describe the numerical scheme more succinctly, we introduce the following convenient notation.

Definition 7.1

(1) Let $x(i,j) = x_i - \Delta t\, v_{j_1}$. Then define $s = s(i,j)$ to be the spatial node such that $x(i,j)$ lies in $[x_s, x_{s+1})$.

(2) The reconstructed distribution function $\widetilde{f}^n_{i,j}$ is defined as

$$\widetilde{f}^n_{i,j} = \frac{x(i,j) - x_s}{\Delta x} f^n_{s+1,j} + \frac{x_{s+1} - x(i,j)}{\Delta x} f^n_{s,j}.$$

We also need to define the discrete ellipsoidal Gaussian:

$$\mathcal{M}_{\nu,j}(f^n_i) = \frac{\rho^n_i}{\sqrt{\det(2\pi \mathcal{T}^n_{\nu,i})}} \exp\Big(-\frac{1}{2}(v_j - U^n_i)\{\mathcal{T}^n_{\nu,i}\}^{-1}(v_j - U^n_i)\Big),$$

and the discrete macroscopic field ρ^n_i, U^n_i, T^n_i, Θ^n_i and $\mathcal{T}^n_{\nu,i}$:

$$\rho^n_i = \sum_j f^n_{i,j}(\Delta v)^3, \quad \rho^n_i U^n_i = \sum_j f^n_{i,j} v_j (\Delta v)^3, \quad 3\rho^n_i T^n_i = \sum_j f^n_{i,j}|v_j - U^n_i|^2(\Delta v)^3,$$

$$\rho^n_i \Theta^n_i = \sum_j f^n_{i,j}(v_j - U^n_i) \otimes (v_j - U^n_i)(\Delta v)^3, \quad \mathcal{T}^n_{\nu,i} = (1-\nu)T^n_i Id + \nu \Theta^n_i.$$

7.2.1 Derivation of the first order scheme

Let $f_j = f(x, v_j, t)$. We rewrite the ES-BGK model (104) in the characteristic formulation:

$$\frac{df_j}{dt} = \frac{1}{\kappa} A_\nu\big(\mathcal{M}_\nu(f_j) - f_j\big), \quad \frac{dx}{dt} = v_j. \tag{105}$$

Using the implicit Euler scheme on (105), we obtain

$$f^{n+1}_j(x_i) - f^n_j(x_i - v_{j_1}\Delta t) = \frac{\Delta t}{\kappa} A_\nu\{\mathcal{M}_{\nu,j}(f^{n+1}) - f^{n+1}_j\}(x_i).$$

We then approximate $f^n_j(x_i)$ by $f^n_{i,j}$ and $f^n_j(x_i - v_{j_1}\Delta t)$ by $\widetilde{f}^n_{i,j}$ to derive

$$\frac{f^{n+1}_{i,j} - \widetilde{f}^n_{i,j}}{\Delta t} = \frac{1}{\kappa} A_\nu\{\mathcal{M}_{\nu,j}(f^{n+1}_i) - f^{n+1}_{i,j}\}. \tag{106}$$

We attempt to convert (106) into a semi-explicit scheme, keeping beneficial features of implicit schemes such as the stability property. We first impose the conservation of mass, momentum, and energy at the discrete level (throughout this subsection $\approx$ means that they are identical up to negligible error):

$$\frac{\sum_j f^{n+1}_{i,j}\phi(v_j)(\Delta v)^3 - \sum_j \widetilde{f}^n_{i,j}\phi(v_j)(\Delta v)^3}{\Delta t}$$

$$= \frac{1}{\kappa} A_\nu \sum_j \{\mathcal{M}_{\nu,j}(f_i^{n+1})\phi(v_j) - f_{i,j}^{n+1}\phi(v_j)\}(\Delta v)^3 \approx 0,$$

for $\phi(v_j) = 1, v_j, \frac{1}{2}|v_j|^2$. Since r.h.s has a spectral accuracy for fast decaying functions, we can legitimately assume that

$$\sum_j f_{i,j}^{n+1}\phi(v_j) \approx \sum_j \tilde{f}_{i,j}^n \phi(v_j). \tag{107}$$

Therefore, if we define

$$\tilde{\rho}_i^n = \sum_j \tilde{f}_{i,j}^n (\Delta v)^3, \quad \tilde{\rho}_i^n \tilde{U}_i^n = \sum_j \tilde{f}_{i,j}^n v_j (\Delta v)^3,$$
$$3\tilde{\rho}_i^n \tilde{T}_i^n = \sum_j \tilde{f}_{i,j}^n |v_j - \tilde{U}_i^n|^2 (\Delta v)^3,$$

then (107) gives the following relation:

$$\rho_i^{n+1} \approx \tilde{\rho}_i^n, \; U_i^{n+1} \approx \tilde{U}_i^n, \; T_i^{n+1} \approx \tilde{T}_i^n. \tag{108}$$

In the case of the original BGK model ($\nu = 0$), identities in (108) are sufficient to conclude that $\mathcal{M}_{0,j}(f_i^{n+1}) \approx \mathcal{M}_{0,j}(\tilde{f}_i^n)$. But this is not the case for the ellipsoidal case, since the ellipsoidal Gaussian contains the temperature tensor $\mathcal{T}_\nu$, which is not a conserved quantity. For this, we introduce

$$\phi_{i,j}^n \equiv (v_j - U_i^n) \otimes (v_j - U_i^n).$$

Multiplying $\phi_{i,j}^{n+1}(\Delta v)^3$ on both sides of (106) and summing over i and j, we get

$$\frac{\sum_j f_{i,j}^{n+1}\phi_{i,j}^{n+1}(\Delta v)^3 - \sum_j \tilde{f}_{i,j}^n \phi_{i,j}^{n+1}(\Delta v)^3}{\Delta t}$$
$$= \frac{1}{\kappa} A_\nu \sum_j \left\{\mathcal{M}_{\nu,j}(f_i^{n+1})\phi_{i,j}^{n+1} - f_{i,j}^{n+1}\phi_{i,j}^{n+1}\right\}(\Delta v)^3. \tag{109}$$

Let us denote the r.h.s of (109) by R and the l.h.s by L for simplicity. We then recall (108) to observe

$$\phi_{i,j}^{n+1} = (v_j - U_i^{n+1}) \otimes (v_j - U_i^{n+1}) \approx (v_j - \tilde{U}_i^n) \otimes (v_j - \tilde{U}_i^n) = \tilde{\phi}_{i,j}^n.$$

Therefore, the second term of L becomes

$$\sum_j \tilde{f}^n_{i,j}\phi^{n+1}_{i,j}(\Delta v)^3 \approx \sum_j \tilde{f}^n_{i,j}\tilde{\phi}^n_{i,j}(\Delta v)^3 = \tilde{\rho}^n_i \widetilde{\Theta}^n_i,$$

where $\widetilde{\Theta}^n_i$ is defined by

$$\tilde{\rho}^n_i \widetilde{\Theta}^n_i = \sum_j \tilde{f}^n_{i,j}(v_j - \widetilde{U}^n_i) \otimes (v_j - \widetilde{U}^n_i)(\Delta v)^3,$$

so that

$$L = \frac{\rho^{n+1}_i \Theta^{n+1}_i - \tilde{\rho}^n_i \widetilde{\Theta}^n_i}{\Delta t}. \tag{110}$$

On the other hand, we find, for the right-hand side,

$$\begin{aligned}
R &\approx \frac{1}{\kappa} A_\nu \left\{ \rho^{n+1}_i \mathcal{T}^{n+1}_{\nu,i} - \rho^{n+1}_i \Theta^{n+1}_i \right\} \\
&= \frac{1}{\kappa} A_\nu \left[\rho^{n+1}_i \{(1-\nu) T^{n+1}_i Id + \nu \Theta^{n+1}_i\} - \rho^{n+1}_i \Theta^{n+1}_i \right] \\
&= \frac{1}{\kappa} A_\nu (1-\nu) \left[\rho^{n+1}_i T^{n+1}_i Id - \rho^{n+1}_i \Theta^{n+1}_i \right] \\
&= \frac{1}{\kappa} \left\{ \rho^{n+1}_i T^{n+1}_i Id - \rho^{n+1}_i \Theta^{n+1}_i \right\},
\end{aligned}$$

and recall (108) to see that

$$R = \frac{1}{\kappa} \left\{ \rho^{n+1}_i T^{n+1}_i Id - \rho^{n+1}_i \Theta^{n+1}_i \right\} = \frac{1}{\kappa} \left\{ \tilde{\rho}^n_i \widetilde{T}^n_i Id - \tilde{\rho}^n_i \Theta^{n+1}_i \right\}. \tag{111}$$

Now, equating (110) and (111), we rewrite (109) as

$$\frac{\rho^{n+1}_i \Theta^{n+1}_i - \tilde{\rho}^n_i \widetilde{\Theta}^n_i}{\Delta t} = \frac{1}{\kappa} \left\{ \tilde{\rho}^n_i \widetilde{T}^n_i Id - \tilde{\rho}^n_i \Theta^{n+1}_i \right\}.$$

Dividing both sides by ρ^{n+1}_i and gathering relevant terms, we get

$$\Theta^{n+1}_i = \frac{\Delta t}{\kappa + \Delta t} \widetilde{T}^n_i Id + \frac{\kappa}{\kappa + \Delta t} \widetilde{\Theta}^n_i.$$

Therefore, $\mathcal{T}^{n+1}_{\nu,i}$ can be expressed as

$$\begin{aligned}
\mathcal{T}_{\nu,i}^{n+1} &= (1-\nu)T_i^{n+1}Id + \nu\Theta_i^{n+1}\\
&\approx (1-\nu)\widetilde{T}_i^n Id + \nu\left\{\frac{\Delta t}{\kappa+\Delta t}\widetilde{T}_i^n Id + \frac{\kappa}{\kappa+\Delta t}\widetilde{\Theta}_i^n\right\}\\
&= \left(1-\frac{\kappa\nu}{\kappa+\Delta t}\right)\widetilde{T}_i^n Id + \left(\frac{\kappa\nu}{\kappa+\Delta t}\right)\widetilde{\Theta}_i^n\\
&\equiv (1-\widetilde{\nu})\widetilde{T}_i^n Id + \widetilde{\nu}\widetilde{\Theta}_i^n\\
&\equiv \widetilde{\mathcal{T}}_{\widetilde{\nu},i}^n,
\end{aligned} \tag{112}$$

where we denoted $\widetilde{\nu} = (\kappa\nu)/(\kappa+\Delta t)$. Using (108) and (112), the implicitly defined discrete ellipsoidal Gaussian can now be rewritten in an explicit way as

$$\begin{aligned}
\mathcal{M}_{\nu,j}(f_i^{n+1}) &=\approx \mathcal{M}_{\nu,j}(\widetilde{\rho}_i^n, \widetilde{U}_i^n, \widetilde{\mathcal{T}}_{\widetilde{\nu},i}^n)\\
&= \frac{\widetilde{\rho}_i^n}{\sqrt{\det(2\pi\widetilde{\mathcal{T}}_{\widetilde{\nu},i}^n)}}\exp\left(-\frac{1}{2}(v_j-\widetilde{U}_i^n)^\top\{\widetilde{\mathcal{T}}_{\widetilde{\nu},i}^n\}^{-1}(v_j-\widetilde{U}_i^n)\right).
\end{aligned}$$

With a slight abuse of notation, we now denote the r.h.s as $\mathcal{M}_{\widetilde{\nu},j}(\widetilde{f}_i^n)$:

$$f_{i,j}^{n+1} = \frac{\kappa}{\kappa + A_\nu\Delta t}\widetilde{f}_{i,j}^n + \frac{A_\nu\Delta t}{\kappa + A_\nu\Delta t}\mathcal{M}_{\widetilde{\nu},j}(\widetilde{f}_i^n).$$

7.2.2 Implicit SL Scheme

Summarizing, our SL scheme for the ES-BGK model (104) reads

$$f_{i,j}^{n+1} = \frac{\kappa}{\kappa + A_\nu\Delta t}\widetilde{f}_{i,j}^n + \frac{A_\nu\Delta t}{\kappa + A_\nu\Delta t}\mathcal{M}_{\widetilde{\nu},j}(\widetilde{f}_i^n), \tag{113}$$

where the discrete ellipsoidal Gaussian $\mathcal{M}_{\widetilde{\nu},j}(\widetilde{f}_i^n)$ is defined as follows:

$$\mathcal{M}_{\widetilde{\nu},j}(\widetilde{f}_i^n) = \frac{\widetilde{\rho}_i^n}{\sqrt{\det(2\pi\widetilde{\mathcal{T}}_{\widetilde{\nu},i}^n)}}\exp\left(-\frac{1}{2}(v_j-\widetilde{U}_i^n)\{\widetilde{\mathcal{T}}_{\widetilde{\nu},i}^n\}^{-1}(v_j-\widetilde{U}_i^n)\right),$$

and the discrete macroscopic fields $\widetilde{\rho}_i^n, \widetilde{U}_i^n, \widetilde{T}_i^n, \widetilde{\theta}_{i,j}^n$, and $\widetilde{\mathcal{T}}_{\widetilde{\nu},i}^n$ $(n \geq 1)$ are given by

$$\widetilde{\rho}_i^n = \sum_j \widetilde{f}_{i,j}^n(\Delta v)^3,\quad \widetilde{\rho}_i^n\widetilde{U}_i^n = \sum_j \widetilde{f}_{i,j}^n v_j(\Delta v)^3,\quad 3\rho_i^n\widetilde{T}_i^n = \sum_j \widetilde{f}_{i,j}^n|v_j-\widetilde{U}_i^n|^2(\Delta v)^3,$$

$$\widetilde{\rho}_i^n\widetilde{\Theta}_i^n = \sum_j \widetilde{f}_{i,j}^n\{(v_j-\widetilde{U}_i^n)\otimes(v_j-\widetilde{U}_i^n)\}(\Delta v)^3,\quad \widetilde{\mathcal{T}}_{\widetilde{\nu},i}^n = (1-\widetilde{\nu})\widetilde{T}_i^n Id + \widetilde{\nu}\widetilde{\Theta}_i^n.$$

In the last line, $\tilde{\nu}$ denotes

$$\tilde{\nu} = \frac{\kappa\nu}{\kappa + \Delta t}.$$

For the initial step ($n = 0$), to ignore the error arising in the discretization of the initial data and simplify the convergence proof, we sample values directly from continuous distribution function and macroscopic fields at $t = 0$:

$$f^0_{i,j} = f_0(x_i, v_j), \quad \widetilde{f}^0_{i,j} = \widetilde{f}_0(x_i, v_j) = f_0(x_i - v_j\Delta t, v_j).$$

7.3 Main Convergence Result

We are now ready to state our main result. We first record the existence result relevant to our convergence proof.

Theorem 7.2 ([118]) *Let $-1/2 < \nu < 1$ and $q > 5$. Let f_0 satisfy $\|f_0\|_{W^{2,\infty}_q} < \infty$. Suppose further that there exist positive constants C^1_0, C^2_0, and α such that*

$$f_0(x, v) \geq C^1_0 e^{-C^2_0|v|^\alpha}.$$

Then, for any final time $T^f > 0$, the ES-BGK model (104) has a unique solution $f \in C([0, T^f], \|\cdot\|_{W^{2,\infty}_q})$ such that:

1. *f is bounded in $\|\cdot\|_{W^{2,\infty}_q}$ for $[0, T^f]$:*

$$\|f(t)\|_{W^{2,\infty}_q} \leq C_1 e^{C_2 t}\left\{\|f_0\|_{W^{1,\infty}_q} + 1\right\}, \quad t \in [0, T^f],$$

   for some constants C_1 and C_2.
2. *The macroscopic fields satisfy the following lower and upper bounds:*

$$\|\rho(t)\|_{L^\infty_x} + \|U(t)\|_{L^\infty_x} + \|T(t)\|_{L^\infty_x} \leq C_q e^{C_q T^f},$$
$$\rho(x, t) \geq C_{N,q} e^{-C_{N,q} t},$$
$$k^\top\{\mathcal{T}_\nu(x, t)\}k \geq C_{N,q} e^{-C_{N,q} t} > 0, \text{ for any } k \in \mathbb{S}^2.$$

Now, we state our main result established in [100].

Theorem 7.3 ([100]) *Let $-1/2 < \nu < 1$. Let f be the unique smooth solution of (104) corresponding to a nonnegative initial datum f_0 satisfying the hypotheses of Theorem 7.2. Let f^n be the approximate solution constructed iteratively by (113). Then, there exists a positive number $r_{\Delta v}$, which is explicitly determined in*

Theorem 7.8, such that, if $\Delta v < r_{\Delta v}$, *then we have*

$$\|f(T^f) - f^{N_t}\|_{L^\infty_q} \leq C\Big\{(\Delta x)^2 + \Delta v + \Delta t + \frac{(\Delta x)^2}{\Delta t}\Big\},$$

where N_t *is defined by* $T^f = N_t \Delta t$ *and* $C = C(T^f, f_0, q, \kappa, \nu) > 0$. *Here,* C *is uniformly bounded in* ν.

Remark 7.4 (1) $\nu = 0$ corresponds to the original BGK model. Therefore, our result holds for the original BGK model too. (2) For the precise definition of $r_{\Delta v}$, see Theorem 7.8. (3) Unfortunately, the critical case $\nu = -1/2$ is not covered. One of the main reasons is that the corresponding existence theory is not established for the continuous problem.

The most important step of the proof is the derivation of the following uniform stability estimate (see Proposition 7.1):

$$C_{0,1} e^{-\frac{A_\nu}{\kappa} T^f} e^{-C_{0,2}|v_j|^\alpha} \leq f^n_{i,j} \leq e^{\frac{(C_{\mathcal{M}}-1)T^f}{\kappa + A_\nu \Delta t}} \|f^0\|_{L^\infty_q} (1+|v_j|)^{-q},$$

which comes from the uniform-in-n control of the discrete ellipsoidal Gaussian in a weighted L^∞ norm (see Lemma 7.11):

$$\|\mathcal{M}_{\tilde{\nu}}(f^n)\|_{L^\infty_q} \leq C_{\mathcal{M}} \|f^n\|_{L^\infty_q}.$$

Two technical issues arise in the process of obtaining the above estimates. First, we need to show that the discrete temperature tensor $\widetilde{\mathcal{T}}^n_{\tilde{\nu},i}$ is strictly positive definite uniformly in n:

$$k^\top \{\widetilde{\mathcal{T}}^n_{\tilde{\nu},i}\} k \geq C_{\nu,q,\kappa,T^f} > 0 \ \text{ for all } \kappa \in \mathbb{S}^2.$$

Otherwise, since the discrete ellipsoidal Gaussian involves the inverse of $\mathcal{T}^n_{\tilde{\nu},i}$ and $\det(\mathcal{T}^n_{\tilde{\nu},i})$, it may blow up as $\mathcal{T}^n_{\tilde{\nu},i}$ approaches arbitrarily close to zero. The second issue is more subtle. It turns out that we need to show that ratios such as

$$\widetilde{\rho}^n_i \widetilde{T}^n_i / \|\widetilde{f}^n\|_{L^\infty_q}, \quad \|\widetilde{f}^n\|_{L^\infty_q} / \widetilde{\rho}^n_i$$

remain strictly larger than Δv, to guarantee the existence of proper decomposition of the macroscopic fields to derive necessary moment estimates. (See Lemma 7.9.) The restriction on the velocity node in Theorem 3.2 mostly comes from this subtle technical issue.

7.4 Stability of the Discrete Distribution Function

In this section, we derive uniform lower and upper bounds for the discrete distribution function $\tilde{f}^n_{i,j}$ and corresponding macroscopic fields. We start with a series of definitions most of which were introduced for technical reasons.

Definition 7.5 (1) We define C_α, $C_{q,\alpha}$, and C_{q-m} $(q \geq m)$ by

$$C_\alpha = \int_{\mathbb{R}^3} e^{-C_{0,2}|v|^\alpha} dv, \quad C_{q,\alpha} = \sup_v \left\{(1+|v|)^q e^{-C_{0,2}|v|^\alpha}\right\},$$

$$C_{q-m} = \int_{\mathbb{R}^3} \frac{1}{(1+|v|)^{q-m}} dv.$$

(2) Throughout this section, we fix C_M, $C_{\mathcal{M}}$ as is defined in Lemma 7.9 and Lemma 7.11, respectively.

Definition 7.6 (1) We say that $f^n_{i,j}$ satisfies the stability estimate E^n_1 for the discrete distribution function if the following statement holds:

$$C_{0,1} e^{-\frac{A_v}{\kappa} T^f} e^{-C_{0,2}|v_j|^\alpha} \leq \tilde{f}^n_{i,j} \leq \frac{e^{\frac{(C_{\mathcal{M}}-1)A_v T^f}{\kappa + A_v \Delta t}} \|f^0\|_{L^\infty_q}}{(1+|v_j|)^q}.$$

(2) We say that $f^n_{i,j}$ satisfies the stability estimate E^n_2 for discrete macroscopic fields if the following statements hold:

$$\tilde{\rho}^n_i \geq \frac{1}{2} C_{0,1} C_\alpha e^{-\frac{A_v}{\kappa} T^f}, \quad \tilde{T}^n_i \geq \left(\frac{C_{0,1}C_\alpha}{2C_M \|f_0\|_{L^\infty_q}}\right)^{2/3} e^{-\frac{2}{3}\left(\frac{1}{\kappa} + \frac{C_{\mathcal{M}}-1}{\kappa + A_v \Delta t}\right) A_v T^f},$$

$$\|\tilde{\rho}^n\|_{L^\infty_x}, \|\tilde{U}^n\|_{L^\infty_x}, \|\tilde{T}^n\|_{L^\infty_x} \leq 2C_q \left\{1 + \left(C_{0,1}C_\alpha\right)^{-1}\right\} e^{\left(\frac{1}{\kappa} + \frac{(C_{\mathcal{M}}-1)}{\kappa + A_v \Delta t}\right) A_v T^f} \|f^0\|_{L^\infty_q}.$$

(3) We define $E^n = E^n_1 \wedge E^n_2$.

Definition 7.7 (1) We define constants a_1, a_2, and a_3 by

$$a_1 = \left(\frac{1}{2}\right)^{\frac{14}{15}} \left(\frac{3}{\pi}\right)^{\frac{1}{5}} \frac{(C_{0,1}C_\alpha)^{\frac{1}{3}}}{C_M^{\frac{2}{15}} \|f_0\|^{\frac{1}{3}}_{L^\infty_q}} e^{-\frac{1}{3}\left(\frac{1}{\kappa} + \frac{(C_{\mathcal{M}}-1)}{\kappa + A_v \Delta t}\right) A_v T^f},$$

$$a_2 = \left(\frac{\pi}{4(q-5)}\right)^{\frac{1}{q-3}} \frac{(C_{0,1}C_{q,\alpha})^{\frac{1}{q-3}}}{C_q^{\frac{1}{q-3}} \left\{1 + \left(C_{0,1}C_\alpha\right)^{-1}\right\}^{\frac{1}{q-3}} \|f_0\|^{\frac{1}{q-3}}_{L^\infty_q}} e^{-\frac{1}{q-3}\left(\frac{2}{\kappa} + \frac{(C_{\mathcal{M}}-1)}{\kappa + A_v \Delta t}\right) A_v T^f},$$

$$a_3 = \left(\frac{1}{2}\right)^{\frac{2(q+6)}{3(q+3)}} \left(\frac{3^q}{\pi}\right)^{\frac{1}{q+3}} \frac{\left(C_{0,1}C_\alpha\right)^{\frac{2q+3}{3(q+3)}}}{\{C_M\}^{\frac{2q}{3(q+3)}} \|f_0\|_{L^\infty_q}^{\frac{2q-3}{3(q+3)}}} e^{-\frac{2q+3}{3(q+3)}\left(\frac{1}{\kappa}+\frac{(C_M-1)}{\kappa+A_\nu\Delta t}\right)A_\nu T^f}.$$

The main goal of this section is the following.

Theorem 7.8 *Choose $\ell > 0$ sufficiently small such that $\Delta v < \ell$ implies*

$$\frac{1}{2}C_\alpha \leq \sum_j e^{-C_{0,2}|v_j|^\alpha}(\Delta v)^3 \leq 2C_\alpha,$$

$$\frac{1}{2}C_{q,\alpha} \leq \sup_j \left\{(1+|v_j|)^q e^{-|v_j|C^\alpha}\right\} \leq 2C_{q,\alpha},$$

$$\frac{1}{2}C_{q-m} \leq \sum_j \frac{(\Delta v)^3}{(1+|v_j|)^{q-m}} \leq 2C_{q-m}.$$

Now, define $r_{\Delta v}$ by

$$r_{\Delta v} = \min\{a_1, a_2, a_3, \ell, 1/2\},$$

and suppose Δv is sufficiently small in the following sense:

$$\Delta v < r_{\Delta v}.$$

Then, $f^n_{i,j}$ satisfies E^n for all $n \geq 0$.

We postpone the proof to the end of this section, after establishing several preliminary results. We begin with the discrete moment estimates:

Lemma 7.9 *Let $q > 5$. Suppose $f^n_{i,j}$ satisfies E^n and Δv satisfies the smallness condition stated in Theorem 7.8. Then there exists a positive constant C_M which depends only on q, such that*

$$\frac{\widetilde{\rho}_i^n}{(\widetilde{T}_i^n)^{\frac{3}{2}}}, \quad \widetilde{\rho}_i^n(\widetilde{T}_i^n + |\widetilde{U}_i^n|^2)^{\frac{q-3}{2}}, \quad \frac{\widetilde{\rho}^n|\widetilde{U}_i^n|^{q+3}}{[(\widetilde{T}_i^n + |\widetilde{U}_i^n|^2)\widetilde{T}_i^n]^{\frac{3}{2}}} \leq C_M\|f^n\|_{L^\infty_q},$$

for all n.

Proof We only present the proof of the first estimate. We split the macroscopic density into the following two parts:

$$\widetilde{\rho}_i^n = \sum_{|v_j-\widetilde{U}_i^n|<r+\Delta v} \widetilde{f}^n_{i,j}(\Delta v)^3 + \sum_{|v_j-\widetilde{U}_i^n|\geq r+\Delta v} \widetilde{f}^n_{i,j}(\Delta v)^3.$$

Then we see that

$$\sum_{|v_j-\widetilde{U}_i^n|\le r+\Delta v} \widetilde{f}_{i,j}^n(\Delta v)^3 \le \|\widetilde{f}^n\|_{L_0^\infty} \sum_{|v_j-\widetilde{U}_i^n|\le r+\Delta v} (\Delta v)^3 \le \pi \|\widetilde{f}^n\|_{L_0^\infty}(r+2\Delta v)^3$$
$$\le 8\pi \|\widetilde{f}^n\|_{L_0^\infty}(r+\Delta v)^3,$$

and

$$\begin{aligned}\sum_{|v_j-\widetilde{U}_i^n|\ge r+\Delta v} \widetilde{f}_{i,j}^n(\Delta v)^3 &= \sum_{|v_j-\widetilde{U}_i^n|\ge r+\Delta v} \widetilde{f}_{i,j}^n \frac{|v_j-\widetilde{U}_i^n|^2}{|v_j-\widetilde{U}_i^n|^2}(\Delta v)^3 \\ &\le \frac{1}{(r+\Delta v)^2}\sum_j \widetilde{f}_{i,j}^n |v_j-\widetilde{U}_i^n|^2(\Delta v)^3 \\ &= \frac{3\widetilde{\rho}_i^n \widetilde{T}_i^n}{(r+\Delta v)^2},\end{aligned}$$

yielding

$$\widetilde{\rho}_i^n \le 8\pi \|\widetilde{f}^n\|_{L_q^\infty}(r+\Delta v)^3 + \frac{3\widetilde{\rho}_i^n \widetilde{T}_i^n}{(r+\Delta v)^2}.$$

We then optimize r by equating the two terms on the right-hand sides, to obtain

$$r+\Delta v = \left(\frac{3\widetilde{\rho}_i^n \widetilde{T}_i^n}{8\pi \|\widetilde{f}^n\|_{L_q^\infty}}\right)^{1/5}. \tag{114}$$

Using the fact that $f_{i,j}^n$ satisfies E^n, it can be easily verified that the r.h.s is greater than a_1, so that

$$\left(\frac{3\widetilde{\rho}_i^n \widetilde{T}_i^n}{8\pi \|\widetilde{f}^n\|_{L_q^\infty}}\right)^{1/5} \ge a_1 > \Delta v.$$

(This is why the quantities a_i, $i = 1, 2, 3$ are introduced in Definition 7.7.) Therefore, we can always find a positive number r satisfying (114). Inserting this, one finds

$$\widetilde{\rho}_i^n \le C_M \{\widetilde{T}_i^n\}^{3/2} \|\widetilde{f}^n\|_{L_q^\infty}.$$

Since the positivity of $\widetilde{T}_i^n$ is guaranteed by (C^n), we can divide both sides by $\{\widetilde{T}_i^n\}^{3/2}$ to get the desired estimate.

We now show that the ellipsoidal Gaussian is controlled by the discrete distribution in L^∞_q. First, we need to establish the following equivalence estimate for the discrete temperature tensor:

Lemma 7.10 *Let $-1/2 < \nu < 1$. Assume $\widetilde{f}^n_{i,j} > 0$ and $\widetilde{\rho}^n_i > 0$. Then the discrete temperature tensor $\mathcal{T}^n_{\tilde{\nu},i}$ and its determinant $\det \mathcal{T}^n_{\tilde{\nu},i}$ satisfy the following equivalence estimates:*

(1) $\min\{1-\tilde{\nu}, 1+2\tilde{\nu}\}T^n_i Id \leq \widetilde{\mathcal{T}}^n_{\tilde{\nu},i} \leq \max\{1-\tilde{\nu}, 1+2\tilde{\nu}\}T^n_i Id.$

(2) $\min\{1-\tilde{\nu}, 1+2\tilde{\nu}\}^3(T^n_i)^3 \leq \det\{\widetilde{\mathcal{T}}^n_{\tilde{\nu},i}\} \leq \max\{1-\tilde{\nu}, 1+2\tilde{\nu}\}^3(T^n_i)^3.$

In the first inequality, $A \geq B$ for 3×3 symmetric matrices A and B means $A-B$ is positive definite.

Proof We only prove (1), since (2) follows immediately from (1). Thanks to (112), we see that

$$
\begin{aligned}
\widetilde{\rho}^n_i \widetilde{\mathcal{T}}^n_{\tilde{\nu},i} &= (1-\tilde{\nu})\widetilde{\rho}^n_i \widetilde{T}^n_i Id + \tilde{\nu}\widetilde{\rho}^n_i \widetilde{\Theta}^n_i \\
&= \frac{(1-\tilde{\nu})}{3}\Big\{\sum_j \widetilde{f}^n_{i,j}|v_j - \widetilde{U}^n_i|^2(\triangle v)^3\Big\} \\
&\quad + \tilde{\nu}\sum_j \widetilde{f}^n_{i,j}(v_j - \widetilde{U}^n_i)\otimes(v_j - \widetilde{U}^n_i)(\triangle v)^3.
\end{aligned}
$$

Then, in view of the identity $k^T\{U\otimes U\}k = (k\cdot U)^2 \ \ (k, U \in \mathbb{R}^3)$, we have

$$
\begin{aligned}
k^\top\{\widetilde{\rho}^n_i \widetilde{\mathcal{T}}^n_{\tilde{\nu},i}\}k &= \frac{(1-\tilde{\nu})}{3}\Big\{\sum_j \widetilde{f}^n_{i,j}|v_j - \widetilde{U}^n_i|^2(\triangle v)^3\Big\}|k|^2 \\
&\quad + \tilde{\nu}\sum_j \widetilde{f}^n_{i,j}\left\{(v_j - \widetilde{U}^n_i)\cdot k\right\}^2(\triangle v)^3.
\end{aligned}
$$

We note from the definition of $\tilde{\nu}$ that $-1/2 < \nu < 1$ implies $-1/2 < \tilde{\nu} < 1$ and divide our estimate into the following two cases:

(a) $0 < \tilde{\nu} < 1$: Since the second term is nonnegative, we have

$$
k^\top\{\widetilde{\rho}^n_i \widetilde{\mathcal{T}}^n_{\tilde{\nu},i}\}k \geq \frac{(1-\tilde{\nu})}{3}\Big\{\sum_j \widetilde{f}^n_{i,j}|v_j - \widetilde{U}^n_i|^2(\triangle v)^3\Big\}|k|^2 = (1-\tilde{\nu})\widetilde{\rho}^n_i \widetilde{T}^n_i|k|^2.
$$

(b) $-\frac{1}{2} < \tilde{\nu} < 0$: By Cauchy-Schwartz inequality, we see that

$$
\begin{aligned}
k^\top\{\widetilde{\rho}_i^n \widetilde{\mathcal{T}}_{\tilde{\nu},i}^n\}k &\geq \frac{(1-\tilde{\nu})}{3}\Big\{\sum_j \tilde{f}_{i,j}^n |v_j - \widetilde{U}_i^n|^2 (\Delta v)^3\Big\}|k|^2 \\
&\quad + \tilde{\nu}\Big\{\sum_j \tilde{f}_{i,j}^n |v_j - \widetilde{U}_i^n|^2 (\Delta v)^3\Big\}|k|^2 \\
&= \frac{(1+2\tilde{\nu})}{3}\Big\{\sum_j \tilde{f}_{i,j}^n |v_j - \widetilde{U}_i^n|^2 (\Delta v)^3\Big\}|k|^2 \\
&= (1+2\tilde{\nu})\widetilde{\rho}_i^n \widetilde{T}_i^n |k|^2.
\end{aligned}
$$

We then combine the above estimates to get

$$
\min\{1-\tilde{\nu}, 1+2\tilde{\nu}\}\widetilde{T}_i^n \leq k^\top\{\widetilde{\mathcal{T}}_{\tilde{\nu},i}^n\}k.
$$

The r.h.s of the inequality follows in a similar manner.

We now prove the main control estimate of the discrete ellipsoidal Gaussian.

Lemma 7.11 *Suppose $f_{i,j}^n$ satisfies E^n, and $\Delta v < r_{\Delta v}$. Then we have*

$$
\|\mathcal{M}_{\tilde{\nu}}(\tilde{f}^n)\|_{L_q^\infty} \leq C_{\mathcal{M}}\|f^n\|_{L_q^\infty},
$$

for some constant $C_{\mathcal{M}}$ which depends only on q and ν.

Proof We divide the proof into the two cases: $q = 0$ and $q \neq 0$.

(1) $q = 0$: Since the exponential part is less than or equal to 1, we see from Lemma 7.10 that

$$
\mathcal{M}_{\tilde{\nu},j}(\tilde{f}_i^n) \leq \frac{\widetilde{\rho}_i^n}{\sqrt{\det(2\pi\widetilde{\mathcal{T}}_{\tilde{\nu},i}^n)}} \leq C_{\tilde{\nu}}\frac{\widetilde{\rho}_i^n}{\{\widetilde{T}_i^n\}^{3/2}}.
$$

Then Lemma 7.9 (1) gives the desired estimate.

(2) $q \neq 0$: We split v_j as

$$
\begin{aligned}
|v_j|^q \mathcal{M}_{\tilde{\nu},j}(\tilde{f}_i^n) &\leq C_q\left\{|\widetilde{U}_i^n|^q + |v_j - \widetilde{U}_i^n|^q\right\}\mathcal{M}_{\tilde{\nu},j}(\tilde{f}_i^n) \\
&\equiv I_1 + I_2.
\end{aligned}
$$

(i) The estimate for I_1: We first bound the exponential part by 1 to get

$$
I_1 \leq C_{\tilde{\nu}}\frac{|\widetilde{U}_i^n|^q \widetilde{\rho}_i^n}{\{\widetilde{T}_i^n\}^{3/2}}.
$$

(a) $|\widetilde{U}_i^n| > \{\widetilde{T}_i^n\}^{\frac{1}{2}}$: Lemma 7.9 (3) gives

$$I_1 \le C\frac{|\widetilde{U}_i^n|^q\widetilde{\rho}_i^n}{\{\widetilde{T}_i^n\}^{\frac{3}{2}}} \le C\frac{|\widetilde{U}_i^n|^{q+3}\widetilde{\rho}_i^n}{|\widetilde{U}_i^n|^3\{\widetilde{T}_i^n\}^{\frac{3}{2}}} \le C\frac{|\widetilde{U}_i^n|^{q+3}\widetilde{\rho}_i^n}{\{\widetilde{T}_i^n+|\widetilde{U}_i^n|^2\}^{\frac{3}{2}}\{\widetilde{T}_i^n\}^{\frac{3}{2}}} \le C_{v,q}\|\tilde{f}^n\|_{L_q^\infty}.$$

(b) $|\widetilde{U}_i^n| \le \{\widetilde{T}_i^n\}^{\frac{1}{2}}$: In this case, we employ Lemma 7.9 (2) as

$$I_1 \le \frac{\{\widetilde{T}_i^n\}^{\frac{q}{2}}\widetilde{\rho}_i^n}{\{\widetilde{T}_i^n\}^{3/2}} \le \widetilde{\rho}_i^n\{\widetilde{T}_i^n\}^{\frac{q-3}{2}} \le \widetilde{\rho}_i^n\{\widetilde{T}_i^n+|\widetilde{U}_i^n|^2\}^{\frac{q-3}{2}} \le C_q\|\tilde{f}^n\|_{L_q^\infty}.$$

(ii) The estimate for I_2: By Lemma 7.10, we have

$$\begin{aligned}
I_2 &\le C_{\tilde{v}}|v_j-\widetilde{U}_i^n|^q\frac{\widetilde{\rho}_i^n}{\{\widetilde{T}_i^n\}^{3/2}}\exp\left(-C_v\frac{|v_j-\widetilde{U}_i^n|^2}{\widetilde{T}_i^n}\right)\\
&= C_{\tilde{v}}\frac{\widetilde{\rho}_i^n}{\{\widetilde{T}_i^n\}^{3/2}}\{\widetilde{T}_i^n\}^{\frac{q}{2}}\left\{\left(\frac{|v_j-\widetilde{U}_i^n|^2}{\widetilde{T}_i^n}\right)^{\frac{q}{2}}\exp\left(-C_v\frac{|v_j-\widetilde{U}_i^n|^2}{\widetilde{T}_i^n}\right)\right\}\\
&\le C_{\tilde{v}}\widetilde{\rho}_i^n\{\widetilde{T}_i^n\}^{\frac{q-3}{2}}.
\end{aligned}$$

In the last line, we used the elementary inequality $|x^ae^{-bx}| \le C_{a,b}$ for some positive $C_{a,b}$ $(a, b, x > 0)$. Then, Lemma 7.9 (2) gives

$$\widetilde{\rho}_i^n\{\widetilde{T}_i^n\}^{\frac{q-3}{2}} \le \widetilde{\rho}_i^n\{\widetilde{T}_i^n+|\widetilde{U}_i^n|^2\}^{\frac{q-3}{2}} \le C_q\|\tilde{f}^n\|_{L_q^\infty}.$$

This completes the proof.

We are not ready to prove our main stability estimate.

Proposition 7.1 *Assume $f_{i,j}^{n-1}$ satisfies E^{n-1}. Then, $f_{i,j}^n$ satisfies the stability estimate E_1^n:*

$$C_{0,1}e^{-\frac{A_v}{\kappa}T^f}e^{-C_{0,2}|v_j|^\alpha} \le \widetilde{f}_{i,j}^n \le \frac{e^{\frac{(C_{\mathcal{M}}-1)A_vT^f}{\kappa+A_v\Delta t}}\|f^0\|_{L_q^\infty}}{(1+|v_j|)^q},$$

for all i, j.

Proof We start with the proof of the uniform lower bound. Since $f_{i,j}^{n-1}$ satisfies E^{n-1}, $\mathcal{M}_{\tilde{v},j}(\widetilde{f}_i^n)$ is strictly positive. Therefore, we have from (113)

$$f_{i,j}^n \ge \frac{\kappa}{\kappa+A_v\Delta t}\widetilde{f}_{i,j}^{n-1},$$

which, in view of the definition of B^{n-1}, gives

$$f_{i,j}^n \geq C_{0,1}\left(\frac{\kappa}{\kappa+A_\nu\Delta t}\right)\left(\frac{\kappa}{\kappa+A_\nu\Delta t}\right)^{n-1} e^{-C_0^2|v_j|^\alpha} = C_{0,1}\left(\frac{\kappa}{\kappa+A_\nu\Delta t}\right) e^{-C_0^2|v_j|^\alpha}.$$

This immediately leads to the same lower bound estimate for $\tilde{f}_{i,j}^n$:

$$\tilde{f}_{i,j}^n = a_{j_1} f_{s(i,j),j}^n + (1-a_{j_1}) f_{s(i,j)+1,j}^n \geq C_{0,1}\left(\frac{\kappa}{\kappa+A_\nu\Delta t}\right)^n e^{-C_{0,2}|v_j|^\alpha}$$

with $a_{j_1} = a_{j_1}^{s(n)}$. We suppressed the dependence on n, which is justified by Lemma 4.4 in [100]. Then we employ the following elementary inequality $(1+x)^n \leq e^{nx}$, $(x \geq 0)$ to derive

$$\left(\frac{\kappa}{\kappa+A_\nu\Delta t}\right)^n = \left(1 + A_\nu\frac{\Delta t}{\kappa}\right)^{-n} \geq e^{-nA_\nu\frac{\Delta t}{\kappa}} \geq e^{-\frac{A_\nu}{\kappa}T^f},$$

where we used $n\Delta t \leq N_t\Delta t = T^f$. The second estimate follows directly from this:

$$\|\tilde{f}_{i,j}^n\|_{L_q^\infty} \geq C_{0,1}e^{-\frac{A_\nu}{\kappa}T^f}\sup_j\left\{(1+|v_j|)^q e^{-C_{0,2}|v_j|^\alpha}\right\} \geq \frac{1}{2}C_{0,1}C_{q,\alpha}e^{-\frac{A_\nu}{\kappa}T^f}.$$

This completes the proof.

We now turn to the upper bound estimate. Applying Lemma 7.11 and to (113), one finds

$$\begin{aligned}
\|f^n\|_{L_q^\infty} &\leq \frac{\kappa}{\kappa+A_\nu\Delta t}\|\tilde{f}^{n-1}\|_{L_q^\infty} + \frac{A_\nu\Delta t}{\kappa+A_\nu\Delta t}\|\mathcal{M}_{\tilde{\nu}}(\tilde{f}^{n-1})\|_{L_q^\infty} \\
&\leq \frac{\kappa}{\kappa+A_\nu\Delta t}\|\tilde{f}^{n-1}\|_{L_q^\infty} + \frac{A_\nu\Delta t}{\kappa+A_\nu\Delta t}C_{\mathcal{M}}\|\tilde{f}^{n-1}\|_{L_q^\infty} \\
&\leq \frac{\kappa+C_{\mathcal{M}}A_\nu\Delta t}{\kappa+A_\nu\Delta t}\|\tilde{f}^{n-1}\|_{L_q^\infty}.
\end{aligned}$$

We then recall A_{n-1} to bound this further by

$$\left(\frac{\kappa+C_{\mathcal{M}}A_\nu\Delta t}{\kappa+A_\nu\Delta t}\right)\left(\frac{\kappa+C_{\mathcal{M}}A_\nu\Delta t}{\kappa+A_\nu\Delta t}\right)^{n-1}\|f^0\|_{L_q^\infty} \leq \left(\frac{\kappa+C_{\mathcal{M}}A_\nu\Delta t}{\kappa+A_\nu\Delta t}\right)^n\|f^0\|_{L_q^\infty}.$$

The second estimate follows from

$$\left(\frac{\kappa+C_{\mathcal{M}}A_\nu\Delta t}{\kappa+A_\nu\Delta t}\right)^n \leq \left(1+\frac{(C_{\mathcal{M}}-1)A_\nu\Delta t}{\kappa+A_\nu\Delta t}\right)^n \leq e^{\frac{(C_{\mathcal{M}}-1)A_\nu n\Delta t}{\kappa+A_\nu\Delta t}} \leq e^{\frac{(C_{\mathcal{M}}-1)A_\nu T^f}{\kappa+A_\nu\Delta t}},$$

where we used $(1+x)^n \leq e^{nx}$ and $n\Delta t \leq N_t\Delta t = T^f$.

Using this, we can prove the uniform lower bound of the macroscopic fields [100]:

Lemma 7.12 *Assume $f^n_{i,j}$ satisfies E^n_1. Then, $f^n_{i,j}$ satisfies E^n_2:*

$$\widetilde{\rho}^n_i \geq \frac{1}{2}C_{0,1}C_\alpha e^{-\frac{A_\nu}{\kappa}T^f}, \quad \widetilde{T}^n_i \geq \left(\frac{C_{0,1}C_\alpha}{2C_M\|f_0\|_{L^\infty_q}}\right)^{2/3} e^{-\frac{2}{3}\left(\frac{1}{\kappa}+\frac{(C_{q,\nu}-1)T^f}{\kappa+A_\nu\Delta t}\right)A_\nu T^f},$$

and

$$\|\widetilde{\rho}^n\|_{L^\infty_x}, \ \|\widetilde{U}^n\|_{L^\infty_x}, \ \|\widetilde{T}^n\|_{L^\infty_x} \leq 2C_q\left\{1+\left(C_{0,1}C_\alpha\right)^{-1}\right\}e^{\left(\frac{1}{\kappa}+\frac{(C_M-1)}{\kappa+A_\nu\Delta t}\right)A_\nu T^f}\|f^0\|_{L^\infty_q}.$$

Note also that Lemma 7.10 then immediately yields the lower bound for $\widetilde{\mathcal{T}}^n_{\widetilde{\nu}}$:

$$\widetilde{\mathcal{T}}^n_{\widetilde{\nu}} \geq C_M C_{0,1}C_\alpha \min\{1-\nu, 1+2\nu\}e^{-\left(\frac{1}{\kappa}+\frac{(C_M-1)T^f}{\kappa+A_\nu\Delta t}\right)A_\nu T^f} Id.$$

Proof of Theorem 7.8
It is clear from our choice of initial data that E_0 holds. Assume E^{n-1} is satisfied. Then Proposition 7.1 and Lemma 7.12 show that $f^n_{i,j}$ satisfies E^n. Therefore, we can conclude that $f^n_{i,j}$ satisfies E^n for all $n \geq 0$ by induction.

7.5 Difference Between Continuous and Discrete Distribution Functions

The main purposes of this section are to establish the estimate on the differences between continuous and discrete distribution functions on the continuity estimate of the discrete ellipsoidal Gaussian (see Proposition 7.2). We start with the difference estimate between distribution functions. The proof can be found in [100].

Lemma 7.13 *Assume $q > 5$ and $\Delta v < r_{\Delta v}$. Let f and f^n denote the solution of (104) and (113), respectively. Then, the difference of $\tilde{f}^n$ and $\tilde{f}(t_n)$ satisfies*

$$\|\tilde{f}^n - \tilde{f}(t_n)\|_{L^\infty_q} \leq \|f^n - f(t_n)\|_{L^\infty_q} + C_{T^f,f_0}(\Delta x)^2,$$

and

$$\Big|\sum_j \widetilde{f}^n_{i,j}\phi(v_j)(\Delta v)^3 - \int_{\mathbb{R}^3}\tilde{f}(x_i,v,t_n)\phi(v)dv\Big| \leq C_{T_f}\|f^n - f(t_n)\|_{L^\infty_q}$$
$$+ C_{T^f,f_0}\left\{(\Delta x)^2 + \Delta v + \Delta v\Delta t\right\},$$

where $\phi(v)$ *denote one of* $1, v, |v|^2, (1-\nu)|v|^2 Id + \nu v \otimes v$ *for* $v \in \mathbb{R}^3$.

This immediately leads to the difference estimates of macroscopic fields [100]:

Lemma 7.14 *Let* $\triangle v < r_\triangle$. *Suppose* f_0 *satisfies the assumptions in Theorem 7.2; then we have*

$$
\begin{aligned}
&|\widetilde{\rho}_i^n - \widetilde{\rho}(x_i, t_n)|,\ |\widetilde{U}_i^n - \widetilde{U}(x_i, t_n)|,\ |\widetilde{\mathcal{T}}_{\nu,i}^n - \widetilde{\mathcal{T}}_{\widetilde{\nu}}(x_i, t_n)| \\
&\qquad\qquad \le C\|f^n - f(t_n)\|_{L_q^\infty} + C_{T^f, f_0}\{(\triangle x)^2 + \triangle v + \triangle v \triangle t\}.
\end{aligned}
$$

We are now ready to prove the main result of this subsection.

Proposition 7.2 *Assume that* $\|f^n\|_{L_q^\infty} < \infty$ *with* $q > 5$. *Then we have*

$$
\|\mathcal{M}_{\widetilde{\nu}}(\widetilde{f}(t_n)) - \mathcal{M}_{\widetilde{\nu}}(\widetilde{f}^n)\|_{L_q^\infty} \le C_{T^f}\|f^n - f(t_n)\|_{L_q^\infty} + C_{T^f}\{(\triangle x)^2 + \triangle v + \triangle v \triangle t\}.
$$

The constants depend only on q, ν, T^f *and* f_0.

Proof We note that

$$
\begin{aligned}
\mathcal{M}_{\widetilde{\nu}}(\widetilde{f}(x_i, t_n))(v_j) - \mathcal{M}_{\widetilde{\nu},j}(\widetilde{f}_i^n) &= \mathcal{M}_{\widetilde{\nu}}\big(\widetilde{\rho}(x_i, t_n)), \widetilde{U}(x_i, t_n), \widetilde{\mathcal{T}}(x_i, t_n)\big)(v_j) \\
&\quad - \mathcal{M}_{\widetilde{\nu}}\big(\widetilde{\rho}_i^n, \widetilde{U}_i^n, \widetilde{\mathcal{T}}_i^n\big)(v_j).
\end{aligned}
$$

Therefore, applying the Taylor series, we expand this as

$$
\begin{aligned}
\mathcal{M}_{\widetilde{\nu}}(\widetilde{f}(x_i, t_n))(v_j) - \mathcal{M}_{\widetilde{\nu},j}(\widetilde{f}_i^n) &= \big\{\widetilde{\rho}(x_i, t_n) - \widetilde{\rho}_i^n\big\}\int_0^1 \frac{\partial \mathcal{M}_{\widetilde{\nu}}(\theta)}{\partial \rho} d\theta \\
&\quad + \big\{\widetilde{U}(x_i, t_n) - \widetilde{U}_i^n\big\}\int_0^1 \frac{\partial \mathcal{M}_{\widetilde{\nu}}(\theta)}{\partial U} d\theta \\
&\quad + \big\{\widetilde{\mathcal{T}}_{\widetilde{\nu}}(x_i, t_n) - \widetilde{\mathcal{T}}_{\widetilde{\nu},i}^n\big\}\int_0^1 \frac{\partial \mathcal{M}_{\widetilde{\nu}}(\theta)}{\partial \mathcal{T}_\nu} d\theta \\
&\equiv I_1 + I_2 + I_3,
\end{aligned} \tag{115}
$$

where

$$
\frac{\partial \mathcal{M}_{\widetilde{\nu}}(\theta)}{\partial X} = \frac{\partial \mathcal{M}_{\widetilde{\nu}}}{\partial X}\bigg|_{(\widetilde{\rho}_{i,\theta}^n, \widetilde{U}_{i,\theta}^n, \widetilde{\mathcal{T}}_{i,\theta}^n)}.
$$

For simplicity of notation, we define transitional macroscopic fields $\widetilde{\rho}_{\theta i}^n$, $\widetilde{U}_{\theta i}^n$, and $\widetilde{\mathcal{T}}_{\theta i}^n$ by

$$(\widetilde{\rho}^n_{\theta i}, \widetilde{U}^n_{\theta i}, \widetilde{\mathcal{T}}^n_{\theta i}) = (1-\theta)\Big(\widetilde{\rho}(x_i,t_n), \widetilde{U}(x_i,t_n), \widetilde{\mathcal{T}}_\nu(x_i,t_n)\Big) + \theta\Big(\widetilde{\rho}^n_i, \widetilde{U}^n_i, \widetilde{\mathcal{T}}^n_{\widetilde{\nu},i}\Big).$$

Since this is a linear combination, we can derive the following estimates for the transitional macroscopic fields:

$$\|\widetilde{\rho}^n_{\theta i}\|_{L^\infty_x}, \|\widetilde{U}^n_{\theta i}\|_{L^\infty_x}, \|\widetilde{\mathcal{T}}^n_{\theta i}\|_{L^\infty_x} \le C_{Tf},$$
$$\widetilde{\rho}^n_{\theta i} \ge C_{Tf} e^{-CT^f}, \quad k^\top\{\widetilde{\mathcal{T}}^n_{\theta i}\}k \ge C_{Tf} e^{-CT^f}|k|^2, \quad k \in \mathbb{R}^3,$$

from the lower and upper bounds of continuous and discrete macroscopic fields given in Theorem 7.2 and Proposition 7.12. On the other hand, Brum-Minkowski inequality implies that

$$\det\{\widetilde{\mathcal{T}}^n_{\theta i}\} \ge \{C_{Tf} e^{-CT^f}\}^{1-\theta}\{C_{Tf} e^{-CT^f}\}^{\theta} \ge C_{Tf} e^{-CT^f}.$$

From these observations we have

$$\begin{aligned}\mathcal{M}_{\widetilde{\nu}}(\theta) &= \frac{\widetilde{\rho}^n_{\theta i}}{\sqrt{\det(2\pi \widetilde{\mathcal{T}}^n_{\theta i})}} \exp\left(-\frac{1}{2}(v_j - \widetilde{U}^n_{\theta i})^\top \{\widetilde{\mathcal{T}}^n_{\theta i}\}^{-1}(v_j - \widetilde{U}^n_{\theta i})\right) \\ &\le C_{Tf,1} e^{-C_{Tf,2}|v_j|^2}.\end{aligned}$$

Similar computations give the following estimates for the integrals in I_i $(i = 1, 2, 3)$:

$$\left|\int_0^1 \frac{\partial \mathcal{M}_{\widetilde{\nu}}(\theta)}{\partial \widetilde{\rho}^n_{\theta i}} ds\right|, \quad \left|\int_0^1 \frac{\partial \mathcal{M}_{\widetilde{\nu}}(\theta)}{\partial \widetilde{U}^n_{\theta i}} ds\right|, \quad \left|\int_0^1 \frac{\partial \mathcal{M}_{\widetilde{\nu}}(\theta)}{\partial \widetilde{\mathcal{T}}_\theta} d\theta\right| \le C_{Tf} e^{-C_{Tf}|v_j|^2}.$$

We now insert all the above estimates into (115) to get

$$\begin{aligned}&|\mathcal{M}^n_{\widetilde{\nu},j}(\widetilde{f}^n_i) - \mathcal{M}_{\widetilde{\nu}}(\widetilde{f}(x_i,t_n))(v_j)| \\ &\quad \le C_{Tf}\Big\{|\widetilde{\rho}^n_i - \widetilde{\rho}(x_i,t_n)| + |\widetilde{U}^n_i - \widetilde{U}(x_i,t_n)| + |\widetilde{\mathcal{T}}^n_{\widetilde{\nu},i} - \widetilde{\mathcal{T}}_{\widetilde{\nu}}(x_i,t_n)|\Big\} e^{-C_{Tf}|v_j|^2}.\end{aligned}$$

Then, Lemma 7.14 gives the desired result.

7.6 *Proof of the Main Convergence Result*

To derive the error estimate, we first find that, under the assumption of Theorem 7.2, the ES-BGK model (104) can be represented in the following form:

$$f(x, v, t + \Delta t) = \frac{\kappa}{\kappa + A_v \Delta t} \widetilde{f}(x, v, t) + \frac{A_v \Delta t}{\kappa + A_v \Delta t} \mathcal{M}_{\tilde{v}}(\widetilde{f})(x, v, t)$$
$$+ \frac{A_v}{\kappa + A_v \Delta t} \big\{ R_1 + R_2 \big\}. \tag{116}$$

Here, we used the following notation $\widetilde{f}(x, v, t) = f(x - v_1 \Delta t, v, t)$ and the remainder terms R_1, R_2 satisfying the following estimate [100]:

$$\|R_1\|_{L_q^\infty} + \|R_2\|_{L_q^\infty} \leq C_{Tf, f_0} (\Delta t)^2. \tag{117}$$

We are now ready to prove our main theorem. Subtracting (113) from (116) and taking L_q^∞ norms, we get

$$\|f(t_{n+1}) - f^{n+1}\|_{L_q^\infty} \leq \frac{\kappa}{\kappa + A_v \Delta t} \|\widetilde{f}(t_n) - \widetilde{f}^n\|_{L_q^\infty}$$
$$+ \frac{A_v \Delta t}{\kappa + A_v \Delta t} \|\mathcal{M}_{\tilde{v}}(\widetilde{f}(t_n)) - \mathcal{M}_{\tilde{v}}(\widetilde{f}^n)\|_{L_q^\infty}$$
$$+ \frac{A_v}{\kappa + A_v \Delta t} \|R_1\|_{L_q^\infty} + \frac{A_v}{\kappa + A_v \Delta t} \|R_2\|_{L_q^\infty}.$$

We then recall the estimates in Lemma 7.13 and Proposition 7.2 and (117) to derive the following recurrence inequality for the numerical error:

$$\|f(t_{n+1}) - f^{n+1}\|_{L_q^\infty} = \Big(1 + \frac{C_{Tf} \Delta t}{\kappa + A_v \Delta t}\Big) \|f(t_n) - f^n\|_{L_q^\infty} + C_{Tf} P(\Delta x, \Delta v, \Delta t), \tag{118}$$

where

$$P(\Delta x, \Delta v, \Delta t) = \frac{(\Delta x)^2 + \Delta t\{(\Delta x)^2 + \Delta v + \Delta v \Delta t\} + (\Delta t)^2}{\kappa + A_v \Delta t}.$$

Put $\Gamma = \frac{C_{Tf} \Delta t}{\kappa + A_v \Delta t}$ for simplicity of notation and iterate (118) until the initial step is reached, to derive

$$\|f(N_t \Delta t) - f^{N_t}\|_{L_q^\infty} \leq (1 + \Gamma)^{N_t} \| f_0 - f^0 \|_{L_q^\infty} + C_v \sum_{i=1}^{N_t} (1 + \Gamma)^{i-1} P. \tag{119}$$

We first note from the definition of $f_{i,j}^0 = f_0(x_i, v_j)$ that

$$\| f_0 - f^0 \|_{L_q^\infty} = \sup_{i,j} |f_0(x_i, v_j) - f_{i,j}^0| (1 + |v_j|)^q = 0.$$

Besides, we use $(1+x)^n \le e^{nx}$ to see that

$$(1+\Gamma)^{N_t} \le e^{N_t\Gamma} \le e^{\frac{C_{Tf} N_t \Delta t}{\kappa + A_\nu \Delta t}} = e^{\frac{C_{Tf} T^f}{\kappa + \Delta t}},$$

so that

$$\sum_{i=1}^{N_t}(1+\Gamma)^{i-1} = \frac{(1+\Gamma)^{N_t}-1}{(1+\Gamma)-1} \le C_{Tf}\left(\frac{\kappa + A_\nu \Delta t}{\Delta t}\right)\left\{e^{\frac{C_{Tf}T^f}{\kappa + A_\nu \Delta t}} - 1\right\},$$

which gives

$$\sum_{i=1}^{N_t}(1+\Gamma)^{i-1}P \le C_{Tf}\left(e^{\frac{C_{Tf}T^f}{\kappa + A_\nu \Delta t}} - 1\right)\left\{\Delta x + \Delta v + \Delta t + \frac{\Delta x}{\Delta t}\right\}.$$

Substituting these estimates into (119), we find

$$\|f(T^f) - f^{N_t}\|_{L^\infty_q} \le C_{Tf}\left(e^{\frac{C_{Tf}T^f}{\kappa + \Delta t}} - 1\right)\left\{(\Delta x)^2 + \Delta v + \Delta t + \frac{(\Delta x)^2}{\Delta t}\right\}.$$

This completes the proof.

7.7 *Further Discussion*

In this section, we briefly reviewed the L^∞ convergence theory of the semi-Lagrangian scheme for the ellipsoidal BGK model. There are a couple of comments to be mentioned. First, when $\kappa = 0$, this error estimate breaks down since the coefficients of the error estimate contain κ^{-1}. Developing a convergence theory that holds uniformly for all ranges of Knudsen numbers seems to be out of reach. The main obstacle is the lack of pertinent uniform estimates for the continuous problem. Deriving regularity estimates that hold uniformly for all ranges of the Knudsen number for arbitrarily large solutions is a major big problem. Second, the result in Theorem 7.2 does not cover the critical case $\nu = -1/2$. One of the main reasons is that the corresponding existence result is missing. Recently, a local in-time existence result for the ES-BGK model with the correct Prandtl number is obtained in [105, 106]. Therefore, we expect that the convergence result can be soon extended to the critical case. Finally, we believe that the argument reviewed in this section is robust and can be extended in many directions such as the SL scheme for BGK models describing multicomponent gases, or high-order SL schemes based on BDF or Runge-Kutta methods.

8 Conclusions

In this chapter, we provide a review on conservative SL methods for several kinetic equations. SL schemes, when coupled with high-order nonlinear interpolation and high-order solvers in time, provide accurate solutions of kinetic equations. At variance with their Eulerian counterpart, they do not suffer from classical CFL-type restriction on the time step, thus achieving higher efficiency. Two causes of loss of conservation are identified, one due to the approximation of the collisional operator and one due to lack of translation invariance in nonlinear reconstructions of the distribution at the feet of the characteristics. The first can be cured by the L^2 projection technique introduced in [63], or its weighted version [18], which is quite general and can be applied to both BGK and Boltzmann equation. The second cause can be eliminated by adopting a suitable conservative reconstruction, introduced in [40].

The technique can be applied to the BGK model, as well as to the full Boltzmann equation. In the case of the BGK model, the accuracy in velocity is spectral. Therefore, for small to moderate Knudsen number, good accuracy is obtained with a relatively small number of grid points in velocity. Likewise, by adopting spectral methods for the approximation of the Boltzmann operator, one can achieve spectral accuracy in velocity even for the full Boltzmann equation. SL methods are very effective also to obtain an accurate solution of Vlasov-type equations. At variance with BTE and BGK equations, in Vlasov equation the characteristics are no longer straight lines; therefore, the conservative reconstruction illustrated in Sect. 3.2 cannot be applied directly. It can be applied as it is by using splitting methods since in each step the characteristics are lines with constant velocity (in each advection step) or constant position (in each drift step). For non-splitting methods the situation is more complicated, and different strategies have to be adopted.

The techniques adopted in this chapter can be extended also to the ellipsoidal BGK model. This model has the advantage of allowing a more accurate description of the gas flow for small Knudsen number, since it can be designed to match the same Navier-Stokes limit of the BTE. Theoretical results are already available for a first-order SL method, as discussed in the previous section. Some of us are currently working on the construction of high-order schemes for the ES-BGK model, which may provide an accurate description of gas flow in intermediate regimes, when macroscopic description is not yet valid.

One of the objectives of this line of research is to provide a unified description of multiscale gas flow, in situations in which a kinetic description is required only in a relatively small region like near boundaries or shocks, while in the rest of the domain standard macroscopic description could be adequate. Asymptotic-preserving schemes for BGK-type models may constitute the appropriate tool to treat seamless transition from full kinetic description to a macroscopic one.

Two major problems remain to be solved. The first one concerns an accurate treatment of boundary conditions, which is necessary to avoid spoiling the effort of the high accurate treatment of the equations in the interior. The second challenge is the development of efficient techniques for the treatment of problems which are very far from equilibrium. Indeed, the procedure described in Sect. 3.4 appears to be very efficient in terms of memory, but it is rather expensive in terms of CPU time. For such a reason alternative approaches are the subject of current investigation.

Acknowledgments G.R. has been partially supported by the following projects: Spoke 1 "FutureHPC & BigData" of the Italian Research Center on High-Performance Computing, Big Data and Quantum Computing (ICSC) and Italian national projects PRIN 2017 (N. 2017KKJP4X entitled "Innovative numerical methods for evolutionary partial differential equations and applications") and PRIN 2022 (N. 2022KA3JBA entitled "Advanced numerical methods for time dependent parametric partial differential equations with applications"), both financed by the Italian Ministry of University and Research (MUR). S. Y. Cho is supported by the Basic Research Program through the National Research Foundation of Korea (NRF) funded by the Ministry of Education and Technology (RS-2022-00166144).

Reviews on CWENO Reconstruction

In the field of conservation laws, the WENO (weighted essentially non-oscillatory) method has been widely adopted for computing non-oscillatory reconstruction to high-order accuracy. The WENO reconstruction reproduces a high-order polynomial from a wide stencil when solutions are smooth (thus providing high accuracy), while it degrades automatically to lower-degree polynomials constructed on smaller stencils when any discontinuity is detected within the wide stencil. By carefully taking nonlinear weights, the method allows to approximate the value of functions at a given particular point such as a cell boundary. However, when a source term is included in equations and one needs to evaluate source terms with high accuracy, it is necessary to compute the source at quadrature points in the interior of the cell. In this case, the WENO reconstruction may become very inefficient because it should be computed with different weights for each quadrature point. This difficulty has been overcome by the central WENO (CWENO) method, originally introduced in [78] (see also [44], where the general description of the CWENO approach is provided). We remark that the CWENO reconstructions achieve arbitrarily high-order uniform accuracy in the interior of the cell, prevent spurious oscillation, and maintain conservation. For such reasons, the CWENO reconstructions are good candidates as a basic reconstruction for the conservative reconstruction [40]. In this section, we briefly review two methods, CWENO23 [78] and CWENO35 [31].

Review on CWENO23 Reconstruction

Here, we review the CWENO23 reconstruction [78]. We start from a polynomial of degree two $P^i_{OPT}(x)$ which interpolates $\bar{u}_{i-1}, \bar{u}_i, \bar{u}_{i+1}$ in the sense of cell averages:

$$\frac{1}{\Delta x}\int_{x_{i+l-\frac{1}{2}}}^{x_{i+l+\frac{1}{2}}} P^i_{OPT}(x)\,dx = \bar{u}_{i+l}, \quad l = -1, 0, 1.$$

Then, this polynomial can be written as $P^i_{OPT}(x) = \tilde{u}_i + \tilde{u}'_i(x - x_i) + \frac{1}{2}\tilde{u}''_i(x - x_i)^2$ with

$$\tilde{u}_i = \bar{u}_i - \frac{1}{24}(\bar{u}_{i+1} - 2\bar{u}_i + \bar{u}_{i-1}), \quad \tilde{u}'_i = \frac{\bar{u}_{i+1} - \bar{u}_{i-1}}{2\Delta x}, \quad \tilde{u}''_i = \frac{\bar{u}_{i+1} - 2\bar{u}_i + \bar{u}_{i-1}}{(\Delta x)^2},$$

and it gives a third-order accurate reconstruction of u in I_i:

$$P^i_{OPT}(x) = u(x) + \mathcal{O}(\Delta x)^3, \quad \forall x \in I_i.$$

In the CWENO23 reconstruction, to avoid oscillations, we use the following convex combination:

$$R_i(x) = \sum_k \omega^i_k P^i_k(x), \quad \sum_k \omega^i_k = 1, \quad \omega^i_k \geq 0, \quad k \in \{L, C, R\} \tag{120}$$

where P^i_L and P^i_R are first-order polynomials such that

$$\int_{x_{i+l-\frac{1}{2}}}^{x_{i+l+\frac{1}{2}}} P^i_L(x)\,dx = \bar{u}_{i+l}, \quad l = -1, 0, \qquad \int_{x_{i+l-\frac{1}{2}}}^{x_{i+l+\frac{1}{2}}} P^i_R(x)\,dx = \bar{u}_{i+l}, \quad l = 0, 1,$$

which gives

$$P^i_L(x) = \bar{u}_i + \frac{\bar{u}_i - \bar{u}_{i-1}}{\Delta x}(x - x_i), \quad P^i_R(x) = \bar{u}_i + \frac{\bar{u}_{i+1} - \bar{u}_i}{\Delta x}(x - x_i).$$

The second-order polynomial $P^i_C(x)$ is obtained from

$$P^i_C(x) := \frac{1}{C_C}\left(P^i_{OPT}(x) - C_L P^i_L(x) - C_R P^i_R(x)\right),$$

with a choice of positive coefficients such that

$$C_L, C_R, C_C \geq 0, \quad C_L = C_R, \quad C_L + C_C + C_R = 1.$$

A common choice is to set $C_L = C_R = 1/4$, $C_C = 1/2$. The nonlinear weights ω_k^i in (120) are chosen as follows:

$$\omega_k^i = \frac{\alpha_k^i}{\sum_\ell \alpha_\ell^i}, \quad \alpha_k^i = \frac{C_i}{(\epsilon + \beta_k^i)^p}, \quad k, \ell \in \{L, C, R\} \tag{121}$$

where the constant ϵ is used to avoid the denominator vanishing and the constant p weights the smoothness indicator. We use $\epsilon = (\Delta x)^2$ or 10^{-6} and $p = 2$ in the numerical tests. An explicit expression of smoothness indicators is the following:

$$\beta_L^i = (\bar{u}_i - \bar{u}_{i-1})^2, \quad \beta_R^i = (\bar{u}_{i+1} - \bar{u}_i)^2,$$
$$\beta_C^i = \frac{13}{3}(\bar{u}_{i+1} - 2\bar{u}_i + \bar{u}_{i-1})^2 + \frac{1}{4}(\bar{u}_{i+1} - \bar{u}_{i-1})^2.$$

We refer to [77] for details on CWENO reconstruction. As a consequence, the reconstruction (120) is third-order accurate in smooth region and automatically becomes second-order accurate in the presence of discontinuity. The final form of the CWENO23 reconstruction $R_i(x)$ is given by

$$R_i(x) = R_i^{(0)} + R_i^{(1)}(x - x_i) + \frac{1}{2}R_i^{(2)}(x - x_i)^2,$$

where

$$R_i^{(0)} = \bar{u}_i - \frac{1}{12}\omega_C^i(\bar{u}_{i+1} - 2\bar{u}_i + \bar{u}_{i-1})$$
$$R_i^{(1)} = \omega_L^i \frac{\bar{u}_i - \bar{u}_{i-1}}{\Delta x} + \omega_R^i \frac{\bar{u}_{i+1} - \bar{u}_i}{\Delta x} + \omega_C^i \frac{\bar{u}_{i+1} - \bar{u}_{i-1}}{2\Delta x}$$
$$R_i^{(2)} = 2\omega_C^i \frac{\bar{u}_{i+1} - 2\bar{u}_i + \bar{u}_{i-1}}{(\Delta x)^2}.$$

Review on CWENO35 Reconstruction

Here we review the CWENO35 [31] reconstruction. We start from a polynomial of degree four, $P_{OPT}^i(x)$, which interpolates $\bar{u}_{i-2}, \bar{u}_{i-1}, \bar{u}_i, \bar{u}_{i+1}, , \bar{u}_{i+2}$ in the sense of cell averages:

$$\frac{1}{\Delta x}\int_{x_{i+l-\frac{1}{2}}}^{x_{i+l+\frac{1}{2}}} P^i_{OPT}(x)\,dx = \bar{u}_{i+l}, \quad l = -2, -1, 0, 1, 2.$$

Then, this polynomial can be written as

$$P^i_{OPT}(x) = \tilde{u}_i + \tilde{u}'_i(x - x_i) + \tilde{u}''_i(x - x_i)^2 + \tilde{u}'''_i(x - x_i)^3 + \tilde{u}''''_i(x - x_i)^4$$

where

$$\begin{aligned}
\tilde{u}_i &= \frac{1067}{960}\bar{u}_i - \frac{29}{480}(\bar{u}_{i+1} + \bar{u}_{i-1}) + \frac{3}{640}(\bar{u}_{i+2} + \bar{u}_{i-2})\\
\tilde{u}'_i &= \frac{17}{24}\frac{\bar{u}_{i+1} - \bar{u}_{i-1}}{\Delta x} + \frac{5}{48}\frac{\bar{u}_{i-2} - \bar{u}_{i+2}}{\Delta x}\\
\tilde{u}''_i &= -\frac{1}{16}\frac{\bar{u}_{i-2} + 22\bar{u}_i + \bar{u}_{i+2}}{(\Delta x)^2} + \frac{3}{4}\frac{\bar{u}_{i+1} + \bar{u}_{i-1}}{(\Delta x)^2}\\
\tilde{u}'''_i &= -\frac{1}{6}\frac{\bar{u}_{i+1} - \bar{u}_{i-1}}{(\Delta x)^3} - \frac{1}{12}\frac{\bar{u}_{i-2} - \bar{u}_{i+2}}{(\Delta x)^3}\\
\tilde{u}''''_i &= \frac{1}{24}\frac{\bar{u}_{i-2} + 6\bar{u}_i + \bar{u}_{i+2}}{(\Delta x)^4} - \frac{1}{6}\frac{\bar{u}_{i+1} + \bar{u}_{i-1}}{(\Delta x)^4}.
\end{aligned}$$

Next, we consider three polynomials $P_1(x)$, $P_2(x)$, $P_3(x)$ which respectively interpolate $\{\bar{u}_{i-2}, \bar{u}_{i-1}, \bar{u}_i\}$, $\{\bar{u}_{i-1}, \bar{u}_i, \bar{u}_{i+1}\}$, $\{\bar{u}_i, \bar{u}_{i+1}, \bar{u}_{i+2}\}$ in the sence of cell average:

$$P^i_\ell(x) = \sum_{k=0}^{2} a_{\ell k}(x - x_i)^k, \quad \ell = 1, 2, 3,$$

where the coefficients $a_{\ell k}$, for $1 \le \ell \le 3$ and $0 \le k \le 2$, are given by

$$\begin{aligned}
a_{10} &= \frac{23}{24}\bar{u}_i + \frac{1}{12}\left(\bar{u}_{i-1} - \frac{1}{2}\bar{u}_{i-2}\right), \quad a_{11} = \frac{3\bar{u}_i - 4\bar{u}_{i-1} + \bar{u}_{i-2}}{2\Delta x}\\
a_{12} &= \frac{\bar{u}_i - 2\bar{u}_{i-1} + \bar{u}_{i-2}}{2(\Delta x)^2}, \quad a_{20} = \frac{13}{12}\bar{u}_i - \frac{1}{24}(\bar{u}_{i-1} + \bar{u}_{i+1})\\
a_{21} &= \frac{\bar{u}_{i+1} - \bar{u}_{i-1}}{2\Delta x}, \quad a_{22} = \frac{\bar{u}_{i+1} - 2\bar{u}_i + \bar{u}_{i-1}}{2(\Delta x)^2}\\
a_{30} &= \frac{23}{24}\bar{u}_i + \frac{1}{12}\left(\bar{u}_{i+1} - \frac{1}{2}\bar{u}_{i+2}\right), \quad a_{31} = \frac{3\bar{u}_i - 4\bar{u}_{i+1} + \bar{u}_{i+2}}{-2\Delta x}\\
a_{32} &= \frac{\bar{u}_i - 2\bar{u}_{i+1} + \bar{u}_{i+2}}{2(\Delta x)^2}.
\end{aligned}$$

Then, we consider a polynomial $P_C(x)$ which satisfies

$$P_C^i(x) = \frac{1}{C_C}\left(P_{OPT}^i(x) - C_1 P_1^i(x) - C_2 P_2^i(x) - C_3 P_3^i(x)\right),$$

where $C_1 = C_3 = \frac{1}{8}$, $C_2 = \frac{1}{4}$, and $C_C = \frac{1}{2}$. Finally, we have

$$P_C^i(x) = \sum_{k=0}^{4} a_{Ck}(x - x_i)^k,$$

where

$$a_{C0} = \frac{1}{C_C}(\tilde{u}_i - C_1 a_{10} - C_2 a_{20} - C_3 a_{30}), \quad a_{C1} = \frac{1}{C_C}(\tilde{u}_i' - C_1 a_{11} - C_2 a_{21} - C_3 a_{31})$$

$$a_{C2} = \frac{1}{C_C}(\tilde{u}_i'' - C_1 a_{12} - C_2 a_{22} - C_3 a_{32}), \quad a_{C3} = \frac{1}{C_C}\tilde{u}_i''', \quad a_{C4} = \frac{1}{C_C}\tilde{u}_i''''.$$

As in (121), we compute nonlinear weights:

$$\omega_k^i = \frac{\alpha_k^i}{\sum_\ell \alpha_\ell^i}, \quad \alpha_k^i = \frac{C_i}{(\epsilon + \beta_k^i)^p}, \quad k, \ell \in \{1, 2, 3, C\}$$

with smoothness indicators:

$$\beta_k^i = \sum_{l \geq 1} \int_{x_{i-\frac{1}{2}}}^{x_{i+\frac{1}{2}}} \Delta x^{2l-1} (\partial_x^{(l)} P_k^i(x))^2 dx, \quad k \in \{1, 2, 3, C\}.$$

Note that we use $\epsilon = (\Delta x)^2$ or 10^{-6} and $p = 2$ in the numerical tests. To sum up, we obtain the CWENO35 reconstruction:

$$R_i(x) = \sum_k \omega_k^i P_k^i(x), \quad \sum_k \omega_k^i = 1, \quad \omega_k^i \geq 0, \quad k \in \{1, 2, 3, C\}.$$

The reconstruction can be represented in form (19), i.e.,

$$R_i(x) = \sum_{\ell=0}^{4} \frac{R_i^{(\ell)}}{\ell!}(x - x_i)^{(\ell)},$$

with

$$
\begin{aligned}
R_i^{(0)} &= \omega_C^i \left(\frac{577}{480}\bar{u}_i - \frac{29}{240}\bar{u}_{i-1} + \frac{19}{960}\bar{u}_{i-2} - \frac{29}{240}\bar{u}_{i+1} + \frac{19}{960}\bar{u}_{i+2} \right) \\
&- \omega_2^i \left(\frac{\bar{u}_{i-1} - 26\bar{u}_i + \bar{u}_{i+1}}{24} \right) + \omega_1^i \left(\frac{23}{24}\bar{u}_i + \frac{1}{12}\bar{u}_{i-1} - \frac{1}{24}\bar{u}_{i-2} \right) \\
&+ \omega_3^i \left(\frac{23}{24}\bar{u}_i + \frac{1}{12}\bar{u}_{i+1} - \frac{1}{24}\bar{u}_{i+2} \right) \\
R_i^{(1)} &= -\omega_C^i \frac{8\bar{u}_{i-1} - \bar{u}_{i-2} - 8\bar{u}_{i+1} + \bar{u}_{i+2}}{12\Delta x} + \omega_1^i \frac{3\bar{u}_i - 4\bar{u}_{i-1} + \bar{u}_{i-2}}{2\Delta x} \\
&- \omega_3^i \frac{3\bar{u}_i - 4\bar{u}_{i+1} + \bar{u}_{i+2}}{2\Delta x} - \omega_2^i \frac{\bar{u}_{i-1} - \bar{u}_{i+1}}{2\Delta x} \\
R_i^{(2)} &= 2\left(\omega_1^i \frac{\bar{u}_i - 2\bar{u}_{i-1} + \bar{u}_{i-2}}{2(\Delta x)^2} + \omega_2^i \frac{\bar{u}_{i-1} - 2\bar{u}_i + \bar{u}_{i+1}}{2(\Delta x)^2} + \omega_3^i \frac{\bar{u}_i - 2\bar{u}_{i+1} + \bar{u}_{i+2}}{2(\Delta x)^2} \right) \\
&- 2\omega_C^i \left(\frac{10\bar{u}_i - 6\bar{u}_{i-1} + \bar{u}_{i-2} - 6\bar{u}_{i+1} + \bar{u}_{i+2}}{4(\Delta x)^2} \right) \\
R_i^{(3)} &= 6\omega_C^i \left(\frac{\bar{u}_{i-1} - \bar{u}_{i+1}}{3(\Delta x)^3} - \frac{\bar{u}_{i-2} - \bar{u}_{i+2}}{6(\Delta x)^3} \right), \\
R_i^{(4)} &= 24\omega_C^i \left(\frac{\bar{u}_{i-2} + 6\bar{u}_i + \bar{u}_{i+2}}{12(\Delta x)^4} - \frac{\bar{u}_{i-1} + \bar{u}_{i+1}}{3(\Delta x)^4} \right).
\end{aligned}
$$

References

1. A. Aimi, M. Diligenti, M. Groppi, C. Guardasoni, On the numerical solution of a BGK-type model for chemical reactions. Eur. J. Mech. B Fluids **26**(4), 455–472 (2007)
2. P. Andries, K. Aoki, B. Perthame, A consistent BGK-type model for gas mixtures. J. Stat. Phys. **106**, 993–1018 (2002)
3. P. Andries, J. Bourgat, P. Le Tallec, B. Perthame, Numerical comparison between the Boltzmann and ES-BGK models for rarefied gases. Comput. Methods Appl. Mech. Eng. **191**(31), 3369–3390 (2002)
4. P. Andries, P. Le Tallec, J. Perlat, B. Perthame, The Gaussian-BGK model of Boltzmann equation with small Prandtl number. Eur. J. Mech. B Fluids **19**(6), 813–830 (2000)
5. V.V. Aristov, F.G. Tcheremissine, The conservative splitting method for solving the Boltzmann equation. USSR Compt. Math. Math. Phys. **1**(20), 208–225 (1980)
6. M.L. Begue, A. Ghizzo, P. Bertrand, E. Sonnendrucker, O. Coulaud, Two-dimensional semi-Lagrangian Vlasov simulations of laser–plasma interaction in the relativistic regime. J. Plasma Phys **62**(04), 367–388 (1999)
7. F. Bernard, A. Iollo, G. Puppo, A local velocity grid approach for BGK equation. Commun. Comput. Phys. **16**(4), 956–982 (2014)
8. N. Besse, E. Sonnendrucker, Semi-Lagrangian schemes for the Vlasov equation on an unstructured mesh of phase space. J. Comput. Phys. **191**(2), 341–376 (2003)

9. P.L. Bhatnagar, E.P. Gross, M. Krook, A model for collision processes in gases. I. Small amplitude processes in charged and neutral one-component systems Phys. Rev. **94**(3), 511 (1954)
10. G.A. Bird, Molecular gas dynamics. NASA STI/Recon Technical Report A **76**, 40225 (1976)
11. M. Bisi, A.V. Bobylev, M. Groppi, G. Spiga, Hydrodynamic equations from a BGK model for inert gas mixtures, in *AIP Conference Proceedings*, vol. 2132 (AIP Publishing LLC, New York, 2019), p. 130010
12. M. Bisi, M. Groppi, G. Martalò, The evaporation–condensation problem for a binary mixture of rarefied gases. Contin. Mech. Thermodyn. **32**, 1109–1126 (2020)
13. M. Bisi, M. Groppi, G. Martalò, Macroscopic equations for inert gas mixtures in different hydrodynamic regimes. J. Phys. A: Math. Theor. **54**(8), 085201 (2021)
14. M. Bisi, M. Groppi, G. Spiga, Kinetic Bhatnagar-Gross-Krook model for fast reactive mixtures and its hydrodynamic limit. Phys. Rev. E **81**(3), 036327 (2010)
15. M. Bisi, G. Spiga, Navier–Stokes hydrodynamic limit of BGK kinetic equations for an inert mixture of polyatomic gases, in *From Particle Systems to Partial Differential Equations: PSPDE V, Braga, Portugal, November 2016 V* (Springer, Berlin, 2018), pp. 13–31
16. A.V. Bobylev, M. Bisi, M. Groppi, G. Spiga, I.F. Potapenko, A general consistent BGK model for gas mixtures. Kinet. Relat. Models **11**(6), 1377–1393 (2018)
17. S. Boscarino, S.Y. Cho, M. Groppi, G. Russo, BGK models for inert mixtures: comparison and applications. Kinet. Relat. Models **14**, 895–928 (2021)
18. S. Boscarino, S.Y. Cho, G. Russo, A local velocity grid conservative semi-Lagrangian schemes for BGK model. J. Comput. Phys. **460**, 111178 (2022)
19. S. Boscarino, S.Y. Cho, G. Russo, A conservative semi-Lagrangian method for inhomogeneous Boltzmann equation. J. Comput. Phys. **498**, 112633 (2024)
20. S. Boscarino, S.Y. Cho, G. Russo, S.-B. Yun, High order conservative semi-Lagrangian scheme for the BGK model of the Boltzmann equation. Commun. Comput. Phys. **29**, 1–56 (2021)
21. S. Boscarino, S.Y. Cho, G. Russo, S.-B. Yun, Convergence estimates of a semi-Lagrangian scheme for the ellipsoidal BGK model for polyatomic molecules. ESAIM: M2AN **56**, 893–942 (2022)
22. W. Boscheri, G. Dimarco, High order finite volume schemes with IMEX time stepping for the Boltzmann model on unstructured meshes. Comput. Methods Appl. Mech. Engrg. **387**, 114180 (2021)
23. W. Boscheri, G. Dimarco, High order modal discontinuous Galerkin implicit–explicit Runge Kutta and linear multistep schemes for the Boltzmann model on general polygonal meshes. Comput. Fluids **233**, 105224 (2022)
24. J.P. Boyd, *Chebyshev and Fourier Spectral Methods* (Courier Dover Publications, New York, 2001)
25. S. Brull, L. Mieussens, Local discrete velocity grids for deterministic rarefied flow simulations. J. Comput. Phys. **266**, 22–46 (2014)
26. S. Brull, C. Prigent, Local discrete velocity grids for multi-species rarefied flow simulations. Commun. Comput. Phys. **28**, 1274–1304 (2020)
27. S. Brull, J. Schneider, A new approach for the ellipsoidal statistical model. Contin. Mech. Thermodyn. **20**, 63–74 (2008)
28. X. Cai, S. Boscarino, J.-M. Qiu, High order semi-Lagrangian discontinuous Galerkin method coupled with Runge-Kutta exponential integrators for nonlinear Vlasov dynamics. J. Comput. Phys. **427**, 110036 (2021)
29. X. Cai, J.-M. Qiu, Y. Yang, An Eulerian-Lagrangian discontinuous Galerkin method for transport problems and its application to nonlinear dynamics. J. Comput. Phys. **439**, 110392 (2021)
30. M. Campos-Pinto, F. Charles, B. Després, Algorithms for positive polynomial approximation. SIAM J. Numer. Anal. **57**(1), 148–172 (2019)
31. G. Capdeville, A central WENO scheme for solving hyperbolic conservation laws on non-uniform meshes. J. Comput. Phys. **227**(5), 2977–3014 (2008)

32. E. Carlini, R. Ferretti, G. Russo, A weighted essentially nonoscillatory, large time-step scheme for Hamilton–Jacobi equations. SIAM J. Sci. Comput. **27**(3), 1071–1091 (2005)
33. J.A. Carrillo, M. Fornasier, G. Toscani, F. Vecil, Particle, kinetic, and hydrodynamic models of swarming, in *Mathematical Modeling of Collective Behavior in Socio-Economic and Life Sciences*, pp. 297–336 (2010)
34. J.A. Carrillo, F. Vecil, Nonoscillatory interpolation methods applied to Vlasov-based models. SIAM J. Sci. Comput. **29**, 1179 (2007)
35. F. Casas, N. Crouseilles, E. Faou, M. Mehrenberger, High-order Hamiltonian splitting for the Vlasov–Poisson equations. Numer. Math. **135**, 769–801 (2017)
36. C. Cercignani, *The Boltzmann Equation and Its Applications* (Springer, New York, 1988)
37. C.Z. Cheng, G. Knorr, The integration of the Vlasov equation in configuration space. J. Comput. Phys. **22**(3), 330–351 (1976)
38. Y. Cheng, A.J. Christlieb, X. Zhong, Energy-conserving discontinuous Galerkin methods for the Vlasov–Ampère system. J. Comput. Phys. **256**, 630–655 (2014)
39. S.Y. Cho, S. Boscarino, M. Groppi, G. Russo, Conservative semi-lagrangian schemes for a general consistent BGK model for inert gas mixtures. Commun. Math. Sci. **20**, 695–725 (2020)
40. S.Y. Cho, S. Boscarino, G. Russo, S.-B. Yun, Conservative semi-Lagrangian schemes for kinetic equations Part I: Reconstruction. J. Comput. Phys. **432**, 110159 (2021)
41. S.Y. Cho, S. Boscarino, G. Russo, S.-B. Yun, Conservative semi-Lagrangian schemes for kinetic equations Part II: applications. J. Comput. Phys. **436**, 110281 (2021)
42. A. Christlieb, W. Guo, M. Morton, J.-M. Qiu, A high order time splitting method based on integral deferred correction for semi-Lagrangian Vlasov simulations. J. Comput. Phys. **267C**, 7–27 (2014)
43. C.K. Chu, Kinetic-theoretic description of the formation of a shock wave. The Phys. Fluids **8**(1), 12–22 (1965)
44. I. Cravero, G. Puppo, M. Semplice, G. Visconti, CWENO: uniformly accurate reconstructions for balance laws. Math. Comp. **87**, 1689–1719 (2018)
45. N. Crouseilles, M. Gutnic, G. Latu, E. Sonnendrücker, Comparison of two Eulerian solvers for the four-dimensional Vlasov equation. II. Commun. Nonlinear Sci. Numer. Simul. **13**(1), 94–99 (2008)
46. N. Crouseilles, M. Mehrenberger, E. Sonnendrücker, Conservative semi-lagrangian schemes for Vlasov equations. J. Comput. Phys. **229**(6), 1927–1953 (2010)
47. G. Dimarco, R. Loubère, Towards an ultra efficient kinetic scheme. Part II: The high order case. J. Comput. Phys. **255**, 699–719 (2013)
48. G. Dimarco, L. Pareschi, Numerical methods for kinetic equations. Acta Numer. **23**, 369–520 (2014)
49. M. Ding, J.-M. Qiu, R. Shu, Accuracy and stability analysis of the semi-Lagrangian method for stiff hyperbolic relaxation systems and kinetic BGK model. Multiscale Model. Simul. **21**(1), 143–167 (2023)
50. M. Falcone, R. Ferretti, *Semi-Lagrangian Approximation Schemes for Linear and Hamilton–Jacobi Equations* (SIAM, New York, 2013)
51. R. Ferretti, Stability of some generalized Godunov schemes with linear high-order reconstructions. J. Sci. Comput. **57**(1), 213–228 (2013)
52. F. Filbet, S. Jin, A class of asymptotic-preserving schemes for kinetic equations and related problems with stiff sources. J. Comput. Phys. **229**, 7625–7648 (2010)
53. F. Filbet, S. Jin, An asymptotic preserving scheme for the ES-BGK model of the Boltzmann equation. J. Sci. Comput. **46**, 204–224 (2011)
54. F. Filbet, C. Mouhot, L. Pareschi, Solving the Boltzmann equation in $N \log 2N$. SIAM J. Sci. Comput. **28**, 1029–1053 (2006)
55. F. Filbet, G. Russo, High order numerical methods for the space non-homogeneous Boltzmann equation. J. Comput. Phys. **186**, 457–480 (2003)
56. F. Filbet, E. Sonnendrücker, Comparison of Eulerian Vlasov solvers. Comput. Phys. Commun. **150**(3), 247–266 (2003)

57. F. Filbet, E. Sonnendrücker, P. Bertrand, Conservative numerical schemes for the Vlasov equation. J. Comput. Phys. **172**(1), 166–187 (2001)
58. A. Frezzotti, Numerical study of the strong evaporation of a binary mixture. Fluid Dyn. Res. **8**(5-6), 175 (1991)
59. J. Friedrich, O. Kolb, Maximum principle satisfying CWENO schemes for nonlocal conservation laws. SIAM J. Sci. Comput. **41**(2), A973–A988 (2019)
60. J. Narski G. Dimarco, R. Loubère, T. Rey, An efficient numerical method for solving the Boltzmann equation in multidimensions. J. Comput. Phys. **353**, 46–81 (2018)
61. V.S. Galkin, N.K. Makashev, Kinetic derivation of the gas-dynamic equation for multicomponent mixtures of light and heavy particles. Fluid Dyn. **29**(1), 140–155 (1994)
62. M.A. Gallis, J.R. Torczynski, Investigation of the ellipsoidal-statistical Bhatnagar–Gross–Krook kinetic model applied to gas-phase transport of heat and tangential momentum between parallel walls. Phys. Fluids **23**(3), 030601 (2011)
63. I.M. Gamba, S.H. Tharkabhushanam, Spectral-lagrangian methods for collisional models of non-equilibrium statistical states. J. Comput. Phys. **228**, 2012–2036 (2009)
64. M. Groppi, G. Russo, G. Stracquadanio, High order semi-Lagrangian methods for the BGK equation. Commun. Math. Sci. **14**(2), 389–414 (2015)
65. M. Groppi, G. Russo, G. Stracquadanio, Semi-Lagrangian approximation of BGK models for inert and reactive gas mixtures, in *From Particle Systems to Partial Differential Equations: PSPDE V, Braga, Portugal, November 2016 V* (Springer, Berlin, 2018), pp. 53–80
66. Y. Güçlü, A.J. Christlieb, W.N.G. Hitchon, Arbitrarily high order convected scheme solution of the Vlasov–Poisson system. J. Comput. Phys. **270**, 711–752 (2014)
67. W. Guo, J.-M. Qiu, Hybrid semi-Lagrangian finite element-finite difference methods for the Vlasov equation. J. Comput. Phys. **234**, 108–132 (2013)
68. E. Hairer, G. Warner, *Solving Ordinary Differential Equations II: Stiff and Differential-Algebraic Problems* (Springer, Berlin, 1996)
69. E. Hairer, G. Warner, S.P. Nórsett, *Solving Ordinary Differential Equations I: Nonstiff Problem* (Springer, Berlin, 1996)
70. S.A. Hollingsworth, R.O. Dror, Molecular dynamics simulation for all. Neuron **99**(6), 1129–1143 (2018)
71. J. Hu, R. Shu, X. Zhang, Asymptotic-preserving and positivity-preserving implicit-explicit schemes for the stiff BGK equation. SIAM J. Numer. Anal. **56**(2), 942–973 (2018)
72. F. Huot, A. Ghizzo, P. Bertrand, E. Sonnendrucker, O. Coulaud, Instability of the time splitting scheme for the one-dimensional and relativistic Vlasov–Maxwell system. J. Comput. Phys. **185**(2), 512–531 (2003)
73. D. Issautier, Convergence of a weighted particle method for solving the Boltzmann (BGK) equation. SIAM J. Numer. Anal. **33**(6), 2099–2119 (1996)
74. L.H. Holway Jr., Kinetic theory of shock structure using an ellipsoidal distribution function. Rarefied Gas Dynamics **1**(1), 193 (1965)
75. J. Kestin, K. Knierim, E.A. Mason, B. Najafi, S.T. Ro, M. Waldman, Equilibrium and transport properties of the noble gases and their mixtures at low density. J. Phys. Chem. Ref. Data. **13**(1), 229–303 (1984)
76. R. Klar, R. Wegener, Kinetic traffic flow models, in *Modeling in Applied Sciences: A Kinetic Theory Approach* (Springer, Berlin, 2000), pp. 263–316
77. D. Levy, G. Puppo, G. Russo, Central WENO schemes for hyperbolic systems of conservation laws. ESAIM Math. Model. Numer. Anal. **33**(3), 547–571 (1999)
78. D. Levy, G. Puppo, G. Russo, Compact central WENO schemes for multidimensional conservation laws. SIAM J. Sci. Comput. **22**(2), 656–672 (2000)
79. D. Levy, G. Puppo, G. Russo, A fourth order central WENO scheme for multidimensional hyperbolic systems of conservation Laws. SIAM J. Sci. Comput. **24**(2), 480–506 (2002)
80. Q. Li, L. Pareschi, Exponential Runge–Kutta for the inhomogeneous Boltzmann equations with high order of accuracy. J. Comput. Phys. **259**, 402–420 (2014)
81. D. Madjarević, S. Simić, Shock structure in helium-argon mixture–a comparison of hyperbolic multi-temperature model with experiment. Europhys. Lett **102**(4), 44002 (2013)

82. J. Meng, L. Wu, J.M. Reese, Y. Zhang, Assessment of the ellipsoidal-statistical Bhatnagar–Gross–Krook model for force-driven Poiseuille flows. J. Comput. Phys. **251**, 383–395 (2013)
83. L. Mieussens, Discrete velocity model and implicit scheme for the BGK equation of rarefied gas dynamics. Math. Models Methods Appl. Sci. **10**(8), 1121–1149 (2000)
84. L. Mieussens, H. Struchtrup, Numerical comparison of Bhatnagar–Gross–Krook models with proper Prandtl number. Phys. Fluids **16**(8), 2797–2813 (2004)
85. C. Mouhot, L. Pareschi, Fast algorithms for computing the Boltzmann collision operator. Math. Comp. **75**, 1833–1852 (2006)
86. D.R. Nicholson, *Introduction to Plasma Theory* (Wiley, New York, 1983)
87. L. Pareschi, B. Perthame, A Fourier spectral method for homogeneous Boltzmann equations. Transport Theor. Stat. Phys. **25**, 369–382 (1996)
88. L. Pareschi, T. Rey, Moment preserving Fourier–Galerkin spectral methods and application to the Boltzmann equation. SIAM J. Numer. Anal. **60**, 3216–3240 (2022)
89. L. Pareschi, G. Russo, Numerical solution of the Boltzmann equation I: Spectrally accurate approximation of the collision operator. SIAM J. Numer. Anal. **37**, 1217–1245 (2000)
90. B. Perthame, An introduction to kinetic schemes for gas dynamics, in *An Introduction to Recent Developments in Theory and Numerics for Conservation Laws: Proceedings of the International School on Theory and Numerics for Conservation Laws, Freiburg/Littenweiler, October 20–24, 1997* (Springer, Berlin, 1999), pp. 1–27
91. S. Pieraccini, G. Puppo, Implicit-explicit schemes for BGK kinetic equations. J. Sci. Comput. **32**, 1–28 (2007)
92. J.-M. Qiu, A. Christlieb, A Conservative high order semi-Lagrangian WENO method for the Vlasov equation. J. Comput. Phys. **229**, 1130–1149 (2010)
93. J.-M. Qiu, G. Russo, A high order multi-dimensional characteristic tracing strategy for the Vlasov–Poisson system. J. Sci. Comput. **71**, 414–434 (2017)
94. J.-M. Qiu, C.-W. Shu, Conservative semi-Lagrangian finite difference WENO formulations with applications to the Vlasov equation. Commun. Comput. Phys. **10**(4), 979–1000 (2011)
95. J.-M. Qiu, C.-W. Shu, Positivity preserving semi-Lagrangian discontinuous Galerkin methods for Vlasov simulations. J. Comput. Phys. **230**, 8386–8409 (2011)
96. J.A. Rossmanith, D.C. Seal, A positivity-preserving high-order semi-Lagrangian discontinuous Galerkin scheme for the Vlasov-Poisson equations. J. Comput. Phys. **230**(16), 6203–6232 (2011)
97. T. Ruggeri, S. Simić, On the hyperbolic system of a mixture of Eulerian fluids: a comparison between single-and multi-temperature models. Math. Methods Appl. Sci. **30**(7), 827–849 (2007)
98. G. Russo, P. Santagati, A new class of large time step methods for the BGK models of the Boltzmann equation. arXiv:1103.5247 (2011)
99. G. Russo, P. Santagati, S.-B. Yun, Convergence of a semi-Lagrangian scheme for the BGK model of the Boltzmann equation. SIAM J. Numer. Anal. **50**(3), 1111–1135 (2012)
100. G. Russo, S.-B. Yun, Convergence of a semi-Lagrangian scheme for the ellipsoidal BGK model of the Boltzmann equation. SIAM J. Numer. Anal. **56**, 3580–3610 (2018)
101. B. Sanderse, A.E. Veldmani, Constraint-consistent Runge–Kutta methods for one-dimensional incompressible multiphase flow. J. Comput. Phys. **384**, 170–199 (2019)
102. J.W. Schmidt, W. Hess, Positivity of cubic polynomials on intervals and positive spline interpolation. BIT **28**(2), 340–352 (1988)
103. C.W. Shu, Essentially non-oscillatory and weighted essentially non-oscillatory schemes for hyperbolic conservation laws, in *Advanced Numerical Approximation of Nonlinear Hyperbolic Equations* (Springer, Berlin, 1998), pp. 325–432
104. S. Simić, M. Pavić-Čolić, D. Madjarević, Non-equilibrium mixtures of gases: Modelling and computation. Riv. Mat. Univ. Parma **6**(1), 135–214 (2015)
105. S.-J. Son, S.-B. Yun, Cauchy problem for the ES-BGK model with the correct Prandtl number. Partial Differ. Equ. Appl. **3**(3), 41 (2022)
106. S.-J. Son, S.-B. Yun, Local-in-time solution to the Ellipsoidal BGK model with the correct Prandtl number. Preprint (2023)

107. E. Sonnendruecker, J. Roche, P. Bertrand, A. Ghizzo, The semi-Lagrangian method for the numerical resolution of the Vlasov equation. J. Comput. Phys. **149**(2), 201–220 (1999)
108. H. Struchtrup, The BGK-model with velocity-dependent collision frequency. Contin. Mech. Thermodyn. **9**, 23–31 (1997)
109. C. Villani, *H-theorem and Beyond: Boltzmann's Entropy in Today's Mathematics* (EMS Publishing House, Zürich, Switzerland, 2008)
110. J. Vranjes, P.S. Krstic, Collisions, magnetization, and transport coefficients in the lower solar atmosphere. Astron. Astrophys **554**, A22 (2013)
111. P. Welander, On the temperature jump in a rarefied gas. Arkiv Fysik **7**, 507–553 (1954)
112. G.B. Whitham, *Linear and Nonlinear Waves*, vol. 42 (Wiley, New York, 2011)
113. L. Wu, J.M. Reese, Y. Zhang, Solving the Boltzmann equation deterministically by the fast spectral method: application to gas microflows. J. Fluid Mech. **746**:53–84, 2014.
114. T. Xiong, G. Russo, J.-M. Qiu, High order multi-dimensional characteristics tracing for the incompressible Euler equation and the guiding-center Vlasov equation. J. Sci. Comput. **77**, 263–282 (2018)
115. T. Xiong, G. Russo, J.-M. Qiu, Conservative multi-dimensional semi-Lagrangian finite difference scheme: stability and applications to the kinetic and fluid simulations. J. Sci. Comput. **79**, 1241–1270 (2019)
116. H. Yoshida, Construction of higher order symplectic integrators. Phys. Lett. A **150**(5–7), 262–268 (1990)
117. S.-B. Yun, Classical solutions for the ellipsoidal BGK model with fixed collision frequency. J. Differ. Equations **259**(11), 6009–6037 (2015)
118. S.-B. Yun, Ellipsoidal BGK model near a global Maxwellian. SIAM J. Math. Anal. **47**(3), 2324–2354 (2015)
119. S.-B. Yun, Entropy production for ellipsoidal BGK model of the Boltzmann equation. Kinet. Relat. Models **9**(3), 605–619 (2016)
120. X. Zhang, C.W. Shu, On maximum-principle-satisfying high order schemes for scalar conservation laws. J. Comput. Phys **229**(9), 3091–3120 (2010)
121. X. Zhang, C.W. Shu, Maximum-principle-satisfying and positivity-preserving high-order schemes for conservation laws: survey and new developments. Proc. R. Soc. A **467**(2134), 2752–2776 (2011)
122. Y. Zheng, H. Struchtrup, Ellipsoidal statistical Bhatnagar–Gross–Krook model with velocity-dependent collision frequency. Phys. Fluids **17**(12), 127103 (2005)

Large Population Limit of Interacting Population Dynamics via Generalized Gradient Structures

Anastasiia Hraivoronska, Jasper Hoeksema, and Oliver Tse

Abstract This chapter focuses on the derivation of a doubly nonlocal Fisher-KPP model, which is a macroscopic nonlocal evolution equation describing population dynamics in the large population limit. The derivation starts from a microscopic individual-based model described as a stochastic process on the space of atomic measures with jump rates that satisfy detailed balance w.r.t. to a reference measure. We make use of the so-called "cosh" generalized gradient structure for the law of the process to pass to the large population limit using evolutionary gamma convergence. In addition to characterizing the large population limit as the solution of the nonlocal Fisher-KPP model, our variational approach further provides a generalized gradient flow structure for the limit equation as well as an entropic propagation of chaos result.

1 Introduction

Population dynamics, a fundamental area of ecology, is concerned with the study of complex phenomena triggered by simple, and often, random changes in a given

A. Hraivoronska
Universite Claude Bernard Lyon 1, CNRS, Ecole Centrale de Lyon, INSA Lyon, Université Jean Monnet, ICJ UMR5208, Villeurbanne, France
e-mail: hraivoronska@math.univ-lyon1.fr
https://hraivoronska.flywheelsites.com/

J. Hoeksema
Department of Mathematics and Computer Science, Eindhoven University of Technology, Eindhoven, The Netherlands
e-mail: j.hoeksema@tue.nl

O. Tse (✉)
Department of Mathematics and Computer Science and Eindhoven Hendrik Casimir Institute, Eindhoven University of Technology, Eindhoven, The Netherlands
e-mail: o.t.c.tse@tue.nl
https://oliver-tse.site

J. A. Carrillo, E. Tadmor (eds.), *Active Particles, Volume 4*, Modeling and Simulation in Science, Engineering and Technology,
https://doi.org/10.1007/978-3-031-73423-6_8

population such as birth, death, immigration, emigration, and mutation that are driven by environmental conditions, resources, etc. and can occur on separate levels of the biological hierarchy—the individual organism, populations of organisms, and communities of populations [4]. In recent years, fueled by the increase in computational power, individual-based models (IBMs) have become a popular tool for ecologists to understand complex phenomena and how they emerge from the behavior of individual organisms. Yet, while they are able to reproduce complex patterns and are computationally tractable, IBMs are more difficult to analyze and require a different set of mathematical tools than their analytical counterpart in classical ecology [19], hence creating the need to derive simplified models that reliably approximate the effects of space and stochasticity in the appropriate scaling regimes [8].

In this chapter, we discuss and prove the large population limit of IBMs for population dynamics to their macroscopic counterparts. In contrast to other works on similar asymptotic limits of stochastic population models [2, 7, 14, 28], our approach makes use of the variational structures associated with the IBMs, which is modeled as jump processes on the space of finite positive measures on a compact Polish space X. To keep the main message of this exposition clear, we choose only to consider the asymptotic limit of birth-death-and-hopping dynamics for a single population type on a compact subset X of $\mathbb{R}^d$ and mention possible generalizations for several instances. In the large population limit, these measure-valued jump processes are shown to converge to the so-called doubly nonlocal Fisher-Kolmogorov-Petrovsky-Piskunov (F-KPP) equation [16]:

$$\partial_t u_t(x) = \mathsf{r}_{\mathsf{b}}(u_t, x) - \mathsf{r}_{\mathsf{d}}(u_t, x)\, u_t(x) + \int_X \big(u_t(y) - u_t(x)\big)\, \mathsf{r}_{\mathsf{h}}(u_t, x, y)\, \pi(\mathrm{d}y), \tag{F-KPP}$$

where π is a given *activity measure* on Ω, the curve $t \mapsto u_t$ describes the evolution of the population density relative to π, and r_j, $j \in \{\mathsf{b}, \mathsf{d}, \mathsf{h}\}$ are density-dependent functions stemming from the birth, death, and hopping rates, respectively, that satisfy an appropriate *detailed-balance* condition. We remark that (F-KPP) is traditionally called doubly nonlocal in the literature because r_j, $j \in \{\mathsf{b}, \mathsf{d}, \mathsf{h}\}$ can depend nonlocally on the density u_t and the equation includes nonlocal hopping dynamic. Since this section is mainly concerned with the derivation of (F-KPP) from IBMs, we will place strong assumptions on r_j, $j \in \{\mathsf{b}, \mathsf{d}, \mathsf{h}\}$, guaranteeing the global well-posedness of the mean-field equation (F-KPP) by standard ODE results in Banach spaces.

Under the detailed-balance condition, the mean-field equation (F-KPP) may be seen as a *generalized gradient flow* on the space of measures in the sense of [39]—here, the term "generalized" refers to having a gradient structure with non-quadratic dissipation potential. Various examples of generalized gradient flows have appeared earlier in rate-independent systems [33], Markov chains, and mass-action kinetics [20, 36, 37]. Our variational approach derives the generalized gradient structure for (F-KPP) from the gradient structure of its underlying microscopic IBM by means of *evolutionary gamma convergence* [34, 38, 42, 43], thus justifying the choice of the gradient structure for the mean-field equation (F-KPP). An alternative convergence technique is based on EDP convergence [9, 29, 34, 35]—with some additional complexity, EDP convergence would provide the *unique* mean-field gradient structure. However, we choose to work with evolutionary gamma convergence to keep the presentation focused on the overall perspective of our approach. Earlier works with a similar strategy can be found in [12, 30], and a setting with pure birth and death with simplified u-dependence was treated in [24] by two of the authors.

Beyond Gradient Dynamics Many systems in nature exhibit (generalized) gradient dynamics, though most do not. *Active* particle systems, which describe a multitude of biological processes, form a large family of intrinsically non-gradient systems. In these systems, each agent can extract energy from their environment and convert it into directed motion or mechanical forces [3, 41]. While this work focuses on gradient dynamics, the models presented here could be generalized to non-gradient systems, where birth-and-death rates and hopping rates do not satisfy the detailed-balance condition [21], as in the case of *active* agents. In these scenarios, the evolutionary gamma-convergence approach used here is not directly applicable but has a potential for adaptation—this adaptation is currently under investigation.

1.1 Approach and Main Results

We start by introducing the space $\mathbb{X}$ of finite configurations over a compact Polish space X. Elements $\boldsymbol{x}$ of $\mathbb{X}$ are symmetric N-tuples that can be identified with atomic measures $\nu \in \Gamma := \mathcal{M}^+(X)$ with finite total variation, i.e.,

$$\mathbb{X} := \left\{\boldsymbol{x} = (x^1, \ldots, x^N) \in X^N,\ N \geq 1\right\} / \sim,$$

with $\sim$ being the equivalence relation $\sigma(\boldsymbol{x}) \sim \boldsymbol{x}$ for every permutation σ, then $\nu = \sum_{i=1}^N \delta_{x^i}$. On $\mathbb{X}$, one defines the so-called Lebesgue-Poisson measure $\lambda \in \Gamma$ (see (4) below for the definition) w.r.t. an activity measure $\pi \in \Gamma$. Under an appropriate scaling corresponding to the large population limit, one obtains a family of scaled and normalized Lebesgue-Poisson measures $\lambda^n \in \mathcal{P}(\mathbb{X})$, $n \geq 1$. Due to the identification $\boldsymbol{x} \leftrightarrow \nu$, one then obtains a *reference measure* $\Pi^n \in \mathcal{P}(\Gamma)$ that is supported on (scaled) atomic measures of the form $\nu = \frac{1}{n}\sum_{i=1}^N \delta_{x^i}$. Note that $N = N(t)$ is the actual number of particles, whereas $n \in \mathbb{N}$ is the scaling parameter corresponding to the average number of particles per unit volume according to π. Here, the space of finite measures $\Gamma = \mathcal{M}^+(X)$ on X is equipped with the narrow topology, i.e., in duality with continuous and bounded functions, making it a Polish space. The total variation norm for any finite measure $\nu \in \Gamma$ is simply denoted by $\|\nu\|$. We also consider the space of signed measures $\mathcal{M}(Z)$ over various Polish spaces Z and the weak-$*$ convergence, i.e., convergence against continuous functions that vanish at infinity.

A large class of (scaled) IBMs for population dynamics can be modeled as Γ-valued stochastic process described via their corresponding (scaled) infinitesimal generator L^n. In our current context, L^n is the sum of two parts consisting of the *birth-and-death generator*:

$$L^n_{\mathsf{bd}} F(\nu) = \int_X n\big[F(\nu + n^{-1}\delta_x) - F(\nu)\big]\, \mathsf{b}_n(\nu, \mathrm{d}x) + \int_X n\big[F(\nu - n^{-1}\delta_x) - F(\nu)\big]\, \mathsf{d}_n(\nu, x)\, \nu(\mathrm{d}x),$$

with the (scaled) measure-dependent birth and death rate b_n, d_n, and the *hopping (or Kawasaki) generator*:

$$L^n_{\mathsf{h}} F(\nu) = \iint_{X\times X} n\big[F(\nu + n^{-1}(\delta_y - \delta_x)) - F(\nu)\big]\, \mathsf{h}_n(\nu, x, \mathrm{d}y)\, \nu(\mathrm{d}x),$$

with the (scaled) measure-dependent hopping rate h_n. The time-marginal law $(\mathsf{P}^n_t)_{t\ge 0}$ of the Γ-valued process then satisfies the *forward Kolmogorov equation*:

$$\partial_t \mathsf{P}^n_t = (L^n)^* \mathsf{P}^n_t. \qquad (\mathsf{FKE}_n)$$

As mentioned, we will restrict ourselves to simple rates that satisfy the detailed balance condition:

Assumption 1.1 (Detailed Balance and Regularity) Let $(\mathsf{b}_n, \mathsf{d}_n, \mathsf{h}_n)_{n\ge 1}$ be a sequence of measurable rates:

$$\mathsf{b}_n : \Gamma\times\mathcal{B}(X) \to [0,+\infty),\ \mathsf{d}_n : \Gamma\times X \to [0,+\infty),\ \mathsf{h}_n : \Gamma\times X\times\mathcal{B}(X) \to [0,+\infty),$$

satisfying the detailed balance condition:

$$\begin{cases} \mathsf{b}_n(\nu, \mathrm{d}x) = \mathsf{d}_n(\nu + n^{-1}\delta_x, x)\, \pi(\mathrm{d}x), \\ \mathsf{h}_n(\nu, x, \mathrm{d}y)\, \pi(\mathrm{d}x) = \mathsf{h}_n(\nu + n^{-1}(\delta_y - \delta_x), y, \mathrm{d}x)\, \pi(\mathrm{d}y), \end{cases} \qquad (\mathsf{DB}_n)$$

and the uniform bounds

$$\sup_{n\ge 1} \sup_{(\nu,x)\in\Gamma\times X} \Big\{ \|\mathsf{b}_n(\nu,\cdot)\| + \mathsf{d}_n(\nu,x) + \|\mathsf{h}_n(\nu,x,\cdot)\| \Big\} < +\infty.$$

We further assume that there exist limit rates $(\mathsf{b}, \mathsf{d}, \mathsf{h})$, such that

$$\sup_{(\nu,x)\in\Gamma\times X} \Big\{ \|\mathsf{b}_n(\nu,\cdot) - \mathsf{b}(\nu,\cdot)\| + |\mathsf{d}_n(\nu,x) - \mathsf{d}(\nu,x)| + \|\mathsf{h}_n(\nu,x,\cdot) - \mathsf{h}(\nu,x,\cdot)\| \Big\} \longrightarrow 0 \quad \text{as } n\to\infty,$$

where the limit rates automatically satisfy the limit detailed balance condition:

$$\left\{\begin{aligned} \mathsf{b}(\nu, \mathrm{d}x) &= \mathsf{d}(\nu, x)\,\pi(\mathrm{d}x), \\ \mathsf{h}(\nu, x, \mathrm{d}y)\,\pi(\mathrm{d}x) &= \mathsf{h}(\nu, y, \mathrm{d}x)\,\pi(\mathrm{d}y). \end{aligned}\right. \tag{DB_∞}$$

For limit passage and the well-posedness of the limit equation, we will assume additional regularity of the limit rates. Namely, we assume:

(1) $\|\cdot\|$-*Lipschitz regularity*: There exist Lipschitz constants ℓ_j, $j \in \{\mathsf{b}, \mathsf{d}, \mathsf{h}\}$ such that

$$\left.\begin{aligned} \|\mathsf{b}(\nu) - \mathsf{b}(\eta)\| \le \ell_{\mathsf{b}}\|\nu - \eta\|, \qquad \sup_{x\in X} |\mathsf{d}(\nu, x) - \mathsf{d}(\eta, x)| \le \ell_{\mathsf{d}}\|\nu - \eta\| \\ \sup_{x\in X} \|\mathsf{h}(\nu, x, \cdot) - \mathsf{h}(\eta, x, \cdot)\| \le \ell_{\mathsf{h}}\|\nu - \eta\|, \end{aligned}\right\}$$

$\forall\, \nu, \eta \in \Gamma.$

(2) *Continuity*: For any $A \in X \times \mathcal{B}(X)$ the functions

$$\nu \mapsto \mathsf{b}(\nu, A), \quad (\nu, x) \mapsto \mathsf{d}(\nu, x), \quad (\nu, x) \mapsto \mathsf{h}(\nu, x, A)$$

are jointly narrowly continuous in $\nu \in \Gamma$ and continuous in $x \in X$.

Example 1.2 An example of rates satisfying Assumption 1.1 is given by

$$\mathsf{b}_n(\nu, \mathrm{d}x) = \psi_{\mathsf{bd}}\left(\int_{y\in X} c(x, y)\,\nu(\mathrm{d}y)\right)\pi(\mathrm{d}x) = \mathsf{d}_n(\nu, x)\pi(\mathrm{d}x), \quad c(x, x) = 0,$$

$$\mathsf{h}_n(\nu, x, \mathrm{d}y) = \psi_{\mathsf{h}}\left(\int_{z\in X} h(x - z, y - z)\,\nu(\mathrm{d}z)\right)\pi(\mathrm{d}y), \; h(x - y, 0) = h(0, y - x),$$

where $\psi_j\colon \mathbb{R} \to [0, \infty)$, $j \in \{\mathsf{bd}, \mathsf{h}\}$ are Lipschitz functions and c, h are bounded continuous functions.

Introducing the (scaled) jump kernel $\kappa^n := \kappa^n_{\mathsf{bd}} + \kappa^n_{\mathsf{h}} : \Gamma \times \mathcal{B}(\Gamma) \to [0, \infty)$ with

$$\kappa^n_{\mathsf{bd}}(\nu, \mathrm{d}\eta) := \int_X \delta_{\nu + n^{-1}\delta_x}(\mathrm{d}\eta)\,\mathsf{b}_n(\nu, \mathrm{d}x) + \int_X \delta_{\nu - n^{-1}\delta_x}(\mathrm{d}\eta)\,\mathsf{d}_n(\nu, x)\nu(\mathrm{d}x),$$

$$\kappa^n_{\mathsf{h}}(\nu, \mathrm{d}\eta) := \iint_{X\times X} \delta_{\nu + n^{-1}(\delta_y - \delta_x)}(\mathrm{d}\eta)\,\mathsf{h}_n(\nu, x, \mathrm{d}y)\nu(\mathrm{d}x),$$

the generator L^n may then be expressed as

$$L^n F(\nu) = \int_\Gamma \overline{\nabla}_n F(\nu, \eta)\,\kappa^n(\nu, \mathrm{d}\eta),$$

where $\overline{\nabla}_n$ is the (scaled) *discrete gradient* defined by $\overline{\nabla}_n F(\nu, \eta) = n[F(\nu) - F(\eta)]$. The previous expression shows that L^n is the generator of a pure jump process on the space of finite measures Γ.

In Sect. 3, we show that under Assumption 1.1, the jump kernel κ^n satisfies the usual detailed-balance condition for jump processes w.r.t. the reference measure Π^n, i.e., the measure

$$\kappa^n(\nu, \mathrm{d}\eta)\,\Pi^n(\mathrm{d}\nu) \quad \text{is symmetric.}$$

Following the approach in [39], one obtains the well-posedness and a generalized gradient flow formulation for the forward Kolmogorov equation (FKE_n) with a generalized gradient structure $(\mathsf{R}_n, \mathsf{R}_n^*, \mathsf{E}_n)$, where R_n, R_n^* are dissipation and dual dissipation potentials, respectively, and E_n is the driving energy for the gradient flow, which is taken to be the (scaled) relative entropy $\mathsf{E}_n := n^{-1}\mathsf{Ent}(\cdot|\Pi^n)\colon \mathcal{P}(\Gamma) \to [0, +\infty]$, where

$$\mathsf{Ent}(\alpha|\beta) := \begin{cases} \displaystyle\int \phi\left(\frac{\mathrm{d}\alpha}{\mathrm{d}\beta}\right)\mathrm{d}\beta & \text{if } \alpha \ll \beta, \\ +\infty & \text{otherwise,} \end{cases} \qquad \phi(s) = s\log s - s + 1, \quad s \geq 0.$$

In this work, we will be adopting the so-called "cosh" generalized gradient structure (cf. Sect. 3.4), as they are the *natural* ones from the large-deviation perspective (see, e.g., [40] for more insights into this structure).

Formally, this means that (FKE_n) may be written as

$$\partial_t \mathsf{P}_t + \overline{\mathrm{div}}_n\, \mathsf{J}_t = 0 \tag{CE_n}$$

$$\mathsf{J}_t = \partial_2 \mathsf{R}_n^*(\mathsf{P}_t, -\overline{\nabla}_n \mathsf{E}_n'(\mathsf{P}_t)), \tag{KE_n}$$

where $\overline{\mathrm{div}}_n$ denotes the (scaled) *discrete divergence*, (CE_n) is the *continuity equation* for the density–flux pair (P, J), and (KE_n) is the *kinetic relation*, which provides the relationship between the force $\mathsf{E}_n'(\mathsf{P})$ induced by the energy E_n at P and the flux J via the dual dissipation potential R_n^*. By *Legendre duality*, one has

$$\mathsf{R}_n(\mathsf{P}, \mathsf{J}) + \mathsf{R}_n^*(\mathsf{P}, \zeta) \geq \langle \mathsf{J}, \zeta \rangle \qquad \forall\, (\mathsf{J}, \zeta), \tag{LD}$$

with equality if and only if $\mathsf{J} = \partial_2 \mathsf{R}_n^*(\mathsf{P}, \zeta)$. Hence, the ($\mathsf{KE}_n$) may be equivalently expressed as

$$\mathsf{R}_n(\mathsf{P}_t, \mathsf{J}_t) + \mathsf{R}_n^*(\mathsf{P}_t, -\overline{\nabla} \mathsf{E}_n'(\mathsf{P}_t)) = \langle \mathsf{J}_t, -\overline{\nabla}_n \mathsf{E}_n'(\mathsf{P}_t) \rangle. \tag{KE_n'}$$

On the other hand, an application of the chain rule gives

$$\frac{\mathrm{d}}{\mathrm{d}t}\mathsf{E}_n(\mathsf{P}_t) = \langle \partial_t \mathsf{P}_t, \mathsf{E}_n(\mathsf{P}_t)\rangle = \langle \mathsf{J}_t, \overline{\nabla}_n \mathsf{E}'_n(\mathsf{P}_t)\rangle, \tag{CR_n}$$

where the second equality follows from the continuity equation (CE_n). Combining the expressions (KE'_n) and (CR_n) and integrating over arbitrary intervals $[s,t] \subset [0,T]$, one obtains the *energy-dissipation balance*:

$$\mathscr{I}_n([s,t];\mathsf{P},\mathsf{J}) := \int_s^t \mathsf{R}_n(\mathsf{P}_r,\mathsf{J}_r) + \mathsf{R}^*_n(\mathsf{P}_r, -\overline{\nabla}_n \mathsf{E}'_n(\mathsf{P}_r))\,\mathrm{d}r + \mathsf{E}_n(\mathsf{P}_t) - \mathsf{E}_n(\mathsf{P}_s) = 0. \tag{EDB_n}$$

In this way, the solution P^n of (FKE_n) can be associated with a density–flux pair $(\mathsf{P}^n,\mathsf{J}^n)$ satisfying both the continuity equation (CE_n) and energy-dissipation balance (EDB_n). Moreover, since $\mathscr{I}_n([s,t];\mathsf{P},\mathsf{J}) \geq 0$ for any density–flux pairs (P,J) satisfying (CE_n), the associated pair $(\mathsf{P}^n,\mathsf{J}^n)$ is in fact a minimizer of the *energy-dissipation functional* $\mathscr{I}_n$, thus providing a variational characterization for the solution of (FKE_n).

Section 4 deals with the large population limit, where we show that a subsequence of generalized gradient flow solutions $(\mathsf{P}^n,\mathsf{J}^n)_{n\geq 1}$ to (FKE_n) converges to some limit pair (P,J) in an appropriate sense as $n \to \infty$, where the limit pair (P,J) satisfies:

(1) The limit continuity equation

$$\partial_t \mathsf{P}_t + \mathrm{div}_\Gamma\, \mathsf{J}_t = 0, \tag{CE_∞}$$

where div_Γ is the Γ-divergence operator associated to the Γ-gradient operator

$$\nabla_\Gamma F(\nu,x) := \sum_{m=1}^{M} \partial_m g\big(\langle f_1,\nu\rangle,\ldots,\langle f_m,\nu\rangle\big)\, f_m(x), \qquad (\nu,x) \in \Gamma \times X,$$

defined for cylindrical functions $F(\nu) = g(\langle f_1,\nu\rangle,\ldots,\langle f_M,\nu\rangle)$ with $g \in C_c^\infty(\mathbb{R}^M)$ and $f_m \in C_b(X)$.

(2) The limit energy-dissipation balance

$$\mathscr{I}_\infty([s,t];\mathsf{P},\mathsf{J}) := \int_s^t \mathsf{R}_\infty(\mathsf{P}_r,\mathsf{J}_r) + \mathsf{R}^*_\infty(\mathsf{P}_r,\quad \nabla_\Gamma \mathsf{E}'_\infty(\mathsf{P}_r))\,\mathrm{d}r + \mathsf{E}_\infty(\mathsf{P}_t) - \mathsf{E}_\infty(\mathsf{P}_s) = 0, \tag{EDB_∞}$$

where $(\mathsf{R}_\infty, \mathsf{R}^*_\infty, \mathsf{E}_\infty)$ is the limit generalized gradient structure with driving energy:

$$\mathsf{E}_\infty(\mathsf{P}) = \int_\Gamma \mathsf{Ent}(\nu|\pi)\,\mathsf{P}(\mathrm{d}\nu).$$

The above convergence is established using evolutionary gamma-convergence techniques, further resulting in *entropic propagation of chaos*, which is an important feature of our approach.

In particular, we show that the limit density–flux pair (P, J) is a generalized gradient flow solution of the so-called Liouville equation:

$$\partial_t \mathsf{P}_t + \mathrm{div}_\Gamma(\mathscr{V}\mathsf{P}_t) = 0, \tag{LiE}$$

where $\mathscr{V} := \kappa_{\mathsf{bd}} + \kappa_{\mathsf{h}} : \Gamma \to \Gamma$ is the measure-valued vector field associated to the limit jump kernels:

$$\begin{aligned} \kappa_{\mathsf{bd}}[\nu](dx) &:= \mathsf{b}(\nu, dx) - \mathsf{d}(\nu, x)\nu(dx), \\ \kappa_{\mathsf{h}}[\nu](dx) &:= \int_X \mathsf{h}(\nu, y, dx)\, \nu(dy) - \nu(dx) \int_X \mathsf{h}(\nu, x, dy). \end{aligned} \tag{1}$$

Observe that the Liouville equation (LiE) resembles a standard transport equation with the velocity field $\mathscr{V}$, whose solution can be obtained by studying its *Lagrangian flow* $\mathscr{S} : [0, T] \times \Gamma \to \Gamma$, associated to

$$\partial_t \nu_t = \mathscr{V}[\nu_t], \qquad \nu \in \Gamma. \tag{MFE}$$

This is the *mean-field equation*. Assuming the well-posedness of (MFE), the unique solution of (LiE) is then given by $\mathsf{P}_t := (\mathscr{S}_t)_\# \mathsf{P}_0$, which can be established utilizing a *superposition principle* (cf. Theorem 4.10).

In the special case $\mathsf{P}_0 = \delta_{\nu_0}$ with $\mathsf{Ent}(\nu_0|\pi) < +\infty$, the well-posedness of (MFE) yields the solution $\mathsf{P}_t = \delta_{\nu_t}$ with $\nu_t := \mathscr{S}_t(\nu_0)$ and $\mathsf{Ent}(\nu_t|\pi) < +\infty$ for every $t \in [0, T]$. Hence, setting $u_t := d\nu_t/d\pi$, one obtains (F-KPP) from (MFE) with the density-dependent functions:

$$\mathsf{r}_{\mathsf{b}}(u, x) = \frac{d\mathsf{b}(u\pi, \cdot)}{d\pi}(x), \quad \mathsf{r}_{\mathsf{d}}(u, x) = \mathsf{d}(u\pi, x), \quad \mathsf{r}_{\mathsf{h}}(u, x, y) = \frac{d\mathsf{h}(u\pi, y, \cdot)}{d\pi}(x), \tag{2}$$

which are well defined due to the detailed balance conditions in Assumption 1.1. As a by-product of the energy-dissipation balance (EDB_∞), we obtain a generalized gradient structure $(\mathscr{R}, \mathscr{R}^*, \mathscr{E})$ for the mean-field equation (MFE) with the driving energy $\mathscr{E} = \mathsf{Ent}(\cdot|\pi) : \Gamma \to [0, +\infty]$ (Fig. 1).

The discussion above can be summarized in the following result, which is a consequence of Theorem 4.11.

Theorem A *Let $\{(\mathsf{P}^n, \mathsf{J}^n)\}_{n\in\mathbb{N}}$ be a sequence of generalized gradient flow solutions to (FKE_n) with initial data $\{\mathsf{P}^n_{in}\}_{n\in\mathbb{N}}$ and suppose there exists $\mathsf{P}_{in} \in dom(\mathsf{E}_\infty)$ such that*

$$\mathsf{P}^n_{in} \rightharpoonup \mathsf{P}_{in}, \qquad \lim_{n\to\infty} \mathsf{E}_n(\mathsf{P}^n_{in}) = \mathsf{E}_\infty(\mathsf{P}_{in}).$$

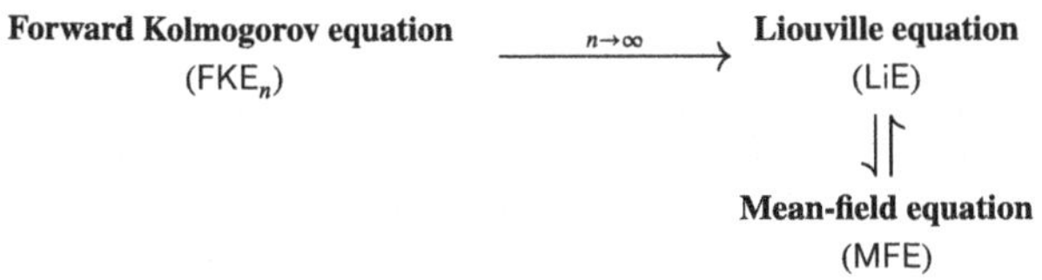

Fig. 1 A depiction of our approach to proving the large-population limit

Then there exists a unique generalized gradient flow solution (P, J) *of* (LiE) *with* $\mathsf{P}_0 = \mathsf{P}_{in}$, *such that:*

- $\mathsf{P}_t^n \rightharpoonup \mathsf{P}_t$ *narrowly for all* $t \in [0, T]$.
- $\lim_{n\to\infty} \mathsf{E}_n(\mathsf{P}_t^n) = \mathsf{E}_\infty(\mathsf{P}_t)$ *for all* $t \in [0, T]$.

If $\mathsf{P}_{in} = \delta_{\bar{\nu}}$, $\bar{\nu} \in dom(\mathscr{E})$, *then* $\mathsf{P}_t = \delta_{\nu_t}$, *where* ν *solves* (MFE) *and satisfies the energy-dissipation balance:*

$$\mathscr{I}([s,t], \nu, \lambda) := \int_s^t \mathcal{R}(\nu_r, \lambda_r) + \mathcal{R}^*(\nu_r, -\mathscr{E}'(\nu_t))\,\mathrm{d}r + \mathscr{E}(\nu_t) - \mathscr{E}(\nu_s) = 0,$$

with an associated flux λ. *If in addition,* $u_t := \mathrm{d}\nu_t/\mathrm{d}\pi \in [a, b]$, $0 < a \le b \in [0, +\infty)$ *for every* $t \in [0, T]$, *then the entropic propagation of chaos holds, i.e.,*

$$\lim_{n\to\infty} \frac{1}{n} \mathsf{Ent}(\mathsf{P}_t^n | \Pi_{\nu_t}^n) = 0 \qquad \textit{for every } t \in [0, T], \tag{3}$$

where Π_ν^n *is the scaled Poisson measure with activity measure* $n\nu$ *(see Sect. 2.2 for details).*

Finally, the curve $t \mapsto u_t$ *solves the doubly nonlocal Fisher-KPP equation* (F-KPP).

Remark 1.3 (Entropic Propagation of Chaos) In the canonical ensemble setting, i.e., when the number of particles $n \in \mathbb{N}$ is fixed and goes to infinity, with the distribution of the particles given by ρ_t^n, there are various notions of propagation of chaos [6]. In this context, the weak convergence of $\mathsf{P}_t^n \to \delta_{\nu_t}$ (where ν_t is now a probability measure) is equivalent to the weak notion of chaoticity (e.g., see [44]). This means that for every $K \in \mathbb{N}$, the K-marginal distributions $\rho_t^{n,K}$ of ρ_t^n converge to $\nu_t^{\otimes K}$. Moreover, global entropic chaoticity, expressed by the convergence

$$\frac{1}{n} \mathsf{Ent}(\rho_t^n | \nu_t^{\otimes n}) \to 0 \qquad \text{as } n \to \infty,$$

implies, through the subadditivity of entropy, the local estimate convergence:

$$\mathsf{Ent}(\rho_t^{n,K} | \nu_t^{\otimes K}) \to 0 \qquad \text{as } n \to \infty.$$

In the grand canonical setting, the situation becomes more complex. While for certain interacting particle systems' birth-and-death processes, the rescaled K-correlation measures do converge to $\nu_t^{\otimes K}$ under suitable strong moment assumptions [15], this does not extend to our setting where we only assume convergence of the rescaled entropies for the initial data. Moreover, for $K > 1$, the assumption of finite entropy relative to Π^n does not even ensure that the K-correlation measures are well defined. Nevertheless, we propose the convergence (3) as a suitable generalization of the global entropic propagation of chaos from the canonical to the grand canonical setting. This convergence suggests that the rescaled entropy of the law of the interacting particle system, relative to suitable Poisson measures, vanishes as $n \to \infty$. It is plausible that this would imply the convergence of appropriately weighted correlation measures, but we leave this to future research.

2 Preliminaries: Configuration Spaces and Large-Population Scaling

2.1 Configuration Space

The basis for the mathematical description of the IBMs we will consider relies on the framework of configuration spaces, which, in our present context, can be found in the works by Kondratiev, Finkelshtein, and others [13, 14, 17]. Throughout, let X be a compact Polish space.

We define the family of *N-point configurations* as the set of N-tuples of indistinguishable particles:

$$\mathbb{X}^N := X^N / \sim, \qquad N \in \mathbb{N}_0 := \mathbb{N} \cup \{0\}, \qquad \mathbb{X}^0 := \emptyset,$$

where $\sim$ is the equivalence relation $(x^1, \dots, x^N) \sim (x^{\sigma_1}, \dots, x^{\sigma_N})$ for any permutation σ. *The space of finite configuration* is then defined as

$$\mathbb{X} := \bigsqcup_{N \in \mathbb{N}_0} \mathbb{X}^N.$$

We equip $\mathbb{X}$ with the disjoint union topology, which is the finest topology for which all the canonical injections $\varphi_N : \mathbb{X}^N \to \mathbb{X}$ are continuous. We also introduce the Borel σ-algebra on $\mathbb{X}^N$ denoted by $\mathcal{B}(\mathbb{X}^N)$ and the corresponding Borel σ-algebra $\mathcal{B}(\mathbb{X})$ on $\mathbb{X}$. The continuous map $\mathsf{N} : \mathbb{X} \to \mathbb{N}_0$, which assigns $\mathsf{N}(\boldsymbol{x}) = N$ for every N-point configuration $\boldsymbol{x} \in \mathbb{X}^N$, is called the *number map*.

The Lebesgue-Poisson measure λ on $\mathbb{X}$ with activity measure π is defined as

$$\lambda := \sum_{N=0}^{\infty} \frac{1}{N!} \pi^{\otimes N} \in \mathcal{M}^+(\mathbb{X}), \tag{4}$$

where $\pi \in \Gamma$ is a given finite Radon measure on X. For example, if X is a compact subset of $\mathbb{R}^d$, one can choose π to be the restriction of the Lebesgue measure to X. It is easy to see that if π is a finite measure, then the Lebesgue-Poisson measure λ is also finite:

$$\lambda(\mathbb{X}) = \sum_{N=0}^{\infty} \frac{1}{N!} \|\pi\|^N = e^{\|\pi\|} < \infty.$$

Together, the triplet $(\mathbb{X}, \mathcal{B}(\mathbb{X}), \lambda)$ forms a measure space.

Remark 2.1 Let us discuss some modifications of the configuration space $\mathbb{X}$ that are relevant for particular mathematical models.

The case when particles have multiple types is relevant for biological models such as tumor growth models, where tissue cells can be either healthy, proliferative, nonproliferative, or necrotic [25]. Other examples include the classical SIR model in epidemiology that includes susceptible, infected, and recovered agents [26] and ecological models of various competing species of plants or animals [18]. Supposing that the system has $K \in \mathbb{N}$ different types of particles and that each particle can change its type, the generalization of $\mathbb{X}^n$ to multi-types can be described with a multi-index $\mathbf{n} = (n_1, \ldots, n_K) \in \mathbb{N}_0^K$, $n_l \in \mathbb{N}_0$, by setting

$$\mathbb{Y}^0 := \emptyset, \qquad \mathbb{Y}^{\mathbf{n}} := \mathbb{X}^{n_1} \times \cdots \times \mathbb{X}^{n_K},$$

and defining the space of finite configurations as

$$\mathbb{Y} := \bigsqcup_{\mathbf{n} \in \mathbb{N}_0^K} \mathbb{Y}^{\mathbf{n}}.$$

Similarly to the one-type case, this space is endowed with the disjoint union topology and the corresponding number map $\mathsf{N} : \mathbb{Y} \to \mathbb{N}_0$, which assigns $\mathsf{N}(\boldsymbol{x}) = n_1 + \cdots + n_K$ for every $\boldsymbol{x} \in \mathbb{Y}^{\mathbf{n}}$.

An important restriction that we need to include in the definition of the configuration space is that each particle can have only one type. Therefore, we consider a subset $\mathbb{Y}_{\neq} \subset \mathbb{Y}$ defined as

$$\mathbb{Y}_{\neq} := \left\{ \boldsymbol{x} \in \mathbb{Y} \;\middle|\; \boldsymbol{x} = (\boldsymbol{x}_1, \ldots, \boldsymbol{x}_K),\ \ x_l^i \neq x_m^j \ \ \forall l, m \in [K],\ l \neq m,\ i \in [n_l], \right.$$
$$\left. j \in [n_m] \right\},$$

where $[n] = \{1, \ldots, n\}$, $n \in \mathbb{N}$.

Here, we have the choice to either restrict the configuration space to $\mathbb{Y}_{\neq}$ or to define a measure on $\mathbb{Y}$ having its support contained in $\mathbb{Y}_{\neq}$—we choose the latter. Following the single-type particle configuration, we define the Lebesgue-Poisson measure λ on $\mathbb{Y}$ as

$$\lambda := \sum_{n_1,\dots,n_K=0}^{\infty} \frac{1}{\prod_{p=1}^{K} n_p!} \pi^{\otimes n_1} \otimes \cdots \otimes \pi^{\otimes n_K}.$$

We then set $\lambda_{\neq} := \mathbf{1}_{\mathbb{Y}_{\neq}} \lambda$, meaning that $\lambda_{\neq}$ is the restriction of λ to $\mathbb{Y}_{\neq}$.

A more general approach to introducing diversity in the particle types is suggested, for instance, in [27] using *marked configuration spaces*. The marked configuration space Ω_X^M over X with marks from a set M is defined as

$$\Omega_X^M := \{(\boldsymbol{x}, m) : \boldsymbol{x} \in \mathbb{X},\ m \in M^{\boldsymbol{x}}\},$$

where $M^{\boldsymbol{x}}$ denotes the set of all maps $[\mathsf{N}(\boldsymbol{x})] \ni i \mapsto m_i \in M$, which is permutation invariant.

2.2 Large-Population Scaling

In performing the large population limit, we consider an appropriate scaling and normalization of the Lebesgue-Poisson measure, i.e., we set, for each $n \geq 1$,

$$\lambda^n := e^{-n\|\pi\|} \sum_{N=0}^{\infty} \frac{n^N}{N!} \pi^{\otimes N} \in \mathcal{P}(\mathbb{X}),$$

i.e., λ^n is a normalized Lebesgue-Poisson measure with activity measure $n\pi$.

In this way, the expected population size scales as n since

$$\int_{\mathbb{X}} \mathsf{N}(\boldsymbol{x})\, \lambda^n(\mathrm{d}\boldsymbol{x}) = e^{-n\|\pi\|} \sum_{N=0}^{\infty} \frac{N n^N}{N!} \|\pi\|^N = n\|\pi\|.$$

We further introduce the (scaled) atomic measure map:

$$\Lambda^n : \mathbb{X} \to \Gamma, \qquad \boldsymbol{x} \mapsto \Lambda^n(\boldsymbol{x}) := \frac{1}{n} \sum_{i=1}^{N} \delta_{x^i}, \quad \boldsymbol{x} = (x^1, \dots, x^N) \in \mathbb{X}^N,$$

which is easily verified to be measurable for each $n \geq 1$. One can then push the scaled Lebesgue-Poisson measure λ^n on $\mathbb{X}$ along the map Λ^n to a probability measure Π^n on Γ, which serves as the reference measure on Γ—we call Π^n the scaled *Poisson measure*. Then by construction, the reference measure Π^n satisfies

$$\int_\Gamma \|\nu\|\, \Pi^n(\mathrm{d}\nu) = \frac{1}{n} \int_{\mathbb{X}} \mathsf{N}(\boldsymbol{x})\, \lambda^n(\mathrm{d}\boldsymbol{x}) = \|\pi\| \qquad \forall n \geq 1,$$

ensuring that the expected mass of scaled atomic measures ν sampled from Π^n equals the mass of the activity measure π. Clearly, the previous equality also provides a uniform finite first-moment estimate for the family $\{\Pi^n\}_{n\geq 1}$. Notice that the support of any measure $\mathsf{P} \in \mathcal{P}(\Gamma)$ satisfying $\mathsf{P} \ll \Pi^n$ contains only atomic measures. Moreover, the reference measure Π^n has exponential moments with

$$\frac{1}{n}\log\int_\Gamma e^{n\|\nu\|}\,\Pi^n(d\nu) = (e-1)\|\pi\| \qquad \forall n \geq 1.$$

In particular, if $\mathsf{P} \in \mathcal{P}(\Gamma)$ has finite energy, i.e., $\mathsf{E}_n(\mathsf{P}) < +\infty$, then P has finite first moment, which follows from a variational representation of exponential integrals (cf. [23, Lemma B.1]), generalizing Varadhan's representation of the relative entropy [45]. More explicitly, for the scaled driving energy $\mathsf{E}_n = n^{-1}\mathsf{Ent}(\cdot|\Pi^n)$, one has that

$$\int_\Gamma \|\nu\|\,\mathsf{P}(d\nu) \leq \mathsf{E}_n(\mathsf{P}) + \frac{1}{n}\log\int_\Gamma e^{n|\nu|}\,\Pi^n(d\nu) = \mathsf{E}_n(\mathsf{P}) + (e-1)\|\pi\|. \tag{5}$$

These facts play an essential role in the asymptotic limit of E_n, which we now illustrate to provide a first glimpse into our approach.

Recall the functional $\mathsf{E}_\infty\colon \Gamma \to [0,+\infty]$ mentioned earlier, given by

$$\mathsf{E}_\infty(\mathsf{P}) := \int_\Gamma \mathsf{Ent}(\nu|\pi)\,\mathsf{P}(d\nu) = \int_\Gamma \mathscr{E}(\nu)\,\mathsf{P}(d\nu).$$

Proposition 2.2 *Let* $(\mathsf{P}^n)_{n\geq 1} \subset \mathcal{P}(\Gamma)$ *be a sequence satisfying*

$$\sup_{n\geq 1}\mathsf{E}_n(\mathsf{P}^n) < +\infty.$$

Then, there exists some $\mathsf{P} \in \mathcal{P}(\Gamma)$ *such that* $\mathsf{P}^n \rightharpoonup \mathsf{P}$ *narrowly in* $\mathcal{P}(\Gamma)$. *In addition,*

$$\mathsf{E}_\infty(\mathsf{P}) \leq \liminf_{n\to\infty}\mathsf{E}_n(\mathsf{P}^n).$$

The proof of Proposition 2.2 is rather elementary but requires a delicate approximation result that can be found in [31]. We include the proof here to highlight the general strategy used to establish the corresponding limits for the other terms in the energy-dissipation functional $\mathscr{I}_n$.

Proof of Proposition 2.2 For any $n \geq 1$, one uses inequality (5) to obtain

$$\int_\Gamma |\nu|\,\mathsf{P}^n(d\nu) \leq \sup_{n\geq 1}\mathsf{E}_n(\mathsf{P}^n) + (e-1)\|\pi\|,$$

therewith implying the tightness of the sequence $(\mathsf{P}^n)_{n\geq 1} \subset \mathcal{P}(\Gamma)$ w.r.t. narrow convergence. One can then extract a (not relabelled) subsequence and a point $\mathsf{P} \in \mathcal{P}(\Gamma)$ for which $\mathsf{P}^n \rightharpoonup \mathsf{P}$ narrowly in $\mathcal{P}(\Gamma)$.

Suppose for the moment that $\mathsf{P} = \delta_\eta$ for some $\eta \in \Gamma$. For cylinder functions of the form $F(\nu) = \langle \varphi, \nu \rangle$, $\varphi \in B_b(X)$, Varadhan's representation [45] then yields

$$\liminf_{n\to\infty} \mathsf{E}_n(\mathsf{P}^n) \geq \langle \varphi, \eta \rangle - \int_X (e^{\varphi(x)} - 1)\, \pi(\mathrm{d}x).$$

Taking the supremum over $\varphi \in B_b(X)$ and using Varadhan's representation again, we arrive at

$$\liminf_{n\to\infty} \mathsf{E}_n(\mathsf{P}^n) \geq \mathsf{Ent}(\eta|\pi) = \int_\Gamma \mathsf{Ent}(\nu|\pi)\, \mathsf{P}(\mathrm{d}\nu).$$

The result for general $\mathsf{P} \in \mathcal{P}(\Gamma)$ follows from [31, Theorem 3.5]. □

Remark 2.3 In general, one can consider reference measures in Gibbs form, i.e.,

$$e^{-\mathcal{F}_n(\nu)}\, \Pi^n(\mathrm{d}\nu), \qquad \mathcal{F}_n \colon \Gamma \to (-\infty, +\infty].$$

The detailed balance condition on the jump kernel κ^n is then specified relative to this measure.

3 The Forward Kolmogorov Equation

We now turn to the forward Kolmogorov equation (FKE_n), which describes the law of the measure-valued process corresponding to the IBM. We will discuss well-posedness, the phrasing of (FKE_n) in terms of a continuity equation, and its variational formulation via the generalized gradient structure outlined previously.

3.1 Continuity Equation

Let $T > 0$ be an arbitrary but fixed time horizon. We consider curves $(\mathsf{P}_t^n)_{t\geq 0} \subset \mathcal{P}(\Gamma)$ that satisfy the continuity equation (CE_n) in the following sense:

Definition 3.1 A tuple $(\mathsf{P}, \mathsf{J}^{\mathsf{bd}}, \mathsf{J}^{\mathsf{h}})$ satisfies the continuity equation, or simply $(\mathsf{P}, \mathsf{J}) \in \mathsf{CE}_n$, if:

(1) The map $[0, T] \ni t \mapsto \mathsf{P}_t \in \mathcal{P}(\Gamma)$ is narrowly continuous, and

$$\sup_{t\in[0,T]} \int_\Gamma \|\nu\| \mathrm{d}\mathsf{P}_t < \infty.$$

(2) The Borel families $\{\mathsf{J}_t^{\mathsf{bd}}\}_{t\geq 0} \subset \mathcal{M}(\Gamma \times X)$, $\{\mathsf{J}_t^{\mathsf{h}}\}_{t\geq 0} \subset \mathcal{M}(\Gamma \times X \times X)$ satisfy

$$\operatorname{supp} \mathsf{J}_t^{\mathsf{h}} \subset \{(\nu, x, y) : x \in \operatorname{supp} \nu\}, \qquad \int_0^T \|\mathsf{J}_t^{\mathsf{bd}}\| + \|\mathsf{J}_t^{\mathsf{h}}\| \, \mathrm{d}t < \infty.$$

(3) For any bounded Borel function $F \in B_b(\Gamma)$ and interval $[s,t] \subset [0,T]$,

$$\int_\Gamma F \mathrm{d}\mathsf{P}_t - \int_\Gamma F \,\mathrm{d}\mathsf{P}_s = \int_s^t \Bigg[\iint_{\Gamma\times X} \overline{\nabla}_{n,\mathsf{bd}} F(\nu, x)\, \mathsf{J}_r^{\mathsf{bd}}(\mathrm{d}\nu \mathrm{d}x) + \iiint_{\Gamma\times X^2} \overline{\nabla}_{n,\mathsf{h}} F(\nu, x, y)\, \mathsf{J}_r^{\mathsf{h}}(\mathrm{d}\nu \mathrm{d}x \mathrm{d}y) \Bigg] \mathrm{d}r,$$

with (scaled) discrete derivatives

$$\overline{\nabla}_{n,\mathsf{bd}} F(\nu, x) := n\big[F(\nu + n^{-1}\delta_x) - F(\nu)\big],$$
$$\overline{\nabla}_{n,\mathsf{h}} F(\nu, x, y) := n\big[F(\nu + n^{-1}(\delta_y - \delta_x)) - F(\nu)\big].$$

Formally, the continuity equation can be viewed in terms of a discrete divergence in the sense that

$$\partial_t \mathsf{P}_t + \overline{\operatorname{div}}_{n,\mathsf{bd}}\, \mathsf{J}_t^{\mathsf{bd}} + \overline{\operatorname{div}}_{n,\mathsf{h}}\, \mathsf{J}_t^{\mathsf{h}} = 0,$$

where $\overline{\operatorname{div}}_{n,\mathsf{bd}}$ and $\overline{\operatorname{div}}_{n,\mathsf{h}}$ are the dual to the discrete derivatives $\overline{\nabla}_{n,\mathsf{bd}}$ and $\overline{\nabla}_{n,\mathsf{h}}$, respectively.

Remark 3.2 Condition (3.1) of Definition 3.1 is satisfied whenever $\mathsf{J}_t^{\mathsf{h}}(\mathrm{d}\nu, \mathrm{d}x, \mathrm{d}y))$ $= \mathsf{h}_t(\nu, x, \mathrm{d}y)\nu(\mathrm{d}x)\mathsf{P}_t(\mathrm{d}\nu$ for some time-and-measure-dependent kernel h. Moreover, the moment Condition (3.1) holds by using similar arguments from [39], since even though the kernel κ^n is unbounded, we have

$$\sup_{t\in[0,T]} \int_\Gamma \kappa^n(\nu, \Gamma)\, \mathsf{P}(\mathrm{d}\nu) < \infty. \tag{6}$$

Remark 3.3 The above form of the continuity equation is suitable for taking large population limits, but to apply the tools of [39] it is useful to bring this into a form of the continuity equation ($\overline{\mathsf{CE}}_n$) similar to the one stated in the introduction. In particular, with a little abuse of notation, setting the *edge flux*

$$\mathsf{J}_t^\circ(\mathrm{d}\nu \mathrm{d}\eta) := \frac{n}{2} \int_{x\in X} \delta_{\nu+n^{-1}\delta_x}(\mathrm{d}\eta)\, \mathsf{J}_t^{\mathsf{bd}}(\mathrm{d}\nu, \mathrm{d}x) - \frac{n}{2} \int_{x\in X} \delta_{\nu-n^{-1}\delta_x}(\mathrm{d}\eta)\, \mathsf{J}_t^{\mathsf{bd}}(\mathrm{d}(\nu - \delta_x), \mathrm{d}x)$$

$$+n \iint_{(x,y)\in X^2} \delta_{\nu+n^{-1}(\delta_y-\delta_x)}(\mathrm{d}\eta)\, \mathsf{J}_t^{\mathsf{h}}(\mathrm{d}\nu, \mathrm{d}x\mathrm{d}y),$$

Conditions (3.1) and (3.1) of Definition 3.1 can then be rephrased as:

(2') The Borel family $\{\mathsf{J}_t^\circ\}_{t\geq 0} \subset \mathcal{M}(\Gamma \times \Gamma)$ has finite total variation over $[0, T]$.
(3') For any bounded Borel function $F \in B_b(\Gamma)$ and interval $[s, t] \subset [0, T]$,

$$\int_\Gamma F\mathrm{d}\mathsf{P}_t - \int_\Gamma F\mathrm{d}\mathsf{P}_s = \int_s^t \iint_{\Gamma^2} (F(\nu) - F(\eta))\, \mathsf{J}_r^\circ(\mathrm{d}\nu\mathrm{d}\eta)\mathrm{d}r.$$

Vice versa, any Borel family $\{\mathsf{J}_t^\circ\}_{t\geq 0}$ such that $\mathsf{J}_t^\circ(\mathrm{d}\nu\mathrm{d}\eta) \ll \kappa^n(\nu, \mathrm{d}\eta)\Pi^n(\mathrm{d}\nu)$ can be decomposed into a birth-and-death $\mathsf{J}_t^{\mathsf{bd}}$ and a hopping $\mathsf{J}_t^{\mathsf{h}}$ part.

In Lemma 3.6, we show the reversibility of the kernel κ^n with respect to Π^n. Together with the above this implies, via a truncation argument of the kernels κ^n, mass moment estimates on P_t, and adaptation from the techniques in [39], which whenever $\mathsf{E}_n(\mathsf{P}_0) < \infty$ with suitable moment estimates, classical solutions to (FKE_n) exist and have a variational representation in terms of generalized gradient flows; see Theorem 3.8.

3.2 *Solutions to the Forward Kolmogorov Equation*

We consider the following family of solutions to the forward Kolmogorov equation (FKE_n).

Definition 3.4 A narrowly continuous curve $\mathsf{P} = (\mathsf{P}_t)_{t\in[0,T]}$ is a solution to (FKE_n) if

$$\sup_{t\in[0,T]} \int_\Gamma \|\nu\|\mathrm{d}\mathsf{P}_t < \infty, \tag{7}$$

and for every bounded Borel function $F \in B_b(\Gamma)$ and interval $[s, t] \subset [0, T]$,

$$\int_\Gamma F\mathrm{d}\mathsf{P}_t - \int_\Gamma F\mathrm{d}\mathsf{P}_s = \int_s^t \int_\Gamma L^n F(\nu)\, \mathsf{P}_r(\mathrm{d}\nu)\mathrm{d}r. \tag{8}$$

To state a solution of (FKE_n) in terms of the tuple $(\mathsf{P}, \mathsf{J}^{\mathsf{bd}}, \mathsf{J}^{\mathsf{h}})$, we define

$$\mathsf{T}^{n,\pm}(\nu, x) := \Big(\nu \pm n^{-1}\delta_x, x\Big), \qquad (\nu, x) \in \Gamma \times X,$$

$$\mathsf{T}^{n,\mathsf{h}}(\nu, x, y) := \Big(\nu + n^{-1}(\delta_y - \delta_x), y, x\Big), \qquad (\nu, x, y) \in \Gamma \times X \times X,$$

describing the birth, death, and hopping transitions and jump intensity measures

$$\vartheta_{\mathsf{P}}^{n,\mathsf{b}}(\mathrm{d}\nu, \mathrm{d}x) := \mathsf{b}_n(\nu, \mathrm{d}x)\mathsf{P}(\mathrm{d}\nu), \qquad \vartheta_{\mathsf{P}}^{n,\mathsf{d}}(\mathrm{d}\nu, \mathrm{d}x) := \mathsf{d}_n(\nu, x)\nu(\mathrm{d}x)\mathsf{P}(\mathrm{d}\nu),$$

$$\vartheta_{\mathsf{P}}^{n,\mathsf{h}}(\mathrm{d}\nu, \mathrm{d}x, \mathrm{d}y) := \mathsf{h}_n(\nu, x, \mathrm{d}y)\nu(\mathrm{d}x)\mathsf{P}(\mathrm{d}\nu).$$

Note that $\mathsf{T}_{\#}^{n,-}\vartheta_{\mathsf{P}}^{n,\mathsf{d}}$ and $\mathsf{T}_{\#}^{n,\mathsf{h}}\vartheta_{\mathsf{P}}^{n,h}$ are also well-defined measures.

Lemma 3.5 *A curve* P *is a solution to* (FKE_n) *if and only if* $(\mathsf{P}, \mathsf{J}) \in \mathsf{CE}_n$ *with*

$$\mathsf{J}_t^{\mathsf{bd}} = \vartheta_{\mathsf{P}_t}^{n,\mathsf{b}} - \mathsf{T}_{\#}^{n,-}\vartheta_{\mathsf{P}_t}^{n,\mathsf{d}}, \qquad \mathsf{J}_t^{\mathsf{h}} = \frac{1}{2}\left(\vartheta_{\mathsf{P}_t}^{n,\mathsf{h}} - \mathsf{T}_{\#}^{n,\mathsf{h}}\vartheta_{\mathsf{P}_t}^{n,\mathsf{h}}\right), \tag{9}$$

for almost every $t \in [0, T]$.

Proof For any P satisfying the mass bound (7) and any $F \in B_b(\Gamma)$, we have that

$$\begin{aligned}
&\iint_{\Gamma\times X} \overline{\nabla}_{n,\mathsf{bd}}F(\nu, x)\mathrm{d}\left(\vartheta_{\mathsf{P}}^{n,b} - \mathsf{T}_{\#}^{n,-}\vartheta_{\mathsf{P}}^{n,d}\right)\\
&= \iint_{\Gamma\times X} n\big[F(\nu + n^{-1}\delta_x) - F(\nu)\big]\mathsf{b}_n(\nu, \mathrm{d}x)\mathsf{P}(\mathrm{d}\nu)\\
&\quad + \iint_{\Gamma\times X} n\big[F(\nu - n^{-1}\delta_x) - F(\nu)\big]\mathsf{d}_n(\nu, x)\nu(\mathrm{d}x)\mathsf{P}(\mathrm{d}\nu)\\
&= \int_\Gamma L_{\mathsf{bd}}^n F(\nu)\, \mathsf{P}(\mathrm{d}\nu).
\end{aligned}$$

Similarly, for the hopping part, we have that

$$\begin{aligned}
&-\iiint_{\Gamma\times X^2} \overline{\nabla}_{n,\mathsf{h}}F(\nu, x, y)\, \mathsf{T}_{\#}^{n,h}\vartheta_{\mathsf{P}}^{n,h}(\mathrm{d}\nu\mathrm{d}x\mathrm{d}y)\\
&= \iiint_{\Gamma\times X^2} \overline{\nabla}_{n,\mathsf{h}}F(\nu, x, y)\, \vartheta_{\mathsf{P}}^{n,h}(\mathrm{d}\nu\mathrm{d}x\mathrm{d}y)\\
&= \iiint_{\Gamma\times X^2} \overline{\nabla}_{n,\mathsf{h}}F(\nu, x, y)\, \mathsf{h}_n(\nu, x, \mathrm{d}y)\nu(\mathrm{d}x)\mathsf{P}(\mathrm{d}\nu)\\
&= \int_\Gamma L_{\mathsf{h}}^n F(\nu)\, \mathsf{P}(\mathrm{d}\nu),
\end{aligned}$$

thereby concluding the proof. □

3.3 Detailed Balance

We now state the detailed balance condition for both the birth-and-death and hopping parts separately, under Assumption 1.1.

Lemma 3.6 *Both jump kernels κ^n_{bd} and κ^n_{h} satisfy the detailed balance condition w.r.t. Π^n. Consequently, the jump kernel κ^n satisfies the detailed balance condition w.r.t. Π^n. In particular, we have that*

$$\mathsf{T}^{n,-}_{\#}\vartheta^{n,\mathsf{d}}_{\Pi^n} = \vartheta^{n,\mathsf{b}}_{\Pi^n}, \qquad \mathsf{T}^{n,\mathsf{h}}_{\#}\vartheta^{n,\mathsf{h}}_{\Pi^n} = \vartheta^{n,\mathsf{h}}_{\Pi^n}.$$

Proof The symmetry of the birth-and-death part $\kappa^n_{\mathsf{bd}}(\nu, \mathrm{d}\eta)\,\Pi^n(\mathrm{d}\nu)$ follows from [22, Lemma 17.12]. As for the hopping part, we show that

$$\int_\Gamma G(\nu, \eta)\,\kappa^n_{\mathsf{h}}(\nu, \mathrm{d}\eta)\,\Pi^n(\mathrm{d}\nu) = \int_\Gamma G(\eta, \nu)\,\kappa^n_{\mathsf{h}}(\nu, \mathrm{d}\eta)\,\Pi^n(\mathrm{d}\nu),$$

which implies the symmetry of the measure $\kappa^n_{\mathsf{h}}(\nu, \mathrm{d}\eta)\,\Pi^n(\mathrm{d}\nu)$. To simplify the notation, we set

$$\widehat{\mathsf{h}}_n(\boldsymbol{x}, x, \mathrm{d}y) := \mathsf{h}_n(\Lambda^n(\boldsymbol{x}), x, \mathrm{d}y), \qquad \widehat{G}_n(\boldsymbol{x}, \sigma) := G(\Lambda^n(\boldsymbol{x}), \Lambda^n(\sigma)),$$

and $\boldsymbol{x}^y_j := (x_1, \ldots, x_{j-1}, y, x_{j+1}, \ldots, x_N)$ for any $\boldsymbol{x} \in \mathbb{X}^N$, $y \in X$. Hence,

$$\begin{aligned}
&\int_\Gamma G(\nu, \eta)\,\kappa^n_{\mathsf{h}}(\nu, \mathrm{d}\eta)\,\Pi^n(\mathrm{d}\nu)\\
&= e^{-n|\pi|}\sum_{N=0}^{\infty}\frac{n^N}{N!}\int_{X^N}\sum_{j=1}^{N}\int_X \widehat{G}_n(\boldsymbol{x}, \boldsymbol{x}^y_j)\,\widehat{\mathsf{h}}_n(\boldsymbol{x}, x_j, \mathrm{d}y)\,\pi^{\otimes N}(\mathrm{d}x_1\cdots\mathrm{d}x_N)\\
&\overset{(1)}{=} e^{-n|\pi|}\sum_{N=0}^{\infty}\frac{n^N}{N!}\int_{X^N}\sum_{j=1}^{N}\int_X \widehat{G}_n(\boldsymbol{x}, \boldsymbol{x}^y_j)\,\widehat{\mathsf{h}}_n(\boldsymbol{x}^y_j, y, \mathrm{d}x_j)\,\pi(\mathrm{d}y)\,\pi^{\otimes(N-1)}(\widehat{\mathrm{d}x_j})\\
&\overset{(2)}{=} e^{-n|\pi|}\sum_{N=0}^{\infty}\frac{n^N}{N!}\int_{X^N}\sum_{j=1}^{N}\int_X \widehat{G}_n(\sigma^y_j, \sigma)\,\widehat{\mathsf{h}}_n(\sigma, x_j, \mathrm{d}y)\,\pi^{\otimes N}(\mathrm{d}x_1\cdots\mathrm{d}x_N)\\
&= \int_\Gamma G(\eta, \nu)\,\kappa^n_{\mathsf{h}}(\nu, \mathrm{d}\eta)\,\Pi^n(\mathrm{d}\nu)
\end{aligned}$$

where (1) follows from Assumption 1.1 and (2) follows from swapping the role of x_j and y. □

3.4 Generalized Gradient Flow Formulation

As mentioned earlier a generalized gradient structure consists of a dissipation potential R_n, a dual dissipation potential R_n^*, and a driving functional E_n.

We begin by recalling the driving functional:

$$\mathcal{P}(\Gamma) \ni \mathsf{P} \mapsto \mathsf{E}_n(\mathsf{P}) = n^{-1}\mathsf{Ent}(\mathsf{P}, \Pi^n).$$

To state the dissipation potentials, we will need to introduce various objects. First, we define

$$\mathbb{R} \times [0, +\infty) \times [0, +\infty) \ni (w, u, v) \mapsto \Upsilon(w, u, v)$$
$$:= \begin{cases} 0 & \text{if } w = 0, \\ \psi\left(w/\sqrt{uv}\right)\sqrt{uv} & \text{if } u, v > 0, \\ +\infty & \text{if } w \neq 0 \text{ and } uv = 0, \end{cases}$$

where ψ is the Legendre dual of

$$\mathbb{R} \ni z \mapsto \psi^*(z) := 2(\cosh(z/2) - 1).$$

It is straightforward to check that Υ is jointly convex, lower semicontinuous, and 1-homogeneous (cf. [39]).

Birth-and-Death Dissipation Potentials We define the birth-and-death dissipation and dual dissipation potentials as

$$\begin{aligned} &\mathcal{P}(\Gamma_n) \times \mathcal{M}(\Gamma \times X) \ni (\mathsf{P}, \mathsf{J}) \mapsto \mathsf{R}_{n,\mathsf{bd}}(\mathsf{P}, \mathsf{J}) \\ &\quad := 2\iint_{\Gamma\times X} \Upsilon(\mathsf{J}/2, \vartheta_{\mathsf{P}}^{n,\mathsf{b}}, \mathsf{T}_{\#}^{n,-}\vartheta_{\mathsf{P}}^{n,\mathsf{d}}), \\ &\mathcal{P}(\Gamma_n) \times B_b(\Gamma \times X) \ni (\mathsf{P}, \zeta) \mapsto \mathsf{R}_{n,\mathsf{bd}}^*(\mathsf{P}, \zeta) \\ &\quad := 2\iint_{\Gamma\times X} \psi^*(\zeta)\mathrm{d}\sqrt{\vartheta_{\mathsf{P}}^{n,\mathsf{b}}\left(\mathsf{T}_{\#}^{n,-}\vartheta_{\mathsf{P}}^{n,\mathsf{d}}\right)}, \end{aligned} \tag{10}$$

where for any two measures μ, ν, the square root of its product is given by $\sqrt{\mu\nu} = \sqrt{\frac{\mathrm{d}\mu}{\mathrm{d}\sigma}\frac{\mathrm{d}\nu}{\mathrm{d}\sigma}}\sigma$ for any common dominating measure σ, i.e., $\mu, \nu \ll \sigma$ (see [39, Section 2.3] for a detailed discussion on concave transformations of measures). We further define the birth-and-death Fisher information as

$$\mathcal{P}(\Gamma_n) \ni \mathsf{P} \mapsto \mathsf{D}_{n,\mathsf{bd}}(\mathsf{P}) := 4\mathcal{H}^2(\vartheta_{\mathsf{P}}^{n,\mathsf{b}}, \mathsf{T}_{\#}^{n,-}\vartheta_{\mathsf{P}}^{n,\mathsf{d}}) \in [0, +\infty],$$

where $\mathcal{H}$ is the Hellinger distance between two measures defined as

$$\mathcal{H}^2(\mu,\nu) = \frac{1}{2}\int\left(\sqrt{\frac{\mathrm{d}\mu}{\mathrm{d}\sigma}} - \sqrt{\frac{\mathrm{d}\nu}{\mathrm{d}\sigma}}\right)^2 \mathrm{d}\sigma$$

for some common dominating measure σ.

Hopping Dissipation Potentials In a similar fashion, we define the hopping dissipation and dual dissipation potentials as

$$\begin{aligned}
\mathcal{P}(\Gamma_n)\times\mathcal{M}(\Gamma\times X^2) \ni (\mathsf{P},\mathsf{J}) &\mapsto \mathsf{R}_{n,\mathsf{h}}(\mathsf{P},\mathsf{J}) \\
&:= \iiint_{\Gamma\times X^2} \Upsilon(\mathsf{J},\vartheta_{\mathsf{P}}^{n,\mathsf{h}},\mathsf{T}_{\#}^{n,\mathsf{h}}\vartheta_{\mathsf{P}}^{n,\mathsf{h}}), \\
\mathcal{P}(\Gamma_n)\times B_b(\Gamma\times X^2) \ni (\mathsf{P},\zeta) &\mapsto \mathsf{R}_{n,\mathsf{h}}^*(\mathsf{P},\zeta) \\
&:= \iiint_{\Gamma\times X^2} \psi^*(\zeta)\mathrm{d}\sqrt{\vartheta_{\mathsf{P}}^{n,\mathsf{h}}\left(\mathsf{T}_{\#}^{n,\mathsf{h}}\vartheta_{\mathsf{P}}^{n,\mathsf{h}}\right)},
\end{aligned} \tag{11}$$

and the hopping Fisher information as

$$\mathcal{P}(\Gamma_n) \ni \mathsf{P} \mapsto \mathsf{D}_{n,\mathsf{h}}(\mathsf{P}) := 2\mathcal{H}^2(\vartheta_{\mathsf{P}}^{n,\mathsf{h}},\mathsf{T}_{\#}^{n,\mathsf{h}}\vartheta_{\mathsf{P}}^{n,\mathsf{h}}) \in [0,+\infty].$$

Remark 3.7 If $\mathsf{P} = U\Pi^n$, then by Lemma 3.6, we have that

$$\sqrt{\vartheta_{\mathsf{P}}^{n,\mathsf{b}}\left(\mathsf{T}_{\#}^{n,-}\vartheta_{\mathsf{P}}^{n,\mathsf{d}}\right)} = \sqrt{U^{\mathsf{bd}}(\nu)U(\nu)}\,\mathsf{b}_n(\nu,\mathrm{d}x)\Pi^n(\mathrm{d}\nu),$$

allowing us to express the Fisher information as

$$\begin{aligned}
\mathsf{D}_{n,\mathsf{bd}}(\mathsf{P}) &= 2\iint_{\Gamma\times X}\left(\sqrt{U^{\mathsf{bd}}(\nu,x)} - \sqrt{U(\nu)}\right)^2 \mathsf{b}_n(\nu,\mathrm{d}x)\Pi^n(\mathrm{d}\nu) \\
&= 2n^{-1}\iint_{\Gamma\times X}\left(\overline{\nabla}_{n,\mathsf{bd}}\sqrt{U}(\nu,x)\right)^2 \mathsf{b}_n(\nu,\mathrm{d}x)\Pi^n(\mathrm{d}\nu),
\end{aligned}$$

where $U^{\mathsf{bd}}(\nu,x) := U(\mathsf{T}^{n,+}(\nu,x))$.

As for the hopping case, if $\mathsf{P} = U\Pi^n$, then

$$\sqrt{\vartheta_{\mathsf{P}}^{n,\mathsf{h}}\left(\mathsf{T}_{\#}^{n,\mathsf{h}}\vartheta_{\mathsf{P}}^{n,\mathsf{h}}\right)} = \sqrt{U^{\mathsf{h}}(\nu)U(\nu)}\,\mathsf{h}_n(\nu,x,\mathrm{d}y)\nu(\mathrm{d}x)\Pi^n(\mathrm{d}\nu),$$

and the Fisher information reads as

$$\mathsf{D}_{n,\mathsf{h}}(\mathsf{P}) = \iiint_{\Gamma\times X^2}\left(\sqrt{U^{\mathsf{h}}(\nu)} - \sqrt{U(\nu)}\right)^2 \mathsf{h}_n(\nu,x,\mathrm{d}y)\nu(\mathrm{d}x)\Pi^n(\mathrm{d}\nu)$$

$$= n^{-1} \iiint_{\Gamma \times X^2} \left(\overline{\nabla}_{n,\mathsf{h}} \sqrt{U}(\nu, x)\right)^2 \mathsf{h}_n(\nu, x, dy)\nu(dx)\Pi^n(d\nu),$$

where $U^{\mathsf{h}}(\nu, x) := U(\mathsf{T}^{n,\mathsf{h}}(\nu, x))$.

Putting the birth-and-death and hopping part together, we obtain the formulation of our action functional:

$$\mathsf{R}_n(\mathsf{P}, \mathsf{J}) := \mathsf{R}_{n,\mathsf{bd}}(\mathsf{P}, \mathsf{J}^{\mathsf{bd}}) + \mathsf{R}_{n,\mathsf{h}}(\mathsf{P}, \mathsf{J}^{\mathsf{h}}),$$

and total Fisher information $\mathsf{D}_n := \mathsf{D}_{n,\mathsf{bd}} + \mathsf{D}_{n,\mathsf{h}}$.

Formally, one can check that (FKE_n) can be written via the kinetic relation as

$$\partial_t \mathsf{P}_t + \overline{\mathrm{div}}_{n,\mathsf{bd}}\, \mathsf{J}^{\mathsf{bd}}_t + \overline{\mathrm{div}}_{n,\mathsf{h}}\, \mathsf{J}^{\mathsf{h}}_t = 0, \qquad \begin{cases} \mathsf{J}^{\mathsf{bd}}_t = \partial_2 \mathsf{R}^*_{n,\mathsf{bd}}(\mathsf{P}_t, -\overline{\nabla}_{n,\mathsf{bd}} \mathsf{E}'_n(\mathsf{P}_t)), \\ \mathsf{J}^{\mathsf{h}}_t = \partial_2 \mathsf{R}^*_{n,\mathsf{h}}(\mathsf{P}_t, -\overline{\nabla}_{n,\mathsf{h}} \mathsf{E}'_n(\mathsf{P}_t)). \end{cases}$$

Indeed, using the shorthand notation U, U^{bd}, and U^{h}, we find

$$\mathsf{E}'_n(\mathsf{P}) = n^{-1} \log U, \; \overline{\nabla}_{n,\mathsf{bd}} \mathsf{E}'_n(\mathsf{P}) = \log U^{\mathsf{bd}} - \log U, \;\; \overline{\nabla}_{n,\mathsf{h}} \mathsf{E}'_n(\mathsf{P}) = \log U^{\mathsf{h}} - \log U,$$

and for the derivatives of the dual dissipation potentials,

$$\begin{aligned} \partial_2 \mathsf{R}^*_{n,\mathsf{bd}}(\mathsf{P}, -\overline{\nabla}_{n,\mathsf{bd}} \mathsf{E}'_n(\mathsf{P})) &= \big(U(\nu) - U^{\mathsf{bd}}(\nu, x)\big)\mathsf{b}(\nu, dx)\Pi^n(d\nu) \\ &= \vartheta^{n,\mathsf{b}}_{\mathsf{P}} - \mathsf{T}^{n,-}_{\#} \vartheta^{n,\mathsf{d}}_{\mathsf{P}}, \\ \partial_2 \mathsf{R}^*_{n,\mathsf{h}}(\mathsf{P}, -\overline{\nabla}_{n,\mathsf{h}} \mathsf{E}'_n(\mathsf{P})) &= \frac{1}{2}\big(U(\nu) - U^{\mathsf{h}}(\nu, x)\big)\mathsf{h}(\nu, x, dy)\nu(dx)\Pi^n(d\nu) \\ &= \frac{1}{2}\left(\vartheta^{n,\mathsf{h}}_{\mathsf{P}} - \mathsf{T}^{n,\mathsf{h}}_{\#} \vartheta^{n,\mathsf{h}}_{\mathsf{P}}\right). \end{aligned}$$

which are the correct fluxes from Lemma 3.5. By a similar calculation, one obtains

$$\mathsf{D}_{n,\mathsf{bd}}(\mathsf{P}) = \mathsf{R}^*_{n,\mathsf{bd}}(\mathsf{P}, -\overline{\nabla}_{n,\mathsf{bd}} \mathsf{E}'_n(\mathsf{P})), \qquad \mathsf{D}_{n,\mathsf{h}}(\mathsf{P}) = \mathsf{R}^*_{n,\mathsf{h}}(\mathsf{P}, -\overline{\nabla}_{n,\mathsf{h}} \mathsf{E}'_n(\mathsf{P})).$$

The rigorous version of the above statements requires the aforementioned variational formulation. Namely, for any tuple $(\mathsf{P}, \mathsf{J}^{\mathsf{bd}}, \mathsf{J}^{\mathsf{h}}) \in \mathsf{CE}_n$, let $\mathscr{I}_n(\mathsf{P}, \mathsf{J}) = \mathscr{I}_n([0, T]; \mathsf{P}, \mathsf{J})$, with

$$\mathscr{I}_n([s, t]; \mathsf{P}, \mathsf{J}) := \int_s^t (\mathsf{R}_n(\mathsf{P}_r, \mathsf{J}_r) + \mathsf{D}_n(\mathsf{P}_r))\, dr + \mathsf{E}_n(\mathsf{P}_t) - \mathsf{E}_n(\mathsf{P}_s) = 0. \quad (12)$$

The functional corresponds to a well-posed gradient system if $\mathscr{I}_n([s,t],\cdot) \geq 0$ for all $s, t \in [0, T]$ and has a unique null minimizer, called the corresponding *generalized gradient flow solution*. For the latter, the energy-dissipation balance holds:

$$\mathscr{I}_n([s,t]; \mathsf{P}, \mathsf{J}) = 0 \qquad \text{for every } [s,t] \subset [0,T].$$

We now have the following result.

Theorem 3.8 *Let* $\mathsf{P}_0 \in dom(\mathsf{E}_n)$. *Then the functional* $\mathscr{I}_n$ *on* (P, J) *defines a well-posed generalized gradient system, and its unique generalized gradient flow solution is such that* P *is the solution to* (FKE_n) *with fluxes given for almost every* $t \in (0,T)$ *by*

$$\mathsf{J}_t^{\mathsf{bd}} = \vartheta_{\mathsf{P}_t}^{n,\mathsf{b}} - \mathsf{T}_{\#}^{n,-}\vartheta_{\mathsf{P}_t}^{n,\mathsf{d}}, \qquad \mathsf{J}_t^{\mathsf{h}} = \frac{1}{2}\left(\vartheta_{\mathsf{P}_t}^{n,\mathsf{h}} - \mathsf{T}_{\#}^{n,\mathsf{h}}\vartheta_{\mathsf{P}_t}^{n,\mathsf{h}}\right).$$

Proof Recall from Remark 3.3 that any $(\mathsf{P}, \mathsf{J}) \in \mathsf{CE}_n$ can also be written in terms of edge fluxes J°. Vice versa, any edge flux $\mathsf{J}^\circ(\mathrm{d}\nu\mathrm{d}\eta) \ll \kappa^n(\nu, \mathrm{d}\eta)\Pi(\mathrm{d}\nu)$ can be decomposed into a birth/death J^{bd} and J^{h} part, with

$$\begin{aligned}
&\mathsf{J}_t^\circ(\mathrm{d}\nu\mathrm{d}\eta)\\
&= \frac{n}{2}\int_{x\in X} \delta_{\nu+n^{-1}\delta_x}(\mathrm{d}\eta)\,\mathsf{J}_t^{\mathsf{bd}}(\mathrm{d}\nu, \mathrm{d}x) - \frac{n}{2}\int_{x\in X} \delta_{\nu-n^{-1}\delta_x}(\mathrm{d}\eta)\left(\mathsf{T}_{\#}^{n,+}\mathsf{J}_t^{\mathsf{bd}}\right)(\mathrm{d}\nu, \mathrm{d}x)\\
&\quad + n\iint_{(x,y)\in X^2} \delta_{\nu+n^{-1}(\delta_y-\delta_x)}(\mathrm{d}\eta)\,\mathsf{J}_t^{\mathsf{h}}(\mathrm{d}\nu, \mathrm{d}x\mathrm{d}y).
\end{aligned}$$

Moreover, note that, for any such P,

$$\sup_{t\in[0,T]} \int_\Gamma \|\nu\| \mathrm{d}\mathsf{P}_t < \infty. \tag{13}$$

On the other hand, solutions to (FKE_n) in the sense of Definition 3.4 are solutions to the forward Kolmogorov equation corresponding to a jump process with jump kernel $n\kappa^n$ satisfying detailed balance condition w.r.t. Π^n and the bound (13). The solution to (FKE_n) can be obtained by an approximation argument, where the jump kernel κ^n is truncated so that the results of [39] apply. Using the property that finite energy implies finite first moment (cf. (5)), one can pass to the limit in the truncation to obtain a curve satisfying (8), a priori for any $C_c(\Gamma)$ functions, and a posteriori for any $B_b(\Gamma)$ functions due to the moment estimate.

Additionally, one can modify the arguments of [39] to show that this solution is the unique generalized gradient flow solution for the corresponding well-posed functional:

$$\mathscr{I}([s,t];\mathsf{P},\mathsf{J}^\circ) := \int_s^t \big(\mathsf{R}(\mathsf{P}_r,\mathsf{J}_r^\circ) + \mathsf{D}(\mathsf{P}_r)\big)\,\mathrm{d}r + \mathsf{Ent}(\mathsf{P}_t|\Pi^n) - \mathsf{Ent}(\mathsf{P}_s|\Pi^n),$$

with Fisher information $\mathsf{D}(\mathsf{P}) := 2\mathcal{H}^2(\mathsf{P}n\kappa^n, n\kappa^n\mathsf{P})$ and dual dissipation potential

$$\mathsf{R}^*(\mathsf{P},\zeta) := \iint_{\Gamma^2} \Psi^*(\zeta)\mathrm{d}\sqrt{(\mathsf{P}n\kappa^n)(n\kappa^n\mathsf{P})},$$

where $(\mathsf{P}\kappa^n)(\mathrm{d}\nu\mathrm{d}\eta) := \kappa^n(\nu,\mathrm{d}\eta)\mathsf{P}(\mathrm{d}\nu)$ and $(\kappa^n\mathsf{P})(\mathrm{d}\nu\mathrm{d}\eta) := \kappa^n(\eta,\mathrm{d}\nu)\mathsf{P}(\mathrm{d}\eta)$.

Note that for birth-and-death cases, we have

$$\kappa^n_{\mathsf{bd}}(\nu,\mathrm{d}\eta)\mathsf{P}(\mathrm{d}\nu) = \int_X \delta_{\nu+n^{-1}\delta_x}(\mathrm{d}\eta)\,\vartheta_{\mathsf{P}}^{n,\mathsf{b}}(\mathrm{d}\nu,\mathrm{d}x) + \int_X \delta_{\nu-n^{-1}\delta_x}(\mathrm{d}\eta)\,\vartheta_{\mathsf{P}}^{n,\mathsf{d}}(\mathrm{d}\nu,\mathrm{d}x),$$

$$\kappa^n_{\mathsf{bd}}(\eta,\mathrm{d}\nu)\mathsf{P}(\mathrm{d}\eta) = \int_X \delta_{\nu+n^{-1}\delta_x}(\mathrm{d}\eta)\left(\mathsf{T}_\#^{n,-}\vartheta_{\mathsf{P}}^{n,\mathsf{d}}\right)(\mathrm{d}\nu,\mathrm{d}x)$$
$$+ \int_X \delta_{\nu-n^{-1}\delta_x}(\mathrm{d}\eta)\left(\mathsf{T}_\#^{n,+}\vartheta_{\mathsf{P}}^{n,\mathsf{b}}\right)(\mathrm{d}\nu,\mathrm{d}x),$$

and for the hopping part

$$\kappa^n_{\mathsf{h}}(\nu,\mathrm{d}\eta)\mathsf{P}(\mathrm{d}\nu) = \iint_{(x,y)\in X^2} \delta_{\nu+n^{-1}(\delta_y-\delta_x)}(\mathrm{d}\eta)\,\vartheta_{\mathsf{P}_t}^{n,\mathsf{h}}(\mathrm{d}\nu,\mathrm{d}x\mathrm{d}y),$$

$$\kappa^n_{\mathsf{h}}(\eta,\mathrm{d}\nu)\mathsf{P}(\mathrm{d}\eta) = \iint_{(x,y)\in X^2} \delta_{\nu+n^{-1}(\delta_y-\delta_x)}(\mathrm{d}\eta)\,(\mathsf{T}_\#^{n,\mathsf{h}}\,\vartheta_{\mathsf{P}}^{n,\mathsf{h}})(\mathrm{d}\nu,\mathrm{d}x\mathrm{d}y).$$

Thus, we establish that in fact

$$\mathscr{I}([s,t];\mathsf{P},\mathsf{J}^\circ) = n\mathscr{I}_n([s,t];\mathsf{P},\mathsf{J}),$$

and therefore our solution in the sense of Definition 3.4 exists and is the unique gradient flow solution corresponding to the functional $\mathscr{I}_n$. □

3.5 Uniform Estimates

Here, we establish the necessary estimates to obtain compactness in Theorem 4.11. Let $\mathfrak{d}$ be the metric on $\mathcal{P}(\Gamma)$ defined by

$$\mathfrak{d}(\mathsf{P},\mathsf{Q}) := \sup_{F\in\mathcal{F}} \int_\Gamma F\mathrm{d}(\mathsf{P}-\mathsf{Q}), \tag{14}$$

with

$$\mathcal{F} := \left\{ F \in \mathrm{Cyl}(\Gamma) : \sup_{(\nu,x)\in\Gamma\times X} |\nabla_\Gamma F(\nu, x)| \le 1 \right\}.$$

Remark 3.9 Using a similar superposition principle as in Sect. 4.3, one can show that $\mathfrak{d}$ is in fact the 1-Wasserstein metric on $\mathcal{P}(\Gamma)$ induced by the total variation metric on Γ.

Note that $\mathfrak{d}$ is narrowly lower semicontinuous, and convergence in $\mathfrak{d}$ implies narrow convergence on narrowly compact subsets by density of $\mathcal{F}$ in $C_b(\Gamma)$. We then have the following uniform bounds.

Lemma 3.10 *Let the family* $(\mathsf{P}_0^n)_{n\ge1} \subset \mathcal{P}(\Gamma)$ *be such that*

$$\limsup_{n\to\infty} \mathsf{E}_n(\mathsf{P}_0^n) < +\infty.$$

Then for the corresponding family $(\mathsf{P}^n, \mathsf{J}^n)$ *of generalized gradient flow solutions,*

$$\limsup_{n\to\infty} \sup_{t\in[0,T]} \mathsf{E}_n(\mathsf{P}_t^n) < \infty, \qquad \limsup_{n\to\infty} \mathfrak{d}(\mathsf{P}_t^n, \mathsf{P}_s^n) \le K|t-s|,$$

for some constant $K > 0$*, independent of* $s, t \in [0, T]$.

Proof The estimates for $\mathsf{E}_n(\mathsf{P}_t^n)$ follow directly from the fact that the driving energy E_n is decreasing along solutions. In particular, we obtain the bound

$$\limsup_{n\to\infty} \sup_{t\in[0,T]} \int_\Gamma \|\nu\|\, \mathsf{P}_t^n(d\nu) < +\infty.$$

Next, take any $F \in \mathcal{F}$. Note that for any $\nu \in \Gamma_n$ and $x, y \in X$,

$$|F(\nu + n^{-1}\delta_x) - F(\nu)| \le n^{-1},$$

$$\int_X |F(\nu - n^{-1}\delta_x) - F(\nu)|\, \nu(dx) \le n^{-1}\|\nu\|,$$

$$\int_X |F(\nu + n^{-1}(\delta_y - \delta_x) - F(\nu)|\nu(dx) \le 2n^{-1}\|\nu\|,$$

which, together with the uniform mass estimates and the uniform bounds on b_n, d_n, and h_n, leads after substituting F in (8) to a constant $K > 0$ such that for every $s, t \in [0, T]$, $F \in \mathcal{F}$,

$$\left| \int_\Gamma F(\nu) d(\mathsf{P}_t^n - \mathsf{P}_s^n) \right| \le K|t-s|.$$

Taking the supremum over $F \in \mathcal{F}$, we derive the asserted uniform estimate for the distance $\mathfrak{d}$. □

4 Large-Population Limit

In this section, we consider the large-population limit for the forward Kolmogorov equation (FKE_n). As indicated in the introduction, we expect the measure-valued jump processes to converge to

$$\partial_t \nu_t = \mathscr{V}[\nu_t], \qquad \mathscr{V} = \kappa_{\mathsf{bd}} + \kappa_{\mathsf{h}} : \Gamma \to \Gamma, \tag{MFE}$$

with limit rates κ_{bd} and κ_{h} given in (1), which is equivalent to (F-KPP) if we take $\nu_t = u_t \pi$ and use the measure-dependent variants of the rates $\mathsf{r_b}$, $\mathsf{r_d}$, and $\mathsf{r_h}$ given in (2). However, we do not work directly with the measure-valued jump processes. Instead, we consider the forward Kolmogorov equation (FKE_n) and the convergence to (MFE) is done in the sense that $\mathsf{P}_t^n \rightharpoonup \delta_{\nu_t}$ narrowly as $n \to \infty$.

On the level of the measure-valued evolution, we expect (FKE_n) to converge to the Liouville equation:

$$\partial_t \mathsf{P}_t + \mathrm{div}_\Gamma\big(\mathscr{V}\mathsf{P}_t\big) = 0. \tag{LiE}$$

It is important to recognize that (MFE) and (LiE) are closely related. If ν is a solution of (MFE), then $\mathsf{P}_t = \delta_{\nu_t}$ solves (LiE). Thus, (LiE) can be seen as the lifting of (MFE) in Γ to $\mathcal{P}(\Gamma)$.

In Sect. 4.1, we show a formal derivation of (LiE) as a limit of (FKE_n) and present its generalized gradient flow formulation. Section 4.2 is dedicated to the well-posedness and variational formulation of (MFE). We explore the connection between (LiE) and (MFE) in Sect. 4.3, where we introduce the superposition principle and finalize the variational characterization of (LiE). In Sect. 4.4, we prove the main convergence theorem, consequently resulting in Theorem A.

4.1 Liouville Equation and Its Generalized Gradient Flow Formulation

First, we give an intuitive explanation of why we expect (LiE) as the limit for (FKE_n). Let $F : \Gamma \to \mathbb{R}$ be a cylindrical function of the form

$$F(\nu) = g\big(\langle f_1, \nu\rangle, \ldots, \langle f_M, \nu\rangle\big), \qquad g \in C_c^\infty(\mathbb{R}^M), \quad f_1 = 1, \; f_m \in C_b(X), \\ m = 2, \ldots, M, \tag{15}$$

and denote by $\mathrm{Cyl}(\Gamma)$ the collection of such functions.

We recall the birth-and-death generator L^n_{bd} and see how it acts on a cylindrical function F. A simple application of the Taylor expansion provides

$$
\begin{aligned}
n\big[F(\nu+n^{-1}\delta_x)-F(\nu)\big] &= \sum_{m=1}^{M} \partial_m g\big(\langle f_1,\nu\rangle,\ldots,\langle f_m,\nu\rangle\big) f_m(x) + O\big(\tfrac{1}{n}\big)\\
&= \nabla_\Gamma F(\nu,x) + O\big(\tfrac{1}{n}\big)
\end{aligned}
$$

$$
\begin{aligned}
n\big[F(\nu-n^{-1}\delta_x)-F(\nu)\big] &= -\sum_{m=1}^{M} \partial_m g\big(\langle f_1,\nu\rangle,\ldots,\langle f_m,\nu\rangle\big) f_m(x) + O\Big(\frac{1}{n}\Big)\\
&= -\nabla_\Gamma F(\nu,x) + O\big(\tfrac{1}{n}\big),
\end{aligned}
$$

where $\nabla_\Gamma F$ is the Γ-gradient operator defined on cylindrical functions as

$$
\nabla_\Gamma F(\nu,x) = \sum_{m=1}^{M} \partial_m g\big(\langle f_1,\nu\rangle,\ldots,\langle f_m,\nu\rangle\big) f_m(x), \qquad (\nu,x)\in\Gamma\times X.
$$

Hence, we can write the action of L^n_{bd} on cylindrical functions as

$$
L^n_{\mathsf{bd}} F(\nu) = \int_X \nabla_\Gamma F(\nu,x)\,\big(\mathsf{b}_n(\nu,\mathrm{d}x) - \mathsf{d}_n(\nu,x)\,\nu(\mathrm{d}x)\big) + O\big(\tfrac{1}{n}\big).
$$

Formally, we expect to obtain the pointwise limit birth-and-death generator under appropriate assumptions on the rate functions:

$$
\begin{aligned}
L^\infty_{\mathsf{bd}} F(\nu) = \lim_{n\to\infty} L^n_{\mathsf{bd}} F(\nu) &= \int_X \nabla_\Gamma F(\nu,x)\,\big(\mathsf{b}(\nu,\mathrm{d}x) - \mathsf{d}(\nu,x)\nu(\mathrm{d}x)\big)\\
&= \int_X \nabla_\Gamma F(\nu,x)\,\kappa_{\mathsf{bd}}[\nu](\mathrm{d}x).
\end{aligned}
$$

For the hopping (Kawasaki) generator L^n_{h}, similar calculations give

$$
L^n_{\mathsf{h}} F(\nu) = \iint_{X\times X} \big(\nabla_\Gamma F(\nu,y) - \nabla_\Gamma F(\nu,x)\big)\,\mathsf{h}_n(\nu,x,\mathrm{d}y)\,\nu(\mathrm{d}x) + O\big(\tfrac{1}{n}\big).
$$

We incorporate the notation $(\overline{\nabla}_\Gamma F)(\nu,x,y) := \nabla_\Gamma F(\nu,y) - \nabla_\Gamma F(\nu,x)$ and formally pass $n\to\infty$. We also employ the detailed balance assumption for h to see the following form of the limit hopping generator:

$$
\begin{aligned}
L^\infty_{\mathsf{h}} F(\nu) &= \iint_{X^2} (\overline{\nabla}_\Gamma) F(\nu,x,y)\,\mathsf{h}(\nu,x,\mathrm{d}y)\nu(\mathrm{d}x)\\
&= \int_X \nabla_\Gamma F(\nu,x)\Big[\int_{y\in X} \mathsf{h}(\nu,y,\mathrm{d}x)\nu(\mathrm{d}y) - \int_{y\in X} \mathsf{h}(\nu,x,\mathrm{d}y)\nu(\mathrm{d}x)\Big]
\end{aligned}
$$

$$= \int_X \nabla_\Gamma F(\nu, x)\, \kappa_{\mathsf{h}}[\nu](\mathrm{d}x).$$

Together, we obtain limit generator $L^\infty = L^\infty_{\mathsf{bd}} + L^\infty_{\mathsf{h}}$, which, by duality, gives the Liouville equation (LiE).

Having formally obtained (LiE) as a reasonable limit for (FKE_n), we defined the weak solutions for it.

Definition 4.1 (Weak Solutions) A narrowly continuous curve $P = (P_t)_{t\in[0,T]}$ is a weak solution to (LiE) if for all $0 \le s < t \le T$ and all $F \in \mathrm{Cyl}(\Gamma)$ it holds that

$$\int_\Gamma F \mathrm{d}\mathsf{P}_t - \int_\Gamma F \mathrm{d}\mathsf{P}_s = \int_s^t \iint_{\Gamma\times X} \nabla_\Gamma F(\nu, x)\, \mathscr{V}[\nu](\mathrm{d}x)\, \mathsf{P}_r(\mathrm{d}\nu)\mathrm{d}r.$$

Remark 4.2 The Liouville equation (LiE) is the transport equation associated with the measure-valued vector field $\mathscr{V}$. Therefore, the existence of weak solutions to the Liouville equation (LiE) can be proven using the well-posedness of the mean-field equation (MFE). More precisely, if we define the flow map $\mathscr{S} : [0, T] \times \Gamma \to \Gamma$ satisfying the Lagrangian flow

$$\partial_t \mathscr{S}_t(\nu) = \mathscr{V}(\mathscr{S}_t(\nu)), \tag{LF}$$

then one can show that $\mathsf{P}_t = (\mathscr{S}_t)_\#\mathsf{P}_0$ is a weak solution to (LiE) for any initial datum $\mathsf{P}_0 \in \Gamma$.

Remark 4.3 The conditions $g \in C_c^\infty(\mathbb{R}^M)$ and $f_1 = 1$ in (15) imply that $\|\nu\|$ is bounded on the support of F for any $F \in \mathrm{Cyl}(\Gamma)$ and in particular $\mathrm{Cyl}(\Gamma) \subset C_c(\Gamma)$. This frees us from having to prove convergence of mass moments when passing to the limit in (FKE_n).

Generalized Gradient Flow Formulation Now we discuss a variational formulation of the Liouville equation (LiE) given by its generalized gradient structure. This structure resembles the generalized gradient structure of the forward Kolmogorov equation (FKE_n) presented in Sect. 3. First, we introduce the appropriate continuity equation:

$$\partial_t \mathsf{P}_t + \mathrm{div}_\Gamma\, \mathsf{J}_t = 0, \tag{CE_∞}$$

and consider arbitrary curves satisfying (CE_∞) in the following sense.

Definition 4.4 (Continuity Equation) A tuple $(\mathsf{P}, \mathsf{J}^{\mathsf{bd}}, \mathsf{J}^{\mathsf{h}})$ satisfies the continuity equation (CE_∞), or we say for short $(\mathsf{P}, \mathsf{J}) \in \mathsf{CE}_\infty$, if:

(1) The curve $[0, T] \ni t \mapsto \mathsf{P}_t \in \mathcal{P}(\Gamma)$ is narrowly continuous with the mass bound:

$$\sup_{t\in[0,T]} \int_\Gamma \|\nu\| \mathrm{d}\mathsf{P}_t < \infty.$$

(2) The Borel families $\{\mathsf{J}_t^{\mathsf{bd}}\}_{t\in[0,T]} \subset \mathcal{M}(\Gamma \times X)$ and $\{\mathsf{J}_t^{\mathsf{h}}\}_{t\in[0,T]} \subset \mathcal{M}(\Gamma \times X^2)$ satisfy

$$\int_0^T \|\mathsf{J}_t^{\mathsf{bd}}\| + \|\mathsf{J}_t^{\mathsf{h}}\| \mathrm{d}t < \infty.$$

(3) For any $0 \le s < t \le T$ and $F \in \mathrm{Cyl}(\Gamma)$, it holds that

$$\int_\Gamma F \mathrm{d}P_t - \int_\Gamma F \mathrm{d}P_s = \int_s^t \Big[\iint_{\Gamma\times X} \nabla_\Gamma F(\nu, x)\, \mathsf{J}_r^{\mathsf{bd}}(\mathrm{d}\nu \mathrm{d}x) + \iiint_{\Gamma\times X^2} \overline{\nabla}_\Gamma F(\nu, x, y)\, \mathsf{J}_r^{\mathsf{h}}(\mathrm{d}\nu \mathrm{d}x \mathrm{d}y) \Big] \mathrm{d}r. \tag{16}$$

To define the energy-dissipation functional (EDB_∞), we specify the gradient structure $(\mathsf{R}_\infty, \mathsf{R}_\infty^*, \mathsf{E}_\infty)$ with driving energy

$$\mathcal{P}(\Gamma) \ni \mathsf{P} \mapsto \mathsf{E}_\infty(\mathsf{P}) := \int_\Gamma \mathsf{Ent}(\nu|\pi)\, \mathsf{P}(\mathrm{d}\nu).$$

Notice that if $\mathsf{E}_\infty(\mathsf{P}) < +\infty$, then $\mathsf{Ent}(\nu|\pi) < +\infty$ for P-almost every $\nu \in \Gamma$, i.e., P is supported on finite measures having finite relative entropy w.r.t. the activity measure π.

In analogy to Sect. 3.2, we denote the jump intensity measures:

$$\vartheta_{\mathsf{P}}^{\mathsf{b}}(\mathrm{d}\nu, \mathrm{d}x) := \mathsf{b}(\nu, \mathrm{d}x)\mathsf{P}(\mathrm{d}\nu), \qquad \vartheta_{\mathsf{P}}^{\mathsf{d}}(\mathrm{d}\nu, \mathrm{d}x) := \mathsf{d}(\nu, x)\nu(\mathrm{d}x)\mathsf{P}(\mathrm{d}\nu),$$
$$\vartheta_{\mathsf{P}}^{\mathsf{h}}(\mathrm{d}\nu, \mathrm{d}x, \mathrm{d}y) := \mathsf{h}(\nu, x, \mathrm{d}y)\nu(\mathrm{d}x)\mathsf{P}(\mathrm{d}\nu),$$

with the limit rates $\mathsf{b}, \mathsf{d}, \mathsf{h}$ defined in Assumption 1.1. As before, the dissipation potential has two components, the birth-and-death dissipation potential R_{bd} and the hopping dissipation potential R_{h}, given by

$$\mathcal{P}(\Gamma) \times \mathcal{M}(\Gamma \times X) \ni (\mathsf{P}, \mathsf{J}) \mapsto \mathsf{R}_{\infty,\mathsf{bd}}(\mathsf{P}, \mathsf{J}) := 2 \iint_{\Gamma\times X} \Upsilon(\mathsf{J}/2, \vartheta_{\mathsf{P}}^{\mathsf{b}}, \vartheta_{\mathsf{P}}^{\mathsf{d}}),$$
$$\mathcal{P}(\Gamma) \times \mathcal{M}(\Gamma \times X^2) \ni (\mathsf{P}, \mathsf{J}) \mapsto \mathsf{R}_{\infty,\mathsf{h}}(\mathsf{P}, \mathsf{J}) := \iiint_{\Gamma\times X^2} \Upsilon(\mathsf{J}, \vartheta_{\mathsf{P}}^{\mathsf{h}}, \mathsf{T}_\#\vartheta_{\mathsf{P}}^{\mathsf{h}}), \tag{17}$$

where $\mathsf{T}(\nu, x, y) := (\nu, y, x)$. The total dissipation potential is then

$$\mathsf{R}_\infty(\mathsf{P}, \mathsf{J}) := \mathsf{R}_{\infty,\mathsf{bd}}(\mathsf{P}, \mathsf{J}^{\mathsf{bd}}) + \mathsf{R}_{\infty,\mathsf{h}}(\mathsf{P}, \mathsf{J}^{\mathsf{h}}).$$

The corresponding dual dissipation potentials are

$$\mathcal{P}(\Gamma) \times B_b(\Gamma \times X) \ni (\mathsf{P}, \zeta) \mapsto \mathsf{R}^*_{\infty,\mathsf{bd}}(\mathsf{P}, \zeta) := 2 \iint_{\Gamma \times X} \Psi^*(\zeta) \mathrm{d}\sqrt{\vartheta^{\mathsf{b}}_{\mathsf{P}} \vartheta^{\mathsf{d}}_{\mathsf{P}}},$$

$$\mathcal{P}(\Gamma) \times B_b(\Gamma \times X^2) \ni (\mathsf{P}, \zeta) \mapsto \mathsf{R}^*_{\infty,\mathsf{h}}(\mathsf{P}, \zeta) := \iiint_{\Gamma \times X^2} \Psi^*(\zeta) \mathrm{d}\sqrt{\vartheta^{\mathsf{h}}_{\mathsf{P}} \mathsf{T}_{\#}\vartheta^{\mathsf{h}}_{\mathsf{P}}}.$$

Furthermore, we define the birth-and-death and hopping Fisher information by means of the Hellinger distance $\mathcal{H}$:

$$\begin{aligned} \mathcal{P}(\Gamma) \ni \mathsf{P} &\mapsto \mathsf{D}_{\infty,\mathsf{bd}}(\mathsf{P}) := 4\mathcal{H}^2(\vartheta^{\mathsf{b}}_{\mathsf{P}}, \vartheta^{\mathsf{d}}_{\mathsf{P}}) \in [0, +\infty], \\ \mathcal{P}(\Gamma) \ni \mathsf{P} &\mapsto \mathsf{D}_{\infty,\mathsf{h}}(\mathsf{P}) := 2\mathcal{H}^2(\vartheta^{\mathsf{h}}_{\mathsf{P}}, \mathsf{T}_{\#}\vartheta^{\mathsf{h}}_{\mathsf{P}}) \in [0, +\infty]. \end{aligned} \tag{18}$$

The total Fisher information is given by $\mathsf{D}(\mathsf{P}) = \mathsf{D}_{\mathsf{bd}}(\mathsf{P}) + \mathsf{D}_{\mathsf{h}}(\mathsf{P})$. Finally, we introduce the energy-dissipation functional for any $(\mathsf{P}, \mathsf{J}) \in \mathsf{CE}_\infty$:

$$\mathscr{I}_\infty([s,t]; \mathsf{P}, \mathsf{J}) := \int_s^t (\mathsf{R}_\infty(\mathsf{P}_r, \mathsf{J}_r) + \mathcal{D}_\infty(\mathsf{P}_r))\,\mathrm{d}r + \mathsf{E}_\infty(\mathsf{P}_t) - \mathsf{E}_\infty(\mathsf{P}_s). \tag{19}$$

The components of the generalized gradient structure involved in $\mathscr{I}_\infty([s,t]; \mathsf{P}, \mathsf{J})$ were introduced in the lower-semicontinuous form, hinting that they can be obtained as gamma limits of the corresponding structure for (FKE_n) defined in Sect. 3.4. To make a clear connection between the structure $(\mathsf{R}_\infty, \mathsf{R}^*_\infty, \mathsf{E}_\infty)$ and the Liouville equation (LiE), we first need to study the mean-field equation (MFE) in more detail. After establishing the variational formulation for (MFE) in Sect. 4.2, we revisit the variational formulation of (LiE) and clarify the connection between the two and why $(\mathsf{R}_\infty, \mathsf{R}^*_\infty, \mathsf{E}_\infty)$ gives the appropriate gradient flow formulation for (LiE).

4.2 Mean-Field Equation

In this section, we take a closer look at the mean-field equation with velocity field $\mathscr{V} = \kappa_{\mathsf{bd}} + \kappa_{\mathsf{h}}$ with the limit rates derived above:

$$\begin{aligned} \kappa_{\mathsf{bd}}[\nu](\mathrm{d}x) &= \mathsf{b}(\nu, \mathrm{d}x) - \mathsf{d}(\nu, x)\nu(\mathrm{d}x), \\ \kappa_{\mathsf{h}}[\nu](\mathrm{d}x) &= \int_{y \in X} \mathsf{h}(\nu, y, \mathrm{d}x)\,\nu(\mathrm{d}y) - \nu(\mathrm{d}x) \int_{y \in X} \mathsf{h}(\nu, x, \mathrm{d}y). \end{aligned}$$

Under Assumption 1.1, the velocity field $\mathscr{V}$ can easily be shown to be locally Lipschitz having linear growth, i.e., there exists constants $\ell_{\mathscr{V}}$ and $c_{\mathscr{V}} > 0$ such that for any $\nu, \eta \in \Gamma$,

$$\|\mathscr{V}[\nu] - \mathscr{V}[\eta]\| \le \ell_{\mathscr{V}}\big(1 + \|\nu\| + \|\eta\|\big)\|\nu - \eta\|,$$
$$\|\mathscr{V}[\nu]\| \le c_{\mathscr{V}}\|\nu\|.$$

Therefore, the standard ODE theory in Banach spaces (cf. [32]) applies and we obtain the following result.

Proposition 4.5 *Let $\overline{u} \in L^1(X, \pi)$. Then there exists a unique differentiable curve $u \in C_b^1((0, T); L^1(X, \pi))$ satisfying the mean-field equation* (F-KPP) *with $u_0 = \overline{u}$.*
Consequently, the curve $t \mapsto \nu_t := u_t\pi$ is a solution of the mean-field equation (MFE) *with $\nu_0 = \overline{u}\pi$. We call this curve a strong solution of* (MFE).

Under the detailed balance assumption (DB_∞), (MFE) has a gradient flow formulation. To make it precise, we consider the curves satisfying the continuity equation:

$$\partial_t \nu_t = \lambda_t^{\mathsf{bd}} + \overline{\mathrm{div}}\lambda_t^{\mathsf{h}}, \qquad (\mathscr{C}\mathscr{E})$$

where the discrete divergence

$$\mathcal{M}(X^2) \ni j \mapsto (\overline{\mathrm{div}}\, j)(\mathrm{d}x) := \int_{y\in X} j(\mathrm{d}x, \mathrm{d}y) - \int_{y\in X} j(\mathrm{d}y, \mathrm{d}x) \in \mathcal{M}(X)$$

is associated to the discrete gradient $\overline{\nabla} f(x, y) = f(y) - f(x)$ by duality.

Definition 4.6 A triple $(\nu, \lambda^{\mathsf{bd}}, \lambda^{\mathsf{h}})$ satisfies $(\mathscr{C}\mathscr{E})$, or we say $(\nu, \lambda) \in \mathscr{C}\mathscr{E}$, if:

(1) The curve $[0, T] \ni t \mapsto \nu_t \in \Gamma$ is $\|\cdot\|$-absolutely continuous.
(2) The Borel families $\{\lambda_t^{\mathsf{bd}}\}_{t\in[0,T]} \subset \mathcal{M}(X)$, $\{\lambda_t^{\mathsf{h}}\}_{t\in[0,T]} \subset \mathcal{M}(X \times X)$ satisfy

$$\int_0^T \|\lambda_t^{\mathsf{bd}}\| + \|\lambda_t^{\mathsf{h}}\|\mathrm{d}t < \infty.$$

(3) For any $0 \le s < t \le T$ and $f \in C_b(X)$, it holds that

$$\int_X f\mathrm{d}\nu_t - \int_X f\mathrm{d}\nu_s = \int_s^t \Big[\int_X f\mathrm{d}\lambda_r^{\mathsf{bd}} + \iint_{X^2} \overline{\nabla} f\mathrm{d}\lambda_r^{\mathsf{h}}\Big]\mathrm{d}r.$$

We claim that (MFE) has a generalized gradient flow formulation $(\mathscr{R}, \mathscr{R}^*, \mathscr{E})$ with driving energy $\mathscr{E}(\nu) = \mathsf{Ent}(\nu|\pi)$. The dissipation potential $\mathscr{R}$ comprises the birth-and-death and the hopping parts defined as

$$\Gamma \times \mathcal{M}(X) \ni (\nu, \lambda^{\mathsf{bd}}) \mapsto \mathscr{R}_{\mathsf{bd}}(\nu, \lambda^{\mathsf{bd}}) := 2\int_X \Upsilon(\lambda^{\mathsf{bd}}/2, \mathsf{b}_\nu, \mathsf{d}_\nu),$$

$$\Gamma \times \mathcal{M}(X^2) \ni (\nu, \lambda^{\mathsf{h}}) \mapsto \mathscr{R}_{\mathsf{h}}(\nu, \lambda^{\mathsf{h}}) := \int_X \Upsilon(\lambda^{\mathsf{h}}, \mathsf{h}_\nu, \mathsf{T}_\#\mathsf{h}_\nu),$$

where we use the shorthand $\mathsf{b}_\nu(dx) := \mathsf{b}(\nu, dx)$, $\mathsf{d}_\nu(dx) := \mathsf{d}(\nu, x)\nu(dx)$ and $\mathsf{h}_\nu(dx, dy) := \mathsf{h}(\nu, x, dy)\nu(dx)$. Recall that T is the swapping map $\mathsf{T}(\nu, x, y) = (\nu, y, x)$.

The dual dissipation potential $\mathscr{R}^*$ is defined as

$$\Gamma \times B_b(X) \ni (\nu, w) \mapsto \mathscr{R}^*_{\mathsf{bd}}(\nu, w) := 2 \int_X \psi^*(w) \mathrm{d}\sqrt{\mathsf{b}_\nu \mathsf{d}_\nu},$$

$$\Gamma \times B_b(X^2) \ni (\nu, w) \mapsto \mathscr{R}^*_{\mathsf{h}}(\nu, w) := \iint_{X^2} \psi^*(w) \mathrm{d}\sqrt{\mathsf{h}_\nu \mathsf{T}_\# \mathsf{h}_\nu},$$

and the Fisher information is given by

$$\Gamma \ni \nu \mapsto \mathscr{D}(\nu) = 4\mathcal{H}^2(\mathsf{b}_\nu, \mathsf{d}_\nu) + 2\mathcal{H}^2(\mathsf{h}_\nu, \mathsf{T}_\# \mathsf{h}_\nu) \in [0, +\infty].$$

Consequently, the energy-dissipation functional $\mathscr{I}$ reads

$$\mathscr{I}([s, t]; \nu, \lambda) := \int_s^t \mathscr{R}(\nu_r, \lambda_r) + \mathscr{D}(\nu_r) \mathrm{d}r + \mathscr{E}(\nu_T) - \mathscr{E}(\nu_0). \tag{20}$$

Remark 4.7 If $\nu = u\pi$, then we have a simplified expression for the Fisher information:

$$\begin{aligned} \mathscr{D}(\nu) = 2 &\int_X \left(\sqrt{u(x)} - 1\right)^2 \mathsf{r}_\mathsf{b}(u, x)\pi(dx) \\ &+ \iint_{X^2} \left(\sqrt{u(y)} - \sqrt{u(x)}\right)^2 \mathsf{r}_\mathsf{h}(u, x, y)\pi(dx)\pi(dy) \end{aligned}$$

with the density-dependent rates r_b and r_h defined in (2). Moreover, a formal calculation shows that

$$\partial_2 \mathscr{R}^* \left(-\mathscr{E}', -\overline{\nabla}\mathscr{E}'\right) = \left(\mathsf{b}_\nu - \mathsf{d}_\nu, \frac{1}{2}(\mathsf{h}_\nu - \mathsf{T}_\# \mathsf{h}_\nu)\right), \quad \mathscr{D}(\nu) = \mathscr{R}^* \left(-\mathscr{E}', -\overline{\nabla}\mathscr{E}'\right).$$

Theorem 4.8 (Variational Characterization of (MFE)) *For any $(\nu, \lambda^{\mathsf{bd}}, \lambda^{\mathsf{h}}) \in \mathscr{CE}$ with $\mathscr{E}(\nu_0) < \infty$, we have $\mathscr{I}([s, t]; \nu, \lambda) \geq 0$ and ν_t is the unique strong solution to (MFE) if and only if $\mathscr{I}([0, T]; \nu, \lambda) = 0$ and*

$$\lambda^{\mathsf{bd}} = \mathsf{b}_\nu - \mathsf{d}_\nu, \qquad \lambda^{\mathsf{h}} = \frac{1}{2}(\mathsf{h}_\nu - \mathsf{T}_\# \mathsf{h}_\nu).$$

Moreover, if $\mathscr{I}([0, T]; \nu, \lambda) < \infty$, then the chain rule for $\mathscr{E}$ holds, i.e.,

$$t \mapsto \mathscr{E}(\nu_t) \text{ is absolutely continuous,}$$

$$\frac{\mathrm{d}}{\mathrm{d}t}\mathscr{E}(\nu_t) = \int_X \log \frac{\mathrm{d}\nu_t}{\mathrm{d}\pi}\mathrm{d}\lambda_t \quad \textit{for almost every } t \in (0, T).$$

Proof The proof follows by using a regularized entropy functional and combining arguments from [39] for the jump setting and [24] for the case of birth and death.

□

4.3 *Superposition Principle*

The superposition principle, which makes the connection between the mean-field equation (MFE) and the Liouville equation (LiE), is an essential tool for proving uniqueness and variational characterization of solutions. Intuitively, it says that a curve satisfying (CE_∞) with finite action can be represented as a superposition of curves satisfying ($\mathscr{C}\mathscr{E}$).

With the variational formulation for (MFE) at hand, we realize that the energy-dissipation functional $\mathscr{I}_\infty$ given in (19) for (LiE) is "lifted" from the corresponding functional (20). First, observe that if $\mathsf{E}_\infty(\mathsf{P}) < \infty$, then the Fisher information has the following form:

$$\begin{aligned}\int_\Gamma \mathscr{D}(\nu)\,\mathsf{P}(\mathrm{d}\nu) &= 4\int_\Gamma \mathcal{H}^2(\mathsf{b}_\nu, \mathsf{d}_\nu)\,\mathsf{P}(\mathrm{d}\nu) + 2\int_\Gamma \mathcal{H}^2(\mathsf{h}_\nu, \mathsf{T}_\#\mathsf{h}_\nu)\,\mathsf{P}(\mathrm{d}\nu)\\ &= 4\mathcal{H}^2(\vartheta^{\mathsf{b}}_{\mathsf{P}}, \vartheta^{\mathsf{d}}_{\mathsf{P}}) + 2\mathcal{H}^2(\vartheta^{\mathsf{h}}_{\mathsf{P}}, \mathsf{T}_\#\vartheta^{\mathsf{h}}_{\mathsf{P}}) = \mathsf{D}_\infty(\mathsf{P}).\end{aligned}$$

It also holds that $\sqrt{\vartheta^{\mathsf{b}}_{\mathsf{P}}\,\vartheta^{\mathsf{d}}_{\mathsf{P}}}(\mathrm{d}\nu\mathrm{d}x) = \theta^{\mathsf{bd}}_\nu(\mathrm{d}x)\mathsf{P}(\mathrm{d}\nu)$ and $\sqrt{\vartheta^{\mathsf{h}}_{\mathsf{P}}\,\mathsf{T}_\#\vartheta^{\mathsf{h}}_{\mathsf{P}}}(\mathrm{d}\nu\mathrm{d}x\mathrm{d}y) = \theta^{\mathsf{h}}_\nu(\mathrm{d}x\mathrm{d}y)\mathsf{P}(\mathrm{d}\nu)$. If $\mathsf{E}_\infty(\mathsf{P}) < \infty$, one can also show the disintegration property for fluxes

$$\mathsf{J}^{\mathsf{bd}}(\mathrm{d}\nu\mathrm{d}x) = \lambda^{\mathsf{bd}}_\nu(\mathrm{d}x)\mathsf{P}(\mathrm{d}\nu) \quad \text{and} \quad \mathsf{J}^{\mathsf{h}}(\mathrm{d}\nu\mathrm{d}x\mathrm{d}y) = \lambda^{\mathsf{h}}_\nu(\mathrm{d}x\mathrm{d}y)\mathsf{P}(\mathrm{d}\nu).$$

Approaches based on superposition principles similar to ours were also employed in [11, 12, 24], where they are applied to transport equations lifted from the Boltzmann equation, mean-field jump dynamics, and mean-field birth-and-death dynamics, respectively. In all these cases, the proof of the superposition principle relies on the superposition principle for solutions of the continuity equation on $\mathbb{R}^{\mathbb{N}}$ established in [1].

Theorem 4.9 (Superposition Principle) *Let* $(\mathsf{P}, \mathsf{J}) \in \mathrm{CE}_\infty$ *with*

$$\int_0^T \mathsf{R}_\infty(\mathsf{P}_t, \mathsf{J}_t)\,\mathrm{d}t < \infty.$$

Then there exists a Borel probability (path) measure $\mathbb{Q} \in \mathcal{P}(C([0,T];\Gamma))$ *such that:*

(1) The time marginals are

$$(e_t)_\#\mathbb{Q} = \mathsf{P}_t \qquad \forall t \in [0,T],$$

where $e_t \colon C([0,T];\Gamma) \to \Gamma$, $e_t(\nu) = \nu_t$ *is the time evaluation map.*

(2) $\mathbb{Q}$ *is concentrated on* $\|\cdot\|$*-absolutely continuous curves* $\nu \in \mathrm{AC}([0,T];(\Gamma,\|\cdot\|))$ *with* $(\nu,\lambda) \in \mathscr{CE}$ *for some Borel family* $\{\lambda_t\}_{t\in[0,T]} \subset \mathcal{M}(X)$ *of fluxes.*

(3) The following representation holds

$$\int_0^T \mathsf{R}_\infty(\mathsf{P}_t, \mathsf{J}_t)\mathrm{d}t = \int_{C([0,T];\Gamma)} \Big(\int_0^T \mathscr{R}(\nu_t, \lambda_t)\mathrm{d}t\Big)\,\mathbb{Q}(\mathrm{d}\nu).$$

Proof Let $\{f_m\}_{m\in\mathbb{N}} \subset C_b(\Omega)$ be a countable dense set with $f_1 = 1$. We define

$$\Gamma \ni \nu \mapsto \mathbb{T}(\nu) := \Big(\int_X f_1 \mathrm{d}\nu, \int_x f_2 \mathrm{d}\nu, \dots\Big) \in \mathbb{R}^{\mathbb{N}}.$$

We further set $\sigma_t := \mathbb{T}_\# \mathsf{P}_t \in \mathcal{P}(\mathbb{R}^{\mathbb{N}})$ and define a vector field $W \colon [0,T]\times\mathbb{R}^n \to \mathbb{R}^n$ via its components, $W_m := W_m^{\mathsf{bd}} + W_m^{\mathsf{h}}$, $m \in \mathbb{N}$, where

$$W_m^{\mathsf{bd}}(t,z) := \int_X f_m(x)\lambda^{\mathsf{bd}}_{t,\mathbb{T}^{-1}(z)}(\mathrm{d}x),$$

$$W_m^{\mathsf{h}}(t,z) := \iint_{X^2} \big(f_m(y) - f_m(x)\big)\lambda^{\mathsf{h}}_{t,\mathbb{T}^{-1}(z)}(\mathrm{d}x\mathrm{d}y).$$

The pair (σ, W) satisfies the continuity equation in the sense that, for all $g \in \mathrm{Cyl}(\mathbb{R}^{\mathbb{N}})$, we have

$$\int_{\mathbb{R}^{\mathbb{N}}} g\mathrm{d}\sigma_t - \int_{\mathbb{R}^{\mathbb{N}}} g\mathrm{d}\sigma_s = \int_s^t \int_{\mathbb{R}^{\mathbb{N}}} \nabla g \cdot W_r\, \sigma_r(\mathrm{d}\nu)\mathrm{d}r.$$

To see this we define $F := g \circ \mathbb{T}$ and use that $(\mathsf{P}, \mathsf{J}) \in \mathrm{CE}_\infty$ to deduce

$$\begin{aligned}
&\int_{\mathbb{R}^{\mathbb{N}}} g\mathrm{d}\sigma_t - \int_{\mathbb{R}^{\mathbb{N}}} g\mathrm{d}\sigma_s \\
&= \int_\Gamma F\mathrm{d}\mathsf{P}_t - \int_\Gamma F\mathrm{d}\mathsf{P}_s \\
&= \int_s^t \Big[\iint_{\Gamma\times X} \nabla_\Gamma F(\nu,x)\,\mathsf{J}^{\mathsf{bd}}_r(\mathrm{d}\nu\mathrm{d}x) + \iiint_{\Gamma\times X^2} \overline{\nabla}_\Gamma F(\nu,x,y)\,\mathsf{J}^{\mathsf{h}}_r(\mathrm{d}\nu\mathrm{d}x\mathrm{d}y)\Big]\mathrm{d}r
\end{aligned}$$

$$= \int_s^t \Big[\int_\Gamma \sum_{m=1}^{M} \partial_m g(\mathbb{T}(\nu)) \int_X f_m(x) \lambda_r^{\mathsf{bd}}(\mathrm{d}x)\, \mathsf{P}_r(\mathrm{d}\nu)$$

$$+ \int_\Gamma \sum_{m=1}^{M} \partial_m g(\mathbb{T}(\nu)) \iint_{X^2} \big(f_m(y) - f_m(x)\big) \lambda_r^{\mathsf{h}}(\mathrm{d}x\mathrm{d}y)\, \mathsf{P}_r(\mathrm{d}\nu) \Big] \mathrm{d}r$$

$$= \int_s^t \Big[\int_{\mathbb{R}^{\mathsf{N}}} \nabla g(z) \cdot W_r^{\mathsf{bd}}(z)\, \sigma_r(\mathrm{d}\nu) + \int_{\mathbb{R}^{\mathsf{N}}} \nabla g(z) \cdot W_r^{\mathsf{h}}(z)\, \sigma_r(\mathrm{d}\nu) \Big] \mathrm{d}r.$$

Moreover, the defined vector field W belongs to $L^1([0,T] \times \mathbb{R}^{\mathsf{N}}, \mathcal{L}|_{[0,T]} \otimes \sigma_t)$, since

$$\int_0^T \int_{\mathbb{R}^{\mathsf{N}}} |W_m(t,z)| \mathrm{d}\sigma_t \mathrm{d}t = \int_0^T \int_\Gamma |W_m(t, \mathbb{T}(\nu))|\, \mathsf{P}_t(\mathrm{d}\nu) \mathrm{d}t \le \int_0^T \mathsf{R}_\infty(\mathsf{P}_t, \mathsf{J}_t)\, \mathrm{d}t) < \infty.$$

We are now in the position to apply [1, Theorem 7.1] to conclude that there exists a Borel probability measure $\mathbb{q}$ in $\mathcal{P}(C([0,T]; \mathbb{R}^{\mathsf{N}}))$ satisfying $(e_t)_\# \mathbb{q} = \nu_t$ for all $t \in [0,T]$ and is concentrated on absolutely continuous curves $z \in AC([0,T]; \mathbb{R}^{\mathsf{N}})$, which are solutions to $\dot{z}_t = W_t(z_t)$ for almost every $t \in [0,T]$. □

Theorem 4.10 (Variational Characterization of (LiE)**)** *For any* $(\mathsf{P}, \mathsf{J}) \in \mathsf{CE}_\infty$ *with* $\mathsf{P}_0 \in dom(\mathsf{E}_\infty)$*, the energy-dissipation functional* $\mathscr{I}_\infty$ *is finite if and only if there exists a Borel probability measure* $\mathbb{Q} \in \mathcal{P}(C([0,T]; \Gamma))$ *such that:*

(1) The time marginals are $(e_t)_\# \mathbb{Q} = \mathsf{P}_t$ *for all* $t \in [0,T]$.
(2) $\mathbb{Q}$ *is concentrated on the family of* $\|\cdot\|$*-absolutely continuous curves* $\nu \in AC([0,T]; (\Gamma, \|\cdot\|))$ *with* $(\nu, \lambda) \in \mathscr{CE}$*, where* $\lambda = \lambda^{\mathsf{bd}} + \lambda^{\mathsf{h}}$ *satisfies* $\mathsf{J}_t^{\mathsf{bd}} = \lambda_t^{\mathsf{bd}} \mathsf{P}_t$ *and* $\mathsf{J}_t^{\mathsf{h}} = \lambda_t^{\mathsf{h}} \mathsf{P}_t$ *for almost every* $t \in [0,T]$.
(3) The following representation holds:

$$\mathscr{I}_\infty([s,t]; \mathsf{P}, \mathsf{J}) = \int_{C([0,T];\Gamma)} \mathscr{I}([s,t]; \nu, \lambda) \mathrm{d}\mathbb{Q}.$$

In particular, $\mathscr{I}_\infty([0,T]; \mathsf{P}, \mathsf{J}) \ge 0$ *and equal to* 0 *if and only if* P_t *is the weak solution to* (LiE) *with*

$$\mathsf{P}_t = (\mathscr{S}_t)_\# \mathsf{P}_0 \qquad \textit{for every } t \in [0,T], \quad \textit{and fluxes}$$

$$\mathsf{J}^{\mathsf{bd}}_{\mathsf{P}_t} = \vartheta^{\mathsf{b}}_{\mathsf{P}_t} - \vartheta^{\mathsf{d}}_{\mathsf{P}_t}, \qquad \mathsf{J}^{\mathsf{h}} = \frac{1}{2}\left(\vartheta^{\mathsf{h}}_{\mathsf{P}_t} - \mathsf{T}_\# \vartheta^{\mathsf{h}}_{\mathsf{P}_t}\right) \quad \textit{for almost every } t \in [0,T].$$

4.4 Convergence Results

The generalized gradient structures introduced in Sects. 3.4 and 4.1 are particularly useful for rigorously proving the convergence of (FKE_n) to (LiE). We employ a form of convergence called evolutionary gamma convergence. It was first introduced in [42] and led to numerous subsequent studies surveyed in [34, 43]. Let us mention that other notions of convergence were introduced in this context known as EDP convergence [10, 35], which requires further limsup estimates and is not considered in this chapter. We follow the strategy of the evolutionary gamma convergence: establishing the liminf estimate for the energy-dissipation functional and deducing the convergence of gradient flow solutions.

Theorem 4.11 *Let* $\{(\mathsf{P}^n, \mathsf{J}^n)\}_{n\in\mathbb{N}}$ *be a sequence of generalized gradient flow solutions of* (FKE_n) *with initial data* $\{\mathsf{P}^n_{in}\}_{n\in\mathbb{N}}$ *and suppose there exists* $\mathsf{P}_{in} \in$ *dom*(E_∞) *such that*

$$\mathsf{P}^n_{in} \rightharpoonup \mathsf{P}_{in}, \qquad \lim_{n\to\infty} \mathsf{E}_n(\mathsf{P}^n_{in}) = \mathsf{E}_\infty(\mathsf{P}_{in}).$$

Then there exist a (not relabelled) subsequence$\{(\mathsf{P}^n, \mathsf{J}^n)\}_{n\in\mathbb{N}}$ *and a limit pair* (P, J) *such that:*

(1) $(\mathsf{P}, \mathsf{J}) \in \mathsf{CE}_\infty$*, where*

- $\mathsf{P}^n_t \rightharpoonup \mathsf{P}_t$ *narrowly for all* $t \in [0, T]$*.*
- $\mathsf{J}^{n,\mathsf{bd}}_t(\mathrm{d}\nu\mathrm{d}x) \rightharpoonup^* \mathsf{J}^{\mathsf{bd}}_t(\mathrm{d}\nu\mathrm{d}x)$ *weakly-* in* $\mathcal{M}(\Gamma \times X)$ *for all* $t \in [0, T]$*.*
- $\mathsf{J}^{n,\mathsf{h}}_t(\mathrm{d}\nu\mathrm{d}x\mathrm{d}y) \rightharpoonup^* \mathsf{J}^{\mathsf{h}}_t(\mathrm{d}\nu\mathrm{d}x\mathrm{d}y)$ *weakly-* in* $\mathcal{M}(\Gamma \times X^2)$ *for all* $t \in [0, T]$*.*

(2) The following lim inf *inequality holds:*

$$\mathscr{I}_\infty([s,t]; \mathsf{P}, \mathsf{J}) \le \liminf_{n\to\infty} \mathscr{I}_n([s,t]; \mathsf{P}^n, \mathsf{J}^n) \qquad \textit{for every } [s,t] \subset [0,T].$$

(3) (P, J) *is the generalized gradient flow solution of the Liouville equation* (LiE) *with initial datum* P_{in}*.*

Proof We first prove the compactness statement of (1). Consider a sequence $\{(\mathsf{P}^n, \mathsf{J}^n)\}_{n\in\mathbb{N}}$ of generalized gradient flow solutions of (FKE_n) with initial data $\{\mathsf{P}^n_{\mathrm{in}}\}_{n\in\mathbb{N}}$ such that $\sup_{n\in\mathbb{N}} \mathsf{E}_n(\mathsf{P}^n_{\mathrm{in}}) < \infty$. By the uniform estimates of Lemma 3.10 we obtain

$$\limsup_{n\to\infty} \sup_{t\in[0,T]} \mathsf{E}_n(\mathsf{P}^n_t) < \infty, \qquad \limsup_{n\to\infty} \mathfrak{d}(\mathsf{P}^n_t, \mathsf{P}^n_s) \le K|t-s|,$$

for some constant $K > 0$, independent of $s, t \in [0, T]$. Recall $\mathfrak{d}$ is the metric defined in (14), which is narrowly lower semicontinuous and whose convergence implies narrow convergence on narrowly compact subsets. Since the family $\{\mathsf{E}_n\}_{n\ge1}$ is narrowly equicoercive (cf. Proposition 2.2), an Arzelà–Ascoli argument gives a (not relabelled) subsequence P^n and a curve P such that $\mathsf{P}^n_t \rightharpoonup \mathsf{P}_t$ narrowly for all $t \in [0, T]$.

The narrow convergence $\mathsf{P}^n_t \rightharpoonup \mathsf{P}_t$ and Assumption 1.1 provide the following limits:

$$\vartheta^{n,\mathsf{b}}_{\mathsf{P}^n_t} \rightharpoonup^* \vartheta^{\mathsf{b}}_{\mathsf{P}_t}, \qquad \vartheta^{n,\mathsf{d}}_{\mathsf{P}^n_t} \rightharpoonup^* \vartheta^{\mathsf{d}}_{\mathsf{P}_t}, \qquad \vartheta^{n,\mathsf{h}}_{\mathsf{P}^n_t} \rightharpoonup^* \vartheta^{\mathsf{h}}_{\mathsf{P}_t}$$
$$\mathsf{T}^{n,-}_{\#}\vartheta^{n,\mathsf{d}}_{\mathsf{P}^n_t} \rightharpoonup^* \vartheta^{\mathsf{d}}_{\mathsf{P}_t}, \qquad \mathsf{T}^{n,\mathsf{h}}_{\#}\vartheta^{n,\mathsf{h}}_{\mathsf{P}^n_t} \rightharpoonup^* \mathsf{T}^{\mathsf{h}}_{\#}\vartheta^{\mathsf{h}}_{\mathsf{P}_t}. \tag{21}$$

Namely, for any $f \in C_c(\Gamma \times X)$, dropping the t-dependence,

$$\begin{aligned}
\lim_{n\to\infty} \iint_{\Gamma\times X} f(\nu, x)\, \vartheta^{n,\mathsf{b}}_{\mathsf{P}^n}(\mathrm{d}\nu \mathrm{d}x) &= \lim_{n\to\infty} \iint_{\Gamma\times X} f(\nu, x)\mathsf{b}(\nu, \mathrm{d}x)\mathsf{P}^n(\mathrm{d}\nu) \\
&\quad + \lim_{n\to\infty} \iint_{\Gamma\times X} f(\nu, x)(\mathsf{b}_n - \mathsf{b})(\nu, \mathrm{d}x)\mathsf{P}^n(\mathrm{d}\nu) \\
&= \iint_{\Gamma\times X} f(\nu, x)\mathsf{b}(\nu, \mathrm{d}x)\mathsf{P}(\mathrm{d}\nu).
\end{aligned}$$

Thus, $\vartheta^{n,\mathsf{b}}_{\mathsf{P}^n} \rightharpoonup^* \vartheta^{\mathsf{b}}_{\mathsf{P}}$ in $\mathcal{M}(\Gamma \times X)$. By a similar argument, we deduce $\vartheta^{n,\mathsf{d}}_{\mathsf{P}^n} \rightharpoonup^* \vartheta^{\mathsf{d}}_{\mathsf{P}}$ in $\mathcal{M}(\Gamma \times X)$ and $\vartheta^{n,\mathsf{h}}_{\mathsf{P}^n} \rightharpoonup^* \vartheta^{\mathsf{h}}_{\mathsf{P}}$ in $\mathcal{M}(\Gamma \times X^2)$. Moreover, the uniform continuity of $f \in C_c(\Gamma \times X)$ provides

$$\begin{aligned}
\lim_{n\to\infty} \iint_{\Gamma\times X} f(\nu, x)\mathsf{T}^{n,-}_{\#}\vartheta^{n,\mathsf{d}}_{\mathsf{P}^n}(\mathrm{d}\nu \mathrm{d}x) &= \lim_{n\to\infty} \iint_{\Gamma\times X} f(\nu - n^{-1}\delta_x, x)\, \vartheta^{n,\mathsf{d}}_{\mathsf{P}^n}(\mathrm{d}\nu \mathrm{d}x) \\
&= \iint_{\Gamma\times X} f(\nu, x)\, \vartheta^{\mathsf{d}}_{\mathsf{P}}(\mathrm{d}\nu \mathrm{d}x) \\
&\quad + \lim_{n\to\infty} \iint_{\Gamma\times X} \big(f(\nu - n^{-1}\delta_x, x) - f(\nu, x)\big)\, \vartheta^{n,\mathsf{d}}_{\mathsf{P}^n}(\mathrm{d}\nu \mathrm{d}x) \\
&= \iint_{\Gamma\times X} f(\nu, x)\, \vartheta^{\mathsf{d}}_{\mathsf{P}}(\mathrm{d}\nu \mathrm{d}x).
\end{aligned}$$

Consequently, we have convergence $\mathsf{T}^{n,-}_{\#}\vartheta^{n,\mathsf{d}}_{\mathsf{P}^n} \rightharpoonup^* \vartheta^{\mathsf{d}}_{\mathsf{P}}$ weakly-* in $\mathcal{M}(\Gamma \times X)$. Similarly, we find

$$\begin{aligned}
&\lim_{n\to\infty} \iiint_{\Gamma\times X^2} f(\nu, x, y)\big(\mathsf{T}^{n,\mathsf{h}}_{\#}\vartheta^{n,\mathsf{h}}_{\mathsf{P}^n}\big)(\mathrm{d}\nu \mathrm{d}x \mathrm{d}y) \\
&\quad = \lim_{n\to\infty} \iiint_{\Gamma\times X^2} f\big(\nu + n^{-1}(\delta_y - \delta_x), y, x\big)\vartheta^{n,\mathsf{h}}_{\mathsf{P}^n}(\mathrm{d}\nu \mathrm{d}x \mathrm{d}y) \\
&\quad = \iiint_{\Gamma\times X^2} f(\nu, y, x)\, \vartheta^{\mathsf{h}}_{\mathsf{P}}(\mathrm{d}\nu \mathrm{d}x \mathrm{d}y) = \iiint_{\Gamma\times X^2} f(\nu, x, y)\,\big(\mathsf{T}_{\#}\vartheta^{\mathsf{h}}_{\mathsf{P}}\big)(\mathrm{d}\nu \mathrm{d}x \mathrm{d}y).
\end{aligned}$$

By the representation of the fluxes $\mathsf{J}^{n,\mathsf{bd}}$ and $\mathsf{J}^{n,h}$ as in Theorem 3.8, we obtain for all $t \in [0, T]$

$$\mathsf{J}^{n,\mathsf{bd}}_{\mathsf{P}^n_t} \rightharpoonup^* \mathsf{J}^{\mathsf{bd}}_{\mathsf{P}} := \vartheta^{\mathsf{b}}_{\mathsf{P}_t} - \vartheta^{\mathsf{d}}_{\mathsf{P}_t}, \qquad \mathsf{J}^{n,\mathsf{h}}_{\mathsf{P}^n_t} \rightharpoonup^* \mathsf{J}^{\mathsf{h}} := \frac{1}{2}\left(\vartheta^{\mathsf{h}}_{\mathsf{P}_t} - \mathsf{T}_{\#}\vartheta^{\mathsf{h}}_{\mathsf{P}_t}\right).$$

Recall from Remark 4.3 that for any $F \in \mathrm{Cyl}(\Gamma)$ the mass $\|\nu\|$ is bounded on the support of F and in particular $F \in C_c(\Gamma)$. It is then clear that $\nabla_\Gamma F(\nu, x) \in C_c(\Gamma \times X)$, $\overline{\nabla}_\Gamma F(\nu, x, y) \in C_c(\Gamma \times X^2)$, and via Taylor expansion as mentioned in Sect. 4.1, we can find a constant c_F such that for all $\nu \in \Gamma$ and $x, y \in X$,

$$\left|\nabla_\Gamma F(\nu, x) - \overline{\nabla}_{n,\mathsf{bd}} F(\nu, x)\right| + \left|\overline{\nabla}_\Gamma F(\nu, x, y) - \overline{\nabla}_{n,\mathsf{h}} F(\nu, x, y)\right| \le c_F n^{-1}.$$

Therefore, one can pass to the limit in the continuity equation (16) to conclude that $(\mathsf{P}, \mathsf{J}^{\mathsf{bd}}, \mathsf{J}^{\mathsf{h}}) \in \mathsf{CE}_\infty$.

We now turn to the lim inf inequality in (2). Assume that the convergences in (1) hold. We then establish separately lim inf inequalities for all the components of the energy-dissipation functional, i.e.,

$$\liminf_{n\to\infty} \int_0^T \mathsf{R}_n(\mathsf{P}^n_t, \mathsf{J}^n_t)\mathrm{d}t \ge \int_0^T \mathsf{R}_\infty(\mathsf{P}_t, \mathsf{J}_t)\mathrm{d}t, \liminf_{n\to\infty} \int_0^T \mathsf{D}_n(\mathsf{P}^n_t)\mathrm{d}t \ge \int_0^T \mathsf{D}_\infty(\mathsf{P}_t)\mathrm{d}t,$$

$$\liminf_{n\to\infty} \mathsf{E}_n(\mathsf{P}^n_t) \ge \mathsf{E}_\infty(\mathsf{P}_t) \qquad \text{for all } t \in [0, T].$$

Note that the lim inf inequality for E_n follows from Proposition 2.2. As for the dissipation potential, recall that R_n is the sum of two components (we omit the subscript t for now):

$$\mathsf{R}_{n,\mathsf{bd}}(\mathsf{P}^n, \mathsf{J}^{n,\mathsf{bd}}) = 2 \iint_{\Gamma\times X} \Upsilon(\mathsf{J}^{n,\mathsf{bd}}/2, \vartheta^{n,\mathsf{b}}_{\mathsf{P}^n}, \mathsf{T}^{n,-}_{\#}\vartheta^{n,\mathsf{d}}_{\mathsf{P}^n}),$$

$$\mathsf{R}_{n,\mathsf{h}}(\mathsf{P}^n, \mathsf{J}^{n,\mathsf{h}}) = \iiint_{\Gamma\times X^2} \Upsilon(\mathsf{J}^{n,\mathsf{h}}, \vartheta^{n,\mathsf{h}}_{\mathsf{P}^n}, \mathsf{T}^{n,\mathsf{h}}_{\#}\vartheta^{n,\mathsf{h}}_{\mathsf{P}^n}).$$

Due to the convergences in (21) and convergence of the fluxes, we can use the convexity and lower semicontinuity of Υ and a standard relaxation of integral functionals argument (see, e.g., [5, Theorem 3.4.3]) to conclude that

$$\liminf_{n\to\infty} \mathsf{R}_{n,\mathsf{bd}}(\mathsf{P}^n, \mathsf{J}^{n,\mathsf{bd}}) \ge 2 \iint_{\Gamma\times X} \Upsilon(\mathsf{J}^{\mathsf{bd}}/2, \vartheta^{\mathsf{b}}_{\mathsf{P}}, \vartheta^{\mathsf{d}}_{\mathsf{P}}),$$

$$\liminf_{n\to\infty} \mathsf{R}_{n,\mathsf{h}}(\mathsf{P}^n, \mathsf{J}^{\mathsf{h}}) \ge \iiint_{\Gamma\times X^2} \Upsilon(\mathsf{J}^{\mathsf{h}}, \vartheta^{\mathsf{h}}_{\mathsf{P}}, \mathsf{T}_{\#}\vartheta^{\mathsf{h}}_{\mathsf{P}}).$$

The lim inf inequality for the Fisher information follows similarly, from the convexity and lower semicontinuity of the squared Hellinger distance.

Finally, to establish (3), note that $\mathscr{I}_n([0, T]; \mathsf{P}^n, \mathsf{J}^n) = 0$. By the lower semicontinuity results of (2) we find that $\mathscr{I}_\infty([0, T]; \mathsf{P}, \mathsf{J}) \le 0$ and conclude by

Theorem 4.10 that (P, J) is the unique generalized gradient flow solution of the Liouville equation (LiE) with initial datum P_{in}. □

With Theorem 4.11 in hand, the main result of this chapter follows after classical compactness and lower semicontinuity argument (c.f. [42, 43]).

Proof of Theorem A Since all converging subsequences of $(\mathsf{P}^n, \mathsf{J}^n)$ converge to the unique generalized flow of the Liouville equation, we obtain by compactness the convergence of the original sequence $(\mathsf{P}^n, \mathsf{J}^n)$. Moreover, from the convergence $\mathsf{E}_n(\mathsf{P}^n_{\text{in}}) \to \mathsf{E}_\infty(\mathsf{P}_{\text{in}})$ and our previously established lower semicontinuity results, we find that for every $t \in [0, T]$,

$$\begin{aligned}\limsup_{n\to\infty} \mathsf{E}_n(\mathsf{P}^n_t) &\leq \lim_{n\to\infty} \mathsf{E}_n(\mathsf{P}^n_{\text{in}}) - \liminf_{n\to\infty} \int_0^t \left(\mathsf{R}_n(\mathsf{P}^n_r, \mathsf{J}^n_r) + \mathsf{D}_n(\mathsf{P}^n_r)\right) dr \\ &\leq \mathsf{E}_\infty(\mathsf{P}_{\text{in}}) - \int_0^t (\mathsf{R}_\infty(\mathsf{P}_r, \mathsf{J}_r) + \mathsf{D}_\infty(\mathsf{P}_r))\, dr = \mathsf{E}_\infty(\mathsf{P}_t) \\ &\leq \liminf_{n\to\infty} \mathsf{E}_n(\mathsf{P}^n_t),\end{aligned}$$

and hence we obtain convergence of the energies E_n. Substituting $\mathsf{P}^n_{\text{in}} = \delta_{\bar{\nu}}$ and noting that Theorem 4.10 ensures that $\mathsf{P}_t = \nu_t$ with ν being the unique solution to (MFE), the convergence of $\mathsf{E}_n(\mathsf{P}^n_t)$ to $\mathsf{E}(\mathsf{P}_t) = \mathscr{E}(\nu_t)$ follows. Using the uniform upper and lower bounds of u_t, one can follow the arguments in the proof of [24, Theorem 5.4] to transform the latter convergence into the desired entropic propagation of chaos result, i.e.,

$$\lim_{n\to\infty} \frac{1}{n} \mathsf{Ent}(\mathsf{P}^n_t | \Pi^n_{\nu_t}) = 0 \qquad \text{for every } t \in [0, T].$$

□

Acknowledgments The authors acknowledge support from NWO Vidi grant 016.Vidi.189.102 on "Dynamical-Variational Transport Costs and Application to Variational Evolution." A.H. acknowledges support from ERC Advanced Grant on "Everything You Always Wanted to Know About the JKO Scheme."

References

1. L. Ambrosio, D. Trevisan, Well-posedness of Lagrangian flows and continuity equations in metric measure spaces. Anal. PDE **7**(5), 1179–1234 (2014)
2. V. Bansaye, S. Méléard, Stochastic models for structured populations, in *Mathematical Biosciences Institute Lecture Series. Stochastics in Biological Systems*, vol. 1.4 (Springer, Cham; MBI Mathematical Biosciences Institute, Ohio State University, Columbus, OH, 2015). Scaling limits and long time behavior.
3. C. Bechinger, R. Di Leonardo, H. Löwen, C. Reichhardt, G. Volpe, G. Volpe, Active particles in complex and crowded environments. Rev. Mod. Phys. **88**, 045006 (2016)
4. M. Begon, C.R. Townsend, *Ecology: From Individuals to Ecosystems* (Wiley, New York, 2021)

5. G. Buttazzo, *Semicontinuity, Relaxation and Integral Representation in the Calculus of Variations*. Pitman Research Notes in Mathematics Series (Longman Scientific and Technical, Harlow, 1998)
6. L.-P. Chaintron, A. Diez, Propagation of chaos: A review of models, methods and applications. I. Models and methods. Kinet. Rel. Models **15**(6), 895–1015 (2022)
7. N. Champagnat, R. Ferrière, S. Méléard, Unifying evolutionary dynamics: from individual stochastic processes to macroscopic models. Theor. Popul. Biol. **69**(3), 297–321 (2006)
8. S.J. Cornell, Y.F. Suprunenko, D. Finkelshtein, P. Somervuo, O. Ovaskainen, A unified framework for analysis of individual-based models in ecology and beyond. Nat. Commun. **10**(1), 4716 (2019)
9. P. Dondl, T. Frenzel, A. Mielke, A gradient system with a wiggly energy and relaxed EDP-convergence. ESAIM Control Optim. Calc. Var. **25**(Paper No. 68), 45 (2019)
10. P. Dondl, T. Frenzel, A. Mielke, A gradient system with a wiggly energy and relaxed edp-convergence. ESAIM. Control. Optim. Calc. Var. **25**, 68 (2019)
11. M. Erbar, A gradient flow approach to the Boltzmann equation. J. Eur. Math. Soc. **26**(11), 4441–4490 (2023)
12. M. Erbar, M. Fathi, V. Laschos, A. Schlichting, Gradient flow structure for McKean-Vlasov equations on discrete spaces. Discrete Contin. Dyn. Syst. **36**(12), 6799–6833 (2016)
13. D. Finkelshtein, Y. Kondratiev, O. Kutoviy, Individual based model with competition in spatial ecology. SIAM J. Math. Anal. **41**(1), 297–317 (2009)
14. D. Finkelshtein, Y. Kondratiev, O. Kutoviy, Vlasov scaling for stochastic dynamics of continuous systems. J. Stat. Phys. **141**, 158–178 (2010)
15. D. Finkelshtein, Y. Kondratiev, O. Kutoviy, Vlasov scaling for stochastic dynamics of continuous systems. J. Stat. Phys. **141**(1), 158–178 (2010)
16. D. Finkelshtein, Y. Kondratiev, P. Tkachov, Doubly nonlocal Fisher–KPP equation: Speeds and uniqueness of traveling waves. J. Math. Anal. Appl. **475**(1), 94–122 (2019)
17. D.L. Finkelshtein, Y.G. Kondratiev, M.J. Oliveira, Markov evolutions and hierarchical equations in the continuum. II: Multicomponent systems. Rep. Math. Phys. **71**(1), 123–148 (2013)
18. N. Fournier, S. Méléard, A microscopic probabilistic description of a locally regulated population and macroscopic approximations. Ann. Appl. Probab. **14**(4), 1880–1919 (2004)
19. V. Grimm, S.F. Railsback, *Individual-Based Modeling and Ecology*. Princeton Series in Theoretical and Computational Biology (Princeton University Press, Princeton, NJ, 2005)
20. M. Grmela, Multiscale equilibrium and nonequilibrium thermodynamics in chemical engineering, in *Advances in Chemical Engineering*, vol. 39 (Elsevier, Amsterdam, 2010), pp. 75–129
21. J. Hoeksema, *Mean-field limits and beyond: Large deviations for singular interacting diffusions and variational convergence for population dynamics*. Phd thesis 1 (research tu/e / graduation tu/e) (Mathematics and Computer Science, Proefschrift, 2023)
22. J. Hoeksema, *Mean-field limits and beyond: Large deviations for singular interacting diffusions and variational convergence for population dynamics*. PhD thesis (Eindhoven University of Technology, Netherlands, 2023)
23. J. Hoeksema, T. Holding, M. Maurelli, O. Tse, Large deviations for singularly interacting diffusions. arXiv preprint arXiv:2002.01295 (2020)
24. J. Hoeksema, O. Tse, Generalized gradient structures for measure-valued population dynamics and their large-population limit. Calc. Var. Partial Differential Equations **62**(5), Paper No. 158, 72 (2023)
25. A.R. Kansal, S. Torquato, G. Harsh Iv, E. Chiocca, T. Deisboeck, Simulated brain tumor growth dynamics using a three-dimensional cellular automaton. J. Theor. Biol. **203**(4), 367–382 (2000)
26. W.O. Kermack, A.G. McKendrick, A contribution to the mathematical theory of epidemics. Proc. R. Soc. London, Ser. A, Containing Papers of a Mathematical and Physical Character **115**(772), 700–721 (1927)
27. Y.G. Kondratiev, E.W. Lytvynov, G.F. Us, Analysis and geometry on marked configuration space. Methods Funct. Anal. Topology **5**(1), 29–64 (1999)
28. M. Lachowicz, General population systems. Macroscopic limit of a class of stochastic semigroups. J. Math. Anal. Appl. **307**(2), 585–605 (2005)

29. M. Liero, A. Mielke, M.A. Peletier, D.R.M. Renger, On microscopic origins of generalized gradient structures. Discrete Contin. Dyn. Syst. Ser. S **10**(1), 1–35 (2017)
30. J. Maas, A. Mielke, Modeling of chemical reaction systems with detailed balance using gradient structures. J. Stat. Phys. **181**(6), 2257–2303 (2020)
31. M. Mariani, A Gamma-convergence approach to large deviations. arXiv preprint arXiv:1204.0640 (2012)
32. R.H. Martin, Jr., Nonlinear operators and differential equations in Banach spaces, in *Pure and Applied Mathematics* (Wiley-Interscience [Wiley], New York-London-Sydney, 1976)
33. A. Mielke, Differential, energetic, and metric formulations for rate-independent processes, in *Nonlinear PDE's and applications*. Lecture Notes in Mathematics, vol. 2028 (Springer, Heidelberg, 2011), pp. 87–170
34. A. Mielke, On evolutionary Γ-convergence for gradient systems, in *Macroscopic and Large Scale Phenomena: Coarse Graining, Mean Field Limits and Ergodicity* (Springer, Berlin, 2016), pp. 187–249
35. A. Mielke, A. Montefusco, M.A. Peletier, Exploring families of energy-dissipation landscapes via tilting: three types of EDP convergence. Contin. Mech. Thermodyn. **33**(3), 611–637 (2021)
36. A. Mielke, R.I.A. Patterson, M.A. Peletier, D.R.M. Renger, Non-equilibrium thermodynamical principles for chemical reactions with mass-action kinetics. SIAM J. Appl. Math. **77**(4), 1562–1585 (2017)
37. A. Mielke, M.A. Peletier, D.R.M. Renger, On the relation between gradient flows and the large-deviation principle, with applications to Markov chains and diffusion. Potential Anal. **41**(4), 1293–1327 (2014)
38. A. Mielke, R. Rossi, G. Savaré, Variational convergence of gradient flows and rate-independent evolutions in metric spaces. Milan J. Math. **80**(2), 381–410 (2012)
39. M.A. Peletier, R. Rossi, G. Savaré, O. Tse, Jump processes as generalized gradient flows. Calc. Var. Partial Differ. Equations **61**(1), Paper No. 33, 85 (2022)
40. M.A. Peletier, A. Schlichting, Cosh gradient systems and tilting. Nonlinear Anal. **231**(Paper No. 113094), 113 (2023)
41. S. Ramaswamy, The mechanics and statistics of active matter. Annu. Rev. Condens. Matter Phys. **1**(2010), 323–345 (2010)
42. E. Sandier, S. Serfaty, Γ-convergence of gradient flows with applications to Ginzburg–Landau. Commun. Pure Appl. Math. **57**(12), 1627–1672 (2004)
43. S. Serfaty, Γ-convergence of gradient flows on Hilbert and metric spaces and applications. Discrete Contin. Dyn. Syst. **31**(4), 1427 (2011)
44. A.-S. Sznitman, Topics in propagation of chaos, in *École d'Été de Probabilités de Saint-Flour XIX-1989*. Lecture Notes in Mathematics, vol. 1464 (Springer, Berlin, 1991), pp. 165–251
45. S.R.S. Varadhan, Large deviations and applications, in *CBMS-NSF Regional Conference Series in Applied Mathematics*, vol. 46 (Society for Industrial and Applied Mathematics (SIAM), Philadelphia, PA, 1984)

Adjoint Monte Carlo Method

Russel Caflisch and Yunan Yang

Abstract This survey explores the development of adjoint Monte Carlo methods for solving optimization problems governed by kinetic equations, a common challenge in areas such as plasma control and device design. These optimization problems are particularly demanding due to the high dimensionality of the phase space and the randomness in evaluating the objective functional, a consequence of using a forward Monte Carlo solver. To overcome these difficulties, a range of "adjoint Monte Carlo methods" have been devised. These methods skillfully combine Monte Carlo gradient estimators with PDE-constrained optimization, introducing innovative solutions tailored for kinetic applications. In this review, we begin by examining three primary strategies for Monte Carlo gradient estimation: the score function approach, the reparameterization trick, and the coupling method. We also delve into the adjoint-state method, an essential element in PDE-constrained optimization. Focusing on applications in the radiative transfer equation and the nonlinear Boltzmann equation, we provide a comprehensive guide on how to integrate Monte Carlo gradient techniques within both the optimize-then-discretize and the discretize-then-optimize frameworks from PDE-constrained optimization. This approach leads to the formulation of effective adjoint Monte Carlo methods, enabling efficient gradient estimation in complex, high-dimensional optimization problems.

R. Caflisch
Courant Institute of Mathematical Sciences, New York University, New York, NY, USA
e-mail: caflisch@courant.nyu.edu

Y. Yang (✉)
Department of Mathematics, Cornell University, Ithaca, NY, USA
e-mail: yunan.yang@cornell.edu

J. A. Carrillo, E. Tadmor (eds.), *Active Particles, Volume 4*, Modeling and Simulation in Science, Engineering and Technology,
https://doi.org/10.1007/978-3-031-73423-6_9

1 Introduction

The Monte Carlo method, as we know it today, was developed by John von Neumann and Stanisław Ulam during World War II [18, 42, 43, 57], aiming to enhance decision-making in situations characterized by uncertainty. Its name is inspired by the famous casino town of Monaco, reflecting the method's fundamental reliance on the concept of randomness, akin to a roulette game. The Monte Carlo method is renowned for its versatility and efficacy in diverse domains ranging from physics to finance and engineering to artificial intelligence. At its core, the Monte Carlo method involves the use of random sampling to approximate solutions to problems that might be deterministic in principle but are practically infeasible to solve directly.

In many optimization problems arising from a wide array of applications where the Monte Carlo method is most effective, there is a strong need to compute the derivative of an objective function with respect to its parameters. This is crucial if one uses local optimization algorithms such as gradient descent and Newton's method to find the optimizer with fewer objective function evaluations [47]. A significant difficulty of directly computing the gradient is that the objective function is approximated by the Monte Carlo method, suffering from intrinsic randomness. Handling the randomness in the objective function gives rise to the field of Monte Carlo gradient estimation.

In computational mathematics, the technique of Monte Carlo gradient estimation emerges as a critical tool, leveraging stochastic processes to tackle complex optimization challenges [23, 24, 44, 52]. Building on the foundational Monte Carlo method, this approach broadens its application to gradient estimation. In particular, Monte Carlo gradient estimation excels in approximating the gradients of expectations, a task often impractical or impossible due to the complexity of the underlying expectations. This method skillfully overcomes such barriers by sampling from relevant distributions and using these samples for gradient estimation.

There are three primary strategies employed in Monte Carlo gradient estimators:

(i) The likelihood ratio method [24], also known as the score function method [52]
(ii) The pathwise derivative method [22, 23], sometimes referred to as the reparameterization trick [33]
(iii) The coupling method [44, 50]

Each of these strategies possesses distinct characteristics and is suited to specific problem contexts. We will revisit these strategies comprehensively in Sect. 2, exploring their unique properties and applicability.

In this survey, our objective is to explore the application of Monte Carlo gradient estimation techniques in addressing optimization problems that emerge from kinetic theory. Notably, the Monte Carlo method has been a prominent choice for numerically solving kinetic equations, especially those with state variables existing in a seven-dimensional phase space [12]. Alongside the development of mesh-based numerical methods, there has been significant interest in the direct simulation Monte Carlo (DSMC) approach and its various extensions [6, 8, 45, 48, 49].

When dealing with design and control tasks modeled by kinetic equations, such as plasma control and devise design [19], the challenge involves mathematically resolving a kinetic partial differential equation (PDE)-constrained optimization problem. This classical approach typically requires adjusting parameters to align with reference data or to achieve specific desired properties, effectively minimizing an objective functional [29]. The iterative update process in these problems involves the computation of gradients, requiring the solution of two PDEs: the original forward equation and its corresponding adjoint PDE, based on the adjoint-state method. In particular, the gradient is derived from either the optimize-then-discretize (OTD) or the discretize-then-optimize (DTO) frameworks (see Sect. 3). However, this may lead to prohibitive computational costs, particularly in high-dimensional PDEs.

To reduce the computational costs, one can consider using Monte Carlo solvers for gradient-based PDE-constrained optimizations. While there are established Monte Carlo methods for solving the complex forward kinetic PDE, such as the Boltzmann equation and the Vlasov–Poisson equation, a key research question arises:

How to effectively solve the corresponding adjoint equation using a Monte Carlo method?

The primary challenge in developing a Monte Carlo numerical solver for the adjoint equation lies in reconciling the Monte Carlo approach with the computation of gradients. In essence, the Monte Carlo forward solver computes the weak form solution of a given PDE, which is a weighted linear combination of Dirac delta measures. However, the gradient calculations for PDE-constrained optimization problems typically involve multiplying two PDE solutions—one forward and one adjoint. Utilizing the same or a similar Monte Carlo solver for the adjoint PDE as used for the forward PDE presents a problem; the gradient requires the multiplication of two sets of Dirac delta measures over the same phase space, which is not well defined since Dirac delta is not technically a function. Therefore, it is crucial to carefully design an appropriate adjoint Monte Carlo method, which also requires a thorough understanding of the nature of adjoint variables. This survey focuses on this exact issue.

In this survey, we present a general framework for the adjoint Monte Carlo method, illustrating its application in two case studies: the radiative transfer equation (RTE) [38] and the Boltzmann equation [11, 59]. The method skillfully combines Monte Carlo gradient estimation techniques, prevalent in mathematical finance and machine learning [23, 44], with established solutions for PDE-constrained optimization [27]. It introduces a novel approach to handling randomness in the discretization and approximation of forward equations, which is beyond the typical study of PDE-constrained optimization, necessitating new theoretical and computational strategies for the adjoint PDE. An overview diagram of this method is also provided in Fig. 1.

We aim to establish a framework for formulating an adjoint Monte Carlo method, showcased through two detailed case studies: the radiative transfer equation

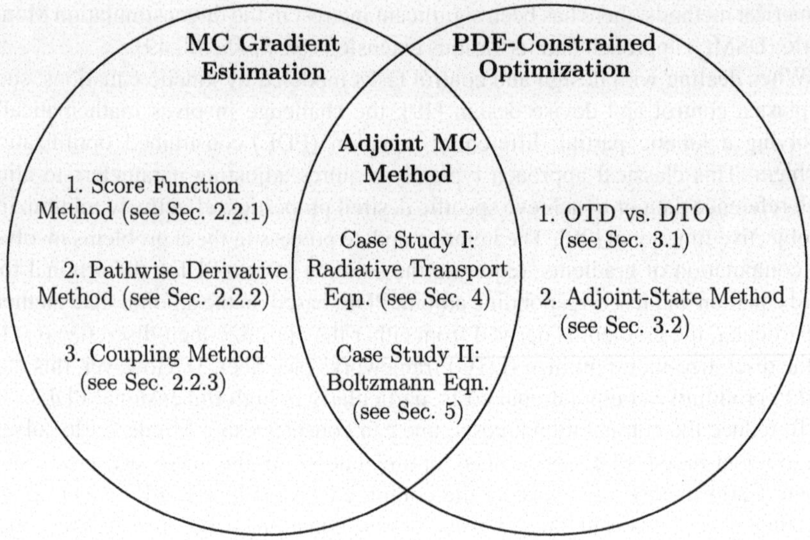

Fig. 1 The diagram illustrates the main structure of this survey chapter. The adjoint Monte Carlo (MC) method integrates effective techniques from Monte Carlo gradient estimation and problem-solving frameworks in PDE-constrained optimization to tackle challenging optimization problems based on various kinetic PDEs

(RTE) [38] and the Boltzmann equation [11, 59]. This exploration will not only address the theoretical aspects but also provide practical insights into its implementation. In a nutshell, the adjoint Monte Carlo method designed for various kinetic PDEs skillfully integrates mature techniques from Monte Carlo gradient estimation, which are widely used in mathematical finance and machine learning [23, 44], with solution frameworks for PDE-constrained optimization problems. Unlike the typical solution procedure in PDE-constrained optimization, the discretization and approximation of the forward (state) equation in the context considered here involve randomness, which requires innovations in both theoretical understandings and computational algorithms for the adjoint PDE. An overview diagram for the adjoint Monte Carlo method is presented in Fig. 1.

Since its first proposal in 2021 [11], the adjoint Monte Carlo method [38, 59] has seen impactful use, notably in topology optimization for rarefied gas flow channel design [25, 60], a spatially inhomogeneous Boltzmann-equation-constrained optimization problem. A key challenge of the adjoint Monte Carlo method is its high memory requirement, which led to the development of a reversible extension for permuted congruential pseudorandom number generators to reduce memory usage [40]. This further influences the work of adjoint sensitivity calculation in plasma edge codes [14]. The adjoint Monte Carlo method for kinetic applications has gathered much research interest in recent years, which warrants the timeliness of this survey.

The rest of the chapter is organized as follows. We first introduce essential mathematical and statistical background in Sect. 2, where we will present an overview of the field of the Monte Carlo gradient and its mainstream strategies. In Sect. 3, we switch gears to revisit PDE-constrained optimization problems and the general mathematical frameworks to tackle them. In particular, we re-derive the adjoint-state method, a key ingredient to guarantee the scalability of PDE-constrained optimization for real-world applications. The central theme of this survey chapter is to showcase how one can combine the Monte Carlo gradient estimation methods in Sect. 2 with strategies that solve the PDE-constrained optimization problem in Sect. 3, two separate research topics, together, and innovate a new class of method, termed as the "adjoint Monte Carlo method," for solving optimization problems regarding kinetic PDEs. In Sect. 4, we give an example of the radiative transport equation (RTE), a linear mean-field equation. To compute the derivative of the objective functional constrained by the RTE, we show how to derive and utilize the adjoint Monte Carlo method based on the OTD and DTO approaches, respectively. In Sect. 5, we look closely at the Boltzmann equation, a nonlinear integrodifferential equation but not a mean-field equation. The binary collision in the Boltzmann equation requires careful treatment, and we obtain the adjoint Monte Carlo method in the DTO approach. With the two examples, we have outlined the steps to construct an adjoint Monte Carlo scheme. Conclusions and more discussions follow in Sect. 6.

2 Monte Carlo Gradient Estimation

The term "Monte Carlo gradient" refers to the estimation of the gradient of a given objective function using Monte Carlo methods. Consider a general objective function in the form of

$$J(\theta) = \int f(x)p(x,\theta)\mathrm{d}x = \mathbb{E}_{X\sim p(x,\theta)}\left[f(X)\right], \tag{1}$$

where $f(x)$ is an observable function, assumed to be independent of θ, and $p(x,\theta)$ is the input density function corresponding to a given distribution that depends on the parameter θ. The optimization problem aims to find an optimal parameter θ that minimizes the objective function $J(\theta)$. If one wants to use optimization algorithms such as steepest descent, conjugate gradient descent, and Newton's method [47], it is essential to have a computationally feasible way of obtaining the gradient of $J(\theta)$.

Often, we do not have direct access to the analytic form of the density function $p(x,\theta)$, or it is challenging to use other deterministic quadrature rules to approximate the integral in (1). In such cases, the objective function (1) can be approximated through the Monte Carlo integration:

$$J(\theta) \approx \widehat{J}(\theta) = \frac{\rho}{N}\sum_{i=1}^{N} f(X_i), \quad X_1,\ldots,X_N \sim p(x,\theta), \tag{2}$$

where $X_1, \ldots, X_N$ are independent and identically distributed (i.i.d.) samples and $\rho = \int p(x, \theta)\mathrm{d}x$ is the total mass and is often independent with respect to the parameter θ. For probability distributions, we always have $\rho = 1$. We remark that $\widehat{J}(\theta)$ given in (2) is an unbiased estimator as it satisfies

$$\mathbb{E}\left[\widehat{J}(\theta)\right] = J(\theta),$$

where the expectation is taken with respect to all the i.i.d. samples $X_1, \ldots, X_N$.

Based on the central limit theorem, we have [13]

$$J(\theta) - \widehat{J}(\theta) \approx \sigma(\theta) N^{-1/2} \mathbf{r},$$

where the random variable $\mathbf{r} \sim N(0, 1)$, the 1D standard normal distribution, and

$$\sigma(\theta) = \sqrt{\int (f - J(\theta))^2 \, p(x, \theta)\mathrm{d}x}$$

is the standard deviation, depending on the parameter θ.

In this survey chapter, the problem of interest is not to estimate $J(\theta)$ but instead to evaluate (or approximate) the gradient of $J(\theta)$ with respect to the parameter θ:

$$G(\theta) := \nabla_\theta J(\theta). \tag{3}$$

There are two main scenarios where the Monte Carlo gradient estimation is useful. First, in some contexts, especially in high-dimensional spaces or complex models, it is challenging or computationally expensive to directly compute the gradient $G(\theta)$. Monte Carlo gradient estimation provides a stochastic approach to approximate these gradients. Second, in many cases, the objective function evaluation can be stochastic in nature due to either the computational complexity of directly evaluating $J(\theta)$ or the lack of access to the full distribution function $p(x, \theta)$. In such scenarios, one has to approximate the objective function $J(\theta)$ through certain Monte Carlo methods. As a result, one only has access to the random variable $\widehat{J}(\theta)$ instead of the deterministic value $J(\theta)$. To examine the sensitivity of the approximated random objective function value $\widehat{J}(\theta)$ to the parameter θ, it is then equivalent to examining the sensitivity of the samples:

$$X_1, X_2, \ldots, X_N \sim p(x; \theta)$$

in the Monte Carlo method with respect to the parameter θ in the underlying distribution. We will use $\widehat{G}(\theta)$ to denote the Monte Carlo gradient estimation of $G(\theta)$.

2.1 Overview

Monte Carlo gradient estimation, sometimes referred to as the "differentiable Monte Carlo," is a topic of interest and an active field of research in a wide range of subjects [44]. In reinforcement learning [32], one research topic is to determine the optimal policy that has a maximum reward. The objective is often an expected reward where the reward is a random variable, similar to the form of $J(\theta)$ in (1). The so-called policy gradient, the gradient of this expectation with respect to the policy parameters in reinforcement learning, is often estimated using Monte Carlo methods. The well-known REINFORCE algorithm is a classic example [58], which utilized the score-function method, which we will discuss in detail later in Sect. 2.2.1.

In variational Bayesian inference, the evidence lower bound (ELBO) is a fundamental concept, which is a method used to approximate complex posterior distributions. The goal of variational inference is to maximize the ELBO with respect to the parameters of the variational distribution. This is equivalent to minimizing the divergence between the true posterior and the variational distribution. The variational problem is often optimized using Monte Carlo gradient estimators [33], especially when the model involves latent variables. The Monte Carlo gradient estimation provides an unbiased estimate of the true gradient. This makes it a powerful tool for optimizing the ELBO, especially when the exact gradient computation is intractable.

Monte Carlo methods are widely used in mathematical finance for valuing complex financial derivatives and risk management. Some financial derivatives have payoffs that are challenging to value using traditional closed-form solutions or numerical methods like finite-difference methods. Monte Carlo simulation can be used to simulate many possible sample paths of the underlying asset(s) and then average the payoffs across these paths to estimate the derivative's value. The gradient (sensitivity) of this value with respect to model parameters can be estimated using the Monte Carlo gradient, e.g., the "Greeks" of financial instruments with respect to underlying parameters [23, Sec. 7.2].

Monte Carlo methods offer a notable advantage in their natural capacity for parallelization. This characteristic stems from the methods' reliance on numerous i.i.d. random samples to approximate solutions. Since these sampling activities are independent of each other, they can be effectively distributed across various processors or computing nodes. This ability to execute in parallel not only speeds up the computational process but also enhances the scalability and efficiency of Monte Carlo methods, making them particularly suitable for tackling large-scale problems in high-performance computing environments.

A significant challenge in utilizing Monte Carlo methods for optimization tasks, as previously noted, lies in calculating the gradients of expectations estimated through these methods, inherently random in nature. Accurately determining these gradients is essential for optimization in fields such as reinforcement learning, Bayesian inference, and simulation-based design. The stochastic nature of Monte

Carlo gradient estimation means that the estimated gradient can have high variance. Techniques like quasi Monte Carlo method [13], importance sampling [54], antithetic variates [36], or control variates [53] can be used to reduce this variance. Depending on the sampling method, there might be bias in the gradient estimates, i.e., $\mathbb{E}[\widehat{G}(\theta)] \neq G(\theta)$. It is crucial to ensure that the estimator is unbiased or that the bias is controlled and understood. This topic has attracted much research endeavor [16, 41, 46, 55].

2.2 Methods

In the following subsections, we will introduce three primary methods of Monte Carlo gradient estimation, each widely employed across diverse fields.

2.2.1 Likelihood Ratio Method/Score Function Method

The Monte Carlo gradient estimator based on the score function method stands as one of the most versatile gradient estimators. It has a wide range of applications. It is known as the score function estimator in [34, 52] and referred to as the likelihood ratio method in [24]. The method is also called the REINFORCE estimator in reinforcement learning [58]. Next, we will briefly introduce its derivation and comment on the properties.

Consider the parameterized density function $p(x, \theta)$. The score function is the derivative of the logarithm of $p(x, \theta)$, i.e.,

$$\nabla_\theta \log p(x, \theta)\,, \tag{4}$$

and it satisfies the relationship

$$\nabla_\theta p(x, \theta) = p(x, \theta) \nabla_\theta \log p(x, \theta)\,. \tag{5}$$

If X is a random variable with its distribution characterized by the density function $p(x, \theta)$, we have another important property regarding the score function:

$$\begin{aligned} \mathbb{E}_{X \sim p(x,\theta)} \left[\nabla_\theta \log p(X, \theta) \right] &= \int \frac{\nabla_\theta p(x, \theta)}{p(x, \theta)} p(x, \theta) \mathrm{d}x \\ &= \nabla_\theta \int p(x, \theta) \mathrm{d}x = \nabla_\theta 1 = 0. \end{aligned} \tag{6}$$

That is, as a random variable, $\nabla_\theta \log p(X, \theta)$, has mean zero. If $\theta \in \mathbb{R}^m$, $\nabla_\theta \log p(X, \theta)$ is a vector-valued random variable. We proceed to compute its covariance matrix $C(\theta)$. The ijth entry of $C(\theta)$ is

$$\begin{aligned} C_{ij}(\theta) &= \int \frac{\partial \log p(x,\theta)}{\partial \theta_i} \frac{\partial \log p(x,\theta)}{\partial \theta_j} p(x,\theta)\mathrm{d}x \\ &= -\mathbb{E}_{X\sim p(x,\theta)}\left[\frac{\partial^2}{\partial \theta_i \partial \theta_j} \log p(X,\theta)\right]. \end{aligned} \tag{7}$$

The last term is the so-called Fisher information. The second equality holds because

$$\nabla_\theta \left(\mathbb{E}_{X\sim p(x,\theta)}\left[\nabla_\theta \log p(X,\theta)\right]\right) = 0\,,$$

as a result of (6).

The score function plays a crucial role in statistical estimation and machine learning, serving as a foundational tool for parameter estimation and model evaluation. It is essentially the gradient of the log-likelihood function with respect to the parameters, providing critical information about how the likelihood changes with respect to these parameters. The importance of the score function lies in its ability to identify the most probable parameters that best fit the data, thereby maximizing the likelihood. In complex models, the score function guides optimization algorithms toward the best parameter values, significantly impacting the accuracy and efficiency of model fitting. Additionally, it is instrumental in constructing various statistical tests and confidence intervals, making it a key component in inferential statistics and model diagnostics.

Given the properties of the score function, we next describe how to use it to estimate the gradient defined in (3). We first reformulate $G(\theta)$ as follows:

$$\begin{aligned} G(\theta) &= \nabla_\theta \left(\int f(x)p(x,\theta)\mathrm{d}x\right) \\ &= \int f(x)\nabla_\theta p(x,\theta)\mathrm{d}x \\ &= \int f(x)(\nabla_\theta \log p(x,\theta))\, p(x,\theta)\mathrm{d}x \\ &= \mathbb{E}_{X\sim p(x,\theta)}\left[f(X)\nabla_\theta \log p(X,\theta)\right], \end{aligned} \tag{8}$$

where we used the identity in (5). As a result, to generate an estimation of $G(\theta)$, denoted by $\widehat{G}$, we can use Monte Carlo integration to approximate the last term in (8):

$$\widehat{G} \approx \frac{1}{N}\sum_{i=1}^{N} f(X_i)\, \nabla_\theta \log p(X_i,\theta)\,, \tag{9}$$

where $X_1, \ldots, X_N$ are i.i.d. samples following the distribution $p(x,\theta)$. Here, we assume that $\rho = \int p(x,\theta)\mathrm{d}x = 1$. The formula outlined in (9) is the score method gradient estimator, one of the most classic techniques for gradient estimation. We can also verify that the estimator $\widehat{G}$ is unbiased:

$$\mathbb{E}_{X_i \sim p(x,\theta),\text{i.i.d.}}[\widehat{G}] = G(\theta)\,.$$

Recall that in the beginning of Sect. 2, we remarked that estimating the Monte Carlo gradient is equivalent to examining the sensitivity of the samples $\{X_i\}$ on the parameter θ. The "score function method" achieves this by first evaluating the score function at those sample positions, i.e., $\nabla_\theta \log p(X_i, \theta)$, and then multiplying the contribution by a weight $f(X_i)$. The final weighted average of the score function values yields the true parameter sensitivity.

Note that we interchanged the order between differentiation and integration in the reformulation of $G(\theta)$ in (8). To justify this step, we need extra conditions on the density function $p(x, \theta)$ and the integrand $f(x)$ [34, 37]. One necessary condition is that $p(x, \theta)$ is continuously differentiable with respect to θ and is also absolutely continuous with respect to θ at the boundary of the support [23, 50].

2.2.2 Pathwise Derivative Method / Reparameterization Trick

The pathwise derivative method [23] is also sometimes referred to as the process derivative [22, 50], stochastic backpropagation [51], and the reparameterization trick [33]. The pathwise derivative method is often used to make the gradient estimation more stable. This method involves reparameterizing the latent variables in a way that allows the gradient to be taken with respect to the parameters of the variational distribution directly rather than through the sampling process.

To begin with, we consider a specific way of sampling the target probability measure μ_θ with density $p(x, \theta)$ where $d\mu(x) = p(x, \theta)dx$. Suppose there exists a parameter θ-independent distribution π such that

$$\mu_\theta = T_\theta \sharp \pi, \qquad T_\theta : \mathbb{R}^n \mapsto \mathbb{R}^n\,. \tag{10}$$

Here, the function T_θ is often referred to as the push-forward map. It turns one measure π to another measure μ and is also mass-preserving:

$$\mu_\theta(A) = \pi\left(T_\theta^{-1}(A)\right)$$

where A is an arbitrary Borel measurable set and $T_\theta^{-1}(A)$ denotes its preimage.

If the relationship (10) is satisfied, we can sample from μ_θ by sampling π. To do so, we first obtain N i.i.d. samples from π, denoted by $Y_1, \ldots, Y_N$, and transform these samples through the deterministic function T_θ:

$$X_i = T_\theta(Y_i), \quad i = 1, \ldots, N\,. \tag{11}$$

Then the resulting samples $X_1, \ldots, X_i, \ldots, X_N \sim \mu_\theta$. If π is also absolutely continuous with respect to the Lebesgue measure in $\mathbb{R}^n$ with the density function $q(x)$, and moreover, T_θ is a bijection, then we have the change of variable formula:

$$q(x) = p\left(T_\theta(x), \theta\right) |\nabla_x T_\theta|,$$

where $|\nabla_x T_\theta|$ represents the determinant of the Jacobian matrix $\nabla_x T_\theta$.

The parameterized push-forward map T_θ is also considered as a path function. For example, when sampling a Gaussian random variable following $N(3, \theta^2)$ with $\theta > 0$, we can first sample from the standard normal distribution $N(0, 1)$, followed by proper translation and dilation through the linear map $T_\theta(x) = \theta\, x + 3$.

A direct consequence of this pathwise sampling is the law of the unconscious statisticians (LOTUS) [20]:

$$\mathbb{E}_{Y\sim\pi}\left[f\left(T_\theta(Y)\right)\right] = \mathbb{E}_{X\sim\mu_\theta}\left[f\left(X\right)\right]. \tag{12}$$

LOTUS tells us that we can perform (Monte Carlo) integration without knowing the underlying probability distribution as long as we are given the corresponding sampling path T_θ and the base distribution π. Using this property, we obtain another reformulation of the gradient G that is different from (8):

$$\begin{aligned} G &= \nabla_\theta \left(\mathbb{E}_{X\sim\mu_\theta}\left[f\left(X\right)\right]\right) \\ &= \nabla_\theta \left(\mathbb{E}_{Y\sim\pi}\left[f\left(T_\theta(Y)\right)\right]\right) \\ &= \mathbb{E}_{Y\sim\pi}\left[\nabla_\theta f\left(T_\theta(Y)\right)\right] \\ &= \mathbb{E}_{Y\sim\pi}\left[\nabla_\theta T_\theta(Y)^\top \nabla_x f(x)\big|_{x=T_\theta(Y)}\right]. \end{aligned} \tag{13}$$

The gradient estimator is then a Monte Carlo approximation of (13), given by

$$\widehat{G} = \frac{1}{N}\sum_{i=1}^{N} \nabla_\theta T_\theta\left(Y_i\right)^\top \nabla_x f\left(T_\theta(Y_i)\right), \quad Y_i \sim \pi\,,\; i = 1, \ldots, N\,. \tag{14}$$

The pathwise gradient estimation is applicable when the function inside the expectation is differentiable with respect to its parameter θ. It involves directly differentiating the simulation, and the formula inherits the structure of a chain rule:

From θ to $T_\theta(X)$ (i.e., $\nabla_\theta T_\theta$), and from $T_\theta(X)$ to $f(T_\theta(X))$ (i.e., $\nabla_x f(x)\,|_{x=T_\theta(X)}$).

Different from the score function method discussed in Sect. 2.2.1, the contribution of the gradient is traced path by path as shown in the diagram in Fig. 2. The dependence of the samples $\{X_i\}_{i=1}^N$ on the parameter θ is evident through the deterministic path function (i.e., the push-forward map) T_θ. This is the fundamental reason why the simple chain rule is applicable here.

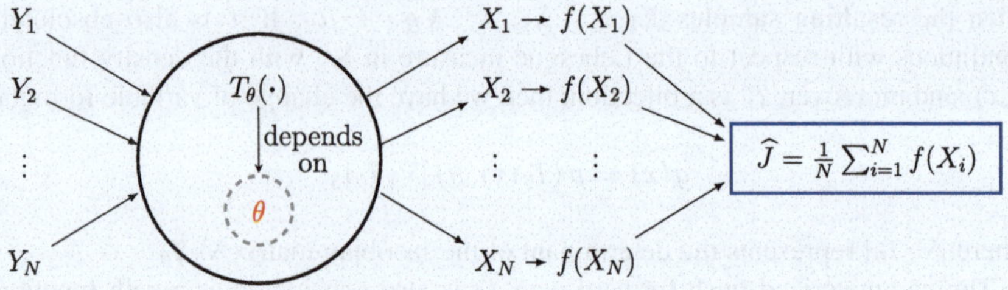

Fig. 2 A diagram showing the dependence of the approximated objective function $\widehat{J}$ on the samples $\{X_i\} \sim \mu_\theta$, which are determined by the parameter θ through the push-forward map $T_\theta(\cdot)$ **path by path** in the "pathwise derivative method" (see Sect. 2.2.2)

The pathwise method is, however, limited to distributions that can be reformulated as (10), the push-forward of a baseline parameter-independent distribution. Moreover, the push-forward map has to be differentiable in terms of the parameters. Often, the process of sampling X from $p(x, \theta)$ and the process of computing the gradient of the samples with respect to the parameter θ, i.e., the computation of the $\nabla_\theta X$, are coupled together. It is possible to decouple these two processes by sampling using a favorite and efficient method and then approximate $\nabla_\theta X$ in another way [44, Sec. 5.3]. Similar to the derivation of the score function estimator, we also used the interchange between differentiation and integration in (13). A sufficient condition requires $f(x)$ to be differentiable. If the interchange is justified, we can again see that the pathwise gradient estimator $\widehat{G}$ given in (14) is unbiased:

$$\mathbb{E}_{Y_i \sim \pi,\text{ i.i.d.}}[\widehat{G}] = G(\theta)\,.$$

2.2.3 Coupling Method

Without loss of generality, we first consider the simple case where θ is a one-dimensional (1D) variable. For a fixed parameter θ, we generate N i.i.d. samples, $X_1, \ldots, X_N$, following $p(x, \theta)$, and then evaluate the Monte Carlo sum $\widehat{J}(\theta)$ following (2). Given a small perturbation $\Delta\theta$ in the parameter θ, we again draw N i.i.d. samples, $\tilde{X}_1, \ldots, \tilde{X}_N$, corresponding to the perturbed density function $p(\tilde{x}, \theta + \Delta\theta)$. We then calculate the new Monte Carlo sum $\widehat{J}(\theta + \Delta\theta)$. In this setup, the sampling procedure of $\{X_i\}$ and $\{\tilde{X}_i\}$ are independent, and we denote the resulting gradient estimation from uncorrelated distributions as $\widehat{G}_{ind}$.

Using the idea of finite-difference approximation to the first-order derivative, we can approximate the gradient $G = \nabla J(\theta) \approx \frac{1}{\Delta\theta}(J(\theta + \Delta\theta) - J(\theta))$ using the Monte Carlo sums:

$$\widehat{G}_{ind} \approx \frac{1}{\Delta\theta}\left(\widehat{J}(\theta + \Delta\theta) - \widehat{J}(\theta)\right) = \frac{1}{N\Delta\theta}\sum_{i=1}^{N}\left(f(\tilde{X}_i) - f(X_i)\right). \tag{15}$$

The variance of $\widehat{G}_{ind}$ can be calculated as follows:

$$\begin{aligned}\mathrm{Var}[\widehat{G}_{ind}] &= \frac{1}{N^2|\Delta\theta|^2}\sum_{i=1}^{N}\left(\mathrm{Var}_{X_i\sim p(x,\theta)}[f(X_i)] + \mathrm{Var}_{\tilde{X}_i\sim p(\tilde{x},\theta+\Delta\theta)}[f(\tilde{X}_i)]\right)\\ &= \frac{1}{N|\Delta\theta|^2}\left(\mathrm{Var}_{X\sim p(x,\theta)}[f(X)] + \mathrm{Var}_{\tilde{X}\sim p(\tilde{x},\theta+\Delta\theta)}[f(\tilde{X})]\right). \end{aligned} \tag{16}$$

Besides the random errors from computing both $\widehat{J}(\theta+\Delta\theta)$ and $\widehat{J}(\theta)$, we address that this gradient estimator is also subject to the finite-difference approximation error $\mathcal{O}(|\Delta\theta|)$. Higher-order difference schemes, such as the central difference method, can be used here. Moreover, reducing the size of $\Delta\theta$ decreases the finite-difference error but, unfortunately, increases the random error as the standard deviation is proportional to $1/|\Delta\theta|$; see Eq. (16). Therefore, to reduce the total error, $|\Delta\theta|$ cannot be chosen to be too large or too small. Certain prior analyses and estimations can be helpful to select a reasonable perturbation $\Delta\theta$; see [11, Sec. 6.1].

The random error not only depends on the size of $\Delta\theta$, but also the concrete sampling schemes for $\{X_i\}$ and $\{\tilde{X}_i\}$. The coupling method in the context of Monte Carlo gradients is a technique used to reduce the variance in the gradient estimation (15) that comes from the two sets of samples. The essential idea behind the coupling method is to *introduce a dependency* between pairs of random variables:

$$X_i \text{ and } \tilde{X}_i, \quad i = 1, \ldots, N,$$

that are used in the Monte Carlo estimation process.

Consider a joint distribution function $P(x, \tilde{x}; \theta, \Delta\theta)$ whose two marginals are $p(x,\theta)$ and $p(\tilde{x}, \theta + \Delta\theta)$, respectively. Let $(X_i, \tilde{X}_i)$ be i.i.d. samples from $P(x, \tilde{x}; \theta, \Delta\theta)$, $i = 1, \ldots, N$. We still use Eq. (15) to estimate the gradient, which is denoted by $\widehat{G}_{cp}$ to highlight the coupled sampling. We proceed to calculate the variance of $\widehat{G}_{cp}$:

$$\begin{aligned}&\mathrm{Var}_{(X_i,\tilde{X}_i)\sim P}[\widehat{G}_{cp}]\\ &= \frac{1}{N^2|\Delta\theta|^2}\sum_{i=1}^{N}\Big(\mathrm{Var}_{X_i\sim p(x,\theta)}[f(X_i)] + \mathrm{Var}_{\tilde{X}_i\sim p(\tilde{x},\theta+\Delta\theta)}[f(\tilde{X}_i)]\\ &\qquad\qquad -2\mathrm{Cov}_{(X_i,\tilde{X}_i)\sim P}[f(X_i), f(\tilde{X}_i)]\Big)\\ &= \mathrm{Var}[\widehat{G}_{ind}] - \frac{2}{N|\Delta\theta|^2}\mathrm{Cov}_{(X,\tilde{X})\sim P}[f(X), f(\tilde{X})]. \end{aligned}$$

Thus, if one can strategically correlate these two sets of random samples such that $\mathrm{Cov}_{(X,\tilde{X})\sim P}[f(X), f(\tilde{X})]$ is as large as possible, the variance of the gradient estimate $\widehat{G}_{cp}$ can be significantly smaller than $\widehat{G}_{ind}$ that is obtained without coupling the two sets of samples. In other words, we want to design correlations so that pairs of samples move together more closely in the sample space, thereby reducing the randomness in the gradient estimation in (15). A simple and practical way to correlate these two sets of samples is to fix the random seed in the random number generator when sampling $\{X_i\}$ and $\{\tilde{X}_i\}$. Other coupling methods exist, and their concrete forms highly depend on the sampling algorithm under consideration.

By reducing the variance of the gradient estimates, the coupling method can lead to more efficient parameter optimization, as it requires fewer samples to achieve a certain level of accuracy. However, it is worth noting that the coupling method does not scale well when the dimension of the parameter θ increases. Consider $\theta \in \mathbb{R}^m$. Then, we need to repeat the procedure in (15) at least m times to obtain the full gradient, as each evaluation following (15) only yields a directional derivative. This involves m different sampling tasks for one gradient evaluation. In contrast, the score function method (Sect. 2.2.1) and the pathwise derivative method (Sect. 2.2.2) require as few as one sampling process to produce a good estimation for the gradient G, especially if combined with the adjoint-state method to be discussed in Sect. 3.2.

We remark that the coupling method can be used to verify the accuracy of the gradient obtained through the adjoint Monte Carlo method [11]. This is because it only requires an efficient sampling algorithm, e.g., forward Monte Carlo solvers for kinetic PDEs, for the target distribution, and does not demand access to the target density $p(x, \theta)$ or the distribution to have a reparameterization structure, which is very hard to ensure when working with applications of kinetic PDEs.

3 PDE-Constrained Optimization and the Adjoint-State Method

Many real-world phenomena are described by PDEs, from fluid dynamics and heat transfer to electromagnetism and elasticity [9]. For example, as the focus of this survey, kinetic equations, which describe the evolution of systems based on their individual particle dynamics, are fundamental in various scientific and engineering disciplines. Optimizing systems described by these phenomena often naturally leads to PDE-constrained optimization problems. Examples include:

- Optimizing the shape of a plasma fusion reactor
- Reconstructing the heat conductivity for a specific material
- Controlling the movements of robots

Most PDE-constrained optimization problems share a common mathematical formulation:

$$\min_{\theta} J(f, \theta) \quad \text{subject to} \quad h(f, \theta) = 0. \tag{17}$$

Here, $J(f, \theta)$ is an objective functional of the state variable f (often the PDE solution) and parameter function θ. The dependence of f on θ is expressed implicitly through the operator h denoting the PDE. Recall that a PDE is an implicit formulation of its corresponding PDE solution. For convenience, we will also denote $f = F(\theta)$ to highlight the dependence between the PDE solution f and the parameter θ. The (nonlinear) operator F is often referred to as the forward operator in inverse problems and the control theory, and the task of solving f given h and θ is regarded as the forward problem.

3.1 OTD and DTO Approaches

When it comes to solving (17), "optimize-then-discretize" (OTD) and "discretize-then-optimize" (DTO) are two fundamental approaches in the context of optimization problems constrained by PDEs. These approaches refer to the sequence in which the optimization problem is tackled with respect to its continuous formulation and discretized counterpart. For the OTD approach, the optimization problem is formulated and solved in its continuous setting. After obtaining the continuous solution (or, more commonly, its optimality condition), the governing PDEs are discretized (e.g., using finite elements, finite differences, and particle methods). For the DTO approach, the governing PDEs and the objective functional are first discretized using proper schemes, which leads to optimization problems with the finite-dimensional unknown and constraints. The optimization is then performed on this discretized problem. We outline these two approaches in a diagram; see Fig. 3.

There have been long-standing discussions on which approach is better [30]. Both OTD and DTO have pros and cons, and the conclusion highly depends on the specific PDE-constrained optimization problem under consideration and the choice of numerical scheme for discretization. OTD often leads to well-structured optimality conditions and can provide insights into the nature of the optimal solution

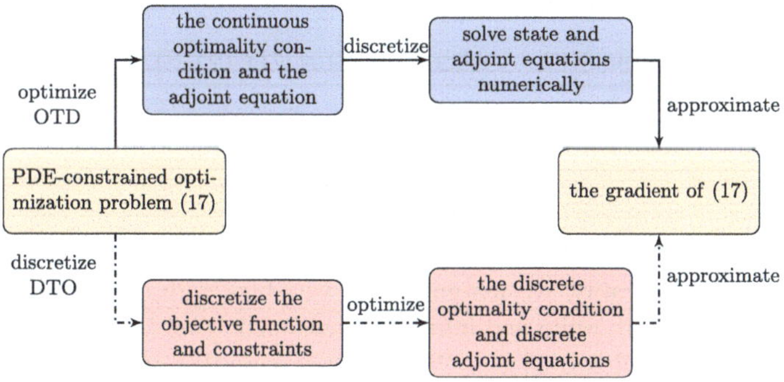

Fig. 3 The OTD approach (solid line) and the DTO approach (dash line) to compute the gradient with respect to the unknown parameter for a general PDE-constrained optimization problem (17)

in terms of analysis. The resulting discrete problem might inherit only some of the properties of the continuous problem. One has to conduct discretization with caution to avoid inconsistency issues that can lead to lower-order methods or even a wrong solution. For the DTO approach, the optimization is performed on a finite-dimensional space, which can be more straightforward and amenable to standard optimization techniques. This approach is often more practical for numerical implementations. The discrete problem might preserve only some of the properties of the continuous problem, as we will see later in Sects. 4–5. The resulting discrete optimization problem can also be large-scale, especially for fine discretizations. The DTO approach is particularly preferred for problems that are otherwise unsolvable under the OTD approach, such as the heat-transfer optimization problem [5, 21]. The OTD approach is very attractive if one wants to utilize the rich geometric structure of the continuous function space [31].

For a given discretization scheme, the gradient obtained from OTD and DTO may not be equivalent, as demonstrated by many examples in the literature [1, 10, 26, 30, 39]. The lack of symmetry in discretization often causes such inconsistency. Examples include the non-Galerkin discretization of the optimality system. Such differences may contribute to (big) errors in optimization for solving nonlinear problems.

3.2 Adjoint-State Method

Both OTD and DTO approaches require deriving the first-order optimality condition, especially the derivative of the (approximated) objective J with respect to the (discretized) parameter θ. Next, we introduce an efficient way to compute the derivative. Under the OTD approach, f and θ are functions, while in the DTO approach, they are both discrete objects as a result of discretization. In the following derivations, we will not distinguish between these two.

Since the PDE constraint h implies that f depends on θ, we will plug in $f = F(\theta)$ into (17). Thus, we deal with the optimization problem:

$$\min_{\theta} \mathfrak{J}(\theta)\,, \quad \text{where} \quad \mathfrak{J}(\theta) = J(F(\theta), \theta)\,. \tag{18}$$

Correspondingly, the PDE constraint becomes

$$h(F(\theta), \theta) = 0. \tag{19}$$

Without loss of generality, we assign all the variables with corresponding linear spaces that can be both finite and infinite dimensional: $f \in U$, $\theta \in V$, $h(f, u) \in W$ and U, V, W are Hilbert spaces. We further assume that operators h and F are smooth enough such that they are Fréchet differentiable.

A first-order variation of h with respect to θ yields

$$h_f(F(\theta), \theta)\, D_\theta F(\theta) + h_\theta(F(\theta), \theta) = 0\,, \tag{20}$$

where h_f and h_θ are Fréchet derivatives of $h(f, \theta)$ with respect to f and θ variables, and $D_\theta F$ is the derivative of $F(\theta)$ with respect to θ. Note that $h_f \in \mathcal{L}(U, W)$, $h_\theta \in \mathcal{L}(V, W)$ and $D_\theta F \in \mathcal{L}(V, U)$, where $\mathcal{L}(A, B)$ denotes the linear operator that maps elements from function space A to function space B.

To calculate the derivative of $\mathfrak{J}$ with respect to θ, we use the chain rule and obtain

$$D_\theta \mathfrak{J} = J_f(f, \theta) D_\theta F(\theta) + J_\theta(f, \theta)\,, \tag{21}$$

where $D_\theta \mathfrak{J} \in V^*$, the dual space of V. Also, with U^* being the dual space of U, $J_f \in U^*$ and $J_\theta \in V^*$ are partial Fréchet derivatives of $J(f, \theta)$ with respect to f and θ variables, respectively.

Using (21) to compute the derivative of $\mathfrak{J}$ requires $D_\theta F$, the linearized operator of the forward map F. Recall that the dependence between θ and f is *implicitly* given through the PDE constraint h. That is, F is defined through (19). As a result, to obtain $D_\theta F$, we have to utilize (20). One may first solve for $D_\theta F(\theta)$ at a particular θ^* through

$$h_f(F(\theta^*), \theta^*)\, D_\theta F(\theta^*) = -h_\theta(F(\theta^*), \theta^*) \tag{22}$$

and then plug the solution into (21). We remark that since the objective function J is the user's choice, J_f and J_θ are usually known and easy to compute based on the form of J. If V, the space of θ, is finite-dimensional, then the cost of this inverse solve (22) increases linearly with respect to the dimension of V (i.e., the intrinsic degrees of freedom in the unknown parameter θ) and can be computationally prohibitive for very-large-scale problems.

Next, we present a more efficient way of obtaining $D_\theta \mathfrak{J}$, called the *adjoint-state method* [9, 29] or the reduced gradient method [28] in the literature. We first derive it using the method of Lagrangian multipliers. Consider the Lagrangian L given by

$$L(f, \lambda, \theta) = J(f, \theta) + \lambda h(f, \theta), \tag{23}$$

where $\lambda \in W^*$, the dual space of W. In the Lagrangian (23), there is no dependence among f, λ and θ, as all of them are independent variables. However, if we plug $f = F(\theta)$ into (23), which is equivalent to restraining the Lagrangian over a particular manifold $\mathcal{M} := \{(f, \theta) : f = F(\theta)\}$, a subset of the full space $U \times V$, we would obtain

$$\begin{aligned} \mathfrak{L}(\lambda, \theta) &:= L(F(\theta), \lambda, \theta) \\ &= J(F(\theta), \theta) + \lambda h(F(\theta), \theta) \\ &= J(F(\theta), \theta) \\ &= \mathfrak{J}(\theta), \end{aligned} \tag{24}$$

where we used (18) and (19). Furthermore, the derivative of $\mathfrak{L}$ with respect to θ yields

$$\mathfrak{L}_\theta(\lambda, \theta) = L_\theta(F(\theta), \lambda, \theta) + L_f(F(\theta), \lambda, \theta)\, D_\theta F(\theta)\,, \tag{25}$$

where $\mathfrak{L}_\theta, L_\theta \in V^*$ and $L_f \in U^*$. We remark that both $\mathfrak{L}$ and $\mathfrak{L}_\theta$ are independent of the variable λ since the last term in (24) is λ-independent.

If we choose a particular $\lambda \in W^*$ such that $L_f(F(\theta), \lambda, \theta) = 0 \in U^*$, then we do not have to involve $D_\theta F(\theta)$ for the calculation of the derivative following the formula (25). **This is the central idea behind the adjoint-state method.** After further simplification for $L_f(F(\theta), \lambda, \theta)$, we obtain

$$L_f(F(\theta), \lambda, \theta) = J_f(F(\theta), \theta) + \lambda\, h_f(F(\theta), \theta))\,.$$

If λ solves the following equation

$$\lambda\, h_f(F(\theta), \theta)) = -J_f(F(\theta), \theta), \tag{26}$$

then λ is θ-dependent and we denote it by $\lambda(\theta)$. Finally, with this particular choice of λ, we have reached a simple form of the derivative following (25):

$$\begin{aligned} D_\theta \mathfrak{J}(\theta) &= D_\theta \mathfrak{L}(F(\theta), \lambda(\theta), \theta) \\ &= L_\theta(F(\theta), \lambda(\theta), \theta) \\ &= J_\theta(F(\theta), \theta) + \lambda(\theta)\, h_\theta(F(\theta), \theta)). \end{aligned} \tag{27}$$

The first equation holds because of (24).

Equation (26) is often referred to as the *adjoint equation*, while solving f based on (19) is referred to as the *state equation*. In the *adjoint-state method*, for a given θ^*, we solve the forward equation once and obtain $f = F(\theta^*)$. It is followed by an adjoint equation solve, which results in λ. Note that λ implicitly depends on θ^* due to Eq. (26). We then obtain the derivative $D_\theta \mathfrak{J}(\theta^*)$ of the constrained problem following the formula (27). If h is a PDE constraint, this amounts to two PDE solves, independent of the dimension of the discretized θ-space V, making it extremely attractive for infinite-dimensional optimization problems. Therefore, the adjoint-state method is widely used for large-scale practical problems with a differentiable equality constraint (19) and a differentiable objective functional J.

Remark 1 We have presented a framework that can apply to both OTD and DTO approaches for a large class of PDE-constrained problems. Note that this requires U (the space for f), V (the space for θ), and W (the space for the range of the equality constraint h) to be vector spaces and, preferably, Hilbert spaces. However, for many kinetic equations, the solution should be regarded as (probability) distribution functions with nonnegativity and total mass conserved. Moreover, for discretization schemes based on particle methods, the relationship between the continuous and discrete variables is not linear, not to mention the further complications originating from the stochasticity in random algorithms (e.g., the Monte Carlo method). In the later sections, we will discuss two specific cases and how to adapt the adjoint-state method to such representative kinetic equation-based optimization problems.

In the following sections, we will study closely two kinetic equation-constrained optimization problems and how to employ and adapt the general framework in Sect. 3 to solve them systemically. Since both the radiative transport equation and the Boltzmann equation can be solved through Monte Carlo particle methods, we will also derive the corresponding adjoint Monte Carlo method for each of the two kinetic equations. Interested readers may refer to [11, 38, 59] for more details in terms of motivation, derivation and convergence proofs.

4 Case Study I: Radiative Transport Equation

The radiative transport equation (RTE), also known as the radiative transfer equation, is a model problem for simulating light propagation in an optical environment. For simplicity, we consider time-dependent RTE with no spatial boundary effect:

$$\begin{cases} \partial_t f + v \cdot \nabla_x f &= \sigma(x)\mathcal{L}[f], \\ f(t=0, x, v) &= f_0(x, v), \end{cases} \qquad x \in \mathbb{R}^{d_x}, \ v \in \Omega. \tag{28}$$

Here, $f(t, x, v)$ is the distribution function of photon particles at time t on the phase space (x, v) and d_x is the dimension for the spatial domain. The left-hand side of the equation describes the photon moving in a straight line in x with velocity v, whereas the right-hand side characterizes the photon particles' interaction with the media characterized by the function $\sigma(x)$. The term $\mathcal{L}[f]$ is defined by

$$\mathcal{L}[f] = \frac{1}{|\Omega|}\langle f \rangle_v - f, \quad \text{with} \quad \langle f \rangle_v = \int_\Omega f(x, v) \mathrm{d}v. \tag{29}$$

The entire right-hand side represents that particles at location x have a probability proportional to $\sigma(x)$ to be scattered, into a new direction uniformly chosen in the v space. As a result, the distribution function in the phase space exhibits a gain on $|\Omega|^{-1}\langle f \rangle_v(x)$ and a loss on $f(x, v)$. We will use $\langle \cdot \rangle_s$ to denote the integration in variable s with respect to the Lebesgue measure.

The forward problem (28), i.e., given σ and solving for f, has a unique solution under very mild conditions on both $\sigma(x)$ and the initial data f_0 [4, 17]. Moreover, the equation preserves mass because $\rho = \langle f \rangle_{xv}$ is a constant in time. Without loss of generality, we set $\rho = 1$. Note that (28) is a linear equation with respect to f.

4.1 Constrained Optimization Problem

We consider a PDE-constrained optimization originated from an inverse problem as an example. Assume we have measurement which is the final-time solution of the RTE:

$$d(x, v) \approx f(t = T, x, v),$$

and the goal is to identify the variable absorption parameter $\sigma(x)$ in the spatial domain given the known initial condition. The objective functional for the corresponding computational inverse problem is the mismatch of the simulated solution and the measured data, and the constraints come from the fact that the simulated solution satisfies a forward RTE. The corresponding PDE-constrained optimization is of the form

$$\min_{\sigma} J\left(f(\cdot;\sigma)\right), \quad \text{s.t. } f(\cdot;\sigma) \text{satisfies (28)}, \tag{30}$$

where $f(\cdot;\sigma)$ is the simulated data that solves (28) with the given initial condition and the absorbing parameter $\sigma(x)$. In this particular application setup, we may consider the objective functional:

$$J\left(f(\cdot;\sigma)\right) = \frac{1}{2}\iint |f(t = T, x, v; \sigma) - d(x, v)|^2 \mathrm{d}x\mathrm{d}v\,.$$

Similar to (19), we can rewrite (28) to match the general form of an equality constraint $h(f, \sigma) = 0$. We then denote $f = F(\sigma)$ with F being the forward operator. Next, we consider all relevant function spaces, U (the space of f), V (the space of σ), W (the space of the constraint) to be L^2 over the respective domain.

4.2 The OTD Approach

In the following, we detail how to employ the OTD approach to design an adjoint Monte Carlo method to tackle (30).

4.2.1 Gradient Derivation

We first apply the method of Lagrange multipliers and consider the Lagrangian following the recipe given in Sect. 3.2:

$$L(f,\lambda,\bar{\lambda},\sigma) = J(f) + \langle \lambda\,, \partial_t f + v\cdot\nabla_x f - \sigma(x)\mathcal{L}[f]\rangle_{txv} \\ +\langle \bar{\lambda}\,, f(t=0,x,v) - f_0(x,v)\rangle_{xv}.$$

Here, functions $\lambda(t,x,v)$ and $\bar{\lambda}(x,v)$ are Lagrange multipliers with respect to the RTE solution $f(t,x,v)$ for $t > 0$ and the initial condition $f(t = 0,x,v)$, respectively.

We remark that the λ and $\bar{\lambda}$ we used here slightly differ from the adjoint variable λ in Sect. 3.2. The former is in the function space W where $h(f,\sigma)$ belongs to, while the latter is considered an element of W^*, i.e., the *dual space* of W. In other words, the λ, $\bar{\lambda}$ used here represent the Riesz representation of the λ in (23) through the Riesz map. Since all the spaces we consider in this subsection are L^2 for the RTE-based constrained optimization, these two coincide, and thus, we will not differentiate between them when referring to the adjoint variable.

With integration by parts, we rewrite the Lagrangian as

$$\begin{aligned} &L(f,\lambda,\bar{\lambda},\sigma) \\ &\quad= J(f) + \langle f\,, -\partial_t\lambda - v\cdot\nabla_x\lambda - \sigma(x)\mathcal{L}[\lambda]\rangle_{txv} \\ &\qquad+ \langle \bar{\lambda}\,, f(t=0,x,v) - f_0(x,v)\rangle_{xv} \\ &\qquad+ \langle \lambda(t=T,x,v)\,, f(t=T,x,v)\rangle_{xv} - \langle \lambda(t=0,x,v)\,, f(t=0,x,v)\rangle_{xv}\,. \end{aligned}$$

Based on (26), we can set the derivative of $L(f,\lambda,\bar{\lambda},\sigma)$ with respect to $f(t,x,v)$ for any $0 \le t \le T$ evaluated at $f = F(\sigma)$ to be zero. We then obtain the adjoint equation for the RTE-constrained optimization problem:

$$\begin{cases} -\partial_t\lambda - v\cdot\nabla_x\lambda &= \sigma(x)\mathcal{L}[\lambda]\,, \\ \lambda(T,x,v) &= -\frac{\delta J}{\delta f(T,x,v)}\,, \end{cases} \tag{31}$$

where the final condition at T comes from the form of J given in (30). Note that the resulting adjoint solution λ is σ-dependent.

Following the general framework in (27), we then have the Fréchet derivative of $\mathfrak{J}(\sigma) := J(F(\sigma))$ as

$$D_\sigma\mathfrak{J}(\sigma) = L_\sigma(F(\sigma),\lambda,\bar{\lambda},\sigma) = -\langle \lambda, \mathcal{L}[f]\rangle_{tv} = \int_0^T \langle \lambda f\rangle_v \mathrm{d}t - \frac{1}{|\Omega|}\int_0^T \langle f\rangle_v\langle\lambda\rangle_v \mathrm{d}t. \tag{32}$$

The derivation above is carried out completely on the function space in the continuous setting, as done in the OTD framework. Here, we consider the function space for the parameter $\sigma(x)$ to be L^2, so the derivative $D_\sigma\mathfrak{J}(\sigma)$ is indeed the L^2 gradient of $\mathfrak{J}$.

4.2.2 Discretization

Upon obtaining (32), the next step is discretization, which involves numerically solving (i) the state equation (28), (ii) the adjoint equation (31), and (iii) a discrete approximation for the gradient (32).

Assume we have a good numerical scheme for the forward state equation (28). It is also quite tempting to solve the adjoint equation (31) using the same scheme, given the great similarity between (28) and (31), which results from the fact that the state equation (28) is linear with respect to f. However, one has to be careful here as we want to construct a consistent numerical scheme for the adjoint equation with respect to the one applied to the forward equation. That is, the *numerical* solution for the adjoint equation (31) is preferably an adjoint of the *numerical* solution for the state equation (28). Failure to preserve the consistency, i.e., the adjoint relationship on the discrete level, may result in a low-order or even wrong gradient approximation [26].

First, we consider the finite-volume method for the state equation (28). We only consider the 1D case for both x and v as an illustrating example. Denote

$$f_{i,j}^m \approx f(t_m, x_i, v_j)\ \lambda_{i,j}^m \approx \lambda(t_m, x_i, v_j)\ , 0 \le m \le M, 1 \le j \le N_v, 1 \le i \le N_x\,,$$

as the numerical approximation, and let Δx, Δt, and Δv be the corresponding mesh size in x, t, and v, respectively. We consider the time domain $[0, T]$, the spatial domain $D = [-2, 2]$, and the velocity domain $\Omega = [-1, 1]$. Then $M\Delta t = T$, $t_m = m\Delta t$, $\Delta x N_x = |D| = 4$, and $\Delta v N_v = |\Omega| = 2$. We have the following discretization for (28):

$$\begin{aligned} f_{i,j}^{m+1} - f_{i,j}^m + \frac{\Delta t}{\Delta x} v_j^+ (f_{i,j}^m - f_{i-1,j}^m) + \frac{\Delta t}{\Delta x} v_j^- (f_{i+1,j}^m - f_{i,j}^m) \\ = \sigma_i \Delta t \left(|\Omega|^{-1} \langle f_{i,j}^m \rangle_v - f_{i,j}^m \right), \end{aligned} \tag{33}$$

for $0 \le m \le M-1$, $1 \le i \le N_x$, and $1 \le j \le N_v$, with the initial condition

$$f_{i,j}^0 = f_0(x_i, v_j)\,.$$

Here, $v_j = -1+(j-1/2)\Delta v$, $j = 0, \ldots, N_v$, representing the cell center, and $v^+ = \max\{v, 0\}$, $v^- = \min\{v, 0\}$. The average in v is computed by a simple midpoint rule: $\langle f_{i,j}^n \rangle_v = \sum_{j=1}^{N_v} f_{i,j}^n \Delta v$. The consistent scheme for λ in the adjoint equation (31) is

$$\begin{aligned} -(\lambda_{i,j}^{m+1} - \lambda_{i,j}^m) - \frac{\Delta t}{\Delta x} v_j^+ (\lambda_{i+1,j}^m - \lambda_{i,j}^m) - \frac{\Delta t}{\Delta x} v_j^- (\lambda_{i,j}^m - \lambda_{i-1,j}^m) \\ = \sigma_i \Delta t (|\Omega|^{-1} \langle \lambda_{i,j}^{m+1} \rangle_v - \lambda_{i,j}^{m+1}), \end{aligned} \tag{34}$$

for $m = M-1, M-2, \ldots, \geq 0$ and $1 \leq j \leq N_v$, with the final condition depending on the objective function J.

Note that v_j^+ is multiplied by $(f_{i,j}^m - f_{i-1,j}^m)$ in (33), but by $(\lambda_{i+1,j}^m - \lambda_{i,j}^m)$ in (34). The right-hand sides of (33) and (34) are also evaluated at different time points. In other words, we cannot simply replace $\{f_{i,j}^m\}$ in (33) with $\{\lambda_{i,j}^m\}$ and flip the sign on the right-hand side as a numerical solver for the adjoint equation (31), despite the fact that the continuous adjoint equation (31) can be derived by applying these steps to the continuous forward equation (28). These are the differences between consistent forward and adjoint numerical schemes and inconsistent ones.

4.2.3 Discretization with Adjoint Particle Method

The RTE in (28) has an efficient Monte Carlo solver with provable convergence [38, Theorem 2.1]. The step-by-step numerical scheme is detailed in Algorithm 1.

Algorithm 1 Monte Carlo Method for Solving the Forward RTE (28)

```
Preparation: N pairs of samples {(x_n^0, v_n^0)}_{n=1}^N from the initial distribution f_0(x, v); the total
    time steps M and the time spacing Δt; and the parameter function σ(x).
for m = 0 to M − 1 do
    Given {(x_n^m, v_n^m)}_{n=1}^N, set x_n^{m+1} = x_n^m + Δt v_n^m, n = 1, ..., N.
    Draw random numbers {p_n^{m+1}}_{n=1}^N from the uniform distribution U([0, 1]).
    if p_n^{m+1} ≥ α_n^{m+1} = exp(−σ(x_n^{m+1})Δt) then v_n^{m+1} = η_n^{m+1} where η_n^{m+1} ~ U(Ω).
    else v_n^{m+1} = v_n^m.
    end if
end for
```

As noted earlier, the adjoint equation (31) can be seen as the forward equation (28) with the sign in time and velocity reversed, which implies that we can also adapt the same Monte Carlo method for solving the adjoint equation. If we follow this approach, an immediate problem will arise as detailed in [38, Section 2.2]: both f and λ are approximated by empirical distributions, i.e., finite sums of delta measures, and consequently the gradient evaluation in (32) will involve the multiplication of two delta measures over the x and v domains, and these two sets of trajectories meet in the (t, x, v) space with probability zero. Therefore, a direct discretization of the adjoint equation (31) using the same Monte Carlo method leads to *severe inconsistency* between the forward and adjoint numerical schemes.

To circumvent this issue, [38] proposed a different adjoint Monte Carlo approach, which is also much more simple than Algorithm 1. First, we observe that, on the continuous level,

$$\partial_t \langle fg \rangle_{xv} \equiv 0.$$

Since f is approximated by an empirical distribution

$$f_N(t, x, v) \approx \frac{1}{N} \sum_{n=1}^{N} \delta\left(x - x_n(t)\right) \delta\left(v - v_n(t)\right), \tag{35}$$

by combining the previous equations we obtain the relation that

$$\sum_{n=1}^{N} \lambda(t_m, x_n^m, v_n^m) = \sum_{n=1}^{N} \lambda(t_{m+1}, x_n^{m+1}, v_n^{m+1}).$$

It is sufficient to require that for $m = M - 1, \ldots, 0$

$$\begin{aligned} \lambda(t_m, x_n^m, v_n^m) &= \lambda(t_{m+1}, x_n^{m+1}, v_n^{m+1}) = \cdots = \lambda(t^M, x_n^M, v_n^M) \\ &= -\frac{\delta J}{\delta f}(T, x_n^M, v_n^M) =: \psi_n. \end{aligned}$$

The final-time condition comes from (31). In other words, we do not even need to "solve" the adjoint equation (31), as $\lambda(t, x, v)$ at time $t = t_m$ is represented by N weighted adjoint particles:

$$\{(x = x_n^m, v = v_n^m, \psi_n)\}_{n=1}^{N}. \tag{36}$$

Each particle is located at $x = x_n^m$ and $v = v_n^m$, with weight ψ_n. We can perform numerical interpolation and obtain a function approximating $\lambda(t, x, v)$ with certain assumptions on the smoothness of $\lambda(t, x, v)$. We plot two trajectories with indices n and $n + 1$ in Fig. 4 as examples, illustrating that the adjoint variable carries the same value ψ_n and ψ_{n+1} propagating backward in time along the same trajectory from the forward Monte Carlo solver.

Moreover, we have obtained all values in $\{(x_n^m, v_n^m, \psi_n)\}$ immediately after solving the forward equation (28) using the Monte Carlo Algorithm 1. In contrast to the finite volume scheme mentioned above, this adjoint Monte Carlo approach requires zero computational cost in backward time evolution but could potentially demand flops if a continuous format of $\lambda(t, x, v)$ is needed.

The approximation for the gradient (32) by using the Monte Carlo solver for f and the weighted particle representation (36) for λ is proved to converge to the true gradient; see [38, Theorem 2.2] for details. In this step, we fully *correlate* the forward and adjoint Monte Carlo solves in terms of their source of randomness. This is analogous to the **coupling method** discussed in Sect. 2.2.3. Unlike the Monte Carlo gradient estimator from the standard coupling method, the gradient obtained here does not suffer from the finite-difference error, and more importantly, the computational cost is **independent** of the size of the discretized θ.

This is our first example of using the adjoint Monte Carlo method as an efficient approach for kinetic equation-constrained optimization problems.

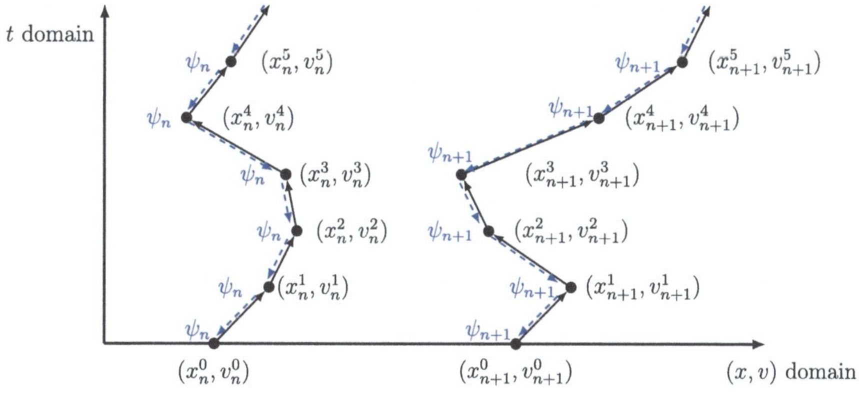

Fig. 4 An illustration of the RTE forward Monte Carlo trajectory (black) and its adjoint Monte Carlo trajectory (blue) for particle indices n and $n + 1$, as examples. The forward and adjoint trajectories coincide since they share the same randomness in the adjoint Monte Carlo method. Moreover, the adjoint variable $\gamma(t, x, v)$ carries a fixed value ψ_n (resp. ψ_{n+1}) along the time trajectory; see (36)

4.3 The DTO Approach

In this subsection, we will employ an alternative approach, the DTO framework, to compute the gradient of the RTE-constrained optimization problem (30) through a *different* adjoint Monte Carlo method. As the first step for the DTO approach, we need to discretize the objective functional and the state equation (28). If we use the finite-volume method, the discrete state equation would be (33), and the corresponding discrete adjoint equation follows (34). Interested reader may refer to [38, Appendix A] for details. In the following, we consider the discretization based on the Monte Carlo Algorithm 1.

Without loss of generality, we consider the objective functional:

$$J = \iint r(x, v) f(T, x, v) \mathrm{d}x \mathrm{d}v = \mathbb{E}_{(X,V)\sim f(T,x,v)} \left[r(X, V) \right] , \tag{37}$$

where J denotes the expectation of r with respect to the final-time RTE solution $f(T, x, v)$. The choice of J is only for notational convenience, and the derivations shall easily apply to general objective functionals. If f is approximated by f_N given in Eq. (35), the value in (37) could be approximated by the Monte Carlo quadrature:

$$J \approx \frac{1}{N} \sum_{n=1}^{N} r \left(x_n^M, v_n^M \right) =: \hat{J} . \tag{38}$$

Next, we assume that for any $m = 0, \ldots M$,

$$(x_n^m, v_n^m) \overset{\text{i.i.d.}}{\sim} f(t_m, x, v), \quad \forall n = 1, \ldots, N.$$

Given any smooth test function $\phi(v)$ and $m = 0, \ldots, M-1$, we have

$$\mathbb{E}_{\eta^{m+1}}\mathbb{E}_{p^{m+1}}[\phi(v)|(x_n^m, v_n^m)] = \alpha_n^{m+1}\phi(v_n^m) + (1-\alpha_n^{m+1})\frac{1}{|\Omega|}\int_\Omega \phi(\eta)d\eta, \tag{39}$$

where

$$\alpha_n^{m+1} = \exp(-\sigma(x_n^{m+1})\Delta t) = \exp(-\sigma(x_n^m + \Delta t v_n^m)\Delta t),$$

while η^{m+1} and p^{m+1} are random variables in Algorithm 1.

Consider a fixed particle index n. For different realizations of $\{(x_n^m, v_n^m)\}_{m=0}^M$, the value $\phi(v_n^m)$ changes with respect to the time index m. As a result, $\phi(v_n^m)$ can be treated as a discrete-time stochastic process on a probability space spanned by random variables $\{p^m, \eta^m\}_{m=1}^M$ and hence naturally generates a filtration.

Moreover, we define the following conditional expectation for $m = 0, 1, \ldots, M-1$:

$$\tilde{\mathbb{E}}^m[\,\cdot\,|(x_n^m, v_n^m)] = \mathbb{E}_{\eta^{m+1}}\mathbb{E}_{p^{m+1}}\cdots\mathbb{E}_{\eta^M}\mathbb{E}_{p^M}[\,\cdot\,|(x_n^m, v_n^m)].$$

Using these notations, the final objective functional (37), after forward Euler time discretization, can be approximated as sums of conditional expectations:

$$J \approx \tilde{J}^m := \frac{1}{N}\sum_{n=1}^N \mathbb{E}_{(x_n^m, v_n^m)\sim f(t_m,x,v)}\left[\tilde{\mathbb{E}}^m[r(x_n^M, v_n^M)|(x_n^m, v_n^m)]\right], \tag{40}$$

as a result of the law of total expectations. Note that the above formula holds for any $0 \le m \le M-1$.

We can further write $\tilde{J}^m$ into $\tilde{J}^m = \sum_{n=1}^N \tilde{J}_n^m$ following (40), where

$$\tilde{J}_n^m = \frac{1}{N}\mathbb{E}_{(x_n^m, v_n^m)\sim f(t_m,x,v)}\left[\mathcal{R}_n^m\left(x_n^m, v_n^m\right)\right], \tag{41}$$

$$\mathcal{R}_n^m\left(x_n^m, v_n^m\right) = \tilde{\mathbb{E}}^m\left[r\left(x_n^M, v_n^M\right)|(x_n^m, v_n^m)\right]. \tag{42}$$

The dependence of $\mathcal{R}_n^m\left(x_n^m, v_n^m\right)$ on the coefficient function $\sigma(x)$ is through the evaluations of $\{\sigma(x_n^{m+1})\}$ where $x_n^{m+1} = x_n^m + \Delta t\, v_n^m$, used in the acceptance-rejection probabilities in each $\mathbb{E}_{p^{m+1}}$; see (39). Note that $\mathcal{R}_n^m$ is conditioned on (x_n^m, v_n^m), so it can be seen as a function of (x_n^m, v_n^m) and consequently a function of $\sigma(x_n^{m+1})$ for a given coefficient $\sigma(x)$. When (x_n^m, v_n^m) is considered a random variable, $\mathcal{R}_n^m$ is also a random variable. Thus, using the **score function method** for computing the Monte Carlo gradient, discussed in Sect. 2.2.1, we can express the derivative of $\mathcal{R}_n^m$ with respect to each $\theta_n^{m+1} := \sigma(x_n^{m+1})$ as

$$\frac{\partial \mathcal{R}_n^m}{\partial \theta_n^{m+1}} = \tilde{\mathbb{E}}^m \left[\frac{\partial \log \kappa_n^{m+1}}{\partial \theta_n^{m+1}} r\left(x_n^M, v_n^M\right)\right] =: G_n^m \,, \tag{43}$$

where the probability for the rejection sampling and its score function are

$$\kappa_n^{m+1} = \begin{cases} \alpha_n^{m+1}, & \text{if } v_n^{m+1} = v_n^m \\ 1-\alpha_n^{m+1}, & \text{otherwise} \end{cases}, \quad \frac{\partial \log \kappa_n^{m+1}}{\partial \theta_n^{m+1}} = \begin{cases} -\Delta t, & \text{if } v_n^{m+1} = v_n^m \\ \Delta t \frac{\alpha_n^{m+1}}{1-\alpha_n^{m+1}}, & \text{otherwise} \end{cases}.$$

In (43), we fix a particular $\left(x_n^m, v_n^m\right)$. When $\left(x_n^m, v_n^m\right)$ is considered a random variable, both G_n^m and $\mathcal{R}_n^m$ are also random variables. Using the same trajectories from the Monte Carlo solver for the forward RTE (see Algorithm 1), we obtain samples $\{\widehat{\mathcal{R}}_n^m\}_{m=0}^{M-1}$ according to (42) and the Monte Carlo gradient $\{\widehat{\mathcal{G}}_n^m\}_{m=0}^{M-1}$ based on (43), for $\mathcal{R}_n^m$ and G_n^m, respectively:

$$\widehat{\mathcal{R}}_n^m = r\left(x_n^M, v_n^M\right), \qquad \widehat{\mathcal{G}}_n^m = \begin{cases} -r\left(x_n^M, v_n^M\right)\Delta t, & \text{if } v_n^{m+1} = v_n^m, \\ r\left(x_n^M, v_n^M\right)\dfrac{\alpha_n^{m+1}\Delta t}{1-\alpha_n^{m+1}}, & \text{otherwise}, \end{cases} \tag{44}$$

with $m = 0, \ldots, M-1$.

Based on (41) and (42), we can also treat $N^{-1}\widehat{\mathcal{R}}_n^m$ as a sample of $\tilde{J}_n^m$. Since all the derivations above are based on the same forward RTE particle trajectories, $\{(x_n^m, v_n^m)\}_{m=0}^M$, $n = 1, \ldots, N$, obtained by running Algorithm 1, the following holds for any $m = 0, \ldots, M-1$:

$$J \overset{\text{time discretization}}{\approx} \sum_{n=1}^N \tilde{J}_n^m \overset{\text{sample (41)}}{\approx} \frac{1}{N}\sum_{n=1}^N \mathcal{R}_n^m \overset{\text{sample (42)}}{\approx} \frac{1}{N}\sum_{n=1}^N \widehat{\mathcal{R}}_n^m,$$

$$\overset{\text{based on (38)}}{\approx} \quad \hat{J} = \frac{1}{N}\sum_{n=1}^N r(x_n^M, v_n^M)\,.$$

Thus, the Monte Carlo gradient of $\hat{J}$ with respect to $\sigma(x_n^{m+1})$ can be approximated by

$$\frac{\partial \hat{J}}{\partial \theta_n^{m+1}} \approx \frac{1}{N}\sum_{i=1}^N \frac{\partial \mathcal{R}_i^m}{\partial \theta_n^{m+1}} = \frac{1}{N}\frac{\partial \mathcal{R}_n^m}{\partial \theta_n^{m+1}} \approx \frac{1}{N}\widehat{\mathcal{G}}_n^m, \quad \forall m\,, \tag{45}$$

since $\mathcal{R}_i^m$ does not depend on $\theta_n^{m+1} = \sigma(x_n^{m+1})$ if $i \neq n$, $\forall m$. Combining (45) with (44), we obtain a Monte Carlo gradient formula for $\frac{\partial \hat{J}}{\partial \theta_n^{m+1}}$.

However, it is worth noting that

$$\frac{\partial \hat{J}}{\partial \theta_n^{m+1}} \neq \frac{\delta J}{\delta \sigma}(x_n^{m+1}), \quad \theta_n^{m+1} = \sigma(x_n^{m+1}). \tag{46}$$

The left-hand side gradient treats $\theta_n^{m+1} = \sigma(x_n^{m+1})$ as a single parameter, and therefore, the effective parameters are MN scalars, $\{\theta_n^{m+1}\}$, where $n = 1, \ldots, N$ and $m = 0, \ldots M-1$. On the other hand, $\frac{\delta J}{\delta \sigma}$ is a functional derivative with respect to the parameter function $\sigma(x)$ (the same with the OTD derivative (32)). Since we assume $\sigma \in L^2$ and the dual space of L^2 is still L^2, the derivative $\frac{\delta J}{\delta \sigma}$ is also the gradient function $G(x)$. The right-hand side of (46) is then function $G(x)$ evaluated at $x = x_n^{m+1}$. We use the following example to show how to relate both sides of (46).

Consider a separate mesh grid $\{\bar{x}_j\}$ in the spatial domain, and Q_j is a small neighborhood of $\bar{x}_j$ for each j. Consider a perturbed parameter function $\sigma(x) + \delta\sigma(x)$, where $\delta\sigma(x) = \mathbb{1}_{x \in Q_j} \epsilon$ for some small constant ϵ. We then have

$$J(\sigma + \delta\sigma) - J(\sigma) \approx \int_x \frac{\delta J}{\delta \sigma}(x)\delta\sigma(x)\mathrm{d}x = \epsilon \int_{Q_j} \frac{\delta J}{\delta \sigma}(x)\mathrm{d}x. \tag{47}$$

On the other hand, with $\hat{J}(\sigma)$ denoting the value (38) calculated with a given parameter function σ, we have the following based on (45):

$$\hat{J}(\sigma + \delta\sigma) - \hat{J}(\sigma) \approx \sum_{n=1}^{N} \sum_{m=0}^{M-1} \frac{\partial \hat{J}}{\partial \sigma(x_n^{m+1})} \delta\sigma(x_n^{m+1}) \approx \epsilon \frac{1}{N} \sum_{n=1}^{N} \sum_{m=0}^{M-1} \mathbb{1}_{x_n^{m+1} \in Q_j} \widehat{\mathcal{G}}_n^m, \tag{48}$$

where we used (45). Combining the last terms in (47)–(48) and assuming the approximated gradient function $\frac{\delta J}{\delta \sigma}$ is piecewise constant on $\{Q_j\}$, we have

$$\frac{\delta J}{\delta \sigma}(\bar{x}_j)\,|Q_j| \approx \int_{Q_j} \frac{\delta J}{\delta \sigma}(x)\mathrm{d}x \approx \frac{1}{N} \sum_{n=1}^{N} \sum_{m=0}^{M-1} \mathbb{1}_{x_n^{m+1} \in Q_j} \widehat{\mathcal{G}}_i^n.$$

Finally, we may approximate the gradient using values of (44) by

$$\begin{aligned} \frac{\delta J}{\delta \sigma}(\bar{x}_j) &\approx \frac{1}{|Q_j|} \frac{1}{N} \sum_{n=1}^{N} \sum_{m=0}^{M-1} \mathbb{1}_{x_n^{m+1} \in Q_j} \widehat{\mathcal{G}}_n^m \\ &= \frac{1}{|Q_j|} \frac{\Delta t}{N} \sum_{n=1}^{N} \sum_{m=1}^{M} \mathbb{1}_{x_n^m \in Q_j}\, r(x_n^M, v_n^M)\, \xi_n^m, \\ \xi_n^m &= \begin{cases} -1, & \text{if } v_n^m = v_n^{m-1}, \\ \frac{\alpha_n^m}{1-\alpha_n^m}, & \text{otherwise.} \end{cases} \end{aligned} \tag{49}$$

Note that sorting is needed in computing (49) for all particles in the long-time horizon.

In this DTO approach, the adjoint particle does not show up explicitly, but is implicitly represented by the particle representation of the gradient, $\mathcal{G}_n^m$, where $1 \le n \le N, 0 \le m \le M-1$. Started from the DTO approach, this is our second example of using an adjoint Monte Carlo method for efficient gradient computation.

4.4 Numerical Examples

Next, we show some numerical examples to demonstrate the adjoint Monte Carlo method for computing the gradient of the objective functional with respect to the parameter. We will also show the result based on the finite volume method (FVM) for comparison.

Consider a 1D spatial domain $[-2, 2]$ and the velocity domain on a unit interval $[-1, 1]$. The time domain is $[0, 0.5]$. The spatial domain is uniformly discretized with the spacing $\Delta x = 0.05$ at which the gradient of the RTE-constrained optimization problem with respect to the scattering coefficient $\sigma(x)$ is evaluated. The objective function is in the form of (37) with $r(x, v) = |v|^2 \mathbb{1}_{x<0}$. The initial distribution $f_0(x, v) = \frac{1}{\sqrt{\pi}}\left(\exp(-4(x-0.5)^2) + \exp(-4(x+0.5)^2)\right)$. The parameter where the gradient is evaluated is $\sigma(x) = 2 + 2\exp(-4x^2)$.

For FVM, we discretize the spatial and time domain with $\Delta x = 0.05$, $\Delta t = 0.01$ (coarse), and $\Delta x = 0.005$, $\Delta t = 0.001$ (fine), respectively. In the fine-grid discretization, we have to reduce the time spacing due to the CFL condition and the use of the forward Euler time integrator. The velocity domain $[-1, 1]$ is uniformly discretized based on the spacing $\Delta v = 0.05$. For the adjoint Monte Carlo method based on the OTD approach (denoted as "P-OTD"), we use $N = 10^6$ particles and time spacing $\Delta t = 0.01$, following the procedure detailed in Sect. 4.2.3. For the adjoint Monte Carlo method based on the DTO approach (denoted as "P-DTO"), we also use $N = 10^6$ particles and time spacing $\Delta t = 0.01$, following the scheme presented in Sect. 4.3. We average the results of adjoint Monte Carlo methods from 64 i.i.d. runs to illustrate the final gradient. We show the approximated gradients from these four methods in Fig. 5a. The values obtained by the adjoint Monte Carlo methods coincide with the ones from FVM with a fine discretization. However, the result obtained from the FVM with a coarse discretization is subject to visible numerical error.

In Fig. 5b, we range the number of particles in the adjoint Monte Carlo methods from $N = 2^{11}$ to 2^{20} and then compute the standard deviation of the L^2 error of the gradient based on the 64 i.i.d. runs. The gradient computed using the FVM with a fine discretization is regarded as the ground truth. Both the P-OTD and P-DTO methods are subject to the standard Monte Carlo error $\mathcal{O}(N^{-1/2})$. It is worth noting that the P-DTO method is subject to a larger random error than the P-OTD method for the RTE-constrained optimization.

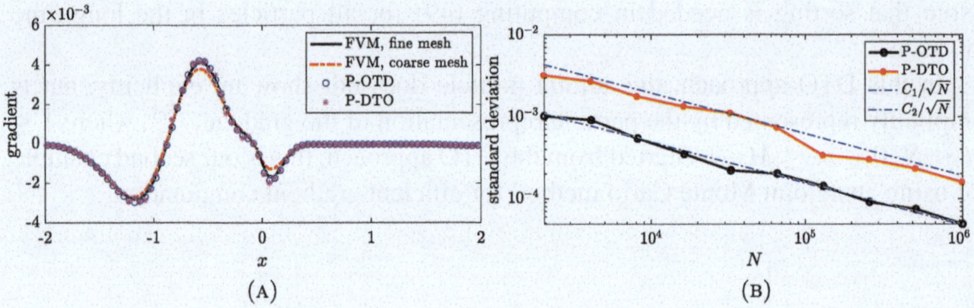

Fig. 5 Illustration of the approximated gradient for an RTE-constrained optimization problem based on various numerical methods (left) and the standard deviation for the adjoint Monte Carlo methods based on the OTD and DTO approaches (right). (**a**) The approximated gradient. (**b**) The standard deviation of adjoint MC methods

5 Case Study II: Boltzmann Equation

In this section, we present an adjoint Monte Carlo method for the Boltzmann equation, a famous kinetic equation with a quadratic nonlinearity. The derivation here are more intricate as we need to incorporate the binary collision feature of the forward problem into the adjoint particle system.

For simplicity, we consider the spatially homogeneous Boltzmann equation:

$$\frac{\partial f}{\partial t} = Q(f, f)\,. \tag{50}$$

The nonlinear collision operator $Q(f, f)$, which describes the binary collisions among particles, is defined as

$$Q(f, f) = \int_{\mathbb{R}^3} \int_{\mathcal{S}^2} q(v - v_1, \sigma)(f(v_1')f(v') - f(v_1)f(v))\mathrm{d}\sigma\,\mathrm{d}v_1,$$

in which (v', v_1') represent the post-collisional velocities associated with the pre-collisional velocities (v, v_1), the collision kernel function $q \geq 0$, and the σ integral is over the surface of the unit sphere $\mathcal{S}^2$.

By conserving the momentum $v + v_1$ and the energy $v^2 + v_1^2$, we have

$$\begin{aligned} v' &= 1/2(v + v_1) + 1/2|v - v_1|\sigma, \\ v_1' &= 1/2(v + v_1) - 1/2|v - v_1|\sigma, \end{aligned} \tag{51}$$

where $\sigma \in \mathcal{S}^2$ is a collision parameter. We will hereafter use the shorthand notation:

$$(f, f_1, f', f_1') = (f(v), f(v_1), f(v'), f(v_1')),$$
$$(\hat{f}, \hat{f}_1, \hat{f}', \hat{f}_1') = \rho^{-1}(f(v), f(v_1), f(v'), f(v_1'))$$

for the values of f in the Boltzmann collision operator Q, as well as the normalized densities $\hat{f}$ with $\rho = \int f(v)\mathrm{d}v$ and $\int \hat{f}(v)dv = 1$.

Physical symmetries imply that

$$q(v - v_1, \sigma) = \tilde{q}(|v - v_1|, \theta), \tag{52}$$

where $\cos\theta = \sigma \cdot \alpha$ with $\alpha = \frac{v-v_1}{|v-v_1|}$ and $\sigma = \frac{v'-v_1'}{|v'-v_1'|} = \frac{v'-v_1'}{|v-v_1|}$. That is, the collision kernel we consider here only depends on the size of the pre-collision relative velocity $|v - v_1|$ and the angle θ between the pre- and post-relative velocity directions.

In [11], the collision rule (51) was rewritten conveniently in terms of matrix-vector multiplications, as

$$\begin{pmatrix} v' \\ v_1' \end{pmatrix} = A(\sigma, \alpha) \begin{pmatrix} v \\ v_1 \end{pmatrix}, \tag{53}$$

where

$$A(\sigma, \alpha) = \frac{1}{2} \begin{pmatrix} I + \sigma\alpha^T & I - \sigma\alpha^T \\ I - \sigma\alpha^T & I + \sigma\alpha^T \end{pmatrix}, \tag{54}$$

in which I is the identity matrix in $\mathbb{R}^3$. It is worth noting that

$$A(\sigma, \alpha)^\top = A(\sigma, \alpha)^{-1} =: B(\sigma, \alpha).$$

We highlight the dependence of A and B on σ and α by writing $A(\sigma, \alpha)$ and $B(\sigma, \alpha)$. Here, for a fixed pair of velocity (v, v_1), we also define the Jacobian matrix $C \in \mathbb{R}^{6\times 6}$ given by

$$C(\sigma, \alpha) = \frac{\partial(v', v_1')}{\partial(v, v_1)}. \tag{55}$$

Even though (53) is a matrix-vector multiplication, C and A do not always coincide since the collision parameter σ can depend on the velocity pair (v, v_1) for certain collision kernel q.

Next, we will assume that the collision kernel q is uniformly bounded from above. For some $\Sigma > 0$, we have

$$q(v - v_1, \sigma) \leq \Sigma \tag{56}$$

for all v, v_1, and σ. In practice, this is not a restriction because one can take Σ to be the largest value of q for the discrete set of velocity and parameter values.

Although the formulation and analysis presented here are valid for the general collision model (52), the following form is often considered:

$$q(v - v_1, \sigma) = \tilde{q}(|v - v_1|, \theta) = C_\kappa(\theta)|v - v_1|^\beta . \tag{57}$$

We refer to [56, Sec. 1.4] for more modeling intuition regarding this type of collision models. The collision model (57) accommodates both the variable hard sphere (VHS) model, when $C_\kappa(\theta)$ is constant, and the variable soft sphere (VSS) model [35] with angular dependence.

With the assumption (56), the Boltzmann equation (50) can be further written as

$$\begin{aligned}
\frac{\partial f}{\partial t} &= \int_{\mathbb{R}^3}\int_{S^2} \left(f' f_1' - f f_1\right) q(v - v_1, \sigma) \mathrm{d}\sigma \mathrm{d}v_1 \\
&= \iint f' f_1' q \mathrm{d}\sigma \mathrm{d}v_1 - \iint f f_1 q \mathrm{d}\sigma \mathrm{d}v_1 \\
&= \iint f' f_1' q \mathrm{d}\sigma \mathrm{d}v_1 + \iint f f_1 (\Sigma - q) \mathrm{d}\sigma \mathrm{d}v_1 - \iint \Sigma f f_1 \mathrm{d}\sigma \mathrm{d}v_1 \\
&= \iint f' f_1' q \mathrm{d}\sigma \mathrm{d}v_1 + f \iint f_1 (\Sigma - q) \mathrm{d}\sigma \mathrm{d}v_1 - \mu f.
\end{aligned} \tag{58}$$

Recall $\rho = \int f_1 \mathrm{d}v_1$ is the total mass. We define $\bar{A} = \int_0^{2\pi} \int_0^{\pi} \sin(\theta)\mathrm{d}\theta \mathrm{d}\varphi = 4\pi$, the surface area of the unit sphere, and then we have

$$\mu = \bar{A}\Sigma\rho. \tag{59}$$

If we multiply both sides of (58) by an arbitrary test function $\phi(v)$, then divide by ρ, and finally integrate over v, we obtain

$$\begin{aligned}
&\frac{\partial(\int \phi \hat{f} \mathrm{d}v)}{\partial t} \\
&= \rho \iiint \phi \hat{f}' \hat{f}_1' q \mathrm{d}\sigma \mathrm{d}v_1 dv + \rho \iiint \phi \hat{f} \hat{f}_1 (\Sigma - q) \mathrm{d}\sigma \mathrm{d}v_1 \mathrm{d}v - \mu \int \phi \hat{f} \mathrm{d}v \\
&= \rho \iiint \phi' \hat{f} \hat{f}_1 q \mathrm{d}\sigma \mathrm{d}v_1 dv + \rho \iiint \phi \hat{f} \hat{f}_1 (\Sigma - q) \mathrm{d}\sigma \mathrm{d}v_1 \mathrm{d}v - \mu \int \phi \hat{f} \mathrm{d}v,
\end{aligned}$$

using $\mathrm{d}v\mathrm{d}v_1 = \mathrm{d}v'\mathrm{d}v_1'$ and then interchanging notation (v, v_1) and (v', v_1') in the first integral. After applying the explicit Euler time integration from t_k to $t_{k+1} = t_k + \Delta t$, we obtain

$$\begin{aligned}
&\int \phi \, \hat{f}(v, t_{k+1}) dv \\
&= (1 - \Delta t \mu) \int \phi \, \hat{f}(v, t_k) \mathrm{d}v \qquad (60) \\
&\quad + \iiint \phi' \, \Delta t \mu \Sigma^{-1} q \, \hat{f}(v, t_k) \hat{f}_1(v_1, t_k) \bar{A}^{-1} \mathrm{d}\sigma \mathrm{d}v_1 \mathrm{d}v \qquad (61)
\end{aligned}$$